C0-AVU-046

Applications of Linear Integrated Circuits

Applications of Linear Integrated Circuits

EUGENE R. HNATEK

DCA Reliability Laboratory Inc.

A Wiley-Interscience Publication

JOHN WILEY & SONS

New York London Sydney Toronto

Library of Congress Cataloging in Publication Data

Hnatek, Eugene R
Applications of linear integrated circuits.

"A Wiley-Interscience publication."
1. Integrated circuits. 2. Electronic apparatus and appliances. I. Title.

TK7874.H53 621.381'73 73-11368
ISBN 0-471-40111-0

Printed in the United States of America

10 9 8 7 6 5 4 3 2 1

To
Susan, Stephen, Jeffrey, and David
and
to the people whose faith and understanding will always be remembered: Frank, Vlasta, Richard, Jenny, Elaine, Dot, Leonard, Gary, Judy, Bob, Linda, Wes, Patti, Ron, and Carol.

Preface

The purpose of this book is to provide applications-oriented information and ideas for linear ICs. The first sections of each chapter will present the reader with general characteristics of various devices including a definition of specifications and their application significance. This is followed by a collection of applications. In many cases these applications show the use of specific IC types. The same basic design considerations, however, apply to any IC of the same general characteristics.

To give more insight into device properties, the first two chapters consist of general information dealing with processes, components, and design principles. This background information will give the user a better understanding of linear IC operation and more specifically will help the user anticipate the result of any nonspecific operating situation in which they might be used. Such a situation is more likely in linear ICs than in other IC forms since a much wider range of functions and operating environments is encountered.

Until recent years, linear ICs have been restricted mostly to building block devices, that is, devices that can be used in a wide variety of functions with the addition of external discrete components. During this period, the typical devices available were operational amplifiers, video amplifiers, comparators, differential amplifiers, etc. The number of devices designed for a specific end use were comparatively few. The last few years have seen a large expansion in the range and performance of linear IC components while current pricing gives them the economic edge over discrete designs. This is shown significantly by the penetration linear ICs are making into new and varied designs such as timers, phase-locked loops, preamplifiers, etc.

This trend will accelerate as the penetration of linear ICs expands into wider markets. New technologies will give an increased range of components so that almost any function feasible in discretes will be possible in monolithic form. The general exception will be when fundamental physical limitations, such as power dissipation, prevail. Outside such physical limitations, the use of ICs in linear functions will become predominant.

A wide variety of functional linear ICs have been developed to provide a powerful array of components for analog circuit design. This book is dedicated to providing application ideas and basic principles which when clearly understood can lead to successful, reliable, and practical circuit designs.

I should like to thank my friends at National Semiconductor Corporation, especially Charles Signor, for permission to use portions of their applications literature. Additionally, I should like to thank my two typists for their painstaking hours in preparing this manuscript—my mother, Mrs. Vlasta Hnatek, and my sister-in-law, Mrs. Robert Banks. Finally, I would like to thank my editor, George Novotny, for his time and Alan Boston and Angelo Pestarino for their encouragement during the preparation of this manuscript. A special thanks also goes to Mike Economy.

Although every effort has been made to provide meaningful applications information, the author will assume no responsibility for the use of the circuits described herein and make no representation as to freedom of patent infringement.

San Jose, California *Eugene R. Hnatek*

Contents

Chapter 1

LINEAR INTEGRATED CIRCUIT TECHNOLOGY 1

Materials Processing 1
Linear Processes 3
Linear Integrated Circuit Components 9

Chapter 2

LINEAR INTEGRATED CIRCUIT DESIGN TECHNIQUES 13

General Considerations 13
Bias Circuits 13
Current-Source Loads 17
Level-Shifting 19
Output Stages 22
Active Component Simulation 25

Chapter 3

THE OPERATIONAL AMPLIFIER 28

Integrated Circuit Operational Amplifiers 28
The Ideal Operational Amplifier 29
Design Considerations for Frequency Response and Gain 31
Operational Amplifier Terminology 34
Operational Amplifier Input-Circuit Design Considerations 38
Factors to Consider in Choosing an Operational Amplifier 49

Chapter 4

OPERATIONAL AMPLIFIER APPLICATIONS 53

Basic Operational Amplifier Connections 53
Operational Amplifier Biasing and Compensating 64

Voltage-Followers, Buffers, and Drivers 95
Precision Measuring and Reference Circuits 109
Sample-and-Hold Circuits 147
Oscillators, Function Generators, and Active Filters 156
Analog-Computing and Servo Amplifiers 203
Operational Amplifier Voltage-Regulators 223
Appendix: Design Criteria for Second-Order Active Filters 235

Chapter 5

MONOLITHIC COMPARATORS AND INTERFACE CIRCUITS **238**

Comparators 238
Interface Circuits 279

Chapter 6

MONOLITHIC VOLTAGE-REGULATORS **331**

Introduction 331
Definition of Terms 332
Basic Regulator Types 333
Integrated Circuit Regulator Families 335
Dissipative Voltage-Regulators 337
Switching Regulators 378
Development of a 5–V, 1–A Monolithic Regulator 409

Chapter 7

THE INTEGRATED CIRCUIT TIMER **421**

Introduction 421
Application Information 423

Chapter 8

AUDIO CIRCUITS **445**

Audio Amplifiers 445
FM Detector-Limiter 491
Phase-Locked Loops (PLL) 496

INDEX **513**

Chapter 1

Linear Integrated Circuit Technology

MATERIALS PROCESSING

The first step in making an IC is to design a circuit which can be integrated by using a set of rules, some of which are described in Chapter 2. Masks are then created which represent the patterns of various layers of material to be diffused. The next step is to "grow" a cylinder of silicon. Various diameters of this cylinder are currently used, but in general the range is between 2 and 3 in. Wafers 10 mils thick are then cut from the cylinder as seen in Figure 1-1.

Since each wafer can contain 250 or more circuits, the surface must be highly uniform. To accomplish the uniformity required, the wafers are polished to produce a mirror-smooth surface. Molecular structure of the surface material as well as other surface characteristics are important to circuit performance. This requirement is satisfied by growing additional semiconductor material, an oxide, on the top surface of the wafer by a method known as *epitaxy*.

A coating of photographic emulsion, called photoresist, is now put onto the wafer. The mask which corresponds to the first layer of material to be diffused is placed in an optical jig along with the wafer. An ultraviolet (uv) light passing through the mask forms a photographic image of the mask on the wafer. Portions of the photoresist exposed to the light polymerize. The wafer is then developed and washed in a way similar to photographic film. Emulsion that is not polymerized becomes dissolved in the developing process. The remaining polymerized emulsion forms a protection against diffusion.

The wafers are then placed into a diffusion furnace and under the conditions of temperature (900°C) and time (1 to 6 hr) certain materials such as phosphorus, arsenic, or boron are diffused into the exposed patterned areas in order to achieve the desired electrical characteristics. Temperature, time, amount and type of diffusants, depth of diffusion, and rate of diffusion are all critical factors. Temperatures in the furnace, for example, are controlled to better than ± 0.5°C.

Figure 1-1. Individual wafers after cutting and polishing.

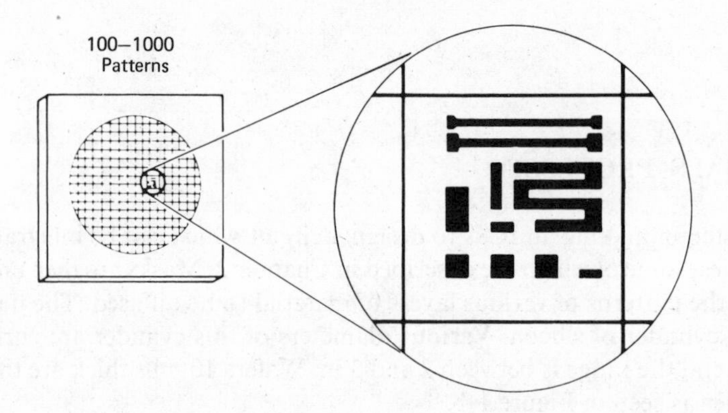

Figure 1-2. More than 250 identical circuits are processed on a single wafer.

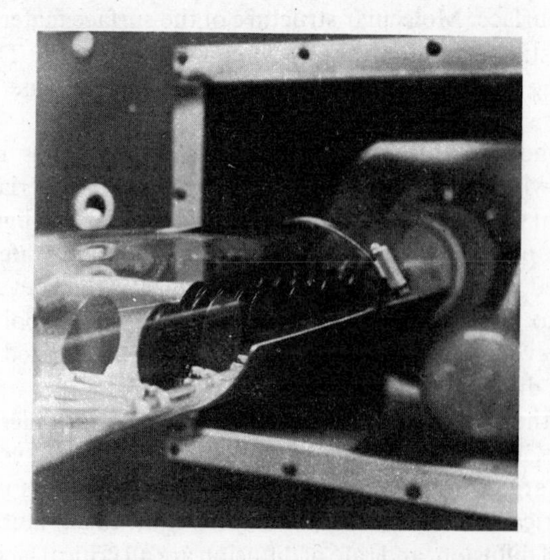

Figure 1-3. A typical diffusion furnace.

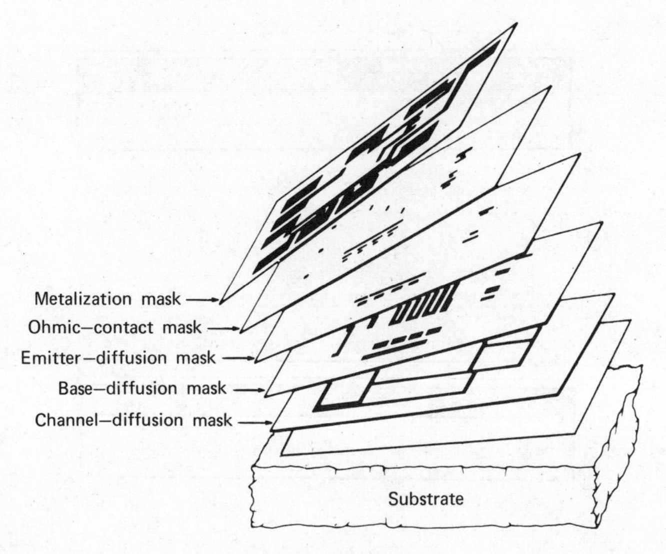

Figure 1-4. The series of overlaying masks used in integrated circuit processing.

These three basic steps,

1. Exposure to light through a mask
2. Development
3. Diffusion

are repeated using different masks resulting in an IC which consists of a series of overlaying patterns, Figure 1-4. A "metal mask" is the final pattern. It serves as the interconnecting links between the individual components on the chip. Figure 1-5 graphically illustrates the major steps in the processing of an IC.

Once the above processes have been completed, each circuit on the wafer is tested under computer control and circuits which fail the test are marked. The wafer is then cut into individual "die" or "chips."

The final major step before testing and shipment is packaging. A variety of package types is available all of which provide three principal characteristics:

1. Protection of the die from a hostile ambient environment such as mechanical abuse, dust, and corrosive elements
2. Thermal dissipation of heat generated during operation
3. A convenient means for electrically connecting the chip to an external circuit

LINEAR PROCESSES

Integrated circuits are divided into three general categories: (1) Linear, (2) digital, and (3) metal oxide semiconductor (MOS). Distinctly different design and process techniques are used for each type. The main difference between linear processes and other IC processes is their diversity. While digital circuits are commonly restricted to low-voltage switching, linear circuits may be fabricated with anything from switching to linear characteristics, high or low voltages, high or low frequency, or

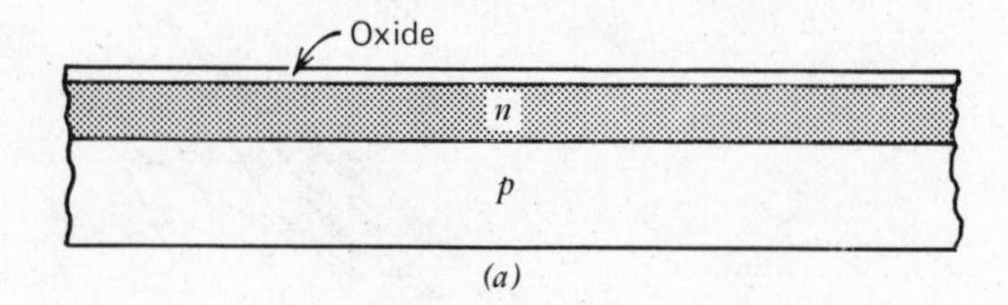

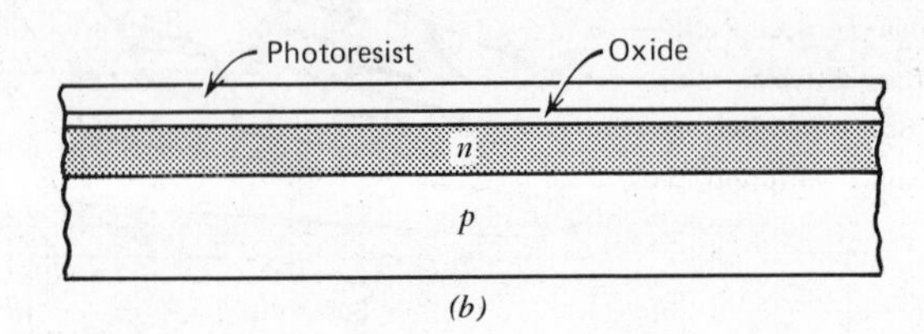

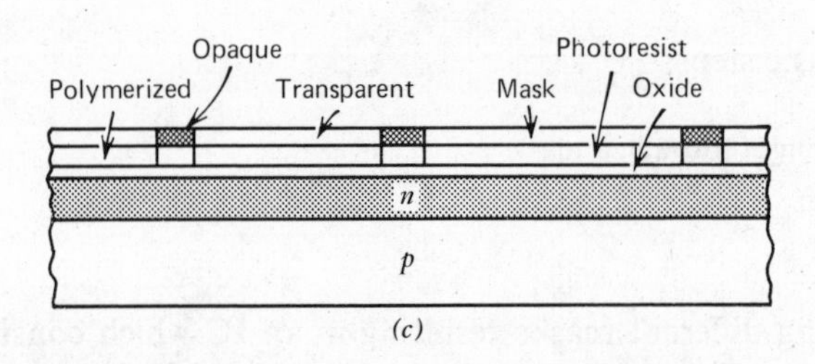

Figure 1-5. Major integrated circuit processing steps: (*a*) Starting material is a slice of *p*-type silicon, between 2 and 3 in. in diameter and a few mils thick. In an epitaxial reactor, a thin layer of *n*-type silicon is grown on the slice. Then an oxide layer is grown. (*b*) The entire top surface is covered with a layer of photoresist. (*c*) The mask containing the isolation pattern is placed on top, and the photoresist is exposed to uv light. The portions of the photoresist exposed to the light polymerize; the rest can be dissolved. (*d*) The oxide that is not protected by the polymerized photoresist is etched away. (*e*) The *p*-type dopant is diffused through the windows. The diffused regions connect with the underlying *p* region (the substrate) and form isolation pockets in the epitaxial layer. The edge of the junction is under the oxide. (*f*) The diffusion windows are closed with a new oxide layer, and the slice is again covered with photoresist. (*g*) The photoresist is exposed through the mask which outlines all shallow *p* regions, and the oxide is again etched away in the unexposed areas. (*h*) The *p*-type dopant is diffused into the unprotected regions, and the slice is covered with another oxide layer. (*i*) Again the slice is covered with photoresist. The resist is exposed through the mask, which outlines all shallow *n* regions, and the oxide is etched away in the unpolymerized areas. (*j*) A shallow layer of high *n*-type dopant concentration is diffused into the unprotected areas. (*k*) Another oxide layer is grown. (*l*) The slice is covered with photoresist for the fourth time. The resist is exposed through a mask in the areas where contact to the devices must be made, and the unprotected oxide is removed. (*m*) The entire slice is covered with a thin metal film. (*n*) With one more series of photolithographic steps, the portions of the metal layer not needed for interconnection are removed.

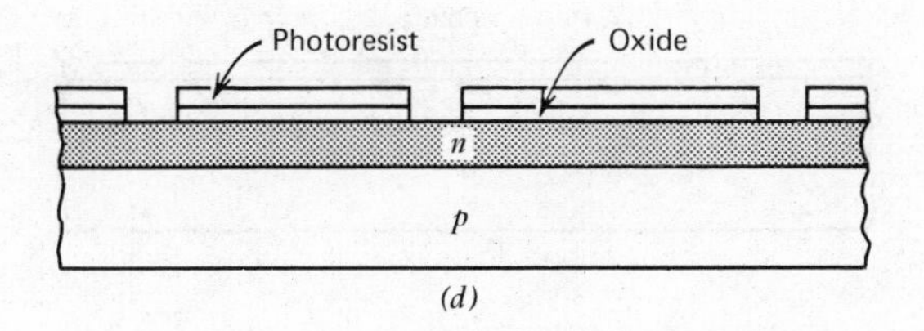

(d)

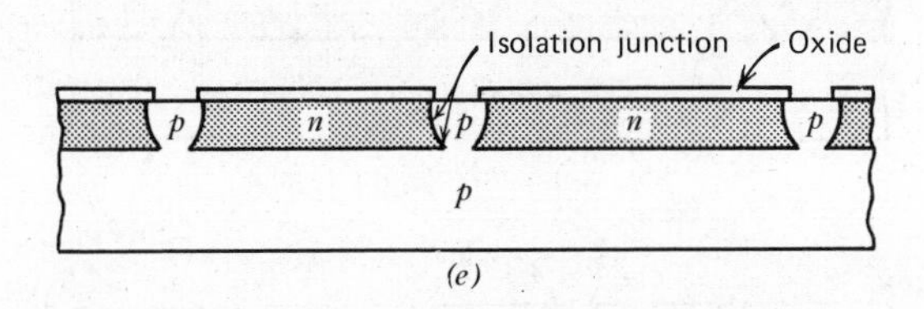

(e)

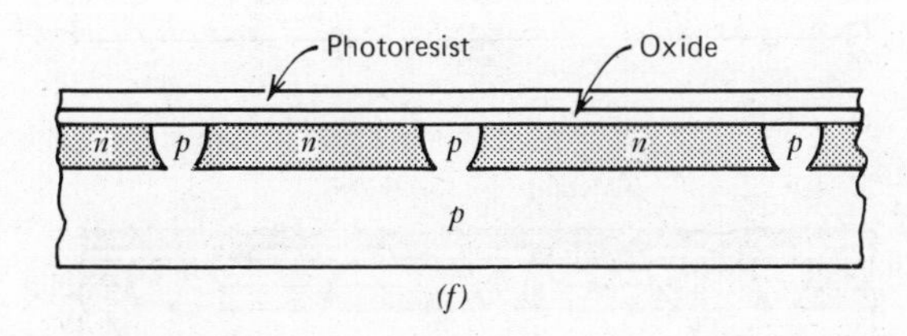

(f)

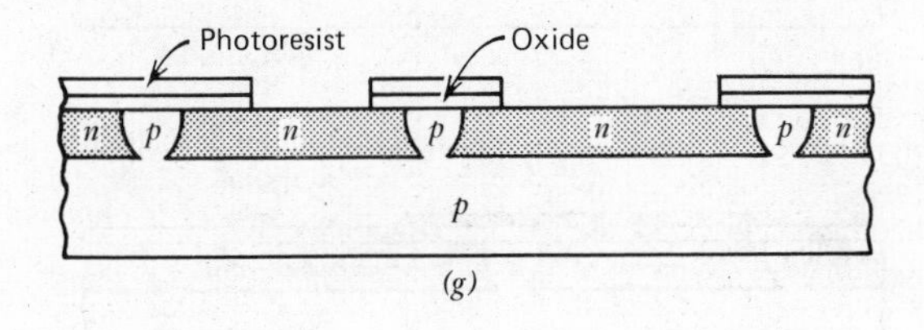

(g)

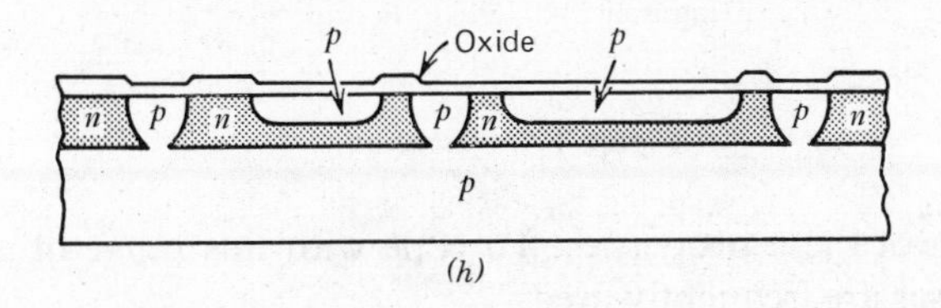

(h)

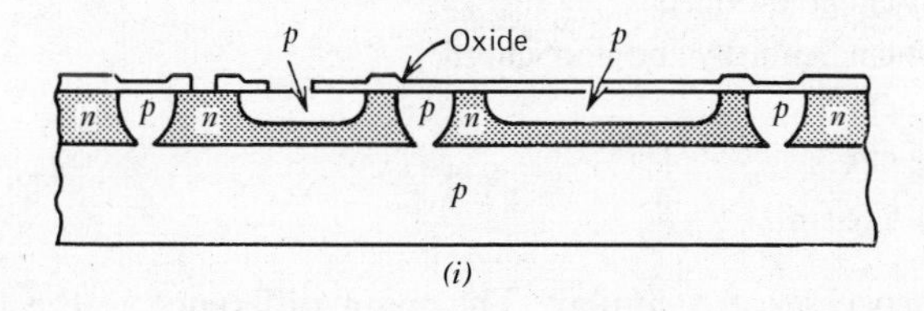

(i)

Figure 1-5. *Continued*

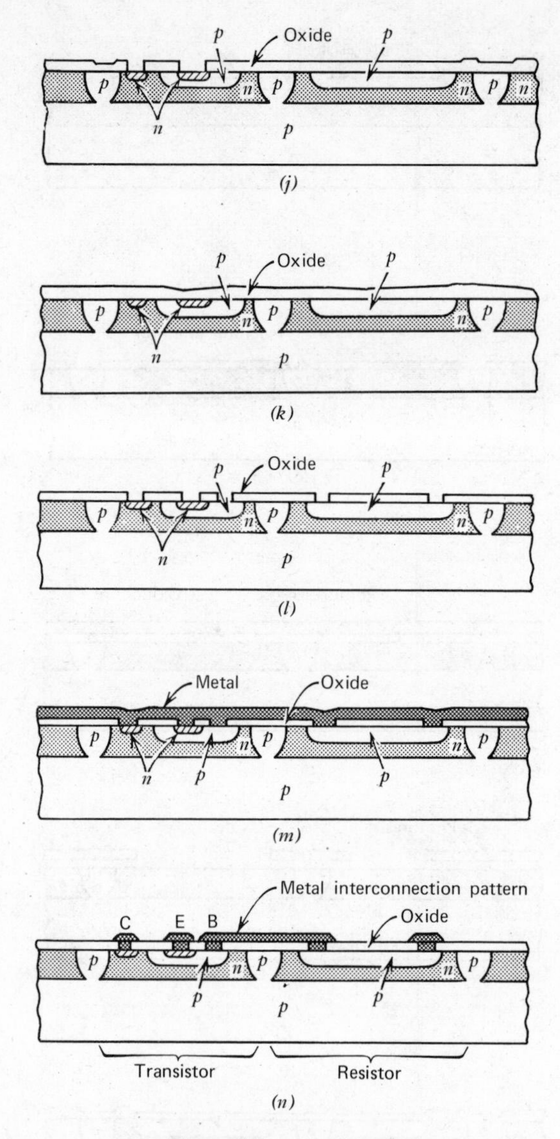

Figure 1-5. *Continued*

any combination of these properties. To cope with this range of applications, the following processes are frequently used:

Epitaxial	0.25 Ω-cm gold-doped and nongold-doped
	0.5 Ω-cm gold-doped
	1.0 Ω-cm Schottky and nonSchottky
	2.5 Ω-cm
	5.0 Ω-cm
Dielectric	2 to 15 Ω-cm

All epitaxial processes are similar. The main difference is the resistivity of the deposited epitaxial layer in which the components are formed. This difference

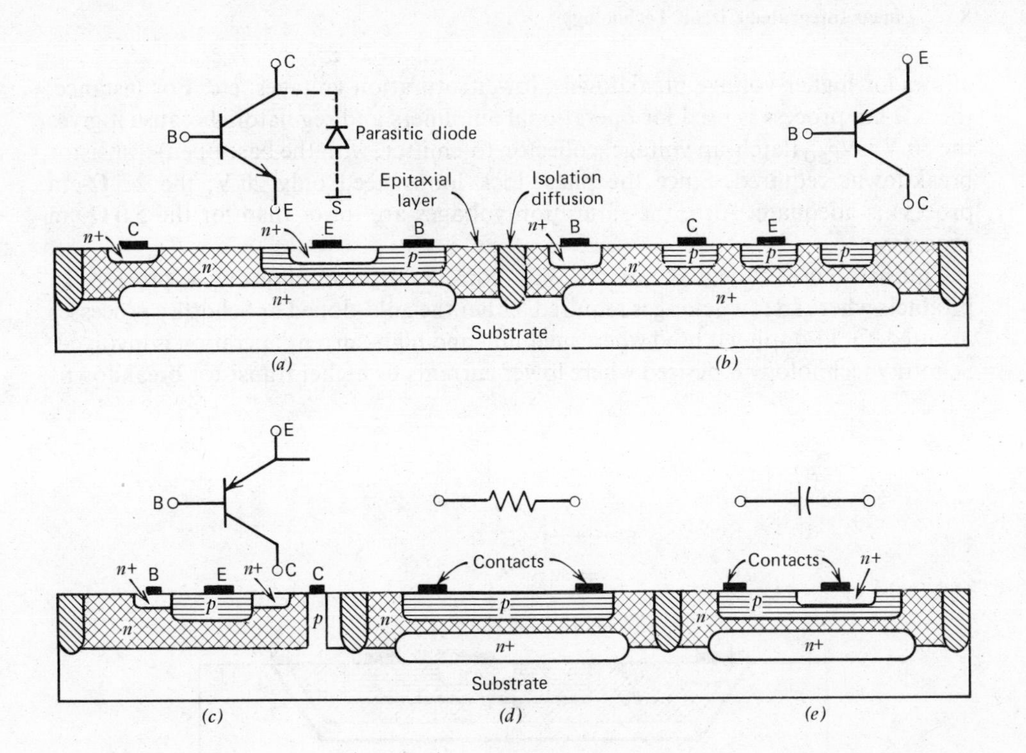

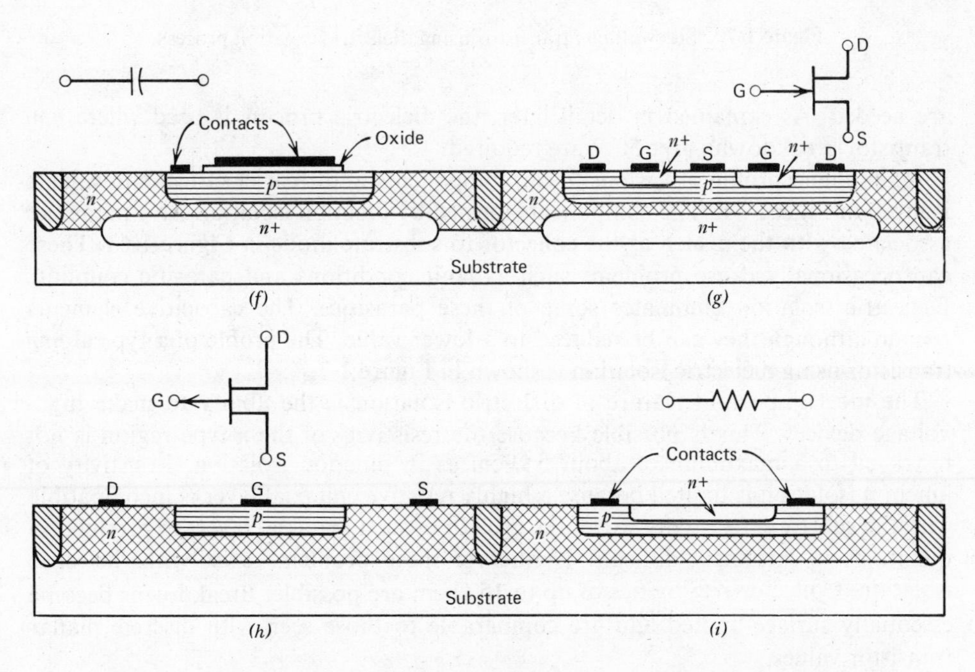

Figure 1-6. (*a*) An *npn* transistor, (*b*) a lateral *pnp* transistor, (*c*) a vertical *pnp* transistor, (*d*) a base resistor, (*e*) a junction capacitor, (*f*) an oxide capacitor, (*g*) a *p*-channel J FET, (*h*) an *n*-channel J FET, and (*i*) a pinch resistor.

allows for higher voltage breakdowns, lower saturation voltages, etc. For instance, the 5 Ω-cm process is used for operational amplifiers and regulators because it gives the 50 V LV_{CEO} (latch-up voltage, collector-to-emitter, with the base open) transistor breakdowns required. Since the phase-lock loops need only 20 V, the 2.5 Ω-cm process is adequate. Also, the saturation voltages are lower than for the 5.0 Ω-cm process.

Transistor breakdown is not the only consideration in choosing a process. In products where fast switching is required, either the gold-doped or Schottky processes are used. Gold-doping is used where medium- and high-current operation is involved. Schottky technology is desired where lower currents or higher transistor breakdowns

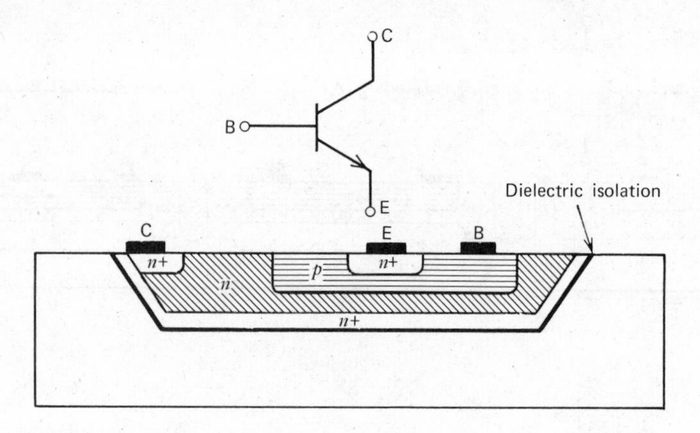

Figure 1-7. Sidewall *npn* transistor using dielectric isolation process.

are needed. As explained in detail later, the dielectric process is used where *npn* transistor breakdowns over 50 V are required.

A simplified profile of various linear components and their electrical equivalents is given in Figure 1-6. The components formed by junction isolation have parasitics associated with them such as the collector to substrate diode in Figure 1-6*a*. These can occasionally cause problems such as fault conditions and parasitic coupling. Dielectric isolation eliminates some of these parasitics. The capacitive elements remain although they can be reduced to a lower value. The profile of a typical *npn* transistor using dielectric isolation is shown in Figure 1-7.

The most important feature of dielectric isolation is the ability to make high-voltage devices. This is possible because the resistivity of the *n*-type region is not restricted to a maximum of about 5 Ω-cm as in junction isolation. Resistivity of junction isolation is limited because a highly resistive epitaxial layer is incompatible with the substrate resistivity. Also, the growth of such epitaxial layers in production quantities is extremely difficult. Neither of these problems exist with dielectric isolation. Collector resistivities of up to 15 Ω-cm are possible. Breakdowns become essentially surface-limited and are comparable to those seen with discrete planar transistor values.

Unless costly extra processing is used, the vertical *pnp* and the *n*-channel FET which are used in junction isolation are not available in dielectric processing. The loss of the vertical *pnp* is serious. Though its use is restricted to the emitter-follower configuration, it has the advantages of good frequency response and good high-current beta. The *n*-channel field-effect transistor (FET) is used primarily in biasing circuits where it can be arranged to supply the bias circuit starting current. The circuit parameters can then be made independent of supply voltages.

LINEAR INTEGRATED CIRCUIT COMPONENTS

Transistors

It is instructive to compare, in a general manner, discrete transistors with those manufactured in ICs. The most important differences for *npn* transistors are the parasitic substrate diode (transistor) and the top contact collector, as seen in Figure 1-8.

The magnitude of the parasitic *pnp* beta depends upon the process and geometry used, but it ranges from about five for high-resistivity processes to very much less than one for gold-doped processes. The parasitic *pnp* only becomes active when the *npn* transistor goes into saturation. Normally such effects are not important, but in some circuit configurations latching effects may be observed. That is, a positive feedback path may be established which is self-sustaining. Alternatively, or perhaps coincidently, this path may cause high currents to flow. These potential problems are easily avoided with judicious layout procedures. The effect of the top contact is to increase the saturation resistance. In small signal devices this is not really significant, but at higher currents (around 500 mA) this becomes an economic factor as the die area must be increased and yields drop.

At the expense of some extra processing, *npn* transistors with very high beta may be made. The processing steps used are the same as for the regular *npn* but the emitter is diffused longer to give a very narrow base width. If regular beta *npn*s are made at the same time, complications in the masking sequence occur.

The *pnp* transistors available, both lateral and vertical, are different than discrete *pnp*s. The names *lateral* and *vertical* are derived from the mode of transistor action that occurs in the two components. Referring to Figures 1-6*b* and 1-6*c*, it can be seen that in the lateral *pnp* current flows laterally from emitter to collector through

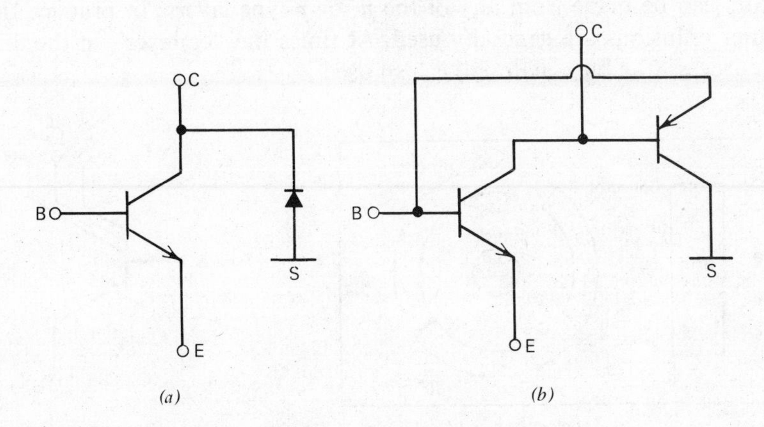

Figure 1-8. *npn* transistor parasitics shown as (*a*) diode and (*b*) *pnp* transistor.

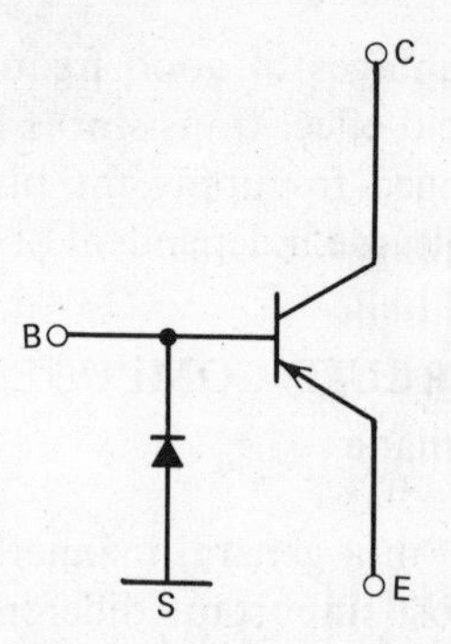

Figure 1-9. Lateral *pnp* transistor parasitic diode.

the *n*-epitaxial region. The presence of the buried-layer diffusion reduces to a comparatively low-level collection by the isolation area. It is not eliminated entirely, however, which gives rise to the parasitic diode shown in Figure 1-9. The vertical *pnp* is similarly constructed but in this case the buried-layer diffusion is omitted resulting in isolation diffusion acting as the collector.

Frequency response is the primary difference between these devices. The lateral *pnp* is restricted to frequencies below 1 to 2 MHz while the vertical *pnp* upper range is around 10 to 20 MHz. Another important feature of the lateral *pnp* is its comparatively low beta range and the low current at which beta peaks. In addition, the lateral *pnp* collector can be split to give multiple collector devices as shown in Figure 1-10. This configuration can also be used to give fairly precise values of beta by tying one of the collectors to the base.

Recent process development allows the addition of Schottky barrier devices to monolithic design (Figure 1-11). The advantage is very fast-switching circuits without gold-doping. With this technique, the properties of nongold-doped devices can be maintained while switching speeds are greatly improved. This is very desirable in devices containing analog and digital circuitry such as voltage comparators.

Resistors

Resistors can be made from any of the *n*- or *p*-type layers. In practice the base and emitter diffusions are generally used. At times the "epilayer" in the dielectric isolation process (the bulk material) is also used.

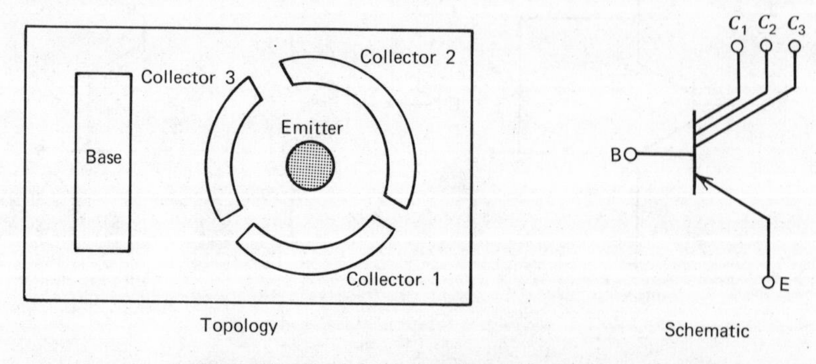

Figure 1-10. Multiple collector lateral *pnp*.

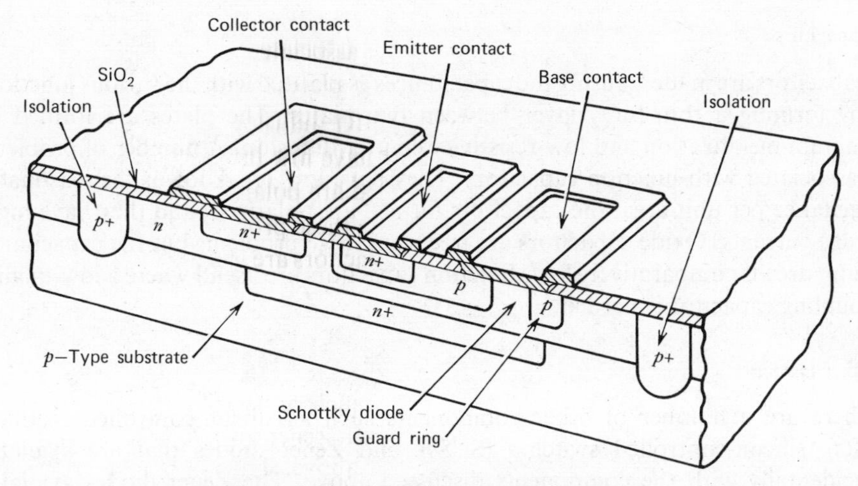

Figure 1-11. Schottky transistor geometry.

Required characteristics determine the material or processes used for a given resistor. The most commonly used is the base-diffused resistor. Size is scaled for the value desired. Base-pinch resistors are used where high values are required and the limitations of accuracy and breakdown are not a problem.

Emitter resistors are useful where low-value, temperature-insensitive resistors are needed. This diffusion is also often used as a "cross-under," that is, a low-resistance connecting path. This can simplify layout design considerably.

A new development in resistors that will offer great flexibility in design is the use of ion implantation. With this technique, resistivities orders of magnitude higher and temperature coefficients orders of magnitude lower than base-diffused resistors are possible. We can expect to see this technology used extensively in high-voltage circuits, low-power circuits and in complex linear functions where die areas would otherwise be excessive.

Junction Field-Effect Transistors

The *n*-channel FET is fabricated in the epitaxial layers and is obtained by pinching-off the epi with isolation diffusion. Because of this construction, the nomenclature FET is something of a misnomer since the gate is not available as an input. Its usefulness is as a bias circuit-starting element. It is much smaller in area than an equivalent value resistor and has a sufficiently high-breakdown voltage for this purpose.

The *p*-channel FET is a more useful general-purpose device. Its most important limitation is the breakdown voltage which is restricted to around 5 V. Processing of both field-effect and punch-through transistors is similar. Changes in the regular process follows for both devices.

Punch-through transistors are usually required at the same time regular *npn* transistors are needed. Therefore, steps must be taken to fabricate transistors with two different base widths; the wider for regular *npn* transistors, and the narrower for the punch-through. This can be achieved by reducing the depth of the base diffusion or by increasing the depth of the emitter diffusion. This requires much closer control of base and emitter diffusions.

Capacitors

Capacitors are made by using the capacitances associated with the various junctions or by forming a thin SiO_2 layer between two plates. The plates are formed by aluminum metalization and low-resistivity emitter diffusion. A number of problems are associated with junction capacitors. They have low breakdowns for reasonable capacitance per unit area, the capacitors formed are polarized, and they have high-leakage currents. Oxide capacitors are free from these problems but the capacitance per unit area is comparatively low. Junction capacitors are useful where a low-quality decoupling capacitor is needed.

Other Devices

There are a number of other components such as silicon-controlled rectifiers (SCRs), silicon-controlled switches (SCSs), and Zener diodes that are available coincidentally with the components discussed above. The Zener diodes available are the emitter-base junction and the emitter-isolation junction. Other junctions have breakdowns that are so high and so variable that they are seldom used.

Chapter 2

Linear Integrated Circuit Design Techniques

GENERAL CONSIDERATIONS

The components available for linear ICs were reviewed in Chapter 1. Some of the differences between IC components and those available to discrete circuit design were discussed. These differences may be conveniently summarized as

1. Limited resistor values and accuracy
2. No inductance
3. Low value of total circuit capacitance
4. Poor *pnp* transistor performance
5. Limited power dissipation

On the other hand, linear IC designs have advantages in the ease with which a wide range of active components can be incorporated in a circuit and in the excellent match that can be achieved between components, both passive and active. To cope with the limitations in linear IC components designers have evolved a number of special techniques, some of which are described in this chapter.

BIAS CIRCUITS

In discrete designs, biasing is almost always done with high-value resistor networks. Because of the die area required, this is impractical in linear ICs. An alternate method is to use the *n*-channel FET as shown in Figure 2-1.

The *n*-channel FET Q_1 supplies current to Zener V_z. A resistor bias chain consisting of R_1 and R_2 is then used to supply the required bias voltage V_{BIAS}. This simple technique can be elaborated upon to temperature compensate the Zener diodes temperature coefficient (TC) by adding a transistor Q_2 as shown in Figure 2-2. Here, the positive Zener TC is added to the negative TC of the transistor forward emitter-base diode. More elaborate techniques, which maintain constant current in the diodes, buffer it from the load, and adjust the composite TC to zero, are commonly used where accurate references are required. Both *npn* and *pnp* current

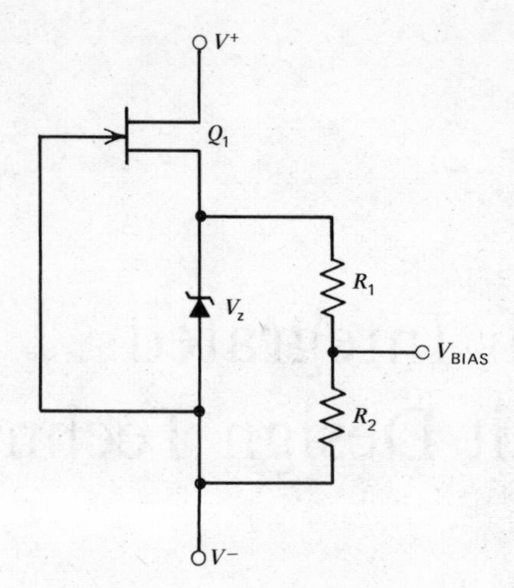

Figure 2-1. Simple bias circuit.

references are easily obtainable as voltage references. Figure 2-3 depicts the circuitry.

If Q_1, Q_2, Q_3, and Q_4 are the same geometries and therefore match very well, it can be shown that

$$I_{npn} = \frac{V_{\mathrm{BIAS}} - V_{\mathrm{BE}}}{R_4} \quad \text{and} \quad I_{pnp} = \left(\frac{V_{\mathrm{BIAS}} - V_{\mathrm{BE}}}{R_3}\right)\frac{R_1}{R_2} \tag{2-1}$$

These equations show that circuit currents can be made independent of external voltage supplies and temperature insensitive as well. Schemes such as these or variations of them are used in the more modern operational amplifiers such as the NE 531. Their use ensures such characteristics as voltage gain, offset voltage, input

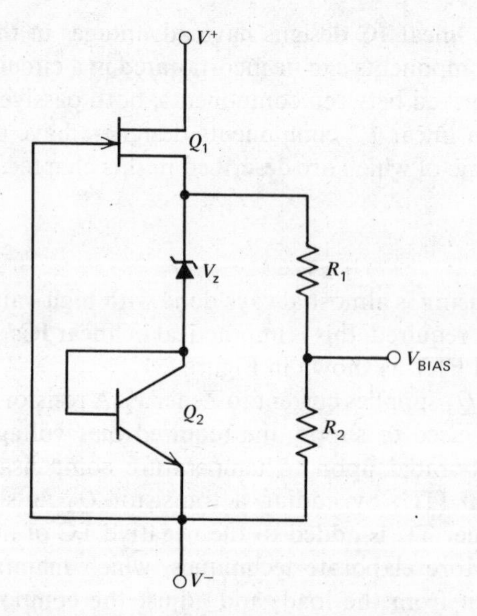

Figure 2-2. Bias circuit with temperature compensation.

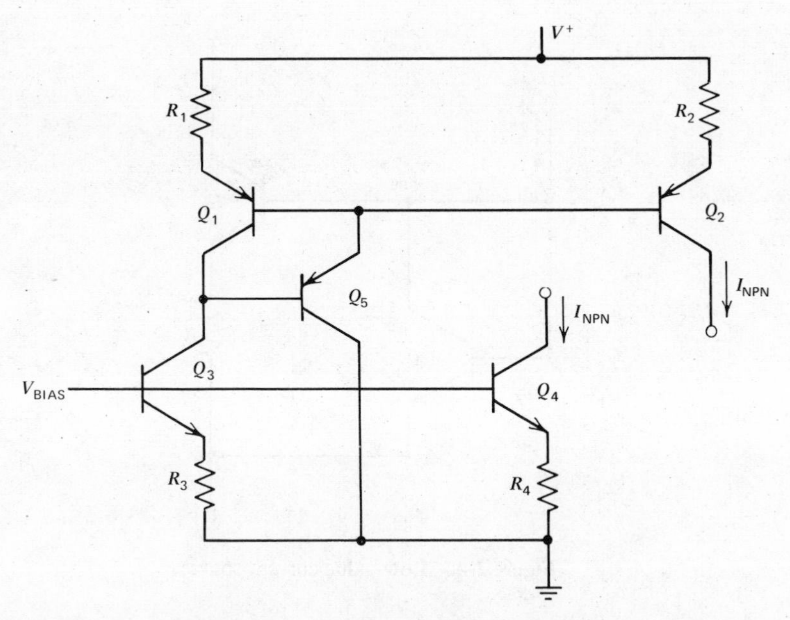

Figure 2-3. *npn* and *pnp* current references.

bias and offset currents and slew rate vary only slightly with operating voltages and temperatures.

The circuits above are valuable for high- and medium-value currents. Low-value currents are supplied by the well-known circuit of Figure 2-4. If the geometries of Q_1 and Q_2 are identical, the following equations hold:

$$V_{BE_1} = V_{BE_2} + I_n R \tag{2-2}$$

$$V_{BE_1} = \frac{kT}{q} \ln \frac{I_b}{I_s} \tag{2-3}$$

$$V_{BE_2} = \frac{kT}{q} \ln \frac{I_n}{I_s} \tag{2-4}$$

$$I_n R = \frac{kT}{q} \ln \frac{I_b}{I_n} \qquad \frac{kT}{q} \approx 2.58 \times 10^{-2} \text{ at } 27^\circ\text{C} \tag{2-5}$$

Where the subscripts refer to their respective transistors, beta is assumed to be large, and the other symbols have their usual meaning.

This is a transcendental equation which can be presented graphically as shown in Figure 2-5. As can be seen, it becomes possible to obtain very low currents with reasonable values of resistance. This circuit is used in the μA709 where I_{BIAS} is set by one resistor and the operating power supplies.

This principle may be extended to provide a particularly useful voltage reference. The circuit in Figure 2-6 illustrates how this might be done. From the same considerations as before the reference voltage can be shown to be

$$V_{REF} = K_2 \frac{kT}{q} \ln K_1 \tag{2-6}$$

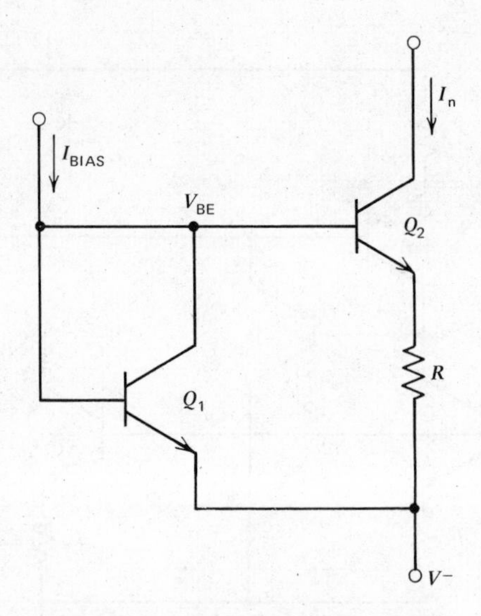

Figure 2-4. Low-value current source.

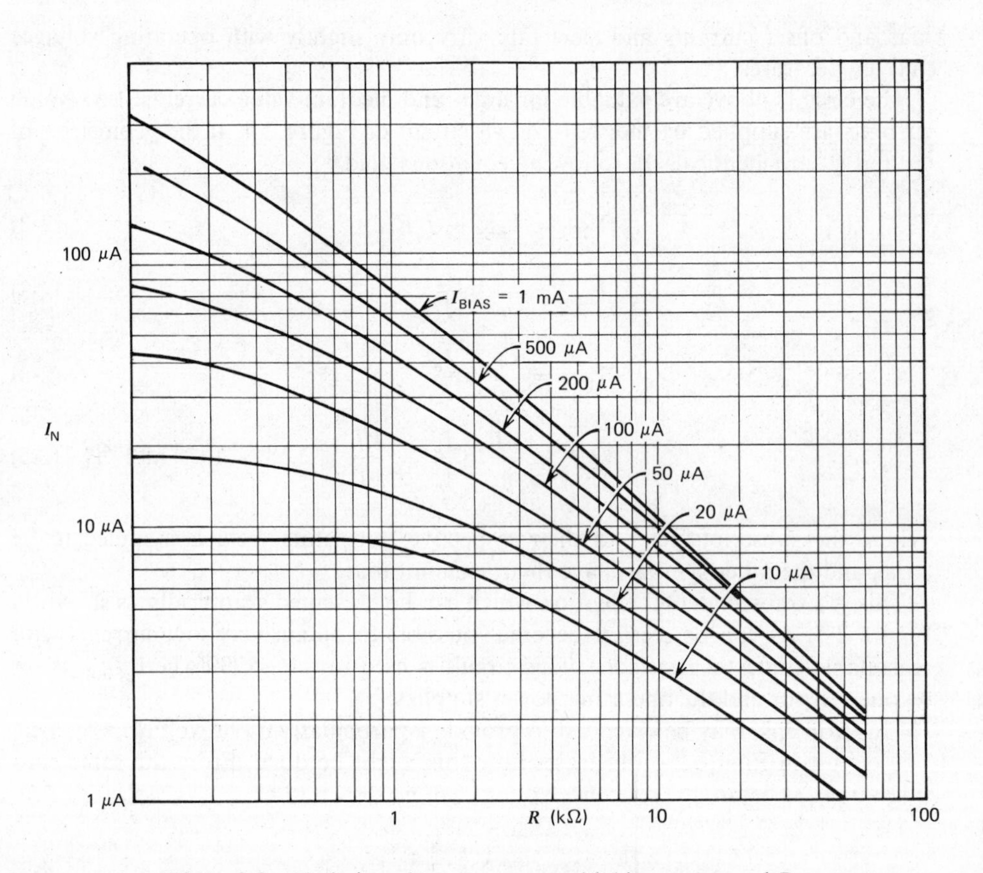

Figure 2-5. Variation of current source with bias current and R.

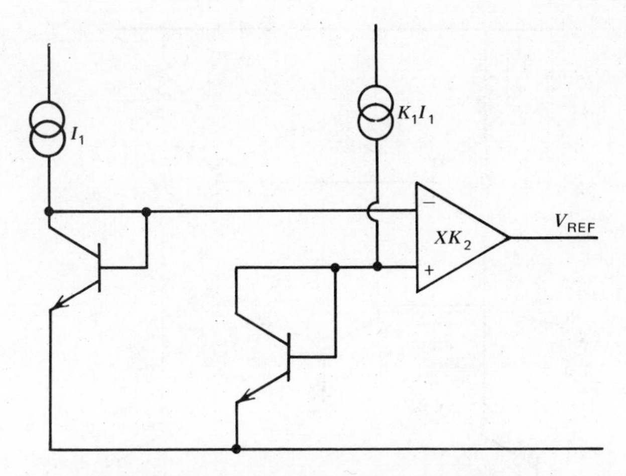

Figure 2-6. Non-Zener voltage reference.

There are a number of interesting and useful properties of this circuit. Compared with the Zener reference, it is much less noisy and can be used at lower-supply voltages. With judicious circuit implementation, the voltage can be controlled to about ±10% and low TCs achieved.

CURRENT-SOURCE LOADS

Using current sources for load resistors is another technique exploited in linear IC designs. The schematics of Figure 2-7 show how this may be done. The circuit shown is the simple differential amplifier. Figure 2-7*a* gives a conventional circuit while Figure 2-7*b* is an IC implementation using *pnp* current sources.

The value of the current source as a load lies in the fact that it is equivalent to a very high-value resistor, but it takes a relatively small die area. This, of course, gives very high-voltage gains. Other less obvious benefits are, first, this gain is very

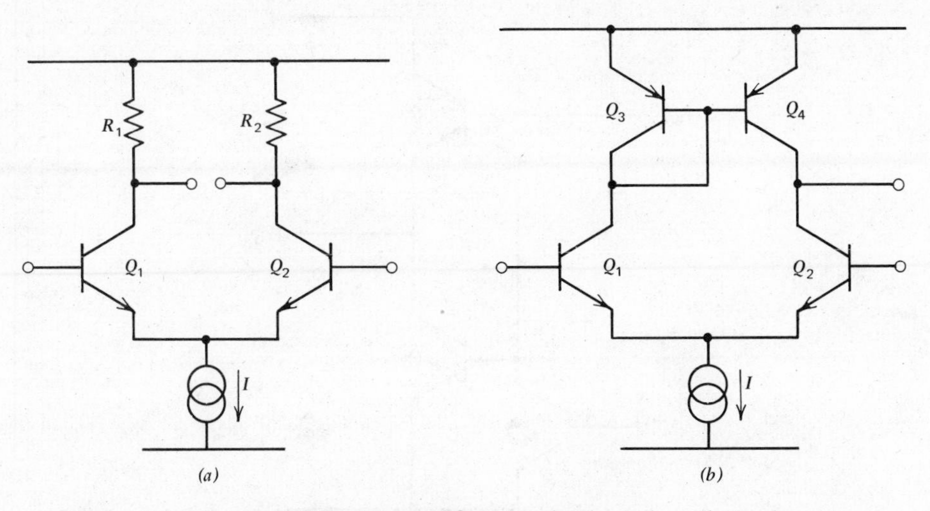

Figure 2-7. Simple differential amplifier: (*a*) simple design, (*b*) load resistors replaced by current-source loads.

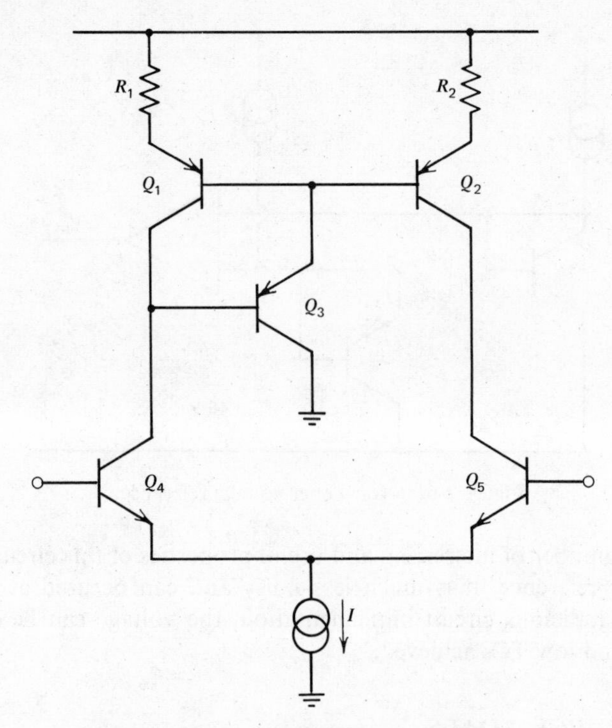

Figure 2-8. Improved differential amplifier.

linear through the output swing, as the output conductance is proportional to the current, and the gain is therefore independent of current; and secondly, a large output swing is available. The active load circuit also has the feature of summing the gain from both output sides. A disadvantage of the circuit in Figure 2-7*b* is that the noise in the *pnps* is summed into the *npn* input noise. Another problem is that the self-biasing scheme used in Figure 2-7*b* can introduce some added offset current if

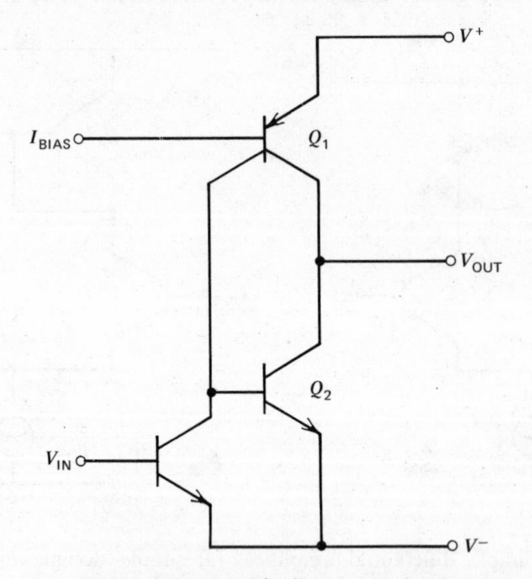

Figure 2-9. Multicollector *pnp* load.

the *pnp* betas are low because then the *pnp*s run at different current levels. A change in the circuit which avoids this problem is shown in Figure 2-8.

The current imbalance between Q_1 and Q_2 is now only

$$I_{B_3} = \frac{2I_{C_1}}{\beta_1 \beta_3} \tag{2-7}$$

where I_{B_3} is Q_3 base current, I_{C_1} is Q_1 collector current, and β_1 and β_3 are betas of Q_1 and Q_3, respectively. This imbalance is now negligible. Resistors have also been added to the *pnp* emitters to increase the output impedance and, therefore, the gain.

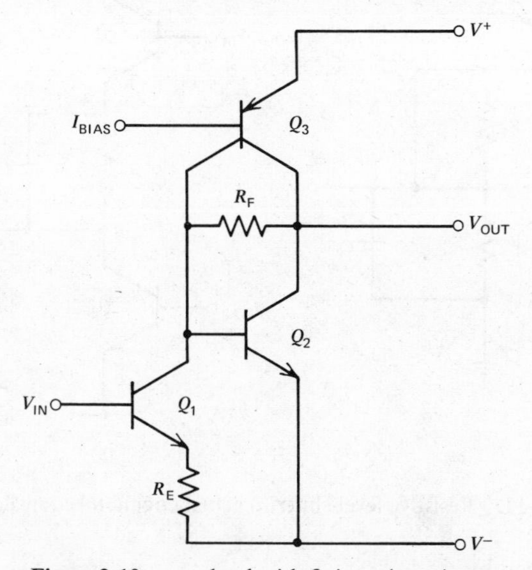

Figure 2-10. *pnp* load with finite gain resistors.

Use of the *pnp* as a load can be extended by using multiple collector *pnp*s as indicated in Figure 2-9. This circuit could give a voltage gain of many millions, which for linear amplification would be generally impractical. A practical realization would incorporate feedback to define the gain, as shown in Figure 2-10, where the voltage gain is given by the ratio

$$A_V \simeq \frac{R_F}{R_E} \tag{2-8}$$

A further practical modification to the circuit of Figure 2-10 is to incorporate a method of overcoming the restricted V_{CE} of Q_1. This illustrates a general problem often encountered in linear IC design—level-shifting.

LEVEL-SHIFTING

The need for level-shifting comes from two general requirements: (1) level-interfacing requirements at the input and output of a circuit, and (2) an internal circuit need to maintain a voltage across transistor collector-bases to avoid clipping. The latter situation can be seen in the circuit of Figure 2-10, where the voltage across Q_1 is limited to a V_{CE} minus an IR drop which limits the voltage swing at the output. Discrete design overcomes this problem by a liberal use of coupling and decoupling capacitors which are not available to linear IC designers, unless incorporated externally.

A method of resistive level-shifting is shown in Figure 2-11. The dc voltage is level-shifted down from point A to point B through resistor R_1 by the current in Q_2. The circuits ac performance can be analyzed by the equivalent circuit of Figure 2-11b.

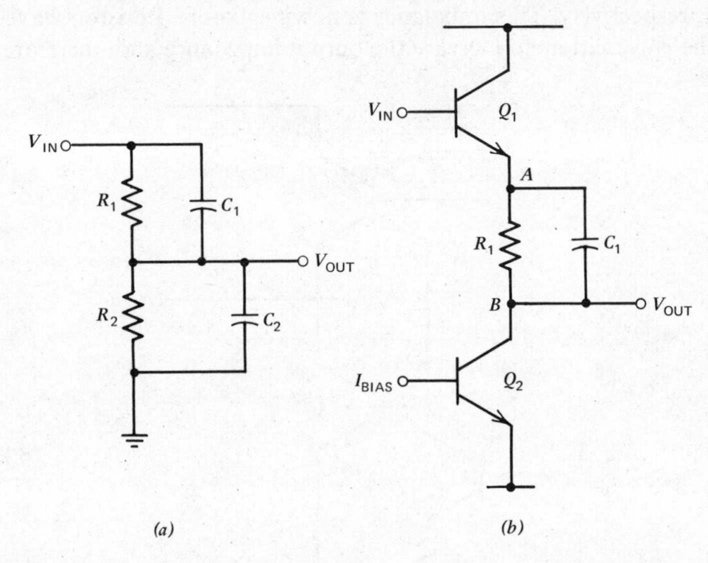

Figure 2-11. Resistive level-shift: (a) actual circuit, (b) equivalent circuit.

Circuit gain is given by

$$A_V = \frac{Z_2}{Z_1 + Z_2} \tag{2-9}$$

where

$$Z_1 = \frac{R_1}{1 + j\omega R_1 C_1} \tag{2-10}$$

$$Z_2 = \frac{R_2}{1 + j\omega R_2 C_2} \tag{2-11}$$

Since maximum gain is desirable, Z_2 should be larger than Z_1. The following conditions must be met for good broad-band operation:

$$R_2 \gg R_1$$

and

$$C_1 \gg C_2$$

then,

$$A_V \approx 1$$

These conditions can generally be met since the output can be fed into an emitter follower with high-input resistance and low-input capacitance. Practical IC values

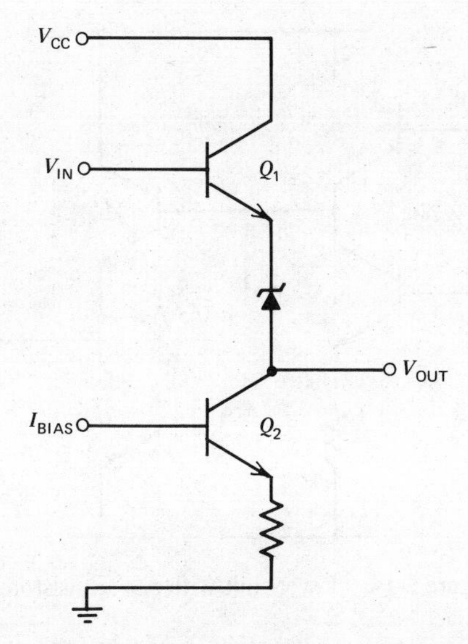

Figure 2-12. Zener diode level-shift circuit.

of R_2 and C_2 would be 5 MΩ and 0.5 pF, respectively. The conditions desired above can be realized, then, with values of R_1 and C_1 10 to 20 kΩ and 15 to 30 pF, respectively. Lower values of R_1 consume excessive current and lower values of C_1 degrade frequency performance which typically might be up to 50 MHz.

Some practical considerations are involved, however, in using this circuit: (1) the values required by R_1 and C_1 require large die sizes; (2) unless the dc bias current through R_1 is made proportional to the supply voltages, the range of voltages over which the circuit will perform is limited, thus creating a supply-voltage incompatibility; and (3) power is consumed without giving any gain.

Another technique of level-shifting used involving a Zener diode is shown in Figure 2-12. The Zener diode commonly used is a transistor-emitter-base junction, giving a voltage drop from 6 to 7 V which is in the range generally required. Multiples

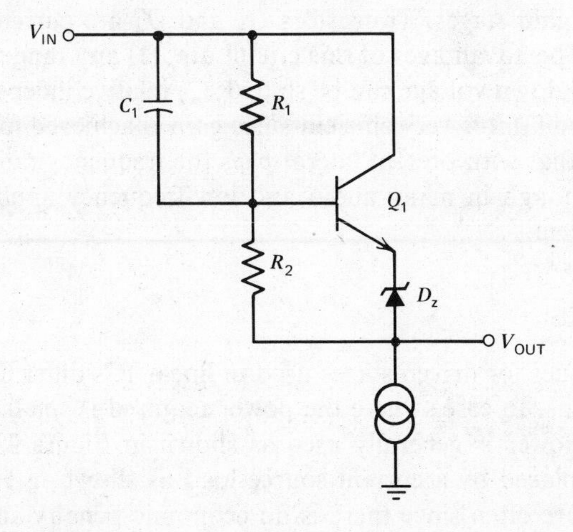

Figure 2-13. Zener-resistive level-shift circuit.

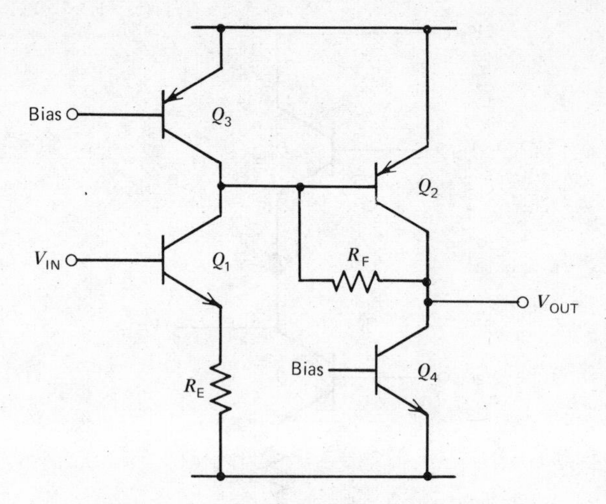

Figure 2-14. Level-shift with *pnp* transistors.

of this value can be obtained by cascading more diodes. Besides the virtue of convenience, this scheme also has the benefit of speed. The inflexibility with respect to power-supply voltages and high noise restrict the use of this method to switching circuits, such as comparators and sense amplifiers.

A useful circuit which uses both techniques discussed above is shown in Figure 2-13. Here a Zener diode is used with a resistive multiplier. The voltage level-shift between input and output is equal to

$$\left(1 + \frac{R_1}{R_2}\right)(V_z + V_{BE})$$

If extended frequency range is needed, capacitor C_1 can be added. With this circuit, intermediate values of level-shift are possible. The most useful level-shift technique involves the use of *pnp* transistors. Implementation of this scheme is illustrated in Figure 2-14.

Here, a similar circuit to Figure 2-10 with *npn* gain stages now has alternate *npn* (Q_1) and *pnp* (Q_2) gain stages. Transistors Q_3 and Q_4 are current-source loads as described earlier. The advantages of this circuit are: (1) any range of voltage within the transistor breakdown voltage can be shifted; (2) relative independence of power-supply changes; and (3) it is very efficient since gain is achieved for the power used. A disadvantage is that with present lateral *pnp*s the frequency range is restricted to below 1 MHz, although in many audio and low-frequency applications this does not present a problem.

OUTPUT STAGES

The design techniques for driven stages used in linear ICs differ little from those of conventional designs. In cases where the power required is small, the conventional class A emitter-follower is generally used as shown in Figure 2-15*a*. The resistor load R_E can be replaced by a current-source load as shown in Figure 2-15*b*. This method is used more often since there is no economic penalty and it can be more easily biased.

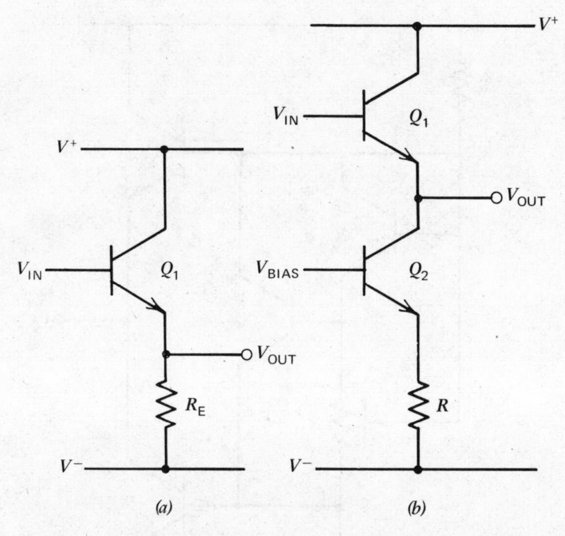

Figure 2-15. Class A emitter-follower output stage (*a*) with a conventional resistor load and (*b*) with a current-source load.

For more efficient output stages with low impedances in both source and sinking current-flow directions, the complementary *npn*/vertical *pnp* design is used, as shown in Figure 2-16. Here Q_1 drives the output transistors Q_2 and Q_3. Diodes D_1 and D_2 bias the output transistors so that some quiescent current is flowing in Q_2 and Q_3. Difficulty is encountered in this scheme in controlling the quiescent current. The circuit of Figure 2-17 is much better from this standpoint since the current through Q_4 and, therefore the voltage across Q_4 and Q_5, can be controlled fairly well. By adjusting the value of R_2 the current flowing through Q_2 and Q_3 is likewise controlled. A further advantage of this scheme is that Q_4, Q_5, and R_2 can be placed in the same isolation tub.

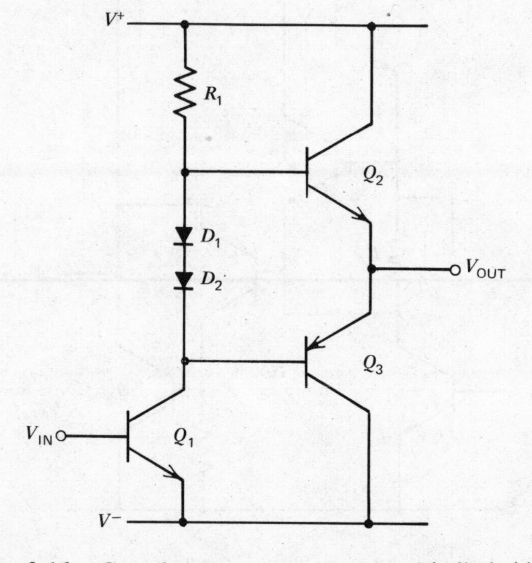

Figure 2-16. Complementary output stage with diode-biasing.

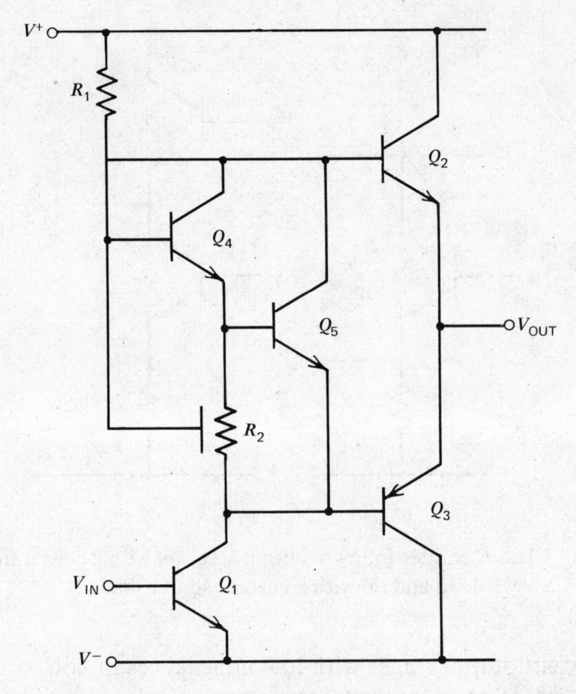

Figure 2-17. Complementary output stage with transistor-biasing.

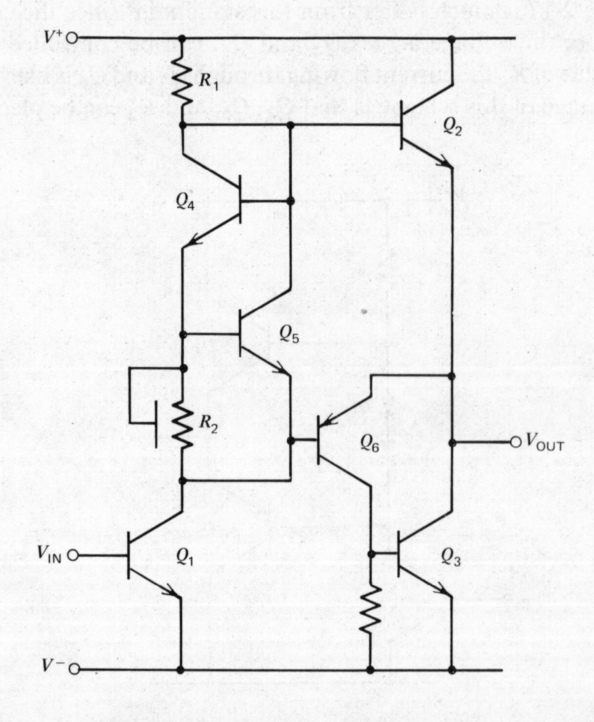

Figure 2-18. Lateral *pnp* output stage.

An alternative to the use of a vertical *pnp* is the compound *npn*/*pnp* circuit which uses a lateral *pnp* transistor as shown in Figure 2-18. A potential problem with this circuit is the local instability of the Q_6/Q_3 loop. Also, the lateral *pnp* Q_6 has poor phase and frequency response.

Protection is commonly added to the output stage so that the maximum current that can be drawn from the output is limited. A circuit that achieves this is shown in Figure 2-19. Looking at Q_1, Q_3, and R_1 one can see that as a large amount of current is drawn through Q_3 into the output, the voltage across R_1 increases until Q_1 is turned ON and removes some of base current driving into Q_3. This prevents the

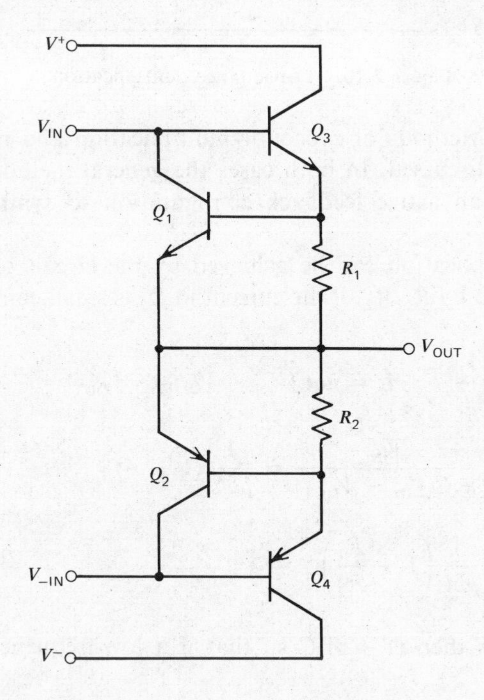

Figure 2-19. Complementary output with short-circuit protection.

current through Q_3 from increasing. The same scheme holds true for the negative direction with Q_2, Q_4, and R_2. However, the amount of base drive to Q_4 is high and, therefore, the amount of current Q_2 must carry is high. Poor current-handling capability and high base-emitter series resistances could make this circuit inoperative or at best make the maximum current through Q_4 indeterminate. If this is the case, the collector of Q_2 is usually taken to some earlier point in the circuit where it can turn the circuit OFF without handling large currents. Variations of this latter circuit are used in almost all of the newer operational amplifier designs.

ACTIVE COMPONENT SIMULATION

Component limitations of linear IC design have been previously discussed. Among these limitations were restrictions in the range of resistor and capacitor values and the nonexistence of inductors. Techniques dealing with resistor-value limitations

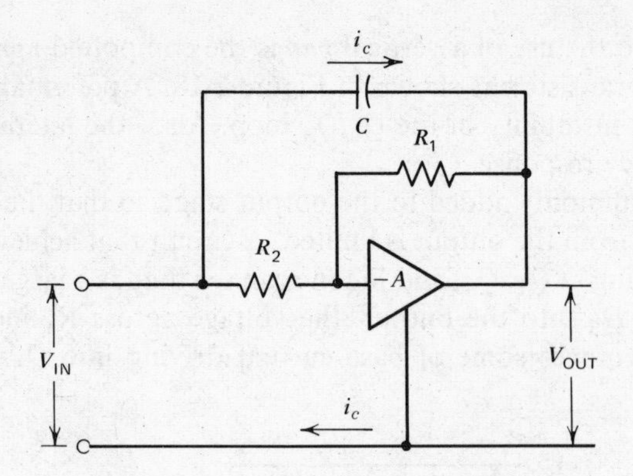

Figure 2-20. Capacitance multiplication.

have been covered. Methods of capacitor multiplication and inductor simulation will now be briefly discussed. In both cases the general method is to incorporate oxide capacitors in an active feedback configuration to synthesize the required impedance.

Capacitance multiplication can be achieved by the circuit of Figure 2-20. The amplifier gain is given by R_1/R_2. If the current in R_2 is small compared to i_c,

$$Z_{IN} = \frac{V_{IN}}{i_c} \qquad i_c = j\omega C(V_{IN} - V_{OUT}), \quad V_{OUT} = -AV_{IN} \tag{2-12}$$

$$= \frac{V_{IN}}{j\omega C(V_{IN} - V_{OUT})} = \frac{1}{j\omega C}(1 + A) \tag{2-13}$$

or

$$= \frac{1}{j\omega C}\left(1 + \frac{R_1}{R_2}\right) \tag{2-14}$$

The capacitance is then $(1 + A)C$ so that if a low-frequency time constant is

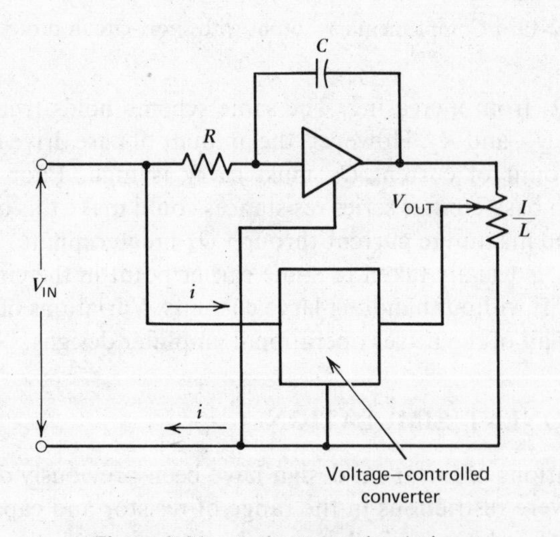

Figure 2-21. Inductance simulation.

required to roll off a high-resistance point R the resultant time constant is given by

$$T = RC(1 + A) \tag{2-15}$$

Inductance simulation is a more complex problem. Its most common need is for filter synthesis. A method of achieving inductance simulation is shown in Figure 2-21. This method is similar to the technique used on analog computers. It solves the general equation for inductors

$$i = \frac{i}{L}\int V\,dt; \qquad V_{OUT} = \int V\,dt \tag{2-16}$$

The purpose of these first two chapters has been to familiarize the reader with IC processes and design techniques. The information presented should be helpful to understand the specific circuits of various linear circuit devices and aid in understanding the limitations of applications which follow in the next few chapters.

Chapter 3

The Operational Amplifier[1]

The basic linear IC building block is a broad-band, high-gain, differential-input amplifier, called the operational amplifier (op amp). The designation *operational amplifier* was originally adopted for a series of high-performance dc amplifiers used in analog computers. These amplifiers performed mathematical operations applicable to analog computation (summation, scaling, subtraction, integration, etc.). Today the availability of inexpensive IC amplifiers has made the packaged op amp useful as a replacement for any low-frequency amplifier.

Operational amplifiers are often the easiest and best way of performing a wide range of linear functions from simple amplification to complex analog computation. Some of the advantages of monolithic op amps over discrete amplifiers are: Low cost, much higher complexities, improved performance, and temperature stabilization of the amplifier chip.

INTEGRATED CIRCUIT OPERATIONAL AMPLIFIERS

A monolithic op amp is a very high-gain dc amplifier, usually with inverting and noninverting inputs and external feedback; it may simply amplify an input signal, or it may operate on a current or voltage in mathematical form. Although it may have one or more inputs, ordinarily there is only a single output, which is internally compensated for both temperature drift and voltage stability. Capacitor-phase compensation is usually applied externally (although some op amps have it built in), as are input, feedback, and any offset-voltage resistors used to establish input impedance, to control loop gain, and to minimize offset voltage that occurs between the two op amp inputs.

The great versatility and many advantages of op amps stems from the use of negative feedback. Negative feedback tends to improve gain stability, reduce output impedance, improve linearity, and in some configurations increase input impedance.

[1] Portions of this chapter were excerpted from E. R. Hnatek, *A User's Handbook of Integrated Circuits*, Wiley, 1973.

Another valuable property of negative feedback, which is the basis for all op amp technology, is that with enough gain, the closed-loop amplifier characteristics become a function of only the feedback components.

Generally, IC op amps use several differential stages in cascade to provide both common-mode rejection and high gain. Therefore, they require both positive- and negative-power supplies. Those ICs which do not require dual-power supplies are rarely (if ever) used as op amps. (Without a differential amplifier, there would be no common-mode rejection.) Most op amp ICs require equal power supplies (such as +15 and −15 V). However, equal power supplies are not required for all ICs.

Unlike most transistor circuits, where it is usual to label one power-supply lead positive and the other negative without specifying which (if either) is common to ground, all IC power-supply voltages should be referenced to ground.

THE IDEAL OPERATIONAL AMPLIFIER

Figure 3-1 is a block diagram of a typical op amp. The ideal op amp offers characteristics approaching the following limits:

1. Infinite voltage amplification A
2. Infinite input impedance Z_{IN}
3. Zero output (source) impedance Z_{OUT}
4. Infinite bandwidth
5. Linear output swing above and below 0 V
6. Symmetrical limiting of output voltage to specified "saturation" levels without damage
7. $V_{OUT} = -AV_{IN}$
8. $V_{OUT} = 0$, when $V_{IN} = 0$
9. Instantaneous recovery from saturation
10. Differential inputs

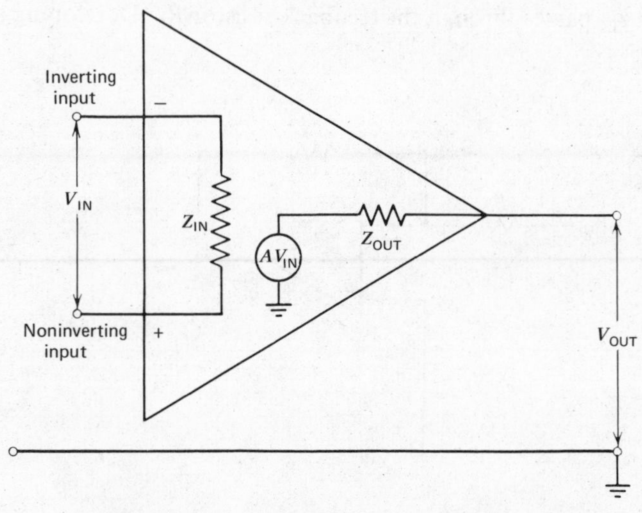

Figure 3-1. Ideal operational amplifier.

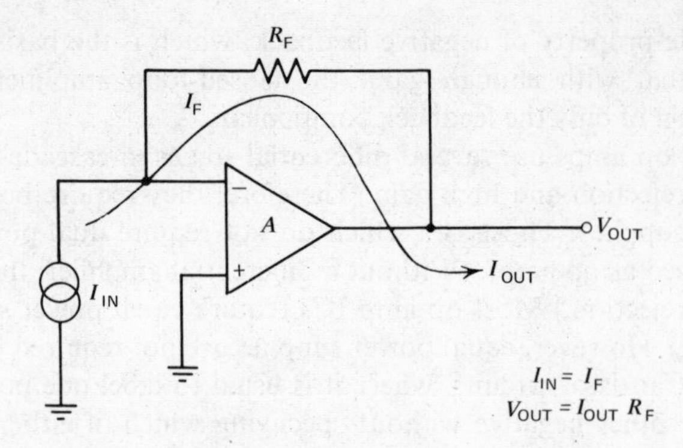

Figure 3-2. Basic current amplifier circuit.

With the ideal op amp, then, the gain A is infinite, the input impedance Z_{IN} is infinite, and the output impedance Z_{OUT} is so low that it approaches zero. This means that almost any signal can turn it on; there is no loading of the preceding stages, and the amplifier itself could drive an infinite number of other devices. In practice, however, Z_{IN} is never infinity and the gain is always finite. Thus there exists a given input impedance and a gain, through a feedback loop, which is usually controlled by a simple resistance. Closed-loop gain depends on the ratio of feedback resistance to input resistance; but loop gain—the product of A (gain) and B (feedback attenuation)—is the ratio between open-loop and closed-loop gain. Loop gain decreases with an increase in frequency and with high values of closed-loop gain. Also, the bandwidth of the amplifier decreases as the closed-loop gain is increased.

Figure 3-2 is the schematic diagram of a simple current amplifier. In this circuit the output voltage is equal to the product of the output current and the feedback-resistor value, whereas the output current is equal to the ratio of the output voltage to the feedback-resistor value. The input and feedback currents are equal.

Since the open-loop input impedance Z_{IN} of Figure 3-2 is virtually infinite, all input current I_{IN} passes through the feedback resistor R_F. Developing an expression

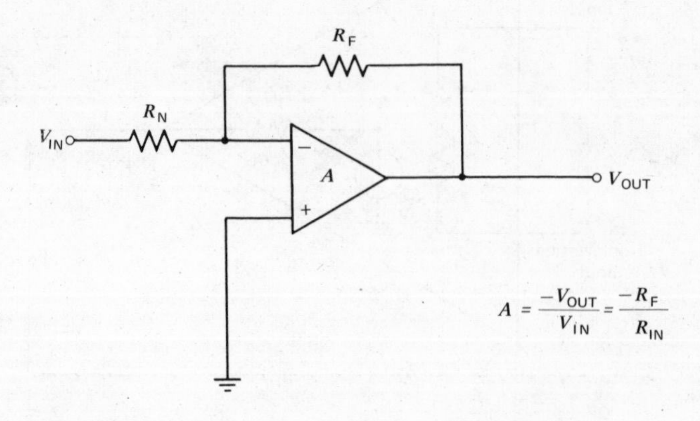

Figure 3-3. Basic amplifier circuit with input resistance added.

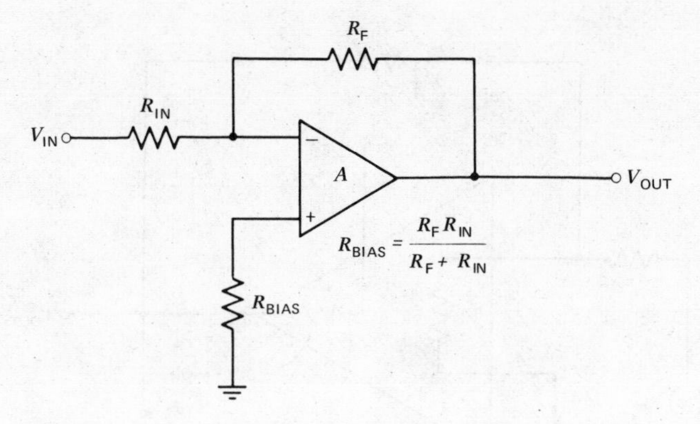

Figure 3-4. Fully developed basic operational amplifier circuit.

for Z_{IN} in terms of R_F, we proceed as follows:

$$Z_{IN} = \frac{V_{IN}}{I_{IN}}$$

$$I_{IN} = \frac{V_{OUT} - V_{IN}}{R_F}$$

$$Z_{IN} = \frac{V_{IN}R_F}{V_{IN} - V_{OUT}}$$

$$= \frac{V_{IN}R_F}{V_{IN} - (-V_{IN}A)}$$

$$= \frac{R_F}{1 + A}$$

If the absolute magnitude of R_F approaches the absolute value of the gain of the device, we no longer have an op amp, and the derived equations no longer hold. For large values of A, Z_{IN} becomes so small that the input is called a virtual ground. This input now becomes a voltage node or null point; therefore, the current from an input source will flow as though the source were returned to ground.

The circuit of Figure 3-3 is similar to that of Figure 3-2 except that an input resistance has been added. This circuit connection facilitates the determination of the gain of the op amp.

Figure 3-4 further develops the circuit of Figure 3-2 by adding a bias resistor R_{BIAS} to minimize the offset voltage at the amplifier's output resulting from input-bias current. The value of the bias resistor is equal to the parallel combination of the feedback resistor and the input resistor. The circuit of Figure 3-4 is the basic op amp circuit used to calculate closed-loop response.

DESIGN CONSIDERATIONS FOR FREQUENCY RESPONSE AND GAIN

Most of the design problems for IC op amps are the result of trade-offs between gain and frequency response (or bandwidth). The open-loop (without feedback) gain and

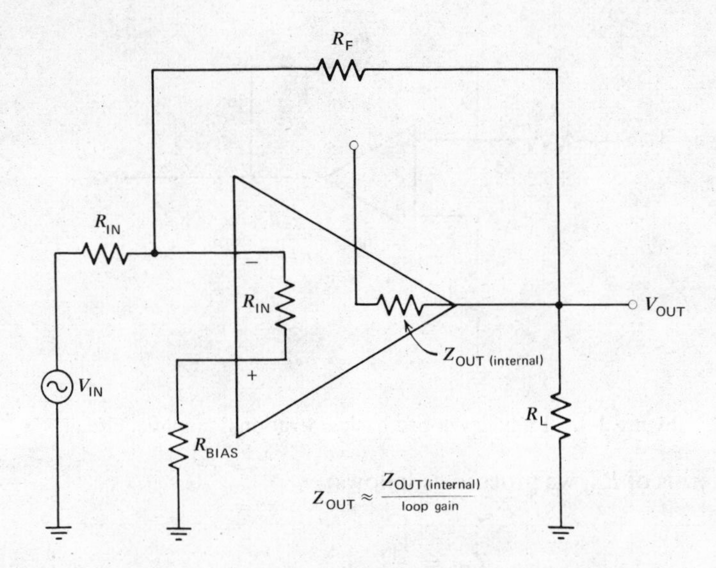

Figure 3-5. Inverting-feedback operational amplifier.

frequency response are characteristics of the basic IC circuit, but these can be modified with external phase compensation networks. The closed-loop (with feedback) gain and frequency response are primarily dependent on external feedback components.

The two basic op amp configurations, inverting feedback and noninverting feedback, appear in Figures 3-5 and 3-6, respectively. Loop gain in these figures is defined as the ratio of open-loop gain to closed-loop gain, as in Figure 3-7. The relationships in Figure 3-7 are based on a theoretical op amp. That is, the open-loop gain rolls off at 6 dB/octave or 20 dB/decade. (The term 6 dB/octave means that the gain drops by 6 dB each time the frequency is doubled. This is the same as a 20-dB drop each time the frequency is increased by a factor of 10.)

If the open-loop gain of an amplifier was as shown in Figure 3-7, any stable closed-loop gain could be produced by the proper selection of feedback components, provided the closed-loop gain was less than the open-loop gain. The only concern

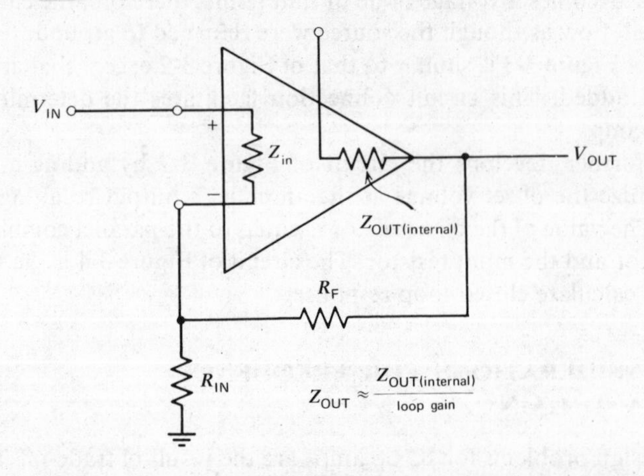

Figure 3-6. Noninverting-feedback operational amplifier.

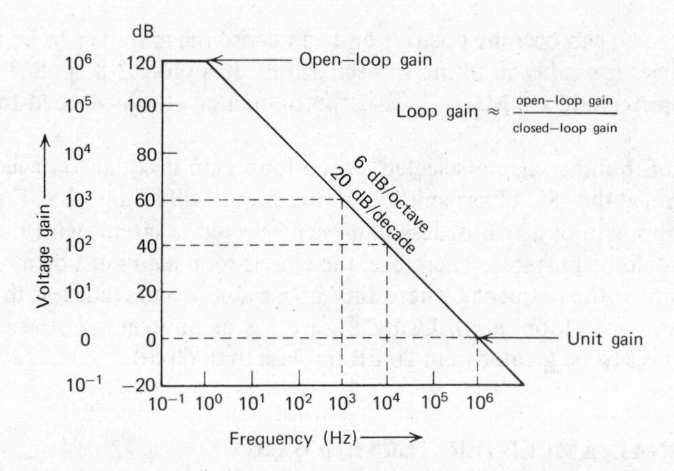

Figure 3-7. Frequency-response curve of a theoretical operational amplifier.

would be a trade-off between gain and frequency response. For example, if a gain of 40 dB (10^2) was desired, a feedback resistance 10^2 times higher than the input resistance would be selected. The gain would then be flat to 10^4 Hz and roll-off at 6 dB/octave to unity gain at 10^6 Hz. If 60-dB (10^3) gain was required, the feedback resistance would be raised to 10^3 times the input resistance. This would reduce the frequency response. Gain would be flat to 10^3 Hz, and then roll-off at 6 dB/octave to unity gain.

The open-loop frequency response curve of a practical amplifier more closely resembles that of Figure 3-8. Here gain is flat at 60 dB to about 200 kHz, then rolls off at 6 dB/octave to 2 MHz. Beyond that point, roll-off continues at 12 dB/octave (40 dB/decade) to 20 MHz, then rolls off at 18 dB/octave (60 dB/decade).

In itself, the sharp roll-off at high frequencies would not be a problem in op amp design (unless the circuit were obliged to operate at a frequency very near the high end). However, the phase response (phase-shift between input and output) changes with frequency. The phase response of Figure 3-8 indicates that a negative feedback

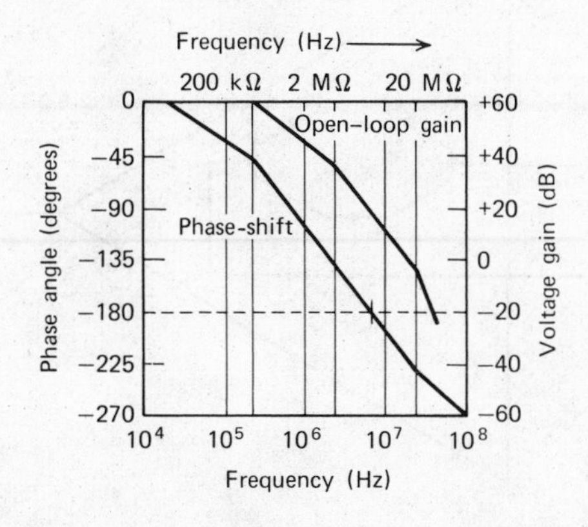

Figure 3-8. Frequency response and phase-shift curve of a practical operational amplifier.

(at low frequency) can become positive and can cause the amplifier to be unstable at high frequencies (possibly resulting in oscillation). In Figure 3-8, a 180° phase shift occurs at approximately 4 MHz. This is the frequency at which open-loop gain is about 20 dB.

As a rule of thumb, when a selected closed-loop gain is equal to or less than the open-loop gain at the 180° phase-shift point, the circuit will be unstable. For example, if a closed-loop gain of 20 dB or less had been selected, a circuit with the curves of Figure 3-8 would be unstable. Therefore, the closed-loop gain must be more than the open-loop gain at the frequency where 180° phase-shift occurs (but less than the flat, low-frequency, open-loop gain). Using Figure 3-8 as an example, the closed-loop gain would have to be greater than 20 dB but less than 60 dB.

OPERATIONAL AMPLIFIER TERMINOLOGY

In order to better understand op amps and to analyze their performance, we must assimilate the associated terminology. The terms presented here are most often used in op amp specifications as measures of amplifier performance.

Input-bias current (I_{BIAS}) is the average of the currents (I_{B_1}, I_{B_2}) flowing into the input terminals when the output is at zero voltage. This is represented in Figure 3-9 and in the equation

$$I_{BIAS} = \frac{I_{B_1} + I_{B_2}}{2}$$

Input-offset current (I_{OS}) is the difference in the currents flowing into the input terminals when the output is at zero voltage. This is expressed as

$$I_{OS} = I_{B_1} - I_{B_2}$$

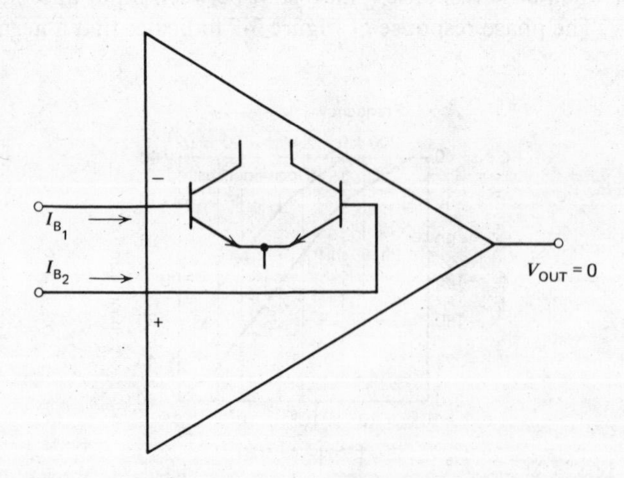

Figure 3-9. An operational amplifier circuit depicting input bias current.

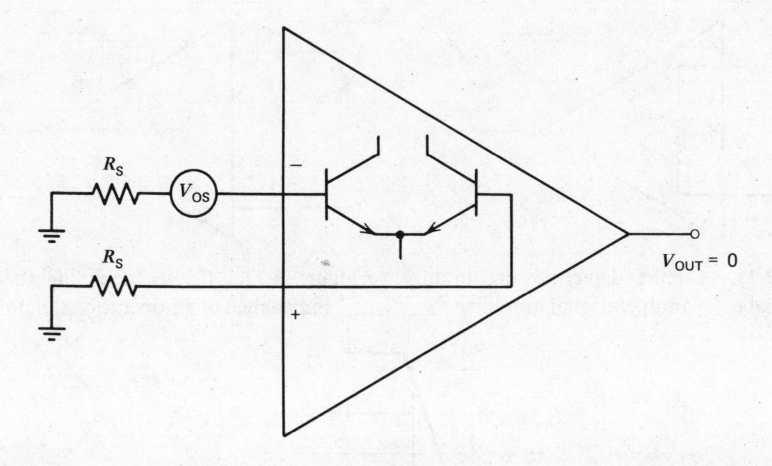

Figure 3-10. In this operational amplifier circuit, showing input offset voltage, $R_S = 10\ k\Omega$ is an industry-defined source resistance.

Typical values of the input-bias and input-offset currents are

$$I_{BIAS} = 0.8 \text{ to } 300 \text{ nA}$$

$$I_{OS} = 0.05 \text{ to } 100 \text{ nA}$$

In an IC op amp, the input devices will not be perfectly matched. Therefore, a delta or difference voltage usually exists between the input terminals.

Input-offset voltage (V_{OS}) is the voltage that must be applied between the input terminals through two equal resistances to force the output voltage to zero (Figure 3-10). Typical values of input-offset voltage are

$$V_{OS} = 0.3 \text{ to } 7.5 \text{ mV}$$

Operational amplifier input impedance (Z_{IN}) is defined as the ratio of the change in input voltage to the change in input current on either input with the other grounded (Figure 3-11). Typical values of input impedance are

$$Z_{IN} = 150 \text{ k}\Omega \text{ to } 30 \text{ M}\Omega$$

Operational amplifier output impedance (Z_{OUT}) is the ratio of the change in output voltage to the change in output current with the output voltage near zero (Figure 3-12). Typical values of output impedance are

$$Z_{OUT} = 0.75 \text{ to } 1000\ \Omega$$

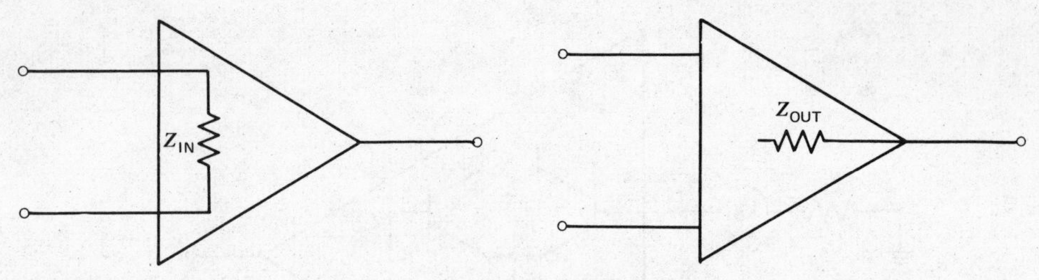

Figure 3-11. Circuit depicting the input impedance of an operational amplifier.

Figure 3-12. Circuit depicting the output impedance of an operational amplifier.

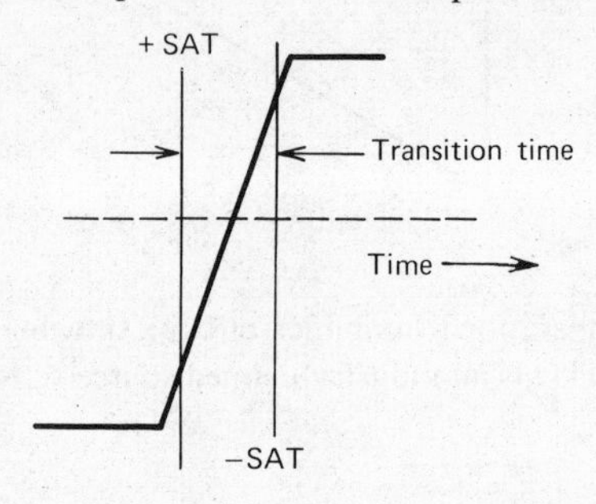

Figure 3-13. Graphical representation of an operational amplifier's transition time.

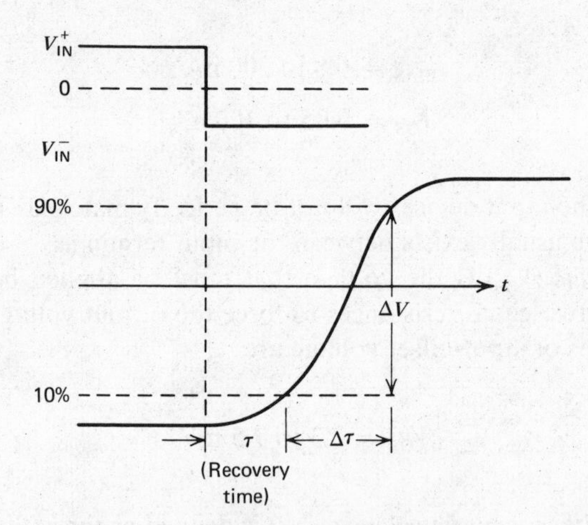

Figure 3-14. A large-amplitude signal drives the operational amplifier from plus to minus saturation.

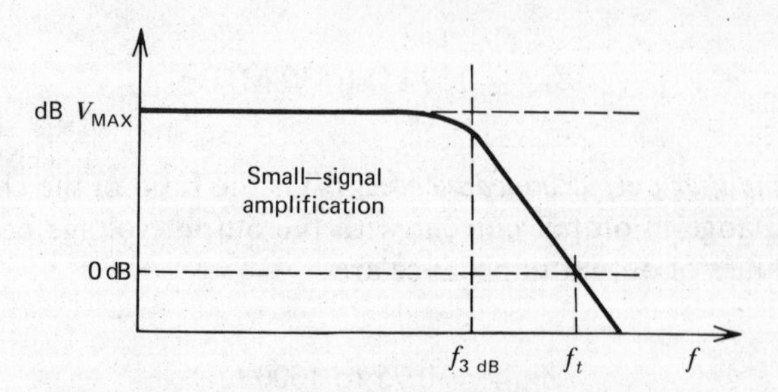

Figure 3-15. Frequency response.

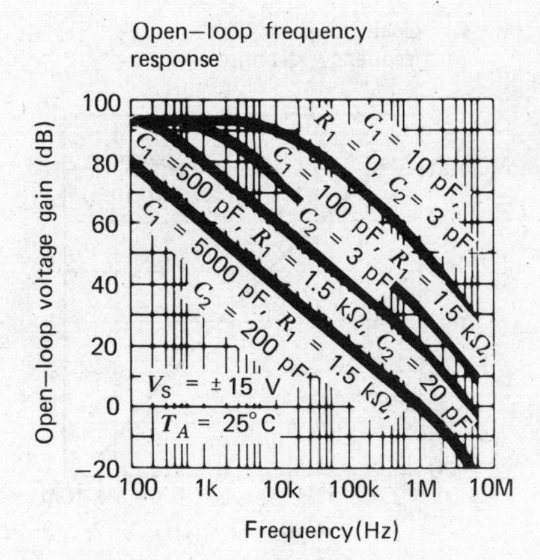

Figure 3-16. Frequency response of a 709 externally compensated operational amplifier.

The common-mode rejection ratio (CMRR) is defined as the ratio of the input-voltage range to the peak-to-peak change in input-offset voltage over this range. Typical values of CMRR are 70 to 120 dB.

Transition Time

Transition time is defined as the time required for a signal to pass through the 10 and 99% points between two specified levels. Worst-case transition time for an amplifier is the time required for the amplifier output to travel between the 10 and 90% points of the swing between positive- and negative-saturation limits (Figure 3-13).

Slew Rate

As can be seen from Figures 3-13 and 3-14, transition time is a function of the voltage swing. A better indication of circuit speed is the rate of voltage change. The slope dv/dt is called the slew rate. Slew rate is defined as the internally limited rate of

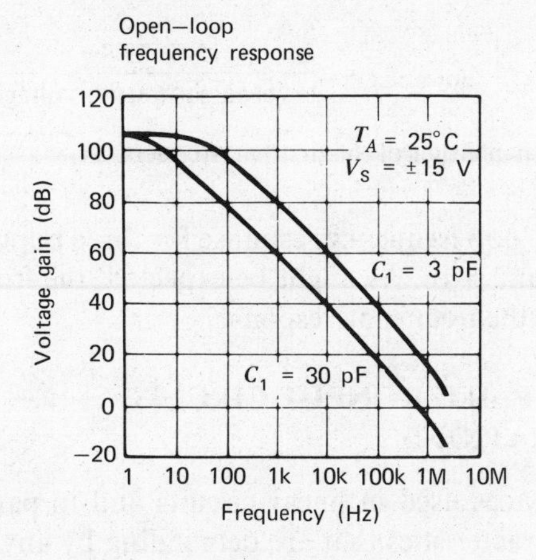

Figure 3-17. Frequency response of a 101/101A externally compensated operational amplifier.

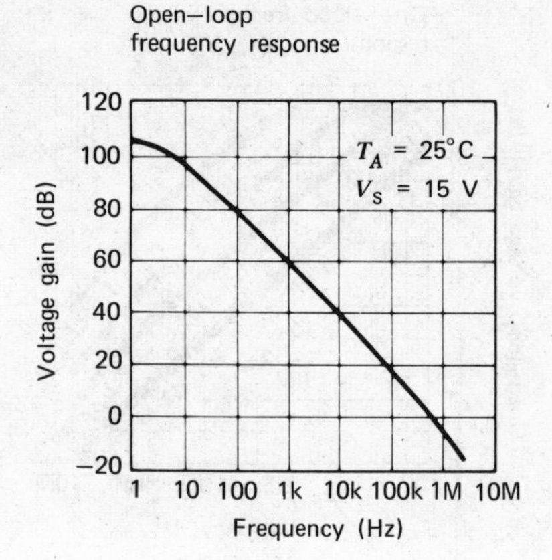

Figure 3-18. Frequency response of a 741 internally compensated operational amplifier.

change in output voltage with a large amplitude step function applied to the input. Typical slew rate values range between 0.1 and 1500 V/μsec.

Frequency Response

The frequency response of an op amp (cf. Figure 3-15) is specified by three frequency measurements:

1. $f_{(3\ \mathrm{dB})}$, the frequency at which the voltage amplification is 3 dB below its maximum value
2. f_t, the frequency at which the amplification factor is unity (A_v dB = 0)
3. $f_{\mathrm{OUT\ MAX}}$, the maximum frequency of full output swing, is not directly related to the curve of Figure 3-15 (frequently omitted from data, it is nevertheless an important parameter)

$$f_{\mathrm{OUT\ MAX}} = \frac{\text{slew rate}}{\pi\ (\text{peak-to-peak saturation voltage})}$$

is also equal to the frequency where we expect slew-rate limiting, or, assuming a symmetrical saturation swing,

$$f_{\mathrm{OUT\ MAX}} = \frac{\text{slew rate}}{2\pi\ (\text{peak saturation voltage})}$$

which is often less than either of the first two frequencies

Typical curves of open-loop frequency response for three popular op amps are given in Figures 3-16, 3-17, and 3-18. As might be expected, the frequency-compensation components determine the frequency response.

OPERATIONAL AMPLIFIER INPUT-CIRCUIT DESIGN CONSIDERATIONS

Monolithic bipolar devices used in linear circuits and in particular, op amps, are required to exhibit characteristics that are demanding by any standard. In order to achieve very low-input bias currents, the input stage transistors need betas on the order of several hundred at microampere current levels. Additionally, the differential amplifier transistor pair has to be closely matched to display minimum offset voltage. This close matching is required of several parameters, especially geometry and beta. As an example, a 1% mismatch in emitter area plus a 2% mismatch in beta will lead to a 1-mV offset voltage. Voltage supply and wide common-mode requirements dictate that the transistor breakdown voltage exceed 40 V for most applications. Other necessary features are low-junction leakages and low noise.

To meet such requirements and still maintain high-production yields, heavy demand is placed on process techniques and control. High betas at low-current levels, good beta-matching, and optimum low-noise performance are all strongly influenced by minimizing recombination centers, both at the silicon–silicon dioxide interface and within the bulk silicon. Most first-generation IC op amps were unsuited for applications involving high-impedance signal sources. The high-input currents of these early devices were just not adaptable as active filters, voltage-followers, integrators, summing amplifiers, sample-and-hold circuits, high-impedance transducer amplifiers, and a few other general-purpose applications.

Three circuit solutions have been advanced to solve the input impedance and offset problem. They are bipolar supergain punch-through transistors, bipolar Darlington pairs and triplers, and the FET.

The effect on slew rate must be considered in making collector current and voltage trade-offs to obtain high-input impedance at low-leakage current. The relation of the R/C slew limitation of input impedance to collector slew rate is as follows:

$$\omega_C = \frac{1}{R_{IN}C_{IN}} = \frac{I_{OUT}}{C_{OUT}V_{OUT}} \tag{3-1}$$

where ω_C is the high-frequency corner, R_{IN} is the effective input resistance, C_{IN} is the effective input capacitance, I_{OUT} is the output current capability, C_{OUT} is the effective output capacitance, and V_{OUT} is the output voltage. Optimized design for bandwidth would require equal weighting of the two right-hand terms. Slew rate is implicitly contained in the relation and can be confirmed by transposing V_{OUT} to the left-hand terms, whence

$$\begin{aligned} \text{slew rate} &= V_{OUT}\omega_C = \frac{V_{OUT}}{R_{IN}C_{IN}} \\ &= \frac{I_{OUT}}{C_{OUT}} \approx \frac{\text{V}}{\text{sec}} \end{aligned} \tag{3-2}$$

For the bipolar transistor, slew rate $= \beta I_B/C_{OUT}$ where β is the transistor current gain, I_B is the base current for the common emitter stage and $C_{IN} = C_{OUT}/\beta$ holds true in the linear range.

Field-effect transistor input stages have long been considered the best way to get low-input currents in an op amp. Low-picoamp input currents can, in fact, be obtained at room temperatures. However, this current, which is the leakage current of the gate junction, doubles every 10°C, so performance is severely degraded at high temperatures. Another disadvantage is the difficulty of matching FETs. Unless expensive selection and trimming techniques are used, typical offset voltages of 50 mV and drifts of 50 μV/°C must be tolerated.

Supergain transistors are now challenging FETs. These devices are standard bipolar transistors which have emitters that have been diffused for extremely high-current gains. Typically, current gains of 5000 can be obtained at 1 μA collector currents which are competitive with FETs. It is also possible to operate these transistors at zero collector-base voltage, eliminating the leakage currents that plague the FET. Therefore they can provide lower error currents at elevated temperatures. Supergain transistors match much better than FETs with typical offset voltages of 1 mV and drifts of 3 μV/°C. Figure 3-19 compares the typical input-offset currents of IC op amps and FET amps.

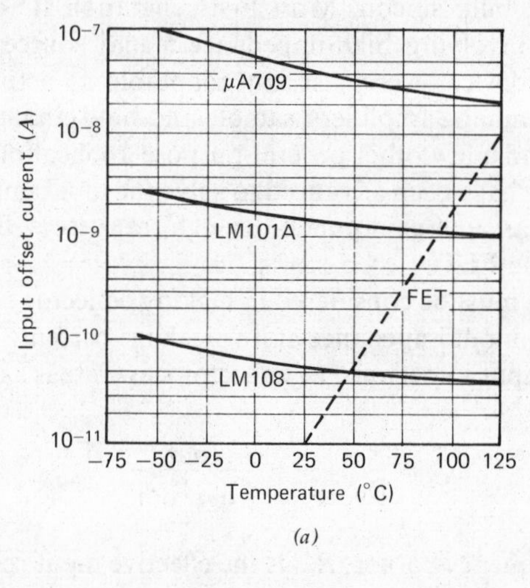

(a)

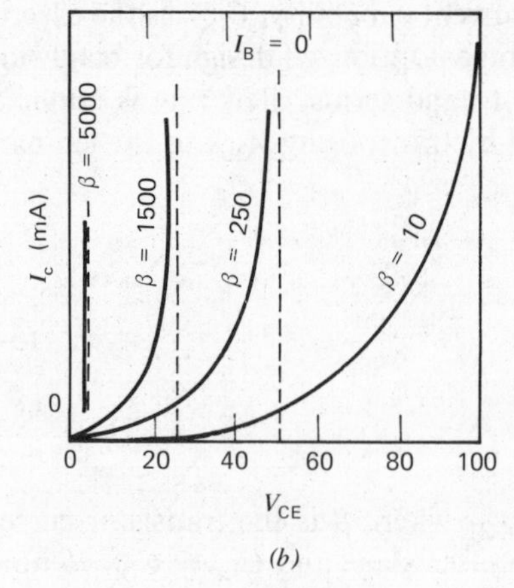

(b)

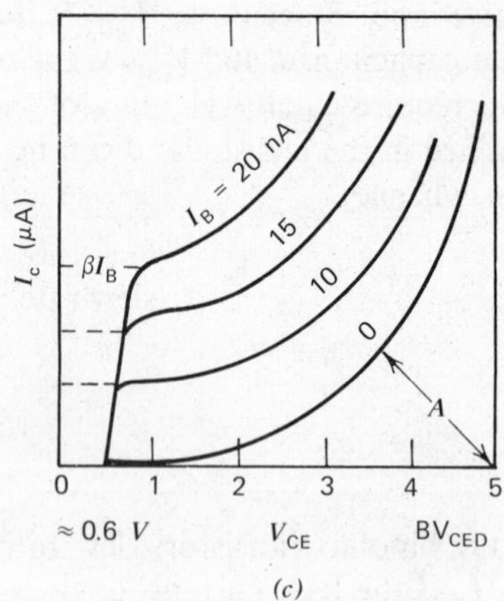

(c)

Figure 3-19. (*a*) Comparison of integrated circuit operational amplifiers; FET input amplifiers are clearly better than the 709s. The LM108 equals typical FET amplifier performance at about 50°C. At 125°C the LM108 is about two orders of magnitude better than FET amplifiers. (*b*, *c*) Collector characteristics of transistors: (*b*) degeneration of breakdown voltage in a double-diffused transistor as it is diffused for higher current gain; (*c*) hypothetical model of a normalized collector characteristic based upon the equilateral hyperbola. Transistor collector characteristics for I_B = 0: (*b*) change of BV_{CEO} (collector-to-emitter breakdown voltage with the base open) with emitter diffusion; (*c*) hypothetical micropower transistor characteristic.

Because supergain input transistors are operated at zero collector-base voltage, high-temperature leakage currents do not show up on the input. Field-effect transistors which in the past have been an obvious choice for the input stage of low input-current op amps, suffer from leakage problems because there is no way to operate them with zero voltage across the gate junction. In applications covering a −55 to 125°C temperature range, supergain transistors, which can give worst-case bias currents of 3 nA and worst-case offset currents of 400 pA, have a distinct advantage over FETs. With existing technology they can equal FETs over a −25 to 85°C temperature range, and it is not difficult to forsee their superiority over a 0 to 70°C temperature range. Furthermore, matched pairs of supergain transistors exhibit typical offset voltages of 0.5 to 1.0 mV with temperature drifts about 2 μV/°C, compared with 40 mV and 50 μV/°C for FETs. Discrete FETs can be compensated or selected for better offset or drift, but at a substantial increase in cost. Metal oxide semiconductor transistors, which do not have leakage problems, are no solution because they exhibit gross instabilities in offset voltage.

Standard bipolar transistors in a Darlington connection have been tried in a number of IC designs to get very low-input currents. First-order calculations indicate that a Darlington transistor connection should be competitive with supergain transistors on input-current specifications. But differential amplifiers, using Darlington-connected transistors, have problems that may not be immediately obvious.

The offset voltage depends not only on the inherent emitter-base voltage match of the transistors but also on the percentage match of current gains. A 10% mismatch in current gains gives a 2.5 mV offset. Within a given process, bias currents drop faster than offset currents as the transistor current gains are raised. Therefore, the better the transistors, the worse the offset voltage.

In addition to being a major contributor to offset voltage, this dependence on current-gain-matching causes other problems. In a simple differential amplifier, the offset-voltage drift can be correlated with offset voltage because of the predictable nature of emitter-base voltage. This is not so with Darlingtons, as bias current-matching is not predictable over temperature. At high temperatures, this effect is aggravated further by leakage currents, so it is impossible to predict performance over a wide temperature range based on room-temperature tests.

Reducing the collector current of double-diffused silicon transistors to very low currents (below 1 μA), as is done in a Darlington connection, does not improve input currents as much as might be expected. Lowering the collector current by a factor of 10 reduces the bias current by about a factor of 7 and the offset current by a factor of 3. These numbers are typical; the results obtained near the edges of a production distribution are significantly worse. In addition, the variation of input currents with temperature goes as the square of the current gain. Since the gain of IC transistors falls off by a factor of 2 to 5 going from 25 down to −55°C, the input currents obtained at the minimum operating temperature are considerably higher than at room temperature.

Other limitations of Darlingtons are that they have higher noise, lower common-mode rejection, and reduced common-mode slew rate. Further, they have one-half the transconductance of a simple differential stage; this doubles the effect of dc offset terms in their output circuitry.

To summarize, Darlingtons can give typical input currents competitive with supergain transistors. However, if the full range of production variables is taken into

Table 3-1 Comparison of operational amplifier categories

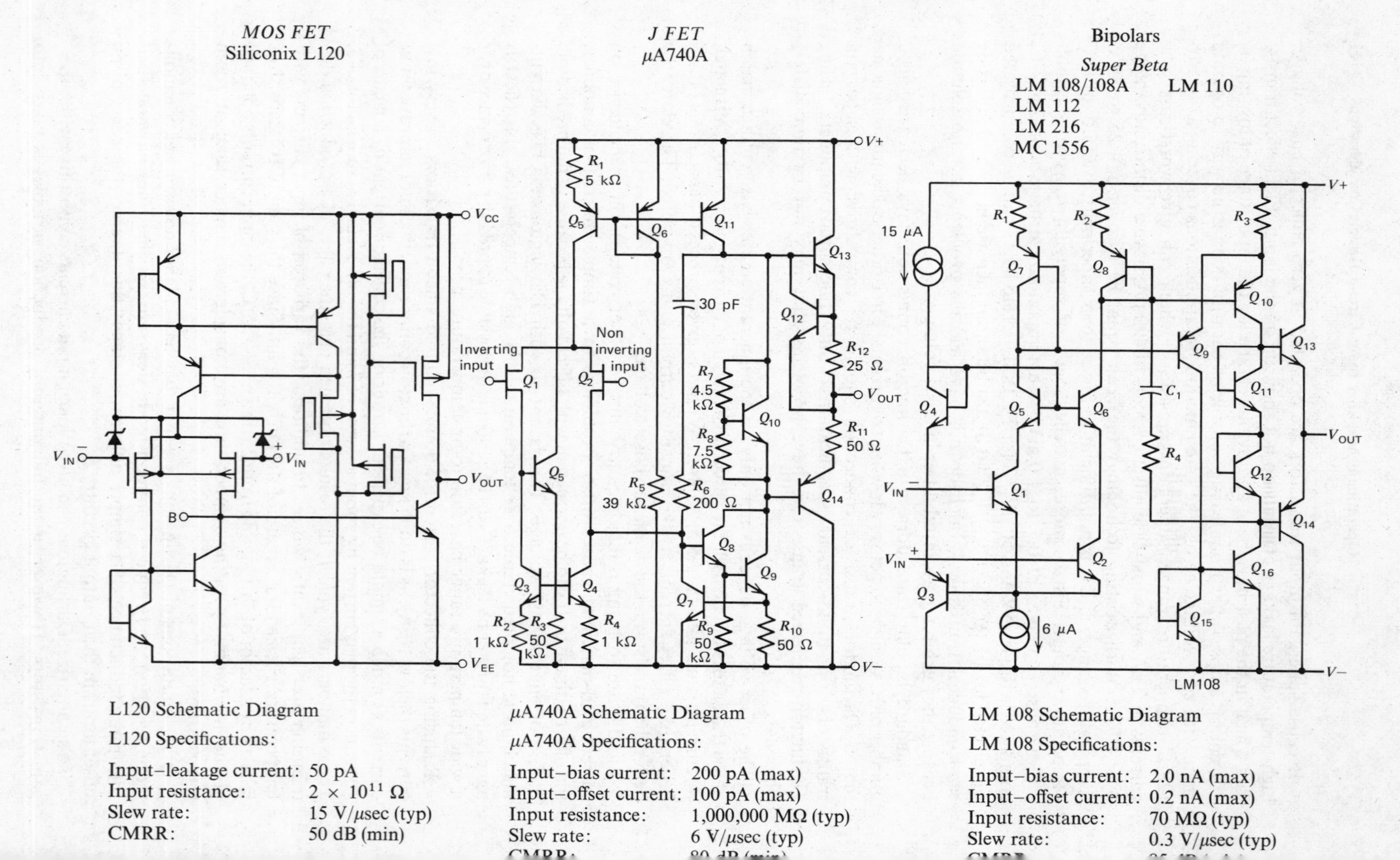

L120 Schematic Diagram

L120 Specifications:

Input–leakage current:	50 pA
Input resistance:	2×10^{11} Ω
Slew rate:	15 V/μsec (typ)
CMRR:	50 dB (min)

μA740A Schematic Diagram

μA740A Specifications:

Input–bias current:	200 pA (max)
Input–offset current:	100 pA (max)
Input resistance:	1,000,000 MΩ (typ)
Slew rate:	6 V/μsec (typ)

LM 108 Schematic Diagram

LM 108 Specifications:

Input–bias current:	2.0 nA (max)
Input–offset current:	0.2 nA (max)
Input resistance:	70 MΩ (typ)
Slew rate:	0.3 V/μsec (typ)

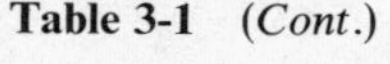

Table 3-1 (*Cont.*)

Bipolars

Darlington	*Regular*	
LM 118	4250	μA747
SE 531	μA709	MC 1558
	μA725	LM 101/101A
	μA741	LM 107

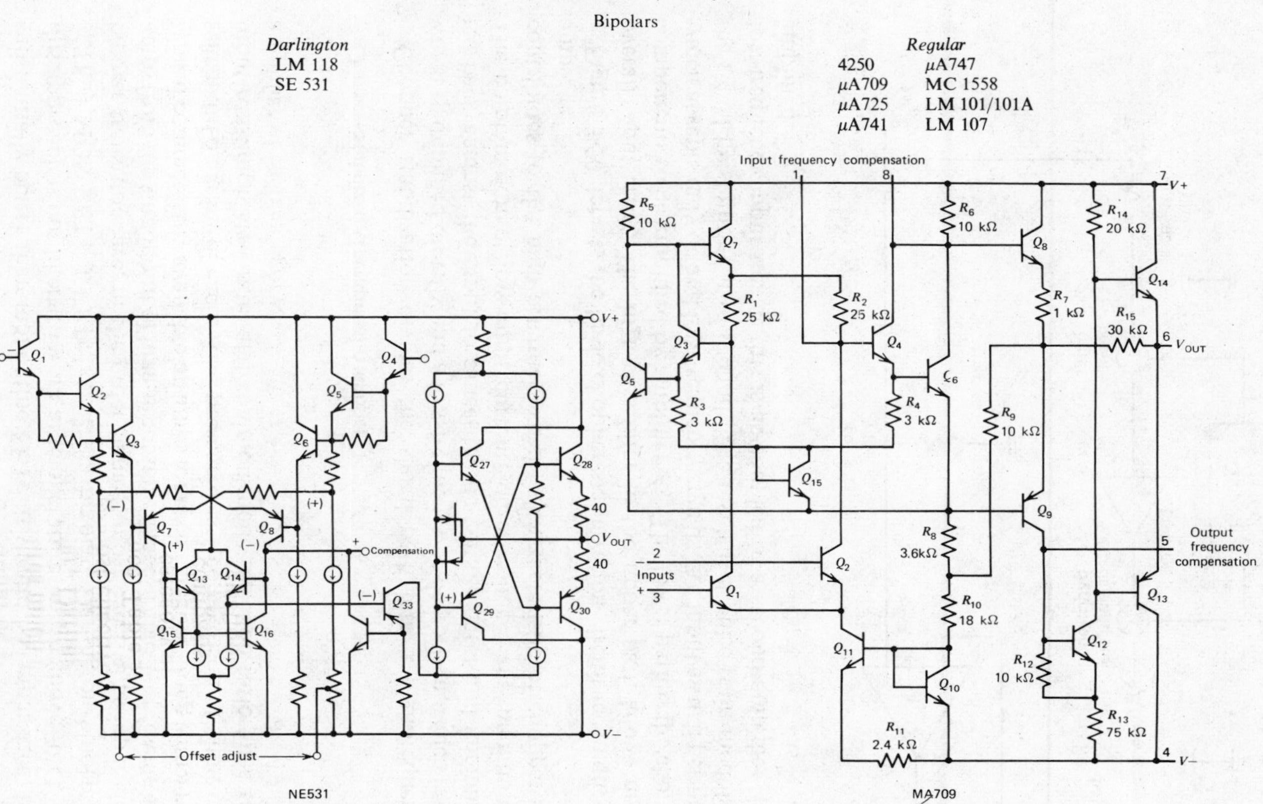

NE531 Schematic Diagram

SE531 Specifications

Input–bias current:	500 nA (max)
Input–offset current:	200 nA (max)
Input resistance:	20 MΩ (typ)
Slew rate:	30–35 V/μs (typ)
CMRR:	70 dB (min)

μA709 Schematic Diagram

μA709 Specifications

Input–bias current:	500 nA (max)
Input–offset current:	200 nA (max)
Input resistance:	400 kΩ (typ)
Slew rate:	0.25 V/μs (typ)
CMRR:	70 dB (min)

[a] All values at 25°C.

account, along with −55 to 125°C operation, the performance is degraded considerably (or the yields reduced) both with respect to offset voltage and input current. Nonetheless, the Darlington connection might become useful to compound supergain transistors with standard transistors. This approach could give input currents less than 50 pA over a 0° to 70°C temperature range, which equals the best FETs. The offset voltage would not be as good as supergain transistors alone, but it would be substantially better than monolithic FETs. It should be emphasized, however, that this approach is not likely to work at temperatures much above 70°C.

Thus MOS FETs, J FETs, superbeta bipolar, Darlington bipolar, and regular bipolar transistors are, in the order named, most effective as input elements in op amps from the standpoint of high-input impedance. Table 3-1 presents a listing of the typical commercially available op amps for each category along with a representative schematic diagram of each. Also presented is a comparison of input-offset current, input resistance, slew rate, and common-mode rejection ratio for a representative amplifier of each category.

Chopper-Stabilized Operational Amplifiers

Chopper stabilization divides the signal path into a high-frequency portion, which is amplified directly and a low-frequency or dc portion, which is chopped and fed back against the input to cancel the dc-voltage offset and minimizes the drift. The combination of low-input current, offset voltage, and minimum offset drift contributes to the high accuracy which is characteristic of chopper-stabilized op amps.

Figure 3-20 depicts the basic chopper-stabilization circuit technique that reduces overall input error. The high-pass filter in the upper path allows high-frequency components to be amplified by amplifier A_1. The lower path is the chopper channel, which accepts only dc and low-frequency signals. An input switch periodically chops the signal by shunting it to ground. The result is a pulse train modulated with the low-frequency input signal; this is ac-coupled to a second amplifier A_2 where it is amplified.

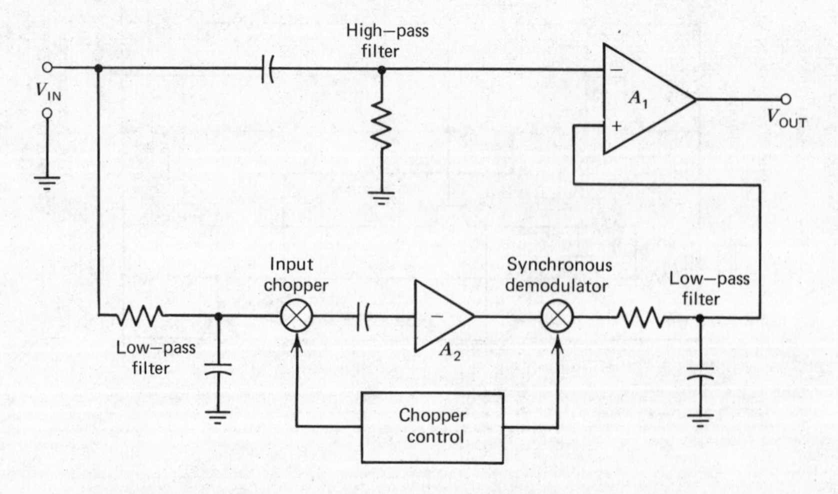

Figure 3-20. Chopper-stabilized operational amplifier block diagram.

To reestablish the proper dc level, the signal is demodulated in synchronism with the input chopper. The demodulated signal has the chopper spikes removed by the output low-pass filter and is summed with the high-frequency components at the noninverting input terminal of A_1. Thus, the low-frequency components are modulated, amplified, demodulated, and summed by the main amplifier. Any dc signals are thus converted to ac and then amplified. Since the dc and low-frequency signals are processed by modulation, both high- and low-frequency amplifiers may be ac-coupled, preventing any dc errors from these amplifiers from appearing at the true input. This in turn means that input current and offset voltage are substantially reduced.

Figure 3-21 serves to further explain the operation of the chopper channel of Figure 3-20. The input to the chopper channel (*a*) consists of a dc and low-frequency component (high-frequency components are rejected by the input low-pass filter). The shorting action of the chopper converts the input signal to that shown in Figure 3-21*b*. The ac signal is capacitively coupled to the inverting amplifier A_2, whose output is shown in Figure 3-21*c*. Since A_2 is ac-coupled, the waveform in Figure 3-21*c* does not have proper ground reference. This dc level is restored as shown in Figure 3-21*d* by a demodulator switch, acting in synchronism with the input chopper. The output low-pass filter removes the high-frequency switching components and passes the dc and low-frequency signals. This amplified inverted version of the input signal, shown in Figure 3-21*e*, is summed at the noninverting terminal of the main amplifier A_1.

Another concept is illustrated in Figure 3-22. This is a circuit diagram for a direct-coupled amplifier, in which the amplifier periodically disconnects itself from the input signal and adjusts its offset voltage to zero. The absence of coupling capacitors provides fast recovery from overdriven conditions—a major problem with traditional chopper-stabilized op amps. With S_1 and S_3 up, the circuit functions as a dc amplifier. When S_1, S_2, and S_3 go down, the amplifier input is grounded and A_2 forces the output of A_1 to ground. Switch S_2 and capacitor C_1 form a sample-and-hold, so that the correction signal for zeroing the offset of A_1 is stored on C_1 and S_2 opens. Here S_3, C_2, and A_5 form a second sample-and-hold; its function is to store the previous output of A_1 while self-zeroing is taking place and so to remove most of the signal discontinuity.

Until 1973, chopper-stabilized amplifiers were available only as modules. Now two companies have introduced chopper-stabilized circuits in monolithic and hybrid form. Texas Instruments came out first with its two-chip circuit, the SN72088/62088, and then Harris Semiconductor announced the monolithic HA2900.

Figure 3-23 depicts the block diagram of the HA2900 where A_1 is the main amplifier, A_2 the auxiliary amplifier, A_3 the sample-and-hold amplifier in the self-zeroing loop of A_2, and A_4 the sample-and-hold amplifier that holds the previous signal during the zeroing interval. The input circuitry of this device is completely symmetrical with respect to the two input lines. This produces a true differential input, in contrast to most stabilized amplifiers which are either inverting or non-inverting devices.

During the period in which A_2 is stabilizing A_1, S_1 and S_4 are closed while S_2 and S_3 are open. The dc and low-frequency components of the input are amplified by A_2 and applied as a correction signal to A_1. The effective input-offset voltage is nearly the same as that of A_2 alone. To keep the offset voltage of A_2 extremely low, it is

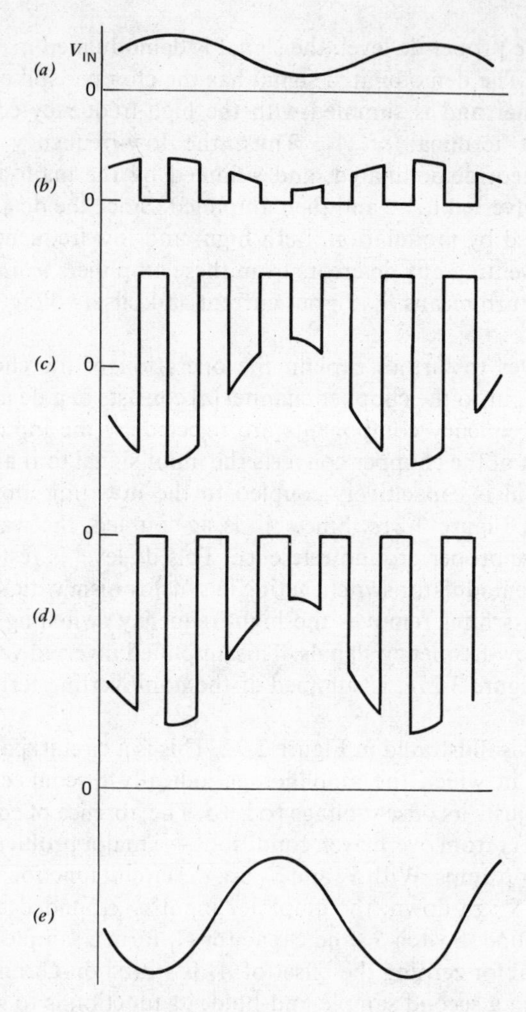

Figure 3-21. The dc and low-frequency input to chopper channel (*a*) is converted to the waveshape (*b*), whose ac signal is capacitively coupled to A_2 (*c*). The dc level is restored in (*d*), while final signal shape (*e*) is summed at the terminal of A_1. (*a*) Input waveform, (*b*) modulated input waveform, (*c*) output of amplifier A_2 prior to demodulation, (*d*) A_2 output after demodulation, (*e*) output after demodulation and low-pass filtering.

periodically zeroed. Switch S_1 opens and S_2 closes, disconnecting A_2 from the input terminals and shorting the inputs of A_2 together—not at ground level, but to a level equal to that of the input common-mode voltage. This results in an extremely high common-mode rejection ratio.

Like most other monolithic op amps, this device does not have a ground terminal; so when S_3 closes, the output of A_2 is forced to equal an internally generated reference voltage, rather than being forced to ground. The same is true of A_4 because it is referenced to the same voltage. The consequence of this is that C_2 changes to a level which will maintain the offset voltage of A_2 at zero. In the meantime, S_3 has opened, so that C_1 maintains its previous level. The offset of A_2 has now been zeroed and can return to its task of stabilizing A_1.

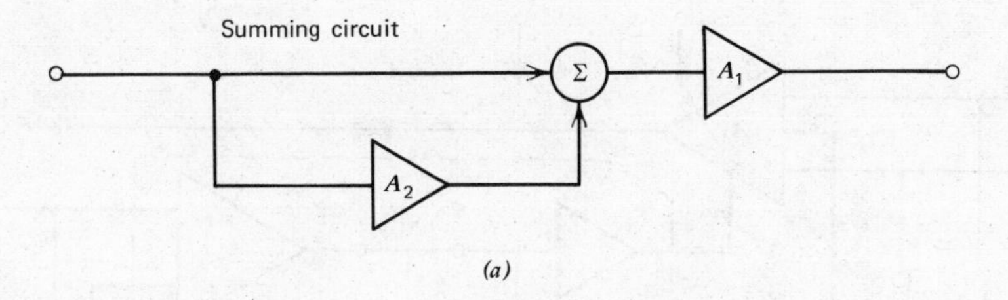

(a)

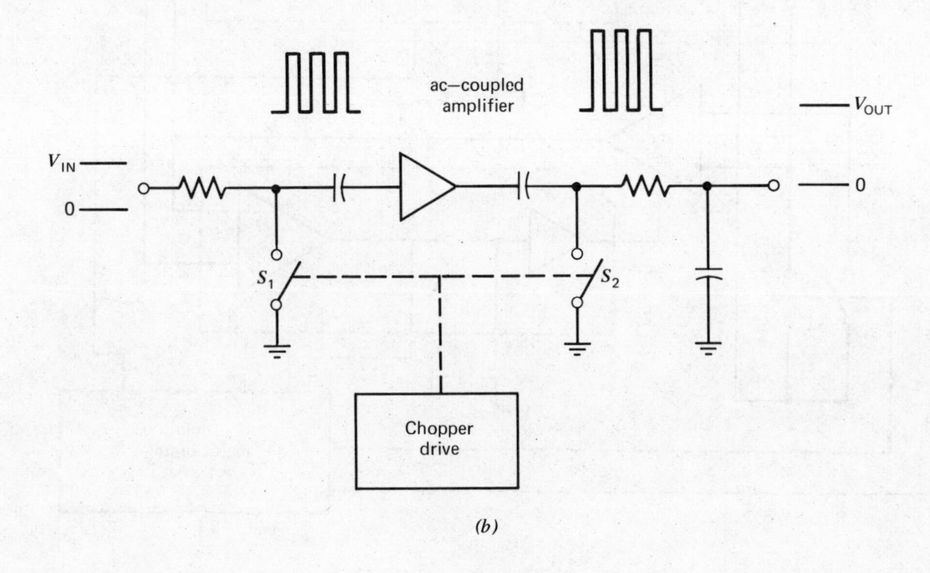

(b)

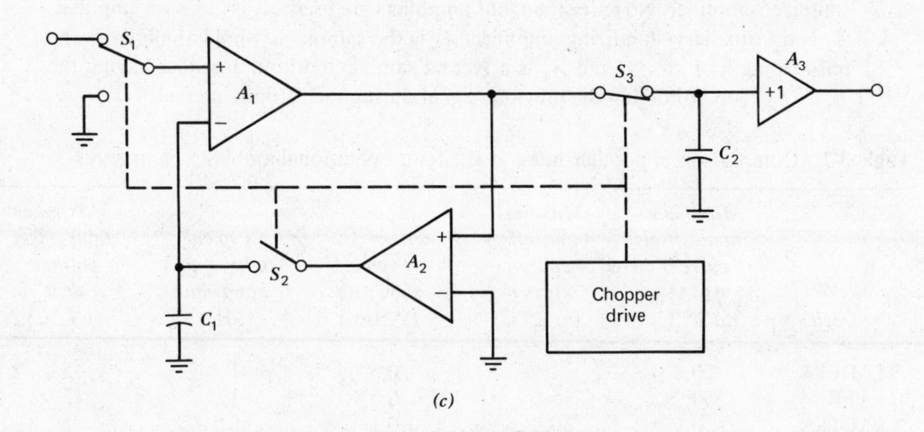

(c)

Figure 3-22. Basic chopper. In any chopper-stabilized configuration (*a*), an auxiliary dc amplifier A_2 chops the input ac waveform (*b*) in such a way that its phase and polarity when added to the input of the main amplifier A_1 produce a minimum dc voltage offset. In the HA2900 (*c*), however, a direct-coupled amplifier periodically disconnects itself from the input signal and adjusts the offset voltage to zero.

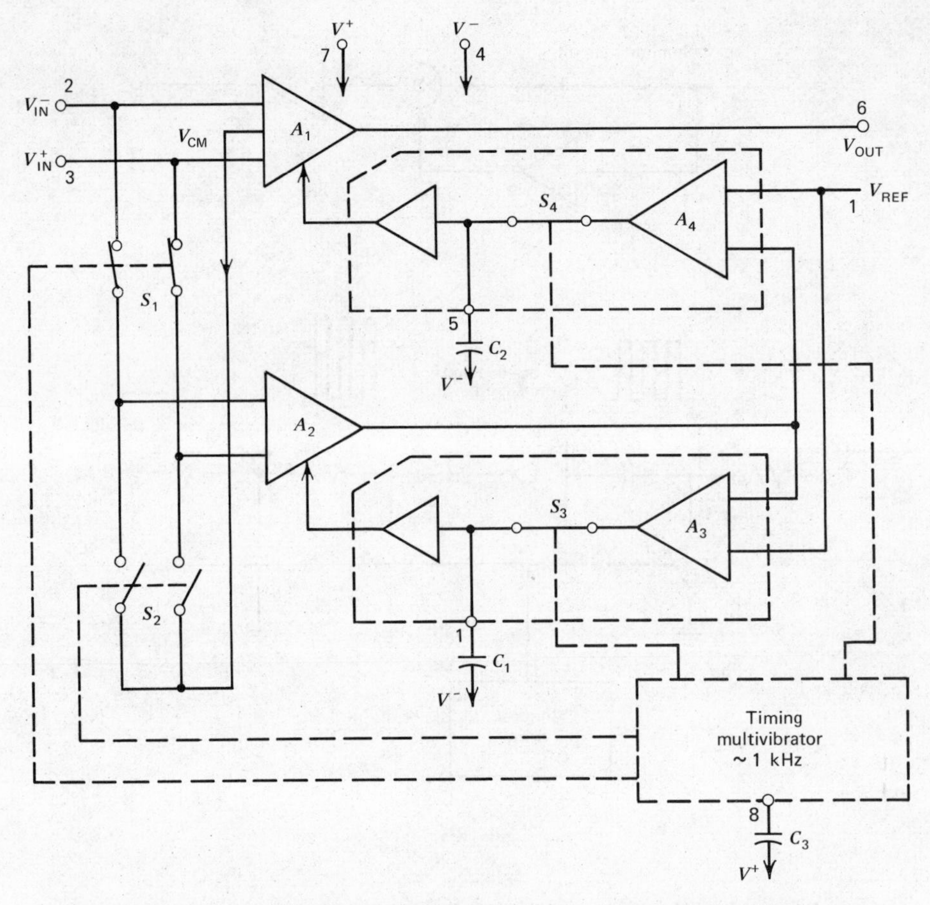

Figure 3-23. Block diagram for the HA2900 shows the complexity of a chopper-stabilized amplifier. No fewer than four amplifiers are used; A_1 is the main amplifier, A_2 is the auxiliary-stabilizing amplifier, A_3 is the sample-and-hold amplifier in the self-zeroing loop of A_2, and A_4 is a second sample-and-hold amplifier having the job of holding the previous signal during the zeroing interval.

Table 3-2 Comparison of popular integrated circuit operational amplifier parameters

Device	*Maximum input-offset current* (n/A) (*at* 25°C)	*Maximum input-offset voltage* (mV) (*at* 25°C)	*Typical slew rate* (V/μsec)	*Typical unity gain bandwidth* (MHz)	*Maximum input-offset voltage drift* (μV/°C)
LM101A	20	2	0.5	1	15
μA741	500	5	0.5	1	15
LM108A	0.4	0.5	0.1	0.5	5
μA725A	40	0.5	0.005	0.05	1
μA740	0.01	20	6.0	1	5[a]
SN62088	0.3	0.075	25.0	3	0.6
HA2900	0.1	0.060	2.5	3	0.6
4250	5	4	0.16	0.25	—

[a] Typical.

Since the opening and closing times of S_1 through S_4 are interleaved, this allows the transient spikes generated when a switch is opened or closed to settle out before other signal paths, which could be affected by these transients, are actuated. The timing multivibrator generates a triangular waveform. Different levels of this triangle are detected by four comparator circuits referenced to different points on a voltage-divider to produce the four desired switch-driving signals. Switches S_1 and S_2 are each composed of two *n*-channel MOS FETs which make excellent no-offset choppers. Switches S_3 and S_4 are complementary bipolar current switches. Amplifiers A_1 and A_2 are each *n*-channel MOS FET input amplifiers which produce extremely low-input currents. Table 3-2 compares the primary characteristics of the popular IC op amps with the Texas Instrument SN62088 and the Harris HA2900 chopper-stabilized op amps.

FACTORS TO CONSIDER IN CHOOSING AN OPERATIONAL AMPLIFIER

In choosing an op amp, finding one that meets the requirements may be only half the job. Often there are many ICs available that could be used. In this case, narrowing down to the one IC best suited for a specific application can be a truly difficult task.

First of all, the specifications on the data sheet do not tell the whole story. Factors often not readily obtainable from the data sheet which can affect one's choice include reliability, acceptability, and compatibility of the package and the supply requirements, capability to withstand the application's environmental stresses, and degree to which the device is accident-proof.

Do Not Overspecify

Decide which specifications are critical and which are not. For example, in building a voltage regulator with op amp ICs where there is a large differential voltage during turn-on, this must be considered to be a very critical parameter. For working with a very high-gain high-source impedance amplifier, offset current would be more critical than offset voltage. If, in such a situation, both offset current and offset voltage were specified very tightly, the cost of the amplifier would be needlessly raised.

There are few applications in which dc-input impedance is important. Input-bias current, which can be easily measured, is the limiting factor in dc applications. Input-offset voltage, input-offset current, and input-bias current measured over the common-mode range of the device will guarantee the integrity of the input stage. Therefore it makes more sense and costs less to test V_{OS}, I_{OS}, and I_{BIAS} at common-mode extremes than to measure dc-input impedance. Where, then, is input impedance important? At frequencies above 1 kHz, the input impedance decreases to the point where it dominates the input-bias current. Therefore, if the application is much above 1 kHz, and it is critical in the application, the ac-input impedance at that frequency may require testing. In these cases, it is still probably less costly to add a voltage-follower such as the LM110 to the amplifier input than to test ac-input impedance.

Trade-Off Cost Versus Performance

Cost, understandably, keeps entering the picture during the evaluation of an op amp. Throughout the elimination process, cost must be continually weighed against performance. Judging cost is complicated because determining the overall system cost involves guesswork. Nevertheless, it should not be too difficult to decide, for example, if using a dual op amp IC instead of two separate units or a high-gain amplifier instead of two low-gain amplifiers are economical moves. At the present time a dual op amp is less costly than two like single op amps and a quad op amp is less costly than four like single op amps or two dual op amps for similar quantities. Thus, the decision of which is most economical is simplified. Although if, as in the case of the dual op amp IC, the close temperature-tracking of two circuits on one chip is especially desirable, the weighing becomes more complex. Of course, when special testing and screening procedures are imposed on the IC manufacturer, the cost will increase. The user must evaluate whether this increased screening (electrical and/or environmental) is worth the added cost and delivery schedule impact. There are several ways to select the most-cost-effective op amp in order to save money. The most-cost-effective means is the designer specifying standard data sheet limits for his design. An off-the-shelf op amp is the best choice if all the user's critical parameters are normally tested by the manufacturer (i.e., if they have a minimum or maximum limit on the data sheet). This ensures that the same part from any manufacturer will function properly in the user's circuit. Also, since tested parameters are typically much better than the data sheet limits, a "guard band" or derating is built into the device. Therefore, the average circuit is better able to perform within specification during unanticipated worst-case conditions.

Note that it must be determined exactly what electrical tests the IC manufacturer actually performs. Because a minimum or maximum limit appears on the manufacturer's data sheet does not mean that this limit is 100% tested. There are some data sheet parameters that are guaranteed by design and not 100% tested. Also, not all manufacturers test the devices over the guaranteed temperature range. This information is vital to both the designer and the buyer when trying to effectively specify and procure the most-cost-effective op amp.

Another way to select the most-cost-effective op amp is to request that the manufacturer select a part to tightened specifications. However, selection to tighter parameters can result in a yield loss and price increase.

If the designer needs a specially selected and tighter parameter, he must determine if the added cost buys him anything. For example, an op amp with a data sheet offset-voltage limit of 2 mV probably can be generally selected for 1 mV or better, and tighter offset-voltage drift can generally be tested by the manufacturer. But this costs money.

Evaluate Operational Amplifier Self-Protection Features

Besides selecting which specifications are needed and which are not, choices regarding safety factors may be required. Problems that will arise under fault conditions should be considered. Will the output become seriously distorted? Will the whole system go up in smoke? For example, ICs op amps can be bought with or without protection against output short-circuit and input overvoltage. Since the emitter-to-base junctions of input transistors are sensitive to damage by large applied voltages, some form of input protection may be desirable. Ungrounded soldering irons, excessive input signals, and static discharges are all apt to challenge the input of the IC.

Evaluate Means of Frequency Compensation

The frequency compensation that will be required differs widely among commercially available op amp ICs. At one time, in fact, opponents of these devices cited the need for frequency compensation networks as the prime reason for not using IC op amps.

The discrete components required for frequency compensation do add to the space requirements and assembly cost. Various arrangements used for commercially available IC op amps can require from zero to seven components, depending on the IC and the application. Certainly, it is of key importance for the designer to consider frequency compensation. He should look at the IC op amp and attempt to determine how prone it is to oscillation if the supply is not bypassed properly. The IC should be evaluated to see how apt it is to oscillate with varying capacitive loads and to ascertain the probability that stray capacitance around the circuit will send it into oscillation. Too much bandwidth in an amplifier can be detrimental. Extra capacitors may be needed across the feedback resistors to prevent oscillation.

Internal frequency compensation reduces the piece-parts count but also limits the performance of the op amp. For a unity-gain noninverting connection, maximum compensation is needed, and this imposes the limits of a certain slew rate and bandwidth. If the op amp is internally compensated, this limitation remains irrespective of the way in which the amplifier is connected in the circuit. For example, in a gain-of-10 circuit, we would have a 0.5 V/μsec slew rate and a power bandwidth of about 5 kHz (0 dB at 1 MHz) for a μA741 internally compensated op amp. If an externally compensated op amp were used, such as the LM101A, we would obtain a higher slew rate and a greater gain-bandwidth product—a factor of 20 better than that obtainable from internally compensated op amps. Furthermore, if used in an inverting mode, the size of the compensation capacitor required for an externally compensated op amp can be cut in half.

Roll-off Curve

The op amp user should know his amplifier's roll-off curve in order to build a circuit with adequate gain stability over the working frequency range. The manufacturer may be perfectly justified in departing from the conventional 6-dB/octave frequency compensation to achieve such desirable features as fast settling time, high-slew rate, fast overload recovery, or increased gain stability over a wide range of frequencies. But to obtain these improved features generally requires fast roll-off characteristics and, therefore, a propensity toward oscillation. The key to preventing instability, of course, is knowing that this kind of amplifier is involved. It is then possible to use one or more well-known circuit techniques to eliminate the oscillations.

Common-Mode Rejection Ratio (CMRR)

Another important parameter is CMRR. An amplifier's CMRR performance can vary with operation conditions, notably with the value of common-mode input voltage. For example, some well-known FET types boast a common-mode voltage range of ± 10 V. But the CMRR figures are specified for a ± 5 V common-mode range. It is possible for the CMRR to degrade by as much as a factor of 10 when the applied common-mode voltage is raised from 5 to 10 V.

Moreover, since CMRR is a nonlinear function of input common-mode voltage, a single specification number can at best give an average value over the best voltage range. For small-input signal variation about some large common-mode voltage, the specified "average" CMRR gives little indication of the actual errors that can be expected due to the steeper error slope at high voltages.

Slew Rate

Another characteristic that is often overlooked is the slew rate. The slew rate usually cannot be calculated from the small-signal response, or vice versa. The designer should evaluate the devices with the particular frequency-compensating network used to determine the small-signal response and the slew rate. These characteristics may not be important if a low-frequency sine wave is all that has to be amplified; however, if a square-wave amplifier is needed or if a sample-and-hold circuit is required, then these characteristics are very important.

Slew rate for the most part is just another way of looking at the rate-limiting of the amplifier's circuitry. The slew-rate specification applies to transient response, whereas full-power response applies to steady-state or continuous response. For a step-function input, slew rate tells how fast the output voltage can swing from one voltage level to another. Fast amplifiers will slew at up to 1500 V/μsec, but amplifiers designed primarily for dc applications often slew at 0.1 V/μsec or less.

Settling Time

Settling time, a parameter of increasing interest, defines the time required for the output to settle within a given percentage of final value in response to a step-function input. Common accuracies of interest are settling times of 0.1 and 0.01%. Therefore, engineers have been forced to use slew rate and unity-gain bandwidth as rough indicators of relative settling time performance when comparing or choosing amplifiers, since no other data was given. As it turns out, these two parameters have little bearing on settling time, particularly to 0.01%. The problem here is that settling time is really a closed-loop parameter (all other op amp specifications being open-loop parameters) and therefore depends on the closed-loop configuration and gain.

Chapter 4

Operational Amplifier Applications

Intensive development of the monolithic op amp has yielded circuits which are quite good engineering approximations of the ideal for finite cost. The low cost and high quality of the amplifiers allows the implementation of equipment and systems functions impractical with discrete components. An example is the low-frequency function generator, which may use 15 to 20 op amp in generation, wave-shaping, triggering, and phase-locking. The availability of the low-cost monolithic op amp makes it mandatory that systems and equipment engineers be familiar with op amp applications. The monolithic op amps presented in this chapter do not generally exhibit the frequency-stabilization components.

BASIC OPERATIONAL AMPLIFIER CONNECTIONS

Inverting Amplifier

The basic inverting op amp circuit appears in Figure 4-1. This circuit gives a closed-loop gain of R_2/R_1 when this ratio is small compared with the amplifier open-loop gain. The input impedance is equal to R_1. The closed-loop bandwidth is equal to the unity-gain frequency divided by 1 plus the closed-loop gain. The only precautions to be observed are: (1) that R_3 should be chosen to be equal to the parallel combination of R_1 and R_2 to minimize the offset-voltage error due to bias current, and (2) that there will be an offset voltage at the amplifier output equal to closed-loop gain times the offset voltage at the amplifier input.

Offset error at the input of an op amp is comprised of two components, which are identified in specifying the amplifier as input-offset voltage and input-bias current. The input-offset voltage is fixed for a particular amplifier; however, the contribution due to input-bias current is dependent on the circuit configuration used. For minimum offset voltage at the amplifier input without circuit adjustment, the source resistances for both inputs should be equal. In this case, the maximum offset voltage would be the algebraic sum of amplifier offset voltage and the voltage drop across the source resistance due to offset current. Amplifier offset voltage is the predominant error

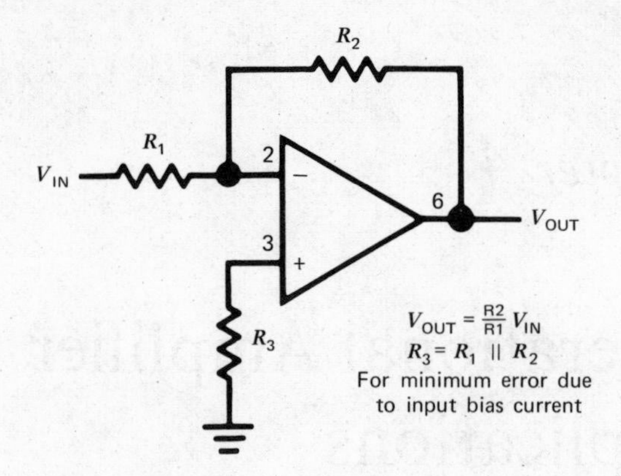

Figure 4-1. Inverting amplifier.

term for low-source resistances, and offset current causes the main error for high-source resistances.

In high-source-resistance applications, offset voltage at the amplifier output may be adjusted by changing the value of R_3 and using the variation in voltage drop across it as an input-offset voltage trim. Offset voltage at the amplifier output is less important in ac-coupled applications. Here the only consideration is that any offset voltage at the output reduces the peak-to-peak linear output swing of the amplifier.

The gain-frequency characteristic of the amplifier and its feedback network must be such that oscillation does not occur. To meet this condition, the phase-shift through amplifier and feedback network must never exceed 180° for any frequency in which the gain of the amplifier and its feedback network is greater than unity. In practical applications, the phase-shift should not approach 180° since this is the situation of conditional stability. Obviously, the most critical case occurs when the attenuation of the feedback network is zero.

Amplifiers that are not internally compensated may be used to achieve increased performance in circuits where feedback network attenuation is high. As an example, the LM101A may be operated at unity gain in the inverting amplifier circuit with a 15-pF compensating capacitor, since the feedback network has an attenuation of 6 dB, whereas it requires 30 pF in the noninverting unity-gain connection when the feedback network has zero attenuation. Since amplifier slew rate is dependent on compensation, the LM101A slew rate in the inverting unity-gain connection will be twice that for noninverting connection, and the inverting gain of 10 connection will yield 11 times the slew rate of the noninverting unity-gain connection. The compensation trade-off for a particular connection is stability versus bandwidth. Larger values of the compensation capacitor yield greater stability and lower bandwidth and vice versa.

Noninverting Amplifier

The high-input-impedance noninverting circuit of Figure 4-2 gives a closed-loop gain equal to the ratio of the sum of R_1 and R_2 to R_1 and a closed-loop 3-dB bandwidth equal to the amplifier unity-gain frequency divided by the closed-loop gain.

The primary difference between this connection and the inverting circuit are that the output is not inverted and that the input impedance is very high and is equal to

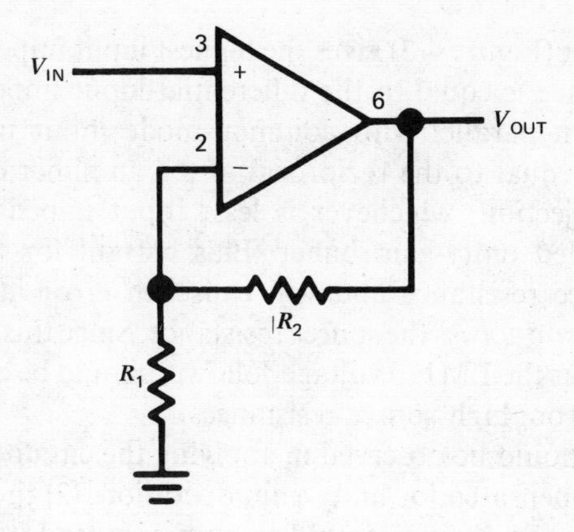

Figure 4-2. Schematic diagram of a noninverting amplifier [*note:* $V_{OUT} = [(R_1 + R_2)/R_1]V_{IN}$; $R_1 \parallel R_2 = R_{SOURCE}$; for minimum error due to input-bias current].

the differential-input impedance multiplied by loop gain. (Open-loop gain/closed-loop gain.) In dc-coupled applications, input impedance is less important than input current and its voltage drop across the source resistance.

Applications precautions are the same for this amplifier as for the inverting amplifier, with one exception. The amplifier output will go into saturation if the input is allowed to float. This may be important if the amplifier must be switched from source to source. The compensation trade-off discussed for the inverting amplifier is also valid for this connection.

Unity-Gain Buffer and Voltage-Follower

The voltage-follower is frequently used as a buffer amplifier to reduce voltage error caused by source-loading and to isolate high-impedance sources from following circuitry. A schematic for the follower is shown in Figure 4-3. The gain of this circuit is unity. The output duplicates, or follows, the input voltage, hence the name *voltage-follower*.

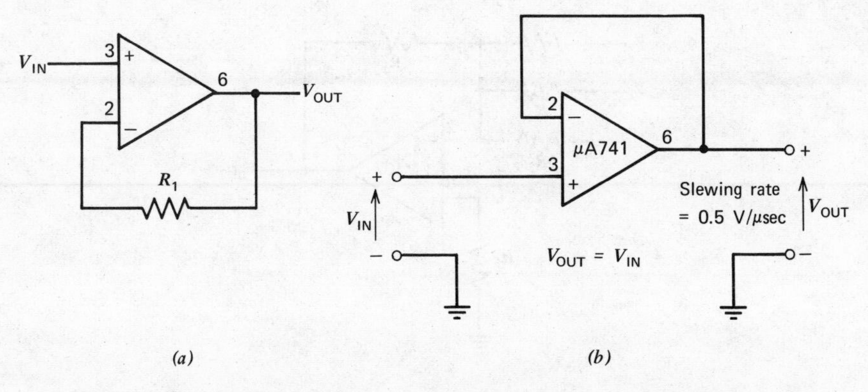

Figure 4-3. (*a*) Schematic diagram of a unity-gain buffer, [*note:* $V_{OUT} = V_{IN}$; $R_1 = R_{SOURCE}$; for minimum error due to input-bias current]. (*b*) Practical unity-gain voltage-follower.

The unity-gain buffer (Figure 4-3) gives the highest input impedance of any op amp circuit. Input impedance is equal to the differential-input impedance multiplied by the open-loop gain, in parallel with common-mode input impedance. The gain error of this circuit is equal to the reciprocal of the amplifier open-loop gain or to the common-mode rejection, whichever is less. Input impedance is a misleading concept in a dc-coupled unity-gain buffer. Bias current for the amplifier will be supplied by the source resistance and will cause an error at the amplifier input because of its voltage drop across the source resistance. Since this is the case, a low-bias current amplifier such as the LM110 voltage-follower should be chosen as a unity-gain buffer when working from high-source resistances.

Three precautions should be observed in applying the circuit of Figure 4-3: (1) the amplifier must be compensated for unity-gain operation, (2) the output swing of the amplifier must be limited by the amplifier common-mode range, and (3) some amplifiers exhibit a latchup mode when the amplifier common-mode range is exceeded. (If the input transistor at the inverting input saturates, then the input to this transistor is fed directly to its collector circuit through the collector-base junction. Thus, the inverting input becomes noninverting if the common-mode limit is exceeded. This results in positive feedback holding the amplifier in saturation, i.e., latchup.)

Summing Amplifier

The summing amplifier, a special case of the inverting amplifier, is shown in Figure 4-4. The circuit gives an inverted output which is equal to the weighted algebraic sum of all three inputs. The gain of any input of this circuit is equal to the ratio of the appropriate input resistor to the feedback resistor, R_4. Amplifier bandwidth may be calculated as in the inverting amplifier shown in Figure 4-1 by assuming the input resistor to be the parallel combination of R_1, R_2, and R_3. Application cautions are the same as for the inverting amplifier. If an uncompensated amplifier is used, compensation is calculated on the basis of this bandwidth as is discussed in the section describing the simple inverting amplifier. The advantage of this circuit is that there is no interaction between inputs, and operations such as summing, and weighted averaging are implemented very easily.

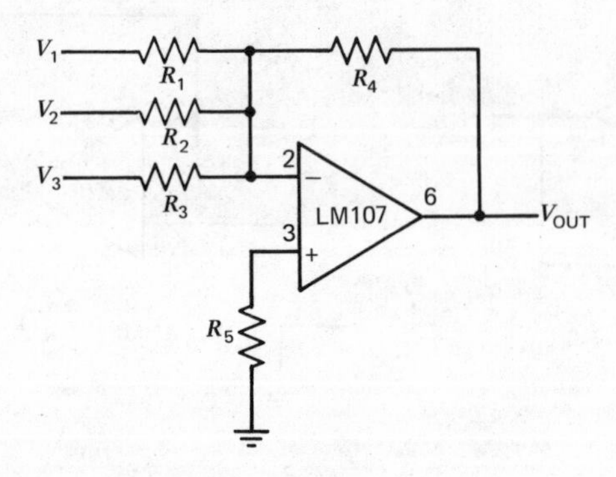

Figure 4-4. Inverting summing amplifier [*note:* $V_{OUT} = -R_4(V_1/R_1 + V_2/R_2 + V_3/R_3)$; $R_5 = R_1 \| R_2 \| R_3 \| R_4$; for minimum offset error due to input-bias current].

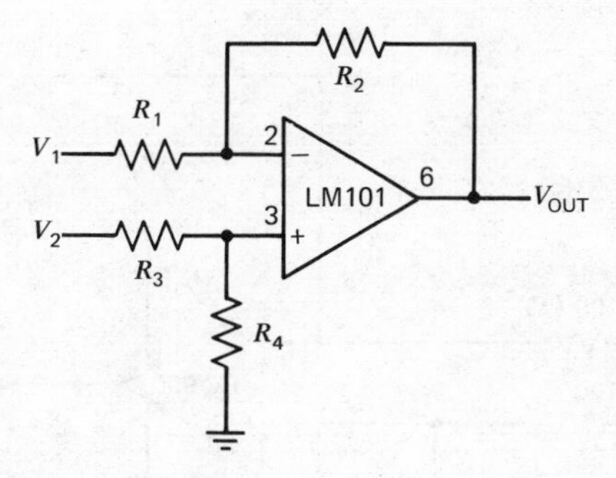

Figure 4-5. Difference amplifier [*note:* $V_{OUT} = \{(R_1 + R_2)/(R_3 + R_4)\}/(R_4/R_1)V_2 - (R_2/R_1)V_1$ for $R_1 = R_3$ and $R_2 = R_4$; $V_{OUT} = R_2/R_1(V_2 - V_1)$; $R_1 \parallel R_2 = R_3 \parallel R_4$; for minimum offset error due to input-bias current].

Difference Amplifier

The difference amplifier is the complement of the summing amplifier and allows the subtraction of two voltages or, as a special case, the cancellation of a signal common to the two inputs. This circuit is shown in Figure 4-5 and is useful as a computational amplifier, in making a differential to single-ended conversion or in rejecting a common-mode signal.

Circuit bandwidth may be calculated in the same manner as for the inverting amplifier, but input impedance is somewhat more complicated. Input impedance for the two inputs is not necessarily equal; inverting input impedance is the same as for the inverting amplifier of Figure 4-1 and the noninverting input impedance is the sum of R_3 and R_4. Gain for either input is the ratio of R_1 to R_2 for the special case of a differential input single-ended output where $R_1 = R_3$ and $R_2 = R_4$. The general expression for gain is given in the figure. Compensation should be chosen on the basis of amplifier bandwidth. Care must be exercised in applying this circuit since input impedances are not equal for minimum bias-current error.

High-Gain Amplifier

Figure 4-6 shows a typical high-gain amplifier application using the HA2900 chopper-stabilized op amp. The gain is 1000 and the bandwidth is about 2 kHz. Either input terminal may be grounded for inverting or noninverting operation, or the inputs may be driven differentially. Symmetrical networks at the device inputs are recommended for any of the three operating modes, to eliminate chopper noise and obtain the best drift characteristics. Total input noise, with $C = 0$, is about 30 μV rms. This noise can be reduced at the expense of bandwidth, by adding capacitors as shown.

Differentiator

The differentiator is shown in Figure 4-7 and, as the name implies, is used to perform the mathematical operation of differentiation. The form shown is not the practical form, it is a true differentiator and is extremely susceptible to high-frequency noise since ac-gain increases at the rate of 6 dB/octave. In addition, the feedback network

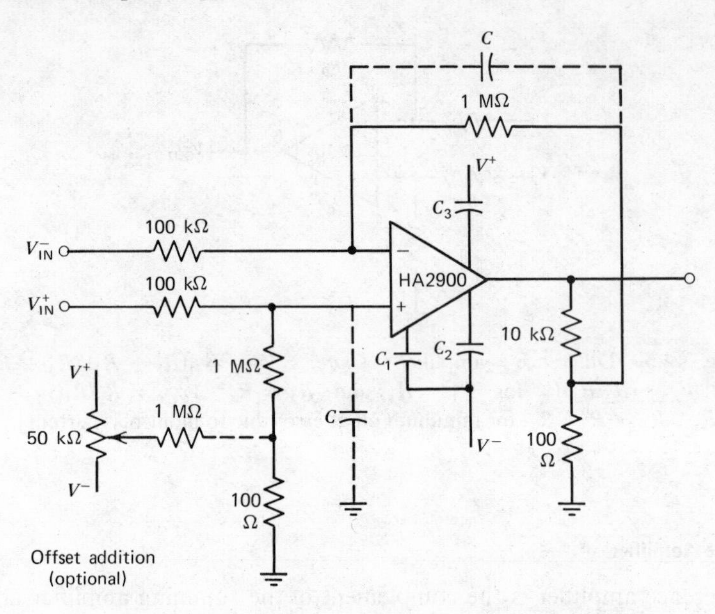

Figure 4-6. High-gain amplifier using the HA2900.

of the differentiator, R_2C_1, is an RC low-pass filter which contributes 90° phase shift to the loop and may cause stability problems even with an amplifier which is compensated for unity gain.

The differentiator circuit of Figure 4-8 provides an output proportional to the derivative of the input signal. The gain function of the differentiator is given by

$$V_{OUT} = -R_2C_1 \frac{dV_1}{dt} \tag{4-1}$$

As the differentiator performs the reverse of the integrator function, a triangular input will produce a square-wave output. For a 2.5 V peak-to-peak triangle wave

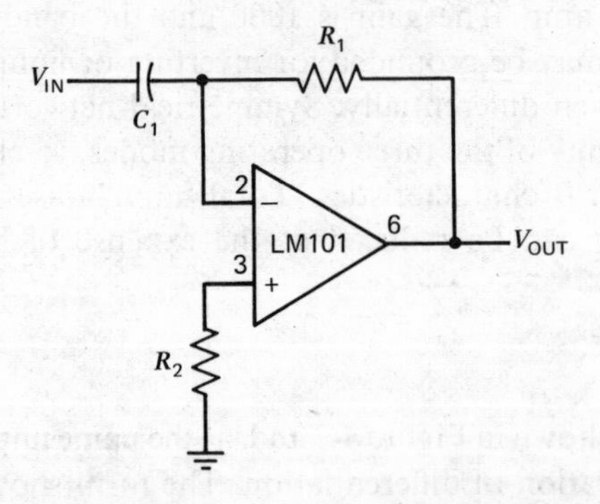

Figure 4-7. Differentiator [*note:* $V_{OUT} = -R_1C_1(d/dt)(V_{IN})$; $R_1 = R_2$; for minimum offset error due to input-bias current].

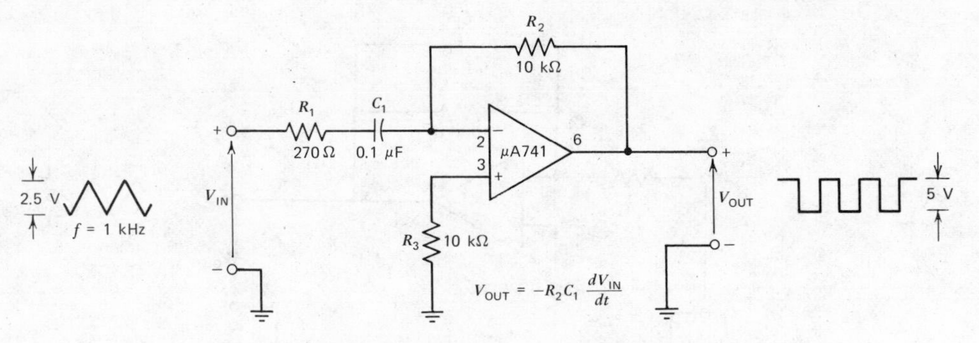

Figure 4-8. Practical differentiator.

with a period of 1 msec we have, for the circuit illustrated,

$$\frac{dV}{dt} = \frac{2.5 \text{ V}}{0.5 \text{ msec}} = 5 \frac{\text{V}}{\text{msec}}$$

$$V_{\text{OUT}} = -(10 \text{ k}\Omega)(0.1 \ \mu\text{F}) = 5 \frac{\text{V}}{\text{msec}}$$

$$= 5 \text{ V}_{\text{p-p}}$$

The resistor R_1 is needed to limit the high-frequency gain of the differentiator. This makes the circuit less susceptible to high-frequency noise and assures dynamic stability. The corner frequency where the gain-limiting comes into effect is given by

$$f = \frac{1}{2\pi R_1 C_1} \tag{4-2}$$

and should be at least ten times the highest input frequency for accurate operation. A maximum value for the corner frequency is determined by stability criteria. In general, it should be no larger than the geometric mean between $1/2\pi R_2 C_1$ and the gain-bandwidth product of the op amp. The μA741 has a gain-bandwidth product of approximately 1 MHz hence the limit on f is given by

$$f < \sqrt{\frac{1 \times 10^6}{2\pi R_2 C_1}} \tag{4-3}$$

The differentiator is subject to damage from fast-rising input signals as is the integrator and, as has been discussed, the circuit is also susceptible to high-frequency instability.

Integrator

The integrator, shown in Figure 4-9, provides an output that is proportional to the time integral of the input signal. The gain function for the integrator is given by

$$V_{\text{OUT}} = -\frac{1}{R_1 C_1} \int V_{\text{IN}} \, dt \tag{4-4}$$

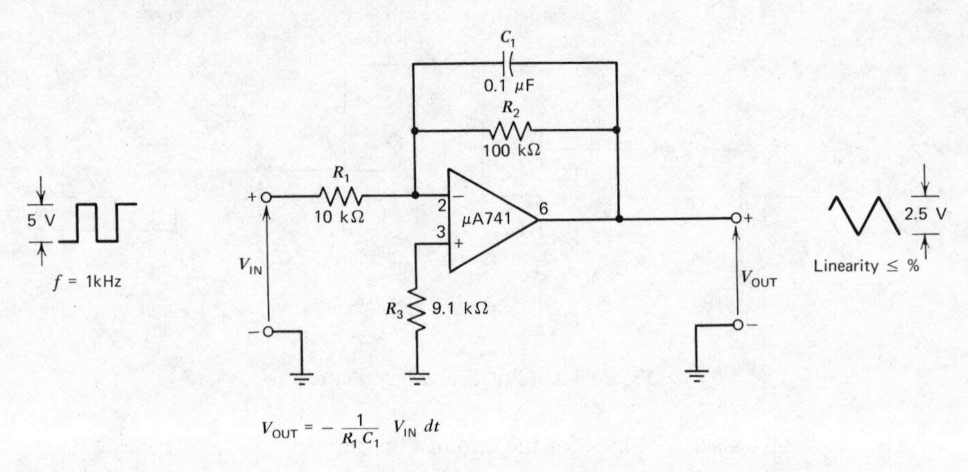

Figure 4-9. Integrator.

As an example, consider the response of the integrator to a symmetrical square-wave input signal with an average value of 0 V. If the input has a peak amplitude of A V, in the period T, then the peak-to-peak output can be calculated by integrating over one-half the input period giving:

$$|V_{\text{OUT p-p}}| = \frac{1}{R_1 C_1} \int_0^{T/2} A\, dt$$

$$= \frac{A}{R_1 C_1}\left(\frac{T}{2}\right) \text{V} \qquad (4\text{-}5)$$

The wave shape will be triangular corresponding to the integral of the square wave. For the component values shown in the figure assuming $A = 5$ V and $T = 1$ msec, we have using equation (4-5),

$$R_1 C_1 = 10^{-3} \text{ sec}$$

$$V_{\text{OUT p-p}} = \left(\frac{5}{10^{-3}}\right)\left(\frac{10^{-3}}{2}\right) = 2.5 \text{ V}_{\text{p-p}}$$

The resistor R_2 is included to provide dc stabilization for the integrator. Its function is to limit the low-frequency gain of the amplifier and thus minimize drift. The frequency above which the circuit will perform as an integrator is given by

$$f = \frac{1}{2\pi R_2 C_1} \text{ Hz} \qquad (4\text{-}6)$$

For best linearity, the frequency of the input signal should be at least ten times the frequency given by (4-6). The linearity of the circuit illustrated is better than 1% with an input frequency of 1 kHz.

Although it is not immediately obvious, the integrator, if it is to operate reliably, requires both a large common-mode and differential-mode input voltage range. There are several ways the input-voltage limits may be inadvertently exceeded. The

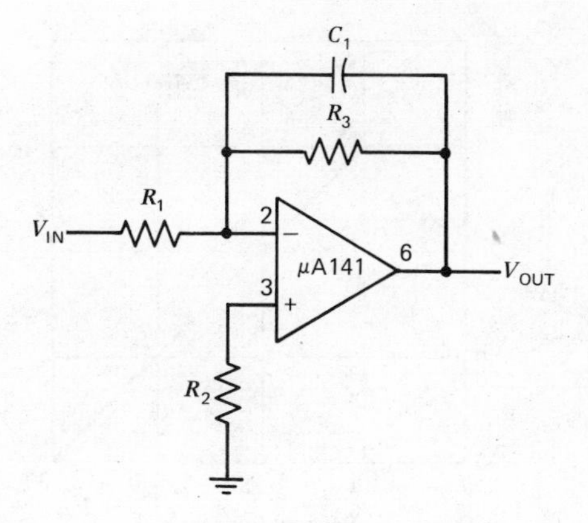

Figure 4-10. Simple low-pass filter [*note:* $f_L = 1/2\pi R_1 C_1$; $f_C = 1/2\pi R_3 C_1$; $A_L = R_3/R_1$].

most obvious is that transients occurring at the output of the amplifier can be coupled back to the input by the integrating capacitor C_1. Thus, either common-mode or differential-mode voltage limits can be exceeded.

Another less obvious problem can occur when the amplifier is driven from fast-rising or fast-falling input signals, such as square waves. The output of the amplifier cannot respond to an input instantaneously. During the short interval before the output reacts, the summing point at *pin* 2 (inverting input) of the amplifier may not be held at ground potential. If the input signal change is large enough, the voltage at the summing point could exceed safe limits for the amplifier.

Simple Low-Pass Filter

The simple low-pass filter is shown in Figure 4-10. This circuit has a 6 dB/octave roll-off after a closed-loop 3 dB point defined by f_c. Gain below this corner frequency is defined by the ratio of R_3 to R_1. The circuit may be considered as an ac integrator at frequencies well above f_c; however, the time domain response is that of a single *RC* rather than an integral. Note that R_2 should be chosen equal to the parallel combination of R_1 and R_3 to minimize errors due to bias current. The amplifier should be compensated for unity-gain, or an internally compensated amplifier can be used (as shown). A gain-frequency plot of circuit response is shown in Figure 4-11 to illustrate the difference between this circuit and the true integrator.

Current-to-Voltage Converter

Current may be measured in two ways with an op amp. The current may be converted into a voltage with a resistor and then amplified or the current may be injected directly into a summing node. Converting into voltage is undesirable for two reasons: First, an impedance is inserted into the measuring line causing an error; second, amplifier offset voltage is also amplified with a subsequent loss of accuracy. The use of a current-to-voltage transducer avoids both of these problems.

The current-to-voltage transducer is shown in Figure 4-12. The input current is fed directly into the summing node and the amplifier output voltage changes to extract the same current from the summing node through R_1. The scale factor of

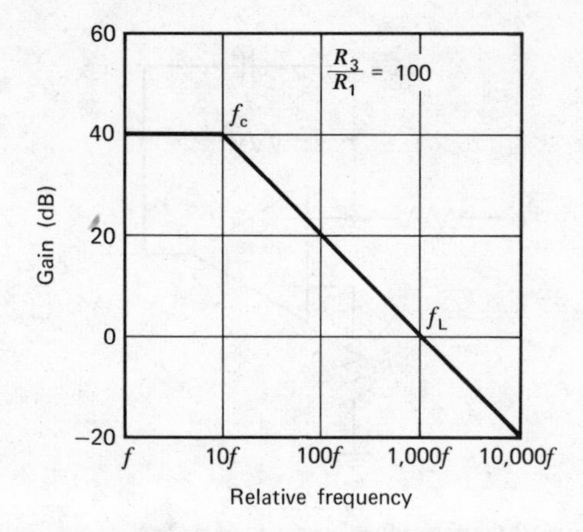

Figure 4-11. Low-pass filter response.

this circuit is R_1 V/A. The only conversion error in this circuit is I_{BIAS} which is summed algebraically with I_{IN}.

The only design constraints are that scale factors must be chosen to minimize errors due to bias current and that, since voltage gain and source impedance are often indeterminate (as with photocells), the amplifier must be compensated for unity-gain operation.

Voltage Comparators

Comparators are amplifiers operated open-loop. They have a variety of applications, including interface circuits, detectors, and sense amplifiers. Integrated comparators, like LM106, μA710, LM111, NE521/522, and NE527/529 are designed for these applications.

General-purpose amplifiers such as the LM101, LM101A, μA741, and LM108 also make good comparators because they have large differential-input voltage ranges and are easily clamped to make their outputs compatible with logic and driver circuits. However, internally compensated, general-purpose amplifiers usually make poor comparators because they switch rather slowly.

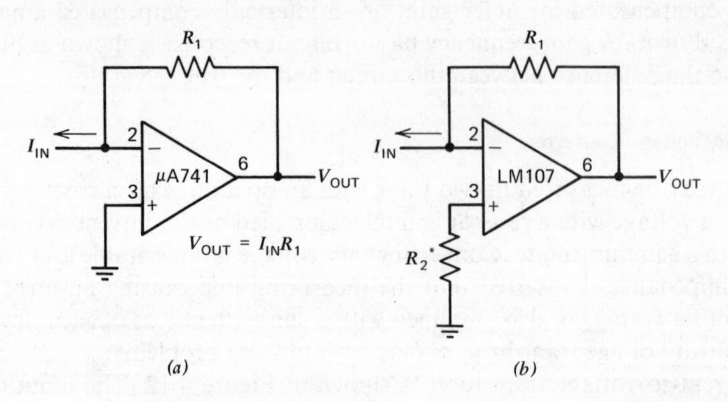

Figure 4-12. Current-to-voltage converters [*note for* (*b*): $V_{OUT} = I_{IN}R_1$; for minimum error due to bias current $R_2 = R_1$].

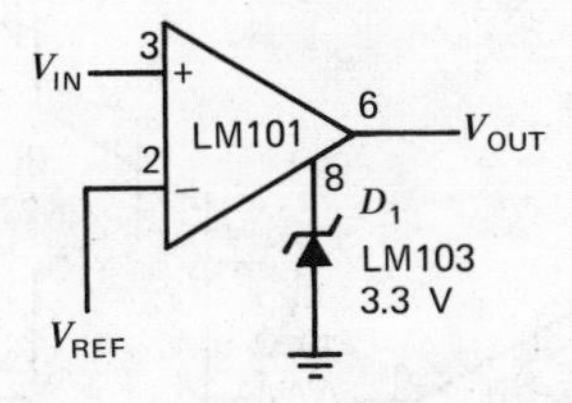

Figure 4-13. Comparator for driving DTL and TTL integrated circuits.

Comparator circuits using the LM101 are shown in Figures 4-13 and 4-14. The one in Figure 4-13 shows a clamping scheme which makes the output signal directly compatible with diode-transistor logic (DTL) or transistor-transistor logic (TTL) ICs. An LM103 breakdown diode clamps the output at 0 V or 4 V in the low or high states, respectively. This particular diode was chosen because it has a sharp breakdown and low equivalent capacitance. When working as a comparator, the amplifier operates open-loop; so normally no frequency compensation is needed. Nonetheless, the stray capacitance between *Pins* 5 and 6 of the amplifier should be minimized to prevent low-level oscillations when the comparator is in the active region. If this becomes a problem, a 3-pF capacitor on the normal compensation terminals will eliminate it.

Figure 4-14 shows the connection of the LM101 as a comparator and lamp driver. Transistor Q_1 switches the lamp, with R_2 limiting the current surge resulting from turning on a cold lamp; R_1 determines the base drive to Q_1 while D_1 keeps the amplifier from putting excessive reverse bias on the emitter-base junction of Q_1 when it turns off. The LM106 and LM111 voltage-comparators can drive lamps directly.

More Output Current Swing

Because almost all monolithic amplifiers use class-B output stages, they have good loaded output-voltage swings, delivering ± 10 V at 5 mA with ± 15 V supplies. Demanding much more current from the IC would require, for one, that the output transistors be made considerably larger. In addition, the increased dissipation could

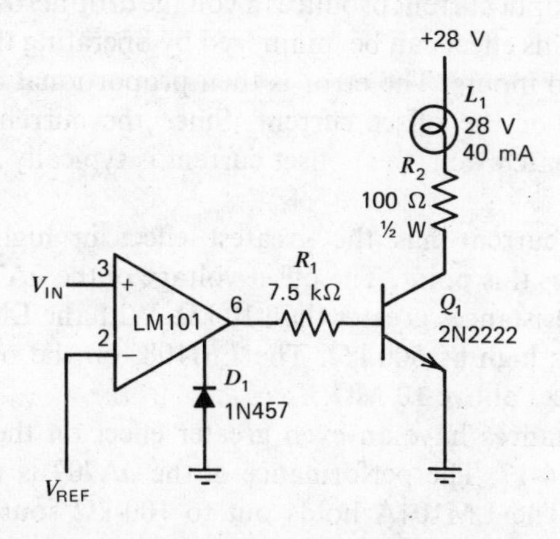

Figure 4-14. Comparator and lamp drive.

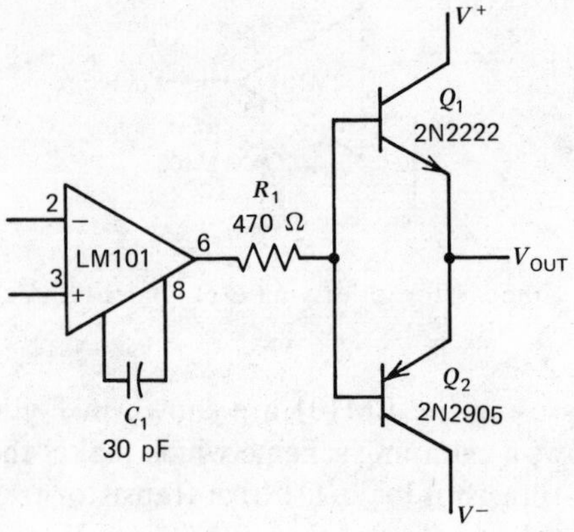

Figure 4-15. High-current output buffer.

give rise to troublesome thermal gradients on the chip as well as excessive package-heating in high-temperature applications. It is therefore advisable to use an external buffer when large output currents are needed.

A simple way of accomplishing this is shown in Figure 4-15. A pair of complementary transistors are used on the output of the LM101 to get the increased current swing. Although this circuit does have a dead zone, it can be neglected at frequencies below 100 Hz because of the high gain of the amplifier; R_1 is included to eliminate parasitic oscillations from the output transistors. In addition, adequate bypassing should be used on the collectors of the output transistors to ensure that the output signal is not coupled back into the amplifier. This circuit does not have current limiting, but it can be added by putting 50-Ω resistors in series with the collectors of Q_1 and Q_2.

OPERATIONAL AMPLIFIER BIASING AND COMPENSATING

Effects of Error Current

In an op amp, the input current produces a voltage drop across the source resistance, causing a dc error. This effect can be minimized by operating the amplifier with equal resistance on the two inputs. The error is then proportional to the difference in the two input currents, or the offset current. Since the current gains of monolithic transistors tend to match well, then offset current is typically a factor of 10 less than the input currents.

Naturally, error current has the greatest effect in high-impedance circuitry. Figure 4-16 illustrates this point. The offset voltage of the μA709 is degraded significantly with source resistances greater than 10 kΩ. With the LM101A this is extended to source resistances high as 500 kΩ. The LM108, on the other hand, works well with source resistances above 10 MΩ.

High-source resistances have an even greater effect on the drift of an amplifier, as shown in Figure 4-17. The performance of the μA709 is worsened with sources greater than 3 kΩ. The LM101A holds out to 100-kΩ sources, while the LM108 still works well at 3 MΩ.

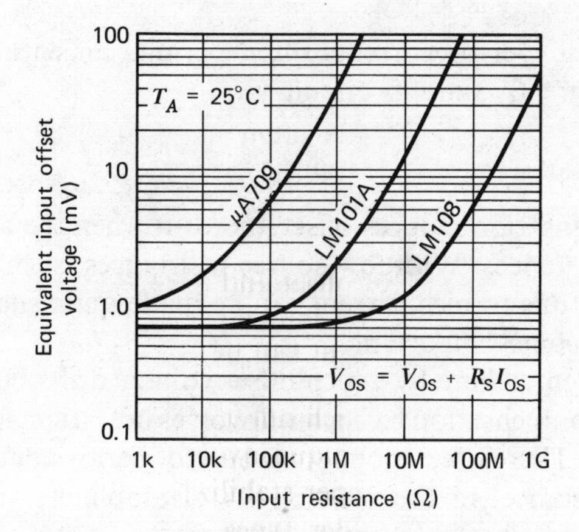

Figure 4-16. Effect of source resistance on typical input-error voltage.

It is difficult to include FET amplifiers in Figure 4-17 because their drift is initially 50 μV/°C, unless they are selected and trimmed. Even though their drift may be well controlled (5 μV/°C) over a limited temperature range, trimmed amplifiers generally exhibit a much higher drift over a −55 to 125°C temperature range. At any rate, their average drift rate would, at best, be like that of the LM101A where 125°C operation is involved.

Applications that require low-error currents include amplifiers for photodiodes or capacitive transducers, as these usually operate at megohm impedance levels. Sample-and-hold circuits, timers, integrators, and analog memories also benefit from low-error currents. For example, with the μA709, worst-case drift rates for these kinds of circuits is in the order of 1.5 V/sec. The LM108 improves this to 3 mV/sec—worst-case over a −55 to 125°C temperature range. Low-input currents are also helpful in oscillators and active filters to get low-frequency operation with reasonable capacitor values. The LM108 can be used at a frequency of 1 Hz with capacitors no larger than 0.01 μF. In logarithmic amplifiers, the dynamic range can be extended by nearly 60 dB by going from the μA709 to the LM108. In other

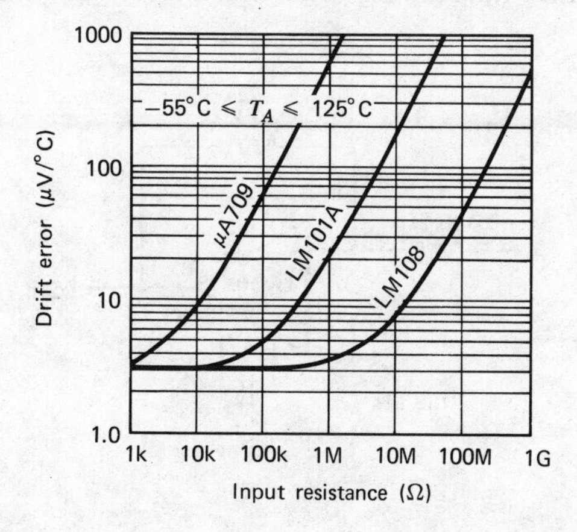

Figure 4-17. Degradation of typical drift characteristics with high-source resistances.

applications, having low-error currents often permits an entirely different design approach which can greatly simplify circuitry.

Bias-Current Compensation

Bias-current compensation reduces offset and drift when the amplifier is operated from high-source resistances. With low-source resistances, such as a thermocouple, the drift contribution due to bias current can be made quite small. In this case, the offset-voltage drift becomes important.

A technique is presented here by which offset voltage drifts better than 0.5 μV/°C can be realized. The compensation technique involves only a single room-temperature balance adjustment. Therefore, chopper-stabilized performance can be realized, with low-source resistance, in the older types of amplifiers such as the LM101, without tedious cut-and-try compensation methods.

The simplest and most effective way of compensating for bias currents is shown in Figure 4-18. Here, the offset produced by the bias current on the inverting input is cancelled by the offset voltage produced across the variable resistor R_3. The main advantage of this scheme, besides its simplicity, is that the bias currents of the two input transistors tend to track well over temperature so that low drift is also achieved. The disadvantage of the method is that a given compensation setting works only with fixed-feedback resistors, and the compensation must be readadjusted if the equivalent parallel resistance of R_1 and R_2 is changed.

Figure 4-19 shows a similar circuit for a noninverting amplifier. The offset voltage produced across the dc resistance of the source due to the input current is cancelled by the drop across R_3. For proper adjustment range R_3 should have a maximum value about three times the source resistance; and the equivalent parallel resistance of R_1 and R_2 should less than one-third the input-source resistance. This circuit has the same advantages as that in Figure 4-18, however, it can only be used when the input source has a fixed-dc resistance.

Figure 4-20 is a compensation scheme for the voltage-follower connection. The compensating current is obtained through a resistor connected across a diode that is bootstrapped to the output. The diode acts as a regulator so that the compensating current does not change appreciably with signal level, giving input impedances of

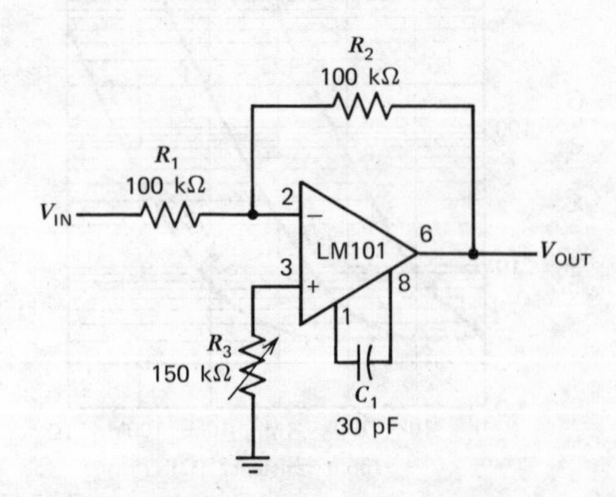

Figure 4-18. Summing amplifier with bias-current compensation for fixed-source resistances.

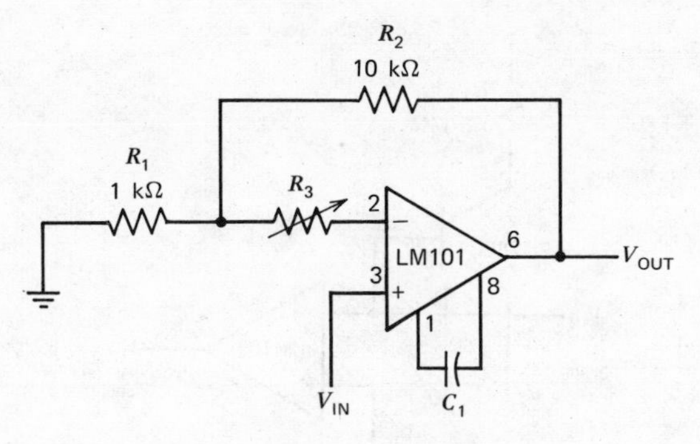

Figure 4-19. Noninverting amplifier with bias-current compensation for fixed-source resistances.

about 1000 MΩ. The negative TC of the diode voltage also provides some temperature compensation. In many applications, such as long-interval integrators, sample-and-hold circuits, switched-gain amplifiers, or voltage-followers operating from an unknown source, the source impedance is not defined. In these cases other compensation schemes must be used.

Compensating with Transistors

Figure 4-21 gives a compensation technique which does not depend upon having a fixed-source resistance. A current is injected into the input terminal from the base of a *pnp* transistor. Since *npn* input transistors are used on the integrated amplifier, the base current of the *pnp* balances out the base current of the *npn*. Further, since a silicon-planar *pnp* transistor has approximately the same current gain versus temperature characteristic as the integrated transistors, an improvement in temperature drift will also be realized. (If the op amp uses a Darlington input stage, however, the drift compensation will not be nearly as good.) However, perfect compensation should not be expected because of unit-to-unit variations in the temperature characteristics of both the *pnp* transistor and the IC.

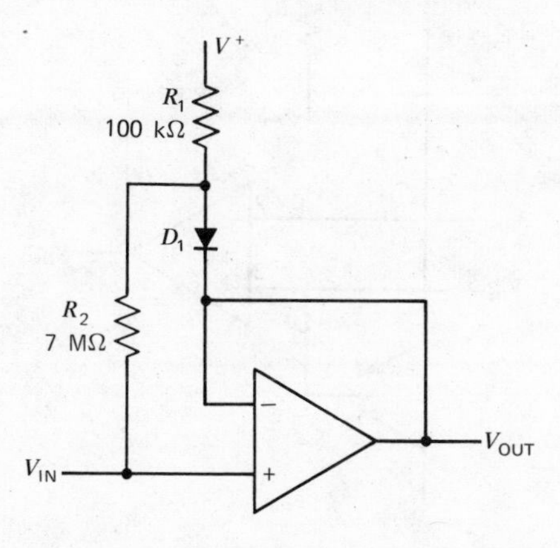

Figure 4-20. Voltage-follower with bias-current compensation.

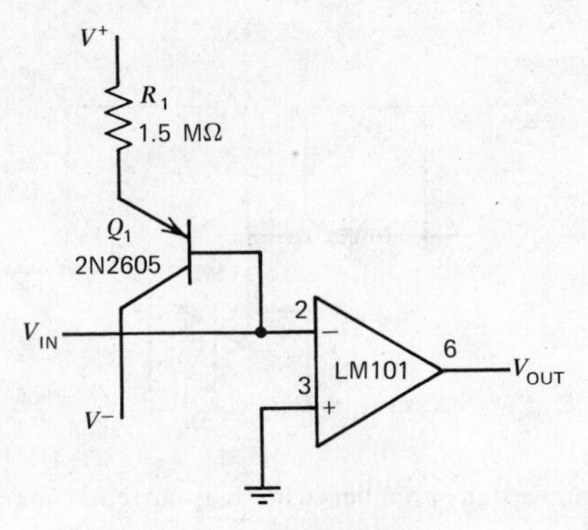

Figure 4-21. Summing amplifier with bias-current compensation.

Although the circuit in Figure 4-21 works well for the summing amplifier connection, it does have limitations in other applications. It could, for example, be used for the voltage-follower configuration by connecting the base of the *pnp* to the noninverting input. However, this would reduce the input impedance (to about 150 MΩ) because the current supplied by the *pnp* will vary with the input-voltage level.

If this characteristic is objectionable, the more complicated circuit shown in Figure 4-22 can be used. The emitter of the *pnp* transistor is fed from a current source

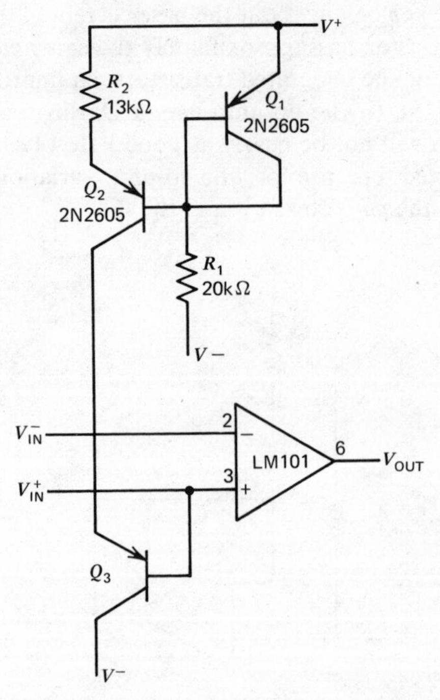

Figure 4-22. Bias-current compensation for noninverting amplifier operated over large common-mode range.

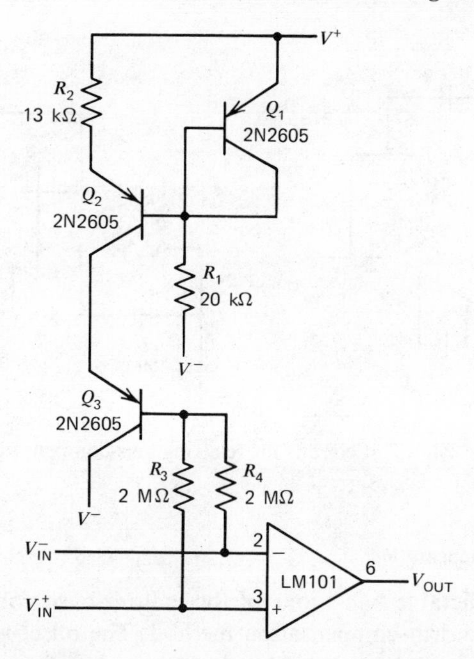

Figure 4-23. Bias-current compensation for differential inputs.

so that the compensating current does not vary with the input-voltage level. The design of the current source is such as to give it about the same characteristics as those on the input stage of the better monolithic amplifiers. It provides closer compensation with changes in temperature and supply voltage. The circuit makes use of the emitter-base voltage differential between two transistors operated at different collector currents. It is not really necessary that these transistors be well matched, since the devices are operated at much different collector currents.

Bias Adjustment

All the circuits discussed thus far have been tailored for particular applications. Figure 4-23 shows a completely general scheme wherein both inputs are current-compensated over the full common-mode range as well as against power supply and temperature variations. This circuit is suitable for use either as a summing amplifier or as a noninverting amplifier. It is not required that the dc impedance seen by both inputs be equal, although lower drift can be expected if they are.

As was mentioned earlier, all the bias-compensation circuits require adjustment. With the circuits in Figures 4-18 and 4-19, this is merely a matter of adjusting the potentiometer for zero output with zero input. It is not so simple with the other circuits, however. For one, it is difficult to use potentiometers because a very wide range of resistance values are required to accommodate expected unit-to-unit variations. Resistor selection must therefore be used. Test circuits for selecting bias-compensation resistors are given in Figure 4-24.

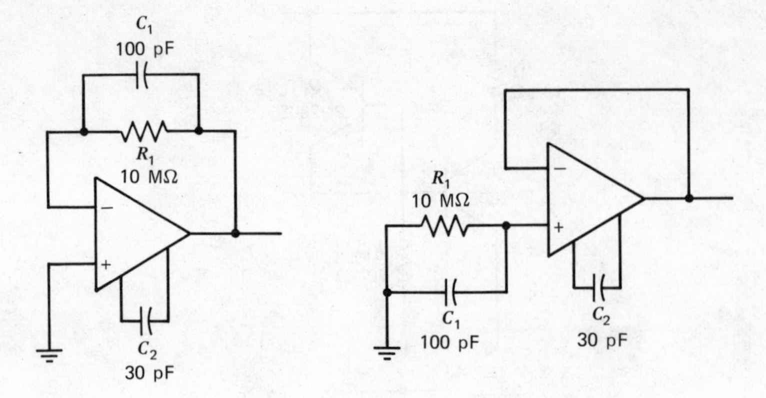

Figure 4-24. Test circuits for selecting bias-compensation resistors.

Offset-Voltage Compensation

The highly predictable behaviour of the emitter-base voltage of transistors has suggested a unique drift-compensation method. The offset-voltage drift of a differential transistor pair can be reduced by about an order of magnitude by unbalancing the collector currents such that the initial offset voltage is zero. The basis for this comes from the equation for the emitter-base voltage differential of two transistors operating at the same temperature

$$\Delta V_{\mathrm{BE}} = \frac{kT}{q} \log_e \frac{\mathrm{I}_{\mathrm{S}_2}}{\mathrm{I}_{\mathrm{S}_1}} - \frac{kT}{q} \log_e \frac{\mathrm{I}_{\mathrm{C}_2}}{\mathrm{I}_{\mathrm{C}_1}} \tag{4-7}$$

where k is Boltzmann's constant, T is the absolute temperature, q is the charge of an electron, $\mathrm{I_S}$ is a constant which depends only on how the transistor is made and $\mathrm{I_C}$ is the collector current.

It is worthwhile noting here that these expressions make no assumptions about the current gain of the transistors. The emitter-base voltage is a function of collector current—not emitter current. Therefore, the balance will not be upset by base current (except for interaction with the dc-source resistance).

The first term in (4-7) is the offset voltage of the two transistors for equal collector currents. It can be seen that this offset voltage is directly proportional to the absolute temperature—a fact which is substantiated by experiment. The second term is the change of offset voltage which arises from operating the transistors at unequal collector currents. For a fixed ratio of collector currents, this is also proportional to absolute temperature. Hence, if the collector currents are unbalanced in a fixed ratio to give a zero emitter-base voltage differential, the temperature drift will also be zero.

For low drift, the transistors must operate from a source resistance that is low enough to render insignificant the voltage drop across the source due to base current (or base-current differential, if both bases see the same resistance). Furthermore, the transistors must be operated at a collector current so low that the emitter-contact and base-spreading resistances are negligible, since (4-7) assumes that they are zero.

In a complete amplifier using this principle (Figure 4-25), a monolithic transistor

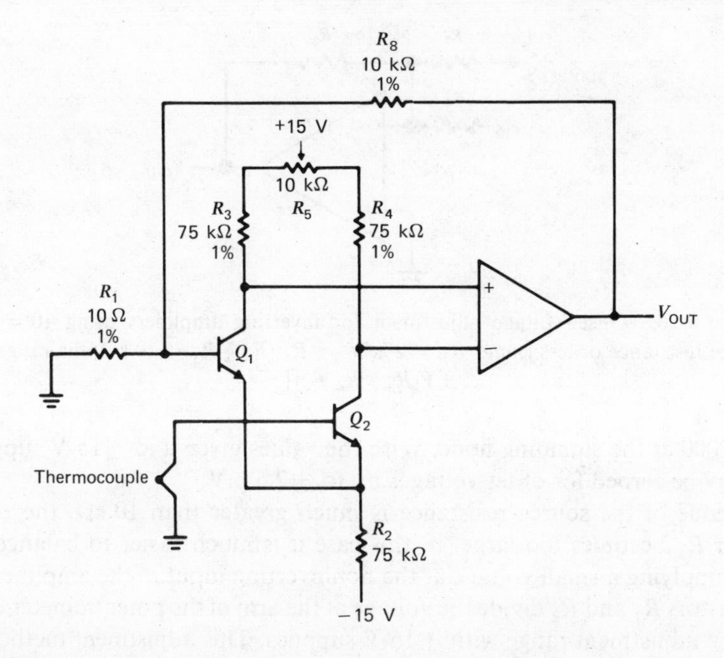

Figure 4-25. Example of a dc amplifier using the drift-compensation techniques.

pair serves as a preamplifier for a conventional op amp. A null potentiometer, which is set for zero output for zero input, unbalances the collector-load resistors of the transistor pair such that the collector currents are unbalanced for zero offset. This gives minimum drift. An interesting feature of the circuit is that the performance is relatively unaffected by supply-voltage variations: A 1-V change in either supply causes an offset voltage change of about 10 μV. This happens because neither term in (4-7) is affected by the magnitude of the collector currents.

In order to get low drift, the gain of the preamplifier must be high enough to prevent the drift of the op amp from degrading performance.

Another important consideration is the matching of the collector-load resistors on the preamplifier stage. A 0.1% imbalance in the load resistors due to thermal mismatches or any other cause will produce a 25-μV shift in offset voltage. This includes the balancing potentiometer, which can introduce an error that will depend on how far it is set off midpoint if it has a TC different from that of the resistors.

Universal Balancing Techniques

The IC op amp is widely accepted as a universal analog component. Although the circuit designs may vary, most devices are functionally interchangeable. However, offset-voltage balancing remains a personality trait of the particular amplifier design. The techniques shown here allow offset voltage balancing without regard to the internal circuitry of the amplifier.

The circuit in Figure 4-26 is used to balance out the offset voltage of inverting amplifiers having a source resistance of 10 kΩ or less. A small current is injected into the summing node of the amplifier through R_1. Since R_1 is 2000 times as large as the source resistance, the voltage at the arm of the potentiometer is attenuated by a

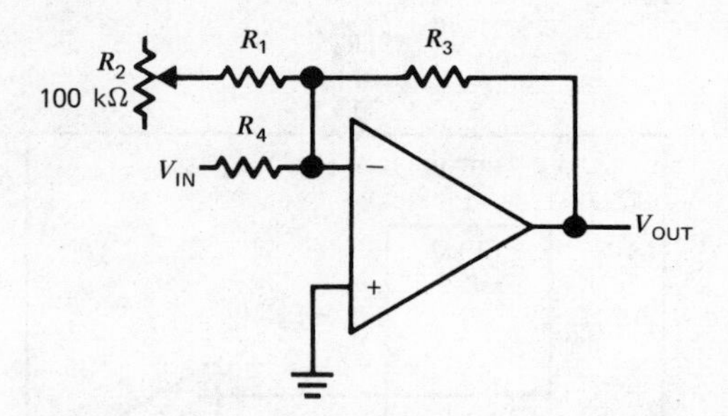

Figure 4-26. Offset-voltage adjustment for inverting amplifiers using 10 = kΩ source resistance or less [*note:* $R_1 = 2000R_3 \parallel R_4$; $R_4 \parallel R_3 \leq 10$ kΩ; the range = $\pm V(R_3 \parallel R_4/R_1)$].

factor of 2000 at the summing node. With the values given and ±15-V supplies, the output may be zeroed for offset voltages up to ±7.5 mV.

If the value of the source resistance is much greater than 10 kΩ, the resistance needed for R_1 becomes too large. In this case it is much easier to balance out the offset by supplying a small voltage at the noninverting input of the amplifier (Figure 4-27). Resistors R_1 and R_2 divide the voltage at the arm of the potentiometer to supply a ±7.5-mV adjustment range with ±15-V supplies. This adjustment method is also useful when the feedback element is a capacitor or nonlinear device.

This technique of supplying a small voltage effectively in series with the input is also used for adjusting noninverting amplifiers. As in Figure 4-28, divider R_1, R_2 reduces the voltage at the arm of the potentiometer to ±7.5 mV for offset adjustment. Since R_2 appears in series with R_4, R_2 should be considered when calculating the gain. If R_4 is greater than 10 kΩ, the error due to R_2 is less than 1%.

A voltage-follower may be balanced by the technique of Figure 4-29, where R_1 injects a current that produces a voltage drop across R_3 to cancel the offset voltage. The addition of the adjustment resistors causes a gain error, increasing the gain by 0.05%. This small error usually causes no problem. The adjustment circuit essentially causes the offset voltage to appear at full output, rather than at low-output levels, where it is a large percentage error.

Differential amplifiers are somewhat more difficult to balance. The offset adjustment used for a differential amplifier can degrade the CMRR. The adjustment

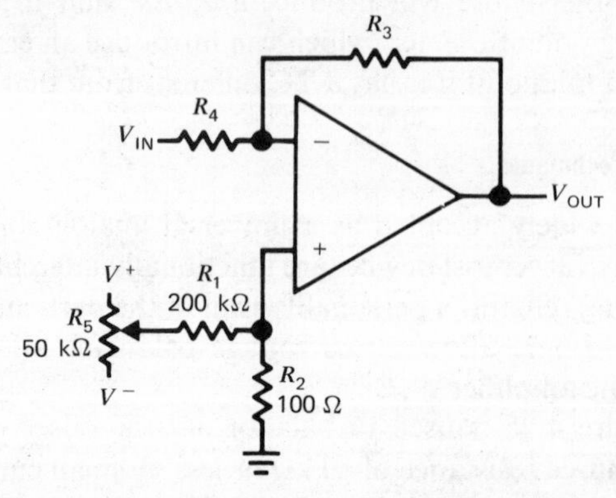

Figure 4-27. Offset-voltage adjustment for inverting amplifiers using any type of feedback element [*note:* range = $\pm V(R_2/R_1)$].

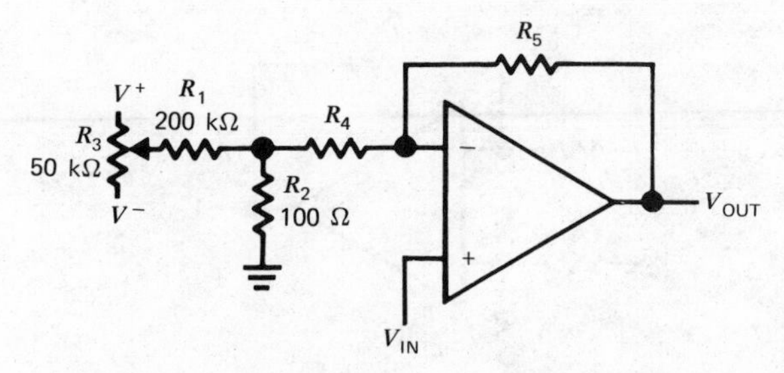

Figure 4-28. Offset-voltage adjustment for noninverting amplifiers [*note:* range = $\pm V(R_2/R_1)$ and the gain = $1 + R_5/(R_4 + R_2)$].

circuit in Figure 4-30 has minimal effect on the common-mode rejection. The voltage at the arm of the potentiometer is divided by R_4 and R_5 to supply an offset correction of ± 7.5 mV. Here R_4 and R_5 are chosen such that the CMRR is limited by the amplifier for values of R_3 greater than 1 kΩ. If R_3 is less than 1 kΩ, the shunting of R_4 by R_5 must be considered when choosing the value of R_3.

The techniques described for balancing the offset voltage at the input of the amplifier offer two main advantages: First, they are universally applicable to all op amps and allow device interchangeability with no modifications to the balance circuitry. Second, they permit balancing without interfering with the internal circuitry of the amplifier. As examples, the input-offset voltage of the μA741 and the μA725 may be easily nulled by connecting a 10-kΩ potentiometer and a 100-kΩ potentiometer, respectively, as shown in Figure 4-31.

Frequency Compensation

One of the most frequent complaints that users had with early op amps was the complexity of their frequency-compensation networks, which can be seen in Figure 4-32. Several manufacturers developed a number of amplifiers that are internally frequency compensated: National's LM107 and LM112, Fairchild's μA741 and μA747, Motorola's 1558, to name the industry standards. In addition, National's general-purpose op amps in the LM101, LM101A, and LM108 series are very simple to compensate for most applications. Each can be compensated by a single 30-pF capacitor, and thus they are interchangeable.

The LM101 and LM101A are generally compensated by a 30-pF capacitor connected between the compensation terminals (the capacitor is connected from the input

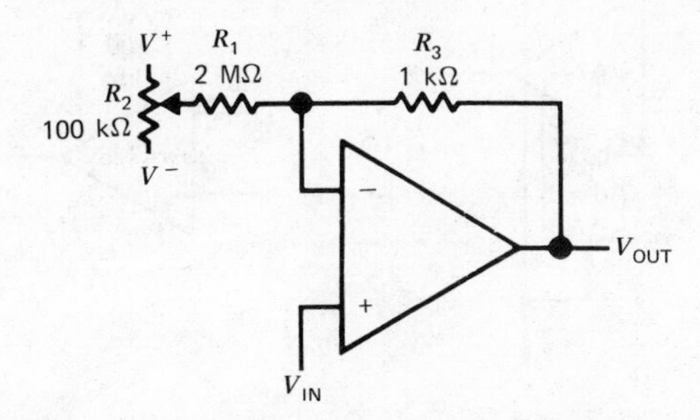

Figure 4-29. Offset-voltage adjustment for voltage followers [*note:* range = $\pm V(R_3/R_1)$].

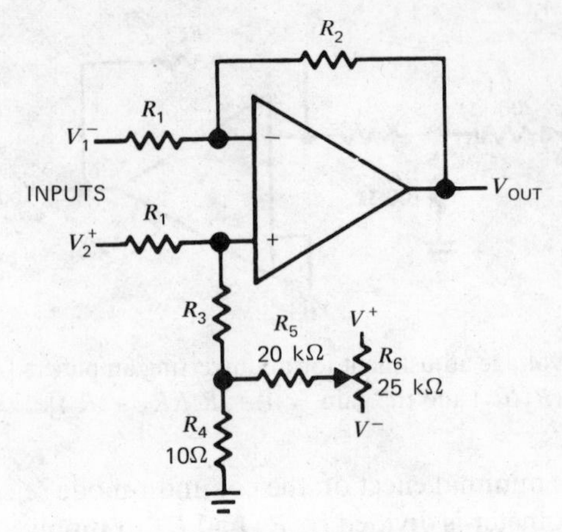

Figure 4-30. Offset-voltage adjustment for differential amplifiers [*note:* $R_2 = R_3 + R_4$; the range $= \pm V(R_5/R_4)(R_1/R_1 + R_3)$; and the gain $= R_2/R_1$].

to the second-stage output). The LM108 can be compensated by this method or by a 100-pF capacitor connected from the output of the second stage to ground, as shown in Figure 4-33. The latter technique improves the high-frequency power-supply rejection of the LM108 by a factor of 10. Another optional technique, called feedforward compensation, can be used on the LM101A to multiply its slew rate approximately 20 times. Still another method, two-pole compensation, extends power bandwidth and reduces gain error.

Frequency-Compensation Requirements

The ease of designing with op amps sometimes obscures some of the rules which must be followed with any feedback amplifier to keep it from oscillating. In general, these problems stem from stray capacitance, excessive capacitive loading, inadequate supply-bypassing, or improper frequency compensation.

In frequency-compensating an op amp, it is best to follow the manufacturer's recommendations. However, if operating speed and frequency response is not a consideration, a greater stability margin can usually be obtained by increasing the

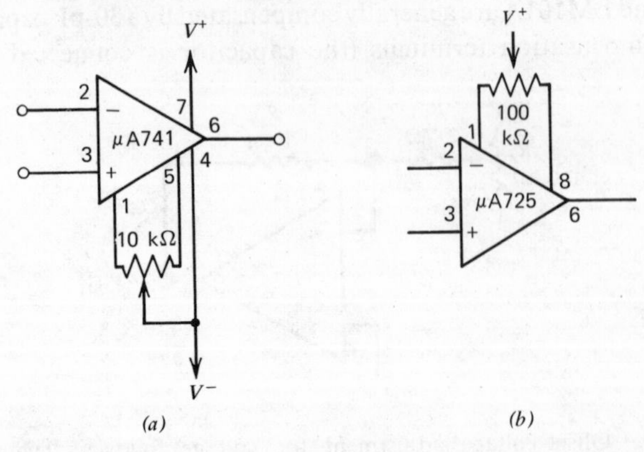

Figure 4-31. Voltage-offset null adjustment circuits for the μA741 and μA725.

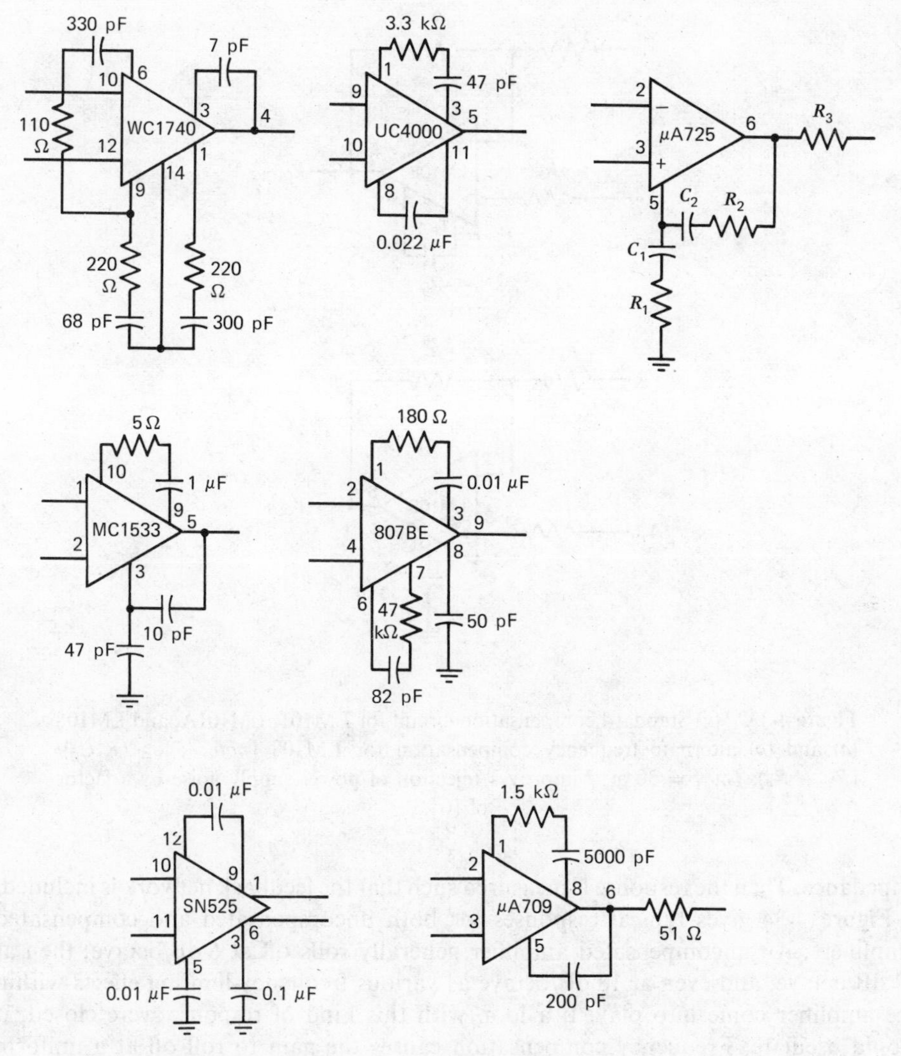

Figure 4-32. Typical frequency-compensation networks used with earlier monolithic operational amplifiers [*note:* use $R_3 = 51\ \Omega$ when the amplifier is operated with capacitive load].

size of the compensation capacitors. For example, replacing the 30-pF compensation capacitor on the LM101 with a 300-pF capacitor will make it ten times less susceptible to oscillation problems in the unity-gain connection. Similarly, on the μA709, using 0.05 μF, 1.5 kΩ, 2000 pF, and 51 Ω components instead of 5000 pF, 1.5 kΩ, 200 pF, and 51 Ω will give 20 dB more stability margin. Capacitor values less than those specified by the manufacturer for a particular gain connection should not be used since they will make the amplifier more sensitive to strays and capacitive loading, or the circuit can even oscillate with worst-case units.

The basic requirement for frequency-compensating a feedback amplifier is to keep the frequency roll-off of the loop gain from exceeding 12 dB/octave when it goes through unity gain. Figure 4-34 shows what is meant by loop gain. The feedback loop is broken at the output, and the input sources are replaced by their equivalent

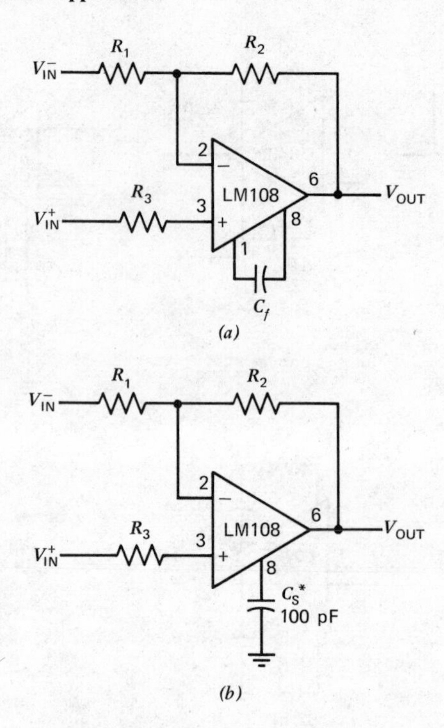

Figure 4-33. (*a*) Standard compensation circuit for LM101, LM101A, and LM108 (*a*) and (*b*) alternate frequency compensation for LM108 [*note:* $C_1 \geq (R_1 C_0)/(R_1 + R_2)$; $C_{OUT} = 30$ pf; * improves rejection of power-supply noise by a factor of 10].

impedance. Then the response is measured such that the feedback network is included.

Figure 4-34 gives typical responses for both uncompensated and compensated amplifiers. An uncompensated amplifier generally rolls off at 6 dB/octave, then at 12 dB/octave, and even at 18 dB/octave as various frequency-limiting effects within the amplifier come into play. If a loop with this kind of response were closed, it would oscillate. Frequency compensation causes the gain to roll off at a uniform

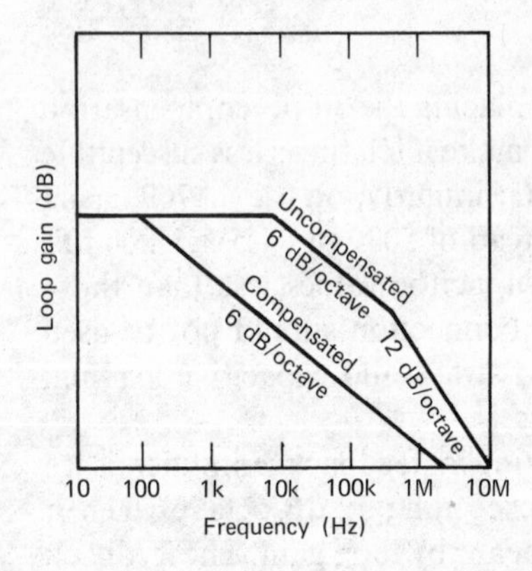

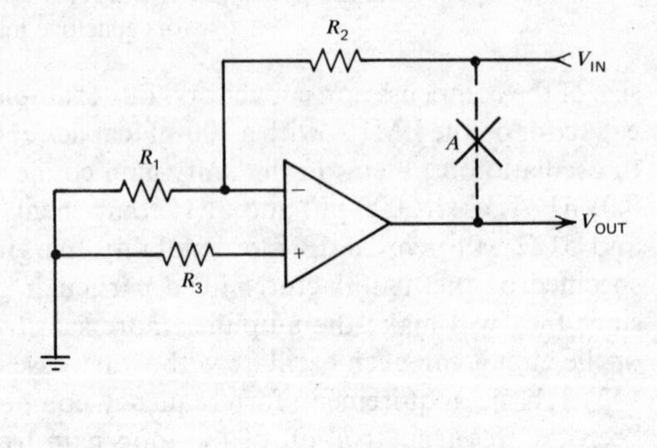

Figure 4-34. Loop gain.

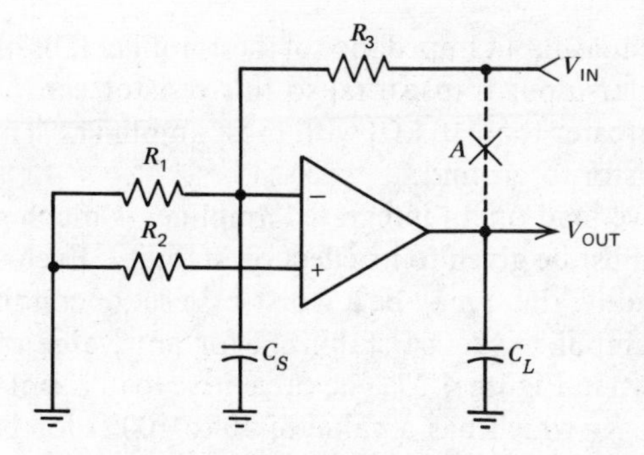

Figure 4-35. External capacitances that affect stability.

6 dB/octave right down through unity gain. This allows some margin for excess roll-off in the external circuitry.

Capacitance Effects on Roll-Off

Some of the external influences which can affect the stability of an op amp are shown in Figure 4-35. One is the load capacitance which can come from wiring, cables, or an actual capacitor on the output. This capacitance works against the output impedance of the amplifier to attenuate high frequencies. If this added roll-off occurs before the loop gain goes through zero, it can cause instability. It should be remembered that this single roll-off point can give more than 6 dB/octave roll-off since the output impedance of the amplifier can be increasing with frequency.

A second source of excess roll-off is stray capacitance on the inverting input. This becomes extremely important with large feedback resistors as might be used with a FET-input amplifier. A relatively simple method of compensating for this stray capacitance is shown in Figure 4-36: A lead capacitor C_1 put across the feedback resistor. Ideally, the ratio of the stray capacitance to the lead capacitor should be equal to the closed-loop gain of the amplifier. However, the lead capacitor can be made larger as long as the amplifier is compensated for unity gain. The only disadvantage of doing this is that it will reduce the bandwidth of the amplifier. Oscillations can also result if there is a large resistance on the noninverting input of the

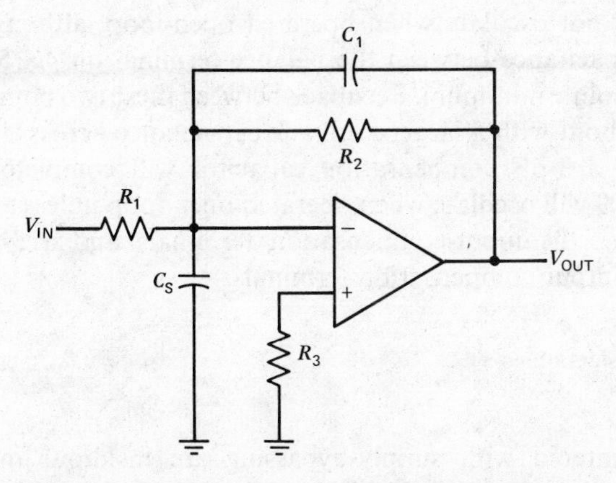

Figure 4-36. Compensating stray-input capacitance.

amplifier. The differential-input impedance of the amplifier falls off at high frequencies (especially with bipolar-input transistors) so this resistor can produce troublesome roll-off if it is much greater than 10 kΩ, with most amplifiers. This is easily corrected by bypassing the resistor to ground.

When the capacitive load on an integrated amplifier is much greater than 100 pF, some consideration must be given to its effect on stability. Even though the amplifier does not oscillate readily, there may be a worst-case set of conditions under which it will. However, the amplifier can be stabilized for any value of capacitive loading using the circuit shown in Figure 4-37. The capacitive load is isolated from the output of the amplifier with R_4 which has a value of 50 to 100 Ω for both the LM101 and the μA709. At high frequencies, the feedback path is through the lead capacitor

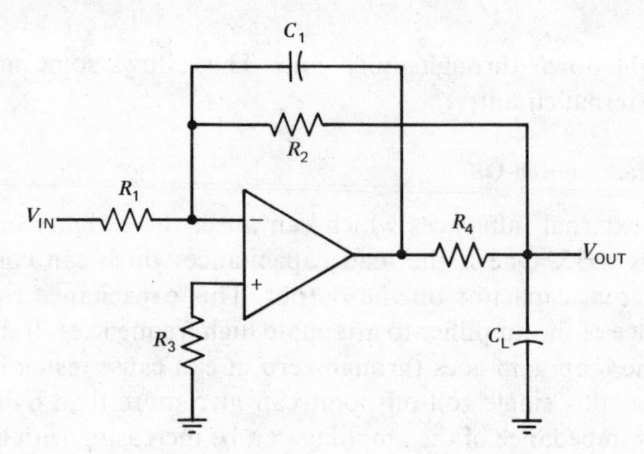

Figure 4-37. Compensating for very large capacitive loads.

C_1 so that the lag produced by the load capacitance does not cause instability. To use this circuit, the amplifier must be compensated for unity gain, regardless of the closed-loop dc gain. The value of C_1 is not too important, but at a minimum its capacitive reactance should be one-tenth the resistance of R_2 at the unity-gain crossover frequency of the amplifier.

When an op amp is operated open-loop, it might appear at first glance that it needs no frequency compensation. However, this is not always the case because the external compensation is sometimes required to stabilize internal-feedback loops.

The LM101 will not oscillate when operated open-loop, although there may be problems if the capacitance between the balance terminal on *pin* 5 and the output is not held to an absolute minimum. Feedback between these two points is regenerative if it is not balanced out with a larger feedback capacitance across the compensation terminals. Usually a 3-pF compensation capacitor will completely eliminate the problem. The μA709 will oscillate when operated open-loop unless a 10-pF capacitor is connected across the input-compensation terminals and a 3-pF capacitor is connected on the output-compensation terminals.

Power-Supply Considerations

BYPASSING

Problems encountered with supply-bypassing are insidious in that they will hardly ever show up in a Nyquist plot. This problem has not really been thoroughly investigated, probably because one sure cure is known: Bypass the positive- and negative-supply terminals of each amplifier to ground with at least a 0.01-μF capacitor.

For example, an LM101 can take over 1-mH inductance in either supply lead without oscillation. This should not suggest that they should be run without bypass capacitors. It has been established that 100 LM101s on a single printed circuit board with common-supply busses will oscillate if the supplies are not bypassed about every fifth device. This happens even though the inputs and outputs are completely isolated.

The μA709, on the other hand, will oscillate under many load conditions with as little as 18 in. of wire between the negative-supply lead and a bypass capacitor. Therefore, it is almost essential to have a set of bypass capacitors for every device.

Operational amplifiers are specified for power-supply rejection at frequencies less than the first break frequency of the open-loop gain. At higher frequencies, the rejection can be reduced depending on how the amplifier is frequency-compensated. For both the LM101 and μA709, the rejection of high-frequency signals on the positive supply is excellent. However, the situation is different for the negative supplies. These two amplifiers have compensation capacitors from the output down to a signal point which is referred to the negative supply, causing the high-frequency rejection for the negative supply to be much reduced. It is therefore important to have sufficient bypassing on the negative supply to remove transients if they can cause trouble from appearing on the output. One fairly large (22 μF) tantalum capacitor on the negative-power lead for each printed-circuit card is usually enough to solve potential problems.

When high-current buffers are used in conjunction with op amps, supply-bypassing and decoupling are even more important since they can feed a considerable amount of signal back into the supply lines. For reference, bypass capacitors of at least 1.0 μF are required for a 50-mA buffer.

When emitter-followers are used to drive long cables, additional precautions are required. An emitter-follower by itself—which is not contained in a feedback loop—will frequently oscillate when connected to a long length of cable. When an emitter-follower is connected to the output of an op amp, it can produce oscillations that will persist no matter how the loop gain is compensated. An analysis of why this happens is not very enlightening, so suffice it to say that these oscillations can usually be eliminated by putting a ferrite bead between the emitter-follower and the cable.

GROUNDING AND LEAD INDUCTANCE

For linear ICs the amount of amplifier sensitivity to power-supply variations usually dictates how "good" a supply must be. As a result, regulation, stability, and noise in the power source contribute little to overall circuit error. To illustrate, assume a 30-μV offset voltage appears at the input of the μA741 op amp for a 1-V change of the power-supply output. Given a $\pm$15-V supply regulated to 0.2% with 0.02%/°C stability, the supply will produce an output variation of 30(0.2% + 0.02%/°C $\times$ 10°C) or 0.12 V for a 10°C ambient change. The resulting amplifier input error is 0.12 V (50 μV/V) or 6 μV, which corresponds to 0.6 μV/°C. Compared to the specified amplifier input-voltage drift of 33 μV/°C, the supply contributes a mere 1.8% of the total input-voltage error.

Operational amplifier input currents and, to a lesser extent, voltage gain and output levels, are all affected by power-supply instability. Similarly, common-mode rejection techniques diminish the circuit response to power-supply noise and ripple.

Current-carrying leads such as ground buses and interconnecting wires can produce side effects that may seriously degrade circuit performance. An especially troublesome problem is the ground loop. Referring to Figure 4-38, resistance in the ground connection simulated by R_1, R_2, and R_3, causes voltage drops to develop due to the various currents in the circuit. The result is increased noise and susceptibility to oscillation.

To minimize such noise, it is best to keep power-return current out of the signal-return leads with a single-point ground (Figure 4-39) that serves the whole system—if it is practical to employ this grounding method throughout the system. If not, the first step is to determine the magnitude of current and the wire (or etch) resistance, and then check the sensitivity of the affected circuit nodes for acceptably low noise.

To eliminate ground loops in more complex systems, it is best to supply power to groups of amplifiers with separate modular supplies (instead of a single supply), returning each common-power lead to a single-point ground where possible. It is in this way that the modularity and low-cost features of these supplies may be used to the best advantage.

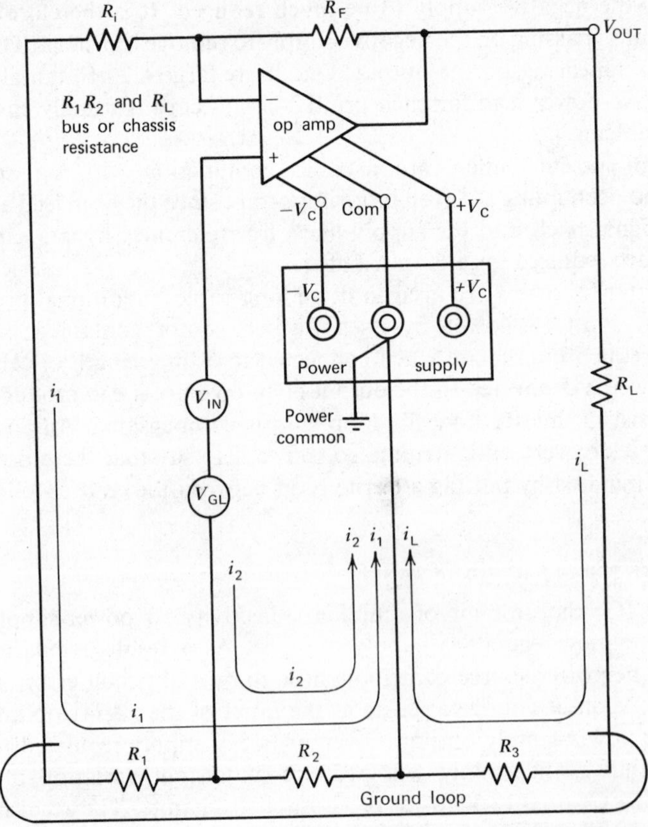

Figure 4-38. Resistors R_1, R_2, and R_3, represent the resistance in the ground circuit. The vector sum of the voltages developed across them is the noise signal that they contribute to the amplifiers. In-phase components can produce oscillation [*note:* V_{GL} is the sum of the voltage drops around the ground loop].

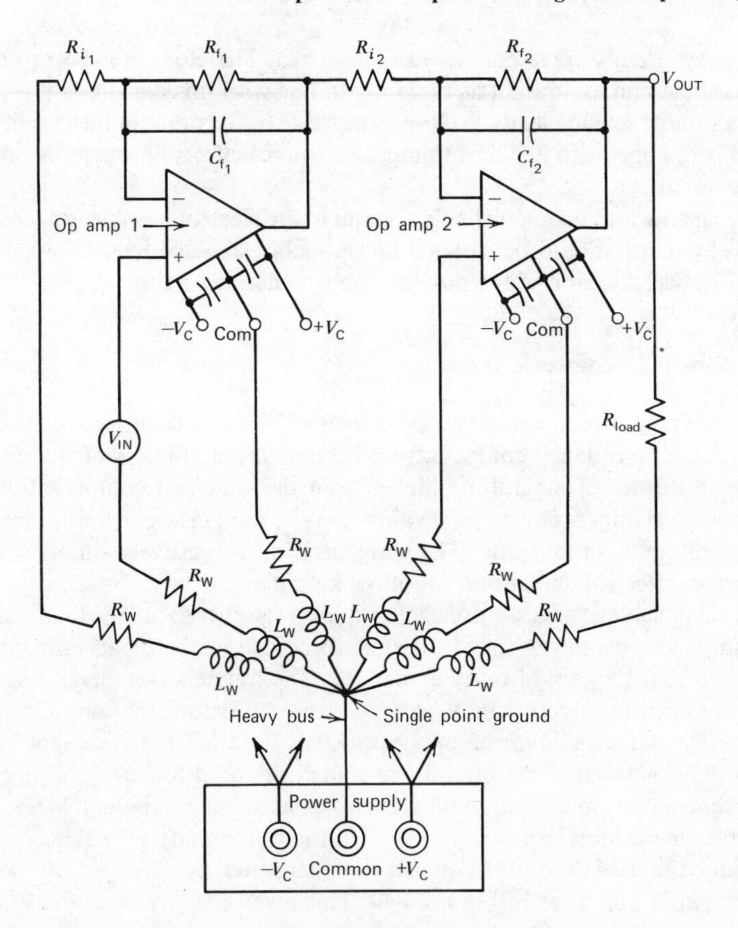

Figure 4-39. Lead inductance degrades systems response and stability [*note:* R_W = wire resistance and L_W = wire inductance].

Proper grounding is especially important in systems in which analog and digital circuits share a common-power supply. Here, the heavy transients that are generated when many digital circuits switch at once can add significantly to the existing ground-loop currents, to greatly affect operation of the analog circuits. Such transient currents are often much larger than the normal ground-loop currents, which might otherwise be tolerable. This problem is often tackled by using separate grounds or employing various ground-isolation schemes between the analog and digital circuits.

Circuit wiring can cause still another adverse effect—that of inductance. The distributed wiring inductance is shown as L_W in Figure 4-39. Lead inductance degrades system transient response and dynamic stability. This inductance can be canceled out by connecting capacitors between the power inputs and common at each amplifier, as illustrated in Figure 4-39.

Each capacitive connection should consist of 1 to 10 μF, shunted by 0.1 μF (when necessary for high-frequency bypass).

Lead inductance in digital circuits is even more critical than in linears because of the ensuing propagation delays. The effects of lead resistance are also greater in digital circuits because this added resistance can cause a significant voltage drop when many circuits are simultaneously switched. Therefore, it is best to keep power leads as short and as straight as possible. In line with this reasoning it is also best to connect remote sensing leads as close as possible to the point in the circuit where that voltage is to appear, to further minimize resistance effects between the power supply and the circuit.

Stray electric and magnetic fields abound in the electronic environment. Therefore, low-level circuits should be isolated and possibly shielded from power-transformer fields. Physical rotation of the power supply sometimes helps.

Feed-Forward Compensation Techniques

The slew rate of an op amp can be increased by using feedforward compensation. This means of frequency compensation extends the usefulness of the device to frequencies an order of magnitude higher than the standard compensation network. With this speed improvement, IC op amps may be used in applications that previously required discrete components. The compensation is relatively simple and does not change the offset voltage or current of the amplifier.

In order to achieve unconditional closed-loop stability for all feedback connections, the gain of an op amp is rolled off at 6 dB/octave, with the accompanying 90° of phase shift, until a gain of unity is reached. The frequency-compensation networks shape the open-loop response to cross unity gain before the amplifier phase-shift exceeds 180°. Unity gain for an op amp such as the LM101A is designed to occur at 1 MHz. This is because the lateral *pnp* transistors used for level-shifting have poor high-frequency response and exhibit excess phase-shift at about 1 MHz. Therefore, the stable closed-loop bandwidth is limited to approximately 1 MHz.

Usually the LM101A is frequency compensated by a single 30-pF capacitor between *pins* 1 and 8, as in Figure 4-40. This gives a slew rate of 0.5 V/μsec. Feed-forward compensation is achieved by connecting a 150-pF capacitor between the inverting input, *pin* 2, and one of the compensation terminals, *pin* 1, as in Figure 4-41. This eliminates the lateral *pnps* from the signal path at high frequencies. Unity-gain bandwidth is 3.5 MHz and the slew rate is 10 V/μsec. Diode D_1 can be added to improve the slew rate with high-speed input pulses.

Figure 4-42 plots the open-loop response using both standard and feedforward

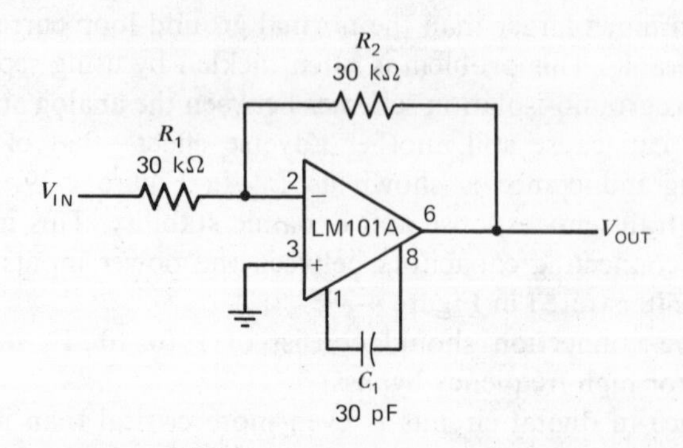

Figure 4-40. Standard frequency compensation.

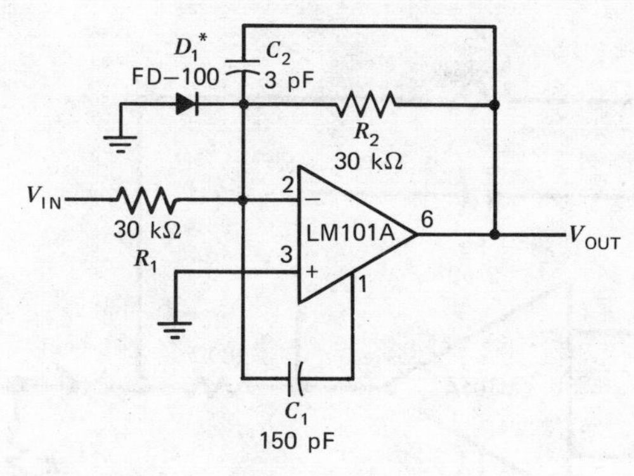

Figure 4-41. Feedforward frequency compensation [*note:* $C_2 \cong (6 \times 10^{-8})/R_2$, * optional to improve response with fast-rising input steps].

compensation techniques. Higher open-loop gain is realized with the fast compensation, since the gain rolls off at about 10 dB/octave until a gain of unity is reached at about 3.5 MHz.

As with all high-frequency, high-gain amplifiers, certain precautions should be taken to ensure stable operation. The power supplies should be bypassed near the amplifier with 0.01-μF disc capacitors. Stray capacitance, such as large lands on *PC* boards, should be avoided at the input, balance, and compensation *pins*. Load capacitance in excess of 75 pF should be decoupled, as in Figure 4-43; however, 500 pF of load capacitance can be tolerated without decoupling at the expense of bandwidth by the addition of 3 pF between *pins* 1 and 8. A small capacitor C_2 is needed as a lead across the feedback resistor to ensure that the roll-off is less than 12 dB/octave at unity gain. The capacitive reactance of C_2 should equal the feedback resistance between 2 and 3 MHz. For integrator applications, the lead capacitor is isolated from the feedback capacitor by a resistor, as in Figure 4-44.

Feedforward compensation may also be used on the LM108, as shown in Figure 4-45. The signal from the inverting input is fed forward around the input stage by a 500-pF capacitor C_1. At high frequencies it provides a phase lead. With this lead, overall phase-shift is reduced and less compensation is needed to keep the amplifier

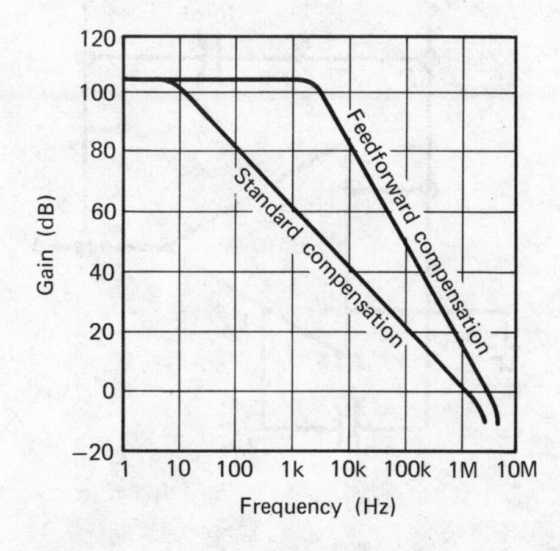

Figure 4-42. Open-loop responses for both frequency-compensation networks.

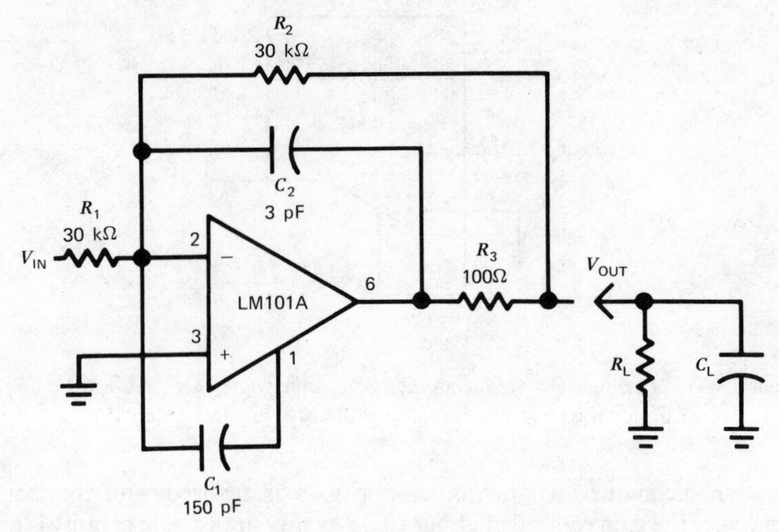

Figure 4-43. Capacitive load isolation.

stable. The C_2–R_1 network provides lag compensation, ensuring that the open-loop gain is below unity before 180° phase-shift occurs. The open-loop gain and phase as a function of frequency is compared with standard compensation in Figure 4-46.

The slew rate is increased from 0.3 V/μsec to about 1.3 V/μsec and the 1 kHz gain is increased from 50 to 10,000. Small-signal bandwidth is extended to 3 MHz. The bandwidth must be limited to 300 MHz because the phase-shift through the lateral *pnp* transistors used in the second stage becomes excessive at higher frequencies.

When the LM108 is used with feedforward compensation, it is less tolerant of capacitive loading and stray capacitance. Precautions must be taken to ensure stability. If load capacitance is greater than about 75 to 100 pF, it must be isolated

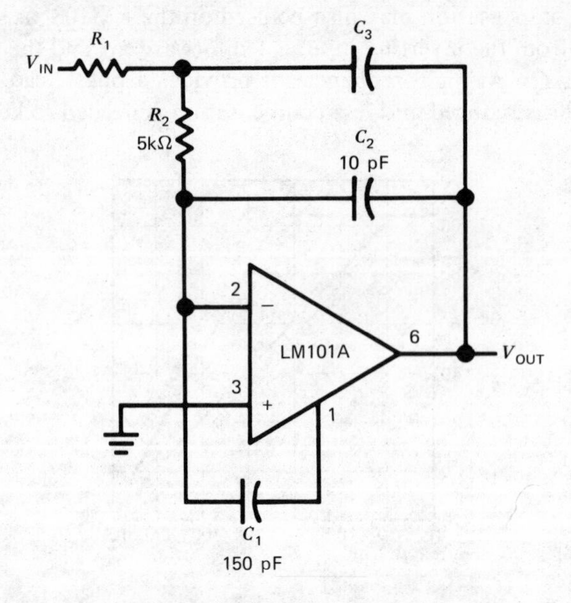

Figure 4-44. Fast integrator.

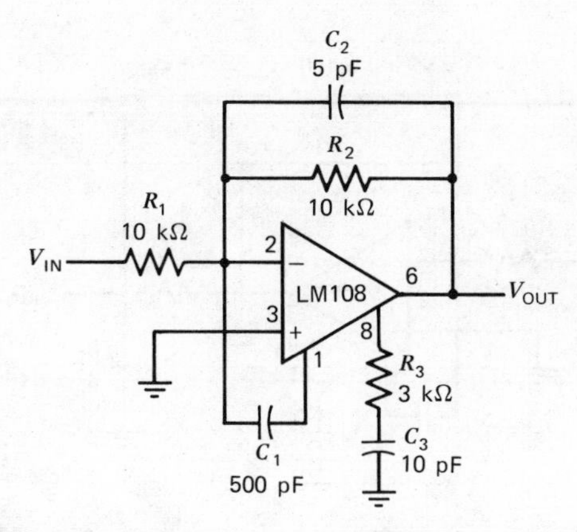

Figure 4-45. LM108 with feedforward compensation.

as shown in Figure 4-47. A small capacitor is always needed to provide a lead across the feedback resistor to compensate for strays at the input. About 3 to 5 pF is the minimum value capacitance at *pins* 1, 2, and 8 when feedforward compensation is used. Additionally, when the source resistance on the noninverting input is greater than 10 kΩ, it should be bypassed with a 0.01-μF capacitor.

As with any externally compensated amplifier, increasing the compensation of the LM108 increased the stability at the expense of slew rate and bandwidth. The circuit shown is for the fastest response. Increasing the size of C_2 to 20 or 30 pF will provide two or three times greater stability and capacitive-load tolerance. Therefore, the size of the compensation capacitor should be optimized for the bandwidth of the particular application.

Feedforward compensation offers a marked improvement over standard compensation. In addition to having higher bandwidth and faster slew rate, there is vanishingly small gain error from dc to 3 kHz, and less than 1% gain error up to 100 kHz as a unity-gain inverter. The power bandwidth is also extended from 6 to 250 kHz. Some applications for this type of amplifier are: Fast-summing amplifier, pulse amplifier, digital-to-analog and analog-to-digital systems, and fast integrator.

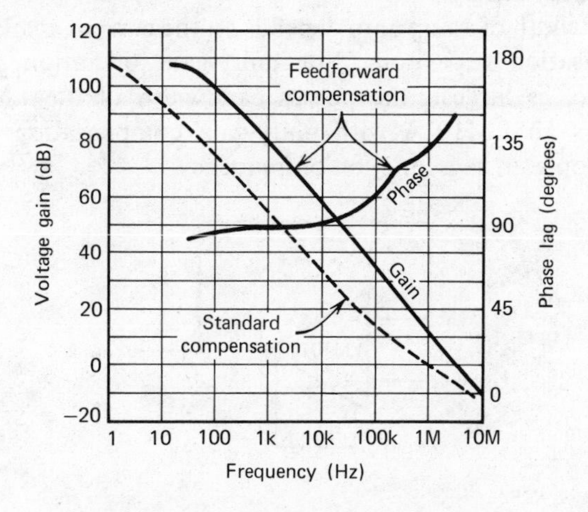

Figure 4-46. Open-loop voltage gain.

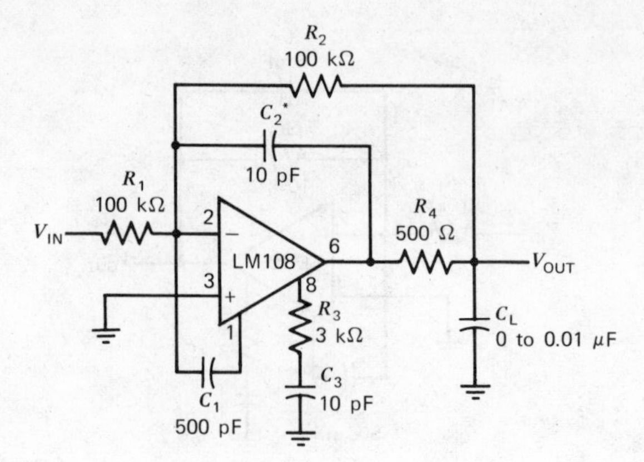

Figure 4-47. Decoupling load capacitance [*note:* * $C_2 > (5 \times 10^5)/R_2$ pF].

Fast Compensation Extends Power Bandwidth

In all IC op amps the power bandwidth depends on the frequency compensation. Normally, compensation for unity-gain operation is accompanied by the lowest power bandwidth. A technique is presented which extends the power bandwidth of the LM101A for noninverting gains of unity to 10, and also reduces the gain error at moderate frequencies.

In order to achieve unconditional stability, an op amp is rolled off at 6 dB/octave, with an accompanying 90° of phase-shift, until a gain of unity is reached. Unity gain in most monolithic op amps is limited to 1 MHz, because the lateral *pnps* used for level-shifting have poor frequency response and exhibit excess phase shift at frequencies above 1 MHz. Hence, for stable operation, the closed-loop bandwidth must be less than 1 MHz where the phase-shift remains below 180°.

For high closed-loop gains, less severe frequency compensation is necessary to roll the open-loop gain off at 6 dB/octave until it crosses the closed-loop gain. The frequency where it crosses must, as previously mentioned, be less than 1 MHz. For closed-loop gains between 1 and 10, more frequency compensation must be used to ensure that the open-loop gain has been rolled off soon enough to cross the closed-loop gain before 1 MHz is reached.

The power bandwidth of an op amp depends on the current available to charge the frequency-compensation capacitors. For unity-gain operation, where the compensation capacitor is largest, the power bandwidth of the LM101A is 6 kHz. Figure 4-48 shows an LM101A with unity-gain compensation and Figure 4-50 shows the open-loop gain as a function of frequency.

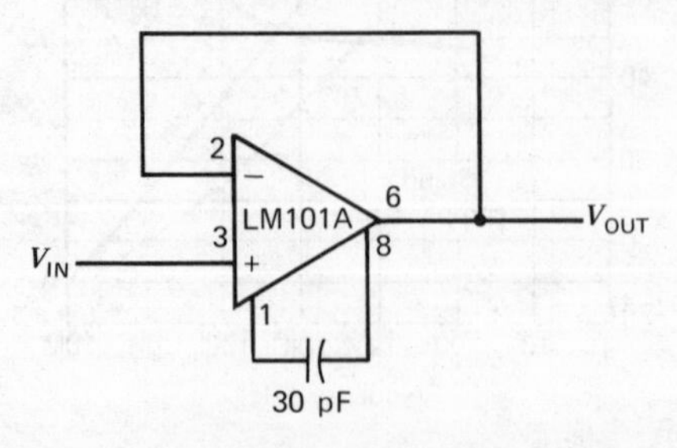

Figure 4-48. LM101A with standard frequency compensation for unity gain.

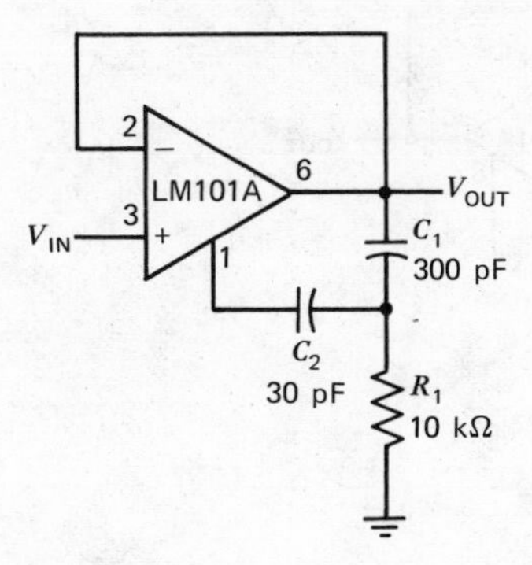

Figure 4-49. LM101A with frequency compensation to extend power bandwidth.

Two-Pole Compensation

A two-pole frequency compensation network, as shown in Figure 4-49 provides more than a factor of 2 in improvement in power bandwidth and reduced gain error at moderate frequencies. The network consists of a 30-pF capacitor, which sets the unity gain frequency at 1 MHz, along with a 300-pF capacitor and a 10 kΩ resistor. By dividing the ac-output voltage with the 10 kΩ resistor and 300-pF capacitor, there is less ac voltage across the 30-pF capacitor and less current is needed for charging. Since the voltage division is frequency-sensitive, the open-loop gain rolls off at 12 dB/octave until a gain of 20 is reached at 50 kHz. From 50 kHz to 1 MHz the 10 kΩ resistor is larger than the impedance of the 300-pF capacitor and the gain rolls off at 6 dB/octave. The open-loop gain plot is shown in Figure 4-50. To ensure sufficient drive to the 300-pF capacitor, it is connected to the output, *pin* 6, rather than *pin* 8. With this frequency-compensation method, the power bandwidth is typically 15 to 20 kHz as a follower, or unity-gain inverter.

This frequency compensation, in addition to extending the power bandwidth, provides an order of magnitude lower gain error at frequencies from dc to 5 kHz. Some applications where it would be helpful to use the compensation are: Differential amplifiers, audio amplifiers, oscillators, and active filters. Figure 4-51 depicts additional examples of *RC* frequency compensation.

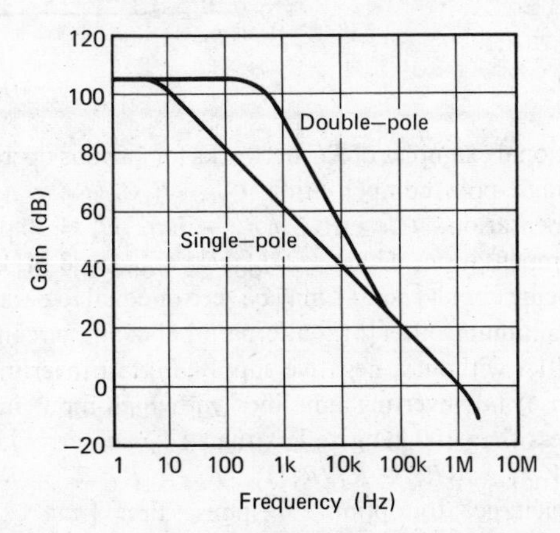

Figure 4-50. Open-loop response for standard and double-pole compensation networks.

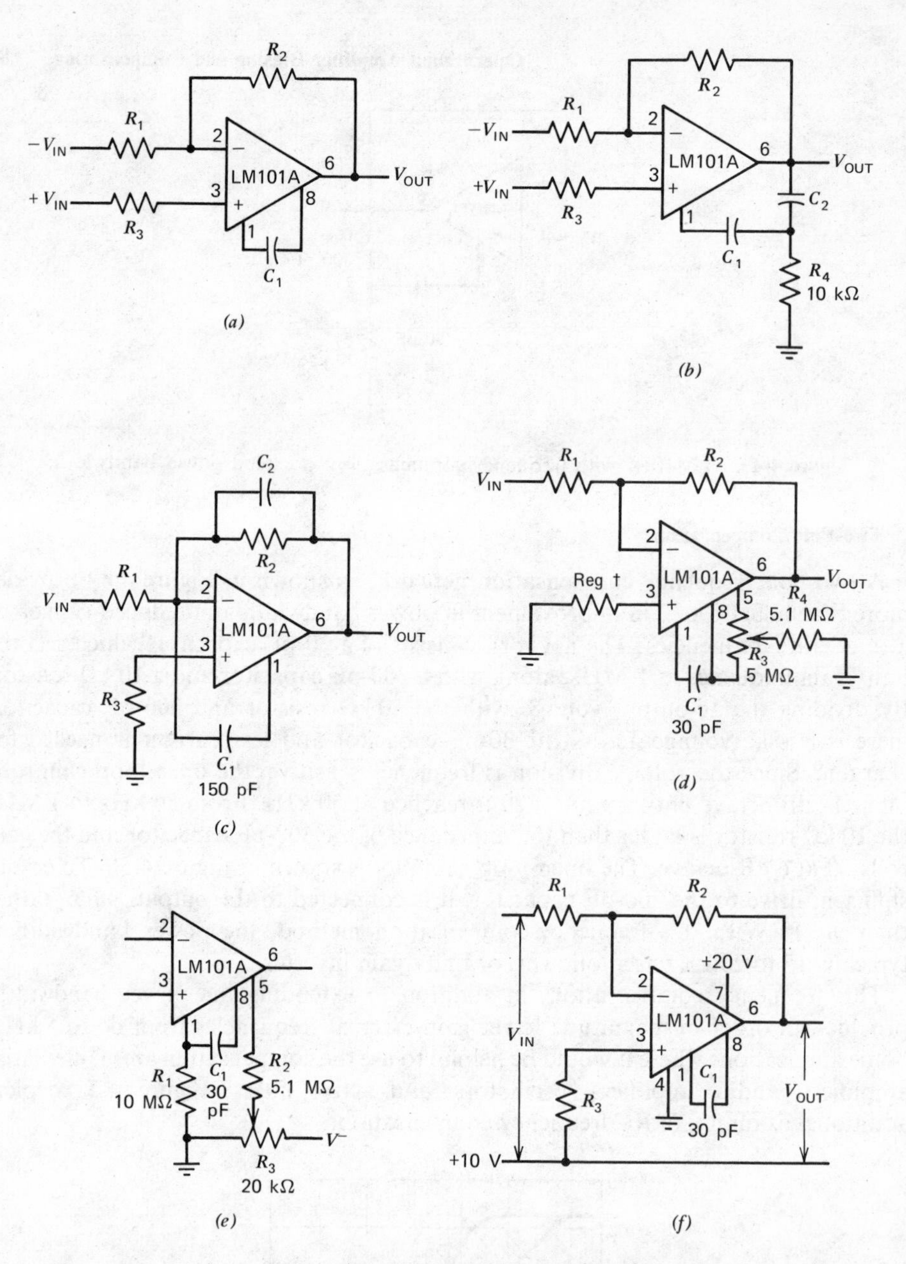

Figure 4-51. Additional examples of *RC* networks for various operational amplifier applications: (*a*) single-pole compensation $C_1 \geq R_1C_S/(R_1 + R_2)$, $C_S = 30$ pF; (*b*) two-pole compensation, $C_1 \geq R_1C_S/(R_1 + R_2)$, $C_S = 30$ pF, $C_2 = 10C_1$; (*c*) feedforward compensation, $C_2 = 1/2\pi f_{OUT}R_2$, $f_{OUT} = 3$ MHz; (*d*) inverting amplifier with balancing circuit [*note:* † may be zero or equal to parallel combination of R_1 and R_2 for minimum offset]; (*e*) alternate balancing circuit; (*f*) circuit for operating the LM101A without a negative supply; (*g*) fast inverting amplifier with high-input impedance; (*h*) inverting amplifier with high-input impedance [*note:* source impedance less than 100 kΩ give less than 1% gain error]; (*i*) noninverting ac amplifier [*note:* $V_{OUT} = \{(R_1 + R_2)/R_1\}V_{IN}$, $R_{IN} = R_3$, $R_3 = R_1 \parallel R_2$]; (*j*) neutralizing input capacitance to optimize response time [*note:* $C_N = R_1/R_2C_S$]; (*k*) offset balancing of μA709; (*l*) unity-gain inverting amplifier [*note:* pin connections shown are for metal can package].

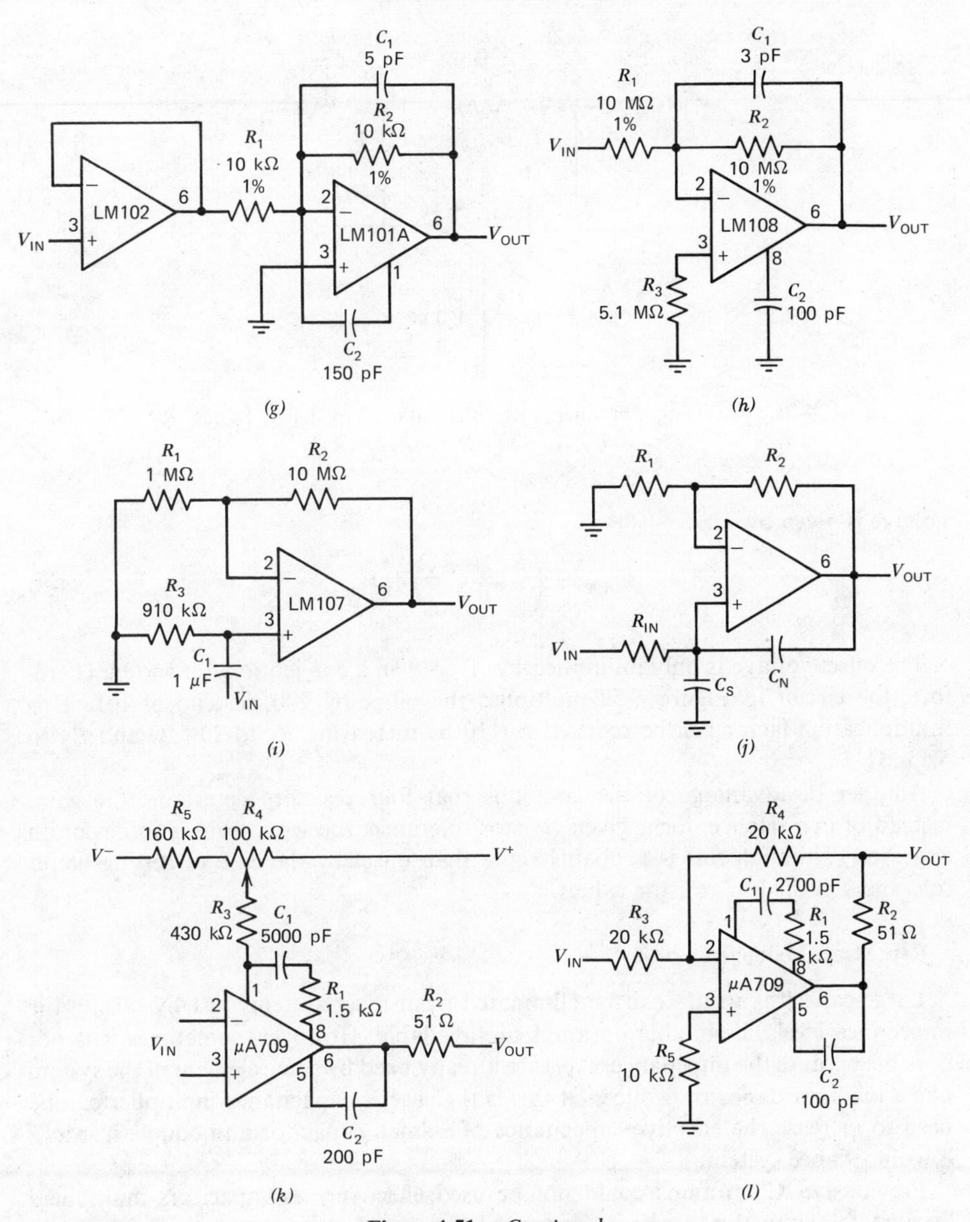

Figure 4-51. *Continued*

Resistance Multiplication

When an inverting op amp must have high-input resistance, the resistor values required can get out of hand. For example, if a 2 MΩ-input resistance is needed for an amplifier with a gain of 100, a 200-MΩ feedback resistor is called for. This resistance can, however, be reduced using the circuit in Figure 4-52. A divider with a ratio of 100 to 1 (R_3 and R_4) is added to the output of the amplifier: Unity-gain feedback is applied from the output of the divider, giving an overall gain of 100 using only 2-MΩ resistors. This circuit does increase the offset voltage somewhat. The output-offset

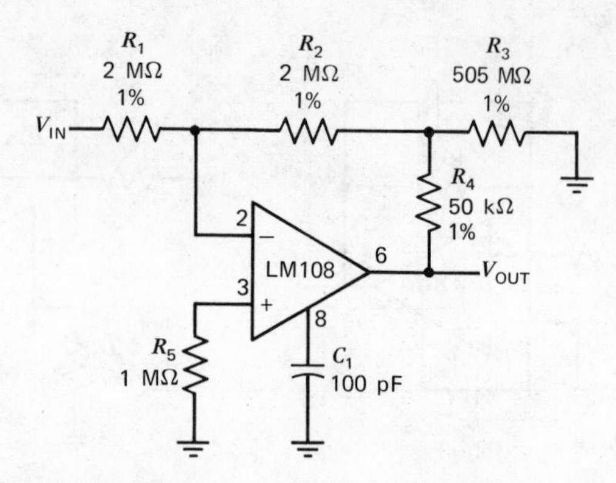

Figure 4-52. Inverting amplifier with high-input impedance [*note:* $R_2 \geq R_1$, $R_2 \gg R_3$, $A_V = R_2(R_3 + R_4)/R_1R_3$].

voltage is given by

$$V_{OUT} = \left(\frac{R_1 + R_2}{R_2}\right)A_V V_{OS} \tag{4-8}$$

The offset voltage is only multiplied by $A_V + 1$ in a conventional inverter. Therefore, the circuit in Figure 4-52 multiplies the offset by 200, instead of 101. This multiplication factor can be reduced to 110 by increasing R_2 to 20 MΩ and R_3 to 5.55 kΩ.

Another disadvantage of the circuit is that four resistors determine the gain, instead of two. Hence, for a given resistor tolerance, the worst-case gain deviation is greater, although this is probably more than offset by the ease of getting better tolerances in the low-resistor values.

Capacitance Multipliers

Large-capacitor values can be eliminated from most systems just by raising the impedance levels, if suitable op amps are available. However, sometimes it is not possible because the impedance levels are already fixed by some element of the system like a low-impedance transducer. If this is the case, a capacitance multiplier can be used to increase the effective capacitance of a small capacitor and couple it into a low-impedance system.

Previously, IC op amps could not be used effectively as capacitors multipliers because the equivalent leakages generated due to offset current were significantly greater than the leakages of large tantalum capacitors. With circuits such as the LM108, this has changed. The circuit shown in Figure 4-53 generates an equivalent capacitance of 100,000 μF with a worst-case leakage of 8 μA—over a −55 to 125°C temperature range.

The performance of the circuit is described by the equations given in Figure 4-53, where C is the effective output capacitance, I_L is the leakage current of this capacitance and R_S is the series resistance of the multiplied capacitance. The series resistance is relatively high, so high-Q capacitors cannot be realized. Hence, such applications as tuned circuits and filters are ruled out. However, the multiplier can still be used in timing circuits or servo compensation networks where some resistance is usually

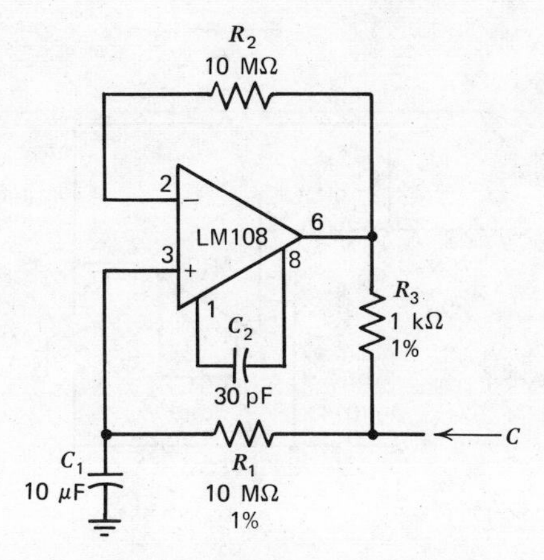

Figure 4-53. Capacitance multiplier [*note:* $C = (R_1/R_3)C_1$, $I_L = (V_{OS} + I_{OS}R_1)/R_3$, $R_S = R_3$].

connected in series with the capacitor or the effect of the resistance can be compensated for.

One final point is that the leakage current of the multiplied capacitance is not a function of the applied voltage. It persists even with no voltage on the output. Therefore, it can generate offset errors in a circuit, rather than the scaling errors caused by conventional capacitors. Figures 4-54 and 4-55 depict two additional capacitance-multiplier circuits.

Design Precautions

Regardless of the type of op amp used, precautions should be taken to protect it from abnormal operating conditions. The majority of damaged ICs are the result of either Zenering the emitter-base junction of *npn* transistors, or excessive current

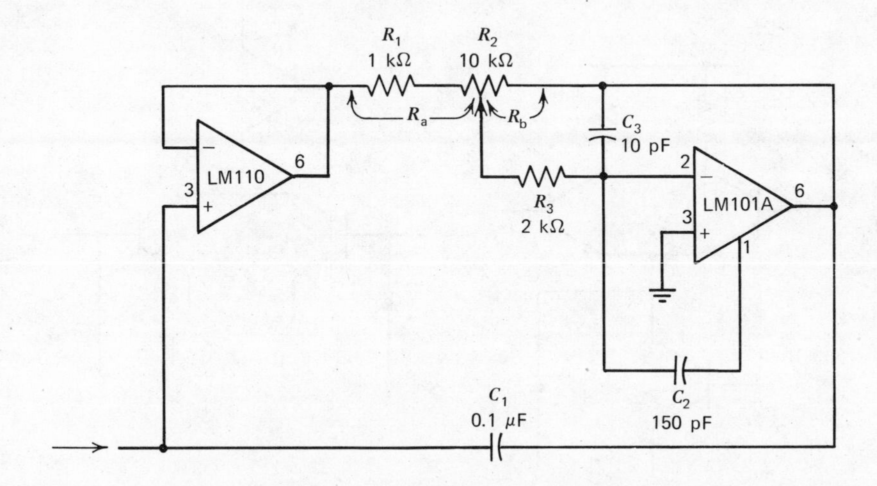

Figure 4-54. Variable-capacitance multiplier [*note:* $C = (1 + R_b/R_a)C_1$].

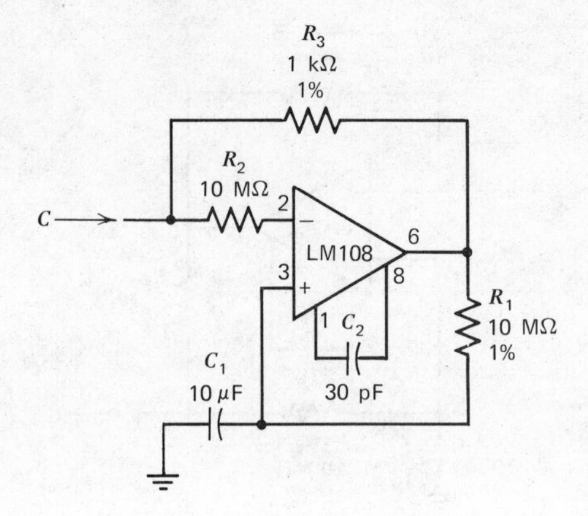

Figure 4-55. Negative capacitance multiplier [*note:* $C = -(R_2/R_3)C_1$, $I_L = (V_{OS} + R_2 I_{OS})/R_3$, $R_S \cong R_3(R_1 + R_{IN})/R_{IN}A_{V\,OUT}$].

fusing the aluminum interconnects on the chip. The emitter-base junction breaks down between 6 and 8 V for most linear processes.

When this junction is Zenered, even for a short time with low current, the beta of the *npn* transistor degrades. If this occurs on the input transistor of an op amp, an increase in bias current and offset current is seen. Metal interconnects on the chip can melt when high currents (400 mA) find a sneak path through isolation diodes.

Figure 4-56 shows two examples of a differential amplifier with a current source, as is used in many amplifier circuits. If the inputs are forced above the positive supply or below the negative supply, damage can occur. For positive-input voltage, the collector-base junction and the isolation junction on the resistors will forward-bias, giving a low-impedance path to the positive supply. If the input is driven more negative than the minus supply, the substrate diode on the current source transistor

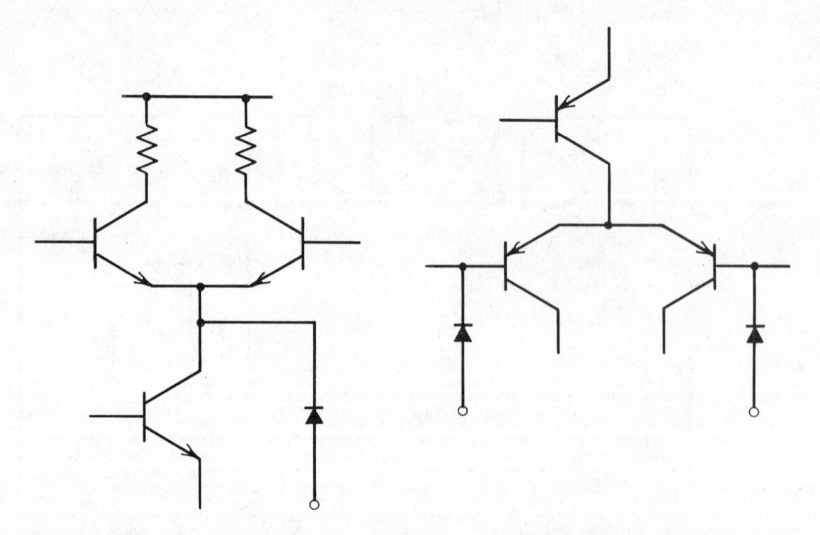

Figure 4-56. Integrated circuit differential amplifiers with substrate diodes.

will forward-bias and Zener the emitter-base junction of the input transistors. Either of these conditions can damage, if not destroy the IC.

When *pnp* transistors are used, positive voltage on the input is usually not dangerous since the emitter-base breakdown is about 60 V. Negative-input voltages will forward-bias the base-substrate diode, and large currents can flow.

Damaging voltages can occur in unsuspected ways. Capacitors connected to an op amp, as in integrators or sample-and-hold circuits, are troublesome. During turn-off when the power-supply voltage decreases, the capacitors can be discharged through the IC.

Power-supply reversal is another problem. During turn-on or turn-off when dual supplies are used, reversal can sometimes occur. This is because loads between the

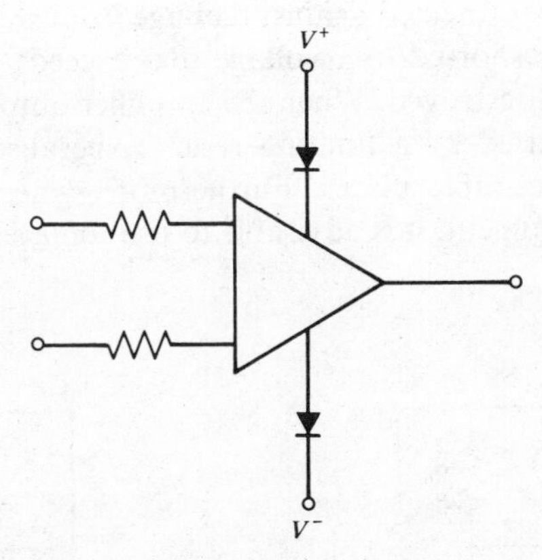

Figure 4-57. Operational amplifier protective circuitry for transients, power-supply reversal, and high common-mode voltages.

supplies and unequal supply-loading cause the output capacitors to discharge at different rates. One supply will then draw current from the other supply, causing its output voltage to reverse.

Thus, in general, maximum voltage ratings must be strictly observed when using op amps. Input and output voltages must not exceed either of the supply voltages during usage. Transients that occur when the power supply is turned ON and OFF often result in a higher input voltage than supply voltage. This condition can cause a catastrophic op amp failure. The minus supply voltage or ground should always be applied first or simultaneously with the positive supply to avoid "latchup" or device destruction. Some op amps can be destroyed when an input voltage is applied to the op amp and both supplies are turned OFF. This condition turns ON the collector-base diode of the input transistors. The op amp will be degraded or destroyed because the positive-power supply, by being OFF, is essentially grounded.

Device inputs can also be degraded if a dc voltage is placed on the output of an integrator or on the input of a sample-and-hold circuit and the positive supply is again OFF. If a large capacitor is used, the inputs may be destroyed.

What is the solution? A high-voltage diode inserted in each power-supply lead will prevent failures caused by transients and power-supply reversals. A resistor inserted in each input-signal line will limit the transient inputs caused by high common-mode voltages. These circuit fixes are diagrammed in Figure 4-57.

The specific techniques given in Figure 4-58 apply to the LM101A, but the advice is applicable to almost any IC op amp, even though the exact reasons may vary with different devices.

When driving either input of the LM101A from a low-impedance source, a limiting resistor should be placed in series with the input lead to limit the peak instantaneous output current of the source of something less than 100 mA. This is especially important when the inputs go outside a piece of equipment, where they could accidentally be connected to high-voltage sources.

Large capacitors on the input (>0.1 μF) should be treated as a low-source impedance and isolated with a resistor. Low-impedance sources do not cause a problem unless their output voltage exceeds the supply voltage. However, since the supplies go to zero when they are turned OFF, the isolation is usually needed.

The output circuitry is protected against damage from shorts to ground on either supply. However, if it is shorted to a voltage that exceeds the positive or negative supplies, the unit can be destroyed. When the amplifier output is connected to a test point, it should be isolated by a limiting resistor, because test points frequently become shorted to undesirable places. Furthermore, when the amplifier drives a load external to the equipment, it is advisable to use some sort of limiting resistance to preclude mishaps.

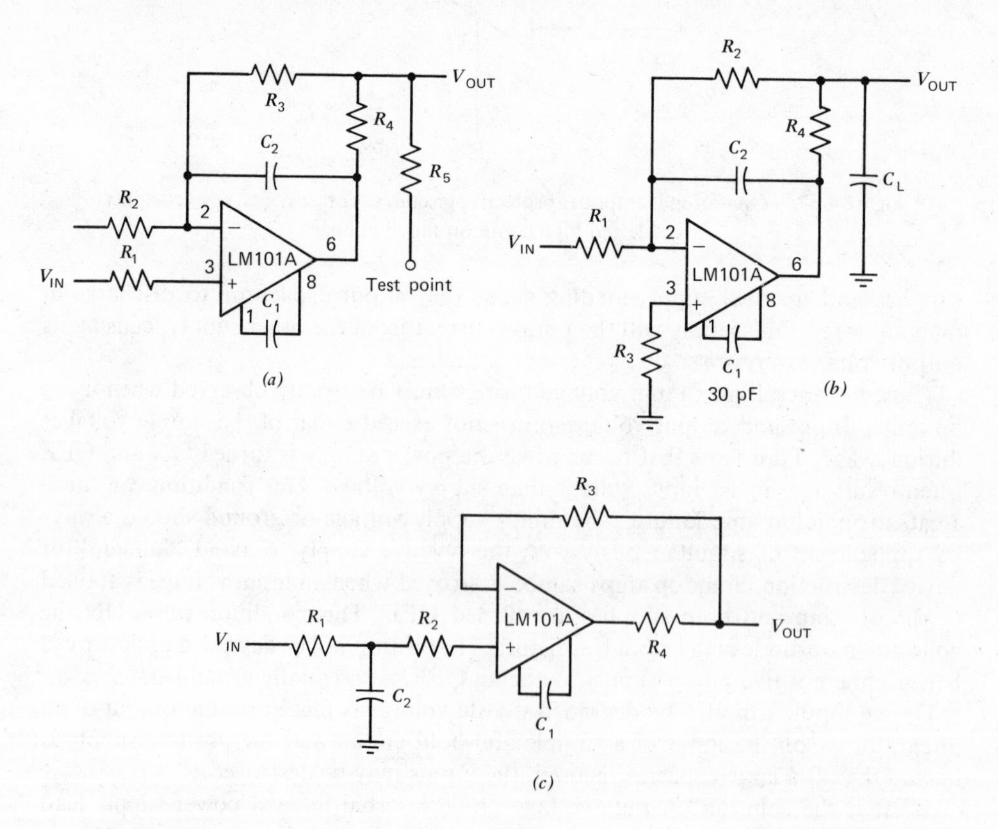

Figure 4-58. Protecting an operational amplifier from abnormal operating conditions: (*a*) protecting against gross fault conditions; (*b*) isolating large capacitive loads; (*c*) current-limiting.

Precautions should be taken to ensure that the power supplies for the IC never become reversed, even under transient conditions. With reverse voltages greater than 1 V, the IC will conduct excessive current, fusing internal aluminum interconnects. If there is a possibility of this happening, clamp diodes with a high peak-current rating should be installed on the supply lines. Reversal of the voltage between V^+ and V^- will always cause a problem, although in many circuits reversals with respect to ground can also lead to difficulties.

The minimum values given for the frequency-compensation capacitor are stable only for source resistances less than 10 kΩ, stray capacitances on the summing junction less than 5 pF, and capacitive loads smaller than 100 pF. If any of these conditions is not met, it becomes necessary to overcompensate the amplifier with a larger compensation capacitor. Alternately, lead capacitors can be used in the feedback network to negate the effect of stray capacitance and large feedback resistors, or an *RC* network can be added to isolate capacitive loads.

Although the LM101A is relatively unaffected by supply-bypassing, this contingency cannot be ignored altogether. Generally it is necessary to bypass the supplies to ground at least once on every circuit card, and more bypass points may be required if more than five amplifiers are used. When feedforward compensation is employed, however, it is advisable to bypass the supply leads of such amplifiers with low-inductance capacitors because of the high frequencies involved.

VOLTAGE-FOLLOWERS, BUFFERS, AND DRIVERS

There are quite a few applications where op amps are used as voltage-followers. These include sample-and-hold circuits and active filters as well as general-purpose buffers for transducers or other high-impedance signal sources. The general usefulness of such an amplifier is particularly enhanced if it is both fast and has a low-input bias current. High speed permits including the buffer in the signal path or within a feedback loop without significantly affecting response or stability. Low-input current prevents loading of high-impedance sources, which is the reason for using a buffer in the first place.

The LM110—An Improved Integrated Circuit Voltage-Follower

The LM102 was designed specifically as a voltage-follower. Therefore, it was possible to optimize performance so that it worked better than general-purpose IC amplifiers in this application. This was particularly true with respect to obtaining low-input currents along with high-speed operation.

One secret of the LM102s performance is that followers do not require level-shifting. Hence, lateral *pnp*s can be eliminated from the gain path. This has been the most significant limitation on the frequency response of general-purpose amplifiers. Second, it was the first IC to use supergain transistors. With these devices, high-speed operation can be realized along with low-input currents.

The LM110 is a voltage-follower that has been designed to supersede the LM102. It is considerably more flexible in its application and offers substantially improved performance. In particular, the LM110 has lower offset-voltage drift, input current, and noise. Further, it is faster, less prone to oscillations, and operates over a wider range of supply voltages. Devices in the LM110 series, including the low-cost LM310,

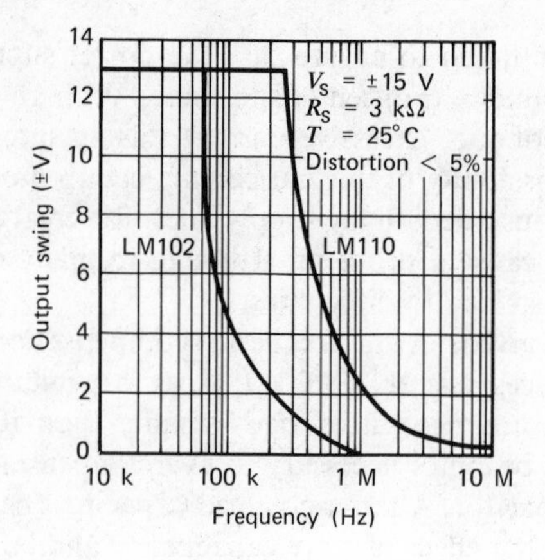

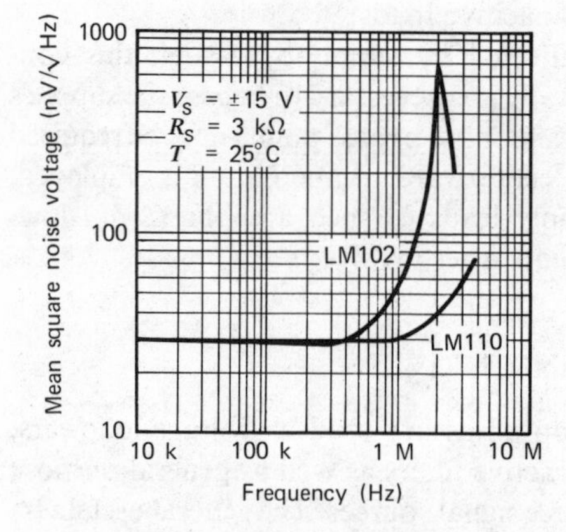

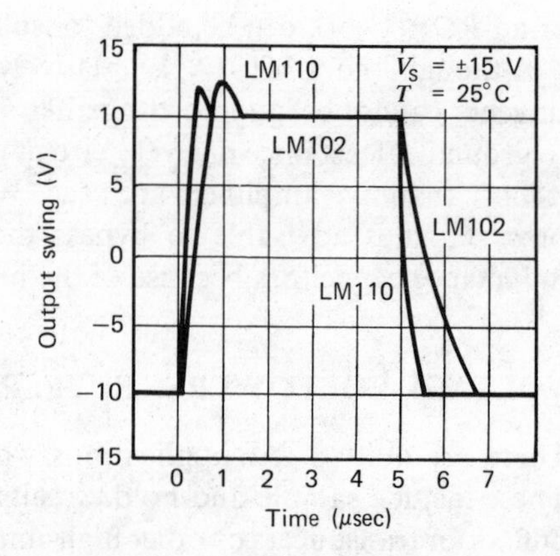

Figure 4-59. Comparisons of LM102 and LM110 performance factors.

slew at 30 V/μsec can operate on 5 V supplies and are therefore compatible with logic circuits.

The advantages of the LM110 over the LM102 are described by the curves in Figure 4-59. Improvements not included are increased output swing under load, larger small-signal bandwidth, and elimination of oscillations with low-impedance sources. The performance of these devices is also compared with general-purpose op amps in Tables 4-1 and 4-2. The advantages of optimizing an IC for this particular slot are clearly demonstrated. Figure 4-60 depicts additional circuits for the LM110.

Operating Hints, LM102

A number of precautions concerning the proper use of the LM102 are worth repeating here:

1. The output is short-circuit-protected; however, the input is clamped to the output to prevent excessive voltage from being developed across the input transistors. If the amplifier is driven from low-source impedances, excessive current can flow through these clamp diodes when the output is shorted. This can be prevented by inserting a resistor larger than 3 kΩ in series with the input.

Table 4-1 Comparing performance of military grade integrated circuit operational amplifiers in the voltage-follower connection

Device	*Offset*[a] *voltage* (mV)	*Bias*[a] *current* (nA)	*Slew*[b] *rate* (V/μsec)	*Bandwidth*[b] (MHz)	*Supply*[c] *current* (mA)
LM110	6.0	10	40	20	5.5
LM102	7.5	100	10	10	5.5
MC1556	6.0	30	2.5	1	1.5
μA715	7.5	4000	20	10	7.0
LM108	3.0	3	0.3	1	0.6
LM108A	1.0	3	0.3	1	0.6
LM101A	3.0	100	0.6	1	3.0
μA741	6.0	1500	0.6	1	3.0

[a] Maximum for $-55°C \leq T_A \leq 125°C$.
[b] Maximum at 25°C.
[c] Typical at 25°C.

2. The circuit cannot deliver its full slew rate into capacitive loads greater than 100 pF unless more sink current is provided on the output with a resistor between *pins* 4 and 5.
3. The amplifier may oscillate when operated with capacitive loads between 200 pF and 0.01 μF.
4. As is the case with any high-frequency amplifier, the power supply leads of the LM102 should be bypassed with capacitors greater than 0.01 μF located as close as possible to the device. This is particularly true if the LM102 is driving capacitive loads.

Figure 4-61*a* gives the connection for getting full-output swing into loads less than 8 kΩ. The external resistor R_1 should not be made less than 100 Ω as this could cause limiting on positive peaks. Figure 4-61*b* shows how to connect a potentiometer to balance out the offset voltage. Figure 4-61*c* gives the placement of a clamp diode which can be used to reduce the overshoot that occurs when the follower is driven with large-input pulses with a leading-edge slope greater than 10 V/μsec. The diode is only needed, however, when the source resistance is less than 30 kΩ since the slope seen by the amplifier will be reduced by the input capacitance with the higher source resistances.

Table 4-2 Comparison of commercial-grade devices

Device	*Offset*[a] *voltage* (mV)	*Bias*[a] *current* (nA)	*Slew*[b] *rate* (V/μsec)	*Bandwidth*[b] (MHz)	*Supply*[a] *current* (mA)
LM310	7.5	7.0	40	20	5.5
LM302	15	30	20	10	5.5
MC1456	10	30	2.5	1	1.5
μA715C	7.5	1500	20	10	10
LM308	7.5	7.0	0.3	1	0.8
LM308A	0.5	7.0	0.3	1	0.8
LM301A	7.5	250	0.6	1	3.0
μA741C	6.0	500	0.6	1	3.0

[a] Maximum at 25°C.
[b] Typical at 25°C.

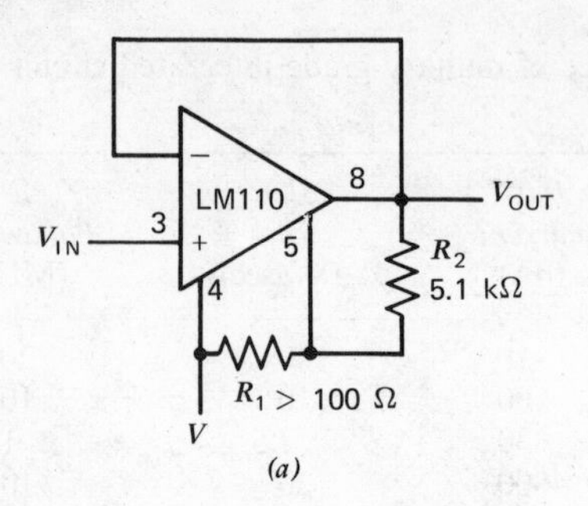

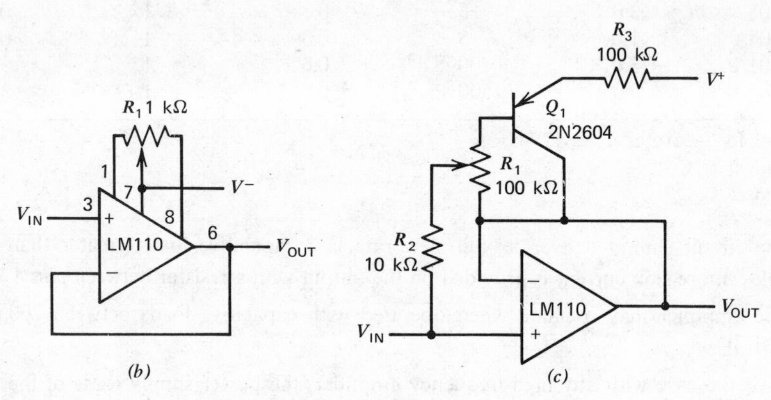

Figure 4-60. Auxiliary circuits for the LM110: (*a*) increasing negative swing under load; (*b*) offset balancing; (*c*) bias-current compensation.

Operational Amplifiers as Voltage-Followers

The comments made earlier in this chapter about the use of op amps as unity-gain buffers must be borne in mind when using standard op amps as voltage-followers. The voltage-follower connection, shown again in Figure 4-62 for the LM101, is its worst-case feedback connection. However, when some special feature of an op amp is wanted, the amplifier may be preferred to a device specifically designed as a voltage-follower. For example, the collection of examples in Figure 4-63 includes an LM101A with the bandwidth-extending compensation described previously.

AC Voltage-Followers

The minimum input resistance of the LM102, which is 10,000 MΩ, can be neglected completely in dc-amplifier applications. However, if the amplifier is ac-coupled, the designer must supply a biasing resistor to supply the input current.

This drastically reduces the input resistance. Figure 4-64 illustrates how to bootstrap the bias resistor of a voltage-follower to get higher input resistance (note that the networks for the LM102 and LM110 are identical). Even though a 200-kΩ bias resistor is used for good dc stability, the input resistance is about 12 MΩ at 100 Hz, increasing to 100 MΩ at 1 kHz.

AC-Coupled High-Input Impedance Amplifier

High-input impedance and low-input capacitance are achieved with a unity-gain buffer amplifier having positive feedback. The circuit in Figure 4-65 has an input

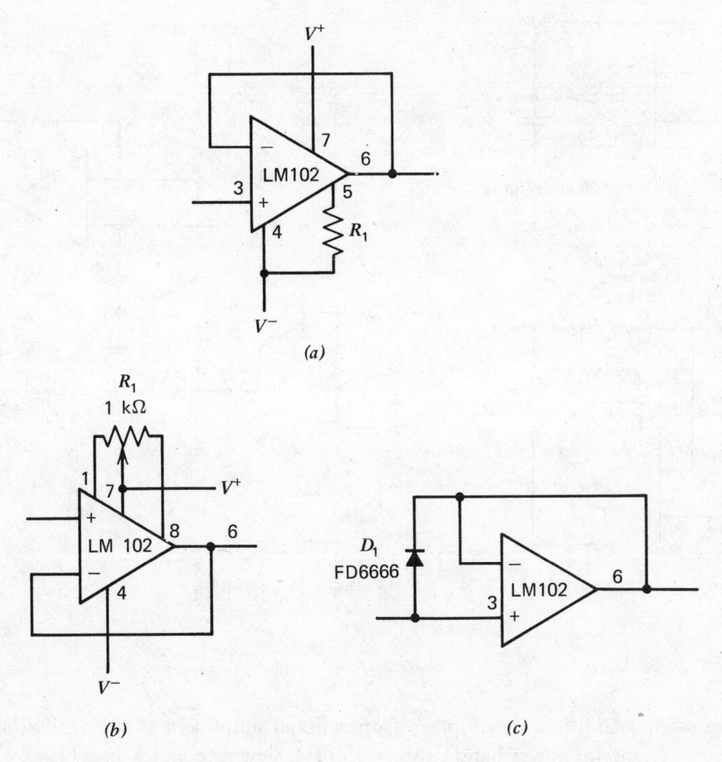

Figure 4-61. Auxiliary circuits for the LM102: (*a*) increased swing; (*b*) offset balancing; (*c*) eliminating overshoot.

impedance of several hundred megohms and an input capacitance that is less than 1.0 pF. The high-input impedance is obtained by positive feedback through C_1 to the positive input of the amplifier. The input capacitance plus the capacitance to ground can be canceled by adding the feedback capacitor C_2 and adjusting R_3.

Since the low-frequency response is determined primarily by C_1, the use of an electrolytic capacitor is advisable. High-frequency response is limited by the op amp. Using an amplifier such as the μA709, the circuit can amplify a 5-μsec wide pulse coupled through a 1.0-pF capacitor.

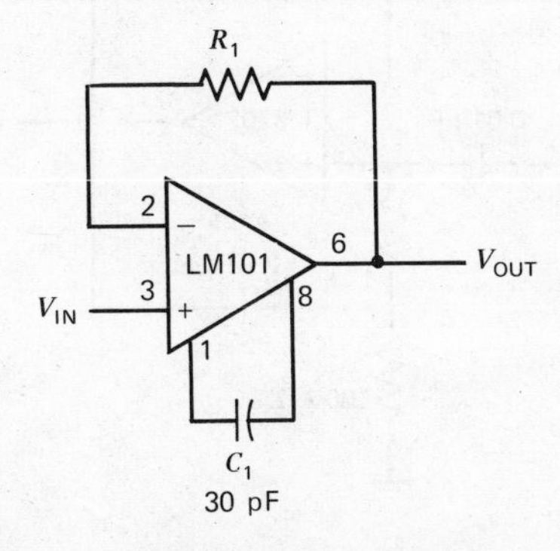

Figure 4-62. Voltage-follower connection of LM101.

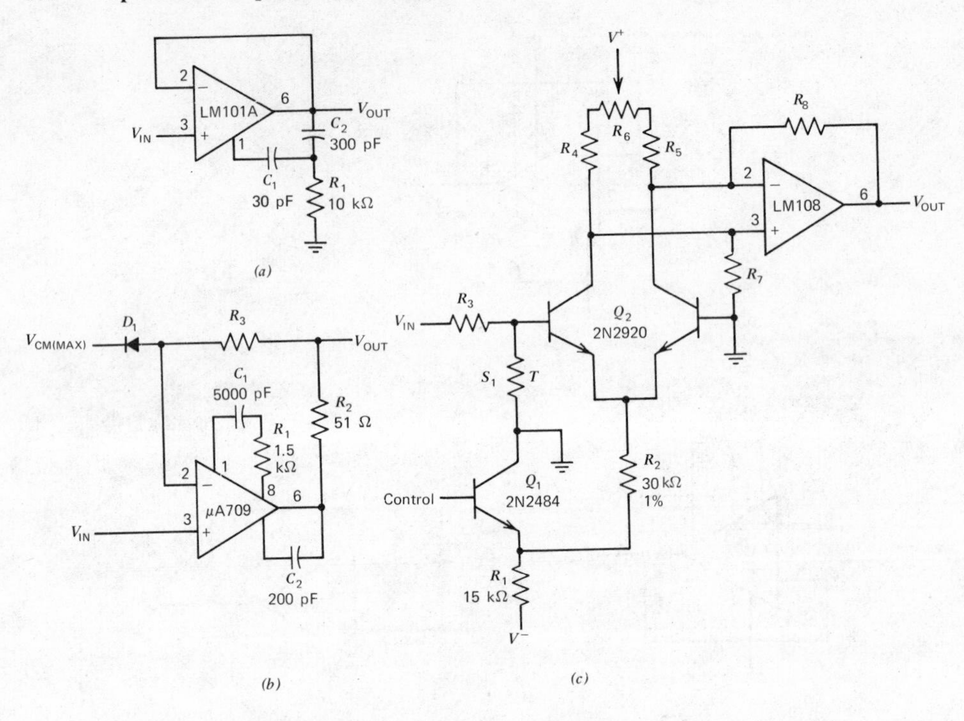

Figure 4-63. Additional examples of operational amplifiers as voltage-followers [*note:* (*a*) power bandwidth = 15 kHz, slew rate = 1 V μsec].

Switching Circuit Drivers

The use of the LM110 or LM102 in a switch circuit to driving the ladder network in an A/D (analog-to-digital) converter is shown in Figure 4-66. Simple transistor switches, connected in the reverse mode for low-saturation voltage, generate the 0 and 5 V levels for the ladder network. The switch output is buffered by A_2 and A_3 to give a low-driving impedance in both the high and low states.

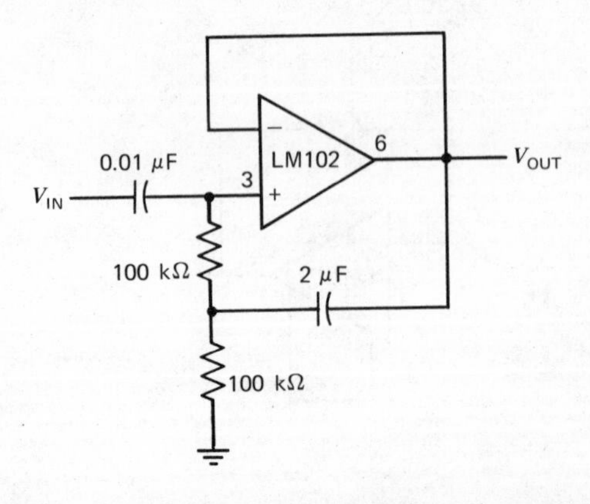

Figure 4-64. ac amplifiers with high-input impedances.

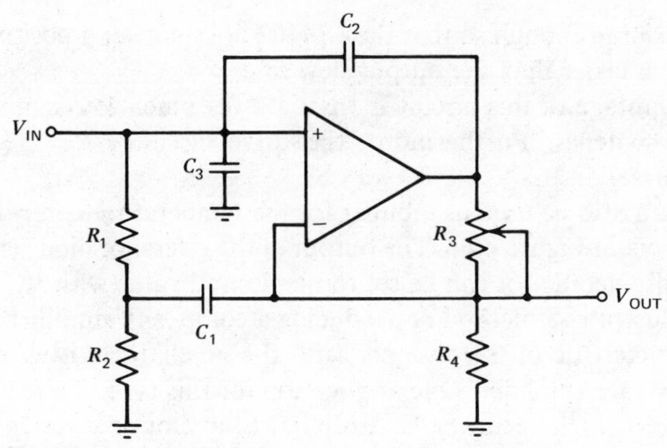

Figure 4-65. High-input impedance amplifier.

The switch transistors can be driven directly from integrated logic circuits. Resistors R_7 and R_8 limit the base drive; the values indicated are for operation with standard TTL or DTL circuits. If necessary, the switching speed of the LM102 can be increased somewhat by bypassing the resistors with 100-pF capacitors.

Even with operation at maximum speed, clamp diodes are not needed on the voltage-followers to reduce overshoot. The pull-up resistors on the switches, R_5 and

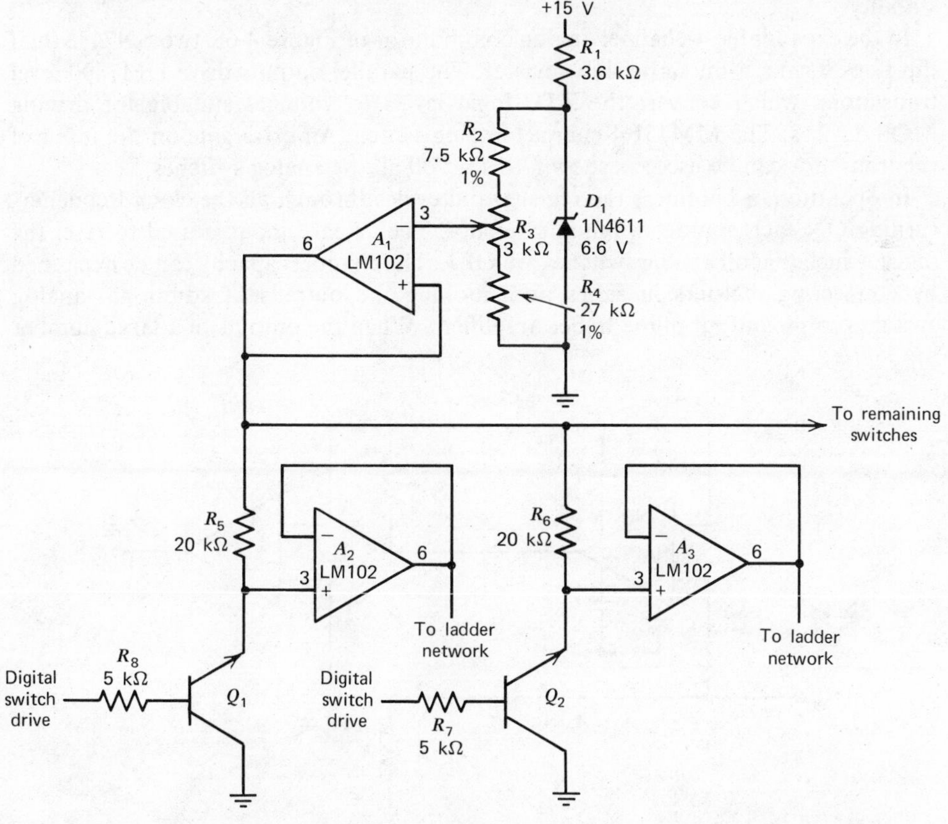

Figure 4-66. Using voltage-followers to drive the ladder network in an A/D converter.

R_6, can be made large enough so that the LM102 does not see a positive-going input pulse that is much faster than the output slew rate.

The main advantage of this circuit is that is gives much lower output resistance than push-pull switches. Furthermore, the drive circuitry for these switches is considerably simpler.

The LM102 can also be used as a buffer for the temperature-compensated voltage reference, as shown in Figure 4-66. The output of the reference diode is divided down with a resistive divider, and it can be set to the desired value with R_3.

Figure 4-67 illustrates a method of producing a composite amplifier that combines the low dc characteristic of the chopper with the ac characteristics of a wideband and/or high-slew-rate amplifier. One application for this type of circuit is as a buffer amplifier for a fast, high-accuracy D/A (digital-to-analog) converter where a large offset voltage would be disastrous. The combined frequency response of the circuit may be tailored for optimum settling time within a very narrow error band.

Analog Commutator Buffers

A voltage-follower's low-input current and fast slewing also make it an excellent buffer amplifier for high-speed analog commutators. The low-input current permits operation with switch resistances even higher than 10 kΩ without affecting the dc stability.

In the expandable 4-channel analog commutator of Figure 4-68, two SN7473 dual flip-flops form a 4-bit static shift register. The parallel outputs drive DM7800 level translators, which convert the TTL logic levels to voltages suitable for driving MOS devices: The MM451 4-channel analog switch. An extra gate on the input of the translator can be used, as shown, to shut off all the analog switches.

In operation, a bit enters the register and cycles through all the clock frequency, turning ON each analog switch in sequence. The "clear" input is used to reset the register such that all analog switches are OFF. The channel capacity can be expanded by connecting registers in series and hooking the outputs of additional analog switches to the output of the buffer amplifiers. When the outputs of a large number

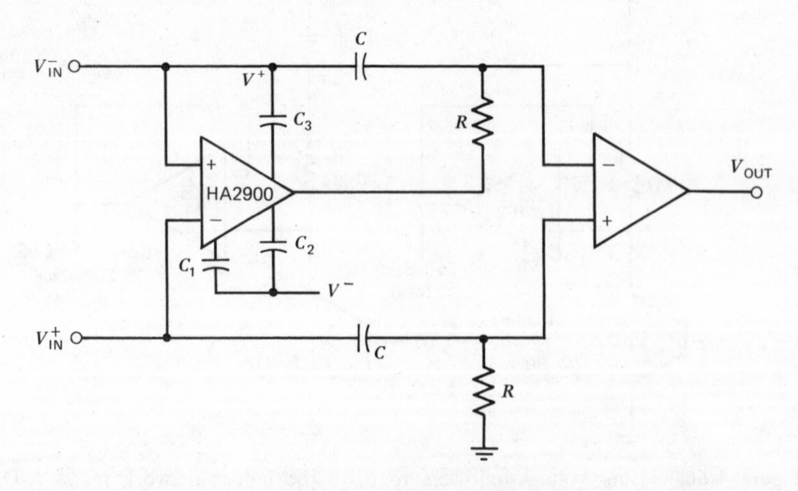

Figure 4-67. Buffer amplifier.

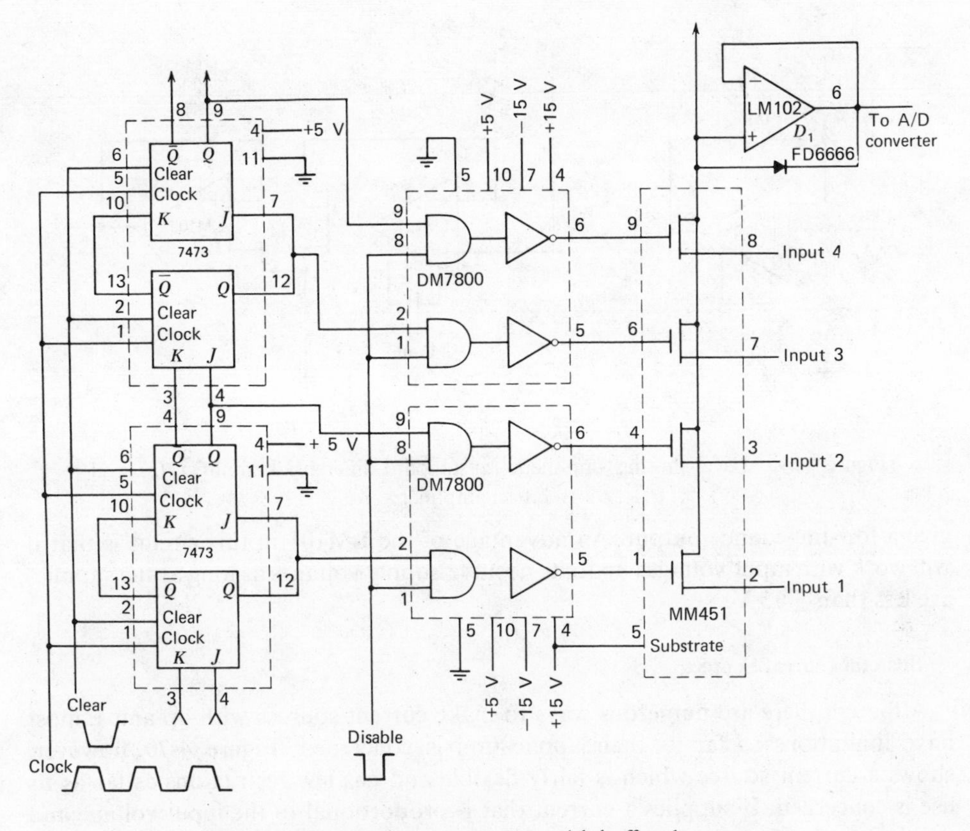

Figure 4-68. Analog commutator with buffered output.

of MOS switches are connected together, the capacitance on the output node can become high enough to reduce accuracy at a given operating speed. This problem can be avoided, however, by breaking up the total number of channels, buffering these segments with voltage-followers, and subcommutating them into the A/D converter.

Level-Shifting Amplifier

Frequently, in the design of linear equipment, it is necessary to take a voltage which is referred to some dc level and produce an amplified output which is referred to ground. The most straightforward way of doing this is to use a differential amplifier similar to that shown in Figure 4-69*a*. This circuit, however, has the disadvantages that the signal source is loaded by current from the input divider, R_3 and R_4, and that the feedback resistors must be very well matched to prevent erroneous outputs from the common-mode input signal.

A circuit which does not have these problems is shown in Figure 4-69*b*. Here, an FET transistor on the output of the op amp produces a voltage drop across the feedback resistor R_1 which is equal to the input voltage. The voltage across R_2 will then be equal to the input voltage multiplied by the ratio, R_2/R_1; and the common-mode rejection will be as good as the basic rejection of the amplifier, independent of the resistor tolerances. This voltage is buffered by an LM102 voltage-follower to

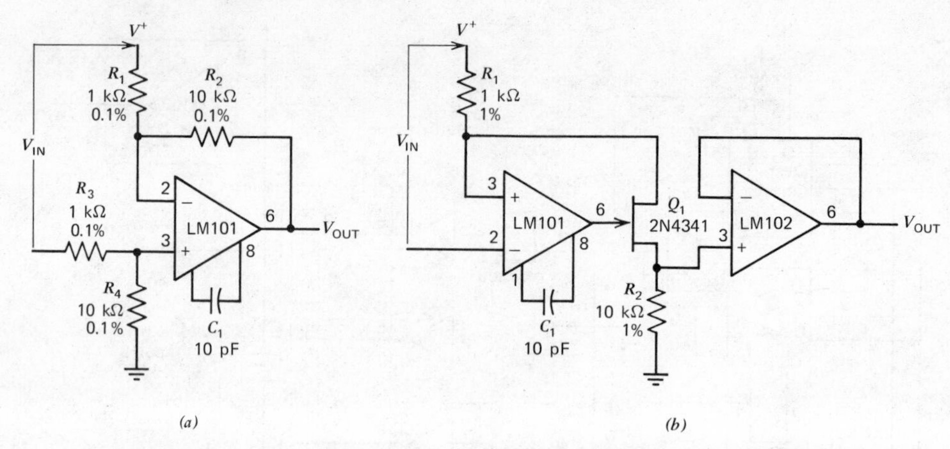

Figure 4-69. Level-shifting amplifiers: (*a*) standard differential amplifier, (*b*) level-isolation amplifier.

give a low-impedance output. An advantage of the LM101 in this circuit is that it will work with input voltages up to its positive-supply voltages as long as the supplies are less than ± 15 V.

Bilateral Current Sources

Although there are numerous ways to make current sources with op amps, most have limitations as far as their application is concerned. Figure 4-70, however, shows a current source which is fairly flexible and has few restrictions as far as its use is concerned. It supplies a current that is proportional to the input voltage and drives a load referred to ground or any voltage within the output swing capability of the amplifier.

With the output grounded, it is relatively obvious that the output current will be determined by R_5 and the gain setting of the op amp, yielding

$$I_{OUT} = -\frac{R_3 V_{IN}}{R_1 R_5} \tag{4-9}$$

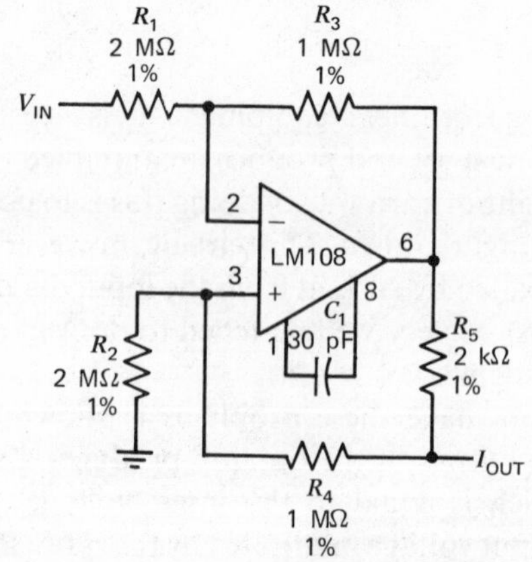

Figure 4-70. Bilateral current sources made with operational amplifiers [*note:* $I_{OUT} = R_3 V_{IN}/R_1 R_5$, $R_3 = R_4 + R_5$, $R_1 = R_2$].

When the output is not at zero, it would seem that the current through R_2 and R_4 would reduce accuracy. Nonetheless, if $R_1 = R_2$ and $R_3 = R_4 + R_5$, the output current will be independent of the output voltage. For $R_1 + R_3 \gg R_5$, the output resistance of the circuit is given by

$$R_{OUT} \cong R_5 \frac{R}{\Delta R} \tag{4-10}$$

where R is any one of the feedback resistors (R_1, R_2, R_3, or R_4) and ΔR is the incremental change in the resistor value from design center. Hence, for the LM108 circuit in Figure 4-70 a 1% deviation in one of the resistor values will drop the output resistance of 200 kΩ. Such errors can be trimmed out by adjusting one of the feedback resistors. In design, it is advisable to make the feedback resistors as large as possible. Otherwise, resistor tolerances become even more critical.

The circuit must be driven from a source resistance which is low by comparison to R_1, since this resistance will imbalance the circuit and affect both gain and output resistance. As shown, the circuit gives a negative-output current for a positive-input voltage. This can be reversed by grounding the input and driving the ground end of R_2. The amplitude of the scale factor will be unchanged as long as $R_4 \gg R_5$. Inserting a voltage-follower into the input signal path, as illustrated in Figure 4-71, will significantly improve the op amps input-impedance capability.

Precision Current Source

Two precision current source configurations are shown in Figures 4-72 and 4-73; they will sink or source conventional current, respectively. Caution must be exercised in applying these circuits. The voltage compliance of the source extends from BV_{CER} (collector-to-emitter breakdown voltage with a resistor between the base and the emitter) of the external transistor to approximately 1 V more negative than V_{IN}. The compliance of the current sink is the same in the positive direction.

The impedance of these current generators is essentially infinite for small currents and they are accurate so long as V_{IN} is much greater than V_{OS} and I_{OUT} is much greater than I_{BIAS}.

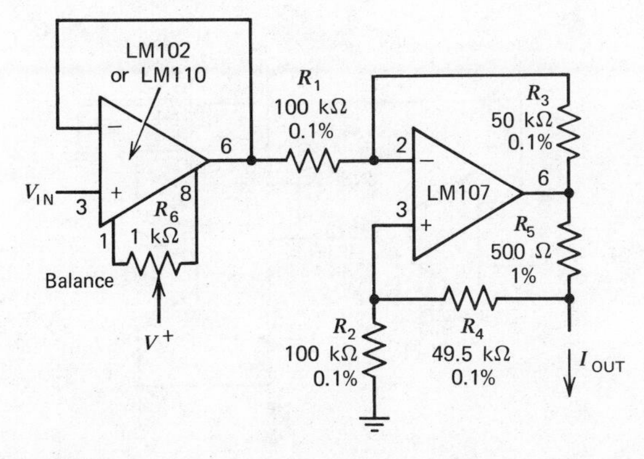

Figure 4-71. Voltage-follower provides bilateral current source with high-input impedance [*note:* $I_{OUT} = R_3 V_{IN}/R_1 R_5$, $R_3 = R_4 + R_5$, $R_1 = R_2$].

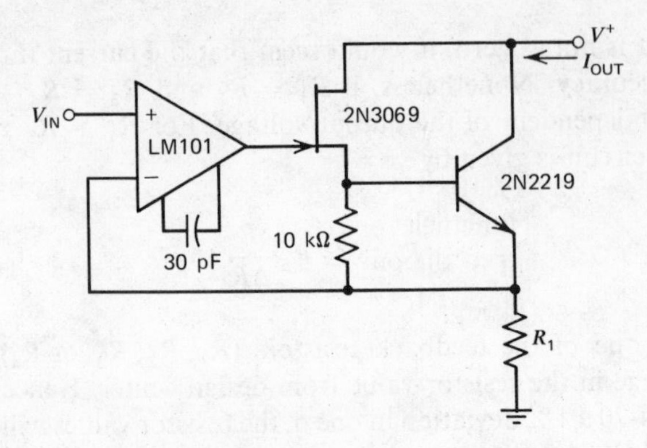

Figure 4-72. Precision current sink [*note:* $I_{OUT} = V_{IN}/R_1$, $V_{IN} > 0V$].

In Figure 4-72, the 2N3069 J FET and 2N2219 bipolar transistor have inherently high-output impedance. Using R_1 as a current-sensing resistor to provide feedback to the LM101 op amp provides a large amount of loop gain for negative feedback to enhance the true current-sink nature of this circuit. For small-current values, the 10 kΩ resistor and 2N2219 may be eliminated if the source of the J FET is connected to R_1.

In Figure 4-73, the 2N3069 J FET and 2N2219 bipolar transistor serve as voltage-isolation devices between the output and the current-sensing resistor, R_1. The LM101 provides a large amount of loop gain to assure that the circuit acts as a current source. For small values of current, the 2N2219 and 10 kΩ resistor may be eliminated with the output appearing at the source of the 2N3069.

It is possible to use a Darlington connection in place of the FET-bipolar combination in cases where the output current is high and the base current of the Darlington input would not cause a significant error.

The amplifiers used must be compensated for unity-gain and additional compensation may be required depending on load reactance and external transistor parameters.

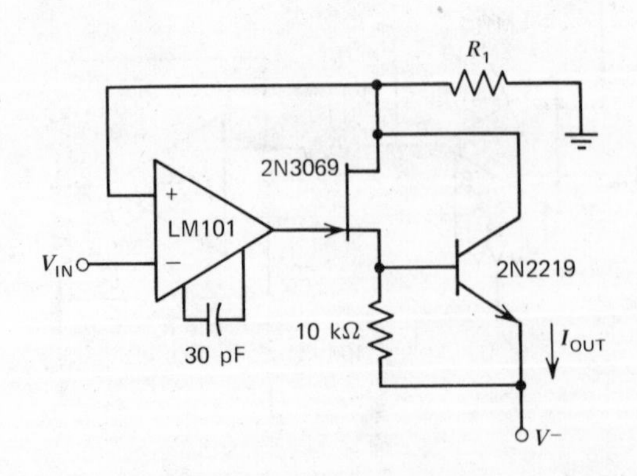

Figure 4-73. Precision current source [*note:* $I_{OUT} = V_{IN}/R_1$, $V_{IN} > 0V$].

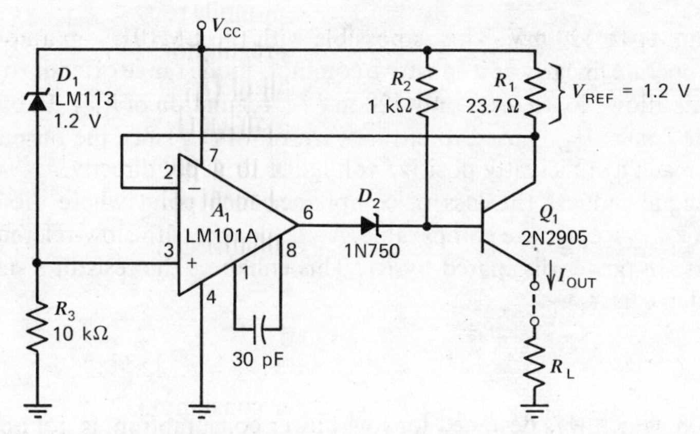

Figure 4-74. High-compliance current source uses a low-voltage Zener to minimize the voltage drop across R_1 [*note:* $I_{OUT} = 1.2\ V/R_1$].

LOW TEMPERATURE COEFFICIENT CURRENT SOURCE

Shown in Figure 4-74 is a basic high-compliance current regulator, arranged to source current and operate from a single supply. The current I_{OUT} supplied by transistor Q_1 is V_{REF} divided by R_1, where V_{REF} is the 1.2-V breakdown voltage of D_1.

The availability of the low-reference voltage allows the voltage compliance of Q_1 to be maximized through an increase in the portion of the supply voltage made available to the load. In this case the compliance is

$$V_{COMP} = V_{CC} - (1.2\ V + V_{CE(SAT)\,Q_1}).$$

If a low-saturation voltage transistor is used for Q_1, the loss of voltage will be negligible in comparison with the 1.2 V dropped across R_1.

An ultrahigh-compliance version of this circuit can be built if the 1.2-V reference is reduced to an even lower potential to decrease further the drop across R_1. This is shown in Figure 4-75. Resistors R_3 and R_4, reduce the reference voltage applied

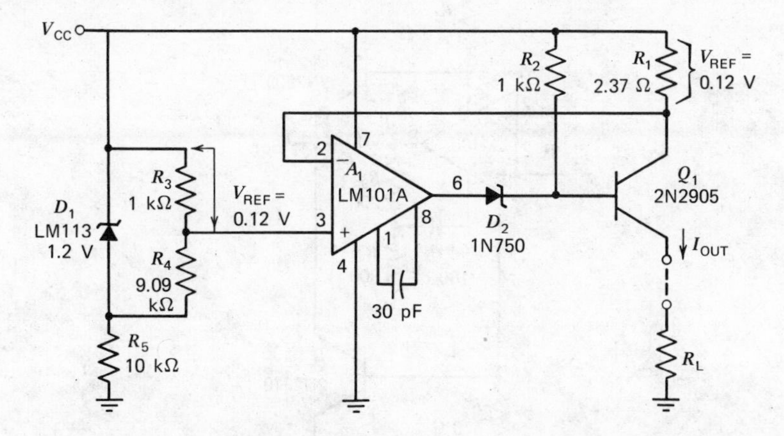

Figure 4-75. Extended compliance circuit uses a resistive divider to lower the apparent Zener voltage, thereby permitting the load voltage to go within a few millivolts of V_{CC} [*note:* $I_{OUT} = 0.12\ V/R_1$].

to the op amp A_1 to 120 mV. This is possible with the LM101A op amp because it continues to operate linearly as its positive common-mode range extends to the supply potential. This allows R_1 to drop only 120 mV for regulation of I_{OUT}. In both circuits the level-shift Zener, D_2, is used to ensure turn-off of Q_1, since the output swing of A_1 does not reach a sufficiently positive voltage to drive Q_1 directly.

This technique reduces the loss of compliance to a point where the regulation drop and the $V_{CE(SAT)}$ of Q_1 are comparable. A side benefit of the low-reference voltage is the decrease in power dissipated by R_1. This enhances the resistor's stability and makes selection easier.

Power Booster

The LM108, which was designed for low-power consumption, is not able to drive heavy loads. However, a relatively simple booster can be added to the output to increase the output current to ± 50 mA. This circuit, shown in Figure 4-76, has the added advantage that it swings the output up to the supplies within a fraction of a volt. The increased voltage swing is particularly helpful in low-voltage circuits.

In Figure 4-76, the output transistors are driven from the supply leads of the op amp. It is important that R_1 and R_2 be made low enough so Q_1 and Q_2 are not turned ON by the worst-case quiescent current of the amplifier. The output of the op amp is loaded heavily to ground with R_3 and R_4.

When the output swings about 0.5 V positive, the increasing positive-supply current will turn on Q_1 which pulls up the load. A similar situation occurs with Q_2 for negative-output swings.

The bootstrapped shunt compensation shown in the figure is the only one that seems to work for all loading conditions. This capacitor C_1 can be made inversely proportional to the closed-loop gain to optimize frequency response. The value given is for a unity-gain follower connection; C_2 is also required for loop stability.

The circuit does have a dead zone in the open-loop transfer characteristic. However,

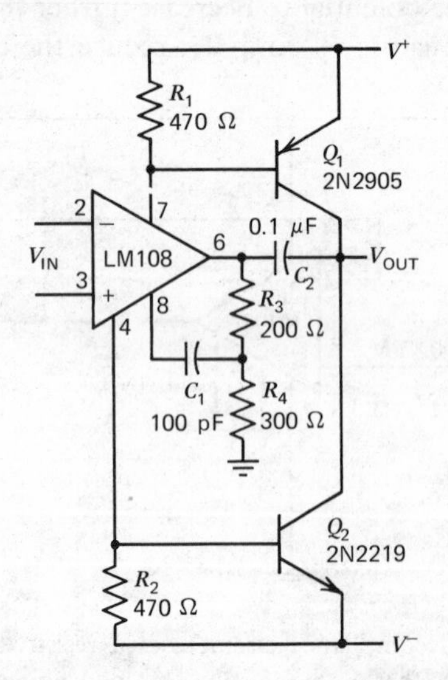

Figure 4-76. Power booster for LM108.

the low-frequency gain is high enough so that it can be neglected. Around 1 kHz, though, the dead zone becomes quite noticeable.

Current-limiting can be incorporated into the circuit by adding resistors in series with the emitters of Q_1 and Q_2 because the short-circuit protection of the LM108 limits the maximum voltage drop across R_1 and R_2.

PRECISION MEASURING AND REFERENCE CIRCUITS

Instrumentation Amplifiers[1]

For measuring low-level signals from remote sources, single-input op amps usually cannot fill the need. These measurements call for adequate CMRR, gain, input impedance, and stability. Usually instrumentation amplifiers—with performance tailored to these applications—can solve the tough measurement problems.

The differential-input single-ended output instrumentation amplifier is one of the most versatile signal-processing amplifiers available. It is used for precision amplification of differential dc or ac signals while rejecting large values of common-mode noise. By using ICs, a high level of performance is obtained at minimum cost.

A single instrumentation amplifier must frequently accept more than 100 multiplexed signals with levels ranging from a few microvolts to several volts. The amplifier must also reject unwanted noise signals from dc to hundreds of megahertz, as well as provide a noise-free, properly scaled, single-ended output signal that can be sampled quickly for processing.

External noise may be separated into two types—common-mode, and normal- or differential-mode signals. Differential-mode noise is the easier to reject. It can be filtered by passive low-pass networks. Common-mode noise is more difficult to reject. The instrumentation amplifier must be able to reject common-mode noise over a large bandwidth.

Special circuits have been designed to combat these problems: IC differential amplifiers offer higher CMRRs than single-ended op amps and can often yield an economical solution. The basic concept of the differential amplifier is shown in Figure 4-77. This circuit is simple and can reject both pick-up and ground-potential interference. Ideally the circuit responds only to the difference signal presented between its two input terminals; it ignores the pick-up and ground voltages that appear in phase on both signal lines.

Common-mode voltage V_{CM} is defined as voltage applied simultaneously to both the inverting and noninverting inputs of an op amp. The CMRR can be considered as the ratio of the differential gain to the common-mode gain. An unbalance of input resistance causes an error signal E_{CM}. But E_{CM} can be expressed as a constant, Θ, times V_{CM}, where

$$\Theta = \frac{Z_2}{Z_2 + R_2} - \frac{Z_1}{Z_1 + (R_1 + R_S)} \tag{4-11}$$

By combining fractions, letting $Z_1 = KZ_2$, and assuming that

$$Z_1 Z_2 \gg Z_1 R_2 + Z_2(R_1 + R_S) + R_2(R_1 + R_2)$$

[1] Portions of this section were excerpted from F. Pouliot, "Simplify Amplifier Selection," *Electronic Design*, **16,** Aug. 2, 1973.

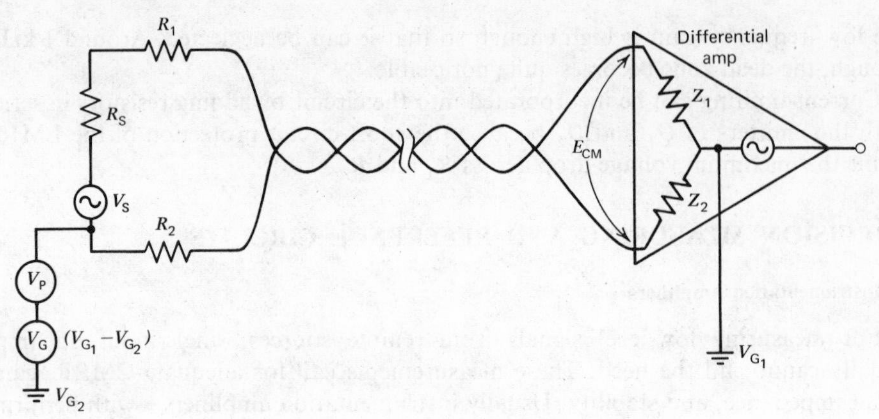

Figure 4-77. The basic differential amplifier can reject all common-mode error voltages [*note:* $V_P + V_6 = V_{CM}$].

we see that E_{CM} becomes

$$E_{CM} = V_{CM}\left(\frac{\text{resistor unbalance}}{KZ_2}\right) \tag{4-12}$$

For most cases $Z_1 = Z_2$; therefore K becomes 1, and the resistor unbalance can be called ΔR. The voltage E_{CM} now simplifies to $(V_{CM})(\Delta R)/Z_2$. To minimize the possibility of one signal line being more susceptible to pick-up, both wires should be twisted together, thereby guaranteeing that they are an equal distance apart and have an equal capacitance to the source of pick-up.

From this analysis, we can see that the amplifier remains relatively immune to ground noise and pick-up—but only if its two input terminals present a high impedance to ground. Often specifications allow for a source unbalance of up to 1000 Ω. Thus if the amplifier's input impedance is 100 MΩ and the V_{CM} is 10 V, the equivalent differential-error input is 0.1 mV. If the circuit is for measuring signals of 100 mV_{p-p}, the error created by the source unbalance is 0.1 mV/100 mV, or 0.1 %.

The relationship between V_{CM}, applied to both input terminals, and the equivalent differential input component is V_{CM} divided by the CMRR. Thus, for a typical differential amplifier with a V_{CM} of 10 V and a CMRR of 10^4, the differential component of V_{CM} becomes 1 mV. If the same amplifier has a differential range of 100 mV, the proportionate common-mode error is the ratio of differential common-mode voltage to actual differential signal (or 1 mV divided by 100 mV, which equals 1 %). That result sets a limit on the accuracy of the signal measurement. Other error sources, such as drift and nonlinearity, would degrade performance still further. The circuit of Figure 4-78 is the well-known single op amp differential configuration.

On examining this circuit and comparing it with that of Figure 4-77, we see that the input impedance of the single op amp is R_1. This immediately leads to a conflict since the higher the value of R_1, the worse the effects of bias-current drift. Additionally, the high gain usually required to bring low-level signals up to, say, 10-V full-scale forces R_2 to become extremely high. For example, if a gain of 100 is needed, R_2 becomes 100 times R_1. Thus if input impedances of 1 MΩ are desired, R_2 would have to be 100 MΩ. To avoid abnormally high resistance values, circuits using potentiometric feedback can be used, but at the penalty of increased noise and drift.

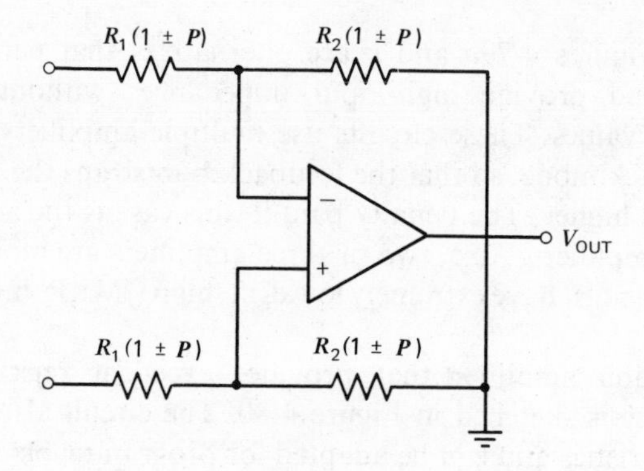

Figure 4-78. Adjusting the resistor ratios to increase gain is extremely difficult when trying to keep the CMRR high.

A more serious problem arises in connection with the circuit's common-mode rejection capability. The CMRR of a single op amp circuit, as shown in Figure 4-77 is given by

$$\text{CMRR} = \frac{1 + G_{CL}}{4P} \tag{4-13}$$

where P is the resistor-matching tolerance of the feedback network and G_{CL} is the closed-loop gain; G_{CL} is simply R_2/R_1 for this circuit. Table 4-3 summarizes the performance of a typical circuit over a three-decade range of gain.

While it is possible to trim resistors to a tighter tolerance than the 0.1% assumed in Table 4-3, any temperature or long-term drift adversely affects the CMRR. The final blow is the difficulty of making gain adjustments—varying the ratios of R_2 to R_1 in both input legs at the same time. It is difficult and uneconomical to buy ganged potentiometers that can track to within 0.1%. So any gain adjustments require subsequent "tweaking" to restore the CMRR to the original level.

Instrumentation amplifiers with large closed-loop gains usually require gain-adjustment potentiometers if a high degree of gain accuracy must be maintained. However, a second-stage design with sufficient gain can eliminate the need for potentiometers without degrading gain accuracy.

Table 4-3 Performance analysis for circuit of Figure 4-78

Gain (G_{CL})	1	10	100	1,000
CMRR = $(1 + G_{CL})/4$ kΩ	500	2750	25,250	250,250
Common-mode error for V_S = 100 mV V_{CM} = 5 V	10%	2%	0.2%	0.02%

Sample calculation for percent-common-mode error in circuit with $G_{CL} = 4$ and measuring 50 mV signals in presence of 10-V common-mode (use 0.1% resistors):

$$\text{common-mode error} \frac{V_{CM}}{(\text{CMRR})(V_S)} \times 100\% = \frac{10\text{ V}}{(1+4)/0.004 \times 0.05} \times 100\% = 16\%$$

Sample calculation for CMRR with a gain of 100 and $P = 0.1\%$:

$$\text{CMRR} = \frac{1 + G_{CL}}{4P} = \frac{1 + 100}{4(0.001)} \approx 25,000$$

The circuits in Figures 4-79*a* and *b* are alternatives that permit single-resistor gain adjustment and provide high-input impedances without excessively high feedback-resistance values. These circuits use multiple amplifiers, connected in the noninverting feedback mode, so that the feedback bootstraps the input resistance to values of $10^9\ \Omega$ and higher. The penalty paid in this case is the additional drift and noise of the extra amplifiers. Also, two or three amplifiers are more costly than one, especially when each must have extremely low drift, high CMRR, high gain, high-input impedance, etc.

An instrumentation amplifier that provides excellent rejection of unwanted common-mode noise is sketched in Figure 4-80. The circuit also offers good gain-bandwidth characteristics and can be adapted for programmable gain selection.

Amplifiers A_1 and A_2 provide the desired input characteristics and signal ranging. A third amplifier A_3 acts as an inverting unity-gain device for differential inputs and operates as a precision broad-bandwidth subtractor for unwanted common-mode signals. To reject common-mode signals over a large bandwidth, A_1 and A_2 must

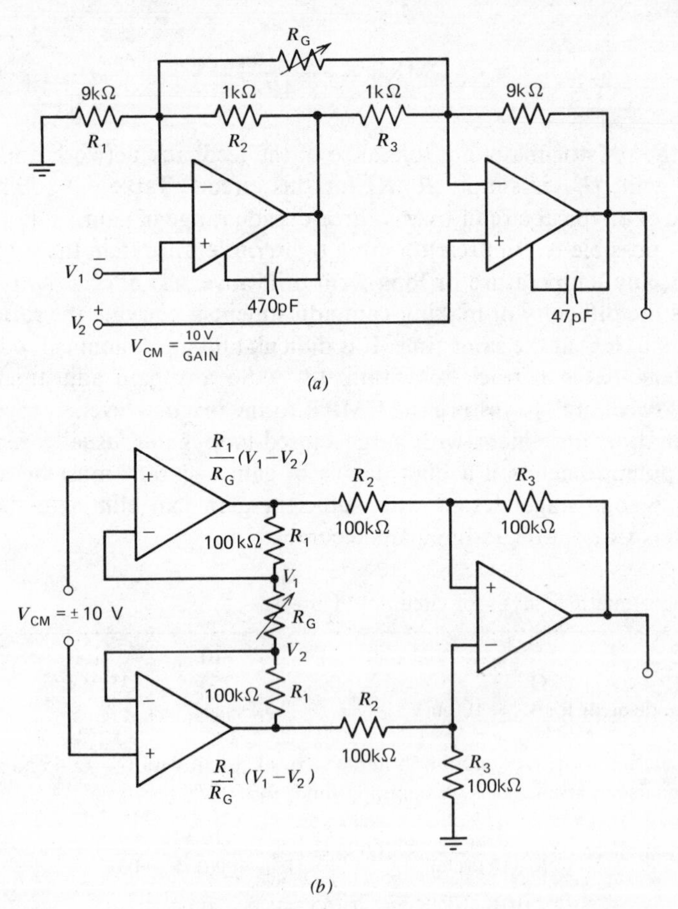

Figure 4-79. Easier gain adjustments are possible with a dual operational amplifier circuit (*a*) and a triple operational amplifier circuit (*b*), but at the expense of higher noise, drift, instability, and cost [*note:* (*a*) $V_{\text{OUT}} = (1 + R_4/R_3)(V_2 - V_1)$ for $R_4/R_3 \times R_2/R_1 = 1$; (*b*) $V_{\text{OUT}} = 2R_1/R_6 \times R_3/R_2(V_1 - V_2)$].

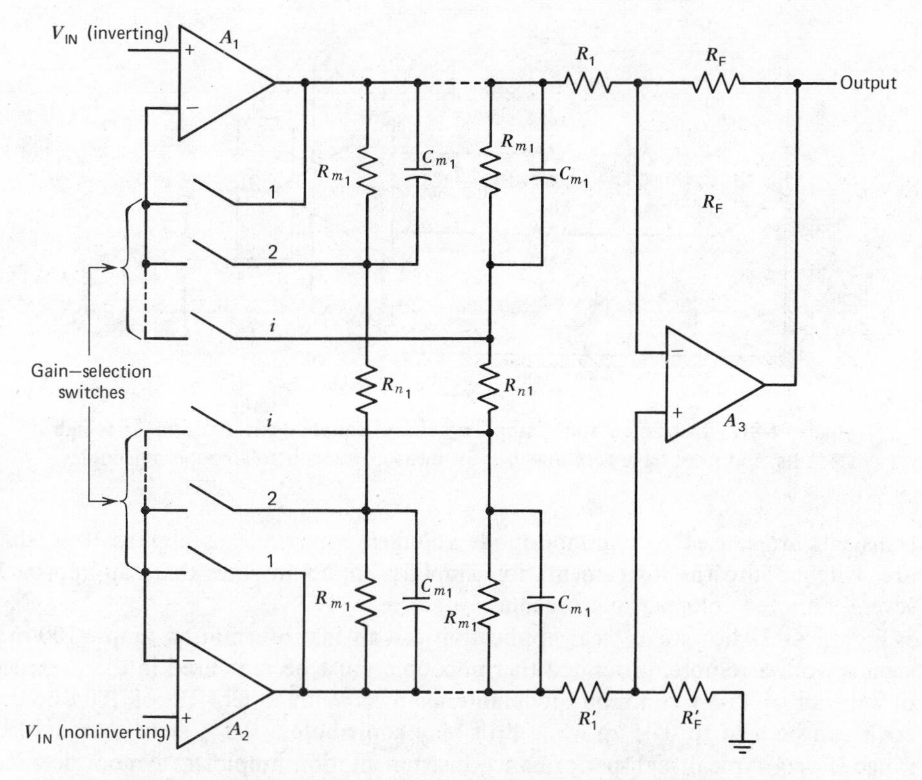

Figure 4-80. Basic amplifier design. Double-amplifier input eases demands on individual amplifiers A_1 and A_2. Amplifier A_3 rejects unwanted common-mode noise signals by subtracting balanced common-mode signals from A_1 and A_2.

perform as broadband unity-gain amplifiers, generating balanced common-mode signals for subtractor amplifier A_3. The CMRR of the overall amplifier can be expressed as

$$\text{CMRR} = \frac{(R_F/R_1)(1 + 2R_m/R_n)}{(1/\text{CMRR}_1 \pm 1/\text{CMRR}_2)(1 + 2R_m/R_n)(R_F/R_1)} + \frac{R_F/\text{CMRR}_3(R_1 + R_F) + R_F/(R_1 + R_F) - R'_F/(R'_1 + R'_F)}{R'_1(R'_1 + R'_F)} \tag{4-14}$$

where CMRR_1, CMRR_2, and CMRR_3 are the individual common-mode rejection ratios of amplifiers A_1, A_2, and A_3, respectively.

Other common-mode input errors, not included in this equation, are caused by the voltage-divider effect between the source impedance of each analog input line and the common-mode input impedance of its amplifier. To prevent common-mode signals from being converted into differential signals from source imbalance at the input of the overall amplifier, amplifiers A_1 and A_2 must maintain a high common-mode input impedance over a large bandwidth.

The high impedance of these two amplifiers also helps the overall amplifier to reject the transients generated when the multiplexer channels are switched. These

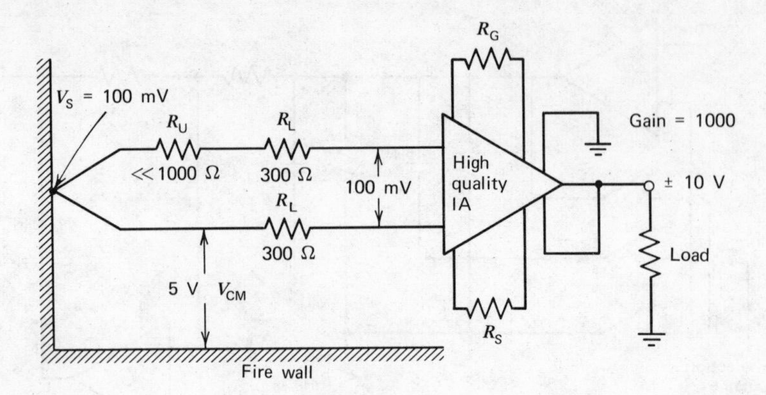

Figure 4-81. Instrumentation amplifiers have simple gain adjustments, high CMRRs, and must have very low drift for measurements involving thermocouples.

transients are caused by common-mode voltages, sometimes as high as 10 V, that are switched into the instrumentation amplifier inputs at rates that can approach several hundred volts per microsecond.

Figure 4-81 shows a typical application for an instrumentation amp—100-mV signals from a remote, grounded thermocouple must be measured in the presence of substantial (5-V) common-mode interference. With a CMRR of 100 dB, the error can be held to 0.05%, while drift may contribute a 0.5% error over a 10°C range. For a typical high-performance instrumentation amplifier, a more detailed outline of the total error contributions is presented in Table 4-4. The overall error for this amplifier is usually less than 1%—a more than tenfold improvement over the circuit of Figure 4-78.

Instrumentation amplifiers usually have two terminals that are not available on op amps. These are the reference and sense lines. They can simplify some types of measurements, as illustrated in Figure 4-82. In the application shown, the chart-recorder pen can be biased so that the amplifier zero and the paper zero coincide—useful if a bipolar recorder uses unipolar paper. Likewise contact potential in medical measurements, and electromechanical emfs in chemical applications, can be compensated very simply.

Table 4-4 Sources of errors in thermocouple circuit

Static errors	
Gain nonlinearity	0.01%
Loading error	Adjusted to zero by gain trim
Initial offsets	Adjusted to zero by trim
Common-mode	V_{CM}/CMRR ÷ V_S = $5/10^5$ ÷ 100 mV = 0.05%
Temperature errors in 25–35°C range	
Gain drift	0.01%/°C × 10 = 0.1%
Current drift	Negligible for R_L = 300 Ω
Voltage drift	0.05%/°C × 10° = 0.5%
Total static error	0.06%
Total drift error	0.6%
Overall error	0.66%

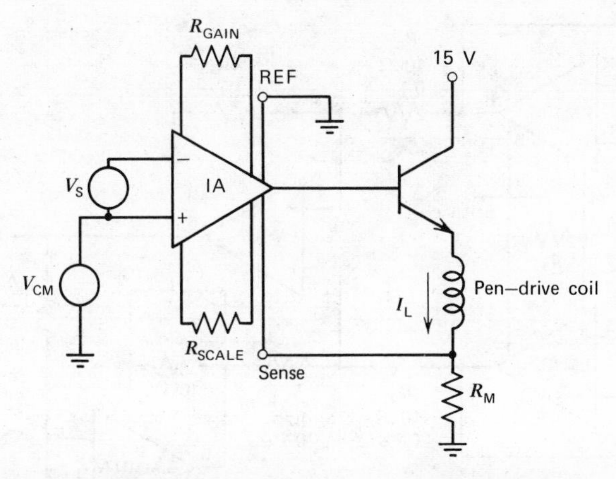

Figure 4-82. A typical general-purpose instrumentation amplifier can be used in chart recorders to cancel out common-mode voltage [*note:* $I_L = R_S/R_6 \times V_S/R_M$].

Figure 4-83 shows a basic instrumentation amplifier which provides a 10-V output for 100-mV input, while rejecting greater than ±11 V of common-mode noise. To obtain good input characteristics, two voltage-followers buffer the input signal. The LM102 is specifically designed for voltage-follower usage and has 10,000-MΩ input impedance with 3-nA input currents. This high of an input impedance provides two benefits: It allows the instrumentation amplifier to be used with high-source resistances and still have low error; and it allows the source resistances to be unbalanced by over 10,000 Ω with no degradation in common-mode rejection. The followers drive a balanced differential amplifier, as shown in Figure 4-83, which provides gain and rejects the common-mode voltage. The gain is set by the ratio of R_4 to R_2 and R_5 to R_3. With the values shown, the gain for differential signals is 100.

Figure 4-84 shows an instrumentation amplifier where the gain is linearly adjustable from 1 to 300 with a single resistor. An LM101A, connected as a fast inverter, is used as an attenuator in the feedback loop. By using an active attenuator, a very low impedance is always presented to the feedback resistors, and common-mode rejection

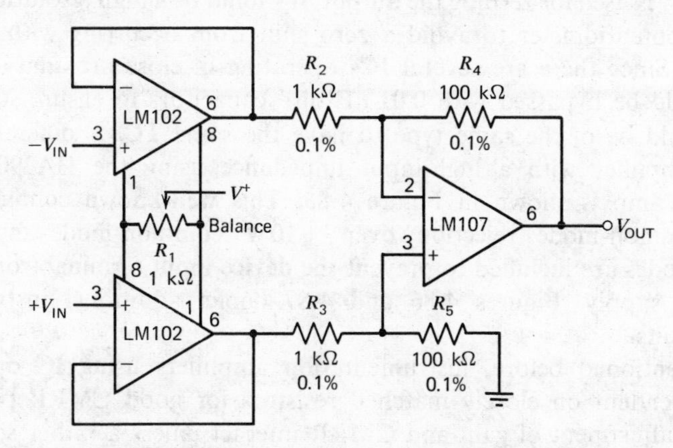

Figure 4-83. Differential instrumentation amplifier with voltage-followers as input buffers [*note:* $R_4 = R_5$, $R_2 = R_3$, $A_V = R_4/R_2$].

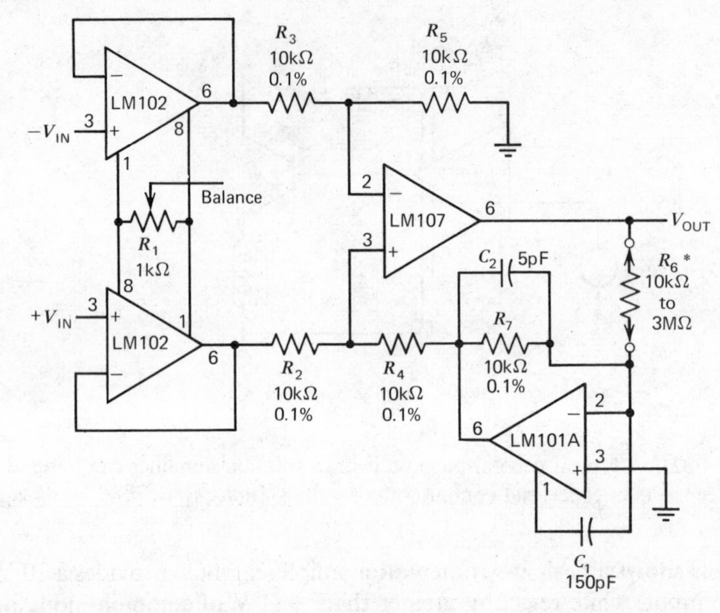

Figure 4-84. Variable-gain, differential-input amplifier [*note:* * gain adjust $A_V = 10^{-4} R_6$].

is unaffected by gain changes. The LM101A, used as shown, has a greater bandwidth than the LM107, and may be used in a feedback network without instability. The gain is linearly dependent on R_6 and is equal to $10^{-4}\ R_6$.

To obtain good CMRRs, it is necessary that the ratio of R_4 to R_2 match the ratio of R_5 to R_3. For example, if the resistors in the circuit shown in Figure 4-83 had a total mismatch of 0.1%, the common-mode rejection would be 60 dB times the closed-loop gain, or 100 dB. The circuit shown in Figure 4-84 would have constant common-mode rejection of 60 dB, independent of gain. In either circuit, it is possible to trim any one of the resistors to obtain CMRRs in excess of 100 dB.

For optimum performance, several items should be considered during construction. The resistor R_1 is used for zeroing the output. It should be a high resolution, mechanically stable potentiometer to avoid a zero shift from occurring with mechanical disturbances. Since there are several ICs operating in close proximity, the power supplies should be bypassed with 0.01-μF disc capacitors to ensure stability. The resistors should be of the same type to have the same TC. A differential instrumentation amplifier with a high-input impedance using the HA2900 chopper-stabilized op amp is shown in Figure 4-85. This well-known configuration has excellent common-mode rejection over ± 10 V common-mode input signals. Protection diodes are included to prevent the device input terminal from exceeding either power supply. Figures 4-86 and 4-87 depict additional instrumentation amplifier circuits.

As was mentioned before, instrumentation amplifiers using IC op amps are normally dependent on closely matched resistors for good CMRR performance. Further, the adjustment of gain and CMRR interact unless a rather sophisticated design approach is taken, and building an instrumentation amplifier with high-input impedance usually requires at least three op amps. By using a μA741 op amp with

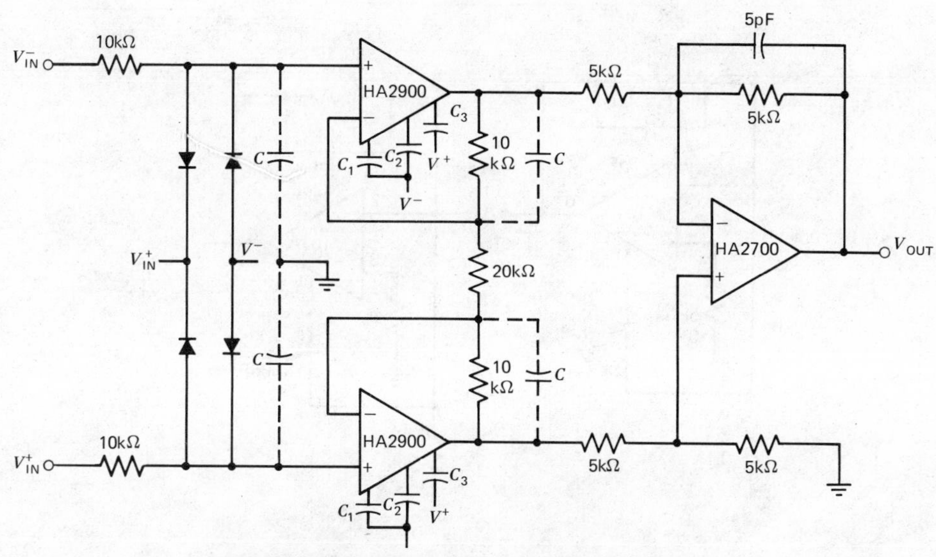

Figure 4-85. Differential instrumentation amplifier.

feedback to the offset adjustment as shown in Figure 4-88,[2] rather than the non-inverting input, many of the above problems can be eliminated.

Some new limitations are encountered though. For example, while most instrumentation amplifiers can be programmed for unity gain, this set-up cannot. To program for unity gain, R_2 would be 2 kΩ, which means that *pin* 5 would have a potential of 2/3 of $-V_{CC}$, which would turn OFF an internal transistor.

Unity gain will also overdrive the input stage. From experimental results, this configuration works best when gain is kept greater than 50. The basic instrumentation

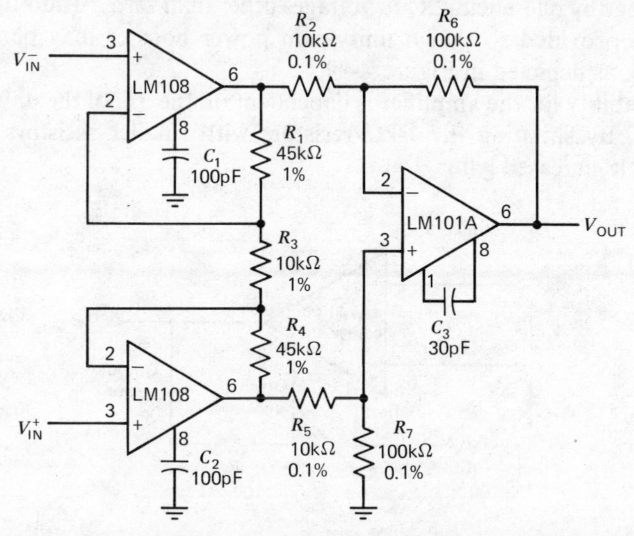

Figure 4-86. Differential input instrumentation amplifier with high common-mode rejection [*note:* $R_1 = R_4$, $R_2 = R_5$, $R_6 = R_7$, $A_V = R_6/R_2(1 + 2R_1/R_3)$; †* matching determines CMRR; $A_V = (R_6/R_2)(1 + 2R_1/R_3)$].

[2] H. Mortensen, "IC Op Amps Make Inexpensive Instrumentation Amplifiers," *EDN*, May 15, 1972.

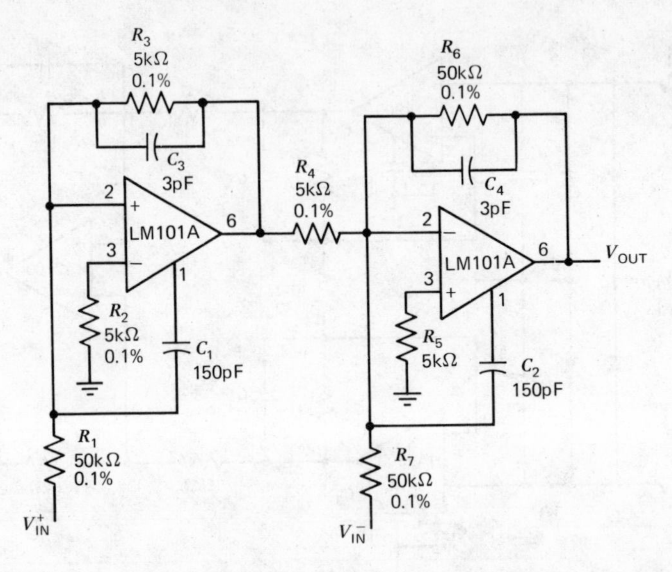

Figure 4-87. Instrumentation amplifier with ±100-V common-mode range [*note:* $R_1 = R_5 = 10R_2$, $R_2 = R_3$, $R_3 = R_4$, $R_1 = R_6 = 10R_3$, $A_V = R_7/R_6$].

amplifier using a μA741, as shown in Figure 4-88, provides 5 V output for 100-mV input. The 100-kΩ resistor (R_2) connected from the output terminal to the offset null provides a feedback path to the internal 1-kΩ resistor, in order to balance the set-up, another 100-kΩ resistor (R_1) is added from ground to *pin* 1.

The gain is given by $R_2/2 \times 1\,\text{k}\Omega$, and the output offset can be varied over the full output range by referencing R_1 to voltages other than zero. In addition, an output sense point is provided so that a unity-gain power booster may be added in the feedback loop, as depicted in Figure 4-89.[3]

The gain stability for the amplifier is dependent on the TC of the diffused resistors (≈0.25%/°C). By shunting the 1-kΩ resistors with smaller resistors this effect is minimized with increased gain.

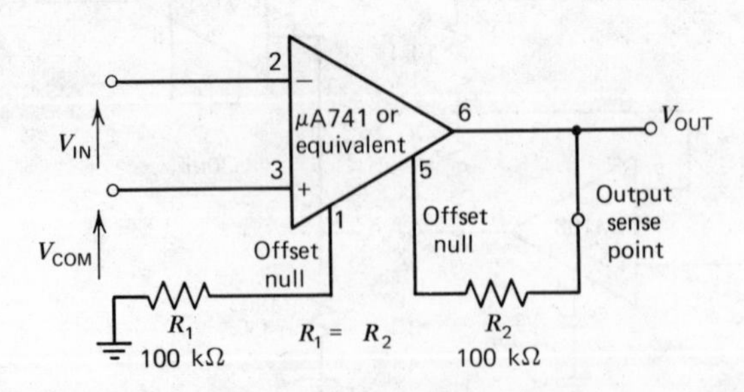

Figure 4-88. Feedback to the offset adjustment of an integrated circuit operational amplifier, rather than the normal feedback to the input, converts the device to an inexpensive instrumentation amplifier. (Courtesy *EDN*.)

[3] *Ibid*.

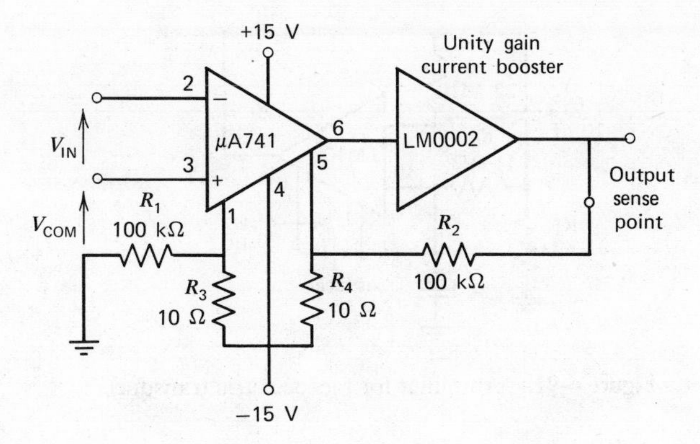

Figure 4-89. Unity-gain power booster added to the basic circuit of Figure 4-88 provides more flexible operation at only a modest increase in cost, and is still less expensive than the more complex disigns. (Courtesy *EDN*.)

Photodiode Sensor

With certain transducers, accuracy depends on the choice of the circuit configuration as much as it does on the quality of the components. The amplifier for photodiode sensors, shown in Figure 4-90, illustrates this point. Normally, photodiodes are operated with reverse voltage across the junction. At high temperatures, the leakage currents can approach the signal current. However, photodiodes deliver a short-circuit output current, unaffected by leakage currents, which is not significantly lower than the output current with reverse bias.

The circuit shown in Figure 4-90 responds to the short-circuit output current of the photodiode. Since the voltage across the diode is only the offset voltage of the amplifier, inherent leakage is reduced by at least two orders of magnitude. Neglecting the offset current of the amplifier, the output current of the sensor is multiplied by R_1 plus R_2 in determining the output voltage.

Figure 4-91 shows an amplifier for high-impedance ac transducers like a piezoelectric accelerometer. These sensors normally require a high input-resistance amplifier. The LM108 can provide input resistances in the range of 10 to 100 MΩ,

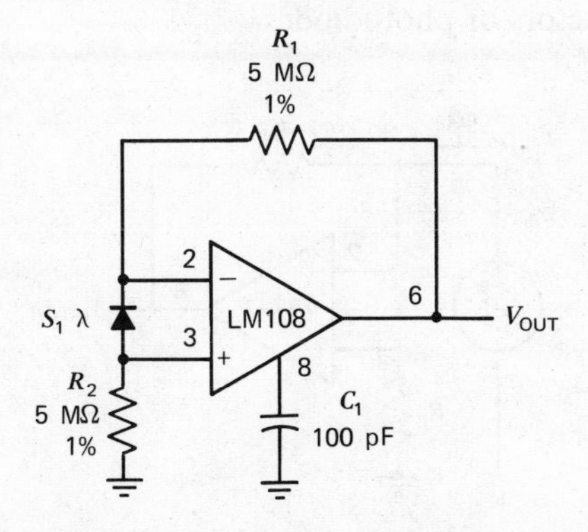

Figure 4-90. Amplifier for photodiode sensor [*note:* V_{OUT} = 10 V/μA].

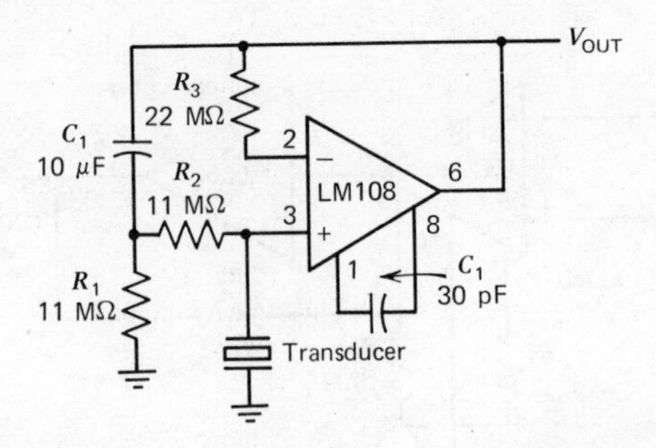

Figure 4-91. Amplifier for piezoelectric transducers.

using conventional circuitry. However, conventional designs are sometimes ruled out either because large resistors cannot be used or because prohibitively large input resistances are needed.

Using the circuit in Figure 4-90, input resistances that are orders of magnitude greater than the values of the dc-return resistors can be obtained. This is accomplished by bootstrapping the resistors to the output. With this arrangement, the lower cut-off frequency of a capacitive transducer is determined more by the RC product of R_1 and C_1 than it is by resistor values and the equivalent capacitance of the transducer.

Photocell Amplifiers

The photocell amplifiers in Figures 4-92, 4-93, and 4-94, are all based upon the current-to-voltage transducer discussed earlier. An internally compensated op amp is also used in these examples to emphasize the similarity, but the basic techniques are adaptable to all general-purpose amplifiers.

All photogenerators display some voltage dependence of both speed and linearity. It is obvious that the current through a photoconductive cell will not display strict proportionality to incident light if the cell terminal voltage is allowed to vary with cell conductance. Somewhat less obvious is the fact that photodiode leakage and photovoltaic cell internal losses are also functions of terminal voltage. The current-to-voltage converter neatly sidesteps gross linearity problems by fixing a constant terminal voltage, zero in the case of photovoltaic cells, and a fixed-bias voltage in the case of photoconductors or photodiodes.

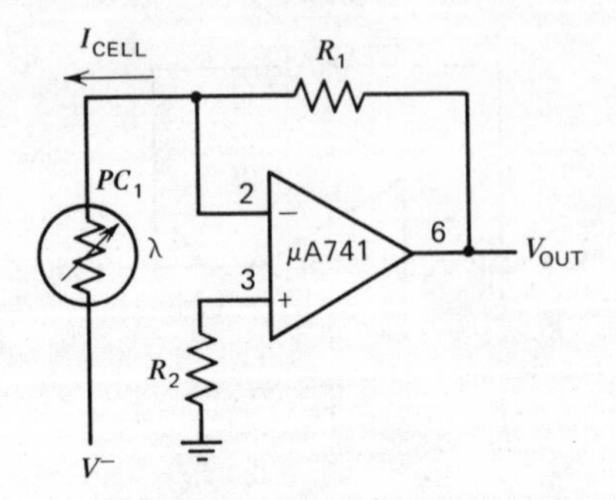

Figure 4-92. Amplifier for photoconductive cell.

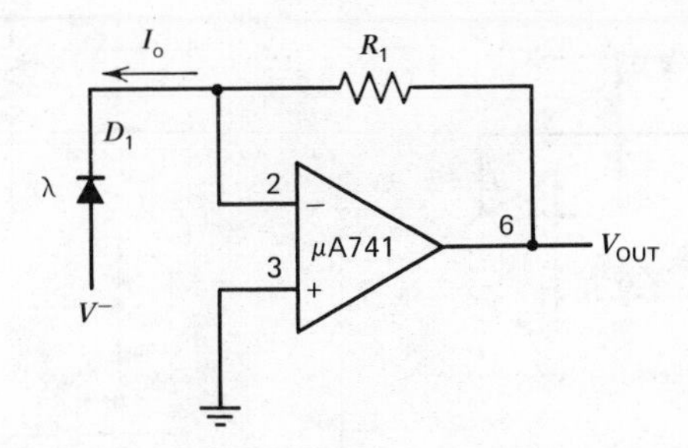

Figure 4-93. Photodiode amplifier [*note:* $V_{OUT} = R_1 I_{OUT}$].

Photodetector speed is optimized by operating into a fixed-low-load impedance. Currently available photovoltaic detectors show response times in the microsecond range at zero load impedance and photoconductors, even though slow, are materially faster at low-load resistances.

The feedback resistance R_1 is dependent on cell sensitivity and should be chosen for either maximum dynamic range or for a desired scale factor; R_2 is elective: In the case of photovoltaic cells or of photodiodes it is not required, in the case of photoconductive cells it should be chosen to minimize bias-current error over the operating range.

Current Monitors

In the current monitors shown in Figure 4-95, R_1 senses the current flow of a power supply. If a J FET such as the 2N3684 is used as a buffer, with drain and source current equal, the output monitor voltage will accurately reflect the power-supply current flow.

High-Impedance Bridge Amplifiers

With high-impedance bridges, the feedback resistances becomes prohibitively large even for the LM108, so the circuit of Figure 4-95 cannot be used. One possible alternative is shown in Figure 4-96. Resistors R_2 and R_3 are chosen so that their equivalent parallel resistance is equal to R_1. Hence, the output of the amplifier will be zero when the bridge is balanced.

When the bridge goes off-balance, the op amp maintains the voltage between its input terminals at zero with current fed back from the output through R_3. This circuit does not act like a true differential amplifier for large imbalances in the bridge.

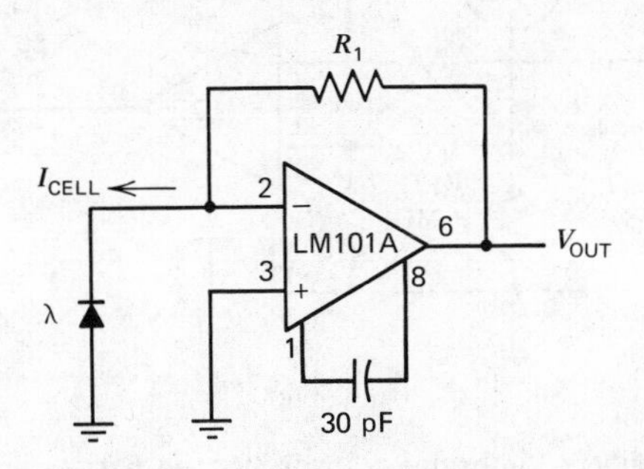

Figure 4-94. Photovoltaic cell amplifier [*note:* $V_{OUT} = I_{CELL} R_1$].

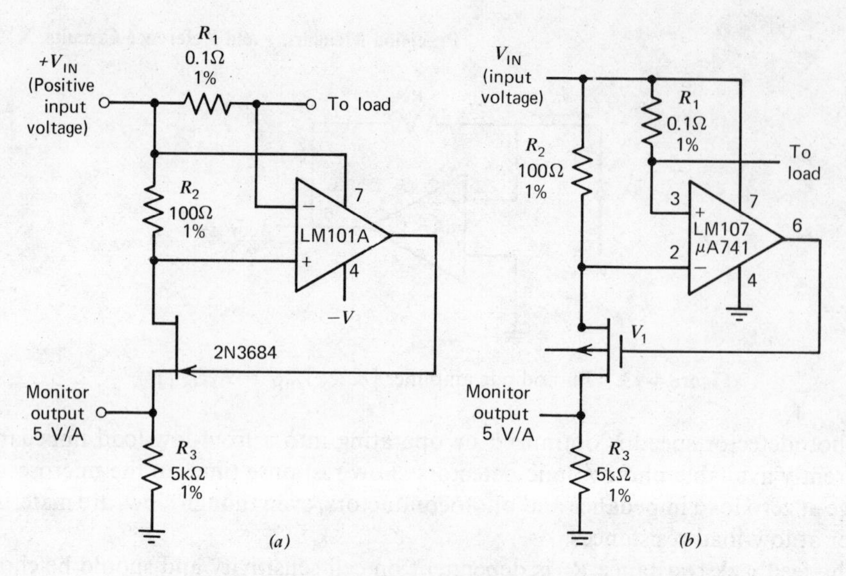

Figure 4-95. Current monitors [*note:* $V_{OUT} = (R_1R_3/R_2)I_L$].

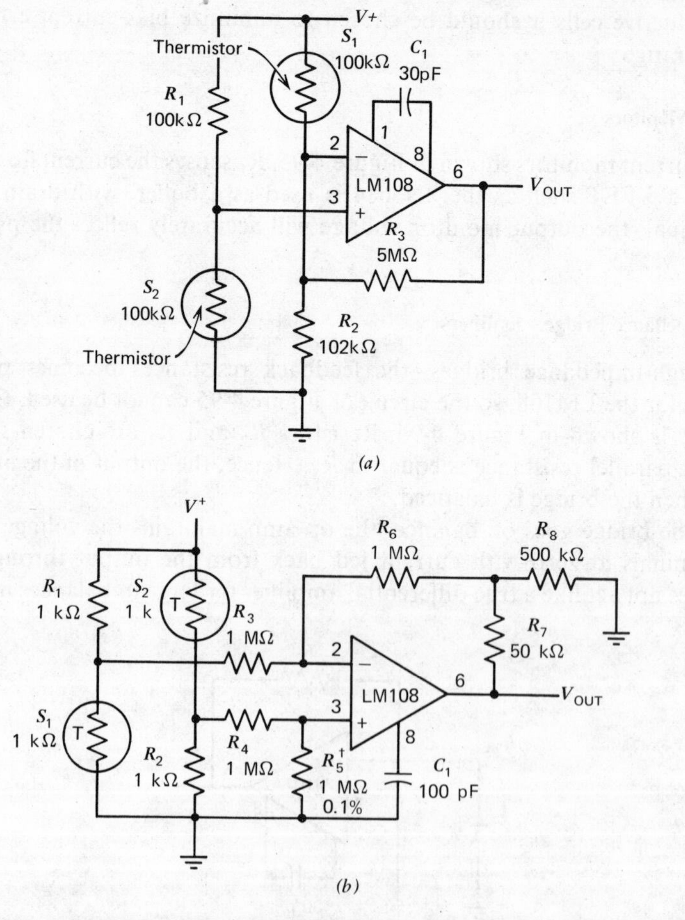

Figure 4-96. Amplifiers for bridge transducers: (*a*) basic amplifier; (*b*) amplifier with low-noise compensation.

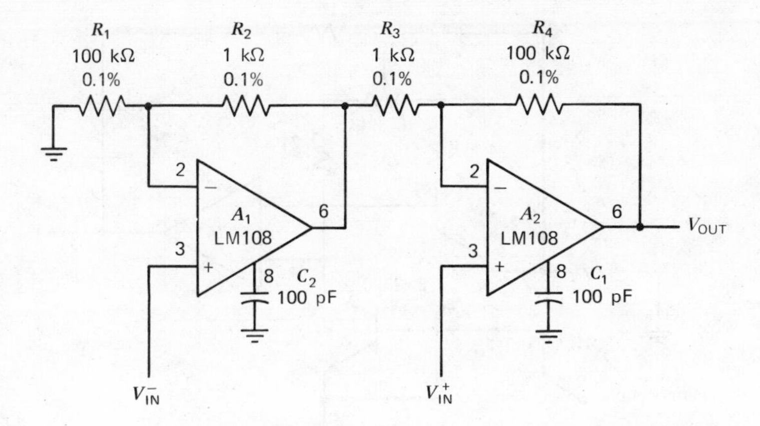

Figure 4-97. Instrumentation amplifier with high-input impedance [*note:* $R_1 = R_4$, $R_2 = R_3$, $A_V = 1 + R_1/R_2$].

The voltage drops across the two sensor resistors, S_1 and S_2, become unequal as the bridge goes off-balance, causing some nonlinearity in the transfer function. However, this is not usually objectionable for small-signal swings.

Figure 4-97 shows a true differential connection that has few of the problems mentioned previously. It has an input resistance greater than 10^{10} Ω, yet it does not need large resistors in the feedback circuitry. With the component values shown, A_1 is connected as a noninverting amplifier with a gain of 1.01; and it feeds into A_2 which has an inverting gain of 100. Hence, the total gain from the input of A_1 to the output of A_2 is 101, which is equal to the noninverting gain of A_2. If all the resistors are matched, the circuit responds only to the differential input signal—not the common-mode voltage.

This circuit has the same sensitivity to resistor-matching as the previous circuits, with a 1% mismatch between two resistors lowering the common-mode rejection to 80 dB. However, matching is more easily accomplished because of the lower resistor values. Further, the high-frequency common-mode rejection is less affected by stray capacitances. The high-frequency rejection is limited, though, by the response of A_1. Several other configurations that can be used with temperature sensors are illustrated in Figure 4-98.

Precision AC/DC Converters

Although semiconductor diodes available today are close to "ideal" devices, they have severe limitations in low-level applications. Silicon diodes have a 0.6-V threshold which must be overcome before appreciable conduction occurs. By placing the diode in the feedback loop of an op amp, the threshold voltage is divided by the open-loop gain of the amplifier. With the threshold virtually eliminated, it is possible to rectify millivolt signals.

Figure 4-99 shows the simplest configuration for eliminating diode-threshold potential. If the voltage at the noninverting input of the amplifier is positive, the output of the LM101A swings positive. When the amplifier output swings 0.6 V positive, D_1 becomes forward-biased; and negative feedback through D_1 forces the inverting

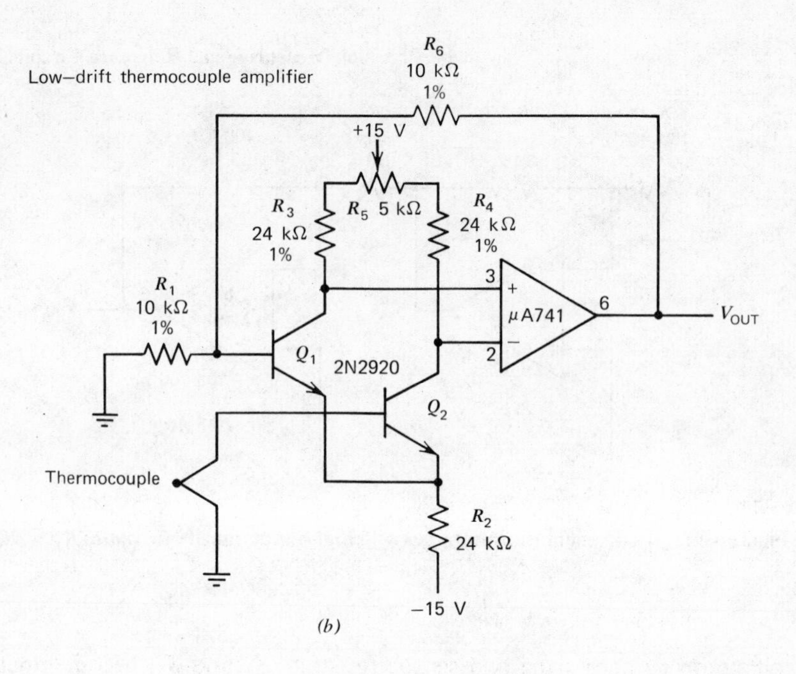

(b)

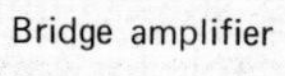

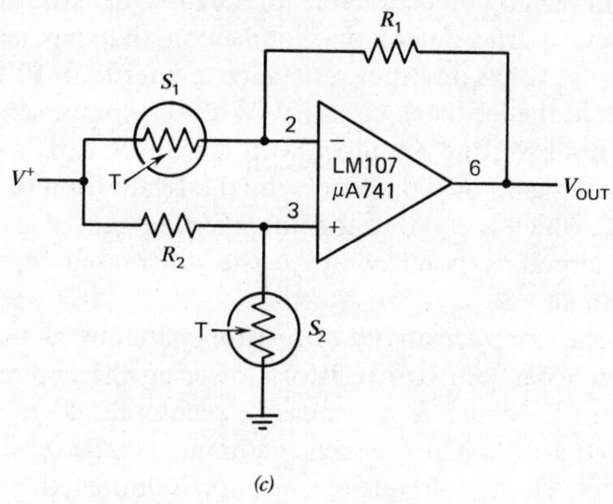

(c)

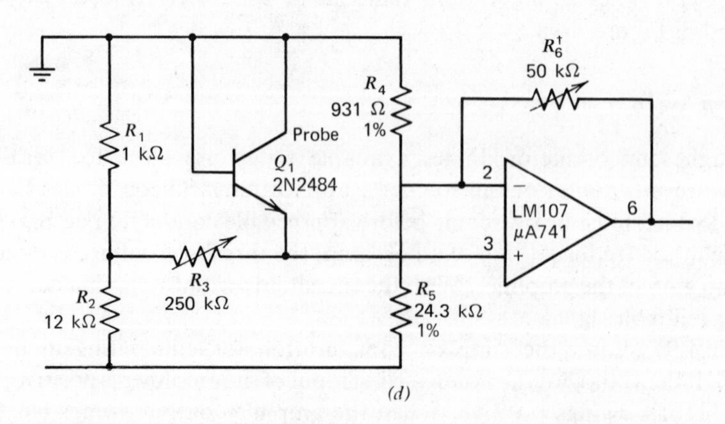

(d)

Figure 4-98. Additional examples of instrumentation amplifiers.

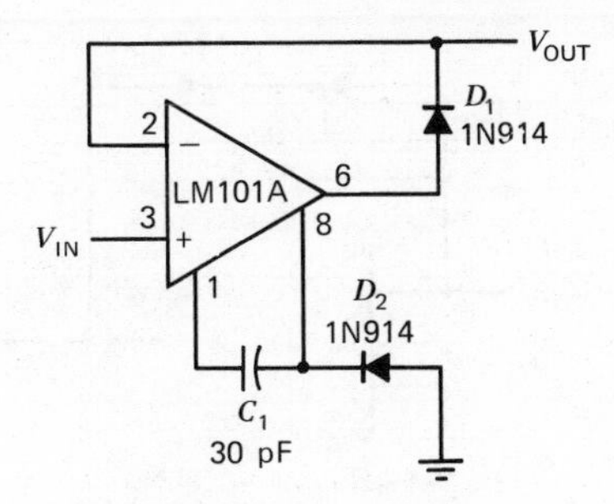

Figure 4-99. Precision diode.

input to follow the noninverting input. Therefore, the circuit acts as a voltage-follower for positive signals. When the input swings negative, the output swings negative and D_1 is cut off. With D_1 cut off, no current flows in the load except the 30-nA bias current of the LM101A. The conduction threshold is very small since less than 100-μV change at the input will cause the output of the LM101A to swing from negative to positive.

Precision Clamp

A useful variation of Figure 4-99 is a precision clamp, as is shown in Figure 4-100. In this circuit the output is precisely clamped from going more positive than the reference voltage. When V_{IN} is more positive than V_{REF}, the LM101A functions as a summing amplifier with the feedback loop closed through D_1. Neglecting offsets, negative feedback keeps the summing node, and therefore the output, within 100 μV of the voltage at the noninverting input. When V_{IN} is about 100 μV more negative than V_{REF}, the output swings positive, reverse-biasing D_1. Since D_1 now prevents negative feedback from controlling the voltage at the inverting input, no clamping action is obtained. On both of the circuits in Figures 4-99 and 4-100 an output clamp diode is added at *pin* 8 to help speed response. The clamp prevents the op amp from saturating when D_1 is reverse-biased.

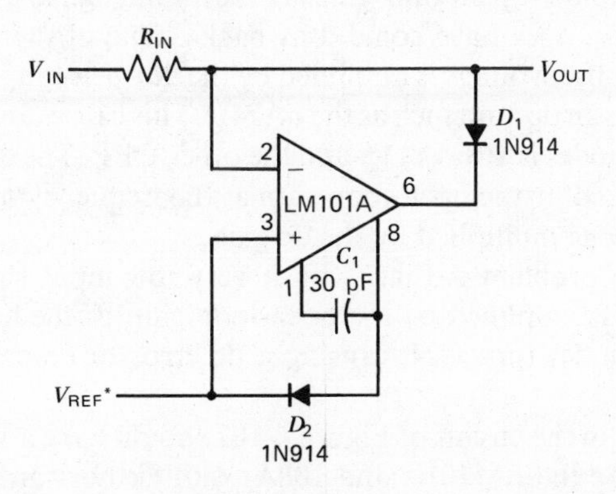

Figure 4-100. Precision clamp [*note:* * V_{REF} must have a source impedance of less than 200 Ω if D_2 is used].

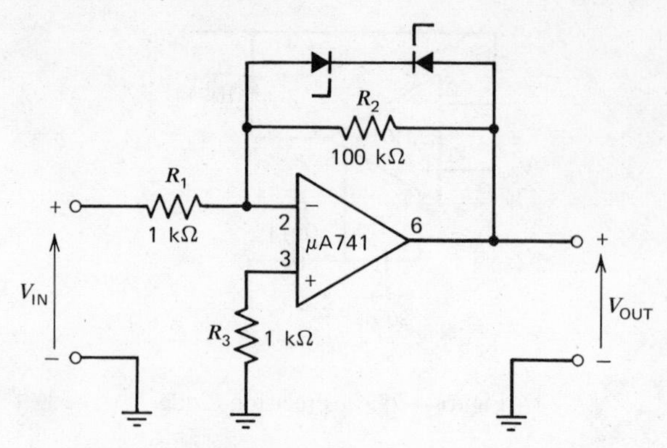

Figure 4-101. Clipping amplifier [*note:* $V_{OUT}/V_{IN} = -R_2/R_1$ if $|V_{OUT}| \leq V_Z +$ 0.7 V where V_Z is the Zener-diode breakdown voltage].

When D_1 is reverse-biased in either circuit, a large differential voltage may appear between the inputs of the LM101A. This is necessary for proper operation and does no damage since the LM101A is designed to withstand large input voltages. These circuits will not work with amplifiers protected with back-to-back diodes across the inputs. Diode protection conducts when the differential-input voltage exceeds 0.6 V and would connect the input and output together. Also, unprotected devices such as the μA709, are damaged by large differential-input signals.

Occasionally, it is desirable to limit the output swing of an amplifier to within specific limits. This can be done by adding nonlinear elements to the feedback network as shown in Figure 4-101.

The Zener diodes quickly reduce the gain of the amplifier if the output tries to exceed the limits set by the Zener voltages. When the Zeners are not conducting, the gain is determined by the feedback resistors R_1 and R_2.

Precision Rectifiers[4]

The basic precision rectifier circuit is shown in Figure 4-102. In this circuit the diodes act as perfect rectifiers because each diode's forward voltage drop is effectively divided by the op amps open-loop gain at the input-signal frequency of interest. The circuit, however, does have some drawbacks. One of the most serious is that the op amp's input-offset voltage is amplified by the dc open-loop gain of the amplifier (50,000 minimum for an op amp such as the μA741). This causes an error output voltage because one of the diodes is always ON and the other OFF. The error occurs when the input signal multiplied by the open-loop gain at the frequency of interest is less than the input-offset voltage multiplied by the dc gain.

A second serious problem is that, with a very low-input signal, the amplifier's input-noise voltage is amplified by the open-loop gain. If the level of input noise is high enough, the diodes turn ON, causing a dc error to be present at $+V_{OUT}$ and $-V_{OUT}$.

The op amp used in the circuit of Figure 4-102 should have a high gain-bandwidth product. Devices like the LM101A and 108A, with feedforward compensation, and the NE531 are ideally suited for this application.

[4] E. R. Hnatek, "Use Integrated Circuits in Transformerless dc-to-dc Converters," *EDN*, Feb. 5, 1973.

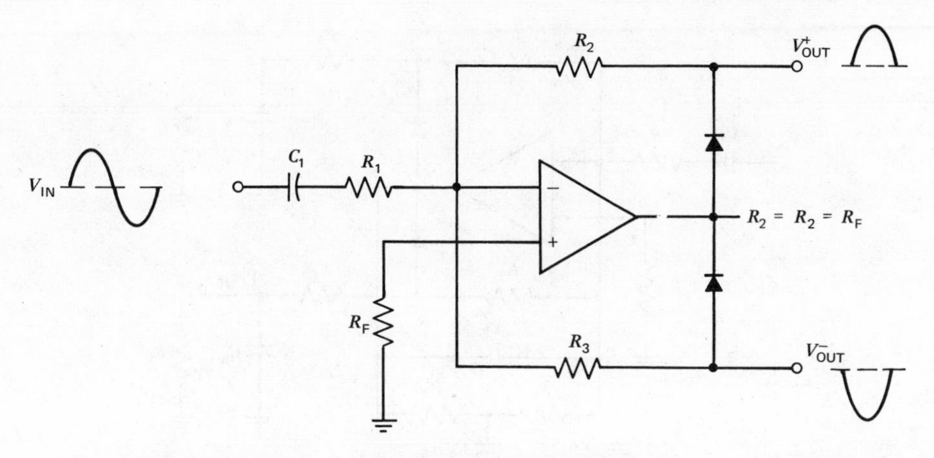

Figure 4-102. The basic precision rectifier circuit suffers from errors in its dc output. (Courtesy *EDN*.)

Figure 4-102 can be modified to overcome the above disadvantages, as shown in Figure 4-103. The addition of C_2, R_4, and R_5 brings the dc closed-loop gain of the circuit to unity. Thus, the op amp's input-offset voltage is only amplified by 1. The addition of C_3 reduces the gain-bandwidth product when the diodes are not conducting. This reduces the gain available to amplify the input noise voltage. However, C_3 should be chosen with care, since it also affects the circuit's gain-bandwidth product with the diodes conducting.

Another variation of the precision rectifier is shown in Figure 4-104. In this circuit, the addition of R_6, C_4 and R_7, C_5 serves two purposes: These components reduce the peak-rectified voltage at $+V_{OUT}$ and $-V_{OUT}$ to an average dc voltage, and they cause the circuit to act as a voltage-doubler.

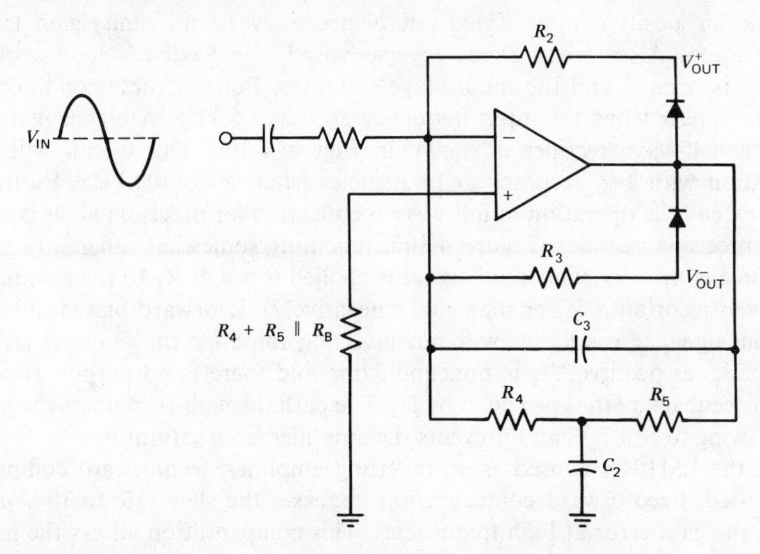

Figure 4-103. Improved performance in the basic precision rectifier is achieved with the addition of R_4, R_5, C_2, and C_3 [*note:* $R_2 = R_3 = R_B$, $R_4 = R_5$]. (Courtesy *EDN*.)

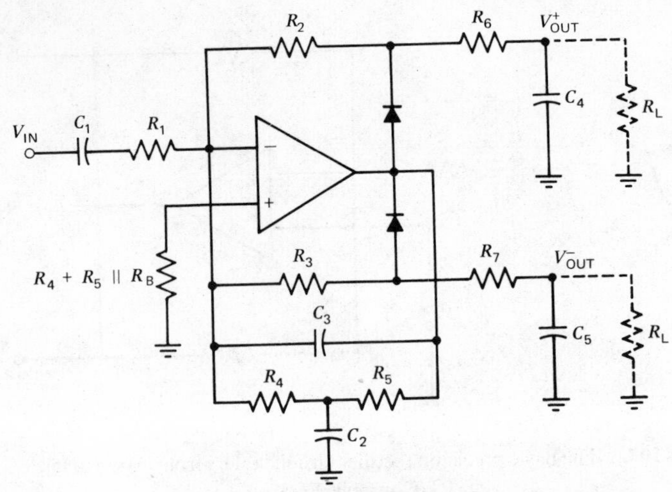

Figure 4-104. This version of the precision rectifier circuit also acts as a voltage-doubler [*note:* $R_2 = R_3 = R_B$, $R_4 = R_5 = R_C$, $R_6 = R_7 = R_A$, $C_3 = C_4$, $10R_B = R_A$, $20R_L \geq R_A$]. (Courtesy *EDN*.)

The time constant of R_6C_4 and R_7C_5 in Figure 4-104 should be set at 1/10 of the single sine-wave period of the highest frequency to be rectified. As a minimum, the values of R_B and R_L are at least ten times R_A; then, the discharge time of C_4 and C_5 is at least ten times as long as the charge time.

Another method of improving circuit performance is to use hot-carrier or Schottky diodes for the rectifiers. These diodes have much lower forward-voltage drops as a function of forward current, which increases the available circuit gain. Figure 4-105 depicts a circuit that provides accurate full-wave rectification.

The circuits in Figures 4-99 and 4-100 are relatively slow. Since there is 100% feedback for positive-input signals, it is necessary to use unity-gain frequency compensation. Also, when D_1 is reverse-biased, the feedback loop around the amplifier is opened and the input stage saturates. Both of these conditions cause errors to appear when the input frequency exceeds 1.5 kHz. A higher performance precision half-wave rectifier is shown in Figure 4-106. This circuit will provide rectification with 1% accuracy at frequencies from dc to 100 kHz. Further, it is easy to extend the operation to full-wave rectification for precision ac/dc converters.

This precision rectifier (Figure 4-106) functions somewhat differently from the circuit in Figure 4-99. The input signal is applied through R_1 to the summing node of an inverting op amp. When the signal is negative, D_1 is forward-biased and develops an output signal across R_2. As with any inverting amplifier, the gain is R_2/R_1. When the signal goes positive, D_1 is nonconducting and there is no output. However, a negative feedback path is provided by D_2. The path through D_2 reduces the negative-output swing to -0.7 V, and prevents the amplifier from saturating.

Since the LM101A is used as an inverting amplifier, feedforward compensation can be used. Feedforward compensation increases the slew rate to 10 V/μsec and reduces the gain error at high frequencies. This compensation allows the half-wave rectifier to operate at higher frequencies than the previous circuits with no loss in accuracy.

The addition of a second amplifier converts the half-wave rectifier to a full-wave

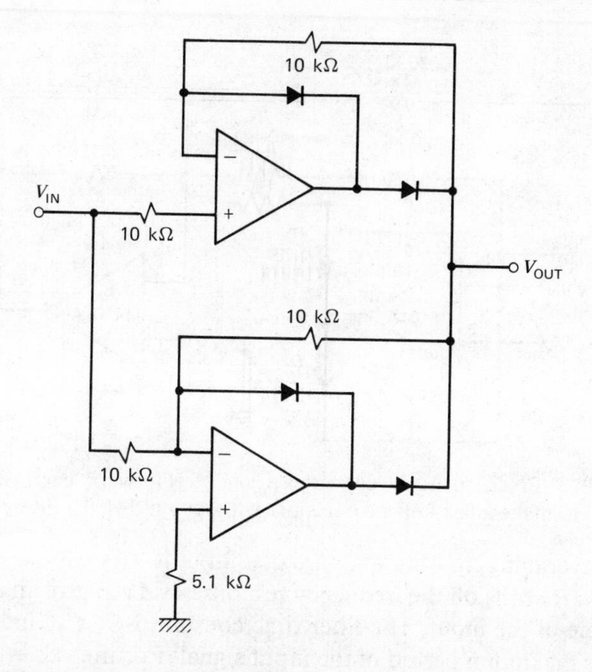

Figure 4-105. Precision full-wave rectifier. (Courtesy *EDN*.)

rectifier. As is shown in Figure 4-107, the half-wave rectifier is connected to inverting amplifier A_2; A_2 sums the half-wave-rectified signal and the input signal to provide a full-wave output. For negative-input signals the output of A_1 is zero and no current flows through R_3. Neglecting for the moment C_2, the output of A_2 is $-(R_7/R_6)V_{IN}$. For positive-input signals, A_2 sums the currents through R_3 and R_6; and

$$V_{OUT} = R_7 \frac{V_{IN}}{R_3} - \frac{V_{IN}}{R_6} \tag{4-15}$$

If R_3 is $(1/2)R_6$, the output is $(R_7/R_6)V_{IN}$. Hence, the output is always the absolute value of the input.

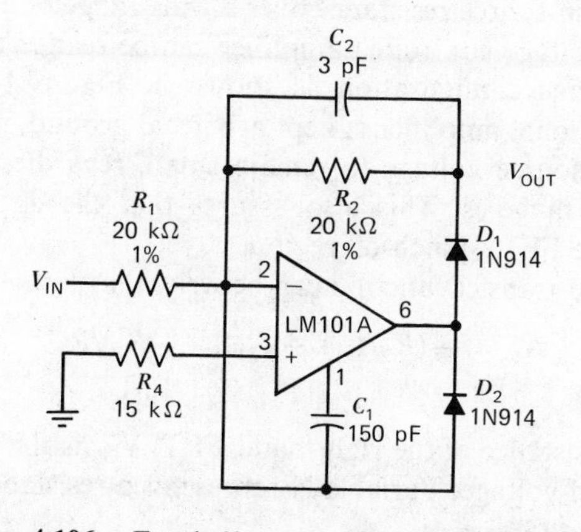

Figure 4-106. Fast half-wave rectifier. (Courtesy *EDN*.)

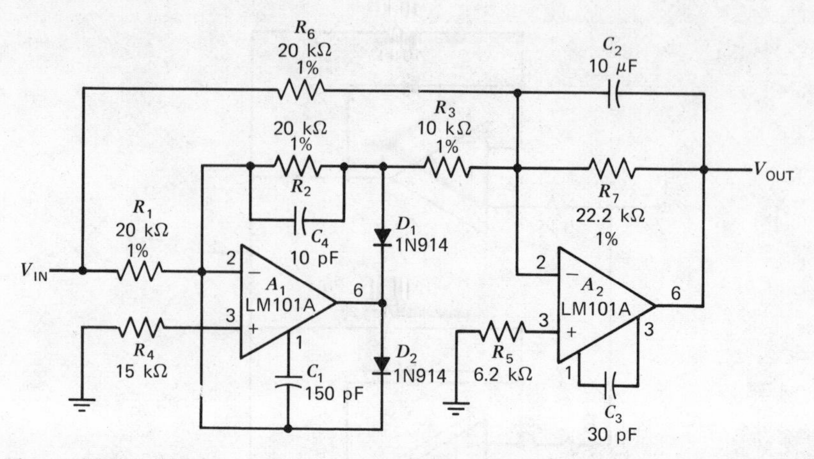

Figure 4-107. Precision ac/dc converter [*note:* feedforward compensation can be used to make a fast full-wave rectifier without a filter]. (Courtesy *EDN*.)

Filtering, or averaging, to obtain a pure dc output is very easy to do. A capacitor C_2 placed across R_7 rolls off the frequency response of A_2 to give an output equal to the average value of the input. The filter time constant is R_7C_2, and must be much greater than the maximum period of the input signal. For the values given in Figure 4-107, the time constant is about 2.0 sec. This converter has better than 1 % conversion accuracy to above 100 kHz and less than 1 % ripple at 20 Hz. The output is calibrated to read the rms value of a sine-wave input.

As with any high-frequency circuit some care must be taken during construction. Leads should be kept short to avoid stray capacitance and power supplies bypassed with 0.01-μF disk ceramic capacitors. Capacitive loading of the fast rectifier circuits must be less than 100 pF or decoupling becomes necessary. The diodes should be reasonably fast and film-type resistors used. Also, the amplifiers must have low-bias currents.

Wide–Range Voltage Controlled Amplifier

When a field-effect transistor is operated as a voltage-controlled resistor, it is usually limited to a relatively small dynamic signal-voltage range. This is due to the nonlinearity of its drain-source resistance over a wide range of drain-source voltage.

But a wide-range voltage-controlled amplifier can be realized if a pair of FETs is connected in the bridge configuration as shown in Figure 4-108. The inverting terminal of the operational amplifier is kept at virtual ground, permitting the range of each FET's drain-source voltage to remain small, regardless of how broad the actual signal-voltage range is. This also assures that the excursions of V_{DS} will remain well within the FET's pinch-off region.

The circuit's voltage-transfer function can be written as:

$$A_V = -(R_2/R_1) + N(R_1 + R_2)/R_1 + NR_2r_{on}[l - (V_{GS}/V_P)] \quad (4\text{-}16)$$

where r_{on} is the on-resistance of the right-hand FET, V_{GS} is the gate-source voltage, and V_P is the pinch-off voltage. Variable N represents a resistance ratio:

$$N = r_{on}/(r_{on} + R_1) \quad (4\text{-}17)$$

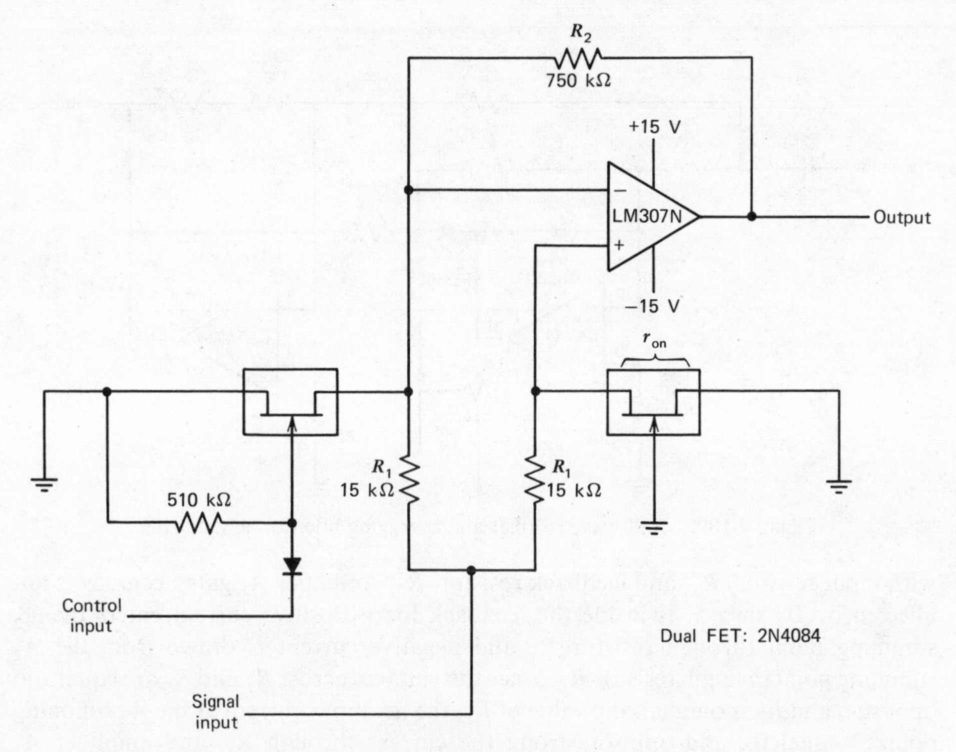

Figure 4-108. Voltage-variable amplifier can operate over a wide range of input-signal voltage.

If N is very small, and r_{on} is much less than R_1, then:

$$AV = -(R_2/R_1)(V_{GS}/V_P) \tag{4-18}$$

Although N must be small, it must, nevertheless, be greater than zero for the circuit to work. The control voltage for the circuit can range from 0 to V_P, and the peak ac input-signal voltage is determined by $I_{DS}R_1$.

Applications for this voltage-controlled amplifier include automatic gain control, true rms conversion, amplitude compression, and signal modulation.

Measurement of Root Mean Square Voltage

The circuit shown in Figure 4-109 is the heart of an average reading, rms-calibrated ac voltmeter. As shown, it is a rectifier and averaging filter. Deletion of C_2 removes the averaging function and provides a precision full-wave rectifier, and deletion of C_1 provides an absolute-value generator.

Circuit operation may be understood by following the signal path for negative and then for positive inputs. For negative signals, the output of amplifier A_1 is clamped to +0.7 V by D_1 and disconnected from the summing point of A_2 by D_2; A_2 then functions as a simple unity-gain inverter with input resistor R_1, and feedback resistor, R_2, giving a positive-going output.

For positive inputs, A_1 operates as a normal amplifier connected to the A_2 summing point through resistor, R_5. Amplifier A_1 then acts as a simple unity-gain inverter

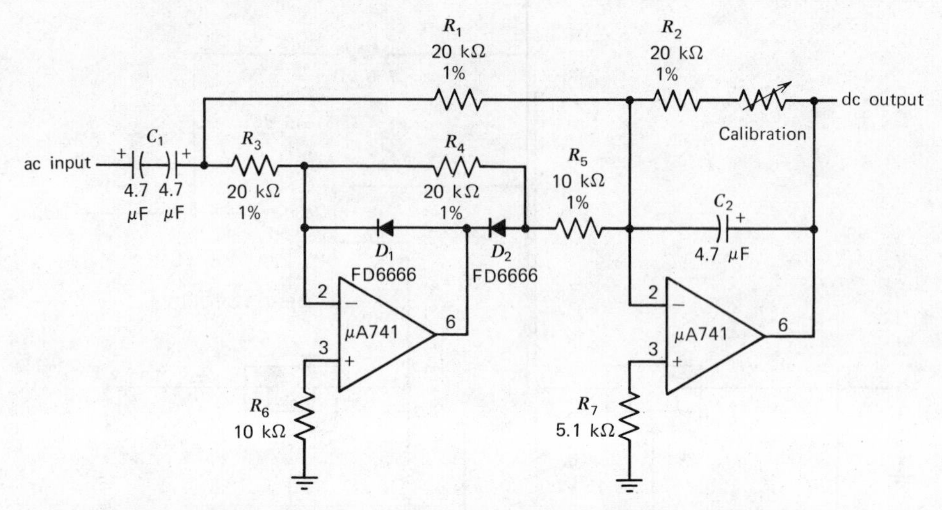

Figure 4-109. Full-wave rectifier and averaging filter for ac voltmeter.

with input resistor, R_3, and feedback resistor, R_5. Amplifier A_1 gain accuracy is not affected by D_2 since it is inside the feedback loop. Positive current enters the A_2 summing point through resistor R_1 and negative current is drawn from the A_2 summing point through resistor R_5. Since the voltages across R_1 and R_5 are equal and opposite, and R_5 is one-half the value of R_1, the net input current at the A_2 summing point is equal to, and opposite from, the current through R_1, and amplifier A_2 operates a summing inverter with unity gain, again giving a positive output.

The circuit becomes an averaging filter when C_2 is connected across R_2. Operation of A_2 then is similar to a simple low-pass filter. The time constant R_2C_2 should be chosen to be much larger than the maximum period of the input voltage which is to be averaged.

Capacitor C_1 may be deleted if the circuit is to be used as an absolute-value generator. When this is done, the circuit output will be the positive absolute value of the input voltage.

The amplifiers chosen must be compensated for unity-gain operation and R_6 and R_7 must be chosen to minimize output errors due to input offset current.

True Root Mean Square Detector[5]

Figure 4-110 depicts an instrumentation amplifier used as a true rms detector. Op amp precision rectifier circuits have greatly eased the problems of ac-to-dc conversion. Inaccuracy due to diode turn-on and nonlinearity is eliminated, and precise rectification of low-level signals is obtained. This makes it possible to measure millivolt ac signals with a dc meter with better than 1% accuracy.

Once the signal is rectified, it is normally filtered to obtain a smooth dc output. The output is proportional to the average value of the ac-input signal, rather than the rms. With known input waveforms such as sine, triangle, or square, this is adequate since there is a known proportionality between rms and average values. However, when the waveform is complex or unknown, a direct readout of the rms value is desirable.

[5] R. C. Dobkin, "True RMS Detector," *Electronic Engineering Times*, May 1973.

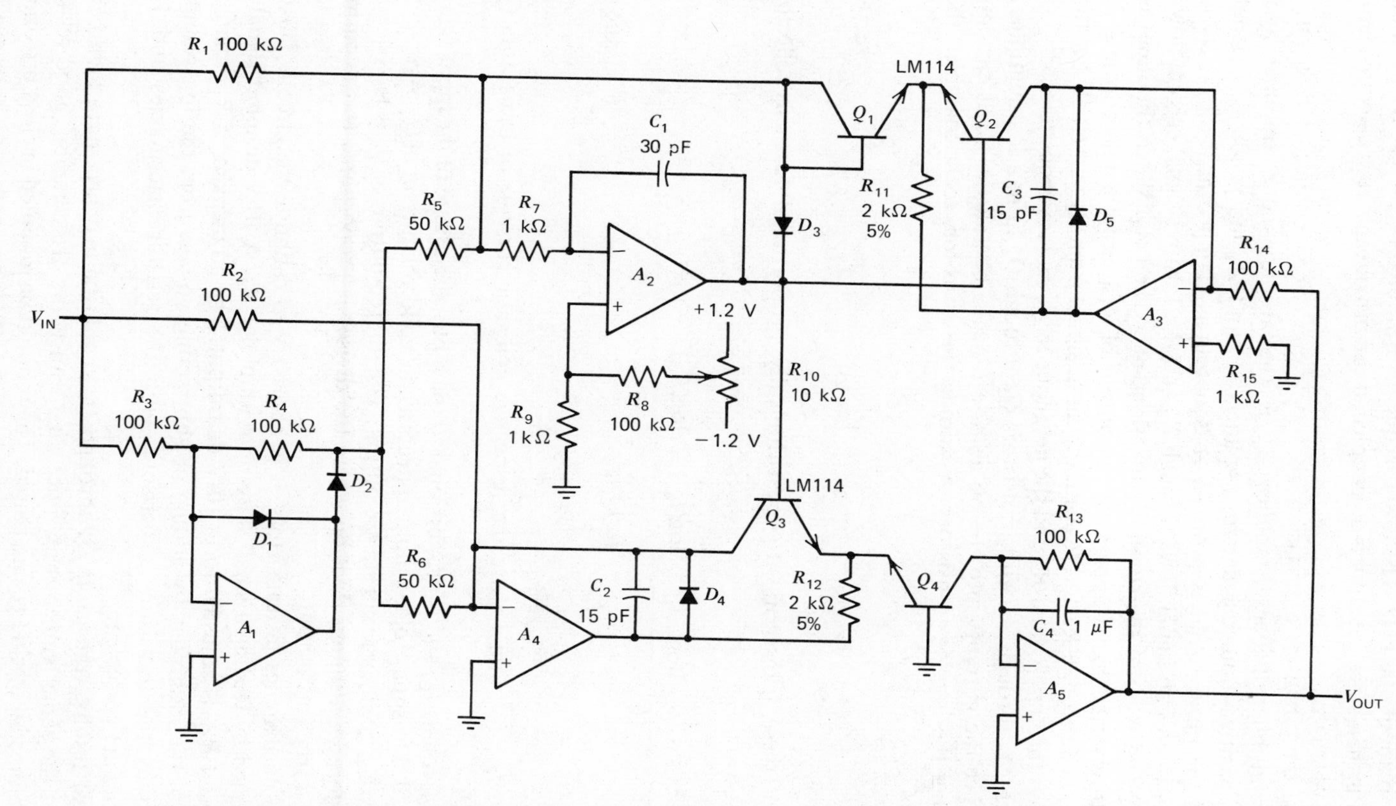

Figure 4-110. True rms detector [*note:* all resistors are 1% unless otherwise specified; all diodes are IN914]. (Courtesy *Electronic Engineering Times.*)

The circuit shown will provide a dc output equal to the rms value of the input. Accuracy is typically 2% for a 20 V_{p-p} input signal from 50 Hz to 100 kHz, although it is usable to about 500 kHz. The lower frequency is limited by the size of the filter capacitor. Further, since the input is dc-coupled, it can provide the true rms equivalent of a dc + ac signal.

Basically, the circuit is a precision absolute-value circuit connected to a one-quadrant multiplier/divider. Amplifier A_1 is the absolute-value amplifier and provides a positive-input current to amplifiers A_2 and A_4 independent of signal polarity. If the input signal is positive, A_1's output is clamped at -0.6 V, D_2 is reverse-biased, and no signal flows through R_5 and R_6. Positive signal current flows through R_1 and R_2 into the summing functions of A_2 and A_4. When the input is negative, an inverted signal appears at the output of A_1 (output is taken from D_2). This is summed through R_5 and R_6 with the input signal from R_1 and R_2. Twice the current flows through R_5 and R_6 and the net input to A_2 and A_4 is positive.

Amplifiers A_2 through A_5 with transistors Q_1 through Q_4 form a log multiplier/divider. Since the currents into the op amps are negligible, all the input currents flow through the logging transistors. Assuming the transitors to be matched, the V_{BE} of Q_4 is

$$V_{BE(Q_4)} = V_{BE(Q_1)} + V_{BE(Q_3)} - V_{BE(Q_2)} \tag{4-19}$$

The V_{BE} of these transistors are logarithmically proportional to their collector currents so

$$\log(I_{C_4}) = \log(I_{C_1}) + \log(I_{C_3}) - \log(I_{C_2}) \tag{4-20}$$

or

$$I_C = \frac{I_{C_1} I_{C_3}}{I_{C_2}} \tag{4-21}$$

where I_{C_1}, I_{C_2}, I_{C_3}, and I_{C_4} are the respective collector currents of transistors Q_1 through Q_4.

Since I_{C_1} equals I_{C_3} and is proportional to the input, the square of the input signal is generated. The square of the input appears as the collector current of Q_4. Averaging is done by C_4, giving a mean square output. The filtered output of Q_4 is fed back to Q_2 to perform continuous division where the divisor is proportional to the output signal for a true rms output.

Due to mismatches in transistors, it is necessary to calibrate the circuit. This is accomplished by feeding a small offset into amplifier A_2. A 10-V dc input signal is applied, and R_{10} is adjusted for a 10-V dc output. The adjustment of R_{10} changes the gain of the multiplier by adding or subtracting voltage from the log voltages generated by the transistors. Therefore, both the resistor inaccuracies and V_{BE} mismatches are corrected.

For best results, transistors Q_1 through Q_4 should be matched, have high beta, and be at the same temperature. Since dual transistors are common, good results can be obtained if Q_1 and Q_2 are paired. They should be mounted in close proximity or on a common heat sink, if possible. As a final note, it is necessary to bypass all op amps with 0.1-μF disk capacitors.

Adjustable Voltage References

Adjustable voltage-reference circuits are shown in Figure 4-111 and 4-112. The two circuits shown have different areas of applicability. The basic difference between the two is that Figure 4-111 illustrates a voltage source which provides a voltage greater than the reference diode while Figure 4-112 illustrates a voltage source which provides a voltage lower than the reference diode. The figures show both positive- and negative-voltage sources.

High-precision extended-temperature applications of the circuit of Figure 4-111 require that the range of adjustment of V_{OUT} be restricted. When this is done, R_1 may be chosen to provide optimum Zener current for minimum Zener TC. Since I_z is not a function of V^+, reference TC will be independent of V^+. The circuit of

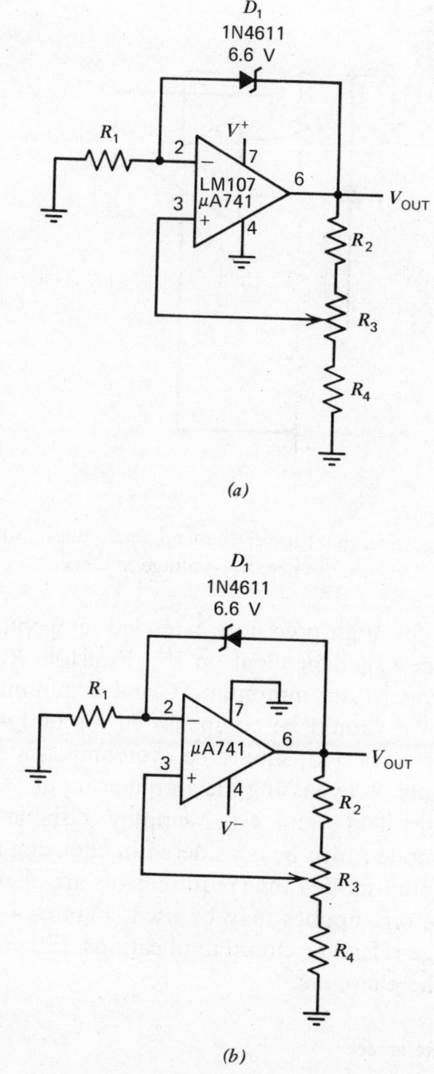

Figure 4-111. Voltage references greater than reference diode: (*a*) positive-voltage reference, (*b*) negative-voltage reference.

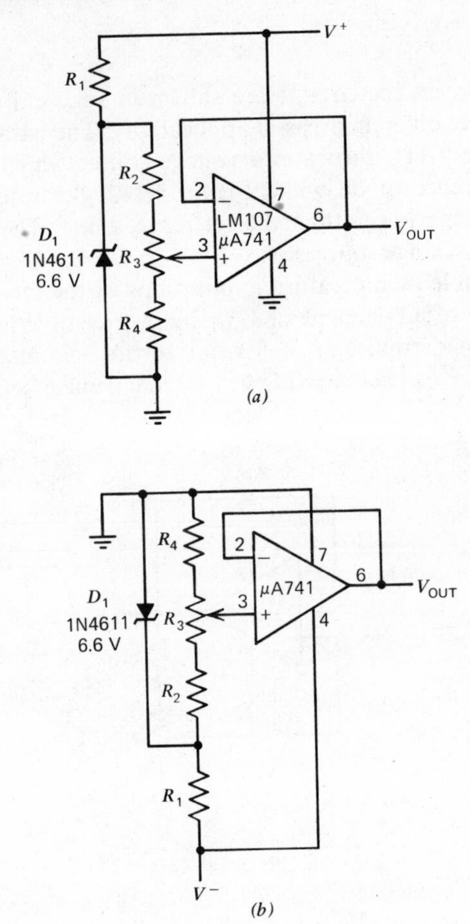

Figure 4-112. Voltage references lower than reference diode: (*a*) positive-voltage reference, (*b*) negative-voltage reference.

Figure 4-112 is suited for high-precision extended-temperature service if V^+ is reasonably constant since I_z is dependent on V^+. Resistors R_1, R_2, R_3, and R_4 are chosen to provide the proper I_z for minimum TC and to minimize errors due to I_{BIAS}.

Both the circuits shown should be compensated for unity-gain operation or, if large capacitive loads are expected, should be overcompensated. Output noise may be reduced in both circuits by bypassing the amplifier input.

The circuits in Figures 4-111 and 4-112 employ a single-power supply which requires that common-mode range be considered in choosing an amplifier for these applications. If the common-mode range requirements are in excess of the capability of the amplifier, two power supplies may be used. Figures 4-113 and 4-114 depict several additional voltage-reference circuit applications. Either the LM107 or μA741 may be used in any of these circuits.

High Stability Voltage Reference

With conventional reference-voltage circuits, the problem is not so much the stability of the temperature-compensated zener used, but the bothersome trimming

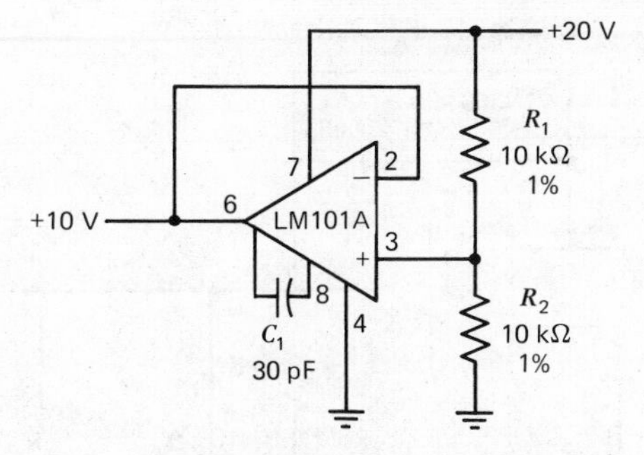

Figure 4-113. Circuit for generating the second positive voltage.

or trial-and-error selection of the components that supply the zener current, and the often nebulous stability of the current-determining circuitry.

With the circuit shown in Figure 4-115, however, the current through the zener diode is truly independent of the power-supply voltage—which may be as low as 10 V. The current is determined by the zener itself, thus one avoids the trimming and component selection needed for other circuits.

To understand how the circuit works, temporarily ignore components D_3, D_2, Q_1, R_4, R_5, and R_6. Let's assume also that zener D_1 is in the breakdown region. Assume a voltage, $-V_1$, at the junction of R_1 and R_2. Then the output of A_1 is $-(V_1 + V_z)$. But resistors R_1 and R_2 form a voltage divider, therefore,

$$-(V_1 + V_Z)\frac{R_2}{R_1 + R_2} = -V_1 \tag{4-22}$$

If we select R_1 equal to $10R_2$, then,

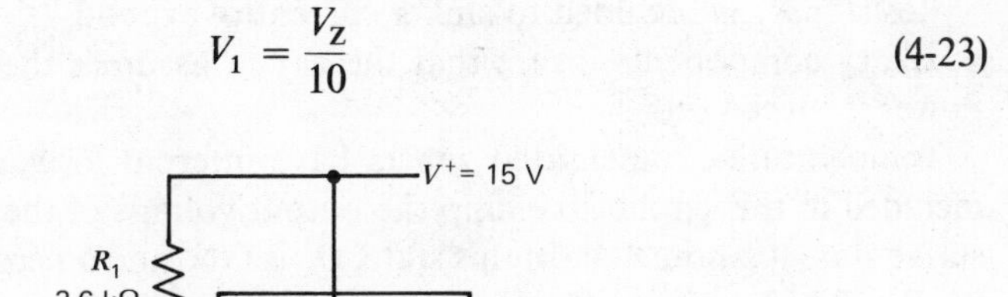

$$V_1 = \frac{V_Z}{10} \tag{4-23}$$

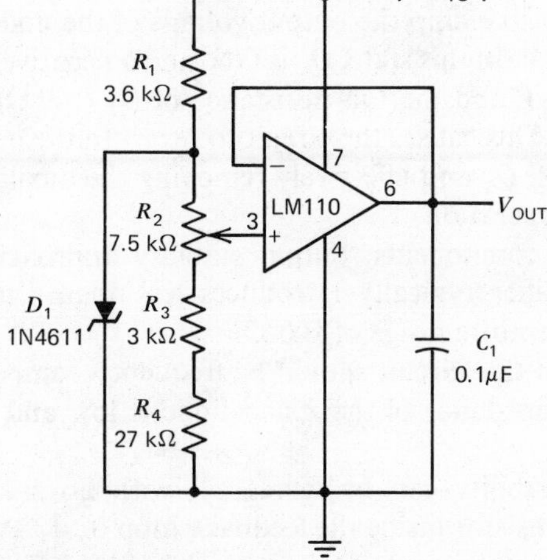

Figure 4-114. Buffered-reference source.

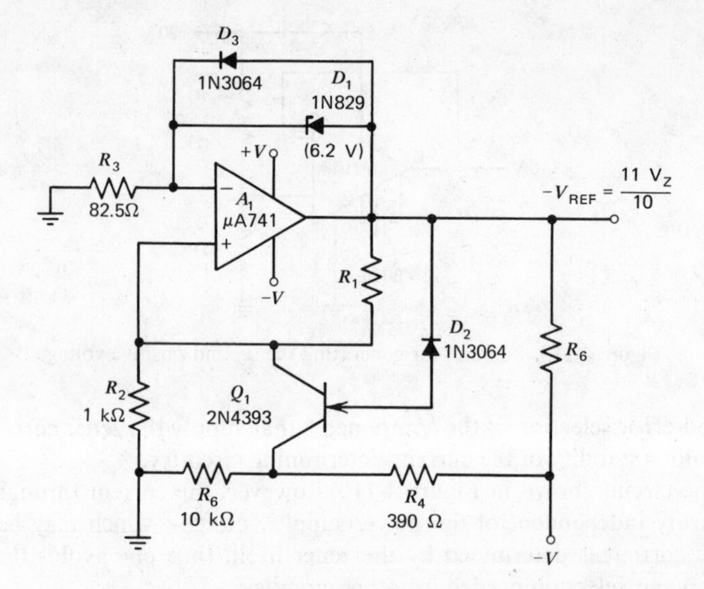

Figure 4-115. Highly stable voltage reference circuit.

Therefore the zener current is given by the following equation:

$$\frac{V_1}{R_3} = \frac{V_Z}{10R_3} \tag{4-24}$$

One can then calculate the value of R_3 needed to supply the specified zener current of 7.5 mA.

Resistor R_6 is included to sink a current of around 7.5 mA. The remainder of the auxiliary components insure that the circuit assumes the current stable state when power is turned on.

Temperature-compensated zeners have internal forward-biased diodes, so D_3 is included in this circuit to clamp the output voltage of the undesired positive-output stable state to about 1 V. In this state, Q_1 is ON, and a negative potential, determined by R_1, R_2, R_4, R_5, $-V$ and the ON resistance of Q_1 (100 Ω), appears at the non-inverting input of A_1. This causes the circuit to revert to the desired negative-output stable state, turning off Q_1 and effectively removing the auxiliary components from the circuit in normal operation.

With the specified components, output stability approaches that of the zener itself. A μA741 amplifier typically introduces less than 1 mV of output-voltage variation over a temperature range of 100°C.

The op amp used in this circuit should be frequency compensated for unity-gain operation since the impedance of the zener diode is low and thus provides almost 100% ac feedback.

Output-current capability can be increased, with no sacrifice in stability, by inserting a booster transistor inside the feedback loop of A_1. Also, the circuit can be built as a positive reference supply by using a *p*-channel FET, reversing the diodes, and changing the current-sinking voltage to $+V$.

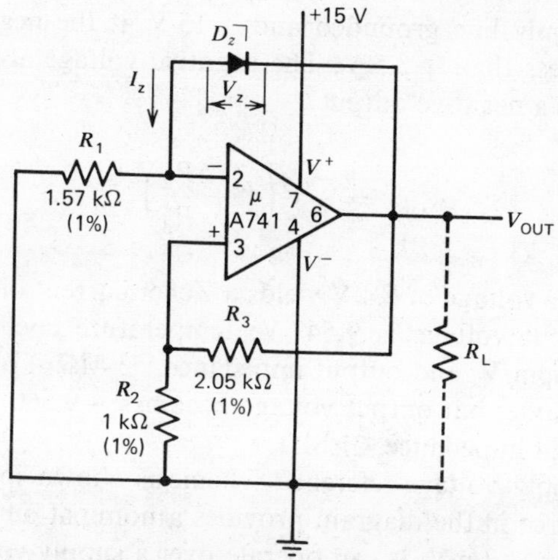

Figure 4-116. Eliminating dual supplies. Voltage-reference source has either positive or negative output, depending on polarity of supply used to bias operational amplifier. Positive-supply results in positive voltage at operational amplifier's non-inverting input and, therefore, positive output. Negative-supply produces negative output. In either case, final output voltage is about 10 V. The Zener diode acts as reference.

Single-Supply Voltage References

When an IC differential op amp is used in a voltage-reference source, two supplies usually are needed—both positive and negative. Moreover, additional circuitry is generally required to establish output polarity. But if one supply line is grounded, only one supply becomes necessary—the one that gives the desired output polarity. So long as the noninverting input of the op amp is maintained at some positive potential, even a small one, the output of the reference source will also be positive. Similarly, small negative potentials at the noninverting input will result in negative outputs.

The voltage-reference source shown in Figure 4-116 employs a single 15-V supply to give a positive output. To produce a negative-output voltage, the positive-supply line is grounded, the negative-supply line is run to −15 V, and Zener polarity must be reversed.

Zener diode D_z establishes the reference voltage (V_z) and the reference current (I_z) for the circuit so that a constant-current source is provided by resistor R_1 and the buffering action of the op amp. Zener current is

$$I_z = \frac{V_z R_2}{R_1 R_3} \tag{4-25}$$

Due to the configuration of the op amp's output stage, circuit output voltage (V_{OUT}) will always be greater than 0.5 V when a positive 15-V supply is applied and the negative-supply line is grounded. Therefore, a few tens of millivolts will always appear at the op amp's noninverting input, making the output positive:

$$V_{OUT} = V_z\left(1 + \frac{R_2}{R_3}\right) \tag{4-26}$$

With the positive-supply line grounded and -15 V at the negative-supply input, V_{OUT} will always be less than -0.5 V. The negative voltage at the op amp's non-inverting input causes a negative output

$$V_{\text{OUT}} = -V_z\left(1 + \frac{R_2}{R_3}\right) \tag{4-27}$$

The reference Zener voltage of 6.4 V yields a Zener current of about 2 mA. For a positive supply, output voltage is 9.547 V; temperature coefficient, 1.9 ppm/°C; voltage stability 9.5 ppm/V; and output impedance, 53 MΩ. For a negative supply, the TC remains the same; but output voltage becomes -9.560 V; voltage stability, 2.6 ppm/V; and output impedance, 21 MΩ.

Another single-supply voltage-reference scheme is shown in Figure 4-117. The voltage-reference source in the diagram provides a nominal output voltage of 10 V that is stable to within ± 7 mV. It can operate over a supply-voltage range of 12 to 18 V. With the output voltage at V_{REF}, the current through the Zener is

$$I_z = V_{\text{REF}} - \frac{V_z}{R_1} \tag{4-28}$$

The current-booster transistor actively keeps the voltage at the inverting input of the op amp at:

$$V_1 = \frac{V_{\text{REF}} R_2}{R_2 + R_3} = I_z R_1 \tag{4-29}$$

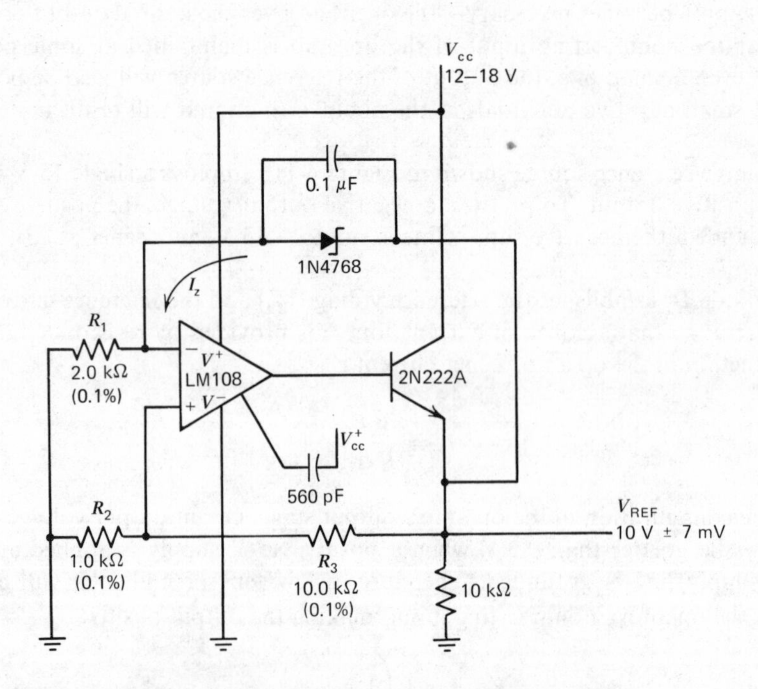

Figure 4-117. Single-supply voltage reference.

Zener current I_z can then be expressed as:

$$I_z = \frac{V_{REF}R_2}{R_1(R_2 + R_3)} \tag{4-30}$$

and reference voltage V_{REF} is held to

$$V_{REF} = \frac{V_z(R_2 + R_3)}{R_3} \tag{4-31}$$

A pitfall in any dc bootstrap scheme like this is the possible existence of stable states other than the desired one. For this circuit, such a state exists when the output voltage is zero. Connecting a capacitor between the power supply and the op amp's compensation input prevents the circuit from being locked into the zero-output-voltage state when it is first turned ON. The capacitor causes the turn-on transient to put the circuit into the desired condition, with the output voltage at V_{REF}.

For the components shown, the output is stable to within ± 7 mV over a temperature range of 0 to 75°C, even with a supply variation of $\pm 10\%$. The output current is 30 mA maximum. The capacitor across the Zener serves to attenuate noise that may be generated by the Zener at high-operating frequencies.

A negative-output voltage can be obtained by reversing the diode, using a *pnp* instead of an *npn* transistor, and grounding the op amp's positive-supply input.

Integrated Nanoammeter Amplifier Circuit[6]

The schematic diagram of a nanoammeter amplifier is presented in Figure 4-118. In this circuit, the meter amplifier is a differential current-to-voltage converter with input protection, zeroing and full-scale adjust provisions, and input-resistor-balancing for minimum-offset voltage. Resistor R'_F (equal in value to R_F for measurements of less than 1 μA) ensures that the input-bias currents for the two input terminals of the amplifier do not contribute significantly to an output-error voltage. The output voltage V_{OUT} for the differential current-to-voltage converter is equal to $-2I_F R_F$, since the floating input current I_{IN} must flow through R_F and R'_F: R'_F may be omitted for R_F values of 500 kΩ or less because a resistance of this value contributes an error of less than 0.1% in output voltage. Potentiometer R_2 provides an electrical meter zero by forcing the input-offset voltage V_{OS} to zero. Full-scale meter deflection is set by R_1. Both R_1 and R_2 only need to be set once for each op amp and meter combination. For a 50-μA, 2-kΩ meter movement, R_1 should be about 4 kΩ to give full-scale meter deflection in response to a 300-mV output voltage. Diodes D_1 and D_2 provide full input protection for overcurrents up to 75 mA.

With an R_F resistor value of 1.5 MΩ, the circuit in Figure 4-117 becomes a nanoammeter with a full-scale reading capability of 100 nA. The voltage drop across the two input terminals is then equal to the output voltage V_{OUT} divided by the open-loop gain. Assuming an open-loop gain of 10,000 gives an input-voltage drop of 30 μV or less.

A unique feature of the 4250-type programmable op amp is the provision for an external master bias-current setting resistor R_8 at *pin* 8. With this resistor set at

[6] M. Vander Kooi, "Simple IC Meter Amplifier Circuit Measures 100 Nanoamps, Full-Scale," *EDN/EEE*, April 15, 1972.

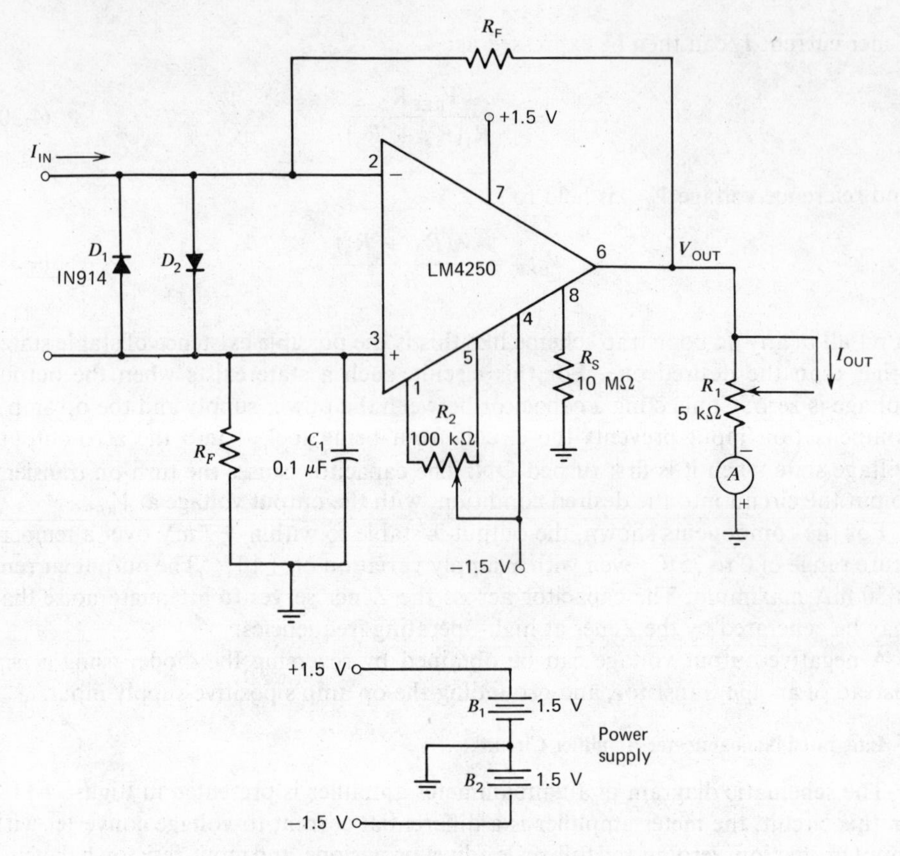

Resistance values for dc nanoammeter and microammeter

I full scale	R_F (Ω)	R'_F (Ω)
100 nA	1.5 M	1.5 M
500 nA	300 k	300 k
1 μA	300 k	0
5 μA	60 k	0
10 μA	30 k	0
50 μA	6 k	0
100 μA	3 k	0

Figure 4-118. Circuit for a dc nonoammeter and microammeter. (Courtesy *EDN/EEE*.)

10 MΩ, the total quiescent current drain of the circuit is 0.6 μA for a total power-supply drain of 1.8 μW. The input-bias current required by the amplifier at this low level of quiescent current is in the range of 600 pA. Notice that this circuit operates on two 1.5-V flashlight batteries and has quiescent power drain so low that an ON-OFF switch (for when the meter is not in use) is not required.

The circuit of Figure 4-118 can be easily modified to handle current readings higher than 100 μA (see Figure 4-119). In the second circuit, resistor R_A develops a voltage drop in response to input current I_A. This voltage is amplified by a factor

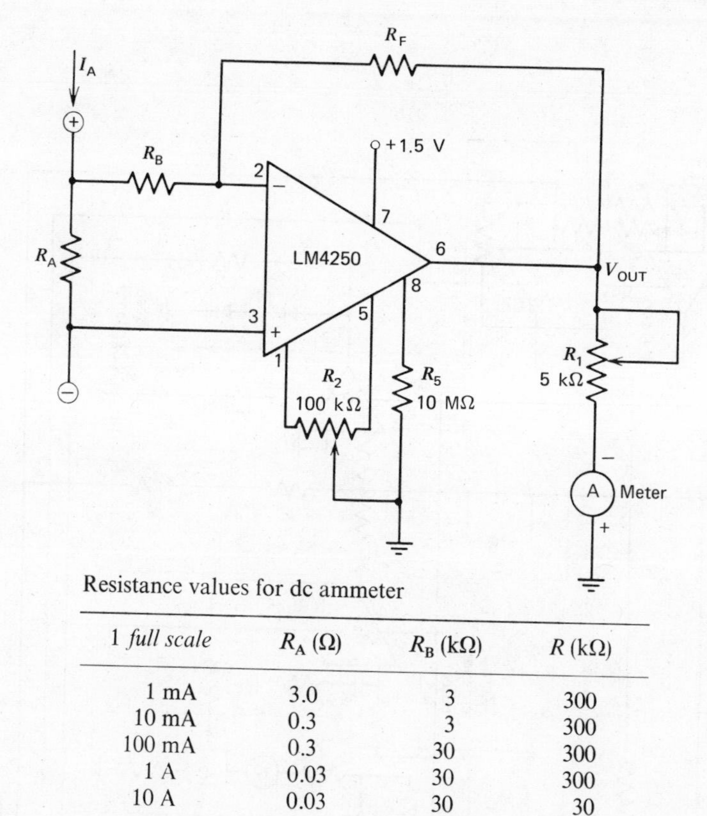

Resistance values for dc ammeter

1 *full scale*	R_A (Ω)	R_B (kΩ)	R (kΩ)
1 mA	3.0	3	300
10 mA	0.3	3	300
100 mA	0.3	30	300
1 A	0.03	30	300
10 A	0.03	30	30

Figure 4-119. Ammeter circuit for measurement of currents from 1 mA to 10 A. (Courtesy *EDN/EEE*.)

equal to the ratio of R_F/R_B: R_B must be sufficiently larger than R_A that the input signal will not be loaded.

Electronic Thermometer

A new and unique use for op amps is the electronic thermometer (Figure 4-120). As shown, the sensor is attached to a bridge whose output goes to both a zero-crossing detector and a sample-and-hold circuit. Upon contacting the mouth, the thermistor in the sensor tip responds, unbalancing the bridge. When the sensor response passes 0 V—the bridge balance point that corresponds to the threshold temperature—the detector switches, turning ON the 15-sec timer and connecting the input of the sample-and-hold circuit to the instrument's meter. Meanwhile, the timer's output closes the switch between the bridge and the sample-and-hold circuit. The sensor output then is connected by way of the bridge and the sample-and-hold circuit to the meter. After 15 sec the timer turns OFF, disconnecting the sample-and-hold circuit from the bridge and turning ON the read lamp, a light-emitting diode. The temperature displayed by the meter equals T_{SS}. The sample-and-hold circuit maintains this reading for at least 7 sec after the read light comes ON.

The battery cut-off circuit continuously monitors the power supply for low output. If either the positive- or negative-battery voltage drops below a predetermined level at which accuracy could be degraded, the thermometer automatically turns OFF and cannot be used again until the batteries are replaced.

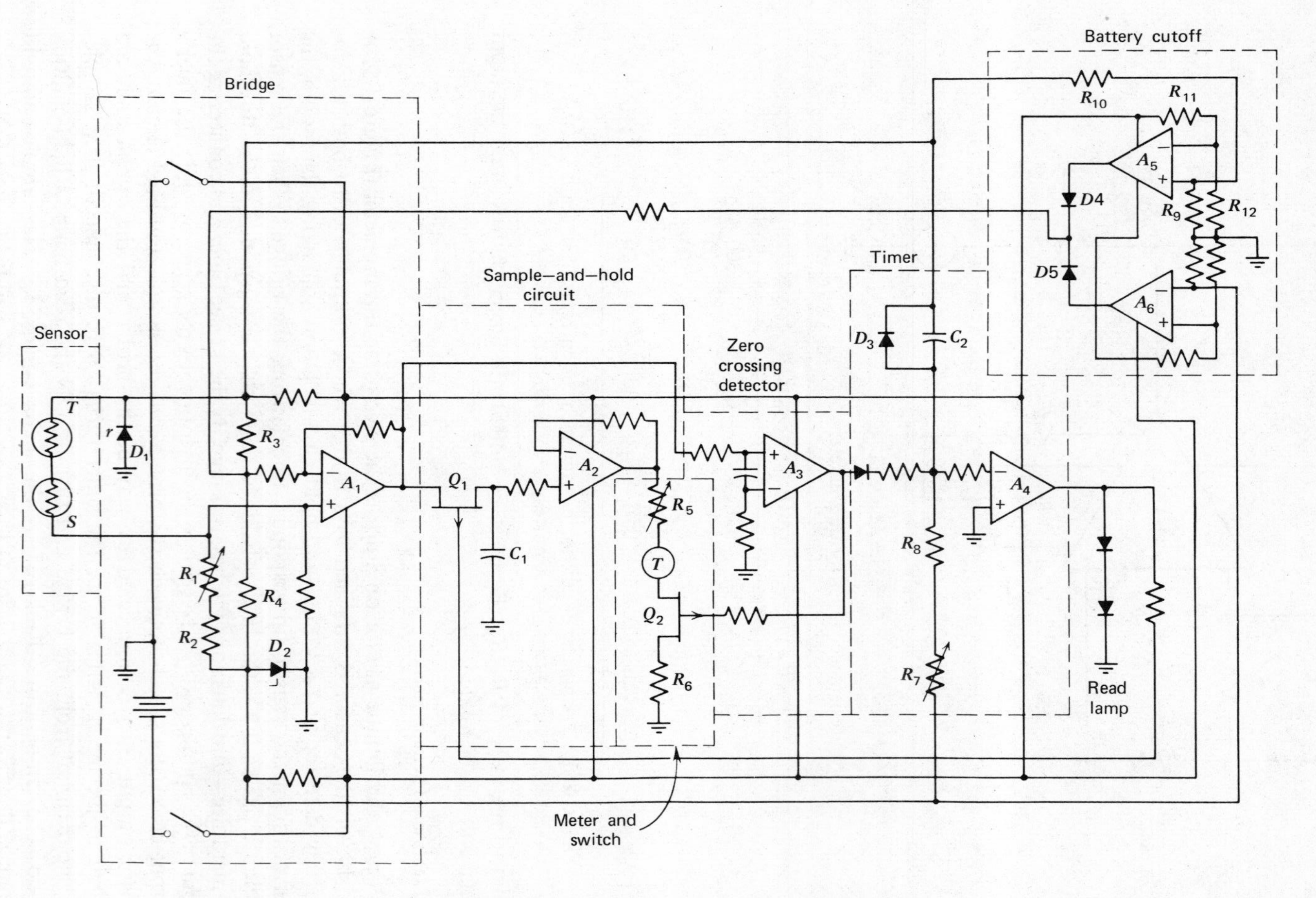

Figure 4-120. Schematic of an electronic thermometer.

The instrument's sensor consists of a thermistor mounted in the tip along with a positive TC resistor. This resistor compensates the total sensor assembly for any lowering of tissue temperature occurring when the thermometer is inserted, as well as the sensor's initial temperature in the thermistor.

The circuit of Figure 4-120 works in the following manner. The bridge consists of a zero-adjustment potentiometer R_1 and resistors R_2, R_3, and R_4. Battery voltage to the bridge is regulated by Zener diodes D_1 and D_2. Operational amplifier A_1, located at the bridge's output, provides a gain of 5 as well as high-input impedance, in order to protect the bridge from loading effects. The output of A_1 passes through through sample switch Q_1 to charge-hold capacitor C_1 at the input of hold amplifier A_2. Like A_3, A_2 is an op amp, but it is connected in a voltage-follower configuration to give very high-input impedance and low-output impedance at a voltage gain of 1. Amplifier A_2 drives a series combination of potentiometer R_6, the meter, the meter switch Q_2, and the scaling resistor R_6.

The output of A_4 also goes to A_3, an open-loop amplifier that provides the zero-crossing detection. Since A_3's positive-input terminal is referenced to ground, its output is positive when A_1's output is below ground and negative when A_1's output is above ground.

When A_3 switches to its negative state, Q_2 turns ON, allowing positive current to flow through the meter. Before it goes negative, A_3's output is more positive than the cathode of diode D_3: therefore, A_3 clamps the voltage across C_2 the timing capacitor. When A_3's output swings negative, C_2 charges through R_7 and R_8. When the junction of C_2 and R_8 reaches the reference voltage of A_1, the amplifier switches to a positive output, lighting the read lamp. The output of A_4 also turns OFF Q_1, maintaining the meter reading: The transistor's charge is sufficient to hold a constant current for a minimum of 7 sec.

Precision Automobile Tachometer[7]

The tachometer circuit shown in Figure 4-121 is composed of three sections: A relaxation oscillator at the input for point-bounce suppression, a monostable multivibrator for pulse generation, and a buffer for driving a meter.

Unijunction transistor Q_1 is operated with an emitter current that is larger than its valley current so that it will not turn OFF after triggering. When the points open, capacitor C_1 charges through resistor R_1 until Q_1 fires (in 0.5 to 0.7 msec) and triggers the monostable. The current through R_1 keeps Q_1 ON and prevents C_1 from charging until the points close. If the points bounce upon closure, they will not be open long enough to allow C_1 to charge and fire Q_1 again.

For every point opening, the monostable multivibrator produces a pulse having a fixed width and amplitude. Normally, the output stage of the op amp produces a negative-saturation voltage. But when a positive trigger from the relaxation oscillator is applied through diode D_1, the op amp's output switches to a positive-saturation voltage, causing capacitor C_2 to charge positively through resistor R_2. Capacitor C_2 stores the charge until V^- is greater than V^+, and the op amp switches back to its stable state.

Diode D_2 clamps the voltage across C_2 to about -0.6 V, while diode D_3 provides TC for changes in D_2's junction-voltage drop. Both of these diodes should be kept

[7] J. Young, "Precision Auto Tachometer Squelches Point Bounce," *Electronics*, December 18, 1972.

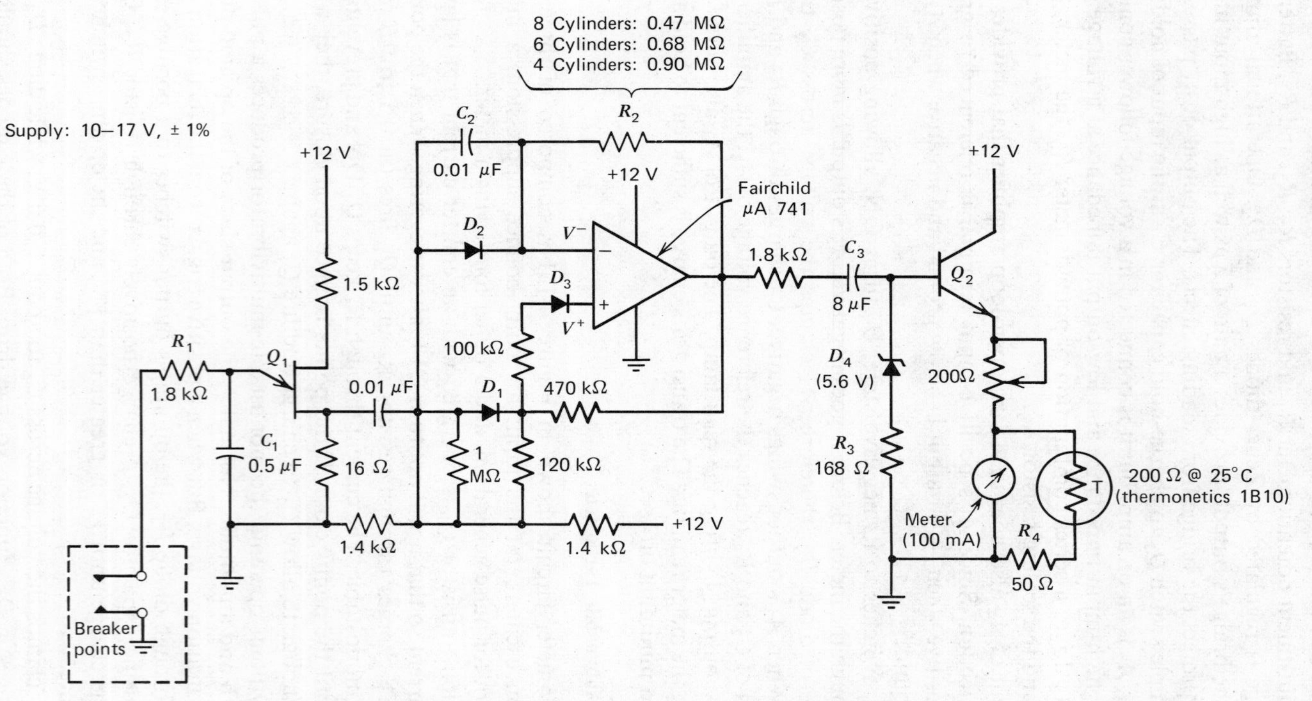

Figure 4-121. Precision auto tachometer. (Courtesy *Electronics*.)

in thermal contact with each other. Since the op amp is left floating so that it can be operated from a car's single-supply voltage, it has a small positive-output voltage when in its untriggered state, making capacitor C_3 necessary to decouple the meter.

Zener diode D_4 and resistor R_3 regulate the output against supply-voltage variations, and the thermistor compensates for temperature variations in the base-emitter voltage of transistor Q_2. If a meter with a full-scale current rating of less than 5 mA is used, the thermistor, as well as transistor Q_2 and resistor R_4 can be omitted.

SAMPLE-AND-HOLD CIRCUITS

Basic Circuits

A sample-and-hold amplifier tracks an analog signal and, when directed by an external digital command, freezes its output at the instantaneous value of the input. But the individual parameters of this performance vary in importance with the particular application.

And the applications are numerous. Sample-and-hold amplifiers are particularly useful where fast-changing signals must be multiplexed in data acquisition systems or where momentary signals must be captured and held. They are frequently used in sampled data systems to decrease system aperature time with A/D converters, and in display systems to provide smooth, glitch-free outputs from D/A converters. Other applications include pulse-stretching, data distribution to multiple readouts, peak-and-valley following, and ratio measurement of time-averaged variables.

The simplest form of sample-and-hold amplifier is the capacitor-switch combination shown in Figure 4-122, with this circuit, the hold command is given when the switch is thrown from position S to position H. But though the circuit works effectively with very slowly changing signals, it causes too much source- and output-loading to be of much practical use.

Most practical modular sample-and-holds are designed for noninverting unity-gain operation, but important new inverting designs are now available. In both types, the control inputs are normally operated at standard logic levels and are usually TTL-compatible. Typically logic "1" is the sample command and logic "0" is the hold command.

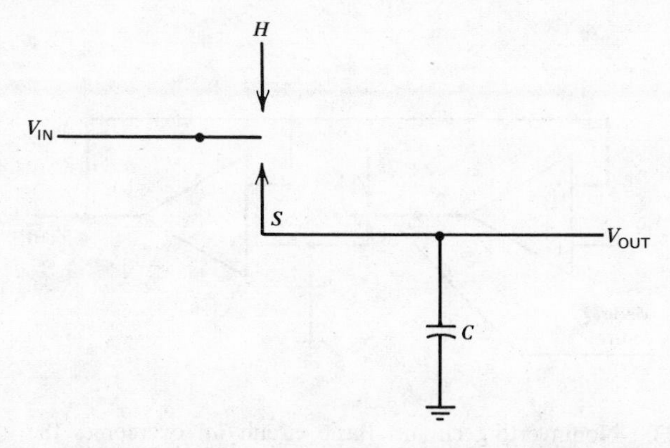

Figure 4-122. Simplest sample-and-hold circuit. If speed and circuit loading are not too important, a capacitor and a switch make an effective sample-and-hold circuit.

The basic noninverting sample-and-hold amplifier consists of a resistor, a switch, a capacitor, and an op amp (Figure 4-123*a*). When the switch is closed, the capacitor charges (or discharges) exponentially to the input voltage just as in the simple capacitor example. The output of the op amp follows the capacitor voltage precisely. Again, when the switch is opened, the capacitor holds the instantaneous value of the input voltage. The advantage of the op amp follower in this circuit is that, once the charge is acquired and the switch is opened, output-loading will not discharge the capacitor.

In such a circuit, a FET switch would most probably be used, and the op amp would have a FET input. However, the storage capacitor still loads the input sources, and this loading, if R is too low, may make the source oscillate or overload it. When the source is overloaded at acquisition time, recovery time normally is long.

Increasing R to prevent these problems will slow the response time, and instead, a buffer amplifier can be added in front of the capacitor (Figure 4-123*b*). Here the input

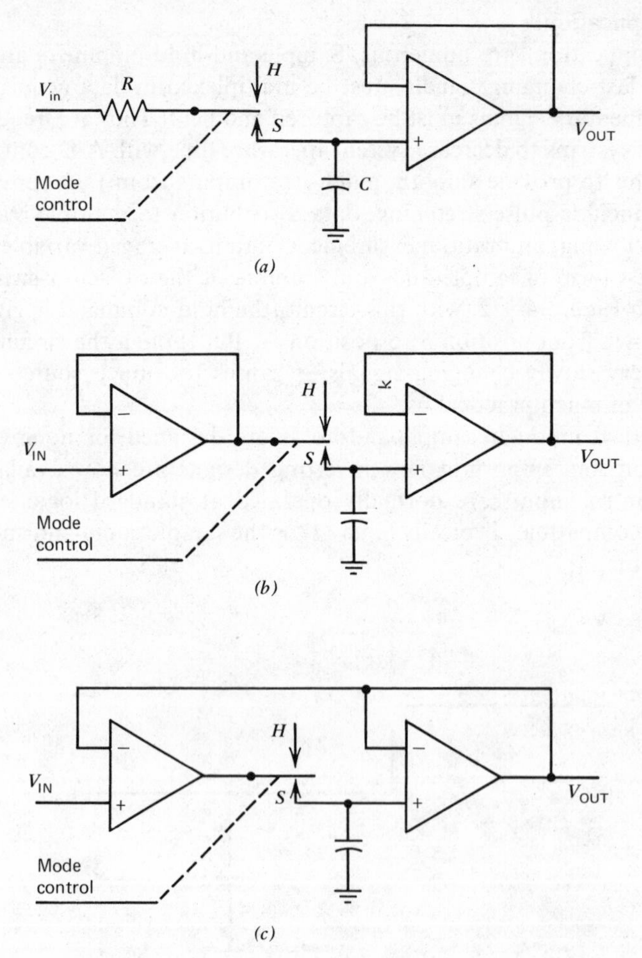

Figure 4-123. Noninverting circuit. Basic circuit (*a*) overcomes the capacitor-switch circuit's sensitivity to output-loading. Source-loading, however, remains a problem unless an input buffer amplifier is added (*b*). Putting a feedback loop around both amplifiers improves low-frequency performance but slows the system (*c*).

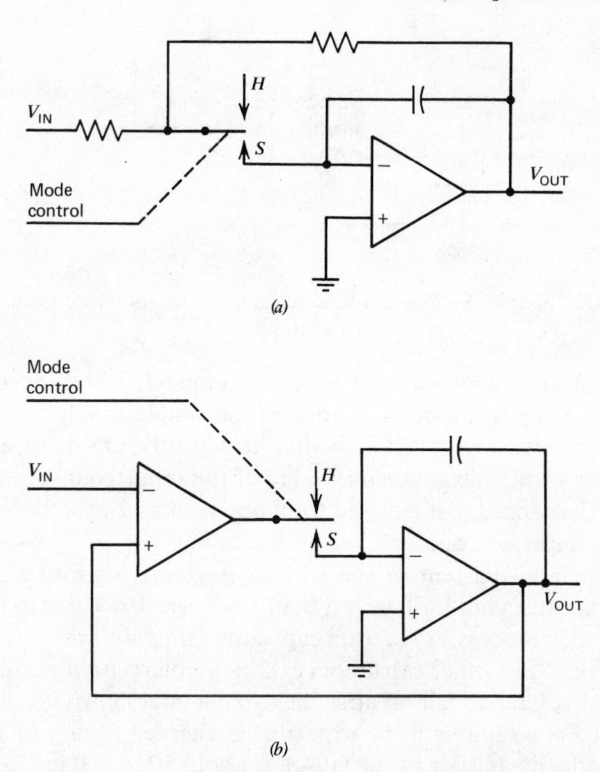

Figure 4-124. Inverting circuit. When a capacitor is inserted in the feedback loop, the inverter effectively isolates the input and minimizes the switch time (*a*). Input impedance is low, however, unless an input buffer amplifier is added (*b*).

is isolated from the holding capacitor, and the buffer amplifier provides the capacitor-charging current.

This circuit is pretty fast, but since the amplifiers work independently a summation of errors results. Consequently, if low-frequency tracking accuracy is more important than speed, the feedback loop can be closed around both amplifiers, forcing both to track as one amplifier (Figure 4-123*c*).

The other basic type of sample-and-hold amplifier—the inverting, or integrating, circuit—is shown in Figure 4-124. Because the capacitor is in the feedback loop, the input is isolated, and the FET switch operates at ground potential, minimizing leakage-current and switching time, while the amplifier is not bothered by a common-mode signal. Although this type of circuit does not require a buffer amplifier to charge the capacitor or isolate the input, its input impedance is significantly lower than in the buffered noninverting types. Therefore, the signal source must have a reasonable drive capability and a low-output impedance. No problem will arise, however, if an op amp is used as a preamp.

The modification of the inverting sample-and-hold amplifier shown in Figure 4-124*b* places an inverting buffer amplifier in front of the switch but within the feedback loop. Since the inverting input is floating, high-input impedance is provided. If a FET buffer is used, the current required to drive the circuit will be in the picoampere range. However, the addition of the input buffer introduces common-mode error, and additional nonlinearity error, besides reducing speed.

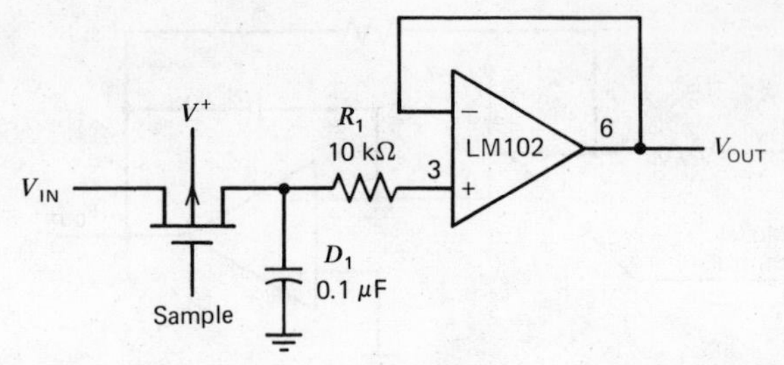

Figure 4-125. Simple sample-and-hold circuit [*note:* polycarbonate dielectric capacitor].

Although there are many ways to make a sample-and-hold device, the circuit shown in Figure 4-125 is undoubtedly one of the simplest. When a negative-going sample pulse is applied to the MOS switch, it will turn ON hard and charge the holding capacitor to the instantaneous value of the input voltage. After the switch is turned OFF, the capacitor is isolated from any loading by the LM102; and it will hold the voltage impressed upon it.

The maximum input current of the LM102 is 10 nA, so with a 10-μF holding capacitor the drift rate in hold will be less than 1 mV/sec. If accuracies of about 1% or better are required, it is necessary to use a capacitor with polycarbonate, polyethylene, or teflon dielectric. Most other capacitors exhibit a polarization phenomenon which causes the stored voltage to fall off after the sample interval with a time constant of several seconds. For example, if the capacitor is charged from 0 to 5 V during the sample interval, the magnitude of the fall-off is about 50 to 100 mV.

Sample-and-Hold Accuracy

The holding accuracy of a sample-and-hold is directly related to the error currents in the components used. During the sample interval, Q_1 is turned ON in Figure 4-126 charging the hold capacitor, C_1, up to the value of the input signal. When Q_1 is turned OFF, C_1 retains this voltage. The output is obtained from an op amp that buffers the capacitor so that it is not discharged by any loading. In the holding mode, an error is generated as the capacitor looses charge to supply-circuit leakages. The accumulation rate for error is given by

$$\frac{dV}{dt} = \frac{I_E}{C_1} \tag{4-32}$$

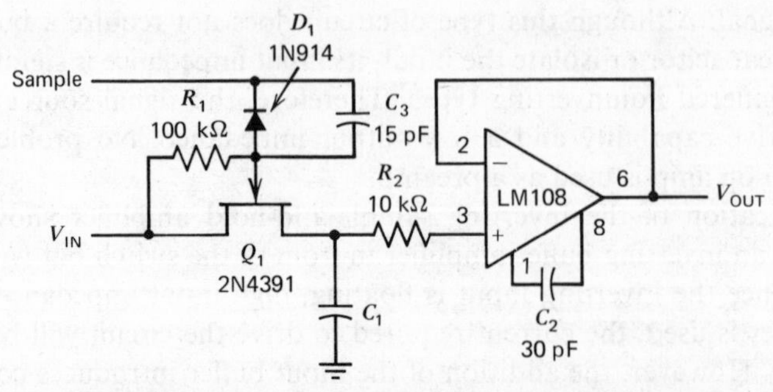

Figure 4-126. Junction FET sample-and-hold circuit.

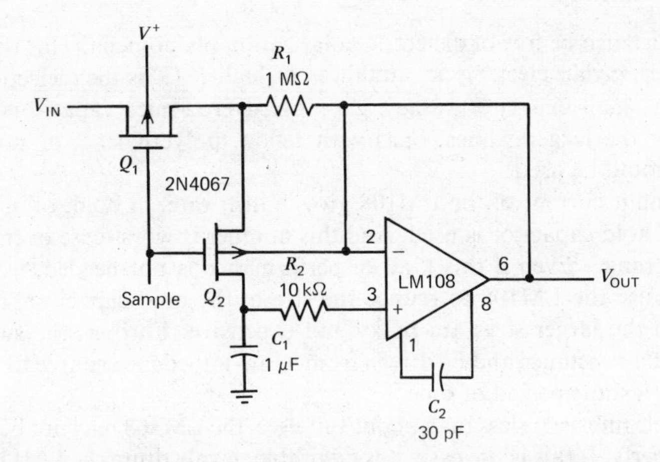

Figure 4-127. Low-leakage sample-and-hold circuit [*note:* teflon, polyethylene, or polycarbonate dielectric capacitor, worst-case drift less than 3 mV/sec].

where dV/dt is the time rate of change in output voltage and I_E is the sum of the input current to the op amp, the leakage current of the holding capacitor, board leakages, and the OFF current of the FET switch.

When high-temperature operation is involved, the FET leakage can limit circuit performance. This can be minimized by using a J FET, as indicated, because commercial J FETs have lower leakage than their MOS counterparts. However, at 125°C even junction devices are a problem. Mechanical switches, such as reed relays, are quite satisfactory from the standpoint of leakage. However, they are often undesirable because they are sensitive to vibration, they are too slow, or they require excessive drive power. If this is the case, the circuit in Figure 4-127 can be used to eliminate the FET leakage.

When using *p*-channel MOS switches, the substrate must be connected to a voltage which is always more positive than the input signal. The source-to-substrate junction becomes forward-biased if this is not done. The troublesome leakage current of a MOS device occurs across the substrate-to-drain junction. In Figure 4-127 this current is routed to the output of the buffer amplifier through R_1 so that it does not contribute to the error current.

The main sample switch is Q_1, while Q_2 isolates the hold capacitor from the leakage of Q_1. When the sample pulse is applied, both FETs turn ON charging C_1 to the input voltage. Removing the pulse shuts OFF both FETs, and the output leakage of Q_1 goes through R_1 to the output. The voltage drop across R_1 is less than 10 mV, so the substrate of Q_2 can be bootstrapped to the output of the LM108. Therefore, the voltage across the substrate-drain junction is equal to the offset voltage of the amplifier. At this low voltage, the leakage of the FET is reduced by about two orders of magnitude.

It is necessary to use MOS switches when bootstrapping the leakages in this fashion. The gate leakage of a MOS device is still negligible at high temperatures; this is not the case with junction FETs. If the MOS transistors have protective diodes on the gates, special arrangements must be made to drive Q_2 so the diode does not become forward-biased.

In selecting the hold capacitor, low leakage is not the only requirement. The capacitor must also be free of dielectric polarization phenomena. This rules out such types as paper, mylar, electrolytic, tantalum, or high-K (K is the dielectric constant) ceramic. For small capacitor values, glass or silvered-mica capacitors are recommended. For the larger values, ones with teflon, polyethylene, or polycarbonate dielectrics should be used.

The low-input current of the LM108 gives a drift rate, in hold, of only 3 mV/sec when a 1-μF hold capacitor is used. And this number is worst-case over the military temperature range. Even if this kind of performance is not needed, it may still be beneficial to use the LM108 to reduce the size of the hold capacitor. High-quality capacitors in the larger sizes are bulky and expensive. Further, the switches must have a low ON resistance and be driven from a low-impedance source to charge large capacitors in a short period of time.

If the sample interval is less than about 100 μsec, the LM108 may not be fast enough to work properly. If this is the case, it is advisable to substitute the LM110 which is a voltage-follower designed for both low-input current and high speed. It has a 30 V/μsec slew rate and will operate with sample intervals as short as 1 μsec.

When the hold capacitor is larger than 0.05 μF, an isolation resistor should be included between the capacitor and the input of the amplifier (R_2 in Figure 4-127). The resistor ensures that the IC will not be damaged by shorting the output or abruptly shutting down the supplies when the capacitor is charged. This precaution is not peculiar to the LM108 and should be observed on any IC op amp.

Sample-and-Hold Stability

A sample-and-hold circuit which combines the low-input current of FETs with the low-offset voltage of monolithic amplifiers is shown in Figure 4-128. The circuit is a unity-gain amplifier employing an op amp and an FET source-follower. In operation, when the sample switch, Q_2, is turned ON, it closes the feedback loop to make the output equal to the input, differing only by the offset voltage of the LM101. When the switch is opened, the charge stored on C_2 holds the output at a level equal to the last value of the input voltage.

Although this circuit does not have a particularly low-output resistance, fixed loads do not upset the accuracy since the loading is automatically compensated for

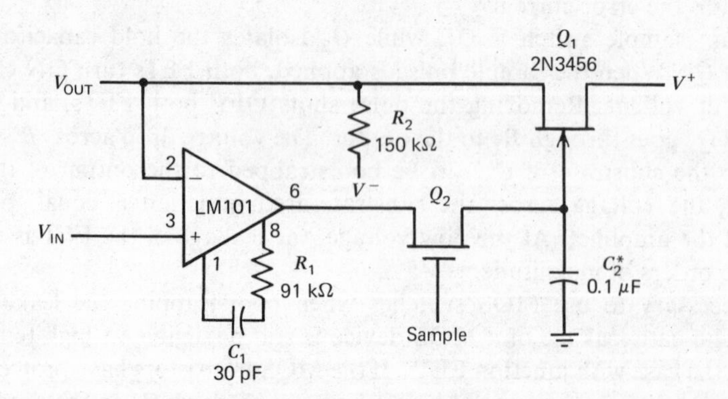

Figure 4-128. Low-drift sample-and-hold circuit [*note:* * polycarbonate dielectric capacitor, ** pin connections shown are for the metal-can package].

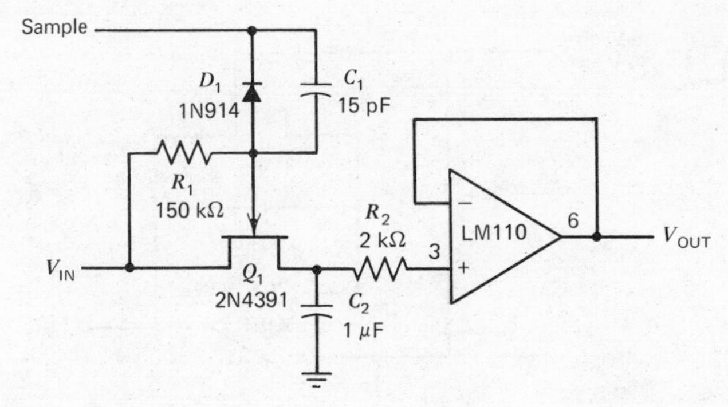

Figure 4-129. Voltage-follower in sample-and-hold circuit [*note:* use capacitors with polycarbonate teflon or polyethylene dielectric].

during the sample interval. However, if the load is expected to change after sampling, a buffer such as the LM102 must be added between the FET and the output.

A second pole is introduced into the loop response of the amplifier by the switch resistance and the holding capacitor, C_2. This can cause problems with overshoot or oscillation if it is not compensated for by adding a resistor, R_1, in series with the LM101 compensation capacitor such that the breakpoint of the R_1C_1 combination is roughly equal to that of the switch and the holding capacitor.

It is possible to use an MOS transistor for Q_1 without worrying about the threshold stability. The threshold voltage is balanced out during every sample interval so only the short-term threshold stability is important. When MOS transistors are used along with mechanical switches, drift rates less than 10 mV/min can be realized.

Additional features of the circuit are that the amplifier acts as a buffer so that the circuit does not load the input signal. Further, gain can also be provided by feeding back to the inverting input of the LM101 through a resistive divider instead of directly.

Further examples are given in the remaining figures. Figure 4-129 is very similar to Figure 4-126 but uses a voltage-follower. In Figure 4-130 the J FETs, Q_1 and Q_2, provide complete buffering to C_1, the sample-and-hold capacitor. During sample, Q_1 is turned ON and provides a path $r_{ds(ON)}$, for charging C_1. During hold, Q_1 is

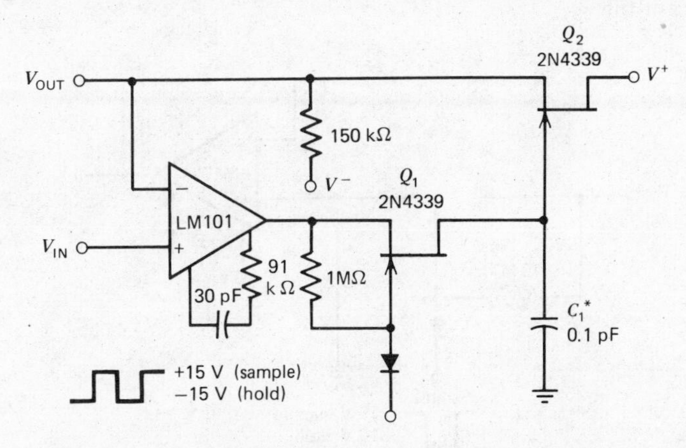

Figure 4-130. Low-drift, J FET-buffered sample-and-hold circuit [*note:* * polycarbonate dielectric capacitor].

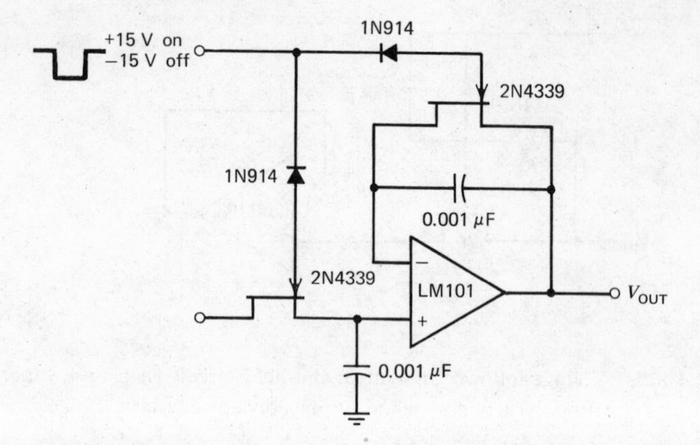

Figure 4-131. Junction J FET sample-and-hold with minimum $r_{ds(ON)}$ error.

turned OFF thus leaving Q_1 $I_{D(OFF)}$ (<50 pA) and Q_2 I_{GSS} (<100 pA) as the only discharge paths; Q_2 serves a buffering function so feedback to the LM101 (or μA748) and output current are supplied from its source.

The logic voltage is applied simultaneously to the sample-and-hold J FETs in Figure 4-131. By matching input impedance and feedback resistance and capacitance, errors due to $r_{ds(ON)}$ of the J FETs is minimized. Figure 4-132 shows a sample-and-hold circuit with offset adjustment.

Digital-to-Analog Converter

Figure 4-133 depicts a D/A converter using a chopper-stabilized op amp. This basic configuration is suitable, whether the binary-weighted resistors (Figure 4-133) or an R-2R ladder network is used. Because the digital information may be processed very quickly, the conversion speed of the total system is normally restricted by the amplifier as it performs a current-to-voltage input transformation. Thus the amplifier must have very low-input currents so that no significant error occurs to the least significant bit.

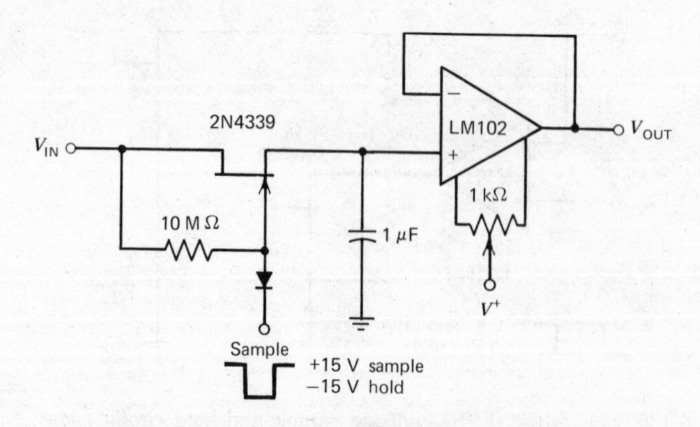

Figure 4-132. Sample-and-hold with offset adjustment.

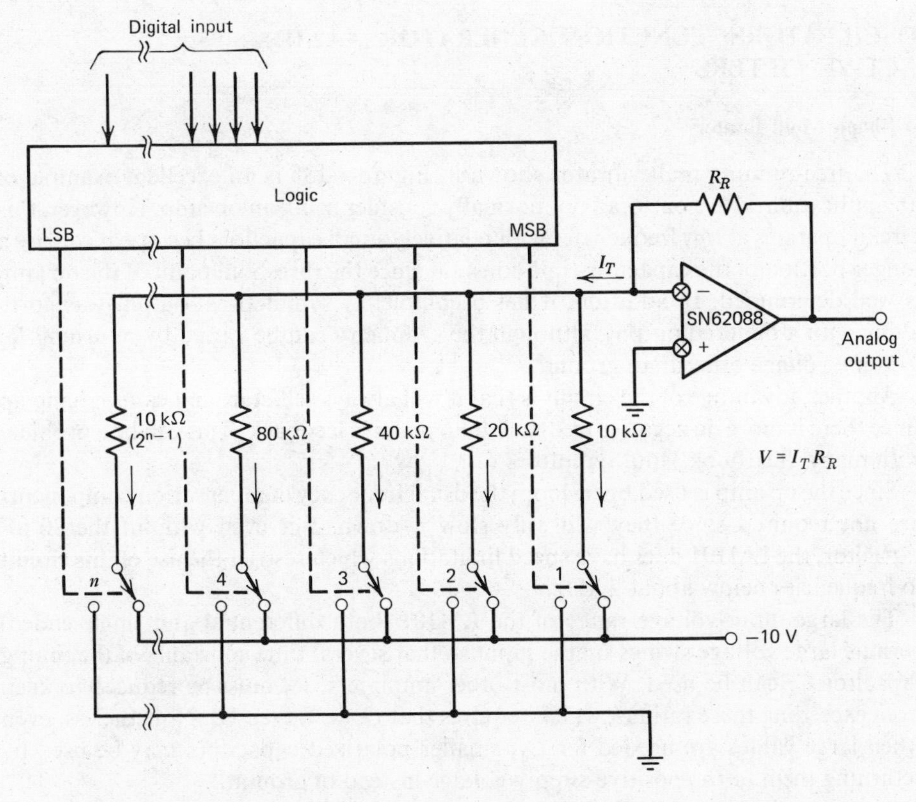

Figure 4-133. Digital-to-analog converter using the 62088 chopper-stabilizer operational amplifier.

Another application for which chopper-stabilized op amps are well suited is the sample-and-hold (Figure 4-134), or peak-detect-and-hold. Requirement of these types of circuits are fast acquisition time, low-input current and low-offset voltage. The circuit shown in Figure 4-134 is in a unity-gain configuration, although it could be generalized with gain resistors in the feedback loop. The amplifier must be able to respond to fast changes at the input and settle to the required system accuracy in a minimum time. The SN62088 provides less than 2-μs settling time for accuracy to 0.1% for a 10-V step.

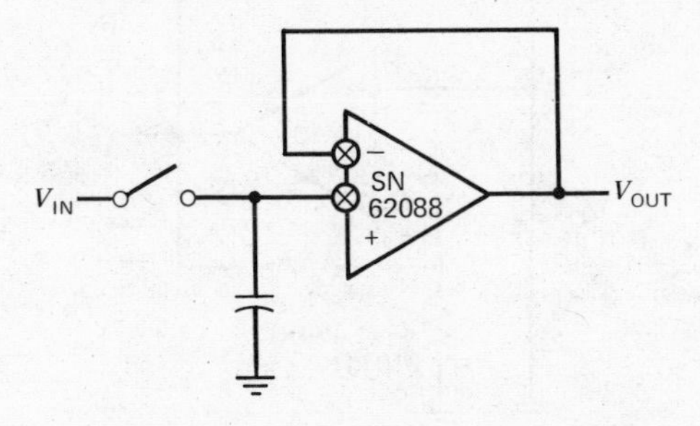

Figure 4-134. Sample-and-hold circuit.

OSCILLATORS, FUNCTION GENERATORS, AND ACTIVE FILTERS

Simple Multivibrator

The free-running multivibrator shown in Figure 4-135 is an excellent example of an application where one does not normally consider using an op amp. However, this circuit operates at low frequencies with relatively small capacitors because it can use a longer portion of the capacitor-time constant since the threshold point of the op amp is well determined. In addition, it has a completely symmetrical-output waveform along with a buffered output, although the symmetry can be varied by returning R_2 to some voltage other than ground.

Another advantage of the circuit is that it will always self-start and cannot hang up since there is more dc negative feedback than positive feedback. This can be a problem with many "textbook" multivibrators.

Since the op amp is used open-loop, the usual frequency compensation components are not required since they will only slow it down. But even without the 30-pF capacitor, the LM101 does have speed limitations which restrict the use of this circuit to frequencies below about 2 kHz.

The large input-voltage range of the LM101 (both differential and single-ended) permits large voltage swings on the input so that several time constants of the timing capacitor C_1 can be used. With most other amplifiers, R_2 must be reduced to keep from exceeding these ratings, which requires that C_1 be increased. Nonetheless, even when large values are needed for C_1, smaller polarized capacitors may be used by returning them to the positive-supply voltage instead of ground.

Several variations of the basic configuration are shown in Figures 4-137 through 4-140. When a square-wave output is needed, a voltage-comparator may be used as shown in Figure 5-31, which depicts a free-running multivibrator using the LM111 comparator.

One-Shot Multivibrator

The circuit for the op amp used to make a temperature-stable one-shot multivibrator (Figure 4-136) provides a highly repeatable trigger point independent of the input-pulse rise time. Also, its output swing can drive FET switches or reed relays directly.

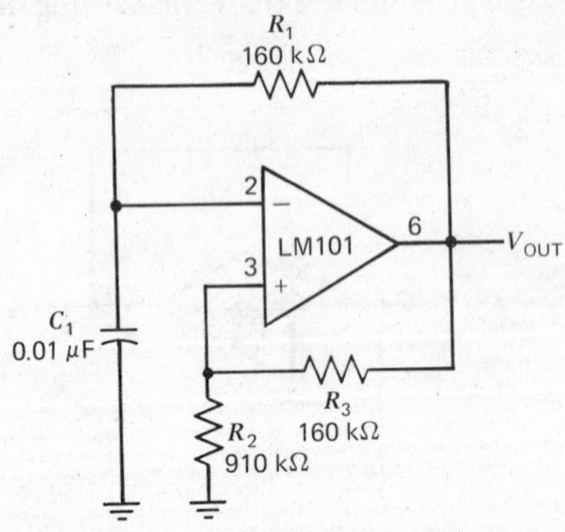

Figure 4-135. Free-running multivibrator.

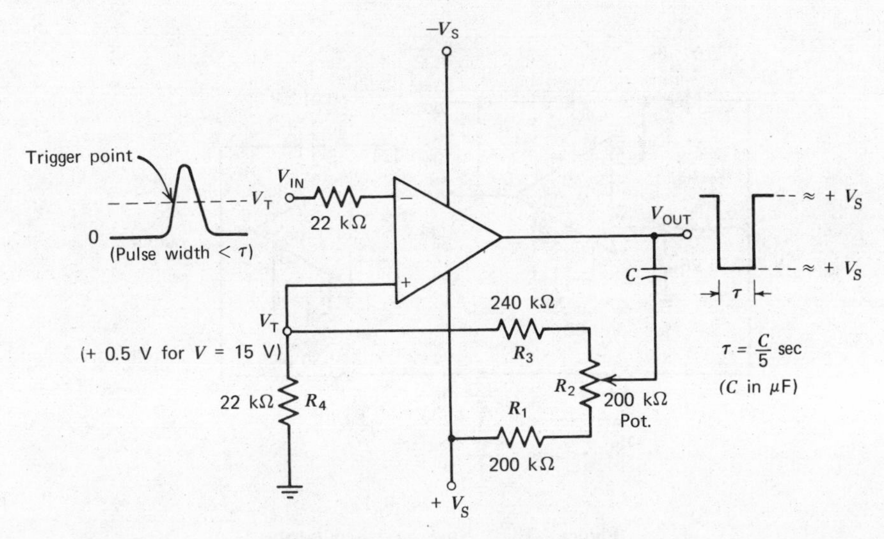

Figure 4-136. One-shot multivibrator.

Voltage-dividers R_1, R_2, R_3, and R_4 hold the input threshold V_T at a positive value. Therefore, when the input is at ground, the output is saturated positively. When the input level exceeds V_T, R_3 provides regeneration to switch the output to negative saturation. Capacitor C then charges exponentially such that V_T returns toward its normal positive value. When V_T passes through 0 V, regenerative action causes the output to rapidly return to positive saturation.

Because the inverting input must return to ground before timeout, an RC differentiator can be used if the input pulse is too long or if its base level is at some level other than ground. Pulse width, variable from microseconds to seconds, is proportional to the value of C: A value of 0.005 μF gives a pulse width of about 1 msec. Resistor R_2 provides a fine adjustment without affecting V_T.

Sine-Wave Oscillators

Although it is comparatively easy to build an oscillator that approximates a sine wave, making one that delivers a high-purity sinusoid with a stable frequency and amplitude is another story. Most satisfactory designs are relatively complicated and will not work without individual trimming and temperature compensation. In addition, they generally take a long time to stabilize to the final output amplitude.

A unique solution to most of these problems is presented in Figure 4-137, where A_1 is connected as a two-pole low-pass active filter and A_2 is connected as an integrator. Since the ultimate phase lag introduced by the amplifiers is 270°, the circuit can be made to oscillate if the loop gain is high enough at the frequency at which the lag is 180°. To ensure starting, the gain is actually made somewhat higher than is required for oscillation. Therefore, the amplitude builds up until it is limited by some nonlinearity in the system.

Amplitude stabilization is accomplished with Zener clamp diodes D_1 and D_2. This does introduce distortion, but the effect is reduced by the subsequent low-pass filters. If D_1 and D_2 have equal breakdown voltages, the resulting symmetrical clipping will virtually eliminate the even-order harmonics. The dominant harmonic

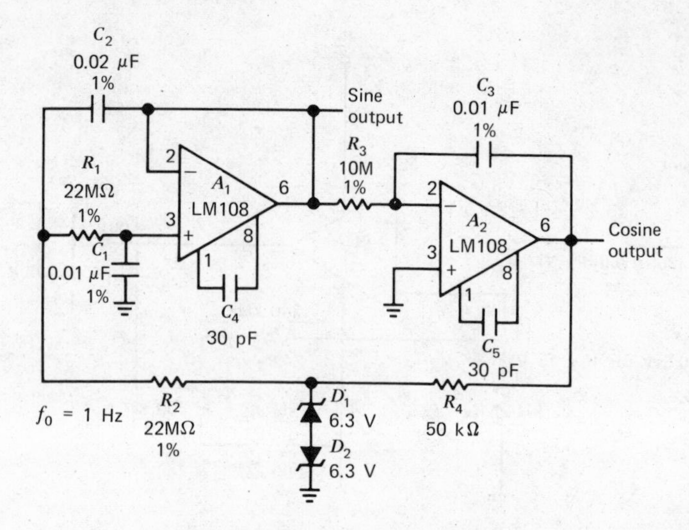

Figure 4-137. Sine-wave oscillator.

is then the third, and this is about 40 dB down at the output of A_1 and about 50 dB down on the output of A_2. This means that the total harmonic distortion of each output is 1 and 0.3%, respectively.

The frequency of oscillation and the oscillation threshold are determined by R_1, R_2, R_3, C_1, and C_3. Therefore, precision components with low TCs should be used. If R_3 is made lower than the value shown, the circuit will accept looser component tolerances before dropping out of oscillation. The start-up will also be quicker. However, the price paid is an increase in distortion. The value of R_4 is not critical, but it should be made much smaller than R_2 so that the effective resistance at R_2 does not drop when the clamp diodes conduct.

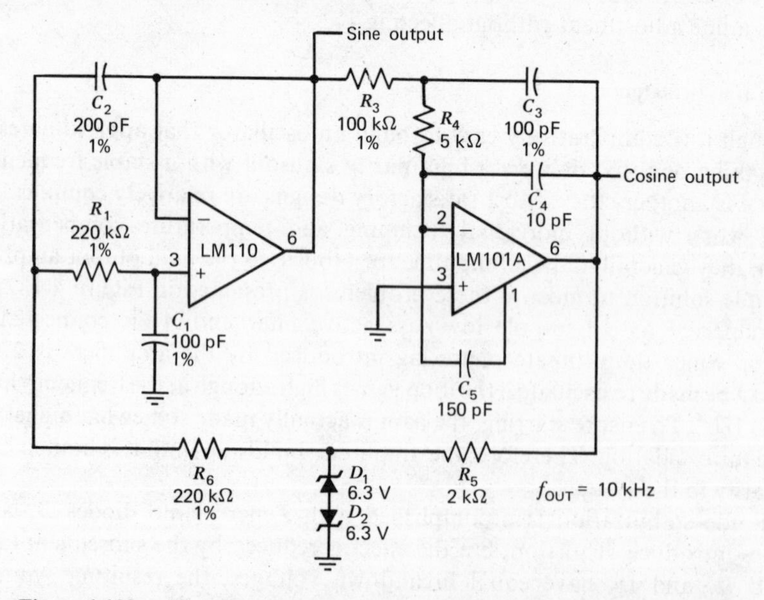

Figure 4-138. High-frequency sine-wave oscillator with quadrature output.

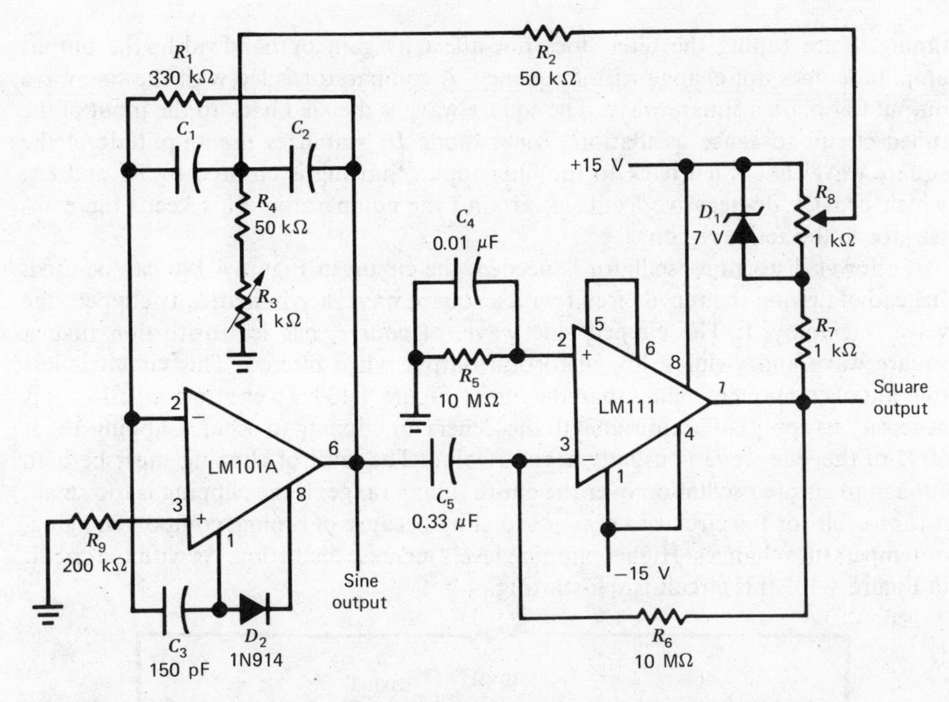

Figure 4-139. Easily tuned sine-wave oscillator [*note:* $C_1 = C_2$, $F_{OUT} = 1/2\pi C_1\sqrt{R_3 R_1}$].

The output amplitude is determined by the breakdown voltages of D_1 and D_2. Therefore, the clamp level should be temperature-compensated for stable operation. Diode-connected (collector shorted to base) *npn* transistors with an emitter-base breakdown of about 6.3 V work well, since the positive TC of the diode in reverse breakdown nearly cancels the negative TC of the forward-biased diode. Added advantages of using transistors are that they have less shunt capacitance and sharper breakdowns than conventional Zeners.

The LM108 is particularly useful in this circuit at low frequencies, since it permits the use of small capacitors. The circuit shown oscillates at 1 Hz, but uses capacitors in the order of 0.01 μF. This makes it much easier to find temperature-stable precision capacitors.

The LM108s are useful in this circuit for output frequencies up to 1 kHz. Beyond that, better performance can be realized by substituting an LM102 or LM110 (Figure 4-138) for A_1 and an LM101A with feedforward compensation for A_2. The improved high-frequency response of these devices extends the operating frequency out to 100 kHz.

Another approach to generating sine waves appears in Figure 4-139. This circuit will provide both sine- and square-wave outputs for frequencies from below 20 Hz to above 20 kHz. The frequency of oscillation is easily turned by varying a single resistor. This is a considerable advantage over Wien bridge circuits, where two elements must be tuned simultaneously to change frequency. Also, the output amplitude is relatively stable when the frequency is changed.

An op amp is used as a tuned circuit, driven by a square wave from a voltage-comparator. Frequency is controlled by R_1, R_2, C_1, C_2, and R_3 with R_3 used for tuning. Since tuning the filter does not affect its gain or bandwidth, the output amplitude does not change with frequency. A comparator is fed with the sine-wave output to obtain a square wave. The square wave is then fed back to the input of the tuned circuit to cause oscillation. Zener diode D_1 stabilizes the amplitude of the square wave that is fed back to the filter input. Starting is ensured by R_6 and C_5, which provide dc-negative feedback around the comparator. This keeps the comparator in the active region.

If a lower distortion oscillator is needed, the circuit in Figure 4-140 can be used. Instead of driving the tuned circuit with a square wave, a symmetrically clipped sine wave is employed. The clipped sine wave, of course, has less distortion than a square wave and yields a low-distortion output when filtered. This circuit is less tolerant of component values than the one in Figure 4-139. To ensure oscillation, it is necessary to apply sufficient signal to the Zeners for clipping to occur. Clipping about 20% of the sine wave is usually a good value. The level of clipping must be high enough to ensure oscillation over the entire tuning range. If the clipping is too small, it is possible for the circuit to cease oscillating because of tuning, component aging, or temperature changes. Higher clipping levels increase distortion. As with the circuit in Figure 4-139 this circuit is self-starting.

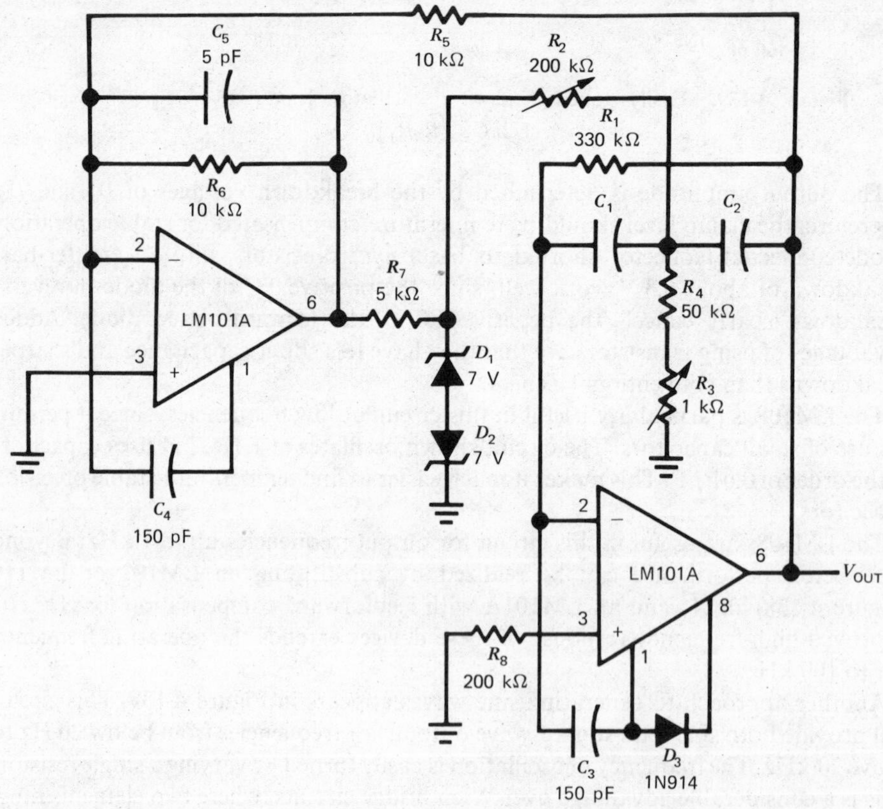

Figure 4-140. Low-distortion sine-wave oscillator [*note:* $C_1 = C_2$, $F_{OUT} = \dfrac{1}{2\pi C_1\sqrt{R_3 R_1}}$].

Table 4-5 Component values for various *frequency ranges*

C_1, C_2 (μF)	*Minimum frequency* (Hz)	*Maximum frequency* (Hz)
0.47	18	80
0.1	80	380
0.022	380	1.7 K
0.0047	1.7 K	8 K
0.002	4.4 K	20 K

Table 4-5 lists the component values for the various frequency ranges. Distortion from the circuit in Figure 4-139 ranges from 0.7 to 2%, depending on the setting of R_3. Although greater tuning range can be accomplished by increasing the size of R_3 beyond 1 kΩ, distortion becomes excessive. Decreasing R_3 lower than 50 Ω can make the filter oscillate by itself. The circuit in Figure 4-140 varies between 0.2 and 0.4% distortion for 20% clipping.

About 20 kHz is the highest usable frequency for these oscillators. At higher frequencies, the tuned circuit is incapable of providing the high-Q band-pass characteristic needed to filter the input into a clean sine wave. The low-frequency end of oscillation is not limited except by capacitor size. In both oscillators, feedforward compensation is used on the LM101A amplifiers to increase their bandwidth.

Figures 4-141 through 4-145 depict other oscillator-type circuits.

Wien Bridge Oscillators

An amplitude-stabilized sine-wave oscillator is shown in Figure 4-146. This circuit provides high-purity sine-wave output down to low frequencies with minimum circuit complexity. An important advantage of this circuit is that the traditional tungsten filament-lamp amplitude-regulator is eliminated along with its time constant and linearity problems. In addition, the reliability problems associated with a lamp are eliminated.

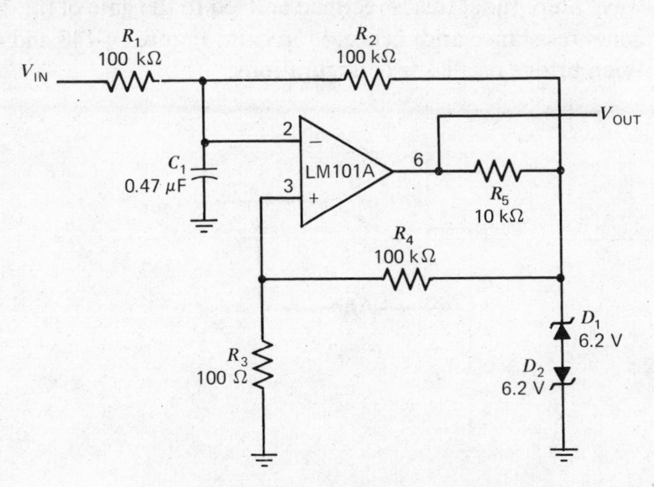

Figure 4-141. Pulse width modulator.

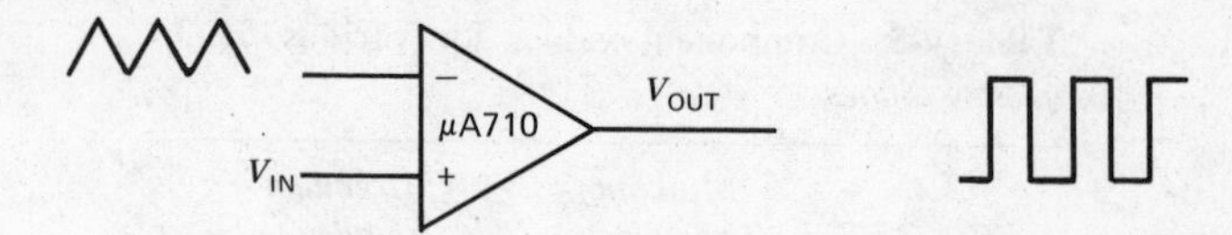

Figure 4-142. Pulse width modulator.

The Wien bridge oscillator is widely used and takes advantage of the fact that the phase of the voltage across the parallel branch of a series network, and a parallel *RC* network connected in series, is the same as the phase of the applied voltage across the two networks at one particular frequency and that the phase lags with increasing frequency and leads with decreasing frequency. When this network—the Wien bridge—is used as a positive-feedback element around an amplifier, oscillation occurs at the frequency at which the phase-shift is zero. Additional negative feedback is provided to set loop gain to unity at the oscillation frequency, to stabilize the frequency of oscillation, and to reduce harmonic distortion.

The circuit presented here differs from the classic usage only in the form of the negative-feedback stabilization scheme. Circuit operation is as follows: Negative peaks in excess of −8.25 V cause D_1 and D_2 to conduct, charging C_4. The charge stored in C_4 provides bias to Q_1, which determines amplifier gain; C_3 is a low-frequency roll-off capacitor in the feedback network and prevents offset voltage and offset-current errors from being multiplied by amplifier gain.

Distortion is determined by amplifier open-loop gain and by the response time of the negative-feedback loop filter, R_5 and C_4. A trade-off is necessary in determining the amplitude-stabilization time constant and oscillator distortion. Resistor R_4 is chosen to adjust the negative-feedback loop so that the FET is operated at a small negative-gate bias. The circuit shown provides optimum values for a general-purpose oscillator.

The major problem in producing a low-distortion constant-amplitude sine wave is getting the amplifier loop gain just right. By using the 2N3069 J FET as a voltage-variable resistor in the amplifier-feedback loop, this can be easily achieved with the circuit in Figure 4-147. The LM103 Zener diode provides the voltage reference for the peak sine-wave amplitude; this is rectified and fed to the gate of the 2N3069, thus varying its channel resistance and, hence, loop gain. Figures 4-148 and 4-149 depict several other Wien bridge oscillator configurations.

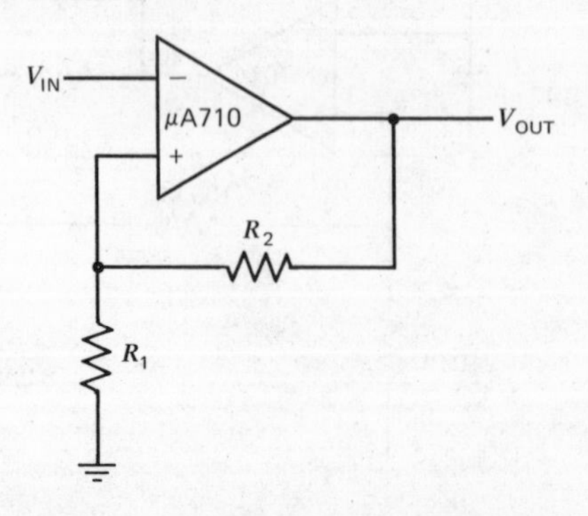

Figure 4-143. Simple Schmitt trigger.

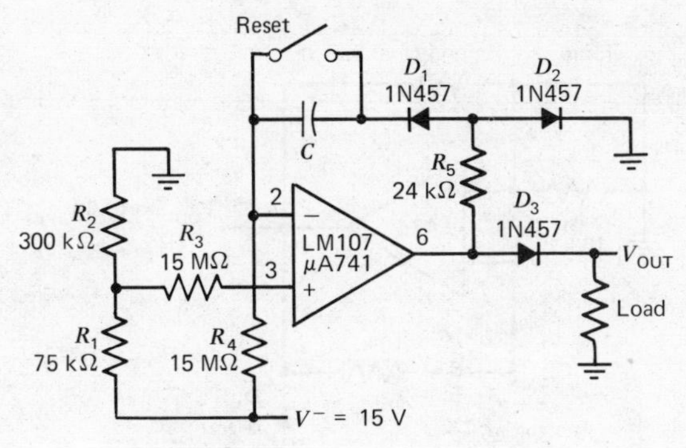

Figure 4-144. Long-interval timer

Triangular-Wave Generator

A constant-amplitude triangular-wave generator is shown in Figure 4-150. This circuit provides a variable-frequency triangular wave whose amplitude is independent of frequency.

The generator embodies an integrator as a ramp generator and a threshold detector with hystersis as a reset circuit. The integrator is described in the section on analog computing and servo amplifiers. The threshold detector is similar to a Schmitt trigger in that it is a latch circuit with a large dead zone. This function is implemented by using positive feedback around an op amp. When the amplifier output is in either the positive- or negative-saturated state, the positive-feedback network provides a voltage at the noninverting input which is determined by the attenuation of the feedback loop and the saturation voltage of the amplifier. To cause the amplifier to change states, the voltage at the input of the amplifier must be caused

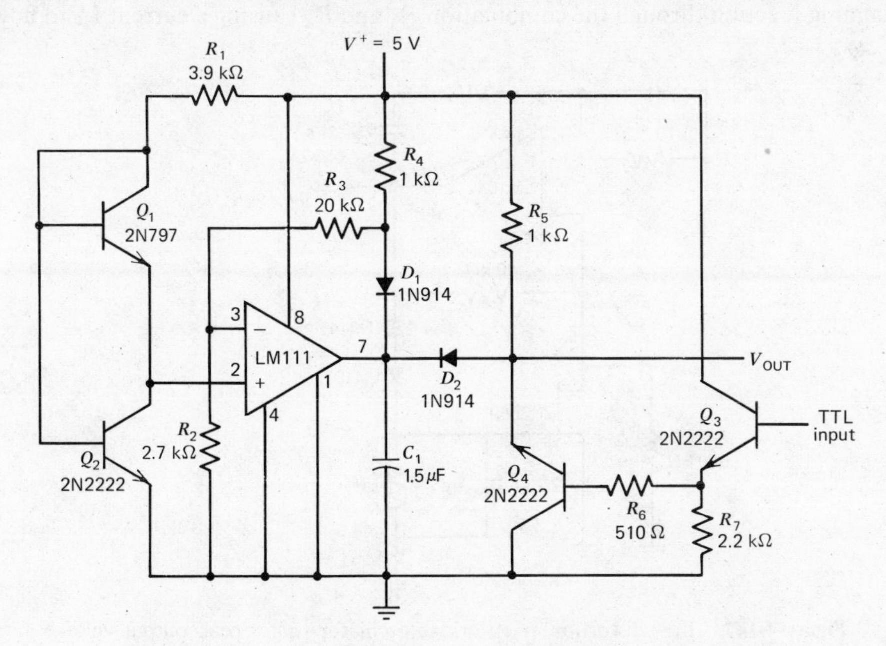

Figure 4-145. Voltage-comparator operating as precision squares.

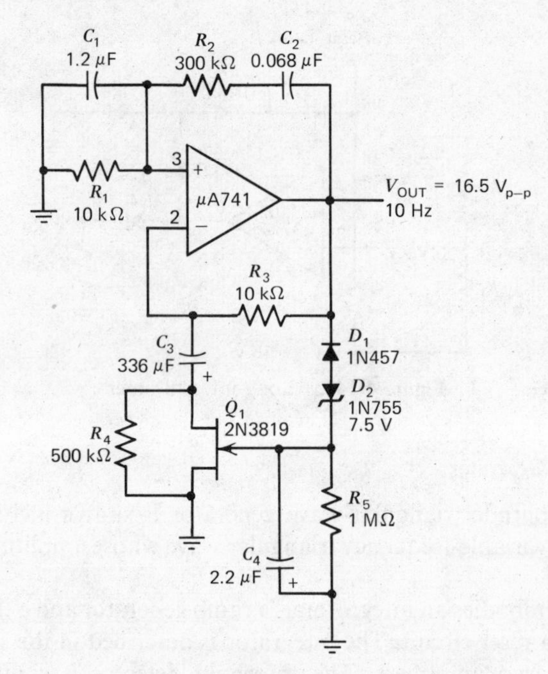

Figure 4-146. Wien bridge sine-wave oscillator.

to change polarity by an amount in excess of the amplifier input-offset voltage. When this is done, the amplifier saturates in the opposite direction and remains in that state until the voltage at its input again reverses. The complete circuit operation may be understood by examining the operation with the output of the threshold detector in the positive state. The detector positive-saturation voltage is applied to the integrator summing junction through the combination R_3 and R_4 causing a current I^+ to flow.

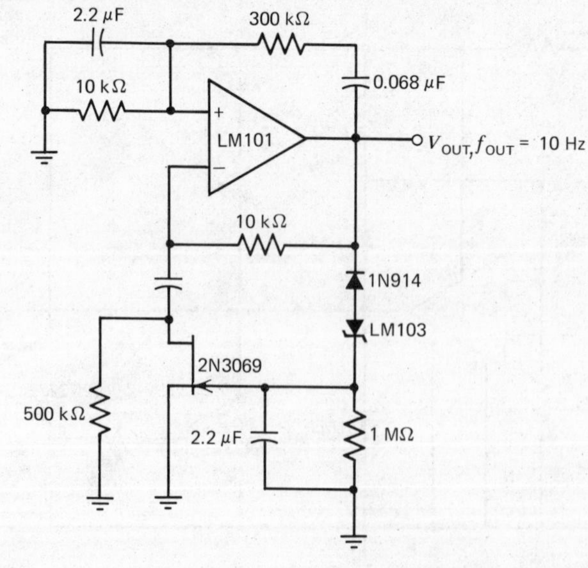

Figure 4-147. Low-distortion Wien bridge oscillator [*note:* peak output voltage $V_p = V_z + 1\ \text{V}$].

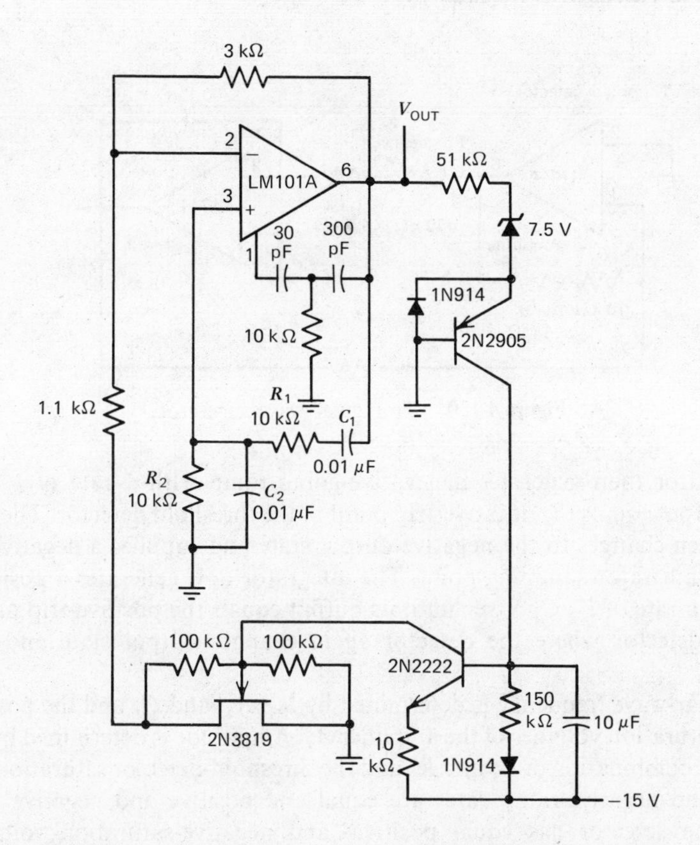

Figure 4-148. Wien bridge oscillator with FET amplitude stabilization [*note:* $R_1 = R_2, C_1 = C_2, f = 1/2\pi R_1C_1$].

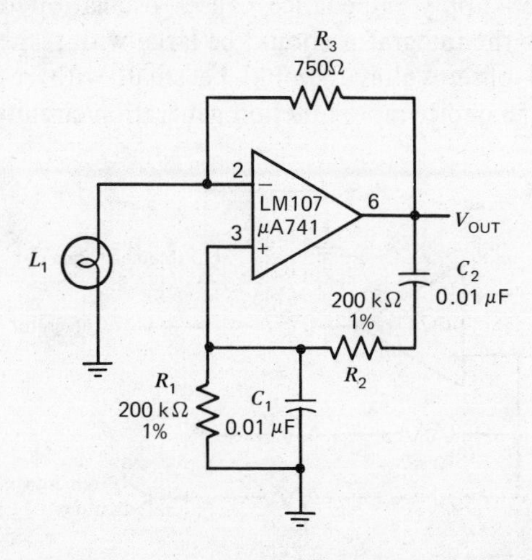

Figure 4-149. Lamp-type Wien bridge sine-wave oscillator [*note:* $R_1 = R_2$, $C_1 = C_2, f = 1/2\pi R_1C_1$, * dialco 8073910].

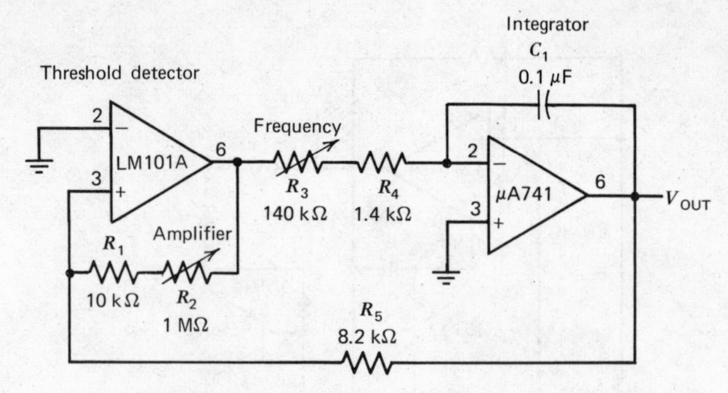

Figure 4-150. Triangular-wave generator.

The integrator then generates a negative-going ramp with a rate of I^+/C_1 V/sec until its output equals the negative-trip point of the threshold detector. The threshold detector then changes to the negative-output state and supplies a negative current, I^-, at the integrator summing point. The integrator now generates a positive-going ramp with a rate of I^-/C_1 V/sec until its output equals the positive-trip point of the threshold detector where the detector again changes output state and the cycle repeats.

Triangular-wave frequency is determined by R_3, R_4, and C_1 and the positive- and negative-saturation voltages of the amplifier A_1. Amplitude is determined by the ratio of R_5 to the combination of R_1 and R_2 and the threshold-detector saturation voltages. Positive- and negative-ramp rates are equal and positive and negative peaks are equal if the detector has equal positive- and negative-saturation voltages. The output waveform may be offset with respect to ground if the inverting input of the threshold detector, A_1, is offset with respect to ground.

The generator may be made independent of temperature and supply voltage if the detector is clamped with matched Zener diodes as shown in Figure 4-151.

The integrator should be compensated for unity gain and the detector may be compensated if power-supply impedance causes oscillation during its transition time. The current into the integrator should be large with respect to I_{BIAS} for maximum symmetry, and offset voltage should be small with respect to V_{OUT} peak. Figures 4-152 and 4-153 depict more function generation circuits.

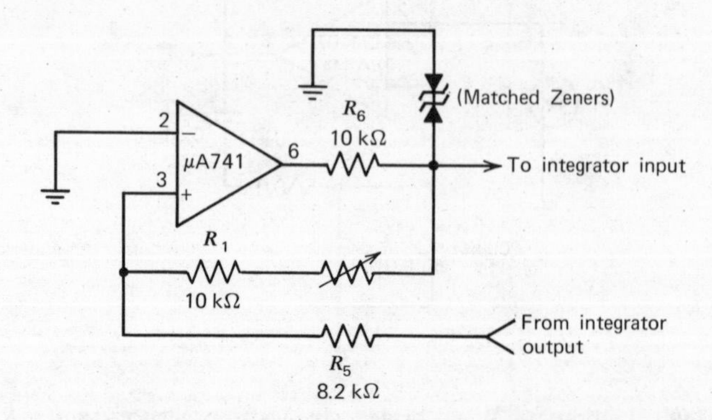

Figure 4-151. Threshold detector with regulated output.

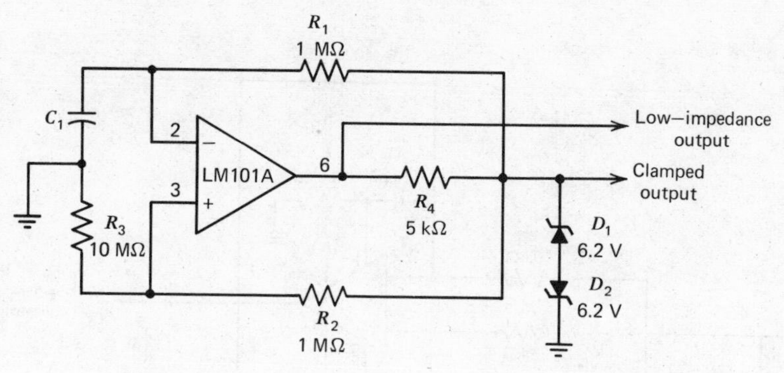

Figure 4-152. Low-frequency square-wave generator.

Active Filters

INTRODUCTION

An active filter is cheaper and lighter than a passive filter at low operating frequencies, specifically in the audio ranges, where an active filter works well. The frequency limitation of an active design lies in the amplifier.

The advent of modern ICs has made it possible to achieve small, high-quality low-frequency filter circuits using active-filter techniques. In this section we examine several basic active filters.

The five most commonly used second-order active filters are low-pass, high-pass, band-pass, band-reject, and all-pass types. Each of these filters uses resistors and capacitors as the passive elements and a high-quality op amp in a positive, fixed-gain configuration as the active element.

To fully characterize each of the five types of filters, the following information is needed:

Voltage-transfer function, $H(s)$
Circuit configuration
Cut-off or center frequency, ω_0
Damping ratio ζ or quality factor Q
Stability functions
Passive-component values

Both the center (or cut-off) frequency and damping ratio (or quality factor) are selected to meet the overall filter requirements. With these two pieces of information determined, the component values can be calculated.

The stability functions provide a measure of the sensitivity of the circuit to changes in component values. If components with tight tolerances are used, the filter performance will be very close to the initial specification. For the sake of economy, of course, component tolerances should be no tighter than the overall filter performance requirements dictate.

All of the transfer functions of these filters have one thing in common—the denominator is second-order. The transfer functions can all be written

$$H(s) = \frac{N(s)}{D(s)} \tag{4-33}$$

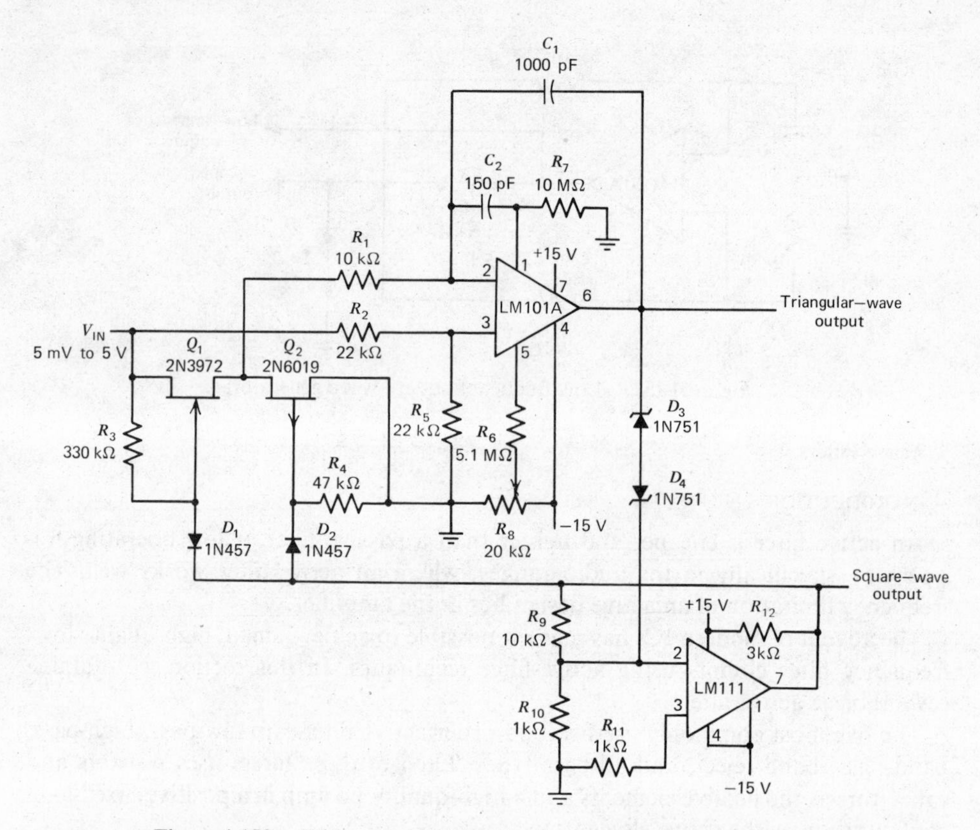

Figure 4-153. Wide-range VCO with square and triangular outputs.

where

$$D(s) = s^2 + 2\zeta\omega_0 s + \omega_0^2 \tag{4-34}$$

If the damping ratio ζ is less than unity, then the roots of $D(s)$ will be complex conjugates and will lie along a circle of constant radius in the s-plane (Figure 4-154). The angle θ (given as $\cos^{-1}\zeta$) and ω_0 determine the polar coordinates of the roots. As ζ varies, these roots will move along the semicircle. By choosing component values the designer can arbitrarily place the poles of his filter anywhere in the left half of the s-plane.

The damping ratio determines the shape of the filter response in the neighborhood of ω_0. The lower the damping ratio the greater the response at the cut-off frequency and the longer it takes for the filter-gain characteristic to approach its asymptotic value of -40 dB/decade.

For the frequency-selective filters, the quality factor Q is often used instead of the damping ratio. This quality factor is given as $Q = (1/2)\zeta$ and is the ratio of the center frequency ω_0 to the -3 dB bandwidth in rad/sec.

The Appendix at the end of this chapter provides the essential design criteria for the previously discussed five types of second-order active filters.

Figure 4-155 is a low-pass filter which is one of the simplest forms of active filters.

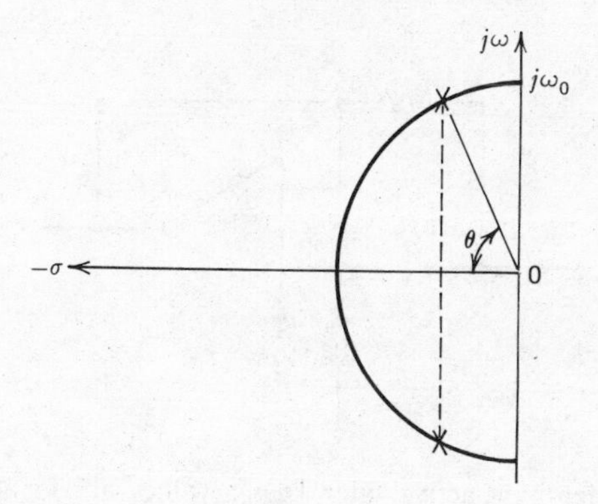

Figure 4-154. The poles of the filter can be arbitrarily placed anywhere in the s-plane in order to achieve the desired value of damping. The angle θ is given as $\cos^{-1}\zeta$, where ζ is determined by the relative values of the passive components chosen for the particular active filter.

The circuit has the filter characteristics of two isolated RC filter sections and also has a buffered, low-impedance output.

The attenuation is roughly 12 dB at twice the cut-off frequency and the ultimate attenuation is 40 dB/decade. A third low-pass RC section can be added on the output of the amplifier for an ultimate attenuation of 60 dB/decade, although this means that the output is no longer buffered.

There are two basic designs for this type of filter. One is the Butterworth filter with maximally flat frequency response. For this characteristic, the component values are determined from

$$C_1 = \frac{R_1 + R_2}{\sqrt{2}\, R_1 R_2 \omega_c} \tag{4-35}$$

$$C_2 = \frac{\sqrt{2}}{(R_1 + R_2)\omega_c} \tag{4-36}$$

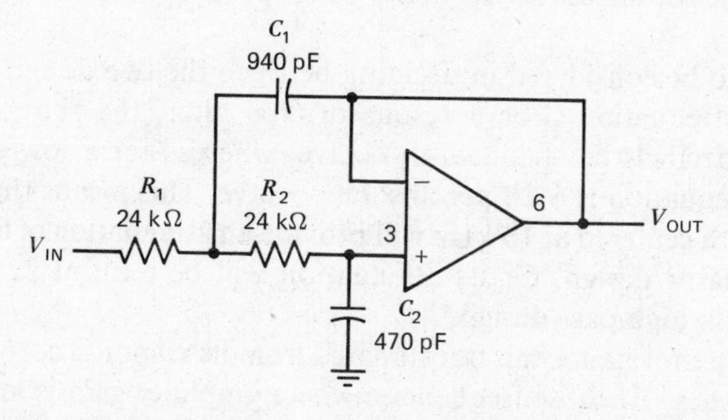

Figure 4-155. Low-pass active filter [*note:* values are for 10 kHz cut-off; use silvered mica capacitors for good temperature stability].

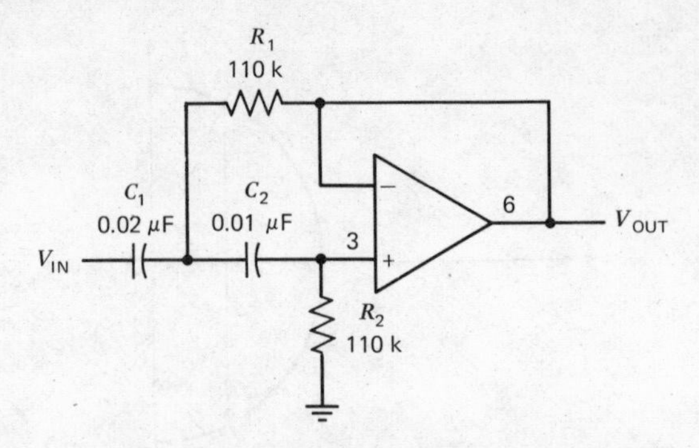

Figure 4-156. High-pass active filter [*note:* values are for 100 Hz cut-off; use metalized polycarbonate capacitors for good temperature stability].

The second kind is the linear-phase filter with minimum settling time for a pulse input. The design equations for this filter are

$$C_1 = \frac{R_1 + R_2}{\sqrt{3}\, R_1 R_2 \omega_c} \tag{4-37}$$

$$C_2 = \frac{\sqrt{3}}{(R_1 + R_2)\omega_c} \tag{4-38}$$

Substituting capacitors for resistors and resistors for capacitors in the circuit in Figure 4-155, a similar high-pass filter is obtained. This is shown in Figure 4-156.

Bandpass Filters

Bandpass filters can be made either by cascading low-pass and high-pass filter sections or by cascading stagger-tuned resonators. The low-pass/high-pass approach is best for wideband applications, while the stagger-tuned-resonator approach is best applied to narrowband filters.

The dividing line between the wide and narrowband approaches is determined by how much gain a section (a pole-pair) can provide and still have excess gain to operate the feedback loop. A two-resonator filter, for example, will have unity gain in the passband because the gain of one section offsets the skirt attenuation of the other. If the bandwidth of the over-all filter is increased by making the center frequency of one section higher or the center frequency of the other section lower, the filter gain will be decreased.

Another point to be considered in deciding between the two design approaches is the filter's skirt attenuation. For a resonator-type filter, the skirt attenuation is approximately 6 decibels per bandwidth octave, whereas for a low-pass/high-pass filter, the skirt attenuation is 6 dB per absolute octave. This means that a filter with a 2-KHz bandwidth centered at 10 kHz will provide an attenuation of 6 dB at 13 kHz if it's a one-resonator design. Or its attenuation will be 6 dB at 22 kHz if it's an equivalent low-pass/high-pass design.

The amplifier's performance can be estimated from its response curve of open-loop gain versus frequency. At those frequencies where amplifier gain is less than 60 dB, filter operation will be poor. And at higher frequencies, if amplifier gain is less than 40 dB, the filter will not work properly.

The low-pass pole existing in the amplifier tends to push the peak filter response to lower frequencies. Although this can be compensated for by trimming, the amplifier's pole frequency typically varies by 3:1 from unit to unit, so that trimming is difficult and rather temperature-sensitive.

Usually, the (Fairchild) 741-type operational amplifier will work well up to about 1 kHz, and the (National Semiconductor) LM101-type op amp, which has feed-forward compensation, will operate satisfactorily at frequencies up to around 10 kHz. Even higher frequencies can be handled by dielectrically isolated op amps such as the (Harris) HA2600-type unit.

The filter-circuit configuration chosen to implement a single filter section will also have some effect on performance. A filter section can contain one, two, or three op amps. Although it's the most expensive to build, the three-amplifier approach has two distinct advantages over the other two circuits—it is the least sensitive to component variations and the easiest to tune. This three-amplifier circuit is commonly referred to as the state-variable filter.

At least three parameters are needed to specify the response of a bandpass filter: center frequency, bandwidth, and gain. These parameters are functions of resistance and capacitance, as well as amplifier gain and bandwidth. For optimum performance, it is best to make the parameters depend on resistor ratios and RC products only. This is because the temperature tracking of resistors and RC products is controllable, but variations in amplifier gain and bandwidth are not.

When the filter's design frequency approaches the amplifier's upper limit (the frequency where the gain is down to 60 dB), the sensitivity of the circuit to amplifier gain and bandwidth becomes increasingly important. High-Q designs require more amplifier gain and are less tolerant of frequency shifts caused by amplifier-bandwidth limitations.

Filter sections can be arranged in any order without affecting the over-all filter response. Also, the gain of one section may be raised, and the gain of another section reduced by a corresponding amount. The Q of a filter section tends to be highest at the band edges, and odd-ordered filters generally contain a low-Q center-frequency section.

These simple considerations can be used to advantage in particular applications. For example, in a three-stage low-level preamplifier filter, the center-tuned section would be first, providing high gain to raise the signal above circuit noise. The remaining sections could then have reduced gain to avoid the undesirable condition of overdrive distortion.

Let us take a look at the detailed design of an active filter that selects a small-amplitude, low-frequency signal in the presence of a high-frequency component that has a considerably greater magnitude. The circuit chosen is a low-frequency bandpass filter having the following frequency response:

$$\frac{V_{\mathrm{OUT}}}{V_{\mathrm{IN}}} = -\frac{K\omega_0 s}{s^2 + 2\,\delta\omega_0 s + \omega_0^2} = -\frac{A'(1/Q)}{1 + (1/Q)(s/\omega_0) + (s/\omega_0)^2} \tag{4-39}$$

characterized by

$$\frac{V_2}{Z_2} = -\frac{Z_2}{Z_2 - (Z_1/K)}$$

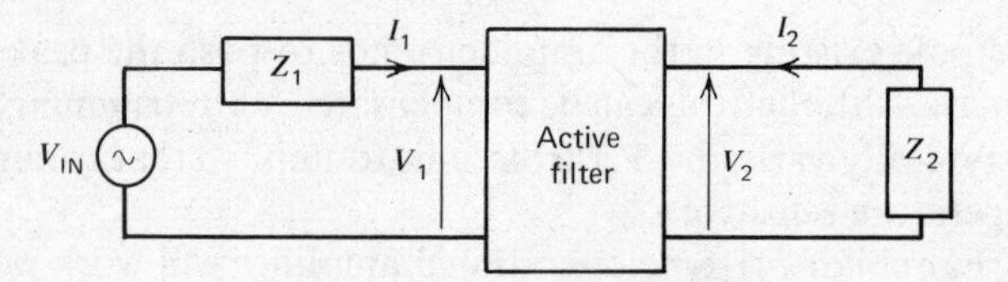

Figure 4-157. Active-filter block diagram.

and shown in Figure 4-157. The defining equations for this circuit are as follows:

$$V_1 = V_2 = i_2 Z_2 \tag{4-40}$$

$$i_2 = K i_1 \tag{4-41}$$

where K is the gain of the active filter.

$$\frac{V_1}{i_1} = -KZ_2 \tag{4-42}$$

$$i_1 = \frac{V_{IN}}{Z_1 - KZ_2} \tag{4-43}$$

Using (4-40) through (4-43), the transfer function is

$$\frac{V_2}{V_{IN}} = -\frac{Z_2}{Z_2 - (Z_1/K)} \tag{4-44}$$

Because we are using a differential-input op amp in this circuit, Figure 4-157 is rearranged as shown in Figure 4-158. The op amp is characterized by zero differential-input voltage and high-input impedances; therefore, the input voltages V_1 and V_2 can be equated, and the input currents i_1 and i_2 will flow through the feedback

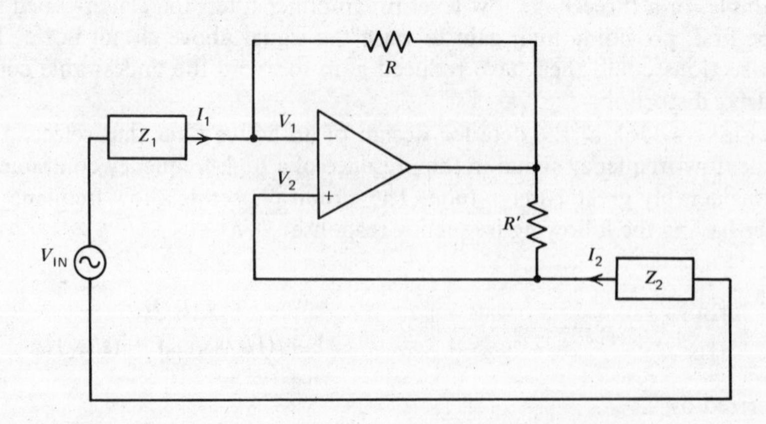

Figure 4-158. Active filter with operational amplifier used to determine the transfer functions.

resistors R and R'. The defining equations are then

$$V_1 = V_2 \tag{4-45}$$

$$i_1 R = i_2 R' \tag{4-46}$$

From (4-41) and (4-46) the gain K is defined as

$$K = \frac{R}{R'} \tag{4-47}$$

The band-pass characteristic is obtained by letting

$$Z_1(s) = \frac{1 + sR_1C_1}{sC_1} \tag{4-48}$$

$$Z_2(s) = \frac{R_2}{1 + sR_2C_2} \tag{4-49}$$

Substitution of (4-48) and (4-49) into (4-44) yields the transfer function

$$A(s) = \frac{-Ks/R_1R_2}{s^2 + s((1/R_2C_2) + (1/R_1C_1) - (K/R_1C_2)) + 1/R_1C_1R_2C_2} \tag{4-50}$$

By letting $R_1 = R_2 = R$ and $C_1 = C_2 = C$, (4-50) becomes

$$A(s) = -\frac{K(s/\omega_0)}{1 + (2 - K)(s/\omega_0) + (s/\omega_0)^2} \tag{4-51}$$

where

$$\omega_0 = \frac{1}{RC} \tag{4-52}$$

$$Q = \frac{1}{2 - K} \tag{4-53}$$

By comparing (4-39) and (4-51) it is shown that

$$2\delta = 2 - K = \frac{1}{Q} \tag{4-54}$$

It is also noted that at $a = j\omega_0$,

$$A(\omega_0) = -KQ \tag{4-55}$$

Equations (4-48) and (4-49) can be synthesized by RC networks such that the band-pass filter can be depicted by Figure 4-159. In this circuit,

$$K = \frac{R_3}{R_4} \tag{4-56}$$

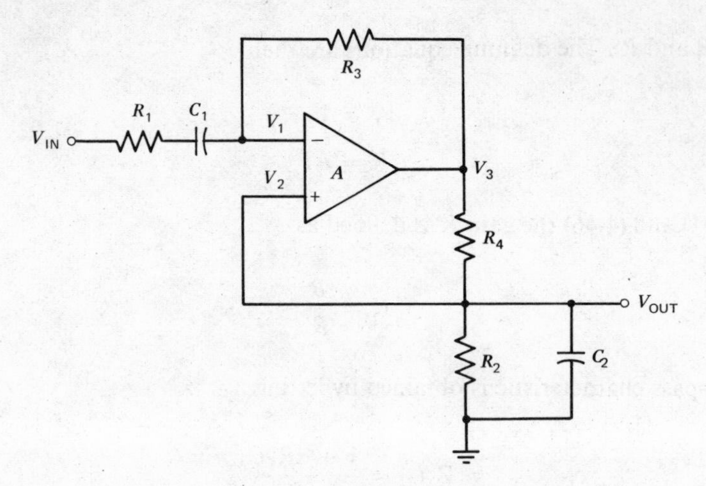

Figure 4-159. *RC* networks are synthesized to replace Z_1 and Z_2 of Figure 4-158 to give the desired band-pass characteristics.

The above equation shows that the center frequency is selected by the proper choice of R and C and that the desired value of the quality factor Q is obtained from the ratio of R_3 and R_4.

OPERATIONAL AMPLIFIER CONSIDERATIONS

Selecting the op amp and determining the magnitude of feedback resistors requires additional investigation. The need for this investigation stems from a consideration of the slew rate and dynamic range requirements, which in turn requires consideration of the expected input signals. A transfer function satisfying (4-51) is desired. However, it may not be attainable because of limitations imposed by the slew-rate capabilities of the op amp and the required output swing, which depends on the gain K and the input signal.

Therefore, an expression for the transfer function of the op amp itself is necessary. Referring to Figure 4-159 we want an equation for the ratio V_3/V_{IN}. For ease of component selection, let $R_1 = R_2 = R$, $C_1 = C_2 = C$, and $\omega_0 = 1/RC$. The transfer function can then be expressed as

$$A(s) = -\frac{s^2(R_3R_4/\omega_0) + s(R_3R_4 + RR_3)}{s^2(RR_4/\omega_0) + s(2RR_4 - RR_3) + (R_4/C)} \tag{4-57}$$

In order to determine the dynamic range required of an op amp in this circuit, it is necessary to know the frequency response of (4-57). This equation can be rearranged and simplified to give

$$A(s) = -\frac{(s/\omega_0)^2(R_3/R) + (s/\omega_0)[(2Q - 1)/Q + R_3/R]}{(s/\omega_0)^2 + (s/\omega_0)(1/Q) + 1} \tag{4-58}$$

The filter will be designed for a given center frequency, ω_0, and quality factor Q. It is apparent that, with ω_0, Q, R, and C determined, the frequency response depends on the choice of R_3. If Q is assumed to be large, the response at the center frequency can be approximated by

$$A(\omega_0) = 2Q\left[\frac{1}{2}\left(\frac{R_3}{R}\right)^2 + \frac{R_3}{R} + 1\right]^{1/2} \tag{4-59}$$

An inspection of (4-58) shows that the gain at high frequencies is equivalent to R_3/R. This can also be seen from an inspection of Figure 4-159.

The most practical method for evaluating (4-57) is by a computer programmed for a Bode plot printout. The frequency response for various values of R_3, is determined to give the maximum gain of the op amp at the center frequency. The peak output can be computed with knowledge of the peak input and maximum gain. This can then be compared to the output-swing limitations of any op amp under consideration. The subsequent limitations on R_3 are: (1) A maximum value that provides operation in the linear region—that is, no saturation and distortion, and (2) a minimum value to ensure that the op amp load impedance is high enough to be properly driven by the IC. Figure 4-159 indicates this load impedance to be a function of R_2, R_3, and R_4, and also of C_2 as a function of frequency.

Additional precautions in the selection of an op amp involve the slew-rate requirements. The slew rate required is determined by the product of the peak value of the op amp output signal and its corresponding frequency in radians per second. The critical areas are then the value of the output at the center frequency where maximum gain occurs, and the output in the higher frequency range. These areas would maximize the slew-rate equation

$$\text{slew rate} = V_{\text{OUT}}(\text{peak}) \times 2\pi f \text{ V/sec} \tag{4-60}$$

APPLYING THE PRINCIPLES

As an example, consider a band-pass filter designed for a center frequency of 82 Hz. The input signal has the waveform shown in Figure 4-160. The dotted lines indicate that the pulse frequency is 50 ± 2 kHz. The rate of deviation is 82 Hz. This results in an 82-Hz signal of 0.18-V peak amplitude at a 4.5 V dc level. For a desired Q of 14.14,

$$K = 2 - \frac{1}{Q} = 1.9293 \tag{4-61}$$

The band-pass frequency of 82 Hz is established by choosing $C = 0.22\ \mu\text{F}$ and $R = 8.87\ \text{k}\Omega$. The value of K given by (4-61) can be obtained by letting $R_3 = 6.24\ \text{k}\Omega$ and $R_4 = 3.24\ \text{k}\Omega$. Since 6.24 kΩ is not a standard value, R_3 will consist of a 6.04-kΩ resistor and a 500-Ω potentiometer.

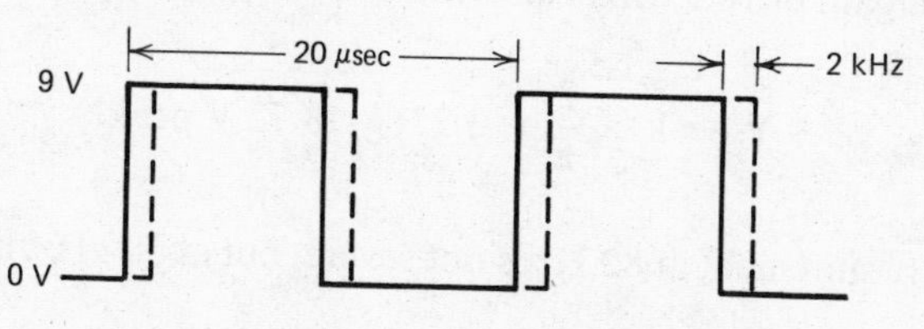

Figure 4-160. Input-signal frequency is 50 ± 2 kHz. The rate of deviation is 82 Hz.

With the above values, the op amp is driving an impedance level of slightly over 2 kΩ. The complete solution shows the maximum gain, which occurs at 82 Hz to be 31.3 dB ($A_V = 36.9$) while an evaluation of (4-59) gives a gain of 31.9 dB ($A_V = 39.4$). Therefore, the output is

$$0.18 \text{ V peak} \times 36.9 = 6.65 \text{ V peak} \tag{4-62}$$

The slew rate is

$$6.65 \text{ V peak} \times 2 \times 3.14 \times 82 \text{ Hz} = 3.42 \times 10^{-3} \text{ V}/\mu\text{sec} \tag{4-63}$$

The values of output swing and slew rate needed to satisfy (4-62) and (4-63) are within the range of the average op amp.

The approximation of (4-58) agrees with the computer-calculated response at 50 kHz—that is,

$$A(\omega \gg \omega_0) = \frac{R_3}{R} = 0.708 \tag{4-64}$$

This corresponds to a gain of −3 dB, which gives an output of

$$9 \text{ V peak} \times \frac{2}{\pi} \times 0.708 = 4.06 \text{ V peak} \tag{4-65}$$

The factor of $2/\pi$ accounts for the peak value of the 50-kHz fundamental. The slew rate is

$$4.06 \text{ V peak} \times 2 \times 3.14 \times 5 \times 10^4 \text{ Hz} = 1.275 \text{ V}/\mu\text{sec} \tag{4-66}$$

A typical device that would appear suited to this application is the internally compensated Fairchild μA741. However, for $A_V = 1$ and $R_1 \geq 2$ kΩ, the slew rate is 0.5 V/μsec, a specification which will not satisfy (4-66). Both the Fairchild μA715C, with a slew rate of 10 V/μsec, and the Motorola MCH1539G, with a typical slew rate of 4.2 V/μsec, will function satisfactorily in this case.

Consider now the case where feedback resistors of higher magnitudes are chosen. Let $R_3 = 28.9$ kΩ and be comprised of a 28-kΩ resistor and a 1-kΩ potentiometer, and $R_4 = 15$ kΩ. In this case, the load resistance is on the order of 10 kΩ. The solution indicates a gain at 82 Hz of 38.4 dB ($A_V = 83.2$). Then the output is

$$0.18 \text{ V peak} \times 83.2 = 14.98 \text{ V peak} \tag{4-67}$$

This output level exceeds the capabilities of several popular op amps. The output at 50 kHz, with a gain of 10.3 dB, is also too great, as (4-68) shows:

$$9 \text{ V peak} \times \frac{2}{\pi} \times 3.27 = 18.72 \text{ V peak} \tag{4-68}$$

The slew-rate requirement at 82 Hz is not severe, but at 50 kHz, it is determined as

$$18.72 \text{ V peak} \times 2 \times 3.14 \times 5 \times 10^4 \text{ Hz} = 5.88 \text{ V}/\mu\text{sec} \tag{4-69}$$

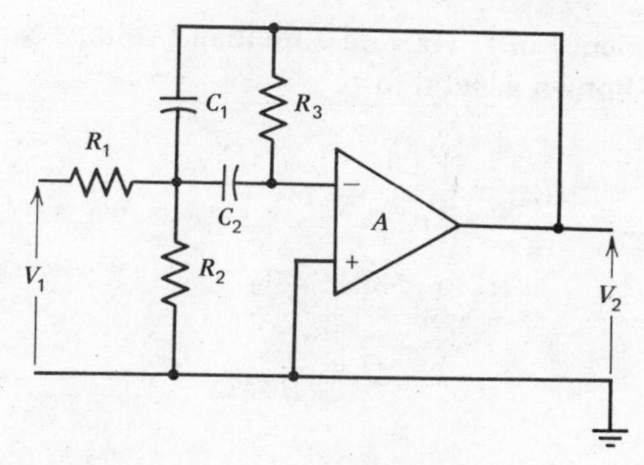

Figure 4-161. Basic network. This active band-pass *RC* filter provides two feedback paths via R_3 and C_1 for precise trade-off control between gain response and phase-shift. The circuit also keeps component count low. When R_2 is very large compared to R_1, it can be omitted altogether.

Here the μA715C meets the slew-rate requirement, but it is still incapable of handling the output-voltage swing requirements imposed by (4-67) and (4-68).

The examples indicate that for proper operation of the filter for this application, the smaller values of the feedback resistors and either a μA715C or MCH1539G op amp must be employed. This does not imply that an R_3 of 6.24 kΩ and an R_4 of 3.24 kΩ represent maximum values that will allow the filter to operate properly, for larger values, up to a limit, will also work. The intent is to show that the choice of the feedback resistors is critical since they determine the gain characteristics and load resistance of the op amp.

Applying these factors to the input signal determines the dynamic range and slew-rate requirements of the circuit, which in turn dictate the selection of the IC. Thus, for the particular band-pass filter described, it is mandatory to consider the output of the op amp itself in addition to the output of the filter.

At least three different active networks can perform the band-pass function. They are the bridged-T, the twin-T, and the multiple-feedback filters. The first two have only one feedback path, while the third has two. That additional path gives the multiple-feedback active band-pass filter (Figure 4-161) several distinct advantages. For example, it uses fewer components. And its response approaches that of an ideal second-order band-pass filter because its op amp operates with nearly infinite gain. Moreover, its output impedance, which is typically 200 to 300 Ω, is really the output impedance of the op amp used. This means that several filters can be cascaded without extreme loading problems between stages. It is also an improvement on the other two active band-pass filters, which usually have a nominal output impedance of 1 kΩ.

The multiple-feedback network is particularly useful in phase-sensitive band-pass systems when precise compromises must be made between frequency selectivity and phase-shift sensitivity. Difficulty in achieving very high values of Q is the single major drawback of the circuit. The maximum Q that can be achieved is approximately 20. When circuit Q is very high, the shunt resistance R_2 must be low in value. This severely attenuates the input signal, and taxes the amplifier and feedback path.

Because of its superior performance, the multiple-feedback filter is the best circuit model for a simplified design procedure. The first step is to consider the voltage-transfer function for an ideal second-order band-pass filter with a -3 dB bandwidth of B Hz, a center frequency of f_0 Hz, and a midband voltage gain of A. The transfer function has the well-known general form

$$H(s) = \frac{V_2(s)}{V_1(s)} = \frac{H\omega_0 s}{s^2 + \zeta\omega_0 s + \omega_0^2} \tag{4-70}$$

where

$$\omega_0 = 2\pi f_0 \tag{4-71}$$

$$Q = \frac{f_0}{B} \tag{4-72}$$

$$\zeta = \frac{1}{Q} \tag{4-73}$$

$$H = \zeta A_0 \tag{4-74}$$

Figure 4-162 shows the filter's magnitude and phase characteristics as a function of normalized frequency, with Q as a third parameter. As can be seen, both the magnitude selectivity and the phase-shift sensitivity increase for higher values of Q. Besides helping to visualize filter response, the graph can be used to determine Q when bandwidth is unknown but frequency rejection and phase-shift are specified with respect to some center frequency.

Now consider the multiple-feedback active band-pass filter. When operating in the inverting mode with infinite gain, its voltage-transfer function is

$$\frac{V_2(s)}{V_1(s)} = -\frac{s/R_1C_1}{s^2 + s(C_1 + C_2)/R_3C_1C_2 + (R_1 + R_2)/R_1R_2R_3C_1C_2} \tag{4-75}$$

Just inspecting (4-70) and (4-75) shows five component values to be determined with only three known constants. Equating the two transfer functions and solving for corresponding coefficients to obtain a new set of equations does not help either. There are still five unknowns and only three equations. The design procedure, therefore, must begin with an assumed relationship between selected components and proceed from there. Let R_{EQ} represent the parallel combination of R_1 and R_2:

$$R_{EQ} = \frac{R_1R_2}{R_1 + R_2} \tag{4-76}$$

Setting (4-70) equal to (4-75) and equating the coefficients of like terms yields

$$R_{EQ}R_3C_1C_2 = \frac{1}{\omega_0^2} \tag{4-77}$$

$$R_{EQ}(C_1 + C_2) = \frac{\zeta}{\omega_0} \tag{4-78}$$

$$\frac{R_{EQ}R_3C_2}{R_1} = \frac{H}{\omega_0} \tag{4-79}$$

Any numerical values satisfying (4-77), (4-78), and (4-79) are an acceptable solution. A simple step-by-step procedure can be used to determine the components and is valid for all values of Q.

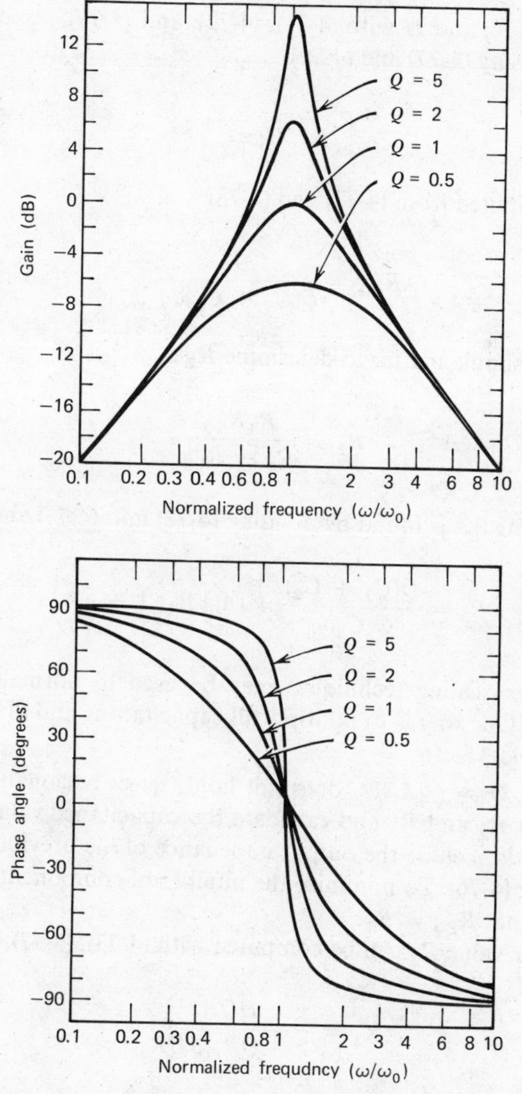

Figure 4-162. Filter curves. Magnitude and phase responses of an ideal second-order band-pass filter show effect of Q on network voltage-transfer function. High Q gives good frequency selectivity at the expense of phase sensitivity. Low Q means large bandwidth and critical phase response. Curves can be used for the multiple-feedback active filter since its performance approximates the ideal.

Since filter bandwidth, center frequency, and midband voltage gain are known desired specifications, Q can be found using the relationship of (4-71). Then the inequality, $Q > \sqrt{|A_0|/2}$, must be checked so that all the components are real—for example, the resistors must be positive numbers.

If the inequality is true, the capacitors are assigned values and the resistors are computed. Set C_1 equal to C_2 and choose an appropriate capacitance value for C_1. This value should be some standard capacitance that will yield resistances in the order of kilohms. The right resistors help to hold noise down and maintain circuit stability.

Next, solve for ω_0, ζ, and H with (4-71), (4-73), and (4-79), respectively. Now R_1 can be determined using (4-77) and (4-79):

$$R_1 = \frac{1}{H\omega_0 C_1} \tag{4-80}$$

and R_{EQ} can be computed from (4-73) and (4-78)

$$R_{EQ} = \frac{1}{Q(C_1 + C_2)\omega_0} \tag{4-81}$$

Using (4-76), it is a simple matter to determine R_2:

$$R_2 = \frac{R_1 R_{EQ}}{R_4 - R_{EQ}} \tag{4-82}$$

The final component, R_3, is found by dividing (4-78) into (4-77) and applying (4-73):

$$R_3 = \frac{Q(C_1 + C_2)}{C_1 C_2 \omega_0} = |A_0|\, R_1\left(1 + \frac{C_1}{C_2}\right) \tag{4-83}$$

Standard impedance-scaling techniques may be used to normalize the computed component values. One way is to multiply all capacitances and divide all resistances by the same numerical factor.

If the inequality, $Q > \sqrt{|A_0|/2}$, does not hold, select reasonable (in the order of kilohms) values for R_1 and R_2 and calculate the capacitance values. The resistance chosen for R_1 should include the output impedance of the previous stage. Now R_{EQ} can be found using (4-76). To minimize the number of components, R_2 can be made infinitely large so that $R_{EQ} = R_1$.

The final resistor value R_3 can be computed with (4-73), (4-77), and (4-79):

$$R_3 = \frac{HQR_1}{1 - QR_{EQ}/HR_1}$$

From (4-78),

$$C_2 = \frac{HR_1}{R_{EQ}R_3\omega_0}$$

Dividing (4-79) into (4-77) yields C_1:

$$C_1 = \frac{1}{HR_1\omega_0}$$

This procedure will in general give unequal, nonstandard capacitance values for C_1 and C_2. It is better to pick standard capacitor values that are close to the computed figures, and then redetermine the resistors from (4-80) through (4-83). Since standard resistor values are graduated in small steps, it is not difficult to select resistors that will do the job.

Sometimes there is an intermediate range of Q values for which either of the two design procedures will work. For this case, the quality ratio falls within the limits established by the condition

$$\sqrt{\frac{C_1\,|A_0|}{C_1 + C_2}} < Q < \sqrt{\frac{R_1\,|A_0|}{R_{EQ}}}$$

A few examples will be presented to illustrate the design technique. First, suppose the problem is to design a band-pass filter with a midband voltage gain of 50 (34 dB), a center frequency of 160 Hz, and a 3-dB bandwidth of 16 Hz. This filter requires a Q of 10 (160/16). The condition, $Q > \sqrt{|A_0|/2}$, is satisfied since

$$10 > \sqrt{50/2} = 5$$

Find the other needed constants:

$$H = \zeta\,|A_0| = \frac{|A_0|}{Q} = \frac{50}{10} = 5$$

$$\omega_0 = 2\pi f_0 = 2\pi(160) = 1000 \text{ rad/sec}$$

Assume $C_1 = C_2 = 0.1\ \mu\text{F}$ and find

$$R_1 = \frac{1}{HC_1\omega_0} = 2\text{ k}\Omega$$

$$R_{EQ} = \frac{1}{Q(C_1 + C_2)\omega_0} = 500\ \Omega$$

$$R_2 = \frac{R_1 R_{EQ}}{R_1 - R_{EQ}} = 667\ \Omega$$

$$R_4 = |A_0|\,R_1\left(1 + \frac{C_1}{C_2}\right) = 200\text{ k}\Omega$$

The complete design and its gain response are shown in Figure 4-163.

The second problem is to design a band-pass filter that has a midband voltage gain of 100 (40 dB), a center frequency of 100 Hz, and no specified bandwidth. However, the circuit must provide at least 20 dB of frequency rejection one decade from the center frequency, and hold phase-shift to $\pm 10°$ maximum for a 10% change in the center frequency.

The gain and phase plots of Figure 4-162 show that the filter must have $Q \approx 1$ to satisfy these last two requirements. The other constants are

$$H = \frac{|A_0|}{Q} = 100$$

$$\omega_0 = 2\pi f_0 = 628 \text{ rad/sec}$$

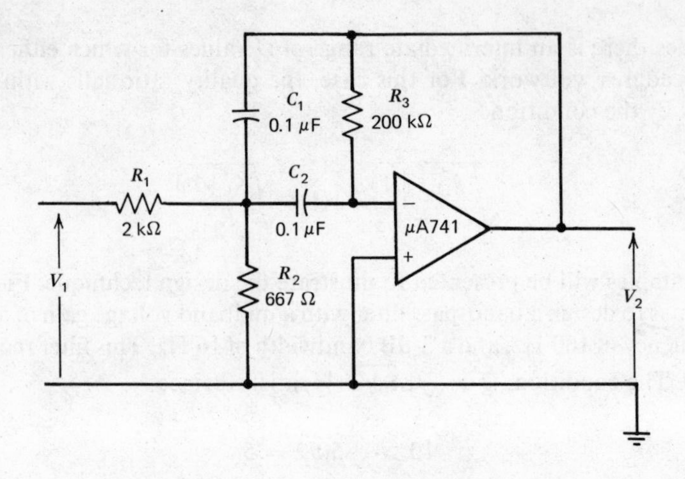

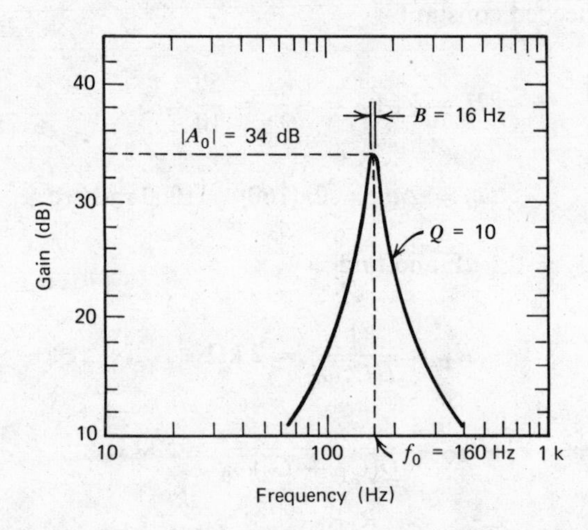

Figure 4-163. High Q. Filter with Q of 10, 16-Hz bandwidth, center frequency of 160 Hz, and gain of 35 dB is designed by selecting the same standard capacitance for C_1 and C_2 and then calculating resistances with handy plug-in formulas. Filter requires only six components, including operational amplifier.

The check, $Q > \sqrt{|A_0|/2}$, does not hold because $1 < \sqrt{50}$. It is now necessary to estimate values for R_1 and R_2. Let $R_1 = 1\ \text{k}\Omega$ and $R_2 = \infty$; then $R_{EQ} = R_1 = 1\ \text{k}\Omega$. Compute the other components:

$$R_3 = \frac{HQR_1}{1 - QR_{EQ}/HR_1} = \frac{HQR_1}{1 - Q/H} = 101\ \text{k}\Omega$$

$$C_2 = \frac{HR_1}{R_{EQ}R_3\omega_0} = \frac{H}{R_3\omega_0} = 1.58\ \mu\text{F}$$

$$C_1 = \frac{1}{HR_1\omega_0} = 0.0159\ \mu\text{F}$$

Standardize the capacitor values by letting $C_1 = 0.01\ \mu\text{F}$ and $C_2 = 1\ \mu\text{F}$, then adjust the resistors:

$$R_1 = \frac{1}{HC_1\omega_0} = 1.59 \text{ k}\Omega$$

$$R_{EQ} = \frac{1}{Q(C_1 + C_2)\omega_0} = 1.58 \text{ k}\Omega$$

$$R_2 = \frac{R_1 R_{EQ}}{R_1 - R_{EQ}} = 250 \text{ k}\Omega$$

$$R_3 = |A_0|\, R_1\left(1 + \frac{C_1}{C_2}\right) = 161 \text{ k}\Omega$$

Note that R_2 is "infinite" compared to R_1. Figure 4-164 illustrates the filter and its gain curve.

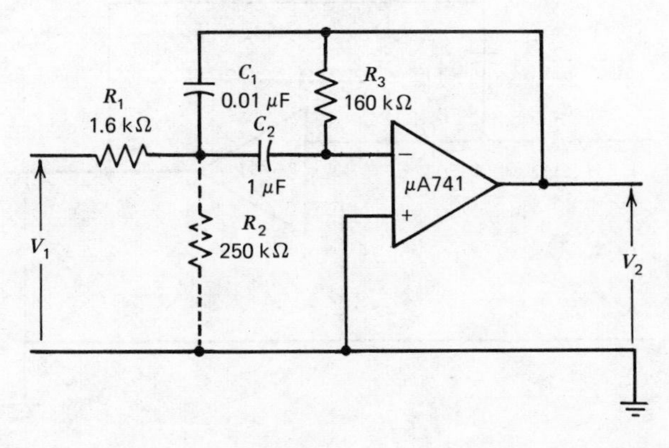

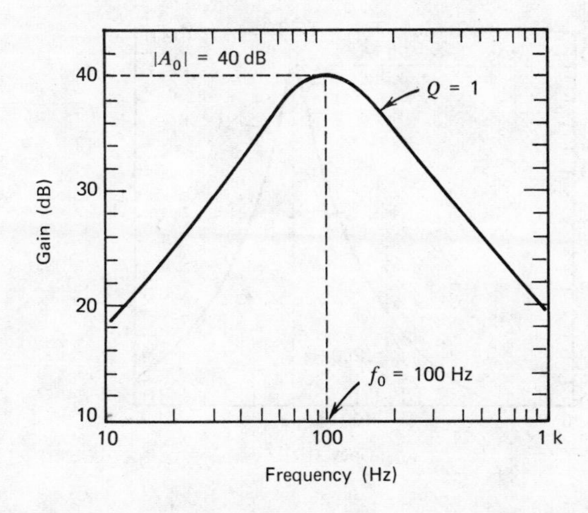

Figure 4-164. Low Q. Five-component filter, which has Q of 1 and gain of 40 dB, rolls off at rate of 20 dB/decade and keeps phase-shift within 10° about the center frequency. Resistor R_2 is not needed since it is "infinitely large" compared to R_1.

Network is designed by choosing R_1 and then computing capacitances.

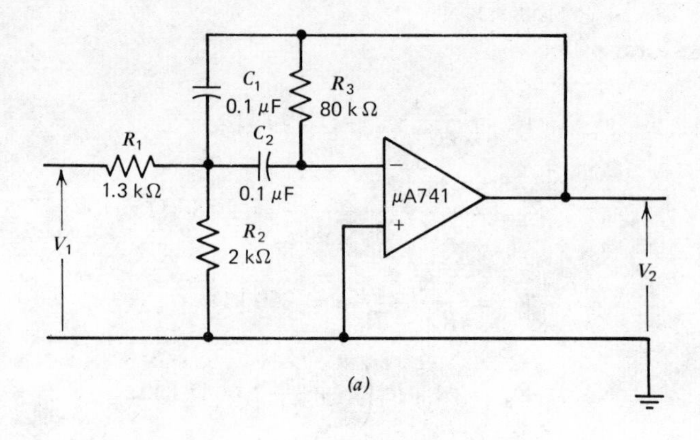

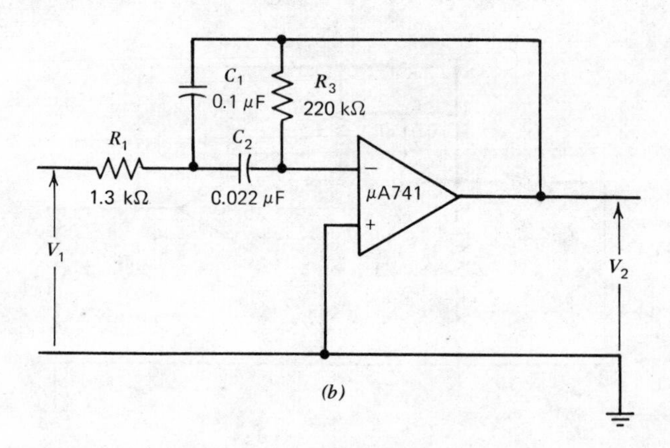

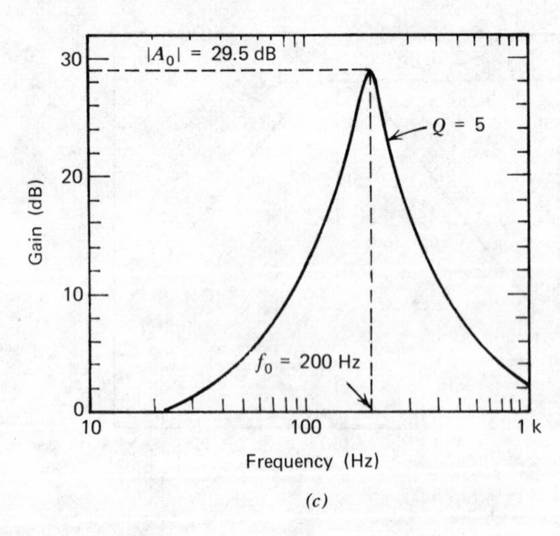

Figure 4-165. Intermediate Q. Two filter designs give the same gain response (c). Though arrived at by different methods, both are valid, but one (a) uses six components, the other (b) only five. Filter gain is 29.5 dB, center frequency is 200 Hz, and Q is 5. For this case, the Q value is intermediate.

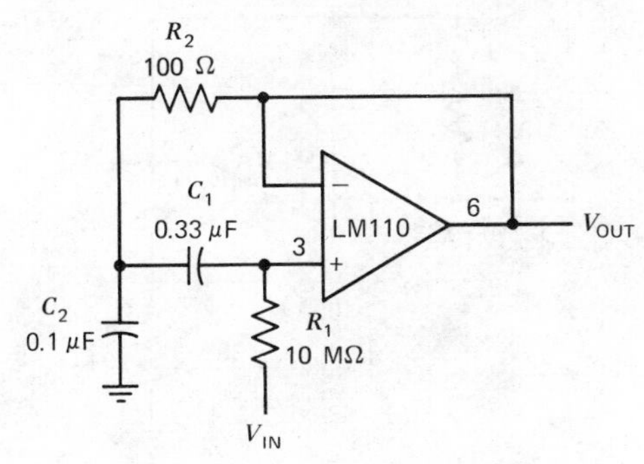

Figure 4-166. Band-pass filter.

For a third problem, design a filter with a midband voltage gain of 30 (29.5 dB), a center frequency of 200 Hz, and a Q of 5. Hence the intermediate Q test, $\sqrt{|A_0|/2}, < Q < \sqrt{|A_0|}$, holds for both design procedures, so that both yield valid results, which is interesting to compare. First choose $C_1 = C_2 = 0.1\ \mu\text{F}$. Then solve for $R_1 = 1.33\ \text{k}\Omega$, $R_{EQ} = 797\ \Omega$, $R_2 = 2\ \text{k}\Omega$, and $R_3 = 80\ \text{k}\Omega$. From these calculations the design in Figure 4-165*a* results.

Now try the other procedure. Let $R_1 = 1\ \text{k}\Omega$ and $R_2 = \infty$. This requires that $R_3 = 180\ \text{k}\Omega$, $C_1 = 0.133\ \mu\text{F}$, and $C_2 = 0.0266\ \mu\text{F}$. Standardize the capacitor values by setting $C_1 = 0.1\ \mu\text{F}$ and $C_2 = 0.022\ \mu\text{F}$. Compute the resistances: $R_1 = 1.33\ \text{k}\Omega$, $R_{EQ} = 1.31\ \text{k}\Omega$, $R_2 \approx \infty$, $R_3 = 222\ \text{k}\Omega$. The filter that results is shown in Figure 4-165*b*. The gain response (Figure 4-165*c*) for each of these filters is identical. Figures 4-166 through 4-169 depict several tuned circuits.

The twin-T network is one of the few *RC* filter networks capable of providing an infinitely deep notch. By combining the twin-T with an LM102 voltage-follower, the usual drawbacks of the network are overcome. The Q is raised from the usual 0.3 to something greater than 50. Further, the voltage-follower acts as a buffer, providing a low-output resistance; and the high-input resistance of the LM102 makes it possible to use large resistance values in the T so that only small capacitors are required, even at low frequencies. The fast response of the follower allows the notch to be used at high frequencies. Neither the depth of the notch nor the frequency of the notch are changed when the follower is added.

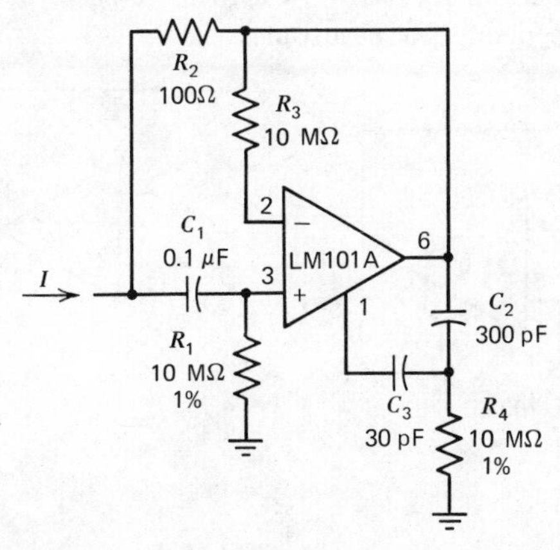

Figure 4-167. Simulated inductors [*note:* $L \cong R_1R_2C_1$, $R_3 = R_2$, $R_p = R_1$].

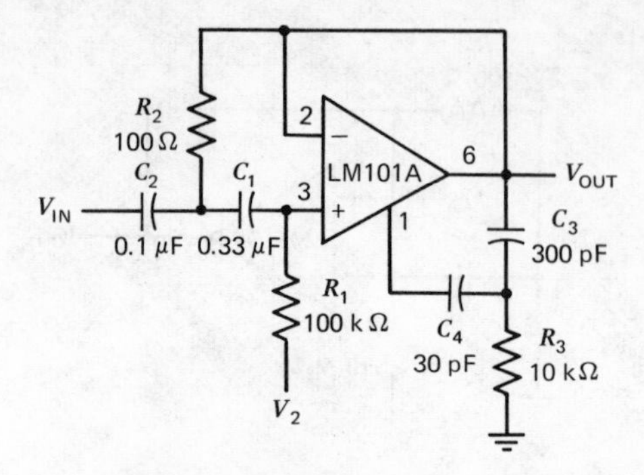

Figure 4-168. Tuned circuit [*note:* $f_{OUT} = 1/2\pi\sqrt{R_1R_2C_1C_2}$].

Figure 4-170 shows a twin-T network connected to an LM102 (or an LM110) to form a high-Q, 60-Hz notch filter. The junction of R_3 and C_3, which is normally connected to ground, is bootstrapped to the output of the follower.

Because the output of the follower is a very low impedance, neither the depth nor the frequency of the notch change, however, the Q is raised in proportion to the amount of signal fed back to R_3 and C_3. Figure 4-171 shows the response of a normal twin-T and the response with the follower added.

In applications where the rejected signal might deviate slightly from the null of the notch network, it is advantageous to lower the Q of the network. This ensures some rejection over a wider range of input frequencies. Figure 4-172 shows a circuit where the Q may be varied from 0.3 to 50. A fraction of the output is fed back to R_3 and C_3 by a second voltage-follower, and the notch Q is dependent on the amount of signal fed back. A second follower is necessary to drive the twin-T from a low-resistance source so that the notch frequency and depth will not change with the potentiometer setting. Depending on the potentiometer setting, the circuit in Figure 4-172 will have a response that falls in the shaded area of Figure 4-171.

An interesting change in the high-Q twin-T occurs when components are not exactly matched in ratio. For example, an increase of 1 to 10% in the value of C_3 will raise the Q, while degrading the depth of the notch. If the value of C_3 is raised by 10 to 20% the network provides voltage gain and acts as a tuned amplifier. A voltage gain of 400 was obtained during testing. Further increases in C_3 cause the circuit to oscillate, giving a clipped sine-wave output.

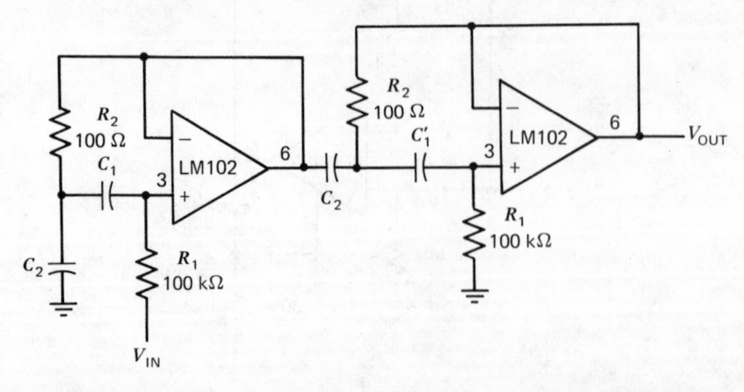

Figure 4-169. Two-stage tuned circuit [*note:* $f_{OUT} = 1/2\pi\sqrt{R_1R_2C_1C_2}$].

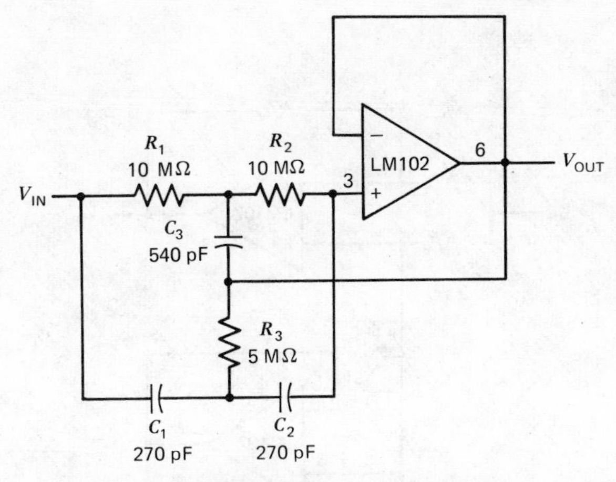

Figure 4-170. High-Q notch filter

$$\left[note: f_{OUT} = 1/2\pi R_1 C_1,\ R_1 = R_2 = 2R_3,\ C_1 = C_2 = \frac{C^3}{2} \right].$$

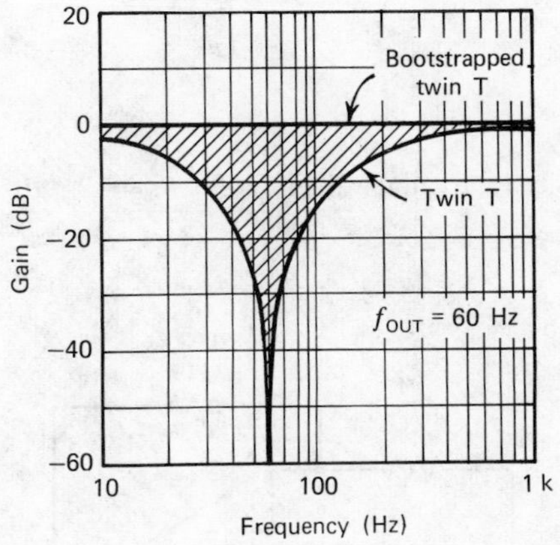

Figure 4-171. Response of filters with high and low Qs.

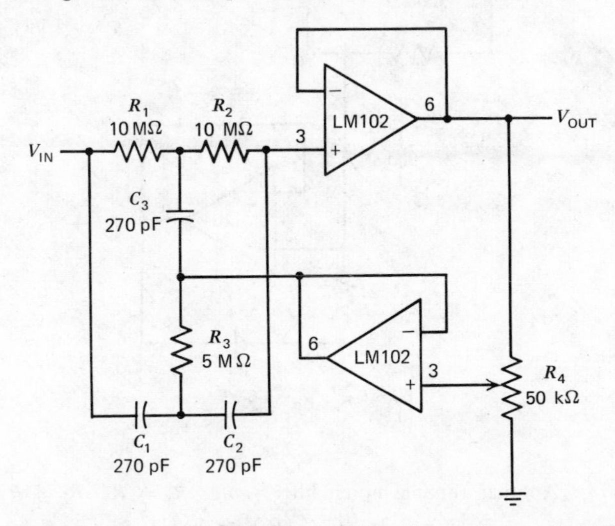

Figure 4-172. Adjustable-Q notch filter [*note:* f_{OUT} = 60 Hz].

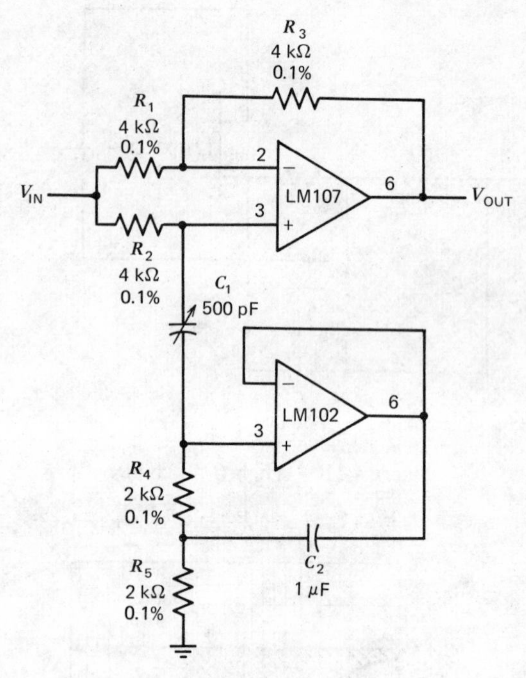

Figure 4-173. Easily tuned notch filter [*note:* $R_4 = R_5$, $R_1 = R_3$, $R_4 = 1/2R_1$, $f_{OUT} = 1/2\pi R_4\sqrt{C_1C_2}$].

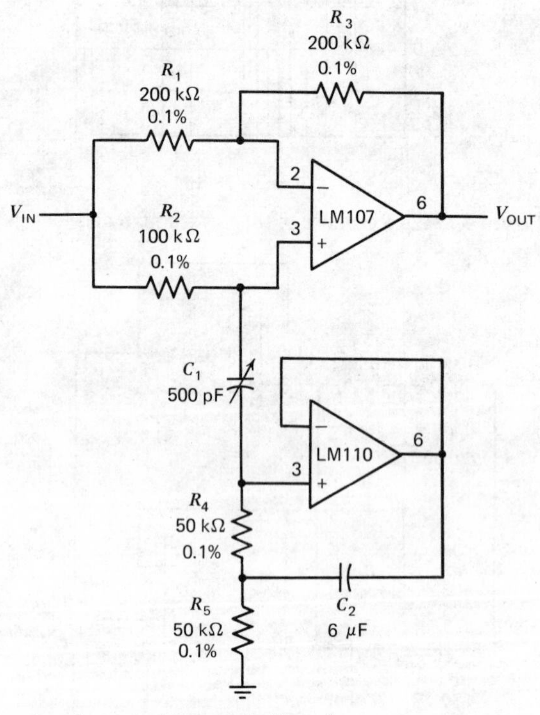

Figure 4-174. Another tunable notch filter [*note:* $R_1 = R_3$, $R_4 = R_5 = R_2/2$, $f_{OUT} = 1/2\pi\sqrt{C_1C_2(R_5)^2} = 60$ Hz].

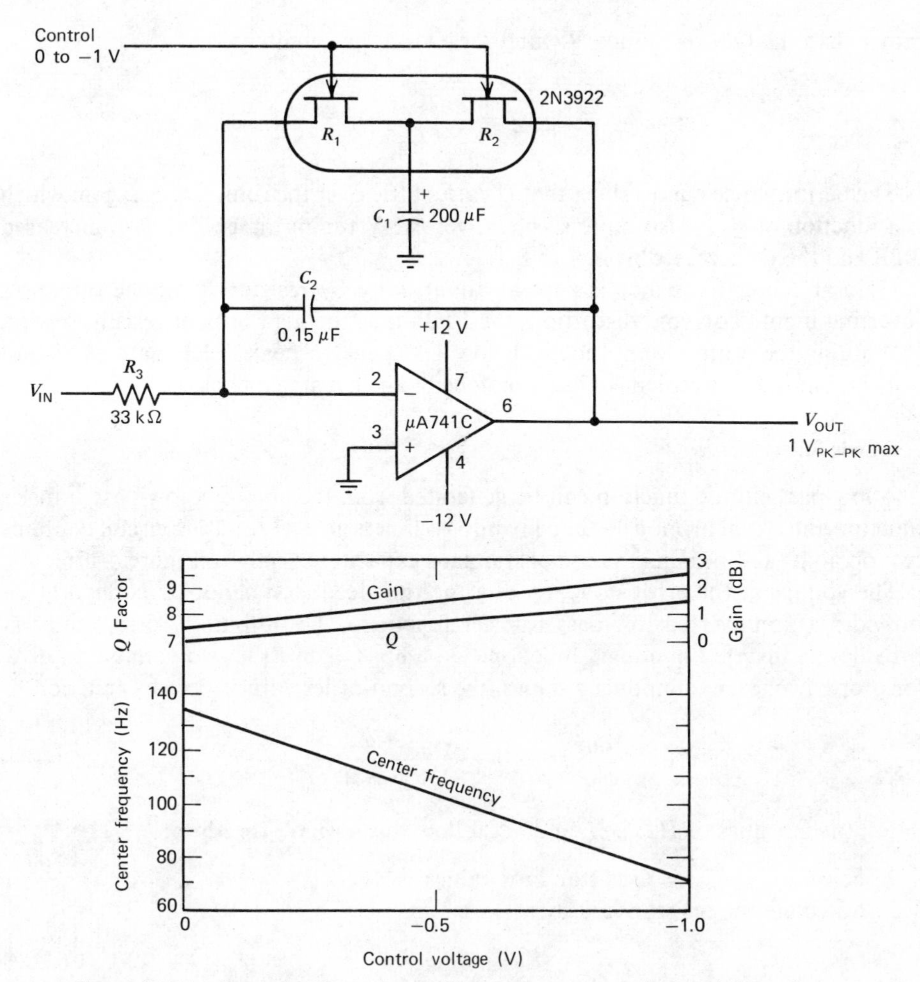

Figure 4-175. Tunable filter using FETs as voltage-controlled variable resistors.

The circuit is easy to use and only a few items need be considered for proper operation. To minimize notch frequency shift with temperature, silver mica, or polycarbonate capacitors should be used with precision resistors. Notch depth depends on the component match, therefore, 0.1% resistors and 1% capacitors are suggested to minimize the trimming needed for a 60-dB notch. To ensure stability of the LM102, the power supplies should be bypassed near the IC package with 0.01-μF disc capacitors. Figures 4-173 and 4-174 depict two additional tuned notched filters.

If the resistors of a conventional bridged-T *RC* network are replaced by the ON resistances of two FETs, the FETs can be made to convert a standard active band-pass filter circuit into an electronically tunable narrow-band filter (Figure 4-175). Since R_{ON} is inversely proportional to the gate voltage, a change in gate voltage linearly varies the center frequency. The center frequency is

$$f = \frac{1}{2\pi\sqrt{R_1 R_2 C_1 C_2}}$$

Furthermore, Q depends primarily on the size of the large shunt capacitor C_1, rather than the ON resistances R_1 and R_2; Q is approximately

$$Q = \frac{1}{2}\sqrt{\frac{C_1}{C_2}}$$

The performance curves show that Q varies little over the tuning range; gain which is a function of Q, is also quite stable. Over a 1-V tuning range, the gain increased 2 dB and the Q increased from 7 to 8.5.

Gain at center frequency is kept at about unity by resistor R_3 at the op amp's inverting input. To avoid distortion, the FETs must be kept in their resistive region by limiting the output amplitude to below 1 V, peak-to-peak; FET matching is not critical, but it is convenient to use a matched pair in a single package.

Elliptic Filter

A low-pass elliptic function can be generated from the sum of a low-pass transfer function and a synchronously tuned bandpass function. The resulting circuit contains two op amps and permits the use of standard capacitance values (Figure 4-176).

The voltage at the error node, v_e, of a multiple-feedback bandpass configuration provides a noninverted low-pass transfer function. The output, V_B of op amp A_1 provides an inverted bandpass function. Op amp A_2 sums the two voltages to give, for properly chosen component values, the second-order elliptic transfer function:

$$\frac{V_{\text{OUT}}}{V_{\text{IN}}} = -H\frac{s^2 + \omega_n^2}{s^2 + \alpha_d\omega_d s + \omega_d^2} \tag{4-84}$$

For given values of H, α_d, ω_n and ω_d, follow this design procedure:

1. Select $C_1 = C_2 = C$ as standard values.
2. Calculate the resistor values

$$R_1 = R_2 = \alpha_d/(\omega_d C)$$

$$R_3 = 2/(\alpha_d\omega_d C)$$

$$R_4 = \frac{2}{\alpha_d\omega_d C}[(\omega_d^2/\omega_n^2) - 1]$$

$$R_5 = \frac{\alpha_d}{\omega_d C}[(\omega_d^2/\omega_n^2) - 1]$$

$$R_6 = H \cdot R_5$$

A catalog of low-pass prototypes ("Filter Design Tables and Graphs," by E. Christian and E. Eisenman, published by John Wiley, New York, 1966) lists elliptic filters by stop-band attenuation, peak-to-peak ripple and order of equation. The elliptic prototype

$$G(p) = \frac{p^2 + 2.270^2}{(p^2 + 0.6226p + 1.138^2)(p + 0.8312)}$$

provides an f_s of 2.0, A_s of 28.6 dB and 0.3-dB ripple.

Frequency and impedance scaling provide the final design (Figure 4-177). In this instance, H is set at 1.667 and the stop-band frequency is scaled to 640 Hz.

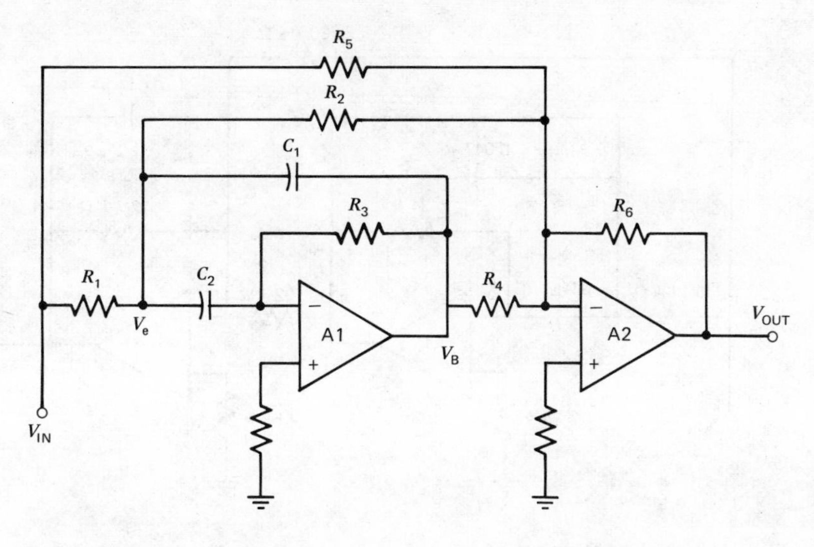

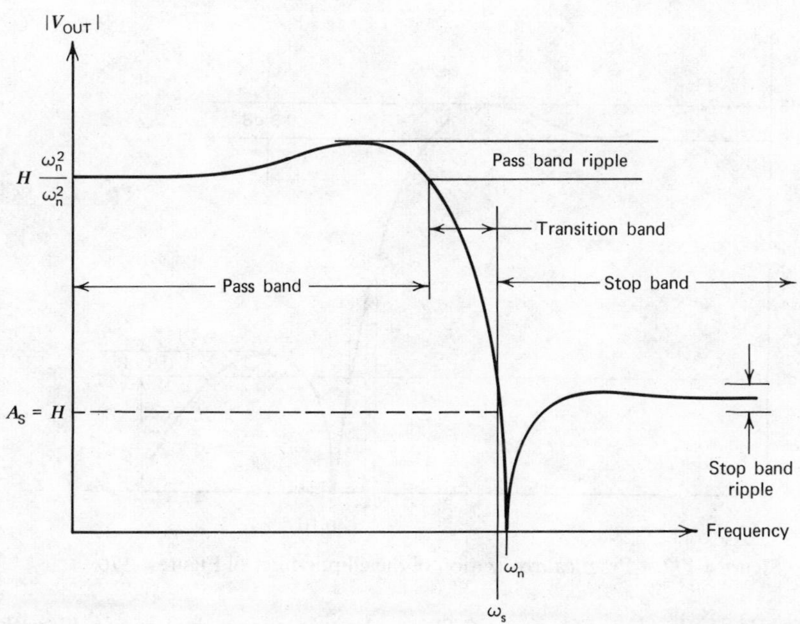

Figure 4-176. Active filter circuit achieves elliptic-filter characteristic through a combination of bandpass V_B and low-pass V_e filter outputs.

Self-Tuned Filter

Audio and instrumentation systems frequently employ band-pass filters to improve the signal-to-noise ratio. If the input-signal frequency varies over a wide range—as it does with many types of vibrating transducers—the bandwidth of the filter usually must be large enough to avoid undue attenuation and phase-shift at the frequency extremes. But a wide-band filter will be less effective in rejecting noise.

One solution is to use a self-tuning filter that automatically adjusts its center

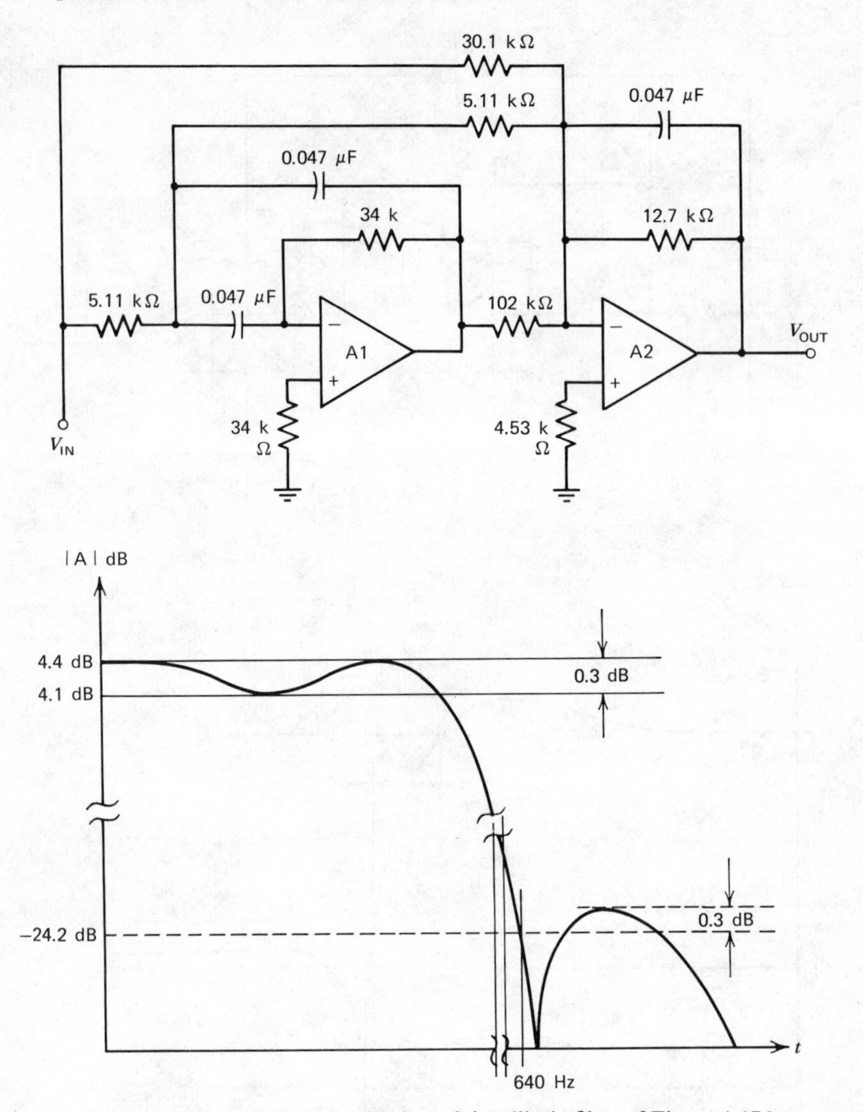

Figure 4-177. Practical realization of the elliptic-filter of Figure 4-176.

frequency to track the signal frequency. This technique allows the use of a filter that has a bandwidth considerably less than the range of input-signal frequencies. The circuit in Figure 4-178 tunes itself over a frequency range of 2 to 20 kHz. It requires no reference frequency other than the input signal, and there is no internal oscillation or synchronization circuitry. The frequency range can be extended in decade steps by capacitor-switching. In this circuit, we have

$$f_{\mathrm{OUT}} = \frac{1}{2\pi C}\left[\frac{1}{R_3}\left(\frac{1}{R_1} + \frac{1}{R_2}\right)\right]^{1/2} \tag{4-85}$$

Thus f_{OUT} can be varied merely by changing R_2 and without affecting the gain or bandwidth. If resistance R_2 can be made to vary approximately as the frequency of the input signal varies, the filter can be made to stay tuned to the signal frequency.

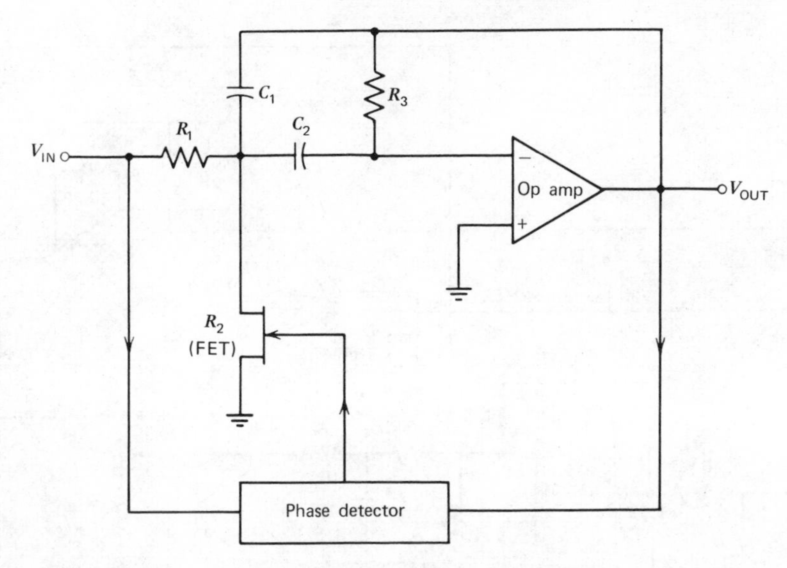

Figure 4-178. Self-tuned filter.

If the filter is properly tuned, there will be 180° of phase-shift between the input and the output of the filter. If, however, the filter is not tuned to the input frequency, then the phase-shift is not 180°, and the phase detector generates an error signal, which is applied to the gate of the FET to control its drain-to-source resistance R_2. The phase detector and the FET form part of a negative-feedback loop around the filter. Because of this configuration, any error in phase resulting from detuning will change the resistance of the FET, thereby retuning the filter. Thus the FET acts as the variable resistance R_2.

Another self-tuned active filter is shown in Figure 4-179, while Figure 4-180 shows the waveforms at various points in the circuit. In Figure 4-179, amplifier A_1, and resistors R_1, R_2, and R_5, FET Q_1, and capacitors C_1, C_2, and C_3 comprise the filter. The FET Q_1 functions as the variable resistor. Resistors R_2 and R_5 limit the amount of variable resistance in the circuit and prevent latchup as the electronic servo hunts for the center frequency when confronted with a noisy input signal. Capacitor C_3 prevents self-oscillation in the op amp A_1. Amplifier A_2 is used as a differentiator, A_3 as an integrator, and A_4 and A_5 as comparators.

In the timing diagram of Figure 4-180 the top waveform is a sinusoidal input signal and the waveform immediately below it is the corresponding output. Since the phase-shift between these two waveforms is not exactly 180°, this indicates that the filter is not tuned to the input-signal frequency. The A_2 output waveform in Figure 4-180 is a differentiated and inverted version of the filter output. The differentiation introduces the additional 90° phase-shift necessary to make the final FET control voltage go both positive and negative on demand. Amplifiers A_4 and A_5 limit the input signal and differentiator output, respectively, resulting in the waveforms illustrated for the output of these amplifiers.

Diodes D_2 and D_3, and transistor Q_2 form a logic circuit, the output of which is the "point-X" waveform in Figure 4-180. When the filter input and output are exactly 180° out of phase (filter in tune), the waveform for point X has a certain average dc level, V_{OUT}. As the phase-shift increases or decreases, the dc level changes

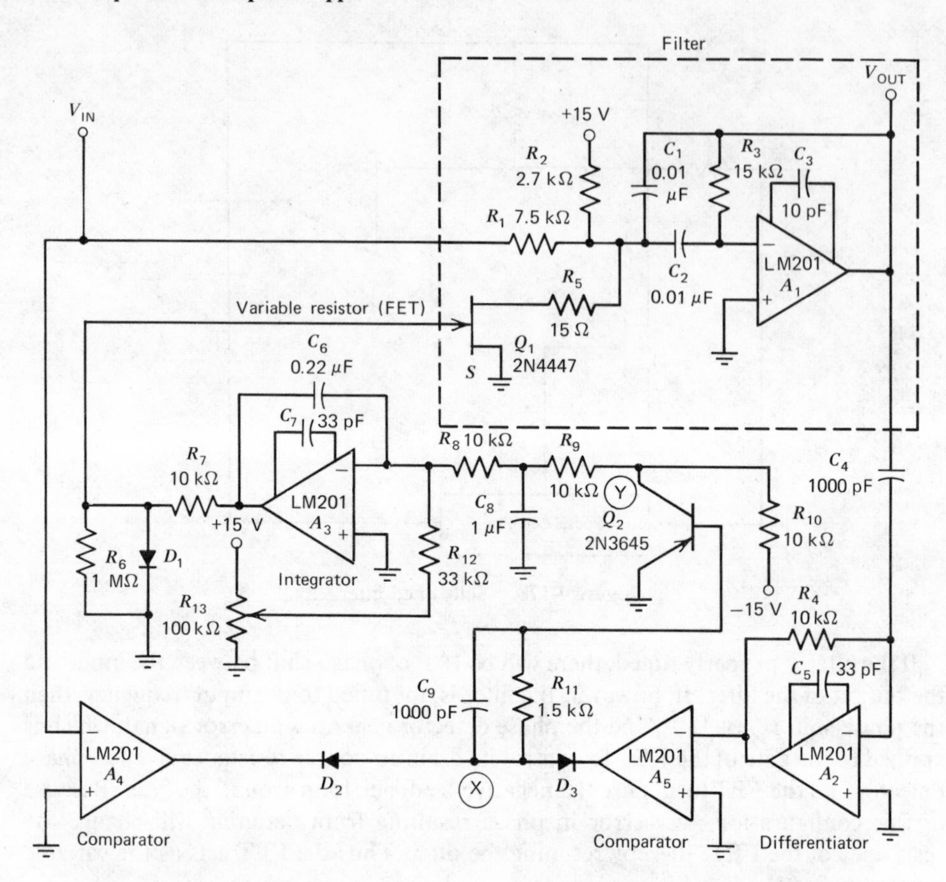

Figure 4-179. Complete self-tuned active band-pass filter uses inexpensive integrated circuit operational amplifiers. The two comparators, and the rectifiers CR_2 and CR_3, comprise the phase detector. Resistor R_{13} nulls the error signal when the filter's input and output signals are exactly 180° out of phase.

to $V_{OUT} \pm \Delta V_{OUT}$, where $\pm \Delta V_{OUT}$ is the error signal produced when the filter is not tuned.

Transistor Q_2 inverts the error signal to produce the bottom waveform in Figure 4-180. The V_{OUT} component is cancelled by an opposing voltage derived from R_{13} and R_{12}. The error signal is amplified by the integrator A_3 to provide a correcting signal to FET Q_1. The integrator has a dc gain of about 10^5. But it has a low ac gain, and this helps remove ripple from the correcting voltage. Resistors R_8, R_9, and R_{10}, and capacitor C_8, provide some prefiltering of the error-signal waveform prior to integration. With the values shown, the filter self-tunes at center frequencies from 2 to 20 kHz. Over this range, the output amplitude remains constant to within $\pm 1\%$.

State-Variable Filters[8]

Another means of configurating active filters is the state variable or biquadratic approach. Although it requires three operational amplifiers, the state-variable active

[8] Portions of the material from this point to the end of this section on "Oscillators, Function Generators, and Active Filters" is excerpted from *The Op Amp and Active Filter Handbook*, Motorola Semiconductor Corporation.

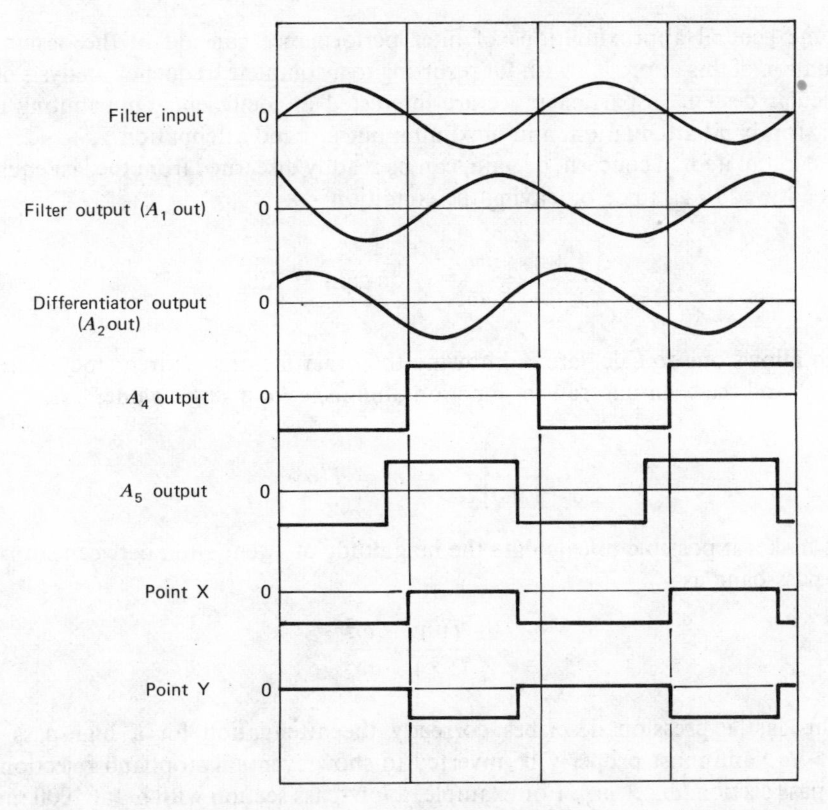

Figure 4-180. Voltage waveforms at various points in the circuit of Figure 4-179. Because the input and output waveforms are not exactly 180° out of phase, an error signal is produced at point X (the output of the phase detector).

filter can operate at fairly high frequencies and can develop large values of Q. Moreover, the operating frequency, gain, and Q of this filter circuit are independent of each other, and the circuit itself is not overly sensitive to variations in component values over the range of operation. The general form of transfer function is

$$\frac{V_2(s)}{V_1(s)} = \frac{ms^2 + cs + d}{s^2 + as + b} \tag{4-86}$$

where the denominator coefficient of s^2 is normalized to unity. This expression has been dubbed a "biquad" because of the quadratic form of both numerator and denominator.

In a control-system approach either the numerator or denominator may be rewritten in a different, more understandable form from a steady-state frequency point of view:

$$s^2 + as + b = s^2 + 2\zeta\omega_0 s + \omega_0^2$$

where ω_0 is the "natural" frequency and ζ is the damping factor. The general form may be rewritten as

$$\frac{V_2(s)}{V_1(s)} = \frac{m(s^2 + 2\zeta_z\omega_z s + \omega_z^2)}{s^2 + 2\zeta_p\omega_p s + \omega_p^2} \tag{4-87}$$

Some general approximations of filter performance can aid in the design and evaluation of this approach without resorting to a complete frequency analysis of any particular design. In particular we are interested in coefficients contributing to dc gain, stop-band attenuation, and maximum out-of-band attenuation.

The initial item of concern, dc gain, can be readily discerned from the last equation if s is allowed to go to zero, leaving the condition

$$\frac{V_2(0)}{V_1(0)} = \frac{m\omega_z^2}{\omega_p^2} = \text{dc gain} = T(0) \tag{4-88}$$

which allows one to calculate m knowing the other factors. As frequency increases $s \to \infty$, and the eventual transfer function diminishes to a single value,

$$\left.\frac{V_2(s)}{V_1(s)}\right|_{s\to\infty} = m = T(\infty) \tag{4-89}$$

This makes it possible to calculate the magnitude of attenuation between stop-band and pass-band as

$$\frac{T(0)}{T(\infty)} = \frac{\omega_z^2}{\omega_p^2} \tag{4-90}$$

This last expression describes correctly the attenuation for a high-pass filter ($\omega_p > \omega_z$) and must properly be inverted to show eventual stopband rejection in a low-pass section ($\omega_z > \omega_p$). For example, a low-pass section with ω_p at 1000 rad/sec and ω_z at 10,000 rad/sec would show an eventual stopband rejection of 40 dB (1/100), which would be expected from the conventional 40 dB/dec roll-off of a second-order low-pass section.

Before values of resistors and capacitors are calculated or selected, the coefficients a, b, c, d, and m should be related to the more relevant parameters ω_z, ω_p, ζ_z, ζ_p, and gain. With the aid of the first two equations, these solutions can be derived:

$$a = 2\zeta_p\omega_p \tag{4-91}$$

$$b = \omega_p^2 \tag{4-92}$$

$$m = \frac{T(0)\omega_p^2}{\omega_z^2} \tag{4-93}$$

$$c = 2m\zeta_z\omega_z \tag{4-94}$$

$$d = m\omega_z^2 \tag{4-95}$$

The general state-variable filter circuit (Figure 4-181) leaves the interconnection of several summing resistors at the mercy of the particular polynomials that have been generated above. There are four distinctly different connections that depend on the relationships of a, b, c, d, and m. The relationships have been tabulated here for easy selection (Table 4-6). Once the case and interconnect have been established, one may go to Table 4-7 for design formulas covering each resistor, given arbitrary values of capacitors.

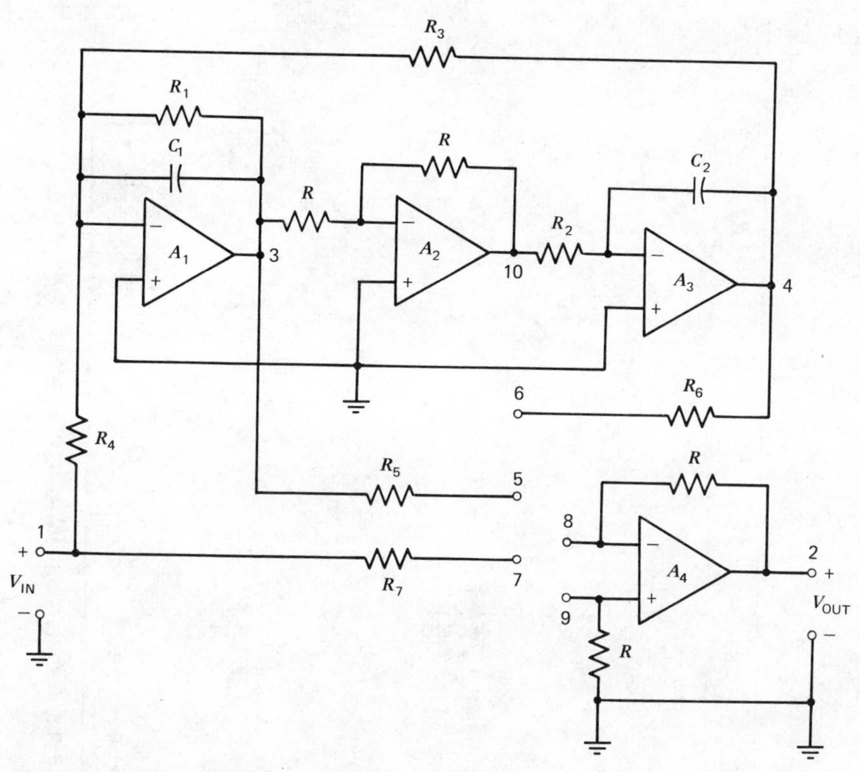

Figure 4-181. Complete biquadratic circuit realization that depends on *pin* connections and component values for filter type and characteristics. (Courtesy Motorola Semiconductor Corporation.)

Table 4-6 Conditions for four cases and their input connections to amplifier A_4 in Figure 4-181

Case	*Subcase*	*Conditions*	*Connections for the summing amplifier*
A	A_1	$ma \geq c$ $mb \geq d$	R_5, R_6, R_7 to the negative-input terminal
	A_2	$ma = c$ $mb \geq d$	R_5, R_6, R_7 to the negative-input terminal
B		$ma < c$ $mb \geq d$	R_5 to the positive-input terminal and R_6, R_7 to the negative-input terminal
C		$ma > c$ $mb < d$	R_6 to the positive-input terminal and R_5, R_7 to the negative-input terminal
D	D_1	$ma < c$ $mb > d$	R_7 to the positive-input terminal and
	D_2	$ma = c$ $mb < d$	R_5, R_6 to the negative-input terminal

Table 4-7 Design formulas for resistors of Figure 4-181

Case		R_4	R_5	R_6	R_7
A	A_1	$\dfrac{1}{k_2(ma-c)C_1}$	k_2R	$\dfrac{k_2}{k_1}\dfrac{ma-c}{mb-d}\sqrt{b}\,R$	$\dfrac{1}{m}R$
	A_2	$\dfrac{\sqrt{b}}{k_2(mb-d)C_1}$	Infinite	$\dfrac{k_2}{k_1}R$	$\dfrac{1}{m}R$
B		$\dfrac{1}{k_2(c-ma)C_1}$	$\left[k_2(1+m)+k_1\dfrac{mb-d}{(c-ma)\sqrt{b}}-1\right]R$	$\dfrac{k_2}{k_1}\dfrac{c-ma}{mb-d}\sqrt{b}\,R$	$\dfrac{1}{m}R$
C		$\dfrac{1}{k_2(ma-c)C_1}$	k_2R	$\left[\dfrac{k_2}{k_1}\dfrac{(ma-c)\sqrt{b}}{d-mb}\left(1+m+\dfrac{1}{k_2}\right)-1\right]R$	$\dfrac{1}{m}R$
D	D_1	$\dfrac{1}{k_2(c-ma)C_1}$	k_2R	$\dfrac{k_2}{k_1}\dfrac{c-ma}{d-mb}\sqrt{b}\,R$	$\left[\dfrac{1}{m}\left(1+\dfrac{1}{k_2}+\dfrac{k_1}{k_2}\dfrac{d-mb}{(c-ma)\sqrt{b}}\right)-1\right]R$
	D_2	$\dfrac{\sqrt{b}}{k_2(d-mb)C_2}$	Infinite	$\dfrac{k_2}{k_1}R$	$\left[\dfrac{1}{m}\left(1+\dfrac{k_1}{k_2}\right)-1\right]R$

Note: $R_1 = (1/aC_1)$, $R_2 = (k_1/\sqrt{b}\,C_2)$, $R_3 = (1/k_1\sqrt{b}\,C_1)$ (where C_1, C_2, and R are arbitrary; k_1 and k_2 are arbitrary and positive).

Let us take an example where we use a biquad high-pass section as an input filter which is to pass 2025 Hz while attenuating 1270 Hz. For the transfer minimum to be at 1270 Hz let the numerator factors assume

$$\omega_z = 1270 \times 2\pi \approx 7960$$
$$\zeta_z = 0.01$$

and for a corner near 2025 Hz the denominator factors may be

$$\omega_p = 1900 \times 2\pi = 11{,}900$$
$$\zeta_p = 0.4$$

and for the final parameter let $T(0) = 1$. Solving the coefficient equations,

$$a = 9.53 \times 10^3$$
$$b = 1.42 \times 10^8$$
$$m = 2.25$$
$$c = 358$$
$$d = 6.3 \times 10^7$$
$$ma = 2.14 \times 10^4$$
$$mb = 3.2 \times 10^8$$

From the calculations its easy to see that $ma > c$ and $mb > d$, which puts this filter in case A_1. At this point we are ready to calculate real resistor values after assuming some standard value of capacitor and letting the scaling factors K_1 and K_2 equal unity. For this instance let $C_1 = C_2 = 0.1\ \mu F$; from Table 4-7,

$$R_1 = \frac{1}{aC_1} = 1050\ \Omega$$

$$R_2 = \frac{1}{\sqrt{b}\,C_2} = \frac{1}{\omega_p C_2} = 840\ \Omega$$

$$R_3 = \frac{1}{\sqrt{b}\,C_1} = \frac{1}{\omega_p C_2} = R_2 = 840\ \Omega$$

$$R_4 = \frac{1}{(ma - c)C_1} = 476\ \Omega$$

$$R_5 = R \qquad \text{(let } R = 10\ \text{k}\Omega\text{)}$$

$$R_5 = 10\ \text{k}\Omega$$

$$R_6 = \frac{ma - c}{mb - d}\sqrt{b}\,R = 9.73\ \text{k}\Omega$$

$$R_7 = \frac{R}{m} = 4.45\ \text{k}\Omega$$

It is obvious from the values of R_1 through R_4 that the capacitor chosen was too large. Since their values are all inversely proportional to C, they may be accurately determined by inspection if C is changed to 0.01 μF:

$$R_1 = 10.5\ \text{k}\Omega$$
$$R_2 = 8.4\ \text{k}\Omega$$
$$R_3 = 8.4\ \text{k}\Omega$$
$$R_4 = 4.76\ \text{k}\Omega$$
$$R_5 = 10\ \text{k}\Omega$$
$$R_6 = 9.73\ \text{k}\Omega$$
$$R_7 = 4.45\ \text{k}\Omega$$

This filter was constructed with components ordinarily available in most laboratories (5% resistors, 10% capacitors) and the results are plotted in Figure 4-182 together with the "ideal" response.

COMPONENT SELECTION

For their state-variable filters, most companies today use the popular μA 741 op amp in combination with metal-film resistors and NPO (nonpolarized) and/or mica capacitors.

RESISTORS

Three types of resistors most often used are carbon-composition, metal-film, and wirewound. Carbon-composition resistors have a rather poor TC of resistance (200 to 500 ppm/°C) and are used for "room-temperature" applications or in filters which may have rather loose performance tolerances with temperature, as in the low-Q stages of two- or three-pole high-pass or low-pass filters. Composition resistors are useful for trimming and padding metal-film or wirewound resistors where the

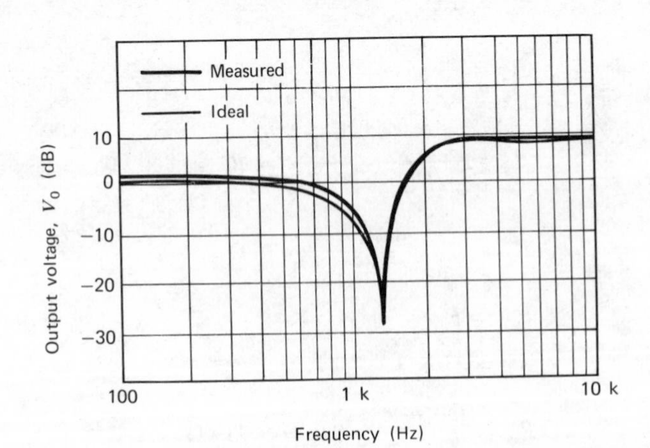

Figure 4-182. Biquad filter response, with 5% resistors and 10% capacitors, compares ideal and measured results [*note:* all amplifiers used were MC1458G]. (Courtesy Motorola Semiconductor Corporation.)

Table 4-8 Typical parameters of integrated circuit resistors

Type	*Range* (Ω)	*Temperature coefficient* (ppm/°C)	*Tolerance* (%)	*Matching* (%)
Based-diffused	100–30 k	500–2,000	±10	1
Emitter-diffused	5–100	900–1,500	±15	2
Base-pinch	5–200 k	4,000–7,000	±50	5
Collector-pinch	10–500 k	4,000–7,000	±50	10
Thin-film (Ta or Ni-Cr)	30–100 k	0 ± 400	±2	0.5
Thick-film	1–10 M	0 ± 500	±20	10

relatively poor TC causes only a small percentage change in the overall value. Carbon resistors are relatively inexpensive and are available in a wide range of values.

Band-pass filters and the high-Q states of a high-pass or low-pass multiple-pole filter require metal-film or even wirewound resistors. Two popular TCs for metal-film resistors are ±100 ppm/°C (T_0) and ±50 ppm/°C (T_2). Metal films can be purchased that have a positive- or negative-only TC and also with lower TCs (±10 ppm/°C). The metal-film resistor is probably the most commonly used resistor for filter applications and is available in a wide range of values. High-Q filters and/or filters which require especially stable parameters with temperature changes may require wirewound resistors with TCs as low as a few ppm/°C. High-frequency applications will require noninductive wound resistors. Integrated circuit technology offers alternatives to discrete resistors. These include diffused, thin- and thick-film resistors.

Base-diffused, emitter-diffused, base-pinch, and collector-pinch resistors are formed simultaneously with the diffusions for the transistors of the circuit. Temperature coefficients and initial tolerances make this type of resistor marginal for active filter applications unless the filter can be designed so that its parameters depend primarily on resistance ratios.

Thin-film resistors are deposited on ceramic or glass substrates. Materials such as nichrome, tantalum, or cermet, may be deposited by evaporation or sputtering. The electrical properties of these resistors are considerably superior to those of diffused silicon resistors. An advantage of thin-film over diffused or thick-film resistors is their superior long-term stability.

Thick-film resistors consist of special resistive inks screened and fired on ceramic substrates. Thick-film resistors are trimmable using sand-blasting or laser techniques. Table 4-8 gives typical (untrimmed) parameters for several IC resistor types.

CAPACITORS

Capacitors present the most severe problem to active-filter designers. Capacitors which have superior characteristics, such as polystyrene, Teflon, NPO ceramic, or mica, are expensive. Large NPO ceramic is available in sizes up to about 0.05 μF; good-quality polystyrene capacitors can be used for the large values (10 μF) in critical applications but they are physically very large. Mica capacitors are available in values up to 0.01 μF, but are larger than Mylar or polycarbonate capacitors of the same value. Physically small ceramic capacitors, such as ceramic disc capacitors and others that have large dielectric constants (from 1200 to 6000), have relatively poor characteristics. Capacitance changes with temperature, frequency, voltage, and

Table 4-9 Typical capacitor parameters for different dielectrics (Burr-Brown Research)

Dielectric	*Power factor*	*Temperature coefficient of capacitance*
Mylar	8×10^{-4} to 14×10^{-4}	+250 ppm/°C, 0 to 70°C, larger at extremes
High-quality polystyrene	1×10^{-4} to 2×10^{-4}	−50 to −100 ppm/°C, −60 to +60°C
High-quality mica	1×10^{-4} to 7×10^{-4}	0 to 70 ppm/°C
NPO ceramic	5×10^{-4} to 20×10^{-4}	0 ± 30 ppm/°C
Polycarbonate	30×10^{-4} to 50×10^{-4}	Nonmonotonic total ±1%
Teflon	0.5×10^{-4} to 1.5×10^{-4}	0 to 70°C, larger at extremes, −250 ppm/°C, −60 to 150°C

time amount to several percent. For high-Q applications, these changes can make a filter stage unstable or have severe amplitude-peaking or attenuation. Such filter stages are usually highly Q-sensitive to element-value changes.

The merit of a capacitor dielectric from the point of view of freedom from losses is expressed in terms of the power factor of the capacitor. The power factor is the sine of the angle by which the current flowing into the capacitor fails to be 90° out of phase with the applied voltage. The tangent of this angle is called the dissipation factor. The reciprocal of the dissipation factor is termed the Q and is the ratio of capacitor reactance to the equivalent series resistance.

With ordinary dielectrics, phase angle is so small that the power factor, the dissipation factor, and the reciprocal of the capacitor Q are, for all practical purposes, equal to each other and to the phase angle expressed in radians. For high-quality capacitors, these are practically independent of capacitance, voltage, and frequency. Although the power factor of a capacitor is determined largely by the type of dielectric, it also is affected by the environment in which it operates: It tends to increase with temperature and is affected by humidity and by the absorption of moisture.

The effect of a capacitor with its power factor can be taken into account by replacing the capacitor with an ideal capacitor associated with a resistance. This resistance may be represented in series or in parallel. For lower power factors ($R_s \ll 1/\omega C$), R_s is given by

$$\text{series resistance} = R_s = \frac{\text{power factor}}{2\pi f C} \tag{4-96}$$

For the parallel resistance, we have approximately

$$\text{parallel resistance} = R_p = \frac{1}{2\pi f C(\text{power factor})} \tag{4-97}$$

A list of dielectric materials and representative performance features are given in Table 4-9.

Integrated circuit capacitors are of three types; *pn* junctions, MOS structures and thin-film types. These capacitors have small values that vary greatly with temperature. The most suitable capacitors for IC filters are those utilized in hybrid construction and are NPO ceramic chips or, for low-frequency work, tantalum capacitor chips.

ANALOG-COMPUTING AND SERVO AMPLIFIERS

Analog-computing amplifiers are divided into two broad classes, linear and non-linear. The amplifiers discussed in the earlier sections were generally linear amplifiers. Operations such as addition and subtraction are easily performed with linear amplifiers. Only a few more examples of summing amplifiers, in Figure 4-183 will be presented here, since the design methods previously discussed apply. Further attention will be paid to differentiators and integrators, which are more difficult to design.

The bulk of this section will be devoted to nonlinear amplifiers. Certain nonlinear amplifiers have important applications in measuring instruments as well as in analog computers. For example, multipliers—a form of logarithmic generator or converter—facilitate measurement of transistor gains as will be discussed shortly.

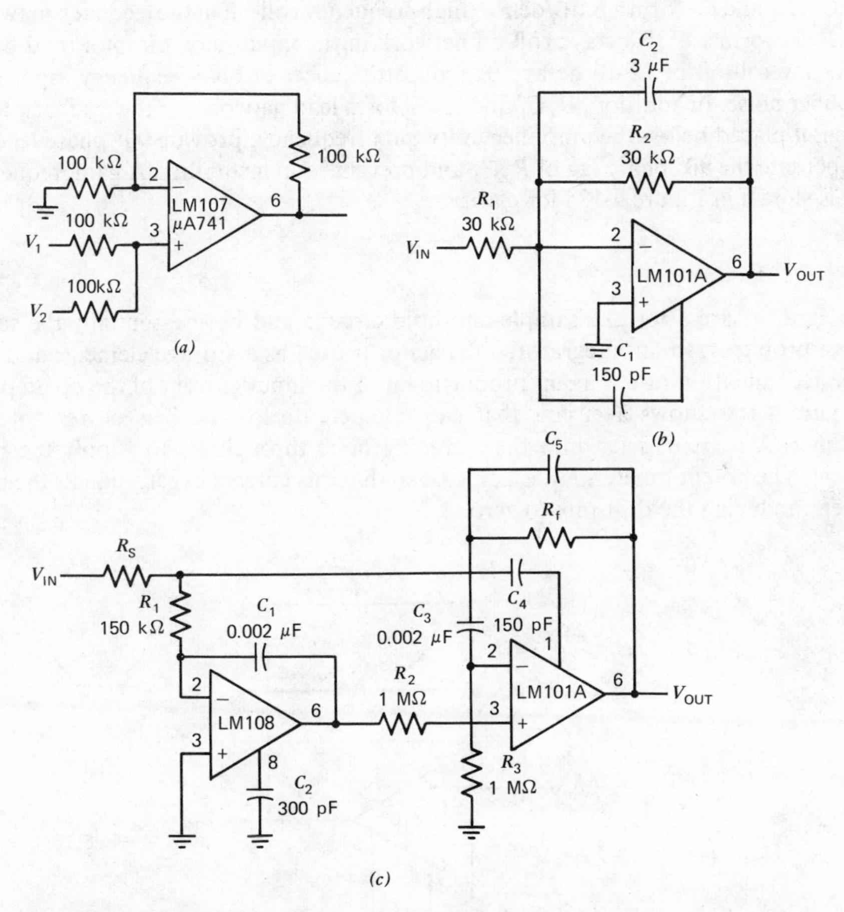

Figure 4-183. Examples of summing amplifiers: (*a*) noninverting ($R_S < 1$ kΩ for 1% accuracy); (*b*) fast-summing (power bandwidth 250 kHz, small-signal bandwidth 3.5 MHz, slew rate 10 V/μsec); (*c*) fast-summing with low-input current (power bandwidth 250 kHz, small-signal bandwidth 3.5 MHz, slew rate 10 V/μsec, $C_5 = 6 \times 10^{-6} R_f$). Note also that in (*c*) the LM101A, in addition to increasing speed, raises high- and low-frequency gain, increases output-drive capability, and eliminates thermal feedback.

In addition to their application in performing functional operations, log generators can provide a significant increase in the dynamic range of signal-processing systems. Also, unlike a linear system, there is no loss in accuracy or resolution when the input signal is small compared to the full scale. Over most of the dynamic range, the accuracy is a percent of the signal rather than a percent of the full scale. For example, using log generators, a simple meter can display signals with 100 dB dynamic range or an oscilloscope can display a 10-mV and 10-V pulse simultaneously. Obviously, without the log generator, the low-level signals are completely lost.

Practical Differentiator

A practical differentiator is shown in Figure 4-184. Both the stability and noise problems mentioned earlier are corrected by the addition of two components, R_1 and C_2; R_2 and C_2 form a 6 dB/octave high-frequency roll-off in the feedback network and R_1C_1 form a 6 dB/octave roll-off network in the input network for a total high-frequency roll-off of 12 dB/octave to reduce the effect of high-frequency input and amplifier noise. In addition, R_1C_1 and R_2C_2 form lead networks in the feedback loop which, if placed below the amplifier unity-gain frequency, provide 90° phase lead to compensate the 90° phase lag of R_2C_1 and prevent loop instability. A gain frequency plot is shown in Figure 4-185 for clarity.

Integrators

Integrators are a lot like sample-and-hold circuits and have essentially the same design problems. In an integrator, a capacitor is used as a storage element; and the error accumulation rate is again proportional to the input current of the op amp.

Figure 4-186 shows a circuit that can compensate for the bias current of the amplifier. A current is fed into the summing node through R_1 to supply the bias current. The potentiometer, R_2, is adjusted so that this current exactly equals the bias current, reducing the drift rate to zero.

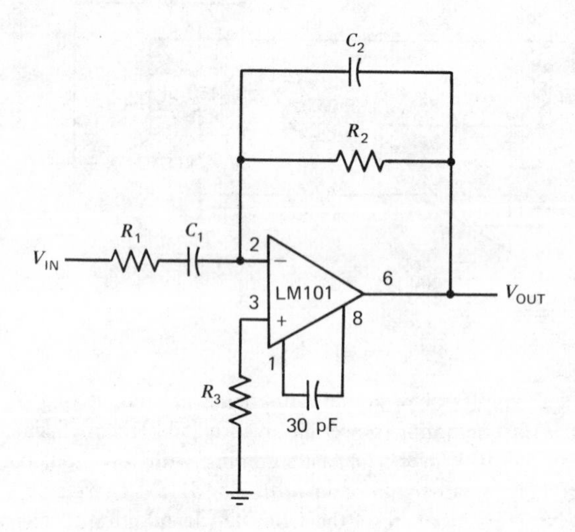

Figure 4-184. Practical differentiator [*note:* $f_c = 1/2\pi R_2C_1$, $f_h = 1/2\pi R_1C_1 = 1/2\pi R_2C_2$, $f_c \ll f_h \ll f_{\text{unity gain}}$].

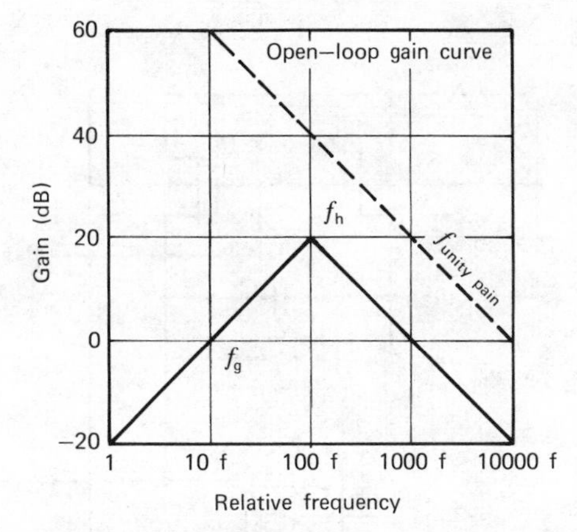

Figure 4-185. Differentiator frequency response.

The diode is used for two reasons. First, it acts as a regulator, making the compensation relatively insensitive to variations in supply voltage. Second, the temperature drift of the diode voltage is approximately the same as the temperature drift of the bias current. Therefore, the compensation is more effective if the temperature changes. Over a 0 to 70°C temperature range, the compensation will give a factor of 10 reduction in input current. Even better results are achieved if the temperature change is less.

Normally, it is necessary to reset an integrator to establish the initial conditions for integration. Resetting to zero is readily accomplished by shorting the integrating capacitor with a suitable switch. However, as with the sample-and-hold circuits, semiconductor switches can cause problems because of high-temperature leakage.

A connection that gets rid of switch leakages is shown in Figure 4-187. A negative-going reset pulse turns ON Q_1 and Q_2, shorting the integrating capacitor. When the switches turn OFF, the leakage current of Q_2 is absorbed by R_2 while Q_1 isolates the output of Q_2 from the summing node; Q_1 has practically no voltage across its junctions because the substrate is grounded; hence, leakage currents are negligible.

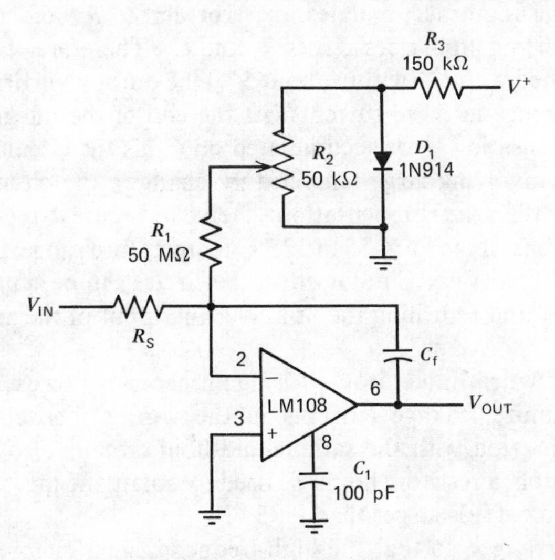

Figure 4-186. Integrator with bias-current compensation.

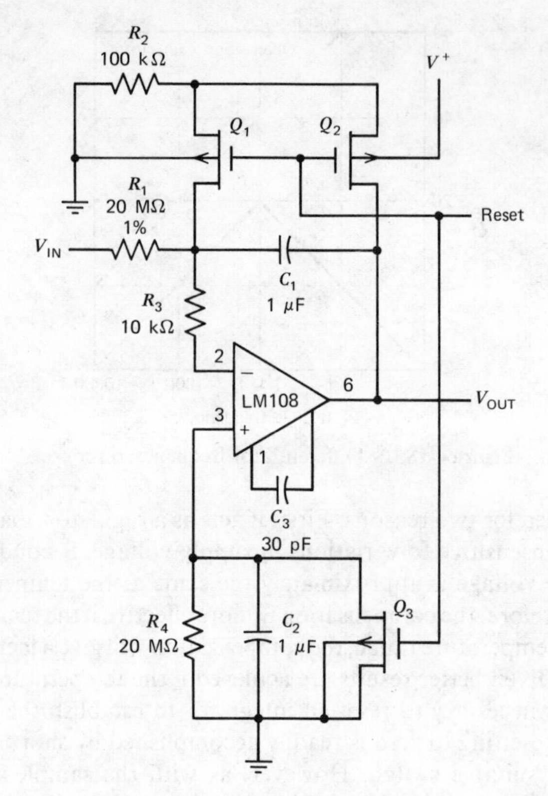

Figure 4-187. Low-drift integrator with reset [*note:* Q_1 and Q_3 should not have internal gate-protection diodes].

The additional circuitry shown in Figure 4-187 makes the error accumulation rate proportional to the offset current, rather than the bias current. Hence, the drift is reduced by roughly a factor of 10. During the integration interval, the bias current of the noninverting input accumulates an error across R_4 and C_2 just as the bias current on the inverting input does across R_1 and C_1. Therefore, if R_4 is matched with R_1 and C_2 is matched with C_1 (within about 5%) the output will drift at a rate proportional to the difference in these currents. At the end of the integration interval, Q_3 removes the compensating error accumulated on C_2 as the circuit is reset.

In applications involving large temperature changes, the circuit in Figure 4-187 gives better results than the compensation scheme in Figure 4-186—especially under worst-case conditions. If over a -55 to 125°C temperature range, the worst-case drift reduced from 3 to 0.5 mV/sec is not needed, the circuit can be simplified by eliminating R_4, C_2, and Q_3 and returning the noninverting input of the amplifier directly to ground.

In fabricating low-drift integrators, it is again necessary to use high-quality components and minimize leakage currents in the wiring. The comments made on capacitors in connection with the sample-and-hold circuits also apply here. As an additional precaution, a resistor should be used to isolate the inverting input from the integrating capacitor if it is larger than 0.05 μF.

The circuit is arranged so that the high-frequency gain characteristics are determined by A_2, while A_1 determines the dc and low-frequency characteristics. The noninverting input of A_1 is connected to the summing node through R_1; A_1 is operated as an integrator, going through unity gain at 500 Hz. Its output drives the noninverting input of A_2. The noninverting input of A_2 is also connected to the summing node through C_3, and C_3 and R_3 are chosen to roll off below 750 Hz. Hence, at frequencies above 750 Hz, the feedback path is directly around A_2, with A_1 contributing little. Below 500 Hz, however, the direct feedback path to A_2 rolls off; and the gain of A_1 is added to that of A_2.

The high-frequency amplifier, A_2, is an LM101A connected with feedforward compensation. It has a 10 MHz equivalent small-signal bandwidth, a 10 V/μsec slew rate and a 250-kHz large-signal bandwidth, so these are the high-frequency characteristics of the complete amplifier. The bias current of A_2 is isolated from the summing node by C_3. Hence, it does not contribute to the dc drift of the integrator. The inverting input of A_1 is the only dc connection to the summing junction. Therefore, the error current of the composite amplifier is equal to the bias current of A_1.

If A_2 is allowed to saturate, A_1 will then start towards saturation. If the output of A_1 gets far off zero, recovery from saturation will be slowed drastically. This can be prevented by putting Zener clamp diodes across the integrating capacitor. A suitable clamping arrangement is shown in Figure 4-188; D_1 and D_2 are included in the clamp circuit along with R_5 to keep the leakage currents of the Zeners from introducing errors. This resistor prevents damage that might occur when the supplies are abruptly shut down while the integrating capacitor is charged.

A good application for the chopper-stabilized op amp is in integrators (Figure 4-189). An op amp for use as a precision integrator should have the high-gain, low-offset voltage, low-bias current, and wide bandwidth that are all available with the chopper amplifier. The gain allows integration over eight decades of frequency. Dual slope A/D converters can now easily be made with 6-digit resolution.

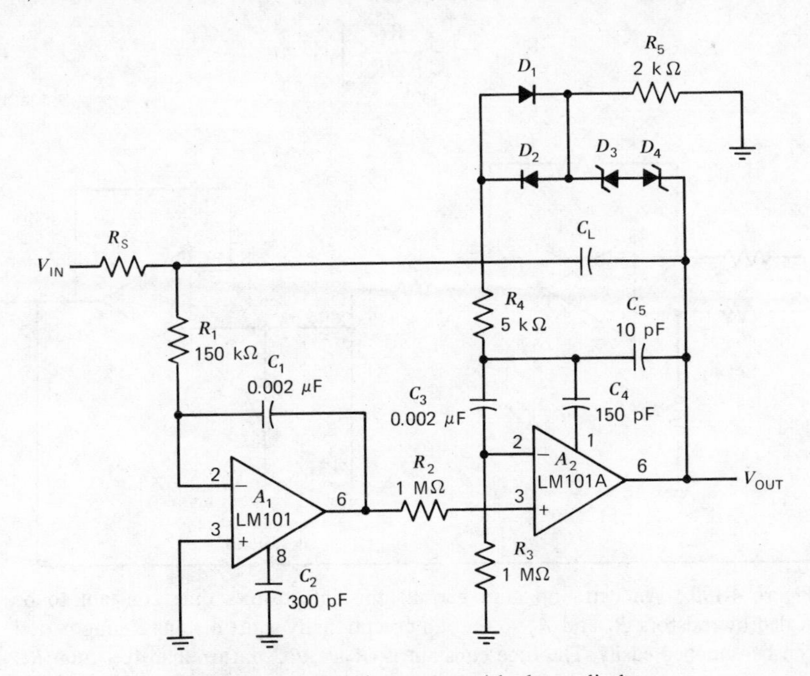

Figure 4-188. Fast integrator with clamp diodes.

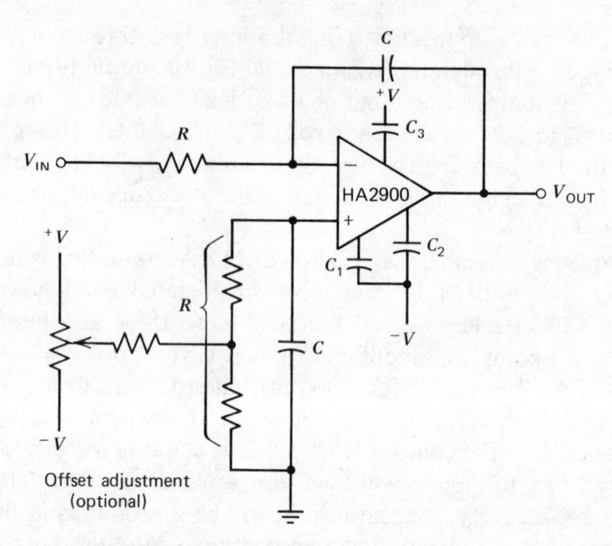

Figure 4-189. Chopper-stabilized operational amplifiers make good integrators.

A simple integrator as shown in Figures 4-9 and 4-186 through 4-189 normally consists of a single operational amplifier and an *RC* network for setting up the desired time constant. Although uncomplicated, this approach can be troublesome if either a very small or a very large time constant is needed.

The integrator in Figure 4-190, however, makes it easy to obtain either short or long timing periods because the values of the timing components are scaled by a straight resistance ratio. The integrator's output voltage is given by:

$$V_{OUT} = -\frac{R_1}{RCR_2} \int V_{IN}\, dt \qquad (4\text{-}98)$$

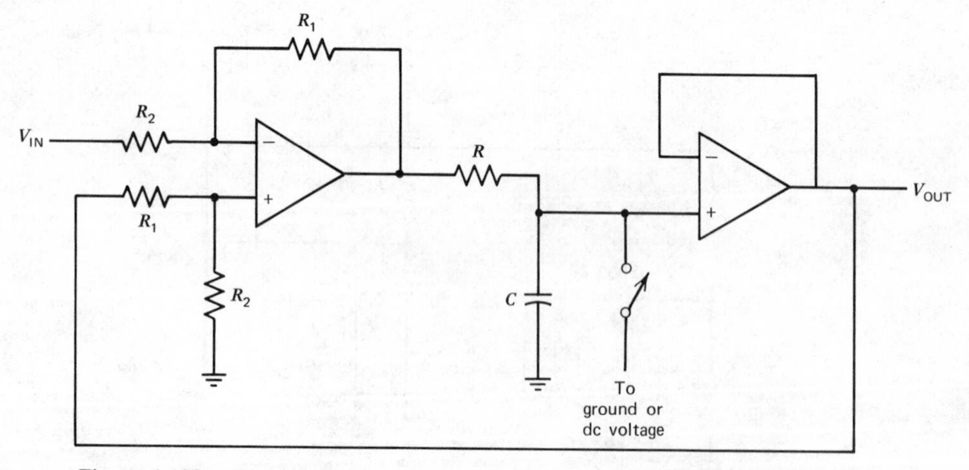

Figure 4-190. An extra op amp permits this integrator's time constant to be scaled by resistors R_1 and R_2 so that an exceptionally short or long timing period can be obtained easily. The time constant is $(R_2/R_1)RC$, rather than the usual *RC* alone.

and its time constant becomes $(R_2/R_1)RC$. The circuit provides very good linearity when precision resistors having a tolerance of $\pm 0.1\,\%$ are used for resistors R_1 and R_2.

Although a second op amp is needed to build the integrator, the circuit offers some additional advantages. For example, it permits initial conditions to be established easily. One of the capacitor's leads goes to ground, and if one end of the switch is connected either to ground or to some dc voltage, the capacitor's initial condition can be set up as either zero or otherwise by simply closing the switch.

Furthermore, when the switch is activated, the integrator's output is not shorted, and the circuit's output op amp operates as a voltage-follower. In a conventional integrator, the initial-condition switch is generally placed across the capacitor, which is in the op amp's feedback loop. With the switch closed, then, the output of a conventional integrator is shorted to the op amp's inverting input.

The integration period of the two-amplifier circuit described here can be as short as 1 ns or as long as 1000 seconds. The bandwidth of the integrator depends on which op amps are used. For high-frequency operation, National's type LM318 op amp and RCA's type CA3100 op amp are recommended.

Fast Integrators

Some integrator applications require both high speed and low-error current. The output amplifiers for photomultiplier tubes or solid-state radiation detectors are examples of this. Although the LM108 is relatively slow, there is a way to speed it up when it is used as an inverting amplifier. This is shown in Figure 4-188.

In addition to increasing speed, this circuit has other advantages. For one, it has the increased output drive capability of the LM101A. Further, thermal feedback is virtually eliminated because the LM108 does not see load variations. Lastly, the open-loop gain is nearly infinite at low frequencies as it is the product of the gains of the two amplifiers. Figures 4-191 and 4-192 show various buffering and compensation techniques for integrators.

Logarithmic Converters

One of the most predictable nonlinear elements commonly available is the bipolar transistor. The relationship between collector current and emitter-base voltage is precisely logarithmic from currents below 1 pA to currents above 1 mA. Using a matched pair of transistors and IC op amp, it is relatively easy to construct a linear to logarithmic converter with a dynamic range in excess of five decades.

The circuit in Figure 4-193 generates a logarithmic output voltage for a linear-input current. Transistor Q_1 is used as the nonlinear-feedback element around an LM108 op amp. Negative feedback is applied to the emitter of Q_1 through divider, R_1 and R_2, and the emitter-base junction of Q_2. This forces the collector current of Q_1 to be exactly equal to the current through the input resistor. Transistor Q_2 is used as the feedback element of an LM101A op amp. Negative feedback forces the collector current of Q_2 to equal the current through R_3. For the values shown, this current is 10 μA. Since the collector current of Q_2 remains constant, the emitter-base voltage also remains constant. Therefore, only the V_{BE} of Q_1 varies with a change of input current. However, the output voltage is a function of the difference in emitter-base

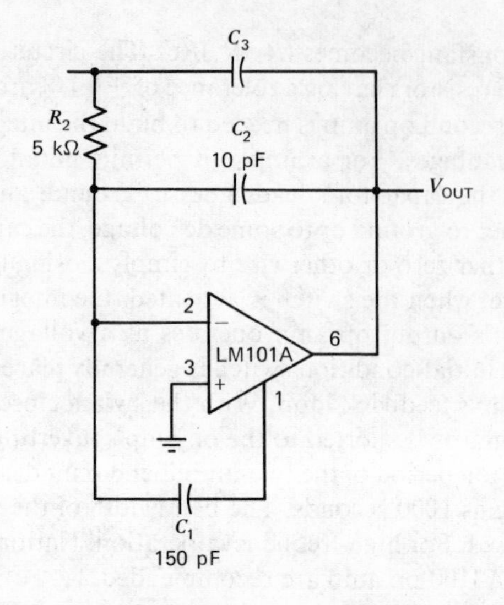

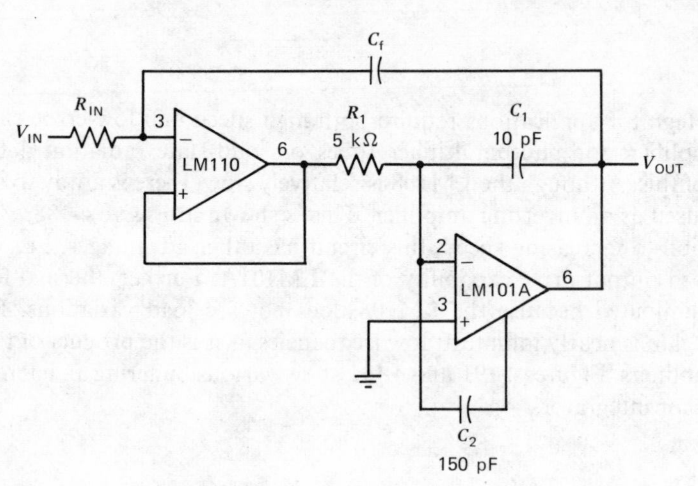

Figure 4-191. Buffer lowers the input current of a fast integrator.

voltage of Q_1 and Q_2

$$V_{OUT} = \frac{R_1 + R_2}{R_2}(V_{BE_2} - V_{BE_1}) \tag{4-99}$$

For matched transistors operating at different collector currents, the emitter-base differential is given by

$$\Delta V_{BE} = \frac{kT}{q} \log_e \frac{I_{C_1}}{I_{C_2}} \tag{4-100}$$

where k is Boltzmann's constant, T is temperature in degrees Kelvin, and q is the charge of an electron. Combining these two equations and writing the expression for

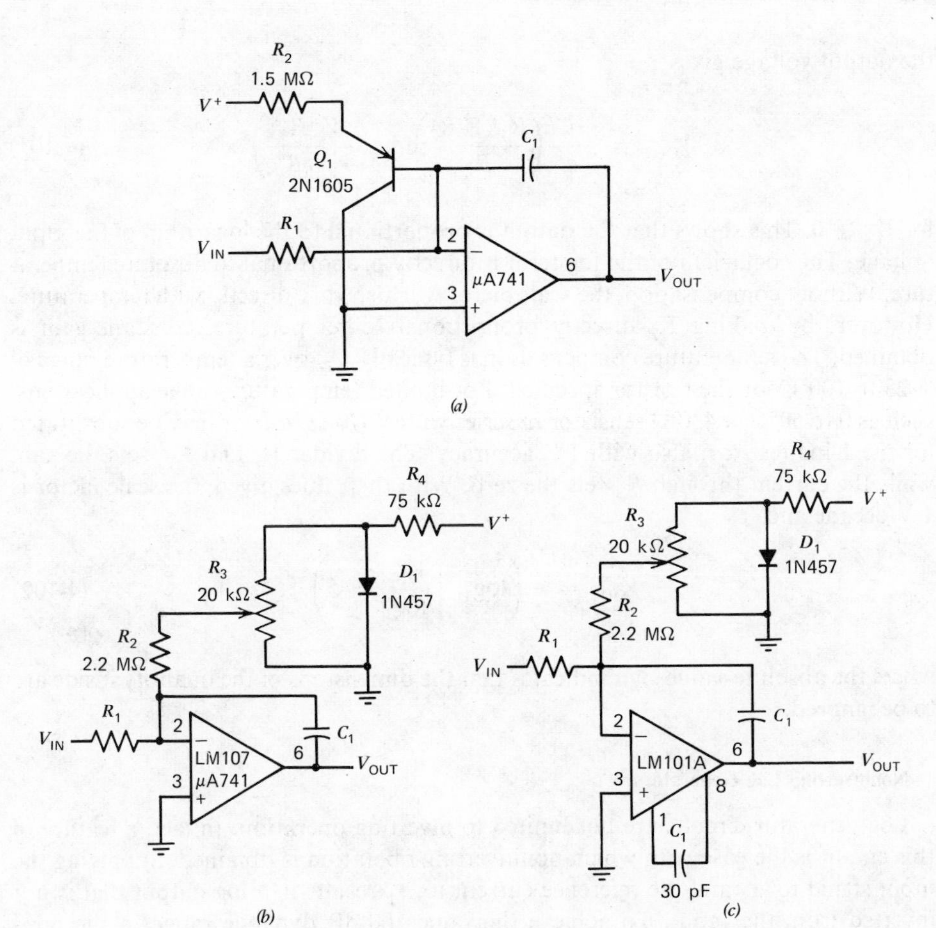

Figure 4-192. Examples of integrators with bias-current compensation.

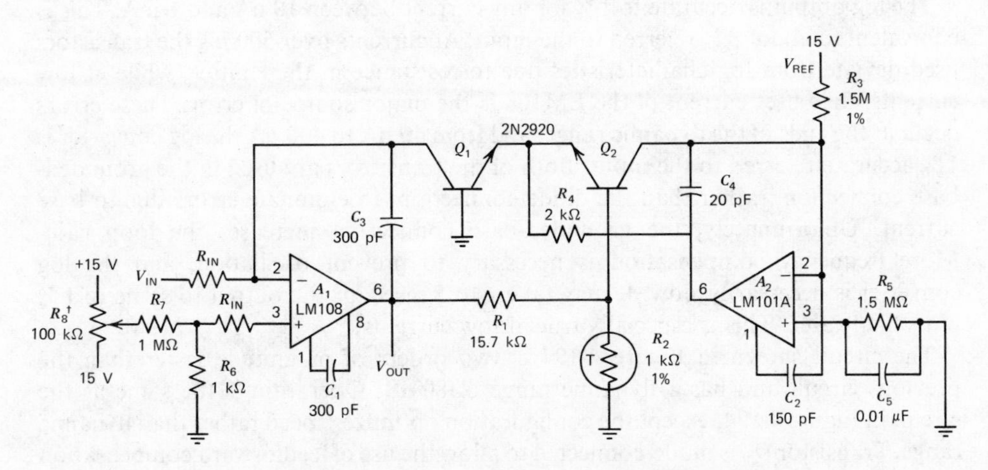

Figure 4-193. Log generator with 100-dB dynamic range.

the output voltage gives

$$V_{\mathrm{OUT}} = \frac{-kT}{q}\left(\frac{R_1 + R_2}{R_2}\right)\log_e\left(\frac{V_{\mathrm{IN}}R_3}{V_{\mathrm{REF}}R_{\mathrm{IN}}}\right) \tag{4-101}$$

for $V_{\mathrm{IN}} \geq 0$. This shows that the output is proportional to the logarithm of the input voltage. The coefficient of the log term is directly proportional to absolute temperature. Without compensation, the scale factor will also vary directly with temperature. However, by making R_2 directly proportional to temperature, constant gain is obtained. The temperature compensation is typically 1% over a temperature range of −25 to 100°C for the resistor specified. For limited temperature-range applications, such as 0 to 50°C, a 430-Ω sensistor in series with a 570-Ω resistor may be substituted for the 1-kΩ resistor, also with 1% accuracy. The divider R_1 and R_2, sets the gain while the current through R_3 sets the zero. With the values given, the scale factor is 1 V/decade and

$$V_{\mathrm{OUT}} = -\left(\log_{10}\left|\frac{V_{\mathrm{IN}}}{R_{\mathrm{IN}}}\right| + 5\right) \tag{4-102}$$

where the absolute-value sign indicates that the dimensions of the quantity inside are to be ignored.

Noninverting Log Generators

Log-generator circuits are not limited to inverting operation. In fact, a feature of this circuit is the ease with which noninverting operation is obtained. Supplying the input signal to A_2 and the reference current to A_1 results in a log output that is not inverted from the input. To achieve the same 100-dB dynamic range in the noninverting configuration, an LM108 should be used for A_2, and an LM101A for A_1. The only other change is the addition of a clamp diode connected from the emitter of Q_1 to ground. This prevents damage to the logging transistors if the input signal should go negative.

The log output is accurate to 1% for any current between 10 nA and 1 mA. This is equivalent to about 3% referred to the input. At currents over 500 μA the transistors used deviate from log characteristics due to resistance in the emitter, while at low currents, the offset current of the LM108 is the major source of error. These errors occur at the ends of the dynamic range, and from 40 nA to 400 μA the log converter is 1% accurate referred to the input. Both of the transistors are used in the grounded-base connection, rather than the diode connection, to eliminate errors due to base current. Unfortunately, the grounded-base connection increases the loop gain. More frequency compensation is necessary to prevent oscillation, and the log converter is necessarily slow. It may take 1 to 5 msec for the output to settle to 1% of its final value. This is especially true at low currents.

The circuit shown in Figure 4-194 is two orders of magnitude faster than the previous circuit and has a dynamic range of 80 dB. Operation is the same as the circuit in Figure 4-194, except the configuration optimizes speed rather than dynamic range. Transistor Q_1 is diode-connected to allow the use of feedforward compensation on an LM101A op amp. To prevent errors due to the finite h_{FE} of Q_1 and the bias

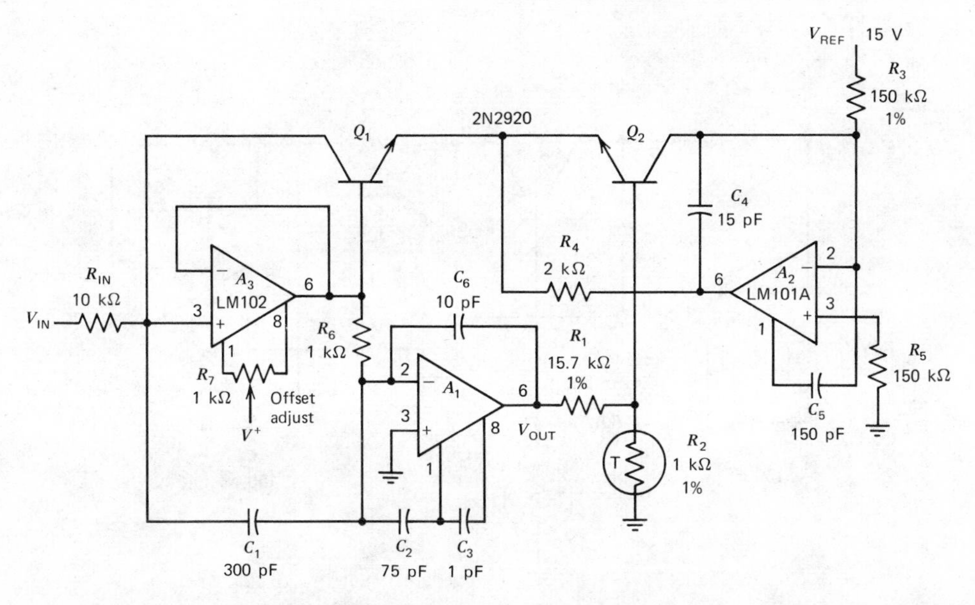

Figure 4-194. Fast-log generator [*note:* * Tel-Labs type Q_{81}, Manchester, N.H.].

current of the LM101A, the LM102 voltage-follower buffers the base current and input current. Although the log circuit will operate without the LM102, accuracy will degrade at low currents. Amplifier A_2 is also compensated for maximum bandwidth. As with the previous log converter, R_1 and R_2 control the sensitivity; and R_3 controls the zero-crossing of the transfer function. With the values shown the scale factor is 1 V/decade and

$$V_{OUT} = -\left(\log_{10}\left|\frac{V_{IN}}{R_{IN}}\right| + 4\right) \tag{4-103}$$

Antilog generator

Antilog or exponential generation is simply a matter of rearranging the circuitry. Figure 4-195 shows the circuitry of the log converter connected to generate an exponential output from a linear input. Amplifier A_1 in conjunction with transistor Q_1 drives the emitter of Q_2 in proportion to the input voltage. The collector current of Q_2 varies exponentially with the emitter-base voltage. This current is converted to a voltage by amplifier A_2. With the values given

$$V_{OUT} = 10^{-V_{IN}} \tag{4-104}$$

Many nonlinear functions such $X^{1/2}, X^2, X^3, 1/X, XY,$ and X/Y are easily generated with the use of logs. Multiplication becomes addition, division becomes subtraction and powers become gain coefficients of log terms. Figure 4-196 shows a circuit whose output is the cube of the input. Actually, any power function is available from this circuit by changing the values of R_9 and R_{10} in accordance with the expression

$$V_{OUT} = V_{IN}^{16.7R_9/R_9+R_{10}} \tag{4-105}$$

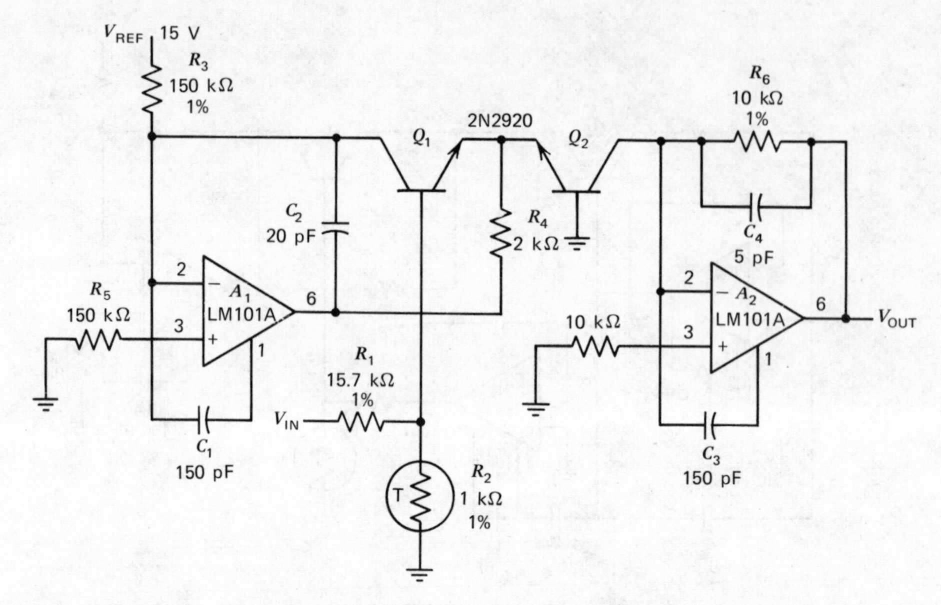

Figure 4-195. Antilog generator [*note:* * Tel-Labs type Q_{81}, Manchester, N.H.].

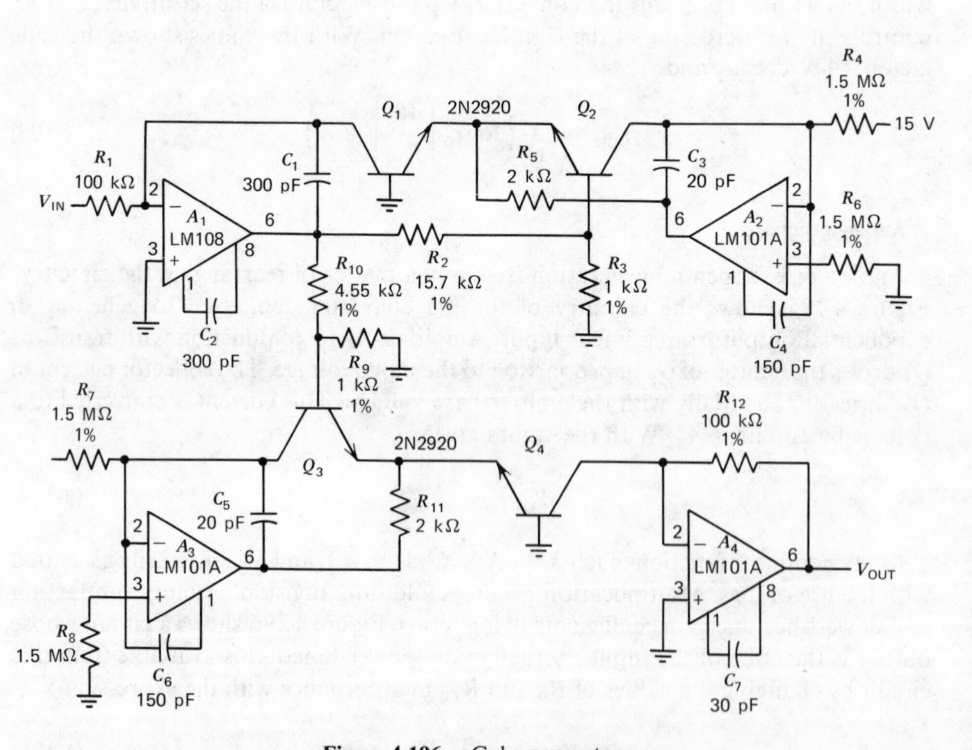

Figure 4-196. Cube generator.

Note that when log and antilog circuits are used to perform an operation with a linear output, no temperature-compensating resistors at all are needed. If the log and antilog transistors are at the same temperature, gain changes with temperature cancel. It is a good idea to use a heat sink which couples the two transistors to minimize thermal gradients. A 1°C temperature difference between the log and antilog transistors results in a 0.3% error. Also, in the log converters, a 1°C difference between the log transistors and the compensating resistor results in a 0.3% error.

Either of the circuits in Figures 4-193 or 4-194 may be used as dividers or reciprocal generators. Equation (4-105) shows the outputs of the log generators are actually the ratio of two currents: The input current and the current through R_3. When used as a log generator, the current through R_3 was held constant by connecting R_3 to a fixed voltage. Hence, the output was just the log of the input. If R_3 is driven by an input voltage, rather than the 15-V reference, the output of the log generator is the log ratio of the input current to the current through R_3. The antilog of this voltage is the quotient. Of course, if the divisor is constant, the output is the reciprocal.

Multiplier-Divider Applications

A complete one-quadrant multiplier-divider is shown in Figure 4-197. It is basically the log generator shown in Figure 4-193 driving the antilog generator shown in Figure 4-195. The log-generator output from A_1 drives the base of Q_3 with a voltage proportional to the log of V_1/V_2. Transistor Q_3 adds a voltage proportional to the log of V_3 and drives the antilog transistor, Q_4. The collector current of Q_4 is converted to an output voltage by A_4 and R_7, with the scale factor set by R_7 at $V_1V_3/10V_2$.

Measurement of transistor-current gains over a wide range of operating currents is an application particularly suited to log multiplier-dividers. Using the circuit in Figure 4-197, *pnp* current gains can be measured at currents from 0.4 μA to 1 mA.

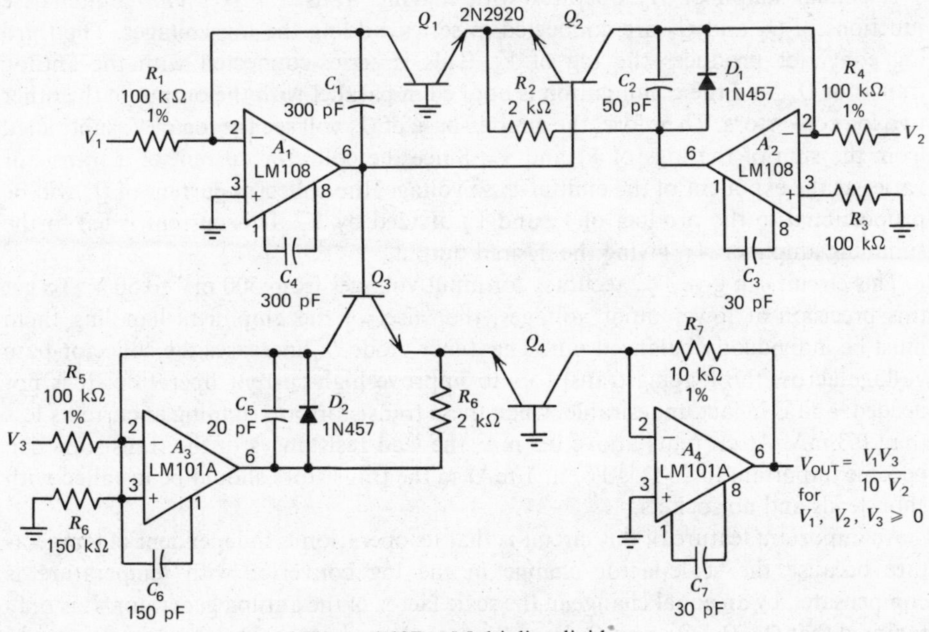

Figure 4-197. Multiplier-divider.

The collector current is the input signal to A_1, the base current is the input signal to A_2, and a fixed voltage to R_5 sets the scale factor. Since A_2 holds the base at ground, a single resistor from the emitter to the positive supply is all that is needed to establish the operating current. The output is proportional to collector current divided by base current, or h_{FE}.

To achieve wide dynamic range with high accuracy, the input op amp necessarily must have low-offset voltage, bias current, and offset current. The LM108 has a maximum bias current of 3 nA and offset current of 400 pA over a -55 to 125°C temperature range. By using equal-source resistors, only the offset current of the LM108 causes an error. The offset current of the LM108 is as low as many FET amplifiers. Further, it has a low and constant TC rather than doubling every 10°C. This results in greater accuracy over temperature than can be achieved with FET amplifiers. The offset voltage may be zeroed, if necessary, to improve accuracy with low-input voltages.

The log converters are low-level circuits and some care should be taken during construction. The input leads should be as short as possible and the input circuitry guarded against leakage currents. Solder residues can easily conduct leakage currents, therefore circuit boards should be cleaned before use. High-quality glass or mica capacitors should be used on the inputs to minimize leakage currents. Also, when the +15-V supply is used as a reference, it must be well regulated.

Another circuit which performs both multiplication and division is shown in Figure 4-198. It gives an output which is proportional to the product of two inputs divided by a third, and it is about the same complexity as a divider alone.

The circuit consists of three log converters and an antilog generator. Taking amplifier A_1, a logging transistor, Q_1, is inserted in the feedback loop such that its collector current is equal to the input voltage divided by the input resistor, R_1. Hence the emitter-base voltage of Q_1 will vary as the log of the input voltage, V_1.

A similar amplifier A_2 operates with logging transistor Q_2. The emitter-base junctions of Q_1 and Q_2 are connected in series, adding the log voltages. The third log converter produces the log of V_3. This is series-connected with the antilog transistor Q_4, and the combination is hooked in parallel with the output of the other two log convertors. Therefore, the emitter-base of Q_4 will see the log of V_3 subtracted from the sum of the logs of V_1 and V_2. Since the collector current of a transistor varies as the exponent of the emitter-base voltage, the collector current of Q_4 will be proportional to the product of V_1 and V_2 divided by V_3. This current is fed to the summing amplifier A_4, giving the desired output.

This circuit can give 1% accuracy for input voltages from 500 mV to 50 V. To get this precision at lower input voltages, the offset of the amplifiers handling them must be individually balanced out. The Zener diode D_4 increases the collector-base voltage across the logging transistors to improve high-current operation. It is not needed, and is in fact undesirable, when these transistors are running at currents less than 0.3 mA. At currents above 0.3 mA, the lead resistances of the transistors can become important (0.25 Ω is 1% at 1 mA) so the transistors should be installed with short leads and no sockets.

An important feature of this circuit is that its operation is independent of temperature because the scale-factor change in the log converter with temperature is compensated by an equal change in the scale factor of the antilog generator. It is only required that Q_1, Q_2, Q_3, and Q_4 be at the same temperature. Dual transistors should

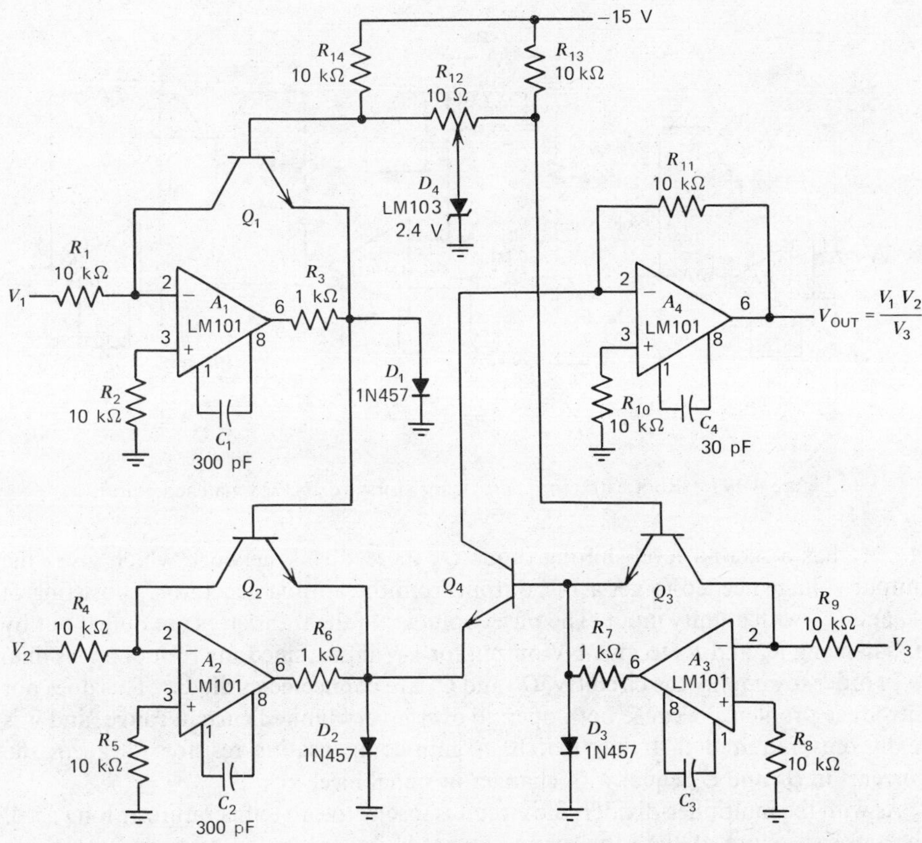

Figure 4-198. Precision analog multiplier-divider [*note:* transistors are 2N3728 matched pairs].

be used and arranged as shown in the figure so that thermal mismatches between cans appear as inaccuracies in scale factor (0.3 %/°C) rather than a balance error (8 %/°C). A balance potentiometer R_{12} nulls out the offset voltages of all the logging transistors; it is adjusted by setting all input voltages equal to 2-V and adjusting for a 2-V output voltage.

The logging transistors provide a gain which is dependent on their operating level, which complicates frequency compensation. Resistors (R_3, R_6, and R_7) are put in the amplifier output to limit the maximum-loop gain, and the compensation capacitor is chosen to correspond with this gain. Finally, clamp diodes D_1 through D_3 prevent exceeding the maximum reverse emitter-base voltage of the logging transistors with negative inputs.

Root Extractor

Taking the root of a number using log converters is a fairly simple matter. All that is needed is to take the log of a voltage, divide it by, say 1/2 for the square root, and then take the antilog. A circuit which accomplishes this is shown in Figure 4-199; A_1 and Q_1 form the log converter for the input signal. This feeds Q_2 which produces a level-shift to give zero voltage into the R_4, R_5 divider for a 1-V input. This divider reduces the log voltage by the ratio for the root desired and drives the buffer amplifier

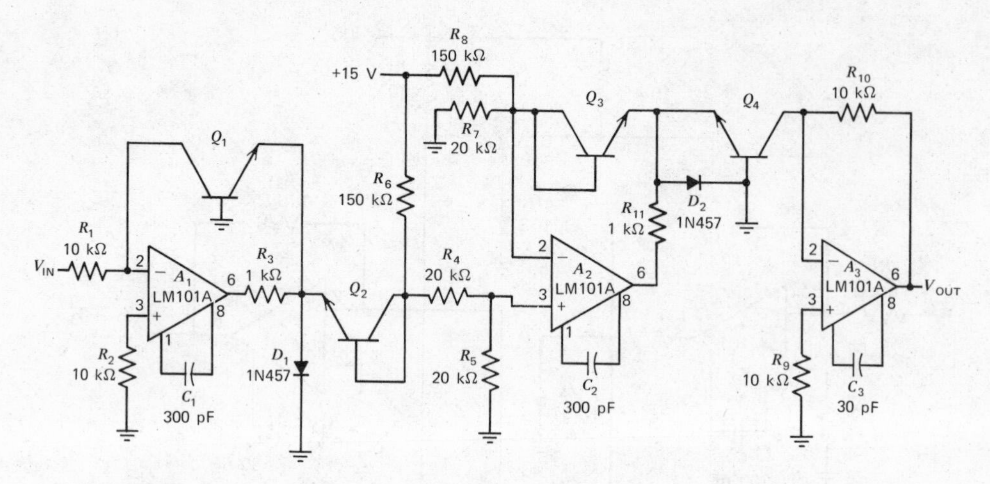

Figure 4-199. Root extractor [*note:* transistors are 2N3728 matched pairs].

A_2; A_2 has a second level-shifting diode Q_3 its feedback network which gives the output voltage needed to get a 1-V output from the antilog generator, consisting of A_3 and Q_4, with a unity input. The offset voltages of the transistors are nulled out by imbalancing R_6 and R_8 to give 1-V output for 1-V input, since any root of one is one.

In order to simplify the circuitry, Q_2 and Q_3 are connected as diodes. This does not introduce problems because both operate over a very limited current range, and it is really only required that they match. A gain-compensating resistor R_7 keeps the currents in Q_2 and Q_3 equal with changes in signal level.

As with the multiplier-divider, the circuit is insensitive to temperature as long as all the transistors are at the same temperature. Using transistor pairs and matching them as shown minimizes the effects of gradients.

Simpler Analog Multiplier

A simple embodiment of the analog multiplier is shown in Figure 4-200. This circuit circumvents many of the problems associated with the log-antilog circuit and provides three-quadrant analog multiplication which is relatively temperature-insensitive and which is not subject to the bias-current errors which plague most multipliers.

Circuit operation may be understood by considering A_2 as a controlled-gain amplifier amplifying V_2, whose gain is dependent on the ratio of the resistance of PC_2 to R_5 and by considering A_1 as a control amplifier which establishes the resistance of PC_2 as a function of V_1. In this way it is seen that V_{OUT} is a function of both V_1 and V_2.

The control amplifier A_1 provides drive for the lamp, L_1. When an input voltage V_1 is present, L_1 is driven by A_1 until the current to the summing junction from the negative supply through PC_1 is equal to the current to the summing junction from V_1 through R_1. Since the negative-supply voltage is fixed, this forces the resistance of PC_1 to a value proportional to R_1 and to the ratio of V_1 to V^-; L_1 also illuminated PC_2 and if the photoconductors are matched, causes PC_2 to have a resistance equal to PC_1.

The controlled-gain amplifier A_2 acts as an inverting amplifier whose gain is equal

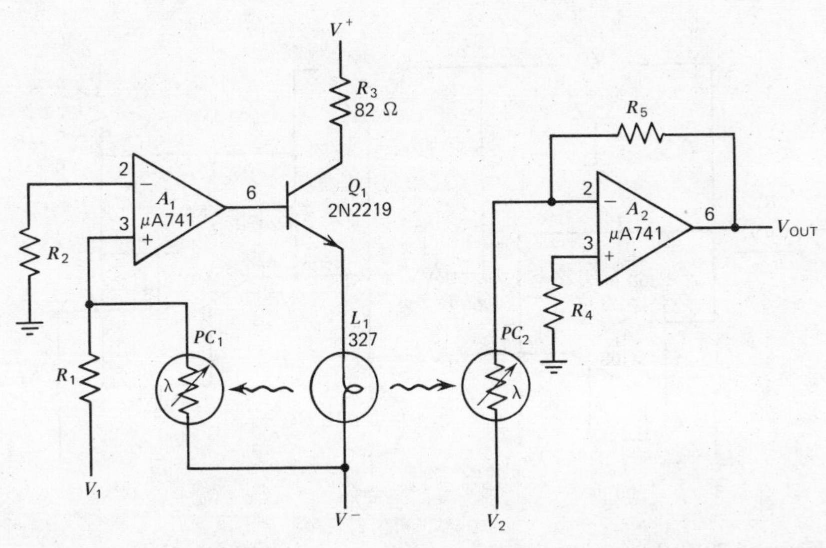

Figure 4-200. Simplified analog multiplier; LM107 may be used in same circuit [*note:* $R_5 = R_1(V^-/10)$, $V_1 > 0$, $V_{OUT} = V_1V_2/10$].

to the ratio of the resistance of PC_2 to R_5. If R_5 is chosen equal to the product of R_1 and V^-, then V_{OUT} becomes simply the product of V_1 and V_2; R_5 may be scaled in powers of 10 to provide any required output scale factor.

For best tracking over temperature, PC_1 and PC_2 should be matched since the TC of resistance is related to resistance-match for cells of the same geometry. Small mismatches may be compensated by varying the value of R_5 as a scale-factor adjustment. The photoconductive cells should receive equal illumination from L_1, a convenient method is to mount the cells in holes in an aluminum block and to mount the lamp midway between them. This mounting method provides controlled spacing and also provides a thermal bridge between the two cells to reduce differences in cell temperature. This technique may be extended to the use of FETs or other devices to meet special resistance or environment requirements.

The circuit as shown gives an inverting output whose magnitude is equal to one-tenth the product of the two analog inputs. Input V_1 is restricted to positive values, but V_2 may assume both positive and negative values. This circuit is restricted to low-frequency operation by the lamp-time constant.

Resistors R_2 and R_4 are chosen to minimize errors due to input offset current as outlined in the section describing the photocell amplifier; R_3 is included to reduce in-rush current when first turning on the lamp L_1.

Temperature Compensation of Logarithmic Converters

When transistors are used as logging elements, the scale factor has a temperature sensitivity of 0.3 %/°C as noted in the preceding sections. The importance of compensating for this characteristic to make log converters accurate over a wide temperature range should be stressed. Therefore, points previously made will be summarized here, using a version of the one-quadrant converter as an example.

In Figure 4-201, Q_1 is the logging transistor, while Q_2 provides a fixed offset to temperature compensate the emitter-base turn on voltage of Q_1. Transistor Q_2 is

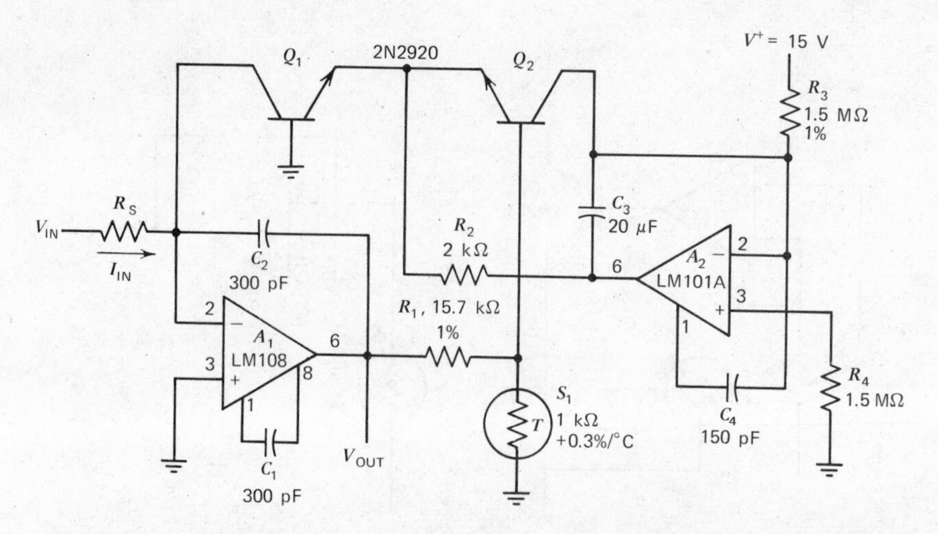

Figure 4-201. Temperature-compensated one-quadrant logarithmic converter [*note:* 10 nA $\leq I_{IN} \leq$ 1 mA].

operated at a fixed collector current of 10 μA by A_2, and its emitter-base voltage is subtracted from that of Q_1 in determining the output voltage of the circuit. The collector current of Q_2 is established by R_3 and V^+ through A_2.

The collector current of Q_1 is proportional to the input current through R_S and, therefore, proportional to the input voltage. The emitter-base voltage of Q_1 varies as the log of the input voltage. The fixed emitter-base voltage of Q_2 subtracts from the voltage on the emitter of Q_1 in determining the voltage on the top end of the temperature-compensating resistor, S_1.

The signal on the top of S_1 will be zero when the input current is equal to the current through R_3 at any temperature. Further, this voltage will vary logarithmically for changes in input current, although the scale factor will have a TC of $-0.3\%/°C$. The output of the converter is essentially multiplied by the ratio of R_1 to S_1. Since S_1 has a positive TC of 0.3%/°C, it compensates for the change in scale factor with temperature.

In this circuit, an LM101A with feedforward compensation is used for A_2 since it is much faster than the LM108 used for A_1. Since both amplifiers are cascaded in the overall feedback loop, the reduced phase-shift through A_2 ensures stability.

Certain things must be considered in designing this circuit. For one, the sensitivity can be changed by varying R_1. But R_1 must be made considerably larger than the resistance of S_1 for effective TC of the scale factor; Q_1 and Q_2 should also be matched devices in the same package, and S_1 should be at the same temperature as these transistors. Accuracy for low-input currents is determined by the error caused by the bias current of A_1. At high currents, the behavior of Q_1 and Q_2 limits accuracy. For input currents approaching 1 mA, the 2N2920 develops logging errors in excess of 1%. If larger input currents are anticipated, bigger transistors must be used; and R_2 should be reduced to ensure that A_2 does not saturate. Similar techniques are used to temperature compensate the two-quadrant multiplier in Figure 4-202.

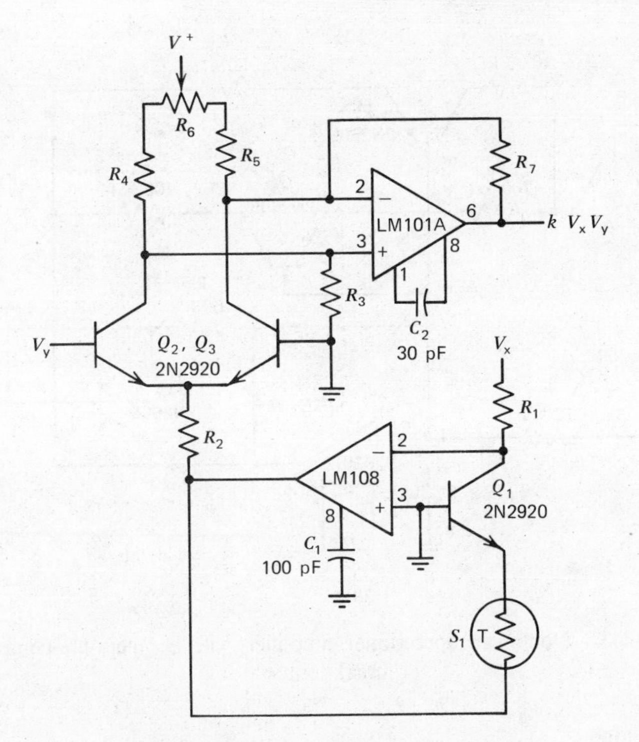

Figure 4-202. Two-quadrant multiplier.

Diode-Breakpoint Nonlinear Amplifiers

When a nonlinear transfer function is needed from an op amp, many methods of obtaining it present themselves. However, they usually require diodes and are therefore difficult to temperature compensate for accurate breakpoints. One way of getting around this is to make the output swing so large that the diode threshold is negligible by comparison, but this is not always practical.

A method of producing very sharp, temperature-stable breakpoints in the transfer function of an op amp is shown in Figure 4-203. For small-input signals, the gain is determined by R_1 and R_2. Both Q_2 and Q_3 are conducting to some degree, but they do not affect the gain because their current gain is high and they do not feed any appreciable current back into the summing node. When the output voltage rises to 2 V (determined by R_3, R_4, and V^-), Q_3 draws enough current to saturate, connecting R_4 in parallel with R_2. This cuts the gain in half. Similarly when the output voltage rises to 4 V, Q_2 will saturate, again halving the gain.

Temperature compensation is achieved in this circuit by including Q_1 and Q_4; Q_4 compensates the emitter-base voltage of Q_2 and Q_3 to keep the voltage across the feedback resistors R_4 and R_6 very nearly equal to the output voltage while Q_1 compensates for the emitter-base voltage of these transistors as they go into saturation, making the voltage across R_3 and R_5 equal to the negative-supply voltage. A detrimental effect of Q_4 is that it causes the output resistance of the amplifier to increase at high-output levels. It may therefore be necessary to use an output buffer if the circuit must drive an appreciable load.

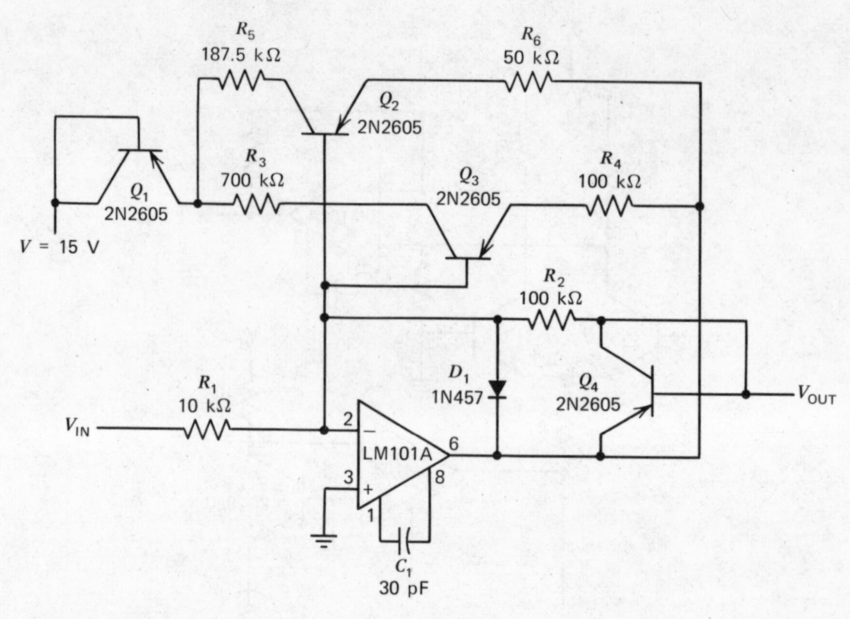

Figure 4-203. Nonlinear operational amplifier with temperature-compensated breakpoints.

Servo Preamplifier

In certain servo systems, it is desirable to get the rate signal required for loop stability from some sort of electrical-lead network. This can, for example, be accomplished with reactive elements in the feedback network of the servo preamplifier.

Many saturating servo amplifiers operate over an extremely wide dynamic range. For example, the maximum error signal could easily be 1000 times the signal required to saturate the system. Cases like this create problems with electrical-rate networks because they cannot be placed in any part of the system which saturates. If the signal into the rate network saturates, a rate signal will only be developed over a narrow

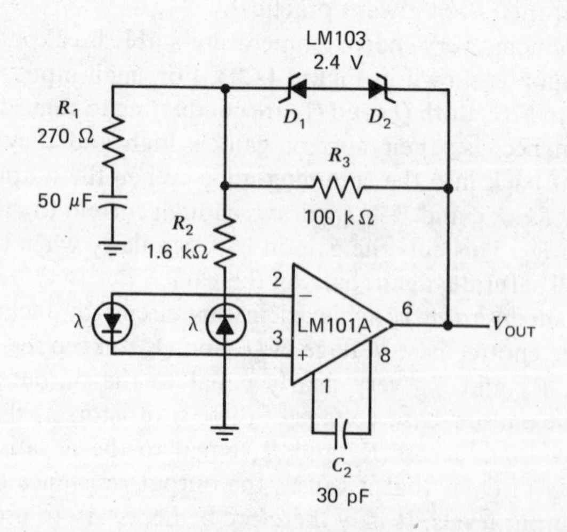

Figure 4-204. Saturating servo preamplifier with rate feedback.

range of system operation, and instability will result when the error becomes large. Attempts to place the rate networks in front of the error amplifier or make the error amplifier linear over the entire range of error signals frequently gives rise to excessive dc error from signal attenuation.

These problems can be largely overcome using the kind of circuit shown in Figure 4-204. This amplifier operates in the linear mode until the output voltage reaches approximately 3 V with 30-μA output current from the solar-cell sensors. At this point the breakdown diodes in the feedback loop begin to conduct, drastically reducing the gain. However, a rate signal will still be developed because current is being fed back into the rate network (R_1, R_2, and C_1) just as it would if the amplifier had remained in the linear-operating region. In fact, the amplifier will not actually saturate until the error current reaches 6 mA, which would be the same as having a linear amplifier with a $\pm$600-V output swing.

OPERATIONAL AMPLIFIER VOLTAGE-REGULATORS

Basic Circuits

The use of an op amp as the precision comparing element in a voltage-regulator is widely known. As in Figure 4-205, a basic regulator circuit employs an op amp to compare a reference voltage with a fraction of the output voltage and to control a series-pass element Q_1 to regulate the output: Q_2 is used to increase the load-current capability of the regulator.

A typical circuit, with variable output voltage, is shown in Figure 4-206. The purpose of the amplifier is to isolate the voltage reference, a Zener diode in Figure 4-206, from changes in loading at the supply output. This results in lower supply-output impedance and hence improved load regulation. Also, because of the high gain of the μA741, the voltage applied to the inverting input of the amplifier from the output-voltage divider is always maintained within a few milivolts of the reference.

Thus, the output voltage may be varied by changing the division ratio of the divider. The 800-kΩ input impedance of the μA741 keeps loading of the reference Zener to a

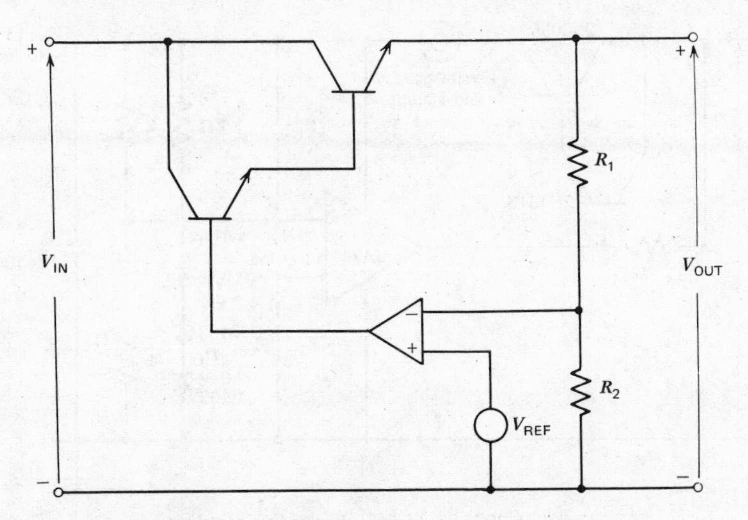

Figure 4-205. Basic operational amplifier regulator circuit.

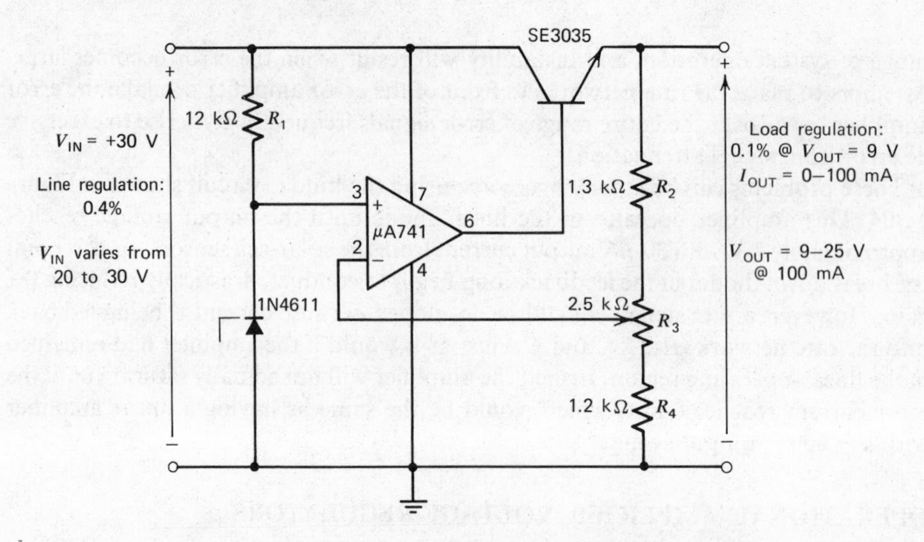

Figure 4-206. Variable voltage-regulated power supply.

minimum. The output impedance of the circuit shown is less than 0.1 Ω and the line regulation is approximately 0.4% for input voltages varying from 20 to 30 V.

Another example of a typical dissipative regulator using an IC as the error amplifier appears in Figure 4-207. The LM107 has built-in output current limiting as protection against burning out. At high-output voltages, however, Q_2 must be chosen so that the current it can draw is held to within the maximum rating of the amplifier. The LM107 op amp also has internal frequency compensation. Capacitor C_4, typically rated in tens of microfarads, will supply the high-frequency components of load demand from its stored energy.

A low-frequency power transistor is chosen as Q_1, using the manufacturer's "safe-area" specification curves to match the regulator's maximum current, voltage, and wattage needs. By choosing a transistor with f_t above 100 MHz for Q_2, the total

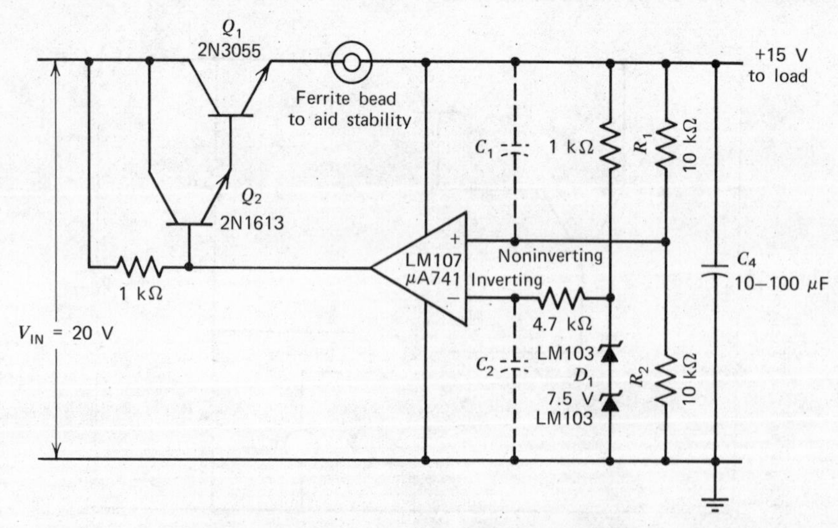

Figure 4-207. A typical dissipative regulator.

phase-shift contribution of the Darlington pair is minimized at gain cross-over, thereby lessening the closed-loop stability problem. Also, since emitter-followers may oscillate because of the effects of inductance in the base circuit, the base-lead lengths should be kept short and, if necessary, a small powdered magnetic bead or toroid can be added to the emitter circuit.

Voltage-dividers R_1 and R_2 form a sampling network by selecting a fraction of the output voltage for comparison with the reference voltage. This fractional load-voltage sample can be adjusted by a trimmer; the sample appears at the noninverting op amp terminal. Stable resistors must be used in the divider—the TC of the voltage-divider is as important as that of the Zener diode D_1. The dashed lines in Figure 4-207 indicate optional circuitry—C_2 for noise suppression, C_1 to aid in loop stabilization, and C_1, R_3, and C_3 to minimize load-voltage overshoot when driving a fast-switching load.

At its input terminals, the op amp is a differential amplifier operated at an emitter current of about 15 μA in order to minimize noise and base current and to ensure high-input impedance. The impedance of the sampling network is a compromise: The current drawn must not be excessive—1 mA of "bleeder" current is typically acceptable. But at the same time, it must be large enough to swamp out any rise in op amp input current with a rise in temperature, thus limiting the regulator's temperature sensitivity. The inverting terminal has a resistance R_4, which is approximately equal to R_1 in parallel with R_2, inserted in series for the same purpose. Other elements are C_2, a noise-filter capacitor, typically 0.01 μF, and C_1, a phase-lead-compensating capacitor, usually between 5 and 20 pF.

Short-circuit load protection for the series-regulating element can be added as fold-back current-limiting, a technique in which load voltage is sampled to sense overcurrent and the series-pass transistor is turned OFF.

Switching Regulator[9]

The switching regulator was primarily developed for use in the aerospace industry because of its higher efficiency, cooler operation, and lower associated weight and size than a conventional linear regulator. For example, for a 100-W output, the dissipation of the switching regulator is 40 W as compared with 200 W for the linear series regulator. The main size reduction stems from operating the switching regulator at between 10 and 25 kHz rather than at 50 or 60 Hz since transformer size is inversely proportional to frequency. However, the switching regulator contains more active components, it costs more, and, most importantly, it generates noise by virtue of its squarewave ON–OFF operation. Recently, the switching regulator has come of age for computer, calculator, CRT and instrumentation applications.

The basic configuration of a switching regulator as shown in Figure 4-208 contains a switch, a comparator, a voltage reference, and a filter. Regulation is achieved by controlling the duty cycle of Q_1 by an operational amplifier that compares the output voltage with a reference voltage. ON time is increased in proportion to OFF time to raise the output level or decreased to lower the output level.

Transistor Q_1 is either ON (saturated) or OFF, so that power dissipation in it is minimum. Freewheeling diode D_1 conducts during the time that Q_1 is cut off, thus maintaining current flow through inductor L_1. It also limits the induced voltage

[9] E. R. Hnatek, "Use Integrated Circuits in Transformerless dc-to-dc Converters," *EDN*, Feb. 5, 1973.

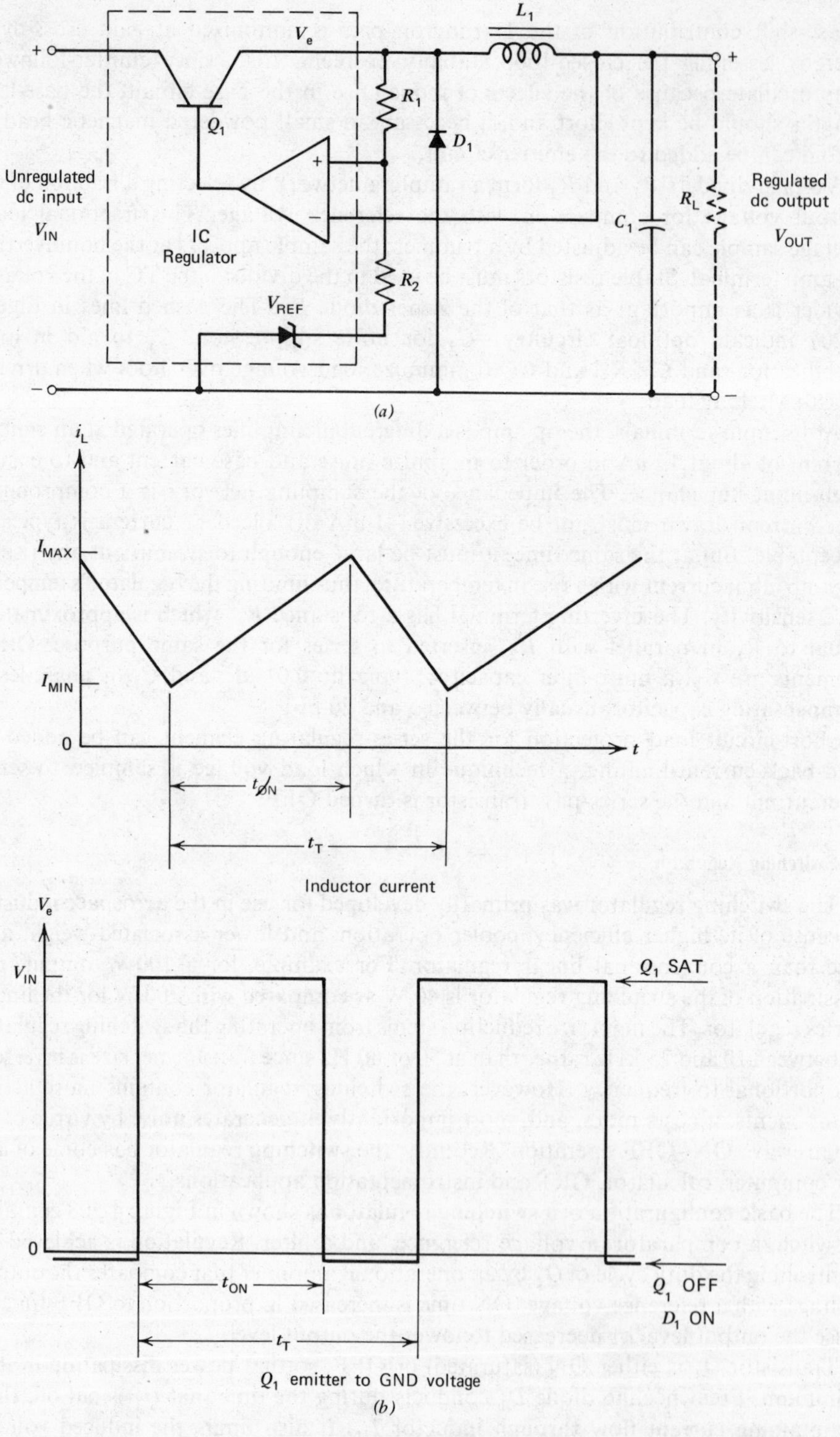

Figure 4-208. Basic switching regulator: (*a*) basic switching regulator schematic; (*b*) switching regulator waveforms.

$[L_1(di/dt)]$ whenever the transistor switch is turned OFF. When Q_1 in ON, the load current i_L through L_1 increases according to the following relationship:

$$V_{IN} - V_{OUT} = L_1\left(\frac{\Delta i_L}{t_{ON}}\right) \tag{4-106}$$

This current flows through the load and charges capacitor C_1.

When V_{OUT} reaches V_{REF}, the voltage comparator turns Q_1 OFF. The current through L_1 then decreases until D_1 is forward biased. At this point the inductor current flows through D_1 decreasing at a rate given by

$$V_{OUT} = L_2\left(\frac{\Delta i_L}{t_{OFF}}\right) \tag{4-107}$$

When the inductor current falls below the load current, C_1 begins to discharge and V_{OUT} decreases. When V_{OUT} decreases to slightly less than V_{REF}, the comparator turns Q_1 back ON and the cycle repeats itself. The output voltage is given by

$$V_{OUT} = \frac{V_{IN} \times t_{ON}}{t_{ON} + t_{OFF}} \tag{4-108}$$

In the circuit of Figure 4-208, both Q_1 and D_1 must be capable of switching fast to minimize losses. Losses occur primarily during transitions between saturation and cutoff, when semiconductor devices are resistive.

However, fast switching increases the amount of noise that must be controlled, so this must be traded off against EMI (electromagnetic interference) requirements. Also, it is necessary to return the catch diode (and any bypass capacitance on the unregulated input) to ground separately from other parts of the circuit. The diode carries large current transients, and large voltage transients can develop across even a short length of wire.

The core of L_1 should have soft saturation characteristics. Cores that saturate abruptly produce excessive peak currents in the switching transistor if the output current goes high enough to run the core close to saturation. A material that has a gradual reduction in permeability with excessive current, yet retains excellent high-frequency characteristics (such as powdered molybdenum-permalloy), is highly desirable.

The filter capacitor ripple rating is extremely important since the ripple current through the regulator filter capacitor will be quite high. Thus, the capacitors should have both low dissipation factors and voltage ratings that are higher than the dc circuit voltages.

Conventional switching regulators are capable of converting from a high-input dc voltage to a lower output voltage at high efficiency. A problem develops, however, if a higher output voltage than the input is desired. Additionally, if an output voltage with opposite polarity to the input voltage is desired, the design gets complicated. Traditionally, both requirements have been satisfied by the use of a transformer. This can be avoided with the type of switching regulator conversion circuit shown in Figure 4-209. It is a switching regulator circuit for a +5 to −15 V converter of the type that has application as a power supply for MOS memories in a logic system where only +5 V is available.

The method by which the regulator of Figure 4-209 generates the opposite polarity

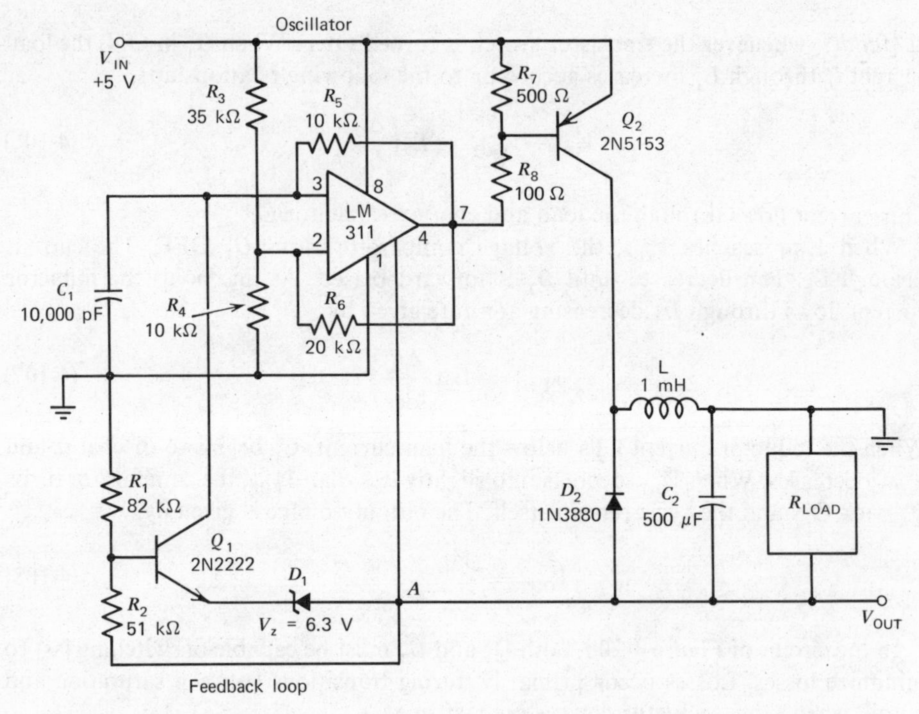

Figure 4-209. Switching regulator dc-to-dc conversion circuit converts a +5 V input to a −15 V output [*note:* $\eta = P_{OUT}/P_{IN} = 75\%$, $F = 6$ kHz 80% duty, $V_{RIPPLE} = 100$ mV @ 200 mA out, $I_L = 200$ mA max, $V_{OUT} = -15$ V, $V_{OUT} = (V_Z + V_{BE})(R_1/R_2 + 1)$]. (Courtesy *EDN*.)

is as follows (see simplified diagram of Figure 4-210). Transistor Q_2 is turned ON and OFF with a given duty cycle. If the base drive is sufficient, the voltage across the inductor is equal to the supply voltage minus V_{SAT}. When the transistor turns OFF, the inductor current has a path through the diode D_2 which builds up a negative voltage across R_L. Capacitor C_2 prevents the voltage from dropping to zero during the transistor ON time.

Referring to the entire circuit, Figure 4-209, the LM311 voltage comparator is chosen because it operates from a single 5-V supply and has high-current capability. An op amp such as the LM301A could be used, but in that case another transistor would be required in conjunction with Q_2, and this would reduce the circuit's efficiency.

The LM311 operates as a high-duty cycle, free-running oscillator. A high-duty cycle is required to start the oscillator (for high current and voltage). Then, once oscillating, the regulator takes over at an approximately 50% duty cycle. The high-duty cycle is also necessary to get the required output voltage. However, with a high-duty cycle, the oscillator has the potential of not starting—that is, stalling in a dc mode. If it stalls for a few seconds, Q_2 gets "wiped out"—there is a dc path to ground due to excessive current flow.

A large amount of dc negative feedback along with R_4 is used to set the output of the LM311 in the center of its active region to ensure starting and reasonable symmetry in the output waveform. Capacitor C_1 reduces the negative feedback at

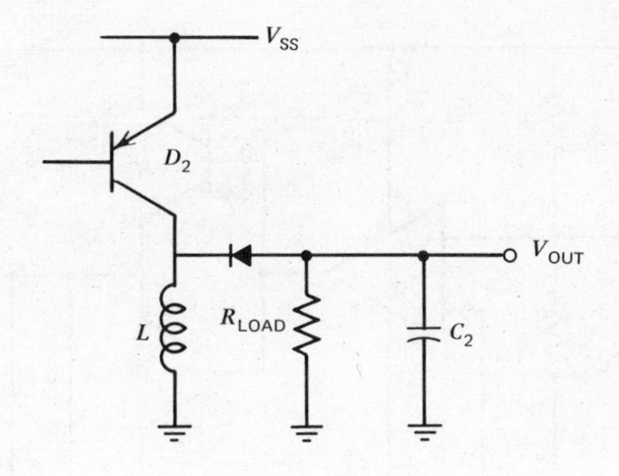

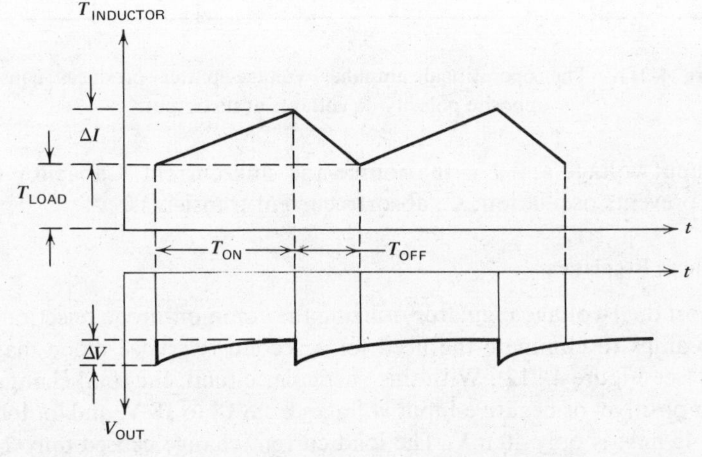

Figure 4-210. The negative-output voltage is developed when Q_2 cuts off and the field around inductor L drives current to ground through R_1 [*note:* $\Delta I = (V_{IN} - V_{SAT})/L \times T_{ON} \approx (V_{IN}/L)T_{ON}$; assuming a large capacitor, this equation becomes: $\Delta I = (V_{OUT} - V_D)/L \times T_{OFF} \approx (V_{OUT}/L)T_{OFF}$; $\Delta V = I_{LOAD}T_{ON}/C_2$]. (Courtesy *EDN*.)

high frequencies, giving net positive feedback and oscillation. A feedback loop is used to regulate the oscillator's duty cycle. The voltage that is sensed at point A is used by means of voltage divider R_1—R_2 to regulate the current through Q_1 and thus control the duty cycle. In the circuit, R_5 provides positive feedback; R_3 and R_4 set the duty cycle; R_3, R_4, and R_6 provide the hysteresis; and D_2 provides drive voltage.

Operational Amplifier Voltage-Splitter

In some instances split voltage is desired when a single-supply voltage is available. A circuit that performs this function is shown in Figure 4-211. Resistors R_1 and R_2 divide the voltage on the floating input. The control circuit essentially is a low-output impedance follower with complementary current boosters in the feedback loop. Current flows into or out of ground when the outputs are unbalanced; the differences in positive and negative feedback drive the outputs back toward balance.

Good balance is maintained because small output variations result from a large current flow. Transistors Q_1 and Q_2 must dissipate power equal to $1/2(IV)$, where

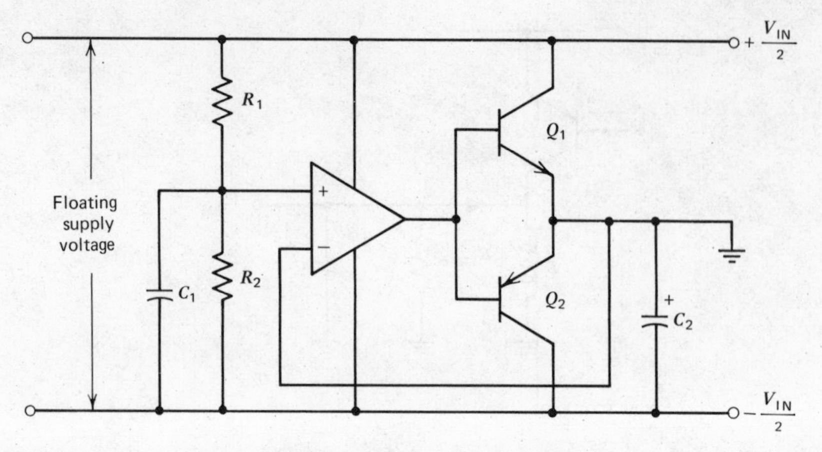

Figure 4-211. The operational amplifier voltage-splitter produces equal but opposite polarity dc voltages at its output.

V is the input voltage and I is the source-and-sink current. Capacitor C_1 reduces noise and prevents oscillation; C_2 absorbs current transients.

Dual-Voltage Regulators

A low-cost dual-voltage regulator utilizing the common-mode rejection properties of two op amps to eliminate the need for a second reference diode may be easily fabricated (see Figure 4-212). With this particular circuit, the total change in output voltage for positive- or negative-input voltages from 14 to 18 V, and for load currents from 0 to 45 mA, is only 40 mV. The load current change caused only 2 mV of the 40-mV output-voltage change.

A variation of the dual-voltage regulator[10] of Figure 4-212 is shown in Figure

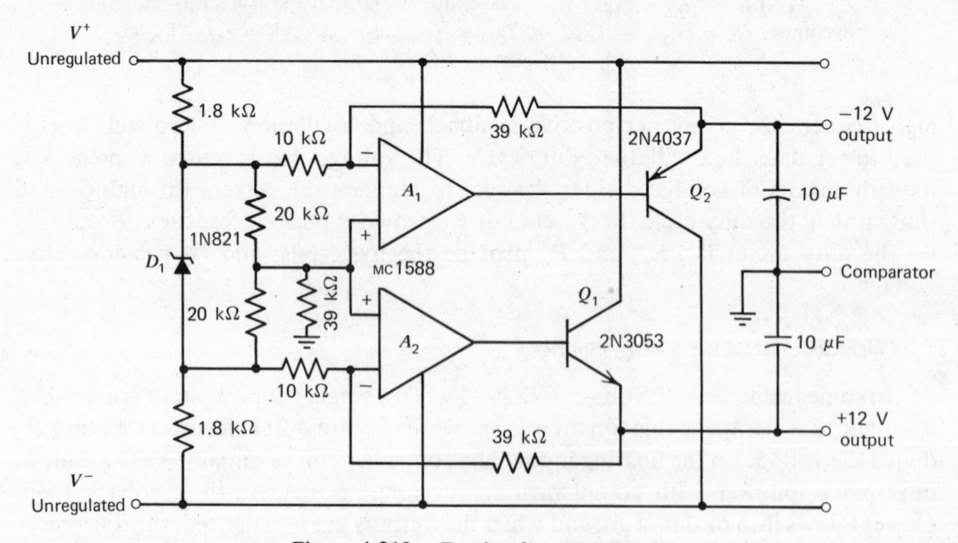

Figure 4-212. Dual-voltage regulator.

[10] NASA Tech. Brief B72-10097, *Precision Voltage Regulator*, 1972.

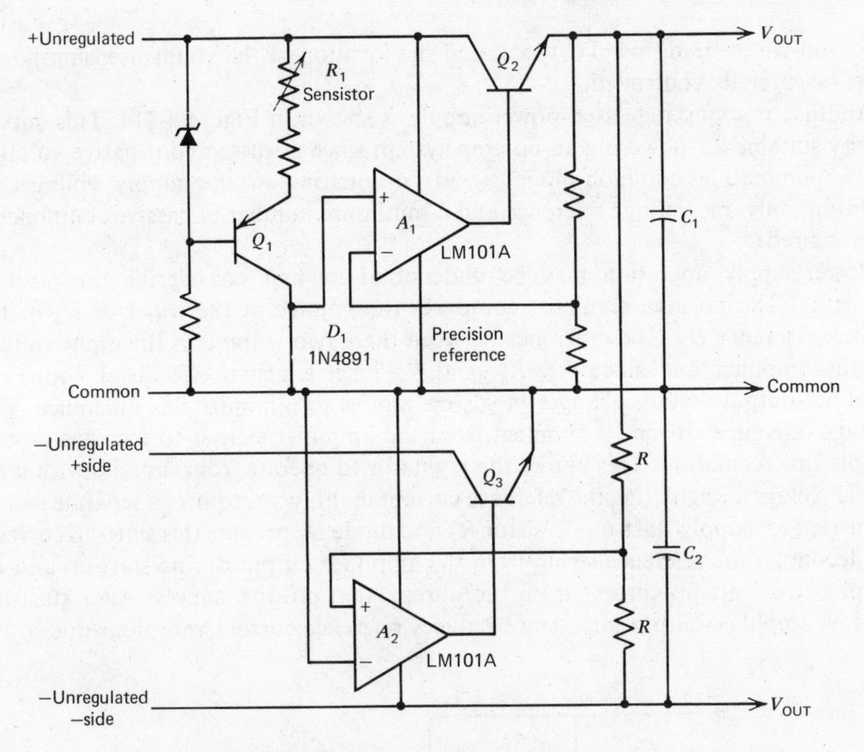

Figure 4-213. Precision dual-voltage regulator.

4-213. In this circuit, precision positive and negative voltages are supplied, which track each other in absolute value and are stable over wide ranges of temperature, input-supply conditions, and output-load conditions.

The dual-voltage regulator of Figure 4-213 operates in the following manner. A constant current is supplied to the precision, aged, temperature-compensated Zener diode D_1 by means of transistor Q_1 and associated components; variations of the Zener current due to the effect of temperature fluctuations on the V_{BE} of Q_1 are automatically compensated for by changes in the value of the silicon resistor R_1 (sensistor). The voltage drop across D_1 is compared by means of the high-gain monolithic op amp A_1 with a similar voltage obtained by means of a voltage divider across the V^+ output and the common terminal. The extremely high gain of the amplifier (typically in excess of 150,000) causes the difference between these voltages to be very small; however, any variation between the compared voltages is supplied as an error signal by A_1 to the series transistor Q_2. The sense of the error signal causes the voltage drop across Q_2 to decrease if the output voltage becomes lower or to increase if the output voltage rises.

Operational amplifier A_2 is identical with A_1 and operates analogously; it compares the difference between the negative- and positive-output voltages with the output common point, thus causing the negative-output voltage to track the positive-output voltage. Error signals from A_2 control the series transistor Q_3 in the same manner that those from A_1 control Q_2.

Capacitors C_1 and C_2 reduce the output impedance at higher frequencies. To obtain maximum performance, all comparison and division resistors must be precision-wirewound, low TC types, and the location of the voltage-sensing points must be carefully controlled.

Another tracking regulated-power supply is shown in Figure 4-214. This supply is very suitable for powering an op amp system since positive and negative voltages track, eliminating common-mode signals originating in the supply voltage. In addition, only one voltage reference and a minimum number of passive components are required.

Power-supply operation may be understood by first considering the positive regulator. The positive regulator compares the voltage at the wiper of R_4 to the voltage reference D_2. The difference between these two voltages is the input voltage for the amplifier and since R_3, R_4, and R_5 form a negative-feedback loop, the amplifier-output voltage changes in such a way as to minimize this difference. The voltage-reference current is supplied from the amplifier output to increase power-supply line regulation. This allows the regulator to operate from supplies with large ripple voltages. Regulating the reference current in this way requires a separate source of current for supply start-up. Resistor R_1 and diode D_1 provide this start-up current; D_1 decouples the reference string from the amplifier output during start-up and R_1 supplies the start-up current from the unregulated positive supply. After start-up, the low amplifier-output impedance reduces reference current variations due to the

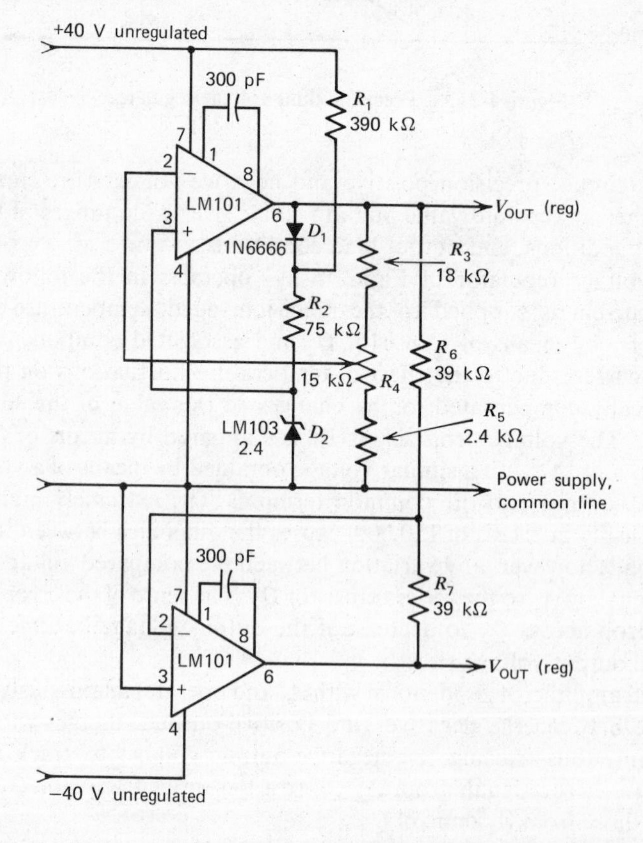

Figure 4-214. Tracking power supply [*note:* output voltage is variable from ±5 to ±35 V; negative output tracks positive output to within the ratio of R_6 to R_7].

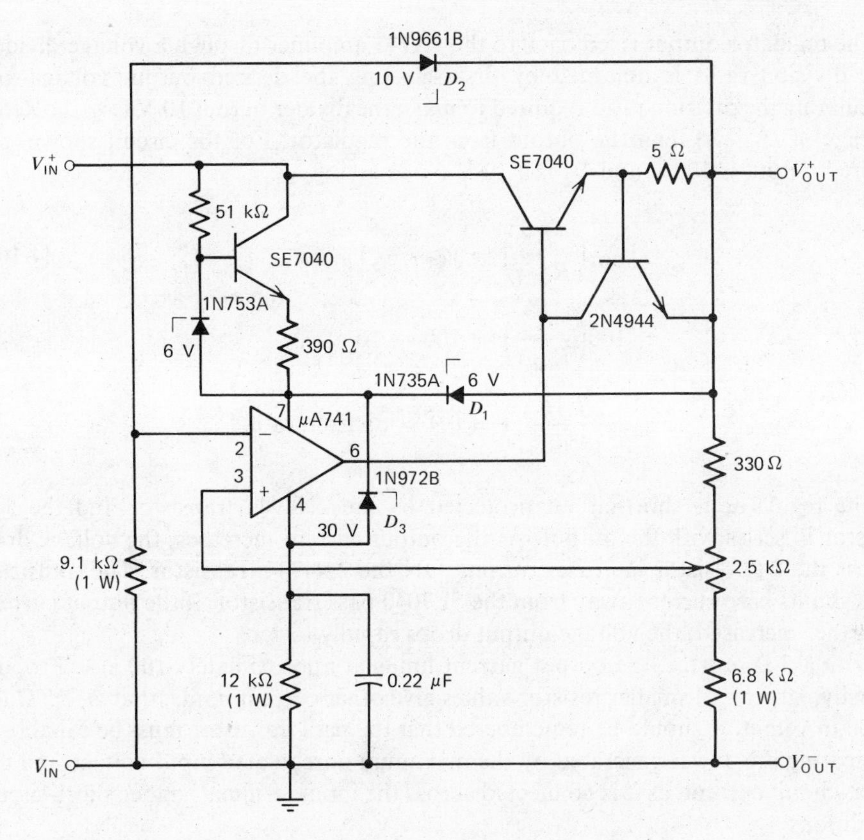

Figure 4-215. High voltage-regulated power supply [*note:* V_{IN} = 120–170 V; line regulation 0.06% V_{IN} = 120–170 V; V_{OUT} = +100 V @ 100 mA; load regulation 0.06%, V_{OUT} = 100 V; I_{OUT} = 0–100 mA].

current through R_1. The negative regulator is simply a unity-gain inverter with an input resistor R_6 and a feedback resistor R_7. The amplifiers must be compensated for unity-gain operation.

The power supply may be modulated by injecting current into the wiper of R_4. In this case, the output-voltage variations will be equal and opposite at the positive and negative outputs. The power-supply voltage may be controlled by replacing D_1, D_2, R_1, and R_2 with a variable-voltage reference.

High Voltage-Regulated Power Supply

Figure 4-215 shows a 100-V regulated power supply using the μA741 as the control amplifier. Zener diodes D_1 through D_3 supply proper operating voltages to the μA741. The diodes reference the amplifier voltages to the power-supply output so the bias levels can follow the output voltage over a wide range of adjustment; D_1 keeps the positive-supply terminal of the μA741 at 6 V above the regulator output. Diode D_2 holds the inverting input at 10 V below the output voltage, while D_3 maintains a 30-V drop across the amplifier's supply terminals.

One of the SE7040 transistors, biased by a Zener-resistor network, acts as a current source supplying operating current to the amplifier and part of the biasing network.

The regulator output is fed back to the μA741 amplifier through a voltage-divider. The division ratio is obtained by first selecting the desired output voltage and calculating the division ratio required to make the divider output 10 V less (the Zener voltage of D_2 less) than the output from the regulator. For the circuit shown, the desired output is 100 V and D_2 is a 10-V diode giving

$$V_{OUT}\left(\frac{R}{R_{total}}\right) = V_{OUT} - V_{D_2} \tag{4-109}$$

$$100\left(\frac{R}{R_{total}}\right) = 100 - 10$$

$$\left(\frac{R}{R_{total}}\right) = 0.9 = \text{division ratio}$$

The regulator is short-circuit protected by the 2N4944 transistor and the 5-Ω resistor in series with the output. As the output current increases, the voltage drop across the 5-Ω resistor increases turning ON the 2N4944 transistor. The transistor thus shunts base current away from the SE7040 pass transistor. If the output current is further increased, the voltage output drops rapidly to zero.

With a 5-Ω resistor, the output current limits at approximately 100 mA. Proportionally, larger and smaller resistor values give other current limits, that is, 2.5 Ω for a 200-mA limit. It should be remembered that the pass transistor must be capable of dissipating the power generated by the maximum unregulated input voltage and the short-circuit current, as it is connected across the input terminals under short-circuit conditions.

The circuit shown provides line and load regulation of approximately 0.06% for input voltages ranging from 120 to 170 V and an output voltage of 100 V at 0 to 100 mA. Higher output currents are possible by using a higher power transistor in place of the SE7040 pass transistor or by excluding the possibility of output short circuits.

APPENDIX
Design Criteria for Second-Order Active Filters[11]

Low-Pass

1. Transfer function

$$H(s) = \frac{\omega_0^2}{s^2 + 2\zeta\omega_0 s + \omega_0^2}$$

2. Circuit configuration:

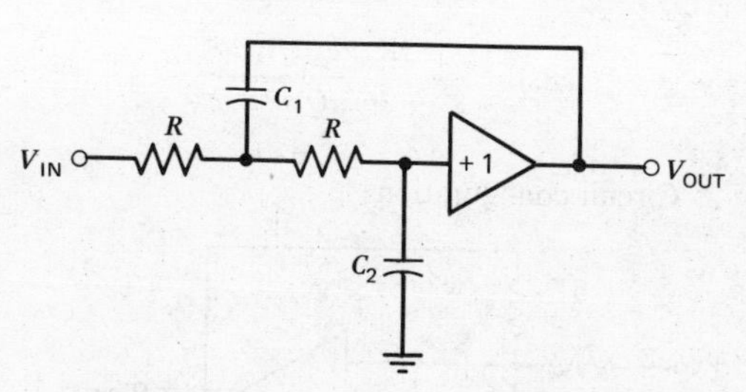

3. Cutoff frequency:

$$\omega_0 = \frac{1}{R\sqrt{C_1 C_2}}$$

4. Damping factor:

$$\zeta = \sqrt{\frac{C_2}{C_1}}$$

5. Stability functions:

(a) $$\frac{\Delta\omega_0}{\omega_0} = -\left(\frac{\Delta R}{R} + \frac{1}{2}\frac{\Delta C_1}{C_1} + \frac{1}{2}\frac{\Delta C_2}{C_2}\right)$$

(b) $$\frac{\Delta\zeta}{\zeta} = \frac{1}{2}\left(\frac{\Delta C_2}{C_2} - \frac{\Delta C_1}{C_1}\right)$$

6. Component values:

$$R = \left(\frac{\zeta}{\omega_0}\right)\frac{1}{C_2}$$

$$C_1 = \left(\frac{1}{\zeta^2}\right)\frac{1}{C_2}$$

High-Pass

1. Transfer function:

$$H(s) = \frac{s^2}{s^2 + 2\zeta\omega_0 s + \omega_0^2}$$

2. Circuit configuration:

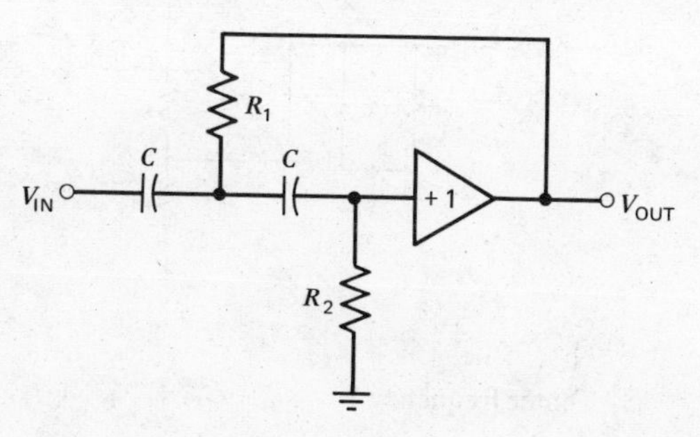

3. Cutoff frequency:

$$\omega_0 = \frac{1}{C\sqrt{R_1 R_2}}$$

4. Damping factor:

$$\zeta = \sqrt{\frac{R_1}{R_2}}$$

5. Stability functions:

(a) $$\frac{\Delta\omega_0}{\omega_0} = -\left(\frac{\Delta C}{C} + \frac{1}{2}\frac{\Delta R_1}{R_1} + \frac{1}{2}\frac{\Delta R_2}{R_2}\right)$$

(b) $$\frac{\Delta\zeta}{\zeta} = \frac{1}{2}\left(\frac{\Delta R_1}{R_1} - \frac{\Delta R_2}{R_2}\right)$$

6. Component values:

$$C = \left(\frac{\zeta}{\omega_0}\right)\frac{1}{R_1}$$

$$R_2 = \left(\frac{1}{\zeta^2}\right)\frac{1}{R_1}$$

[11] H. T. Russell, "Design Active Filters with Less Effect," *Electronics*, June 7, 1971.

Band Pass A

1. Transfer function:

$$H(s) = \frac{K_0(\omega_0/Q)s}{s^2 + (\omega_0/Q)s + \omega_0^2}$$

2. Circuit configuration:

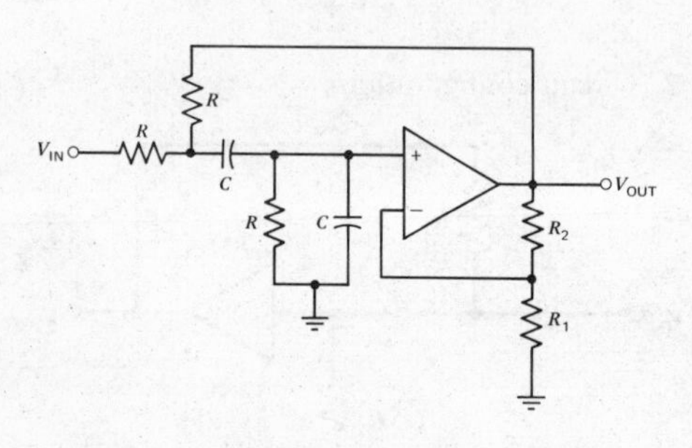

3. Center frequency:

$$\omega_0 = \frac{\sqrt{2}}{RC}$$

4. Quality factor and center frequency gain:

$$Q = \frac{\sqrt{2}}{5 - K}$$

$$K_0 = \frac{K}{5 - K}$$

where

$$K = 1 + \frac{R_2}{R_1}$$

5. Stability functions:

(a) $$\frac{\Delta\omega_0}{\omega_0} = -\sqrt{2}\left(\frac{\Delta R}{R} + \frac{\Delta C}{C}\right)$$

(b) $$\frac{\Delta Q}{Q} = (2\sqrt{2}Q - 1)\left(\frac{\Delta R_2}{R_2} - \frac{\Delta R_1}{R_1}\right)$$

(c) $$\frac{\Delta K_0}{K_0} = (3.54Q)\left(\frac{2.84Q - 1}{3.54Q - 1}\right) \times \left(\frac{\Delta R_2}{R_2} - \frac{\Delta R_1}{R_1}\right)$$

6. Component values:

$$R = \frac{\sqrt{2}}{\omega_0 C}$$

$$R_2 = \left(4 - \frac{\sqrt{4}}{Q}\right)R_1 = (K - 1)R_1$$

Band Pass B

1. Transfer function:

$$H(s) = \frac{K_0(\omega_0/Q)s}{s^2 + (\omega_0/Q)s + \omega_0^2}$$

2. Circuit configuration:

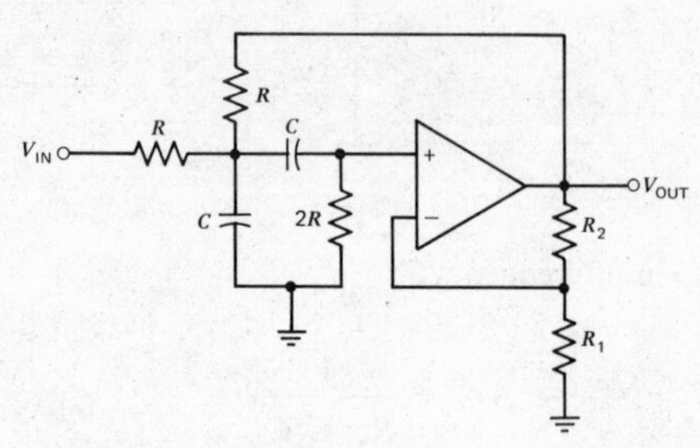

3. Center frequency:

$$\omega_0 = \frac{1}{RC}$$

4. Quality factor and center frequency gain:

$$Q = \frac{1}{3 - K}$$

$$K_0 = \frac{K}{3 - K}$$

where

$$K = 1 + \frac{R_2}{R_1}$$

5. Stability functions:

(a) $$\frac{\Delta\omega_0}{\omega_0} = -\left(\frac{\Delta R}{R} + \frac{\Delta C}{C}\right)$$

(b) $$\frac{\Delta Q}{Q} = (2Q - 1)\left(\frac{\Delta R_2}{R_2} - \frac{\Delta R_1}{R_1}\right)$$

(c) $$\frac{\Delta K_0}{K_0} = 3Q\left(\frac{2Q - 1}{3Q - 1}\right) \times \left(\frac{\Delta R_2}{R_2} - \frac{\Delta R_1}{R_1}\right)$$

6. Component values:

$$R = \frac{1}{\omega_0 C}$$

$$R_2 = \left(2 - \frac{1}{Q}\right)R_1 = (K - 1)R_1$$

Band-Reject

1. Transfer function:

$$H(s) = \frac{s^2 + \omega_0^2}{s^2 + (\omega_0/Q)s + \omega_0^2}$$

2. Circuit configuration:

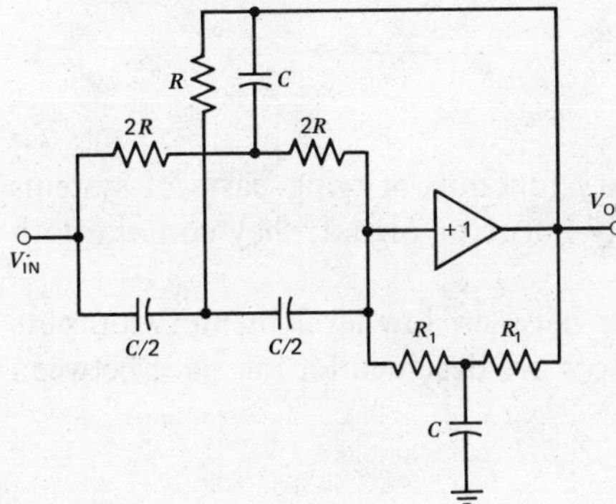

3. Center frequency:

$$\omega_0 = \frac{1}{RC}$$

4. Quality factor:

$$Q = \frac{R_1}{4R}$$

5. Stability functions:

(a) $$\frac{\Delta\omega_0}{\omega_0} = -\left(\frac{\Delta R}{R} + \frac{\Delta C}{C}\right)$$

(b) $$\frac{\Delta Q}{Q} = \left(\frac{\Delta R_1}{R_1} - \frac{\Delta R}{R}\right)$$

6. Component values:

$$R = \frac{1}{\omega_0 C}$$

$$R_1 = \left(\frac{4Q}{\omega_0}\right)\frac{1}{C}$$

$$C_1 = \left(\frac{1}{2Q}\right)C$$

All-Pass (360°)

1. Transfer function:

$$H(s) = \frac{s^2 - (\omega_0/Q)s + \omega_0^2}{s^2 + (\omega_0/Q)s + \omega_0^2}$$

2. Circuit configuration:

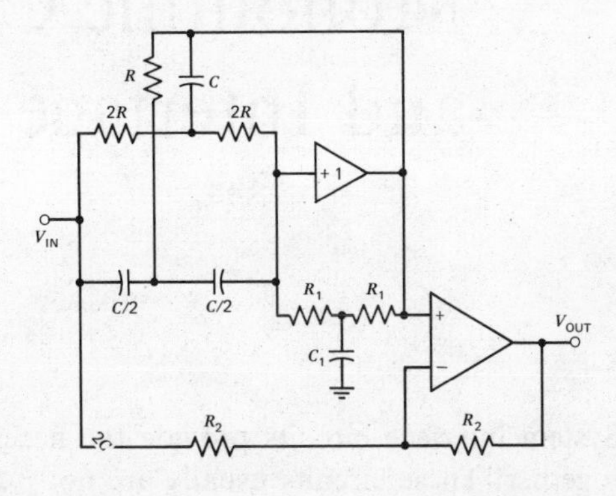

3. Center frequency:

$$\omega_0 = \frac{1}{RC}$$

4. Quality factor:

$$Q = \frac{R_1}{4R}$$

5. Stability functions:

(a) $$\frac{\Delta\omega_0}{\omega_0} = -\left(\frac{\Delta R}{R} + \frac{\Delta C}{C}\right)$$

(b) $$\frac{\Delta\omega}{Q} = \left(\frac{\Delta R_1}{R_1} - \frac{\Delta R}{R}\right)$$

6. Component values:

$$R = \frac{1}{\omega_0 C}$$

$$R_1 = \left(\frac{4Q}{\omega_0}\right)\frac{1}{C}$$

$$C_1 = \left(\frac{1}{2Q}\right)C$$

Chapter 5

Monolithic Comparators and Interface Circuits

System interface circuits provide the necessary function of tying parts of systems together. These circuits usually are not purely linear or digital; they contain both types of circuit functions.

Sense amplifiers are designed for interface between low-level memory outputs and bipolar logic levels; differential comparators are designed for interface between analog systems and TTL/DTL systems.

COMPARATORS

A differential-voltage comparator is a high-gain, differential-input, single-ended output amplifier. The function of the device is to compare a signal voltage on one input with a reference voltage on the other and produce a logic "1" or "0" at the output when one input is higher than the other. Threshold detection is accomplished by putting a reference voltage on one input and the signal on the other. Clearly, an op amp can be used as a comparator, except that its response time is in the tens of microseconds which is often too slow for many applications. Additionally, the large output-voltage swing desired for op amps is often a disadvantage when the comparator is used to drive low-level logic circuits. A comparator is useful as a low-hysteresis, variable-threshold Schmitt trigger, a pulse-height discriminator, a voltage comparator in A/D converters, a zero-crossing detector, a threshold detector, an oscillator, a high noise-immunity digital line receiver, or in pulse-width modulator applications.

Comparator Specifications

Before pursing detailed discussions of comparator specifications and characteristics, it is best to define the basic comparator-specification terms.

Input-Offset Voltage is the voltage between the input terminals required to make the output voltage greater than or less than specified voltages.

Input-Offset Current is the difference between the two input currents for which the output will be driven higher than or lower than specified voltages.

Input-Bias Current is the average of the two input currents.

Input-Voltage Range is the range of voltage on the input terminals (common-mode) over which the offset specifications apply.

Voltage Gain is defined as the ratio of the change in output voltage to the change in voltage between the input terminals producing it.

Response Time is the interval between the application of an input-step function and the time when the output crosses the logic threshold voltage. The input step drives the comparator from some initial, saturated-input voltage to an input level just barely in excess of that required to bring the output from saturation to the logic threshold voltage. This excess is referred to as the voltage overdrive.

Saturation Voltage is the low output-voltage level with the input drive equal to or greater than a specified value.

Strobe on Current is defined as the current that must be drawn out of the strobe terminal to disable the comparator.

Output-Leakage Current is the current into the output terminal with a specified output voltage relative to ground and the input drive equal to or greater than a given value.

Supply Current is the current required from the positive or negative supply to operate the comparator with no output load. The power will vary with input voltage, but is specified as a maximum for the entire range of input-voltage conditions.

The typical transfer functions of two IC comparators, which describe their operation, are shown in Figures 5-1 and 5-2. The transfer function of Figure 5-1 is for the first industry standard comparator, the μA710. That of Figure 5-2 is for the LM111 precision comparator which operates from a single 5-V supply.

A set of comparator specifications must provide the following information in addition to the basic supply voltage, power dissipation, and absolute maximum rating limits: The accuracy with which the comparator can distinguish between

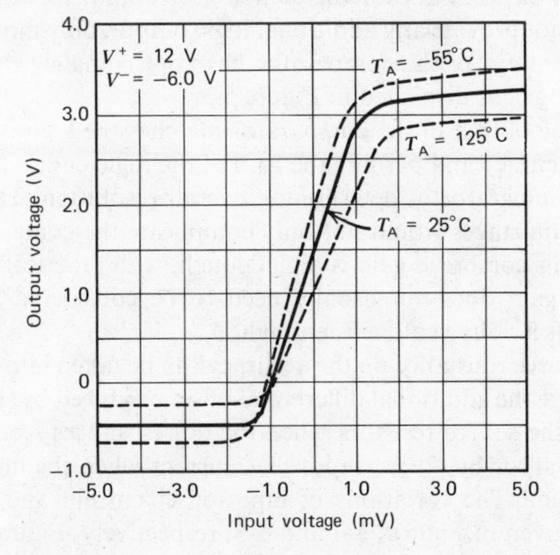

Figure 5-1. μA710 voltage transfer characteristic.

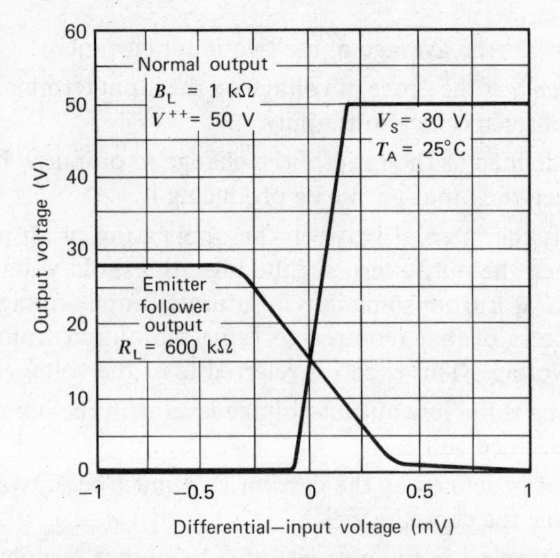

Figure 5-2. LM111 transfer function.

the input signal and the reference signal, and the effect of source resistance on accuracy; the rapidity with which the comparator can make this distinction; the influence of the speed of a decision on accuracy; and the ability of the comparator to drive different types of logic loads. Let us discuss each of these considerations in greater detail.

DC ACCURACY

For a comparator, as stated before, the input-offset voltage is defined as the voltage required between the comparator input terminals to drive the output to the threshold voltage of the logic.

This is in contrast to the definition for an op amp, which is designed to have zero output for zero input. When the output is at the logic threshold voltage, a very small change in input voltage will cause the output of the logic circuit to swing from a 0 to a 1 state. It is not necessary for the comparator output to swing from the 0 to 1 state as this does not provide any additional logic output, only more noise immunity. The input-offset voltage of the comparator has approximately the same TC as the threshold of the logic, as displayed in Figure 5-3.

The fact that the output of the comparator matches the logic allows a lower gain with no loss in accuracy and permits the gain of the logic circuit to be multiplied by the gain of the comparator in determining overall resolution. This is important in that additional gain stages would not only complicate the comparator, but make it slower as well. This combined gain is high enough, with practically all logic circuits, that the voltage gain does not usually need to be considered in determining dc accuracy—the offset voltage alone is enough.

The effect of source resistance on the accuracy can be taken into account by adding to the offset voltage the additional differential offset produced by (1) the offset current flowing through the source resistors when the dc resistances seen by the two input terminals are equal, or by (2) the input-bias current when the input resistances are very much different. The variations of input-offset current and bias current with temperature are given in Figures 5-4 and 5-5, respectively. Figure 5-6 depicts these variations for the NE527 High-Speed Schottky Comparator. To eliminate the

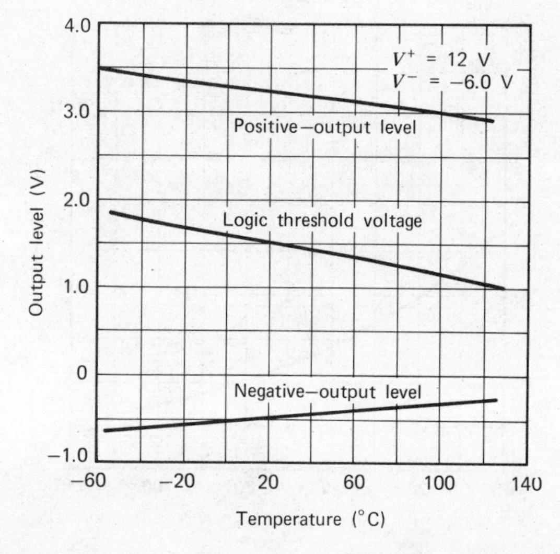

Figure 5-3. Output-voltage levels as a function of ambient temperature.

necessity of using worst-case values for offset voltage and offset current, the offset voltage is specified on the data sheet for the case where the resistances on each input are equal to or less than a given amount.

The use of input resistance to determine loading is not valid with a comparator. The input impedance varies rapidly as the differential-input voltage goes through zero, and is in excess of 1 MΩ for input voltages greater than a couple hundred millivolts. What should be considered is the effect of the input current, which alternately switches between the two input terminals for large differential-input voltages.

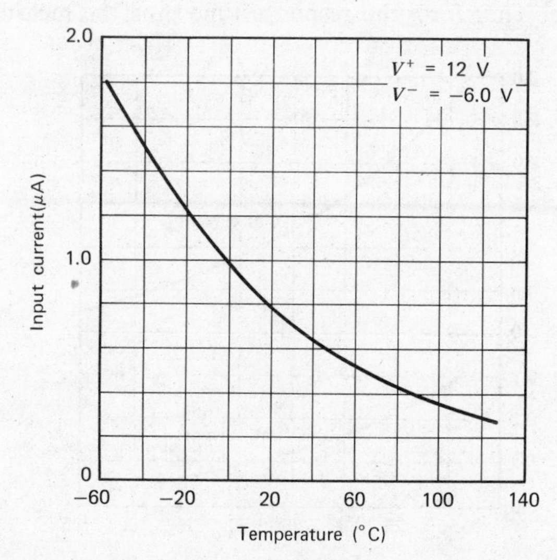

Figure 5-4. Input-offset current as a function of ambient temperature for the μA710.

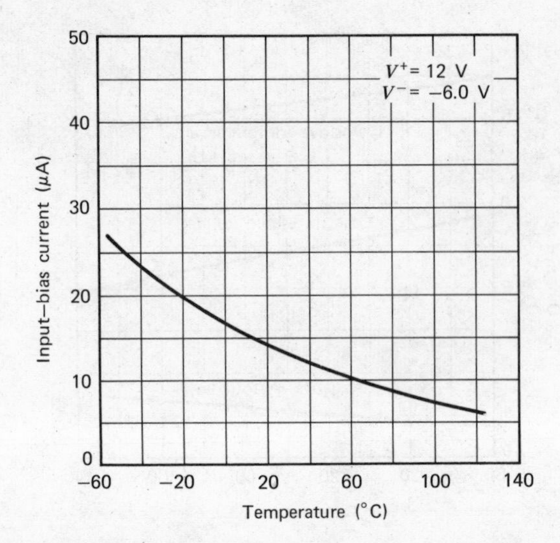

Figure 5-5. Input-bias current as a function of ambient temperature for the μA710.

RESPONSE TIME

The response time of a comparator is a more difficult parameter to define. A worst-case situation would be where the comparator is used to determine the height of a large pulse. A good example is the situation in which a dc reference signal (say 0.1 V) is applied to one comparator input and a step function is applied to the other.

Before the step function is applied, the comparator is completely saturated in one direction by the reference voltage (the comparator will actually reach complete saturation with less than 100 mV across the input terminals). Some time after the step function arrives, the comparator output will reach the logic threshold voltage if the step height differs from the reference voltage only by the offset voltage. In the absence of overshoot, however, this length of time would supposedly be infinite, or at least difficult to determine. Therefore, the response time must be measured with some

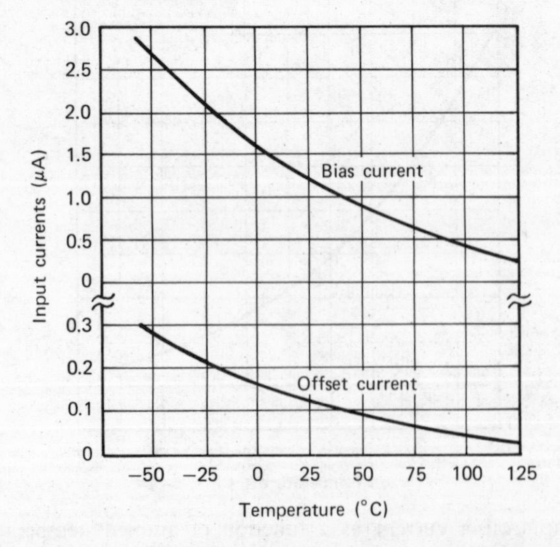

Figure 5-6. Input currents versus temperature for the NE527 comparator.

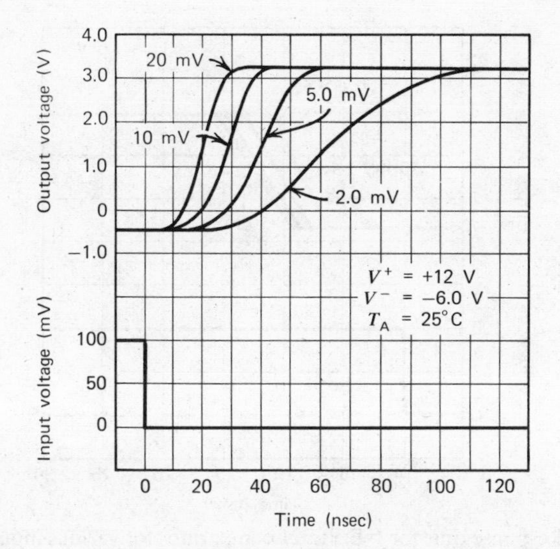

Figure 5-7. Response time for various input overdrives for the μA710.

additional voltage overdrive in order to have any real meaning. That is, the input-step height must be increased by some amount over that required to bring the output from saturation to the logic threshold voltage. The response time can then be defined as the interval between the application of the input step and the time when the output crosses the logic threshold voltage. The error encountered in making the decision within this period of time is then the amount of overdrive required. This definition eliminates offset voltage from the speed measurement in order to provide more flexibility in the interpretation of a specified parameter. This response time will be relatively independent of the step height and reference voltage as long as they are in excess of 100 mV. Smaller step heights will give faster response times since the comparator will not be thoroughly saturated.

Figure 5-7 gives the response time of the μA710 for a 100-mV input step and overdrives of 2-, 5-, 10-, and 20-mV with a positive-going output. The response times are 65, 40, 28, and 20 nsec, respectively, for these overdrives. For smaller input steps, the response time will be somewhat faster for the same overdrive since the internal circuitry of the comparator will not be completely saturated before the pulse is applied. Larger input steps do not materially affect response time since the comparator does not go any deeper into saturation with input voltages larger than 100 mV. Measurement of response time in the above manner permits a determination of how much input-referred error is encountered when the comparison is made in a short time interval.

Figure 5-8 depicts the response time for the NE/SE521 for 5-, 10-, 20-, and 100-mV overdrives for a 100-mV input-step function. The response times are 12.5, 12, 10.5, and 8 nsec, respectively.

With a dual comparator such as the μA711, the speed of the strobe circuitry is important. Strobe release time can be defined as the time required for the output to rise to the logic threshold voltage after a positive-step function has been applied to the strobe terminal (the strobe signal does not actually have to be a step function as long as it is faster than the measured strobe-release time). This definition assumes a dc input signal that will give a positive output when the strobe is released. When the

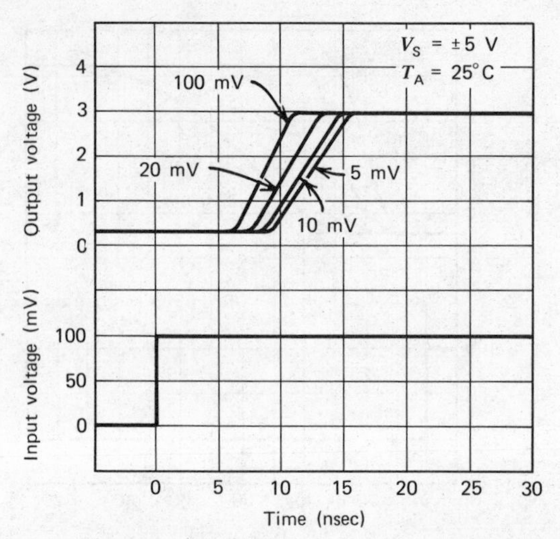

Figure 5-8. Response time for NE/SE521 comparator for various input overdrives.

strobe voltage returns to zero, it will drop the voltage on the base of the output emitter-follower at a speed which is limited only by the fall time of the strobe voltage. Therefore, the strobe-fall time is not pertinent.

The strobe-release time of the μA711 is shown for various input conditions in Figure 5-9. The 0-mV curve is for an input condition that would put the output at the logic threshold voltage under dc conditions. The curve below that is for 1-mV less input voltage, and gives an indication of the feedthrough of the strobe pulse when the comparator is barely turned OFF (it is necessary to wire the comparator into the circuit minimizing the stray capacitance between the strobe terminal and the inputs in order to keep this feedthrough small). The 2-mV and 5-mV curves represent actual operating conditions and give strobe-release times of 14 nsec and 10 nsec, respectively. In these curves, the output does not rise to the maximum output level because of the clamping action of the strobe circuitry when a low-amplitude strobe pulse is used.

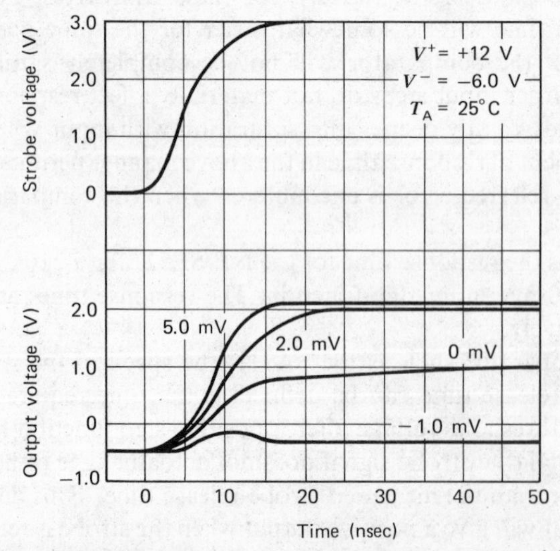

Figure 5-9. Strobe release time for various input overdrives.

When time-strobing is used in the design of sense amplifiers using the μA711, it often becomes necessary to stretch the strobed-output pulse of the sense amplifier in order to make it wide enough to be transmitted over a reasonable line length and trigger the storage circuits. The most straightforward way of accomplishing this is to use a one-shot multivibrator. However, the accuracy requirements for the stretched pulse do not warrant going to the complexity of a one-shot, and thus much simpler methods can be used. A possible alternative is to load the comparator with a capacitor on the output. This will stretch the output pulse. It occurs because the output emitter-follower can charge the load capacitance rapidly, but the output current sink gives a limited rate of discharge. Normally, the strobe width is sufficient to trigger the logic circuits, the only fear being that the pulse will be degraded by cable capacitance.

It is worthwhile to point out that the response time and the strobe-release time for negative-going outputs were not described previously because they are limited by this pulse-stretching characteristic. For a negative-going output, the storage time of the second stage is eliminated; additionally, the second-stage transistor is able to discharge its output capacitance faster than the collector-load resistor can charge it. Similarly, the strobe pull-down time is shorter than the release time on the base of the output emitter-follower, since a negative-going strobe pulse into the Zener diode will forcibly pull down that point. Therefore, the response time for a negative-going output will always be faster than that for a positive output, which depends on the delay time of the input stage and the storage time of the second stage, as well as the feedback and output capacitances of the second stage. Thus, the positive-response time is the one that is specified since it is most indicative of overall device performance; the negative-going response times are limited more by the characteristics of the load than by those of the circuit.

LOGIC COMPATIBILITY

Specification of logic compatibility can be approached by ensuring that the comparator can drive the logic into full saturation under worst-case conditions. After this is done, it is necessary to take into account only the difference between the threshold voltage of the logic used and the logic threshold specified for the comparator to determine the influence on offset voltage. With most standard logic circuits the effect of differing logic threshold voltages on offset is less than 1 mV, so even this can be neglected in the majority of applications.

Figure 5-10 shows how the reduced error currents of the LM111 improve circuit performance. With the μA710 or LM106, the offset voltage is degraded for source resistances above 200 Ω. The LM111, however, works well with source resistances in excess of 30 kΩ. Figure 5-10 applies for equal source resistances on the two inputs. If they are unequal, the degradation will become pronounced at lower resistance levels. Table 5-1 gives the important electrical characteristics of the popular IC comparators.

Comparator Applications

BASIC CIRCUIT

The basic circuit for using the comparator is shown in Figure 5-11*a*. A reference voltage between ±5 V is inserted on one input and the signal is inserted on the other. When the input exceeds the reference voltage, the output switches either positive

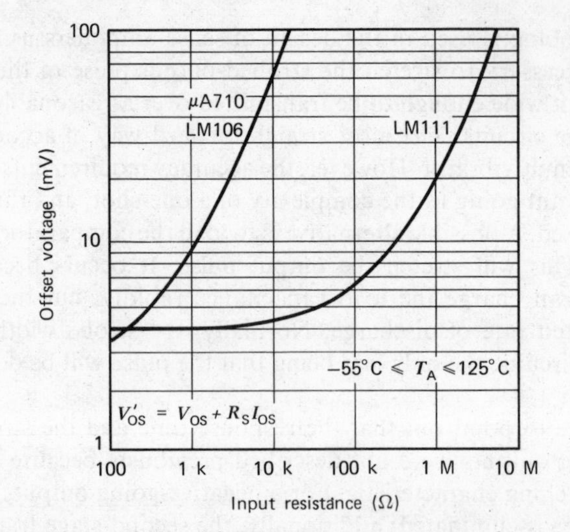

Figure 5-10. Influence of source resistances on worst-case, equivalent input-offset voltages of comparators.

or negative, depending on how the inputs are connected. Figure 5-11*b* shows the general input/output characteristic.

The circuit has a variety of uses. It can be employed as a voltage comparator in A/D converters, where one input is driven by the analog-input signal and the other by a ladder network. It also has use as a tape-drum memory sense amplifier and threshold detector. Other applications include a high-noise-immunity buffer for going from high-level logic into low-level ICs, or a pulse-shape restorer in digital circuits.

In order to achieve maximum accuracy in the circuit of Figure 5-11, the dc resistance of the signal source (R_S) should equal the source resistance of the reference voltage

Table 5-1 Comparing the LM111 with other integrated circuit comparators; values given are worst-case over a −55 to 125°C temperature range, except as noted

Parameter	*LM111*	*LM106*	*μA710*	*SE529*	*NE522*[a]	*Units*
Input-offset voltage	4	3	3	6	10	mV
Input-offset current	0.02	7	7	9	12	μA
Input-bias current	0.15	45	45	36	40	μA
Common-mode range	±14	±5	±5	±6	±3	V
Differential-input voltage range	±30	±5	±5	±5	±6	V
Voltage gain[b]	200	40	1.7	5	5	V/mV
Response time[b]	200	40	40	10	5	nsec
Output drive						
voltage	50	24	2.5	2.5	—	V
current	50	100	1.6	—	—	mA
Fan out (DTL/TTL)	8	16	1	—	—	
Power consumption	80	145	160	—	—	mW

[a] Only 0 to 70°C.
[b] Typical at 25°C.

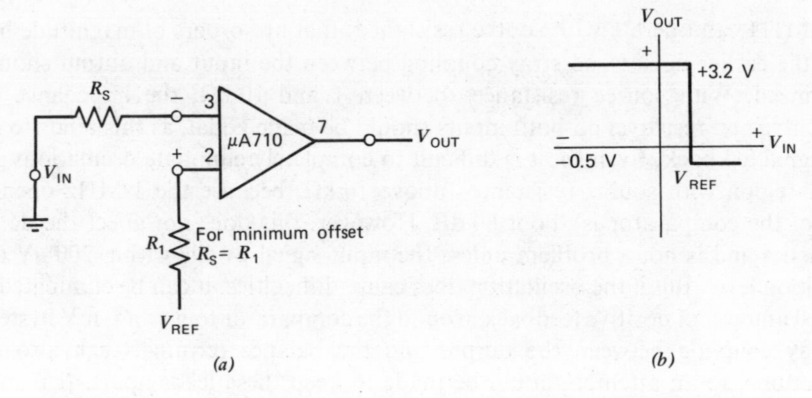

Figure 5-11. Basic level detector circuit: (*a*) circuit; (*b*) transfer function.

(R_1). When this is done, the bias currents of the two inputs produce nearly equal drops across these source resistances, which should be made as low as possible, preferably less than 200 Ω, for best performance. Any increase in offset is then due only to the offset current, which is usually more than an order of magnitude less than the bias current.

COMPARATOR USAGE GUIDELINES

Although the input-voltage range of the μA710 and μA711 is ±5 V, the maximum voltage between the inputs is also ±5 V. If one of the inputs is at +5 V, for example, the other can only be driven as low as ground potential without exceeding the differential-input voltage limit. It is important to observe this maximum rating since exceeding the differential-input voltage limit and drawing excessive current in breaking down the emitter-base junctions of the input transistors could cause gross degradation in the input-offset current and input-bias current.

Exceeding the absolute maximum positive-input voltage limit of the device (+7 V) will saturate the input transistor and possibly cause damage through excessive current. However, even if the current is limited to a reasonable value so that the device is not damaged, erratic operation can result.

Some attention to power-supply-bypassing is required with both the μA710 and μA711. These devices are multistage amplifiers with gain to several hundred megahertz. Long, unbypassed supply leads or sloppy layouts can therefore cause oscillation problems. Bypassing with electrolytic or tubular-paper capacitors is ineffective. What is recommended is that both the positive and negative supplies be bypassed to ground using low-inductance, ceramic-disc capacitors (0.01 μF) located as close as practical to the device. A neat physical layout, keeping the input away from the output, is also important. Additionally, the strobe of the μA711 should be kept away from the inputs if fast-rise strobe pulses are used. The strobe terminal, if unused, should not be left floating in high-speed circuits because of stray capacitive effects.

A comparator such as the LM111 is less susceptible to spurious oscillations both because of its lower speed (200-nsec response time for the LM111 versus 40 nsec for the μA710), and because of its better power-supply rejection. Feedback between the output and the input is a lesser problem with a given source resistance. However, the LM111 can operate with source resistances that are orders of magnitude higher than the earlier devices, so stray coupling between the input and output should be minimized. With source resistances between 1 and 10 kΩ, the impedance (both capacitive and resistive) on both inputs should be made equal, as this tends to reject the signal fed back. Even so, it is difficult to completely eliminate oscillations in the linear region with source resistances above 10 kΩ, because the 1-MHz open-loop gain of the comparator is about 80 dB. However, this does not affect the dc characteristics and is not a problem unless the input signal dwells within 200 μV of the transition level. But if the oscillation does cause difficulties, it can be eliminated with a small amount of positive feedback around the comparator to give a 1-mV hysteresis.

Stray coupling between the output and the balance terminals can also cause oscillations, so an attempt should be made to keep these leads apart. It is usually advisable to tie the balance pins together to minimize the effect of this feedback. If balancing is used, the same result can be accomplished by connecting a 0.1 μF capacitor between these pins.

Normally, individual supply-bypasses on every device are unnecessary, although long leads between the comparator and the bypass capacitors are definitely not recommended. If large current spikes are injected into the supplies in switching the output, bypass capacitors should be included at these points.

When driving the inputs from a low-impedance source, a limiting resistor should be placed in series with the input lead to limit the peak current to something less than 100 mA. This is especially important when the inputs go outside a piece of equipment where they could accidentally be connected to high-voltage sources. Low-impedance sources do not cause a problem unless their output voltage exceeds the negative-supply voltage. However, the supplies go to zero when they are turned off, so the isolation is usually needed.

Large capacitors on the input (greater than 0.1 μF) should be treated as a low-source impedance and isolated with a resistor. A charged capacitor can hold the inputs outside the supply voltage if the supplies are abruptly shut off.

Precautions should be taken to ensure that the power supplies for any IC comparator never become reversed—even under transient conditions. With reverse voltages greater than 1 V, the IC can conduct excessive current, fusing internal aluminum interconnects. This usually takes more than 0.5 A. If there is a possibility of reversal, clamp diodes with an adequate peak-current rating should be installed across the supply bus.

No attempt should be made to operate the circuit with the ground terminal at a voltage exceeding either supply voltage. Further, the 50 V output-voltage rating applies to the potential between the output and the V^- terminal. Therefore, if the comparator is operated from a negative supply, the maximum output voltage must be reduced by an amount equal to the voltage on the V^- terminal.

The output circuitry is protected for shorts across the load. It will not, for example, withstand a short to a voltage more negative than the ground terminal. Additionally, with a sustained short, power dissipation can become excessive if the voltage across the output transistor exceeds about 10 V.

The input terminals can exceed the positive-supply voltage without causing damage. However, the 30 V maximum rating between the inputs and the V^- terminal must be observed. As mentioned earlier, the inputs should not be driven more negative than the V^- terminal.

ZERO-CROSSING DETECTOR

The simplest comparator is the zero-crossing detector which answers the question: Is the input signal greater than or less than zero? Often in the processing of electrical analog data, a power spectrum analysis of some particular signal waveform must be performed. This procedure requires that the frequency content of the signal waveform of interest, within a specified pass-band, be extracted from the broad-band signal and noise. A zero-crossing detector provides a simple and effective means of performing such an operation.

A zero-crossing detector is a device that changes state each time the analog-input signal passes through zero (or through its average reference level). The input signal

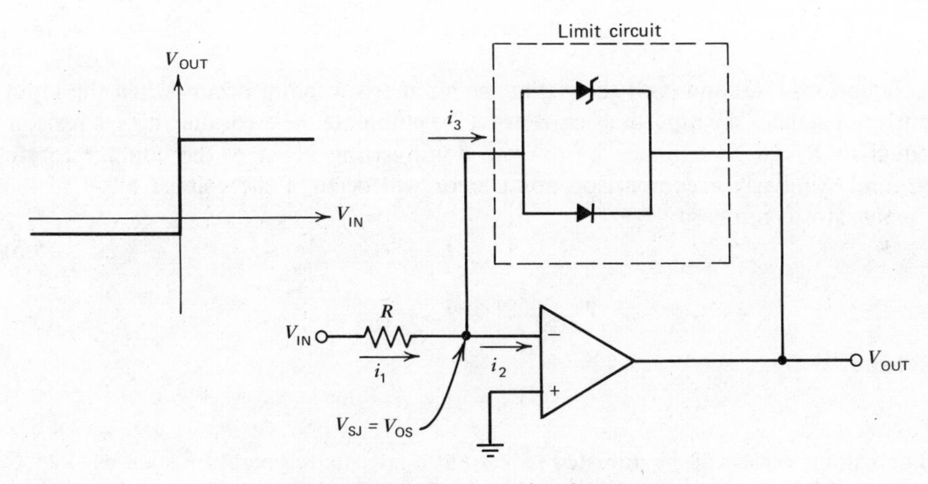

Figure 5-12. Zero-crossing detector.

is thus converted into a train of frequency-dependent pulse widths, and the resultant zero-crossing intervals may then be examined for frequency content.

This "infinite" clipping of the signal virtually eliminates distortion caused by amplitude fluctuations, circuit variations, and noise, and permits simplification of further data-processing through the use of digital techniques. These features are especially valuable for information-processing systems where signal storage and correlation processes are to be utilized.

The commonly used Schmitt trigger and multistage limiter-clipping circuits can exhibit large errors in zero-crossing information. The Schmitt trigger introduces distortion due to its large "on-off" hysteresis characteristic. The circuit does not change state when the input signal passes through zero, but at different levels for positive-going or negative-going signals. Distortion in multistage limiters is caused by a shift in the average reference level after each successive stage of limiting, arising because the signal cannot be fully clipped in one operation. Large peaks (which may be of one polarity only) are clipped by the first stages of the limiter, thus altering the waveform to produce a different average level. The following limiter stage then clips the resultant signal about the new average level, which can completely change the character of the original signal.

A typical circuit for a zero-crossing detector is shown in Figure 5-12. The limit circuit shown in the figure produces one output level when i_3 is positive and a different output level when i_3 is negative. Since the limit circuit changes state when i_3 changes sign, the comparison point occurs when $i_3 = 0$ (assume the summing junction potential is zero):

$$i_1 = I_2 + i_3 \tag{5-1}$$

$$\frac{V_{IN}}{R} = I_2 + i_3 \tag{5-2}$$

When $i_3 = 0$,

$$\text{current comparison: } \frac{V_{IN}}{R} = i_1 = I_2 \tag{5-3}$$

or

$$\text{voltage comparison: } V_{IN} = I_2 R \tag{5-4}$$

Equations (5-3) and (5-4) show that the comparison point occurs when the input current balances the input-bias current I_2. To eliminate the error due to I_2, a resistor equal to R can be connected from the noninverting input of the comparator to ground. Similarly a comparison-point error will occur if the voltage offset of the comparator is not zero:

$$i_1 = I_2 + i_3 \tag{5-5}$$

$$\frac{V_{IN} - V_{OS}}{R} = I_2 + i_3 \tag{5-6}$$

When $i_3 = 0$,

$$V_{IN} = V_{OS} + I_2 R \tag{5-7}$$

The voltage offset can be adjusted to zero at a specific temperature such as +25°C but should be considered at other temperatures because of the unavoidable voltage drift. The error caused by the bias current is I_2R. As discussed above, this error can be eliminated by connecting a resistance of value R between the noninverting input and ground, provided the bias currents of the noninverting and inverting inputs are equal. If they are not equal, the difference is called the differential-bias current (current offset), and (5-7) can still be used except I_2 is now the differential-bias current.

The input impedance of the zero-crossing comparator is R. The circuit has the disadvantage that noise on the input signal will cause i_3 to be "noisy" and therefore V_{OUT} will "chatter" when i_3 is changing sign.

A circuit that will detect zero-crossing in the output of a magnetic transducer within a fraction of a millivolt is shown in Figure 5-13. The magnetic pick-up is connected between the two inputs of the comparator. The resistive divider R_1 and R_2 biases the inputs 0.5 V above ground, within the common-mode range of the IC. The output will directly drive the DTL or TTL. The exact value of the pull-up resistor R_5 is determined by the speed required from the circuit since it must drive any capacitive loading for positive-going output signals. An optional offset-balancing circuit using R_3 and R_4 is included in the schematic.

A zero-crossing detector that drives the data input of MOS logic is shown in Figure 5-14. Here, both a positive supply and the −10-V supply for MOS circuits are used. Both supplies are required for the circuit to work with zero common-mode voltage. An alternate balancing scheme is also shown in the schematic. It differs from the circuit in Figure 5-13 in that it raises the input-stage current by a factor of 3.

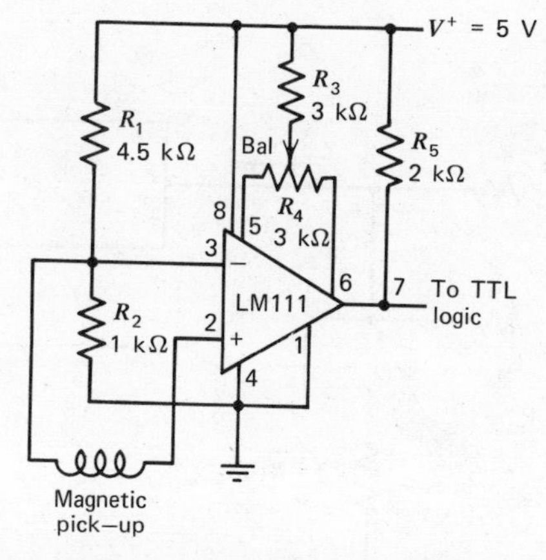

Figure 5-13. Zero-crossing detector for magnetic transducer.

This increases the rate at which the input voltage follows rapidly changing signals from 7 to 18 V/μsec. This increased common-mode slew can be obtained without the balancing potentiometer by shorting both balance terminals to the positive-supply terminal. Increased input-bias current is the price that must be paid for the faster operation.

LEVEL DETECTOR

To make a comparison at some level other than 0 V, the circuit of Figure 5-12 can be modified in one of the ways shown in Figure 5-15. In Figure 5-15*a* two summing resistors are required, but the reference voltage can conveniently be the positive- or negative-power supply, if the supplies are well regulated. The comparison point is scaled by the ratio R_1/R_2. Voltage offset causes the error shown below. When $i_3 = 0$,

$$V_1 = -\frac{R_1}{R_2} V_{\text{REF}} + \overbrace{V_{\text{OS}}\left(1 + \frac{R_1}{R_2}\right)}^{\text{error term}} \qquad (5\text{-}8)$$

Even though R_3 is used to provide bias-current compensation, the differential-bias current (offset current) will cause an error in the comparison point equal to $I_{\text{OS}}R_1$, where I_{OS} is the offset current.

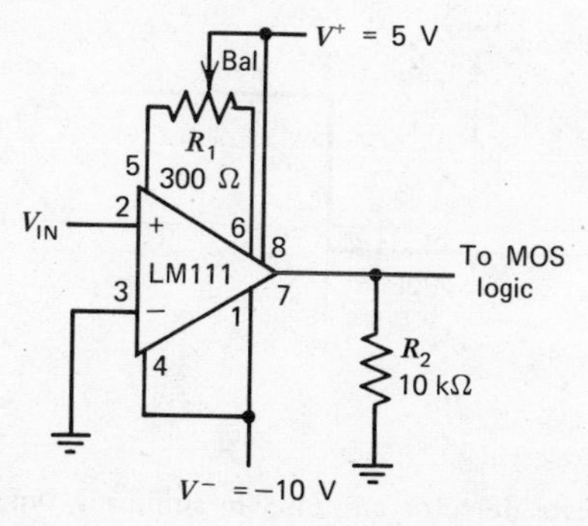

Figure 5-14. Zero-crossing detector driving MOS logic.

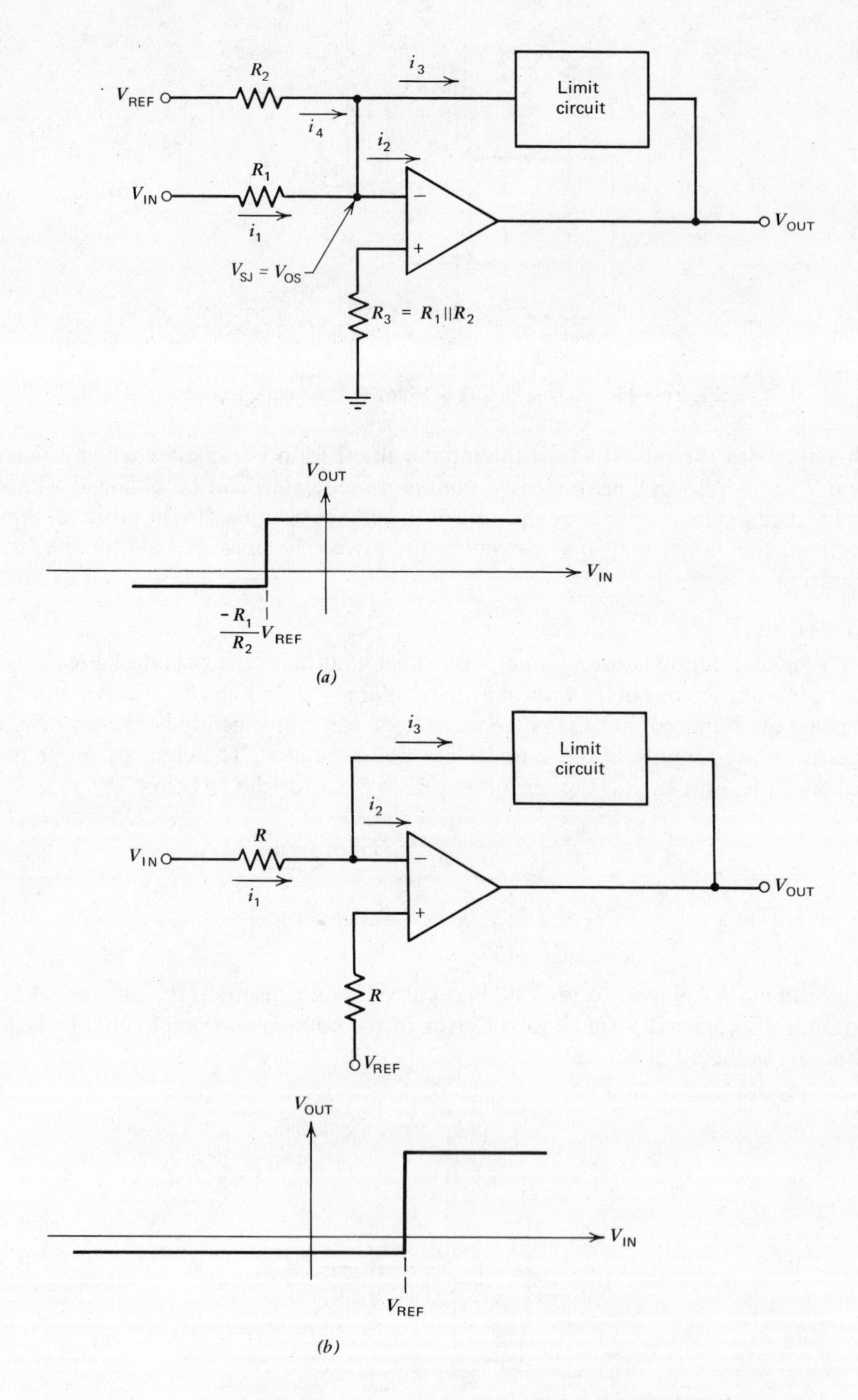

Figure 5-15. Two-level detector circuits: (*a*) summing type (I_2 can be neglected when R_3 is used); (*b*) differential type.

In Figure 5-15*b* only one summing resistor is required, but V_{REF} must be equal to the desired comparison level. The voltage on the inverting input has two major error components: The voltage offset of the amplifier and another error voltage resulting from the finite common-mode rejection of the comparator. Resistor R is needed at the noninverting input if the bias current I_2 causes a significant error.

In some applications where there are large amounts of noise on the signal when it is passing through the threshold of a level detector, it is desirable to have some hysteresis in the transfer characteristic. The hysteresis is generally made somewhat greater than the maximum expected noise. A circuit using external positive feedback to produce hysteresis is shown in Figure 5-16. It operates in the following manner.

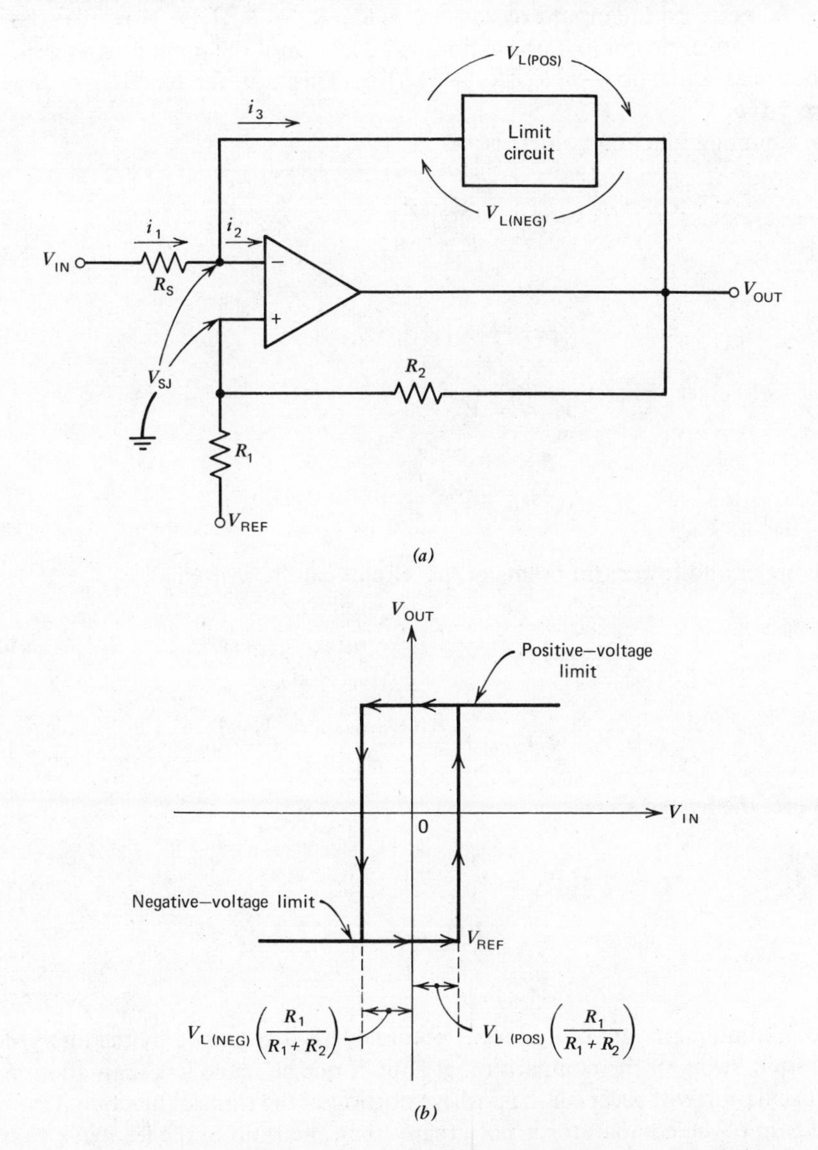

Figure 5-16. Zero-crossing detector with hysteresis: (*a*) circuit diagram ($R_S = R_1 \parallel R_2$ to minimize offsets due to input-bias currents); (*b*) transfer curve.

Assume that the limit circuit contains two 6-V Zener diodes. When the input signal is negative and is approaching zero, the output voltage, V_{OUT}, will be positive since the currents, i_1 and i_3, are negative. The voltage at the two inputs of the op amp will be $+6[R_1/(R_1 + R_2)]$ V; therefore, i_3 will not go to zero until the input voltage is also equal to $+6[R_1/(R_1 + R_2)]$ V. The limit circuit then begins changing to the opposite state (-6 V) and the voltage on the inputs will become $-6[R_1/(R_1 + R_2)]$ V. This action is regenerative because of the positive feedback to the noninverting input, thus increasing the switching speed. The switching can be speeded up even more by placing a small capacitor (10 to 100 pF) in parallel with R_2. Any noise on the input signal at the time this switching occurs will not trigger the circuit to its original state unless the noise on the input exceeds $-2\{6[R_1/(R_1 + R_2)]\}$ V. Now that the input signal is positive, no comparator action will occur until the input crosses zero again and becomes equal to $-6[R_1/(R_1 + R_2)]$ V. This transfer function is shown in Figure 5-16*b*.

The summing junction voltage is

$$V_{SJ} = \frac{V_{OUT}R_1}{R_1 + R_2} \tag{5-9}$$

and

$$V_{OUT} = \pm V_L\left(1 + \frac{R_1}{R_2}\right) \tag{5-10}$$

but

$$V_{OUT} = V_{SJ} \pm V_L \tag{5-11}$$

Thus,

$$V_{SJ} = \pm V_L \frac{R_1}{R_2} \text{ (hysteresis)} \tag{5-12}$$

The upper and lower trip points of this circuit can be written as

$$V_{UT} = V_{REF} + \frac{R_1(V_{OUT(MAX)} - V_{REF})}{R_1 + R_2} \tag{5-13}$$

$$V_{LT} = V_{REF} + \frac{R_1(V_{OUT(MIN)} - V_{REF})}{R_1 + R_2} \tag{5-14}$$

Therefore, the hysteresis is

$$V_{HYST} = V_{UT} - V_{LT}$$

$$= \frac{R_1(V_{OUT(MAX)} - V_{1(MIN)})}{R_1 + R_2} \tag{5-15}$$

The minimum amount of hysteresis obtainable is determined by the forward gain and output swing of the comparator. It should not be made less than about 5 mV, since oscillation will occur on the positive portion of the transfer function if the small signal gain of the comparator is not greater than the ratio of the feedback resistors. When the comparator is used with logic circuits, reduced hysteresis can usually be obtained by taking the positive feedback from the output of the logic.

With the connection in Figure 5-16, a device like the μA710 can be substituted for a Schmitt trigger. It has an advantage in that nonzero trip points can be obtained, and both the upper and lower trip points are easily and independently adjustable over a wide range of positive and negative voltages.

If hysteresis is added to the circuits of Figures 5-15*a* and *b*, the hysteresis can still be calculated by using the above equations, except that the hysteresis will be centered about $-(R_1/R_2)V_{REF}$ in Figure 5-15*a* and about V_{REF} in Figure 5-15*b*. The disadvantage of using hysteresis is that the comparison points do not occur at the zero reference level. Therefore, a trade-off must be made between the degree of noise immunity required and the error at the comparison points. Prefiltering the input signal to reduce the input noise may be helpful. In order to have symmetrical comparison points about zero, the output-voltage limits must be equal in magnitude. In some applications the positive and negative limits will not be the same magnitude, resulting in asymmetry about the nominal comparison point.

Figure 5-17 shows the configuration for a level detector (using the NE/SE521 comparator) employing a 10-mV positive and negative hysteresis loop. Since only one TTL output is available from the 521, an additional inverting gate is necessary to provide hysteresis below the threshold level.

Hysteresis occurs because a small portion of the "one" level output voltage is fed back in phase and added to the input signal. This feedback aids the signal in crossing the threshold. When the signal returns to threshold the positive feedback must be overcome by the signal before switching can occur. The switching process is then assured and oscillations cannot occur. The threshold "dead zone" created by this method prevents output chatter with signals having slow and erratic zero crossings. As is shown in Figure 5-17 the voltage feedback is calculated from the expression

$$V_{HYST} = \frac{V_{OUT} \cdot R_{IN}}{R_{IN} + R_F} \tag{5-16}$$

where V_{OUT} is the gate output voltage. The hysteresis voltage is bounded by the common-mode range (± 3 V) and the ability of the gate to source the current required by the feedback network.

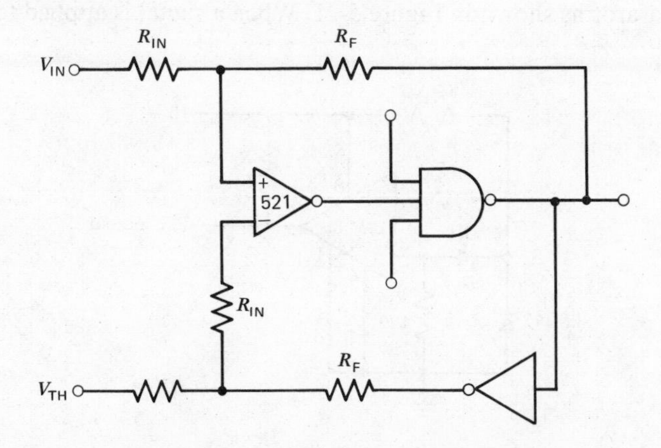

Figure 5-17. Level detector with hysteresis.

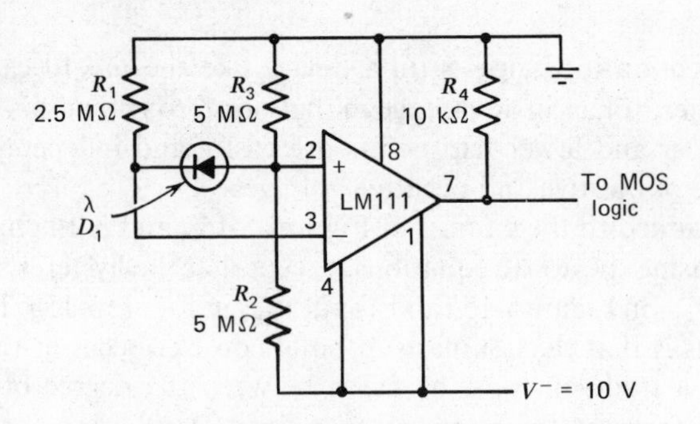

Figure 5-18. Level detector for photodiode.

Figure 5-18 shows a connection for operating with MOS logic. This is a level detector for a photodiode that operates off a −10-V supply. The output changes state when the diode current reaches 1 μA. Even at this low current, the error contributed by the comparator is less than 1%.

Higher threshold currents can be obtained by reducing R_1, R_2, and R_3 proportionally. At the switching point, the voltage across the photodiode is nearly zero, so its leakage current does not cause an error. The output switches between ground and −10 V, driving the data inputs of MOS logic directly.

The circuit in Figure 5-18 can be adapted to work with a 5-V supply. At any rate, the accuracy of the circuit will depend on the supply-voltage regulation, since the reference is derived from the supply. Figure 5-19 shows a method of making performance independent of supply voltage; D_1 is a temperature-compensated reference diode with a 1.23-V breakdown voltage. It acts as a shunt regulator and delivers a stable voltage to the comparator. When the diode current is large enough (about 10 μA) to make the voltage drop across R_3 equal to the breakdown voltage of D_1, the output will change state; R_2 has been added to make the threshold error proportional to the offset current of the comparator, rather than the bias current. It can be eliminated if the bias-current error is not considered significant. Figure 5-20 depicts a photodiode-detector circuit using the NE/SE527 comparator.

PEAK DETECTOR

Peak detection of a signal may be quickly obtained by connecting the NE/SE521 voltage comparator as shown in Figure 5-21. When a signal is applied to the positive

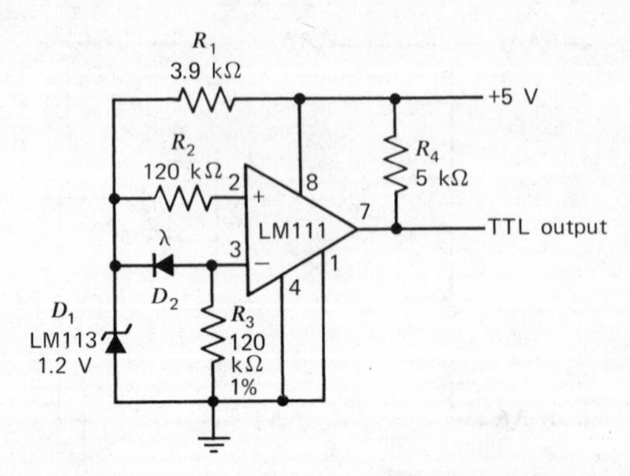

Figure 5-19. Precision level detector for photodiode.

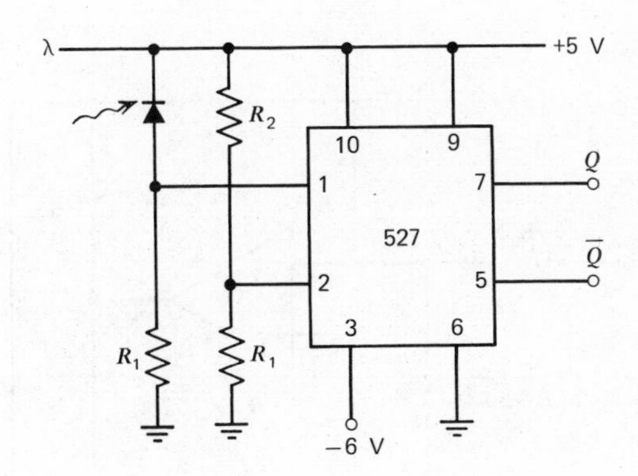

Figure 5-20. Photodiode detector.

input, the negative input is charged to the peak value of that signal through the diode. Some stored charge then provides input-bias current, discharging the inverting input to slightly below the peak signal.

At the peak of the following cycle the input exceeds the stored charge causing the output to go high. With the output high the diode furnishes an additional charge to the inverting input which causes the output to go low again. Thus, the output is a pulse beginning at the peak of the input waveform and having a duration equal to the inverting input.

The 1N457 diode was selected for its relatively slow turn on time to increase the feedback-delay time. If desired, a resistor may be added in series with the diode to provide an additional *RC* time constant to increase the pulse width even further.

WINDOW COMPARATOR

Figure 5-22 shows the circuit diagram and the transfer function for a window comparator. The center of the window is set by the negative of the input V_2, and the window width is twice the input ΔV. Thus, the window can be shifted while maintaining constant window width by varying only one voltage (V_2). Similarly, the window width can be varied by a single voltage, without affecting the center of the window.

The window comparator operates in the following way. When $V_1 + V_2 < 0$, D_2 of A_1 is conducting and D_1 is reverse-biased. Therefore, the output of A_1 does not

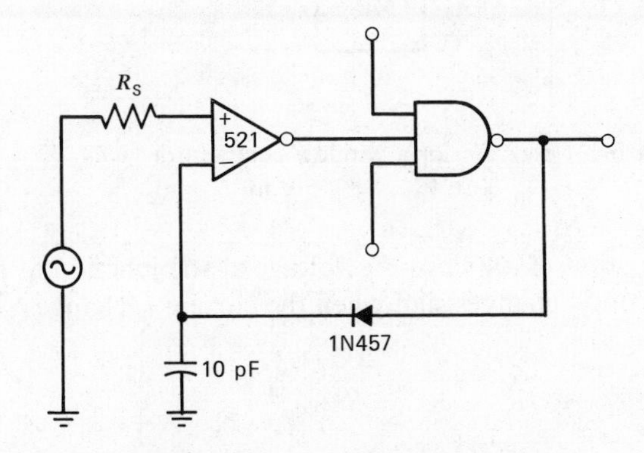

Figure 5-21. Pulse-peak detector.

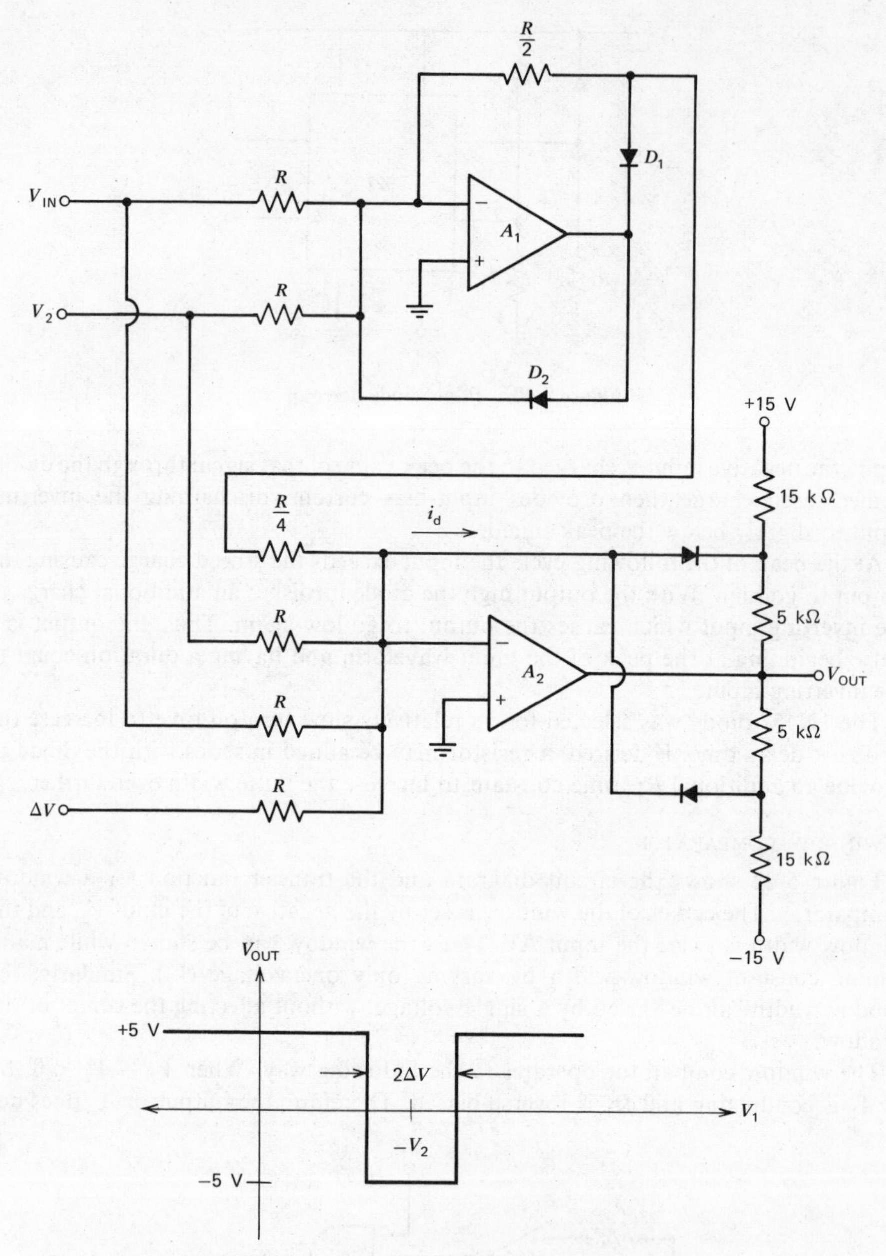

Figure 5-22. Circuit diagram for a window comparator [*note:* $V_{OUT} = -5$ V for $i_D > 0$; $V_{OUT} = +5$ V for $i_D < 0$].

contribute to the output of A_2 since the voltage at the junction of $R/2$ and $R/4$ is 0. The limiter circuit of A_2 changes sign when the current i_d changes sign, or

$$\frac{V_1}{R} + \frac{V_2}{R} + \frac{\Delta V}{R} = i_d = 0 \tag{5-17}$$

so that

$$V_1 = -V_2 - \Delta V \tag{5-18}$$

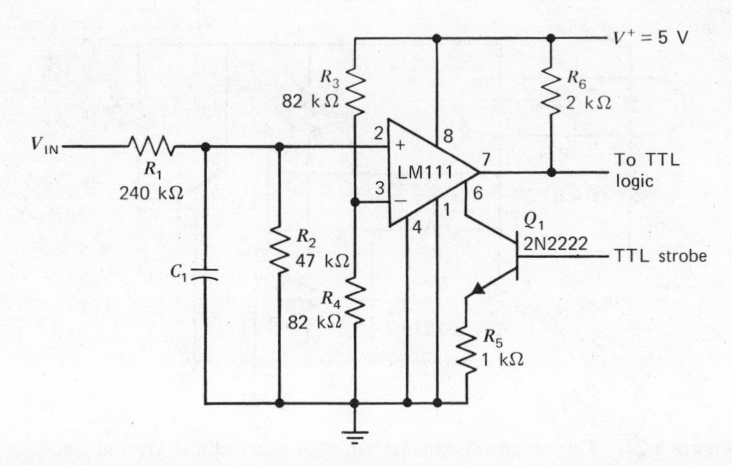

Figure 5-23. Circuit for transmitting data between high-level logic and TTL.

Equation (5-18) gives the lower comparison point of the window. When $V_1 + V_2 > 0$, D_1 will be conducting so that the output of A_1 will be $-1/2(V_1 + V_2)$. Another comparison point will be given by

$$\frac{V_1}{R} + \frac{V_2}{R} + \frac{\Delta V}{R} - \frac{1/2(V_1 + V_2)}{R/4} = i_d = 0 \tag{5-19}$$

so that

$$V_1 = -V_2 + \Delta V \tag{5-20}$$

which is the upper comparison point.

By adding the appropriate logic gates at the outputs of A_1 and A_2 a window comparator can be given three logic outputs called the GO, HIGH, and LOW outputs. Whenever the input signal is inside the window, the GO output will be a logical "1" and the other two outputs will be at a logical "0". If the input signal drops to a value below the window, the LOW output will switch to a logical "1" and the GO output will drop to a logical "0". When the input signal exceeds the window, the HIGH output will be a logical "1" and the other outputs at a logical "0".

DIGITAL INTERFACE CIRCUITS

Figure 5-23 shows an interface between high-level logic and DTL or TTL. The input signal, with 0-V and 30-V logic states is attenuated to 0 and 5 V by R_1 and R_2; R_3 and R_4 set up a 2.5-V threshold level for the comparator so that it switches when the input goes through 15 V. The response time of the circuit can be controlled with C_1, if desired, to make it insensitive to fast noise spikes. Because of the low-error currents of the LM111, it is possible to get input impedances even higher than the 300 kΩ obtained with the indicated resistor values.

The comparator can be strobed, as shown in Figure 5-23, by the addition of Q_1 and R_5. With a logic one on the base of Q_1, approximately 2.5 mA is drawn out of the strobe terminal of the LM111, making the output high independent of the input signal.

Sometimes it is necessary to transmit data between digital equipments, yet maintain a high degree of electrical isolation. Normally, this is done with a transformer.

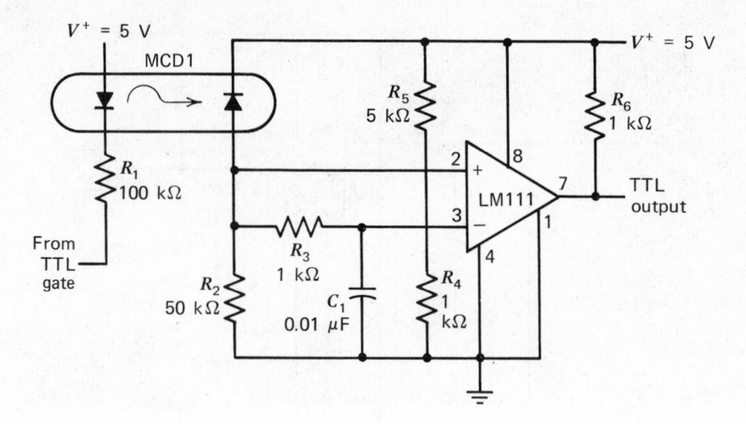

Figure 5-24. Data transmission system with near-infinite ground isolation.

However, transformers have problems with low duty-cycle pulses since they do not preserve the dc level. The circuit in Figure 5-24 is a more satisfactory method of obtaining isolation. At the transmitting end, a TTL gate drives a gallium arsenide light-emitting diode. The light output is optically coupled to a silicon photodiode, and the comparator detects the photodiode output. The optical coupling makes possible electrical isolation in the thousands of megohms at potentials in the thousands of volts. The maximum data rate of this circuit is 1 MHz. At lower rates (~200 kHz) R_3 and C_1 can be eliminated. Figures 5-25 and 5-26 depict ECL (emitter-coupled-logic) to TTL and TTL to ECL interface circuits using the NE/SE527 comparator.

Retrieving signals which have been transmitted over long cables in the presence of high electrical noise is a perfect application for differential comparators. Such systems as automated production lines and large computer systems must transmit high-frequency digital signals over long distances.

If the twisted pair of the system is driven differentially from ground, the signals can be reclaimed easily via a differential-line receiver.

Since the electrical noise imposed upon a pair of wires takes the form of a common-mode signal, the very high common-mode rejection of the NE/SE521 makes the unit ideal for differential-line receivers. Figure 5-27 depicts the simple schematic arrangement. The 521 is used as a differential amplifier having a logic level output. Because

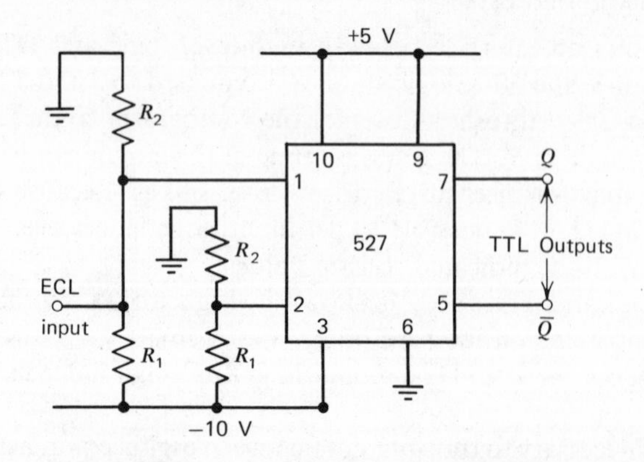

Figure 5-25. Circuit with ECL to TTL interface.

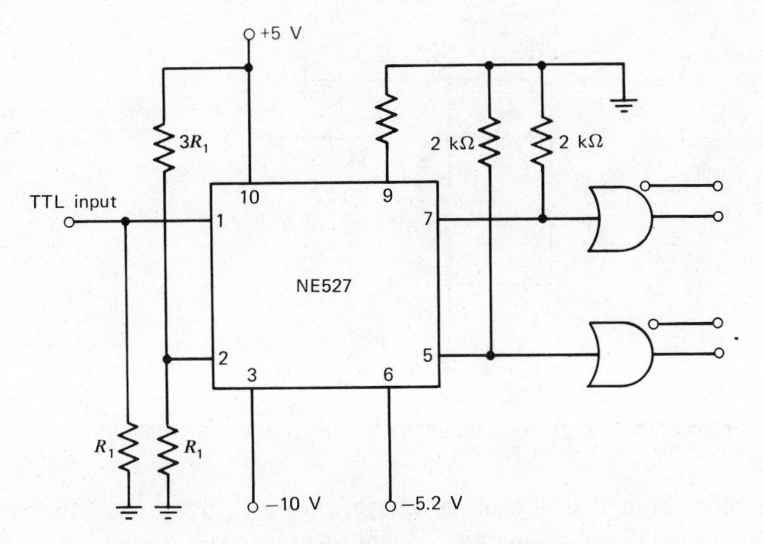

Figure 5-26. Circuit with TTL to ECL interface.

common-mode signals are rejected, noise on the cable disappears and only the desired differential signal remains.

Figure 5-28 shows a line receiver using a μA710 comparator. The resistive divider on the input of the comparator permits higher level logic signals than are possible with the μA710 alone (up to ± 14 V in this case). It is also possible to put a capacitor at the comparator input (C_1) to make the circuit insensitive to fast noise spikes. Because of the high gain of the comparator, fast-rise output pulses can be obtained even with this integrating capacitor.

SENSE AMPLIFIER

Current outputs of the Intel 1103 MOS Memory, require sense amplifiers for conversion to standard DTL/TTL logic. The NE/SE521 comparator is ideal in this

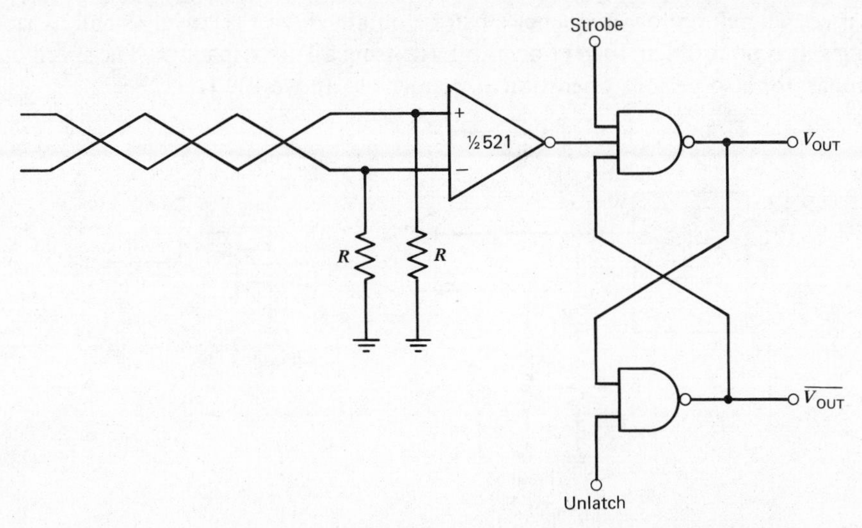

Figure 5-27. Line receiver [*note:* R = 1/2 characteristic line impedance].

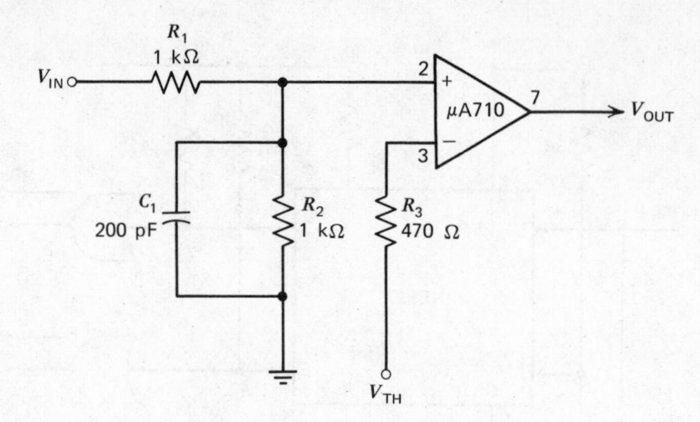

Figure 5-28. High noise-immunity line receiver for slow, high-level logic.

application because of the low-offset voltage, low-offset current, and high speed. Use of the Schottky clamped comparators, as shown in Figure 5-29, significantly increases the total access time of the semiconductor memory. In large memory systems, the NE522 can be used for the "wire-ORed" memory data lines. Figure 5-30 shows the use of the NE527 comparator as an 1103RAM sense amplifier.

MULTIVIBRATORS AND OSCILLATORS

Figure 5-31 depicts a free-running multivibrator using the LM111 comparator. The inputs are biased within the common-mode range by R_1 and R_2; dc stability, which ensures starting, is provided by negative feedback through R_3. The negative feedback is reduced at high frequencies by C_1. At some frequency, the positive feedback through R_4 will be greater than the negative feedback; and the circuit will oscillate. For the component values shown, the circuit delivers a 100-kHz square-wave output. The frequency can be changed by varying C_1 or by adjusting R_1 through R_4 while keeping their ratios constant.

Because of the low-input current of the comparator, large circuit impedances can be used. Therefore, low frequencies can be obtained with relatively small capacitor values. It is no problem to get down to 1 Hz using a 1-μF capacitor. The speed of the comparator also permits operation at frequencies above 100 KHz.

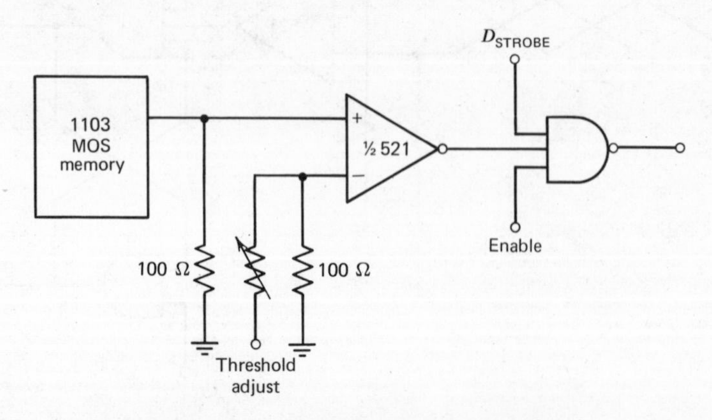

Figure 5-29. 1103 memory sense amplifier.

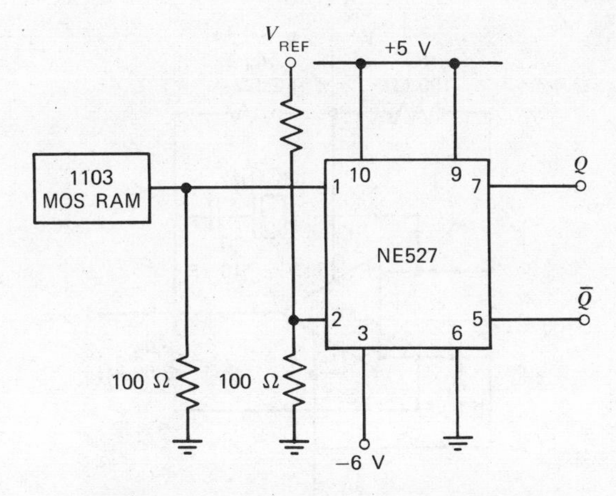

Figure 5-30. MOS memory sense amplifier.

The frequency of oscillation depends almost entirely on the resistance and capacitor values because of the precision of the comparator. Further, the frequency changes by only 1% for a 10% change in supply voltage. Waveform symmetry is also good, but the symmetry can be varied by changing the ratio of R_1 to R_2.

A crystal-controlled oscillator that can be used to generate the clock in slower digital systems is shown in Figure 5-32. It is similar to the free-running multivibrator, except that the positive feedback is obtained through a quartz crystal. The circuit oscillates when transmission through the crystal is at a maximum, so the crystal operates in its shunt-resonant mode. The high-input impedance of the comparator and the isolating capacitor C_2 minimize loading of the crystal and contribute to frequency stability. As shown, the oscillator delivers a 100-kHz square-wave output.

Figure 5-33 shows a typical crystal oscillator circuit using the NE521 comparator. The crystal is operated in its series resonant mode, providing the necessary feedback through the capacitor to the input of the 521. Resistor R_{ADJ} is used to control the amount of feedback for symmetry. Oscillations will start whenever a circuit disturbance such as turning on the power supplies occurs.

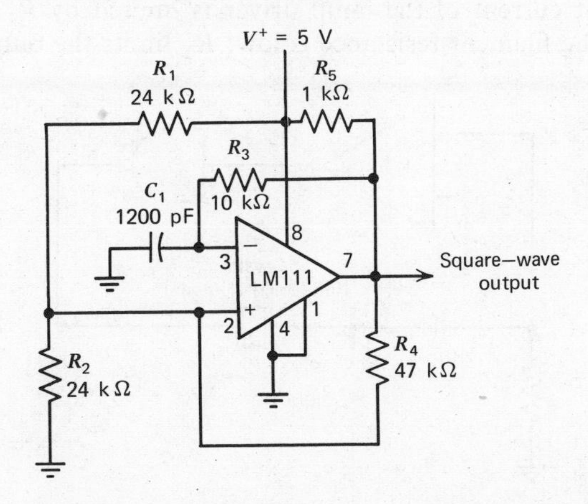

Figure 5-31. Free-running multivibrator [*note:* * TTL or DTL fanout of two].

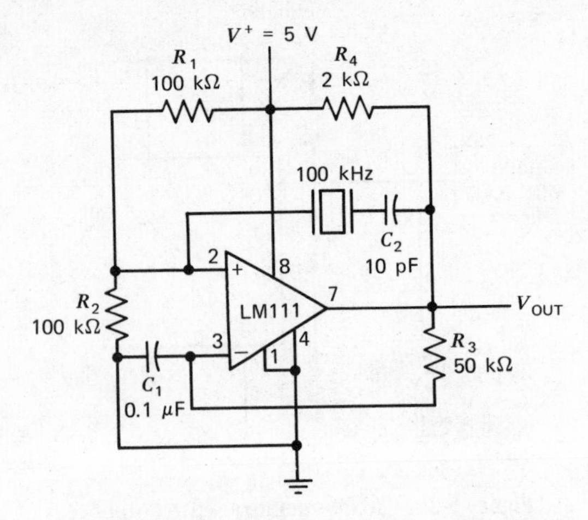

Figure 5-32. Crystal-controlled oscillator.

The NE521 will oscillate up to 70 MHz. However, crystals with frequencies higher than about 20 MHz are usually operated in one of their overtones. To build an oscillator for a specific overtone requires tuned circuits in addition to the crystal to provide the necessary mode suppression. If the spurious modes are not tuned out, the crystal will tend to oscillate at the fundamental frequency.

DOUBLE-ENDED LIMIT DETECTORS

The design of test equipment frequently calls for a circuit that will indicate when a voltage goes outside some preset tolerance limits. The circuit in Figure 5-34, using a μA711 dual comparator, will accomplish this function. A lower limit voltage (V_{LT}) and an upper limit voltage (V_{UT}) are supplied to the dual comparator. When the input voltage exceeds the upper limit or drops below the lower limit, the output of the comparator swings positive and turns ON the lamp driver. A feature of the circuit is that the limit detector can be disabled when it is not being used by grounding the strobe terminals. In addition, up to eight dual comparators can be wired with common outputs and used to feed a single lamp driver.

The peak-output current of the lamp driver is limited by R_2 while the bulb is turning ON and the filament resistance is low; R_1 limits the output current of the

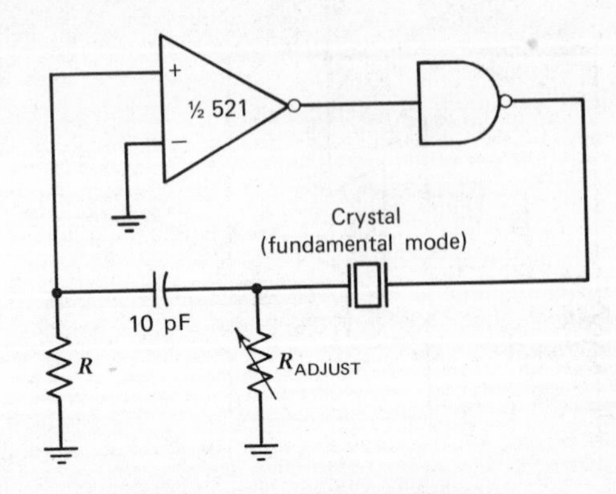

Figure 5-33. Crystal oscillator.

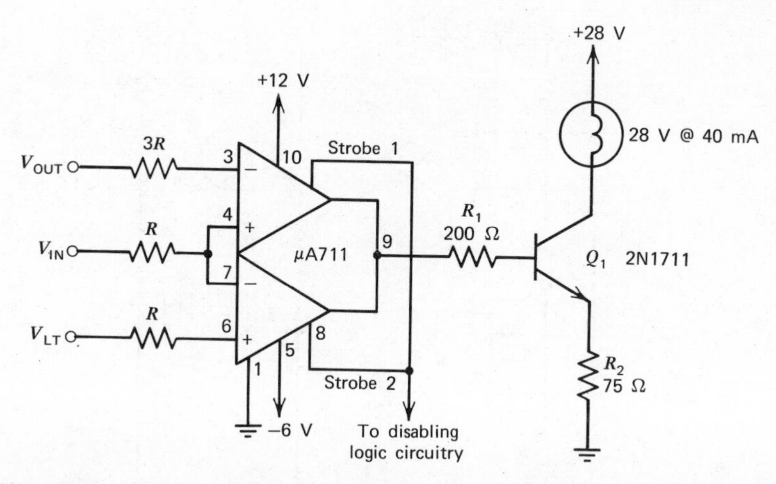

Figure 5-34. Double-ended limit detector for automatic go/no-go test equipment.

comparator after the lamp driver saturates. To make the accuracy dependent on offset currents rather than bias currents, the relative values of the source resistances for the signal and reference voltages should be as indicated on the schematic, and should be as low as possible. Figure 5-35 shows the schematic diagram of a double-ended limit detector using the NE522 comparator.

ANALOG-TO-DIGITAL CONVERTERS

There are many types of A/D converter designs. Each has its own merits. However, where speed of conversion is of prime interest the multithreshold parallel conversion type is used exclusively. It is apparent from Figure 5-35 that the conversion speed of this design is the sum of the delay through the comparator and the decoding gates.

The sacrifices which must be made to obtain speed are the number of components, bit accuracy, and cost. Although the NE521 provides two comparators per package, the length of parallel converters is usually limited to less than 4 bits. Accuracy of multithreshold A/D converters also suffers since the integrity of each bit is dependent upon comparator-threshold accuracy.

The implementation of a 3-bit parallel A/D converter is shown in Figure 5-36. Reference voltages for each bit are developed from a precision resistor ladder network. Values of R and 2 R are chosen so that the threshold is one-half of the least significant bit. This assures maximum accuracy of $\pm 1/2$ bit.

It is apparent from the schematic that the individual strobe line and duality features on the NE521 has greatly reduced the cost and complexity of the design. It should be noted that a parallel converter usually exhibits different delay times for different bits. The output is, therefore, usually strobed into a register only after a certain time has elapsed to ensure that all data has arrived.

COMPARATOR-FOLLOWER COMBINATIONS

Monolithic voltage-comparators are available today which are both fast and accurate. They can detect the height of a pulse with a 5-mV accuracy within 40 nsec. However, these devices have relatively high-input currents and low-input impedances, which reduces their accuracy and speed when operating from high-source resistances. This is probably a basic limitation since the input transistors of the IC must be

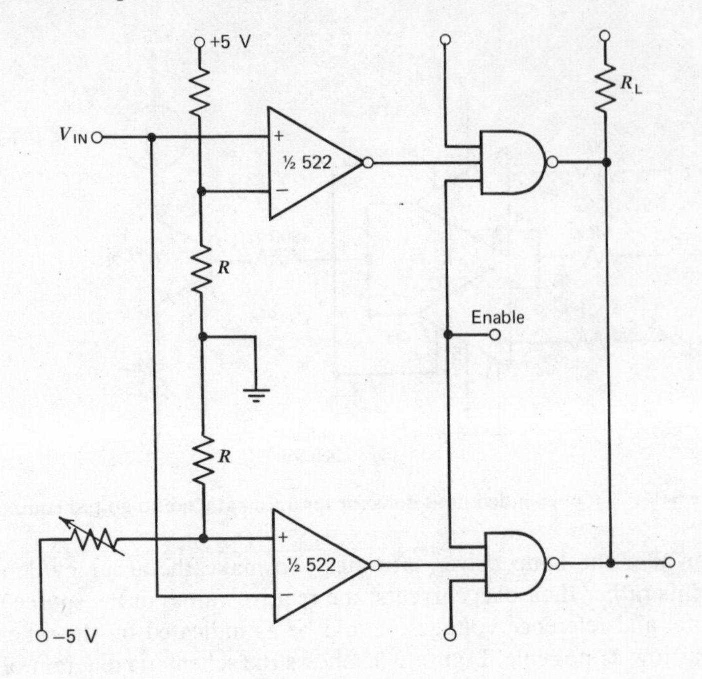

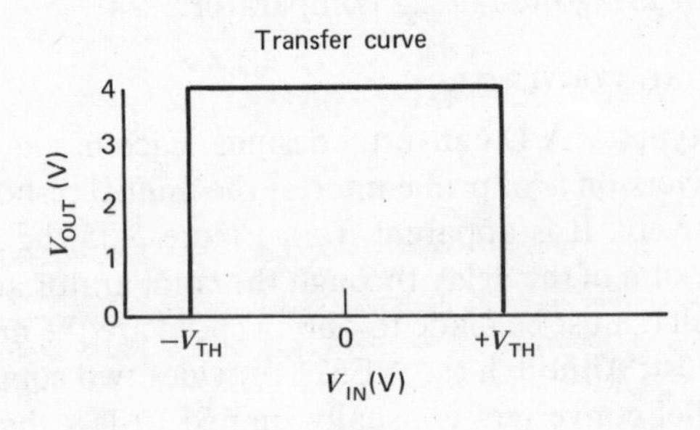

Figure 5-35. Double-ended limit detector.

operated at a relatively high current to get fast operation. Further, the circuit must be gold-doped to reduce storage time, and this limits the current gain that can be obtained in the transistors. High-gain transistors operating at low-collector currents are necessary to get good input characteristics.

One way of overcoming this difficulty is to buffer the input of the comparator. A voltage-follower such as the LM102 or the new LM110 is ideally suited for this job because it is fast and has low-input currents.

At the time the circuits in Figure 5-37 and 5-38 were designed, the LM102 was the best follower available. It reduced the input current of the comparator by more than three orders of magnitude without greatly reducing speed. Better performance is provided by the LM110, as indicated in Tables 4-1 and 4-2.

FAST A/D CONVERTERS

In Figure 5-37*a*, the LM110 voltage-follower buffers the output of a ladder network and drives one input of the comparator. The analog signal is fed to the other input

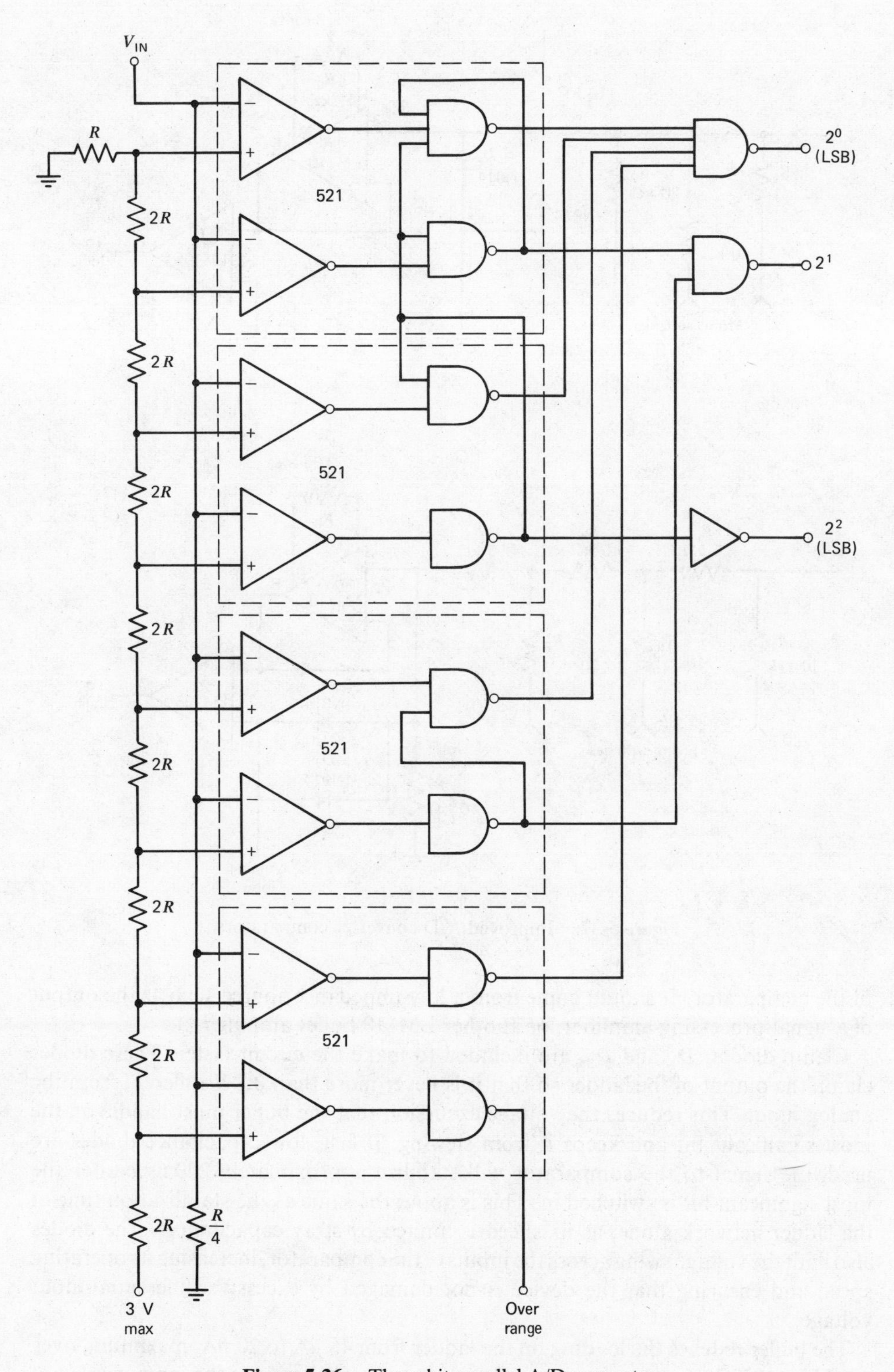

Figure 5-36. Three-bit parallel A/D converter.

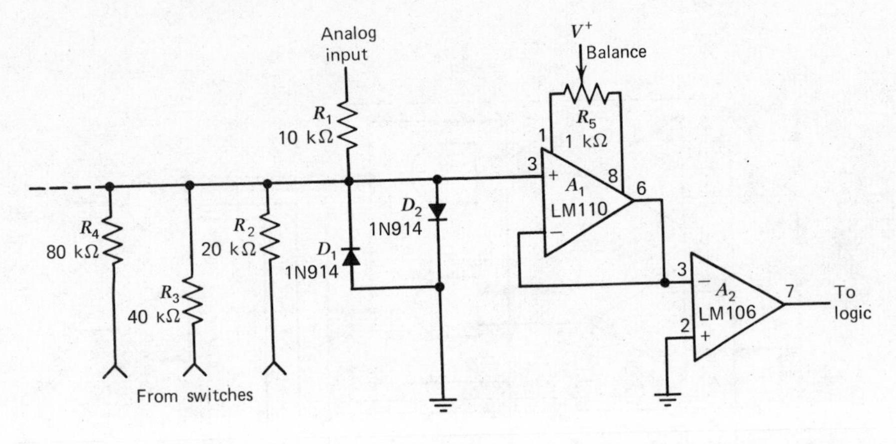

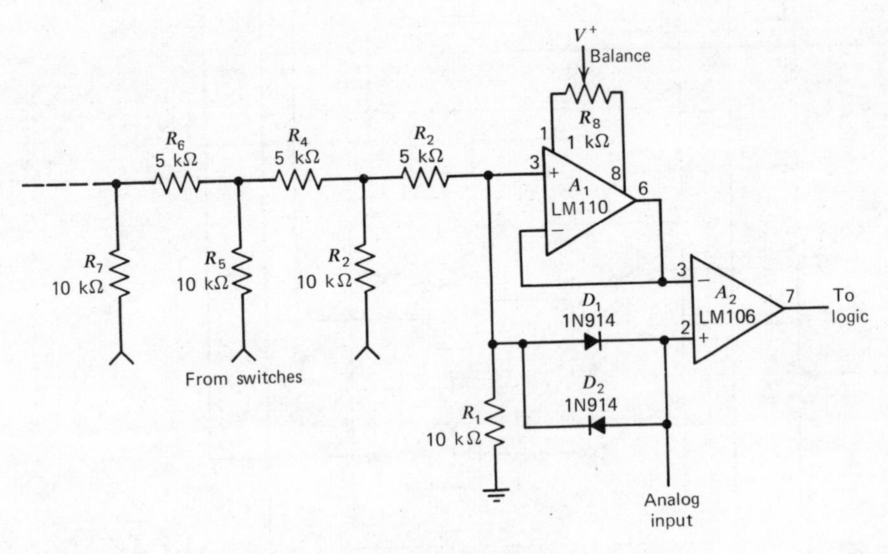

Figure 5-37. Improved A/D converter comparators.

of the comparator. It should come from a low-impedance source such as the output of a signal-processing amplifier, or another LM110 buffer amplifier.

Clamp diodes, D_1 and D_2, are included to make the circuit faster. These diodes clamp the output of the ladder so that it is never more than 0.7 V different from the analog input. This reduces the voltage excursion that the buffer must handle on the most significant bit and keeps it from slewing. If fast, low-capacitance diodes are used, the signal to the comparator will stabilize approximately 200 nsec after the most significant bit is switched in. This is about the same as the stabilization time of the ladder network alone, as its speed is limited by stray capacitances. The diodes also limit the voltage swing across the inputs of the comparator, increasing its operating speed and ensuring that the device is not damaged by excessive differential-input voltage.

The buffer reduces the loading on the ladder from 45 μA to 20 nA, maximum, over a -55 to 125°C temperature range. Hence, in most applications the input current of

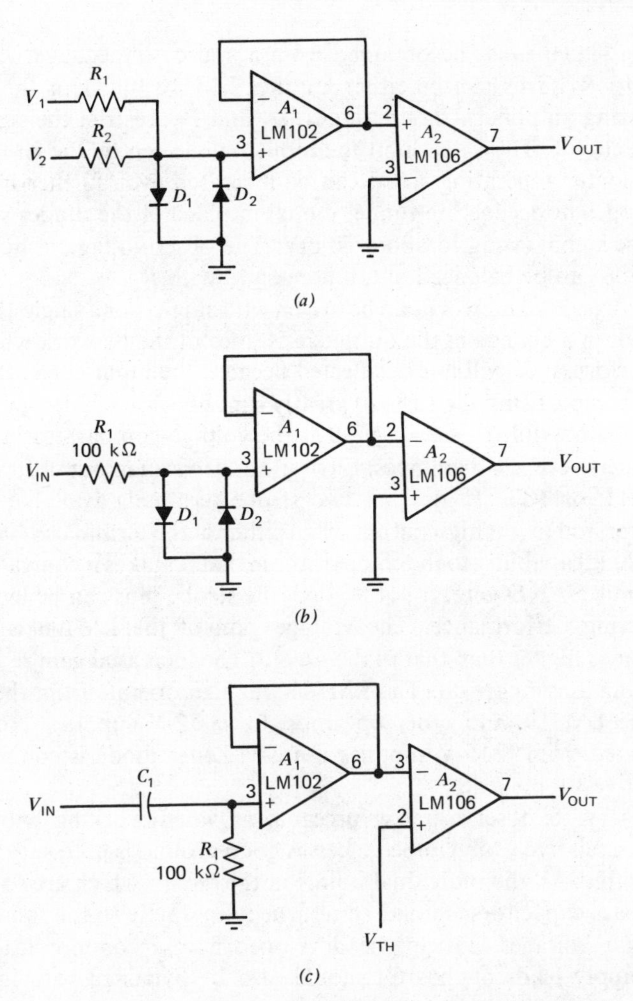

Figure 5-38. Applications requiring comparators with low-input currents: (*a*) comparator for signals of opposite polarity; (*b*) zero-crossing detector: (*c*) comparator for ac-coupled signals.

the buffer is totally insignificant. This low current will often permit the use of larger resistances in the ladder which simplifies design of the switches driving it.

It is possible to balance out the offset of the LM110 with an external 1-kΩ potentiometer, R_9. The adjustment range of this balance control is large enough so that it can be used to null out the offset of both the buffer and the comparator. A 10-kΩ resistor should be installed in series with the input to the LM110, as shown. This is required to make the short-circuit protection of the device effective and to ensure that it will not oscillate. This resistor should be located close to the IC.

A similar technique can be used with A/D converters employing a binary-weighted resistor network. This is shown in Figure 5-37*b*. The analog input is fed into a scaling resistor, R_1. This resistor is selected so that the input voltage to the LM110 is zero when the output of the D/A network corresponds to the analog-input voltage. Hence, if the D/A output is too low, the output of the LM106 will be a logical 0; and the output will change to a logical 1 as the D/A output exceeds the analog signal.

The analog signal must be obtained from a source impedance which is low by comparison to R_1. This can be either another LM110 buffer or the output of the signal-processing amplifier. Clamp diodes, D_1 and D_2, restrict the signal swing and speed up the circuit. They also limit the input signal seen by the LM106 to protect it from over loads. Operating speed can be increased even further by using silicon backward diodes (a degenerate tunnel diode) in place of the diodes shown, as they will clamp the signal swing to about 50 mV. The offset voltage of both the LM110 and the LM106 can be balanced out, if necessary, with R_6.

The binary-weighted network can be driven with single-pole, single-throw switches. This will result in a change in the output resistance of the network when it switches, but circuit performance will not be affected because the input current of the LM110 is negligible. Hence, using the LM110 greatly simplifies switch design.

Although it is possible to use a μA710 as the voltage-comparator in these circuits, the LM106 offers several advantages. First, it can drive a fan out of 10 with standard, integrated DTL or TTL. It also has two strobe terminals available which disable the comparator and give a high output when either of the terminals is held at a logical 0. This adds logic capability to the comparator in that it makes it equivalent to a μA710 and a two-input NAND gate. If not needed, the strobe pins can be left unconnected without affecting performance. The voltage gain of the LM106 is about 45,000 which is 30 times higher than that of the μA710. The increased gain reduces the error band in making a comparison. The LM106 will also operate from the same supply voltage as the LM110, and other op amps, for $\pm$12-V supplies. However, it can also be operated from $\pm$15-V supplies if a 3-V Zener diode is connected in series with the positive-supply lead.

It is necessary to observe a few precautions when working with fast circuits operating from relatively high impedances. A good ground is necessary, and a ground plane is advisable. All the individual points in the circuit which are to be grounded, including bypass capacitors, should be returned separately to the same point on the ground so that voltages will not be developed across common-lead inductance. The power-supply leads of the ICs should also be bypassed with low-inductance 0.01-μF capacitors. These capacitors, preferably disc ceramic, should be installed with short leads and located close to the devices. Lastly, the output of the comparator should be shielded from the circuitry on the input of the buffer, as stray coupling can also cause oscillation.

OTHER FAST COMPARATORS

Although the circuits shown so far were designed for use in A/D converters, the same techniques apply to a number of other applications. Figure 5-38 gives examples of circuits which can put stringent input-current requirements on the comparator. The first is a comparator for signals of opposite polarity. Resistors R_1 and R_2 are required to isolate the two signal sources. Frequently, these resistors must be relatively large so that the signal sources are not loaded. Hence, the input current of the comparator must be reduced to prevent inaccuracies. Another example is the zero-crossing detector in Figure 5-38*b*. When the input signal can exceed the common-mode range of the comparator ($\pm$5 V for the LM106), clamp diodes must be used. It is then necessary to isolate the comparator from the input with a relatively large resistance to prevent loading. Again, bias currents should be reduced. A third example, in Figure 5-38*c* is a comparator with an ac-coupled input. An LM106 will draw an input current which is twice the specified bias current when the signal is above the comparison threshold. Yet, it draws no current when the signal is below the threshold. This asymmetrical current drain will charge any coupling capacitor on the input and produce an error. This problem can be eliminated by using a buffer, as the input current will be both low and constant.

LONG-TIME COMPARATOR

With the exception of the long-time comparator, the circuits in Figure 5-39 can be considered routine applications. In the long-time comparator, the 2N4393 is operated as a Miller integrator. The high Y_{fs} of the 2N4393 (over 12,000 μmhos at 5 mA) yields a stage gain of about 60. Since the equivalent capacitance looking into the gate is C times gain and the gate source resistance can be as high as 10 MΩ, time constants as long as a minute can be achieved.

FREQUENCY DOUBLER

In a digital system, it is a relatively simple matter to divide by any integer. However, multiplying by an integer is quite another story especially if operation over a wide frequency range and waveform symmetry are required.

A frequency doubler that satisfies the above requirements is shown in Figure 5-40. A comparator is used to shape the input signal and feed it to an integrator. The shaping is required because the input to the integrator must swing between the supply voltage and ground to preserve symmetry in the output waveform. An LM108 op amp, that works from the 5-V logic supply, serves as the integrator. This feeds a triangular waveform to a second comparator that detects when the waveform goes through a voltage equal to its average value. Hence, as shown in Figure 5-41, the output of the second comparator is delayed by half the duration of the input pulse. The two comparator outputs can then be combined through an exclusive-OR gate to produce the double frequency output.

With the component values shown, the circuit operates at frequencies from 5 to 50 kHz. Lower frequency operation can be secured by increasing both C_1 and C_2.

SWITCHING REGULATOR

A switching regulator that uses digital current limiting provides a current regulation of better than 1%, while offering good frequency stability and fast response to transients. Essentially, the regulator uses two NAND gates as a pulse-width modulator to keep current from exceeding the present limit.

The circuit of Figure 5-42 employs a zener diode, D_1, to provide both the circuit reference voltage and supply power for the logic. A hot-carrier diode, D_2, is used for efficient elimination of spikes in the output.

As long as the current limit, I_L, is not exceeded, transistor Q_1 remains OFF; I_L is determined by the value of resistor R_1. For the resistor shown, I_L approximately equals 10 amperes.

The signal at Q_1's collector is used to inhibit the output pulses from comparator A_1. When Q_1 is OFF, all pulses pass through NAND gates G_1 and G_2. However, if I_L exceeds the preset 10 A, Q_1 conducts, and its collector voltage goes low. This inhibits the comparator pulses until the current falls below I_L.

A triangular wave is applied to the noninverting comparator input. It is obtained from a multivibrator whose pulse output is integrated and written on the zener

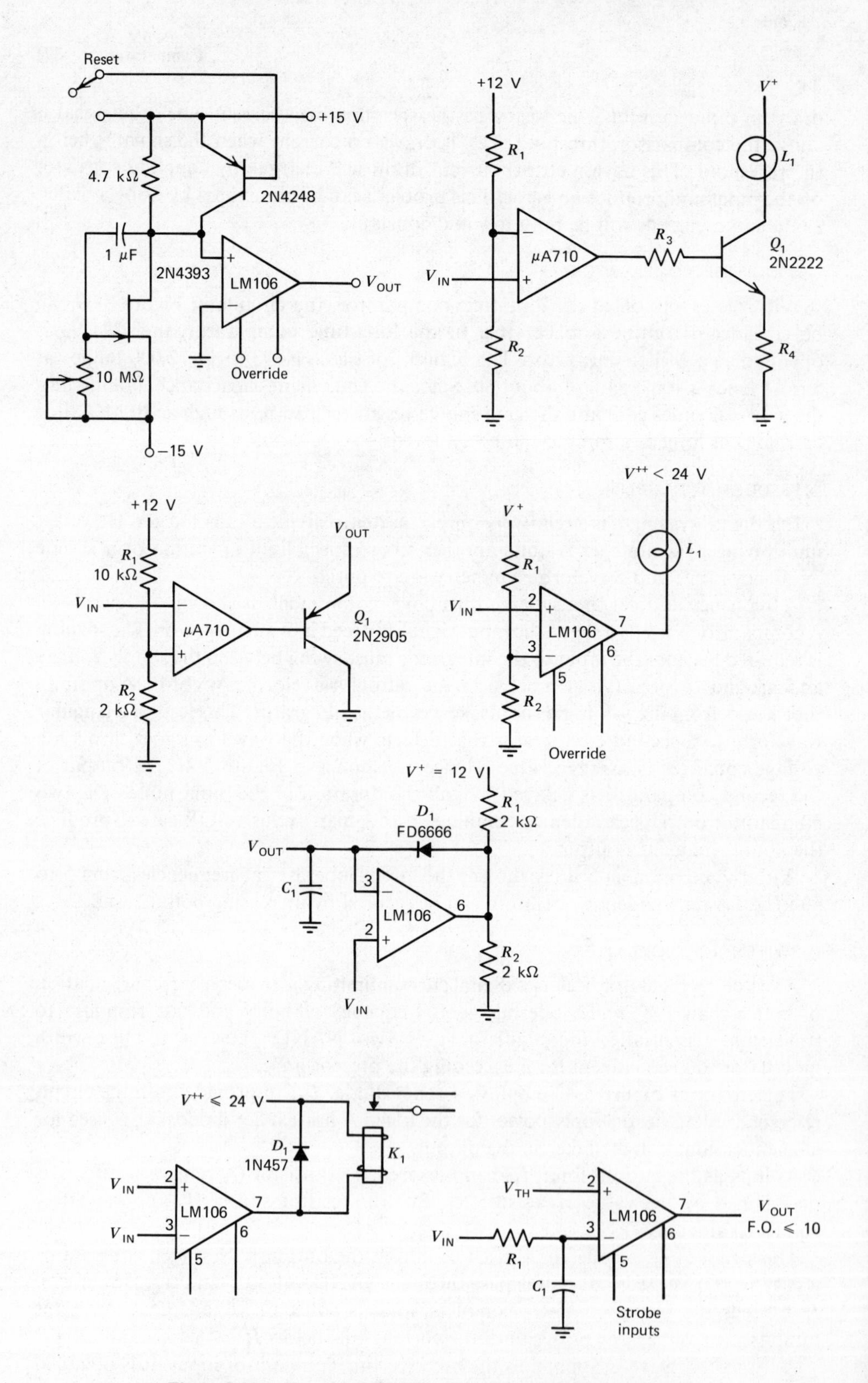

Figure 5-39. Examples of applications for other comparators.

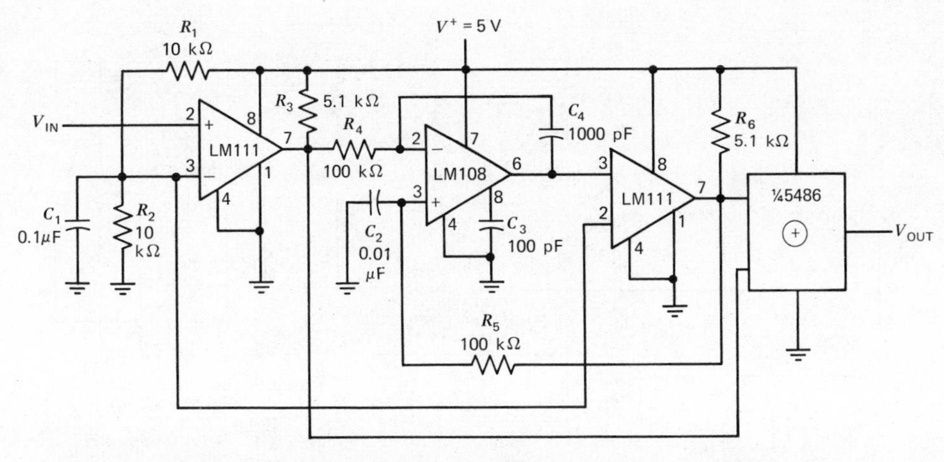

Figure 5-40. Frequency doubler [*note:* frequency range, V_{IN} = 5–50 kHz; V_{OUT} = 10–100 kHz].

reference voltage. The comparator switches one time during each triangle cycle, when the wave's slope is positive; the switching point is determined by setting the 1-kilohm potentiometer. Operating frequency of the switching regulator is the frequency of the triangular wave.

OPERATIONAL AMPLIFIERS AS COMPARATORS

General-purpose amplifiers such as the LM101A also make good comparators because they have differential-input voltage ranges and are easily clamped to make their outputs compatible with logic and driver circuits. Internally compensated op amps are generally poor comparators because they switch very slowly.

A comparator circuit using the LM101A appears in Figure 5-43. This clamping scheme makes the output signal directly compatible with DTL or TTL ICs. A breakdown diode D_1 clamps the output at 0 or 4 V in the low or high states, respectively. This particular diode was chosen because it has a sharp breakdown and a low equivalent capacitance. When working as a comparator, the amplifier operates open-loop; thus frequency compensation is not normally needed. Nonetheless, the stray capacitance between *pins* 5 and 6 of the amplifier should be minimized to prevent low-level oscillations when the comparator is in the active region. If this becomes a problem, a 3-pF capacitor on the normal compensation terminals will eliminate it.

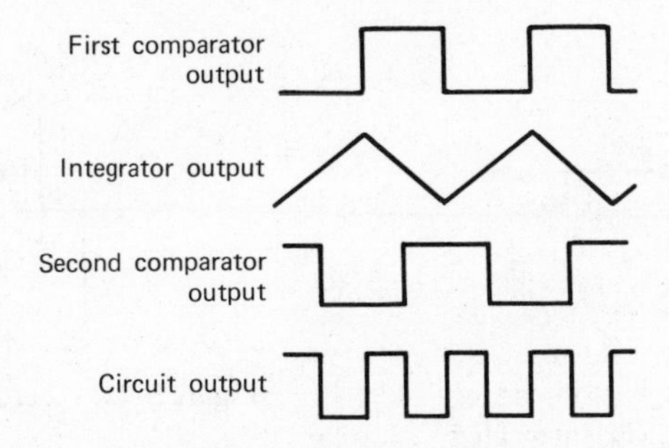

Figure 5-41. Waveforms for the frequency doubler.

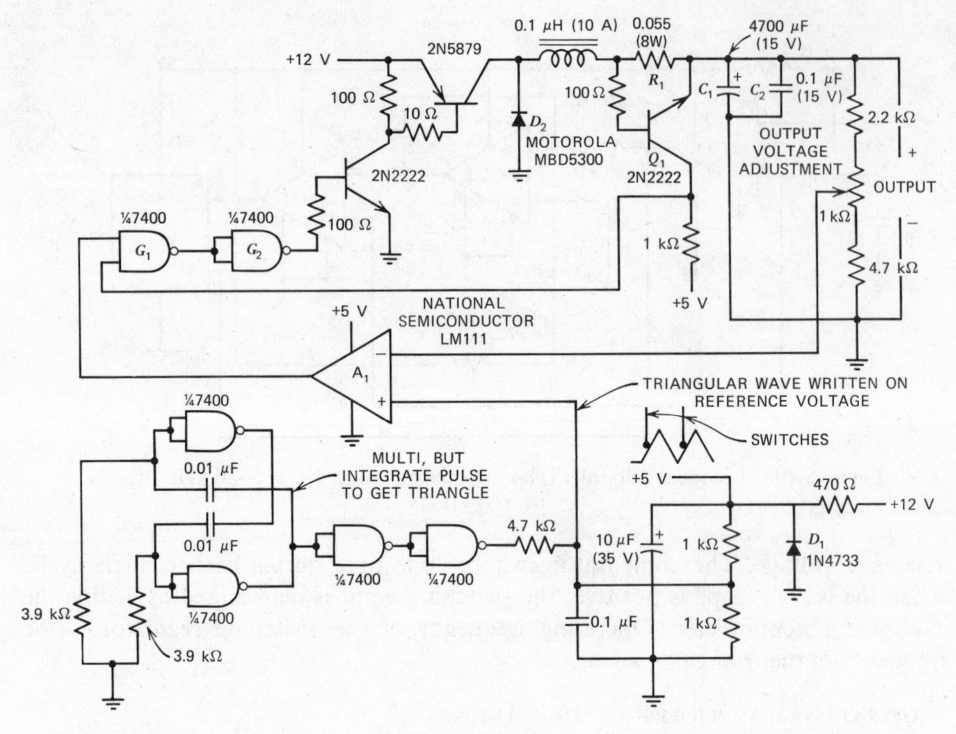

Figure 5-42. Output current switching regulator is held to desired limit by digital comparator.

Like most op amps, it is possible to use the LM108 as a voltage-comparator. Figure 5-44 shows the device used as a simple zero-crossing detector. The inputs of the IC are protected internally by back-to-back diodes connected between them, therefore, voltages in excess of 1 V cannot be impressed directly across the inputs. This problem is taken care of by R_1 which limits the current so that input voltages in excess of 1 kV can be tolerated. If absolute accuracy is required or if R_1 is made much larger than 1 MΩ, a compensating resistor of equal value should be inserted in series with the other input.

In Figure 5-44, the output of the op amp is clamped so that it can drive DTL or TTL directly. This is accomplished with a clamp diode on *pin* 8. When the output swings positive, it is clamped at the breakdown voltage of the Zener. When it swings negative, it is clamped at a diode drop below ground. If the 5-V logic supply is used

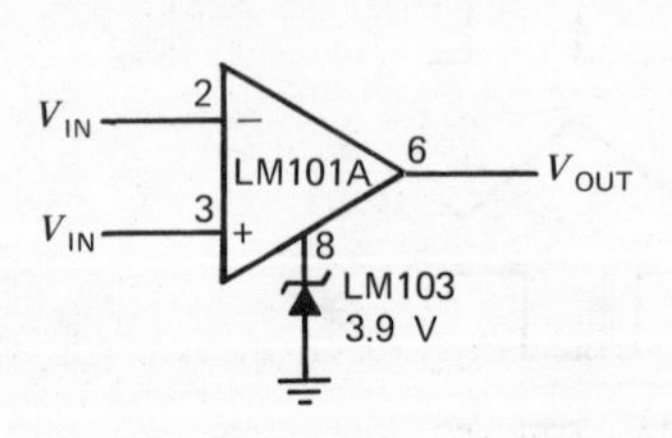

Figure 5-43. Voltage comparator for driving DTL or TTL integrated circuits.

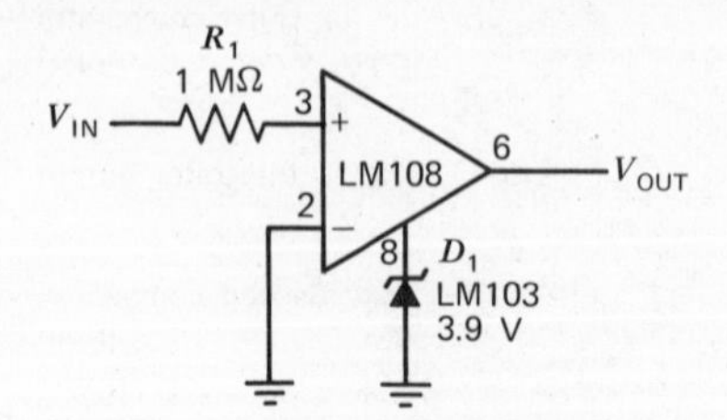

Figure 5-44. Zero-crossing detector.

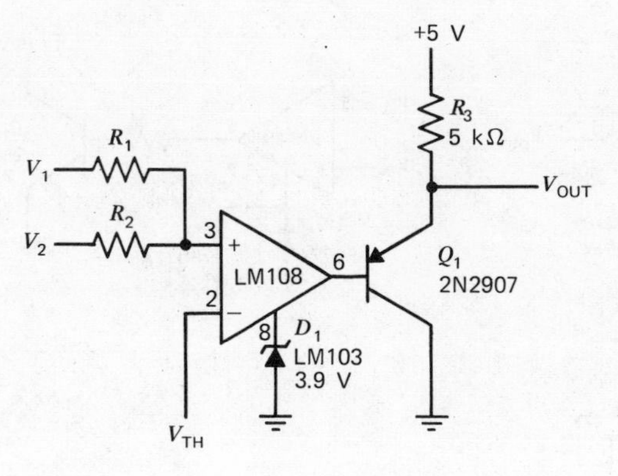

Figure 5-45. Voltage-comparator with output buffer.

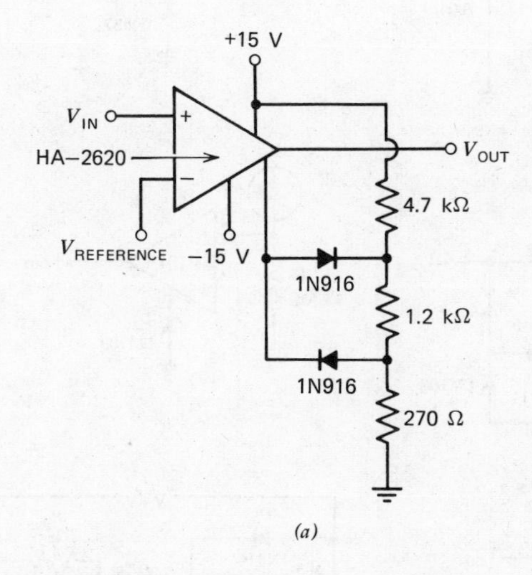

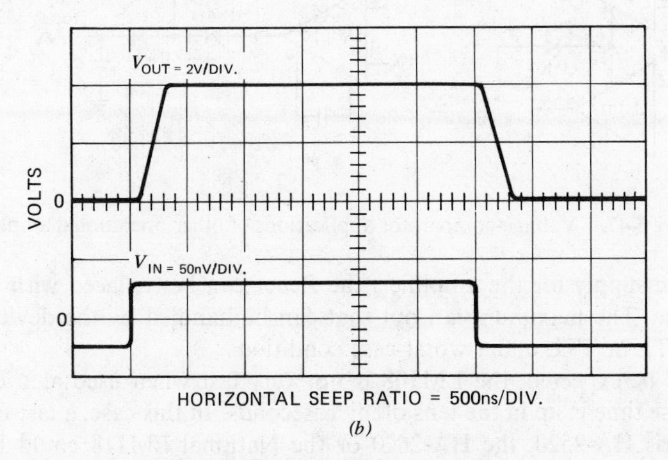

Figure 5-46. High impedance comparator: (a) schematic; (b) waveforms for HA-2620 comparator.

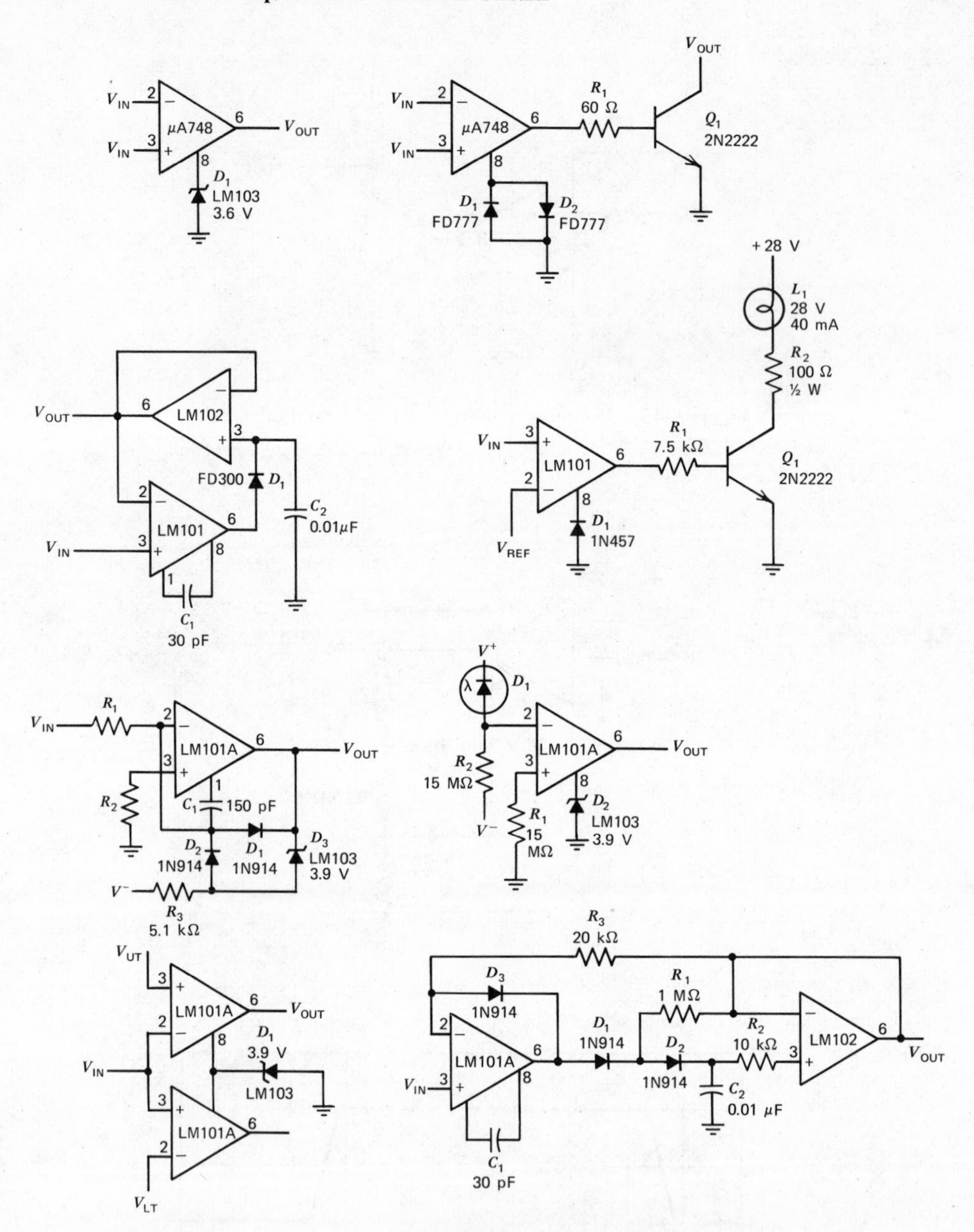

Figure 5-47. Voltage-comparator applications of other operational amplifiers.

as a positive supply for the amplifier, the Zener can be replaced with an ordinary silicon diode. The maximum fan out that can be handled by the device is one for standard DTL or TTL under worst-case conditions.

As might be expected, the LM108 is not very fast when used as a comparator. The response time is up in the tens of microseconds. In this case, a fast op amp such as the Harris HA-2520, the HA-2620 or the National LM118 could be used. An LM103 is recommended for D_1, rather than a conventional alloy Zener, because it has lower capacitance and will not slow the circuit further. The sharp breakdown

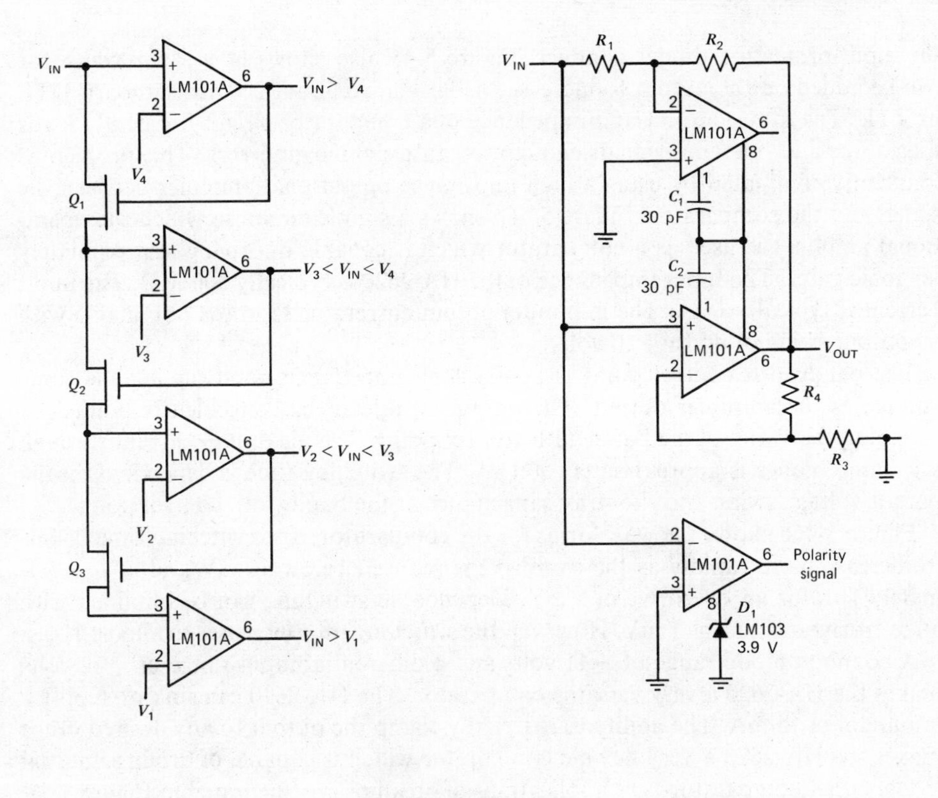

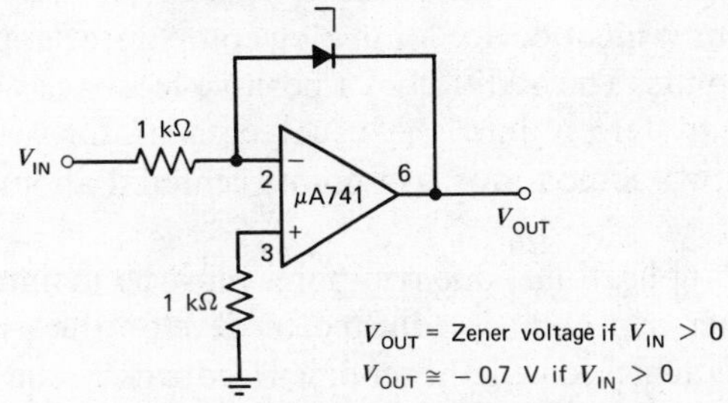

Figure 5-47. *(continued)*

of the LM103 at low currents is also an advantage as the current through the diode in clamp is only 10 μA.

Figure 5-45 shows a comparator for voltages of opposite polarity. The output changes state when the voltage on the junction of R_1 and R_2 is equal to V_{TH} (threshold voltage). Mathematically, this is expressed by

$$V_{TH} = V_2 + \frac{R_2(V_1 - V_2)}{R_1 + R_2}. \tag{5-21}$$

The LM108 can also be used as a differential comparator, going through a transition when two input voltages are equal. However, resistors must be inserted in series with the inputs to limit current and minimize loading on the signal sources when the input-protection diodes conduct. Figure 5-45 also shows how a *pnp* transistor can be added on the output to increase the fan out to about 20 with standard DTL or TTL. The input current and impedance of a comparator circuit frequently loads the source and reference signals enough to cause significant errors. This problem is frequently eliminated by using a high impedance operational amplifier between the signal and the comparator. Figure 5-46*a* shows a simple circuit in which the operational amplifier is used as a comparator which is capable of driving approximately ten logic gates. The input impedance of the HA-2620 is typically 500 MΩ. The input current is typically 1 nA. The minimum output current of 15 mA is obtainable with an output swing of up to ± 10 volts.

The bandwidth control point is a very high impedance point having the same voltage as the amplifier output. The output swing can be conveniently limited by clamping the swing of the bandwidth control point. The maximum current through the clamp diodes is approximately 300 μA. The switching time is dependent on the output voltage swing and the stray capacitance at the bandwidth control point.

Figure 5-46*b* shows the waveforms for the comparator. The switching time begins to increase more rapidly as the overdrive is reduced below 10 mV and is approximately 1 μs for an overdrive of 5 mV. Dependable switching can be obtained with an overdrive as small as 1 mV. However, the switching time increases to almost 12 μs.

A common mode range of ± 11 volts and a differential input range of ± 12 volts makes the HA-2620 a very versatile comparator. The HA-2620 can sink or supply a minimum of 15 mA. The ability to externally clamp the output to any desired range makes the HA-2620 a very flexible comparator which is capable of driving unusual loads. Other comparator-type applications of op amps are illustrated in Figure 5-47.

An op amp is a convenient device for analog comparator applications that require two different trip points. The addition of a positive-feedback network will introduce a precise variable hysteresis into the usual comparator switching action. Such feedback develops two comparator trip points centered about the initial trip point or reference point.

In some control applications, one trip point must be maintained at the reference level, while the other trip point is adjusted to develop the hysteresis. This type of comparator action is achieved with the modified feedback circuit shown in Figure 5-48.

Signal diode D_1 interrupts only one polarity of the positive feedback supplied through resistor R_2. Hysteresis, then, is developed for only one comparator state, and one trip point remains at the original level set by the reference voltage V_{REF}. The second trip point, the one added by hysteresis, is removed from the original trip point by:

$$\Delta V = \frac{R_1(V_Z - V_{REF})}{(R_1 + R_2)} \tag{5-22}$$

where V_Z, the Zener voltage, is greater than reference voltage V_{REF}. Varying resistor R_2 will adjust the hysteresis without disturbing the trip point at V_{REF}.

The circuit's other performance characteristics are similar to the common op amp comparator circuit. The accuracy of both trip points is determined by the op amps input-offset voltage, input-bias current, and finite gain. Resistor R_3 limits the current drain through the Zener diode, and resistor R_4 provides a discharge path for the capacitance of diode D_2.

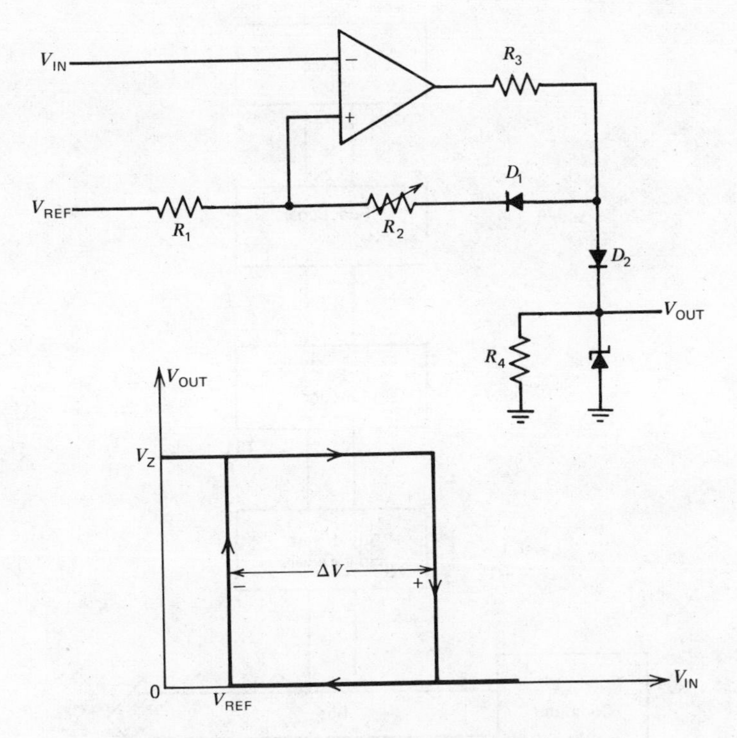

Figure 5-48. Comparator with controllable hysteresis.

The output signal can be taken either directly from the op amp output or from the Zener diode, as shown. With the latter hookup, the output-signal voltage alternates between zero and Zener voltage V_Z, which might be desirable for interfacing with digital logic circuits. It should be noted, however, this output cannot sink current in the 0-V state.

Switching speed is determined by the op amp's slewing-rate limit for high-level input-drive signals. When the input drive is a low-level signal, the output rate of change is limited by the gain available to multiply the input signal's rate of change. Both the slew-rate-limiting and the gain-limiting of switching time are eased if phase compensation is removed from the op amp.

INTERFACE CIRCUITS

Interface circuits are used whenever data must be passed between two points—between computer and nearby terminal (all digital), between computer and remote terminal (front-end digital, output analog), or between sensing and recording equipment, such as thermocouple and recorder (all analog).

In going from a computer to another device, some basic stages must be considered (Figure 5-49). Computers are connected to the outside world through their I/O bus structures. To meet the ever increasing demands for throughout and to keep pace with internal processor speeds, the I/O bus structure has become an important component of the computer architecture and must be well understood before starting on interfacing. To meet the speed requirements demanded for efficient computer use, most bus-structured I/O is based on transmission line technology. This involves

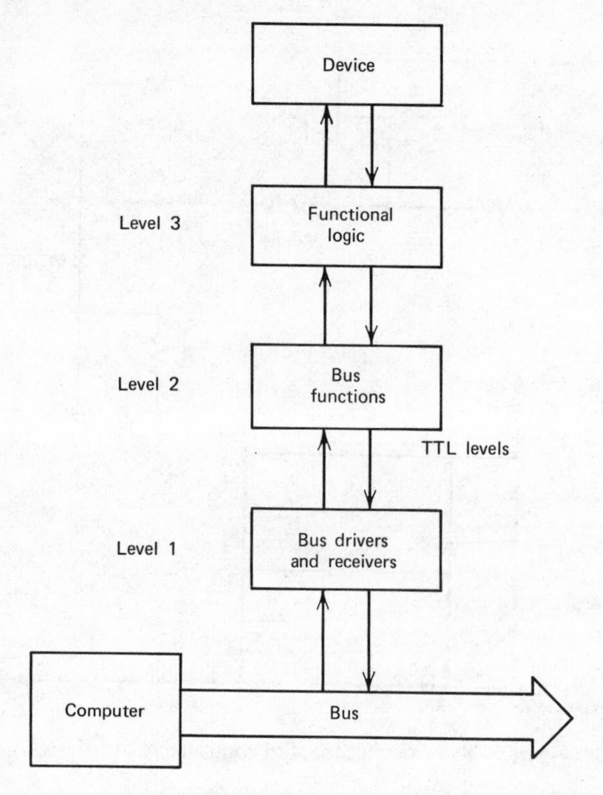

Figure 5-49. Three basic levels are needed when interfacing a computer to an instrument for total compatibility.

using a cable structure with a known characteristic impedance, terminating the bus in its characteristic impedance, and minimizing crosstalk between adjacent signal lines. Because of this controlled environment, care must be exercised in connecting "foreign" devices (usually ICs) to the bus. By adding circuitry carelessly, many of the bus parameters, such as signal-propagation time, logic threshold levels, crosstalk, and signal reflections, are modified to a point where system performance is degraded. This degradation results in faulty and unreliable data transfer between the computer and the device. Such seemingly innocent items as the length and proximity of wires (or etch lines) used to tap signals off the bus may produce undesirable effects. Following the manufacturer's recommendations is of utmost importance.

Since a bus signal line is used both to send and receive data, bus drivers and receivers are required. One of the most important parameters affecting a bus system is the loading placed on the signal lines by these drivers and receivers. Since a bus is terminated (usually by a resistive network), the need for active pull-up is eliminated (Figure 5-50).

The ideal receiver would have zero leakage current and infinite input impedance, and an ideal driver would have an infinite output impedance when OFF and zero output impedance when ON. Unfortunately, ideal drivers and receivers do not exist, and compromises are necessary. Some of these compromises are: To limit the number of drivers and receivers placed on the signal line, to limit the length of the bus itself, and to specify the parameters for receivers and drivers tightly.

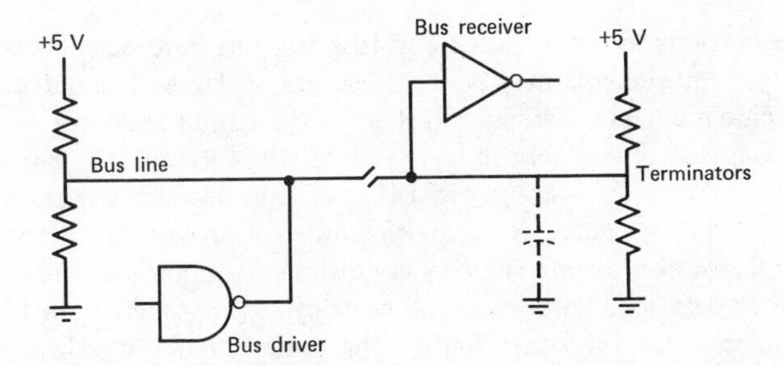

Figure 5-50. A simplified bus line and the interfacing of signal lines to it.

A bus driver must be able to sink enough current to maintain a voltage on the signal line below the ON threshold of any receiver. This current is the result of the terminator network, plus the leakage of all the receivers. In addition, there are transient currents produced by crosstalk and reflections that the driver must be able to handle while still maintaining a voltage below the low threshold of the receiver. The difference between the minimum threshold of the receiver and the voltage maintained by the driver is a measure of the noise margin of the bus. Operating the drivers and receivers within the specifications laid down by the manufacturer provides the greatest noise margin under worst-case conditions.

While this may be an oversimplified version of a bus structure, it points out some of the parameters that should be considered in interfacing at the bus level. If an engineer is not familiar with transmission lines, he should not attempt to work with bus lines. Most mini manufacturers provide proper driver and receiver modules based on their special structure (Figure 5-50). These modules translate bus data into TTL, which is easier to use. Special ICs designed specifically for conversion between various information sources, logic levels, and the drivers required for information displays are referred to as interface circuits. Interface ICs contain both digital and linear functions and have generally been accepted as one segment of the linear IC hierarchy. Computer interface circuits have become major product lines with many IC manufacturers. There are eight[1] primary categories of device types, each with a particular interface function.

Core memory drivers (TTL to core stack)

Core memory sense amplifiers (core stack to TTL)

MOS drivers (TTL to MOS)

MOS sense amplifiers (MOS to TTL)

Light-emitting diode drivers (MOS to LED)

Line drivers (TTL to transmission line)

Line receivers (transmission line to TTL)

Peripheral drivers (multipurpose interface)

CORE CIRCUITS

Most core memory stacks use the coincidence of two currents through one core to write into that core or to read information from it. A selection matrix is necessary

[1] "Special Report: Linear IC's, *Electronic Products*, May 21, 1973.

to control the currents in a core stack and to select one core out of a large matrix. Total core drive requirements may be from 200 mA to 1.2 A. The half current drive levels for coincident current systems will then be from 100 to 600 mA.

Evolved through several generations, the SN55329 represents the latest in a series of core memory drivers. A three- to eight-line built-in decoder serves as the system address register (SAR). A package enable (PE) control is provided to allow just 30 mW device power dissipation during standby conditions. Another input selects a source or sink mode of operation. Total read and write cycle time is less than 1 μsec.

Sense amplifiers form the other half of the core memory interface. A memory sense line output may be positive or negative, and determination of a logic 1 or 0 level is by the absolute value of its amplitude. The core memory sense amplifier must respond to amplitude levels above a set, absolute threshold, regardless of signal polarity.

One levels can be from 4 to 30 or 40 mV, depending on the core size and speed requirements. Zero level signals can be 50% of the 1 level, so good sensitivity is required to distinguish between a 1 and a 0.

Most of the SN55/7520 series sense amplifiers are designed to sense signal levels from 15 to 40 mV, while the SN75236 and SN75237 will sense signals from 4 to 25 mV. Sense amplifiers vary mainly in the type of outputs used (complementary, open-collector, TTL, inverted, buffered, etc.)

Sense Amplifiers. The SN752X series of sense amplifiers provides a stable narrow input threshold, an adjustable reference, and output logic functions to fulfill the several logic variations required in different core-memory systems. Before discussing the application of sense amplifiers, it is helpful to look at the basic organization and performance of ferrite-core memories.

Ferrite-Core Memories. Figure 5-51 shows a simplified block diagram of a memory organization in a computer. The command to the address register selects the appropriate X and Y address lines. Two wires are threaded through each core; one is common to all cores in that row and the other is common to all cores in that column. This arranges a matrix in which any X-address line and any Y-address line have one core in common.

Both reading and writing are accomplished by driving current through the wires linking the cores. If this current exceeds a minimum value, I_m, the core is reset to a 0. Therefore, by driving $I_m/2$ through one of the X-address lines and through one of the Y-address lines, any one core may be reset; only the core common to both lines will receive the sum of the currents and will be reset. The rest of the cores receive $I_m/2$ and will be half-selected. For a core to be set to the 1 state, the current may be reversed in the two selecting wires or additional wires may be provided that link the cores in the opposite direction.

A sense line linking all the cores in the plane together accomplishes the reading. In order to read a specific core a 0 is written into that core; if its initial state was at 1, it is reset and generates a pulse on the sense line. If the core is in the 0 state, no signal results. It is necessary to restart a 1 into the core. The sum of the read and write time is defined as memory-cycle time, a measure of the speed of the memory.

The sense amplifier must be able to detect the difference between the disturbed 0 voltage and disturbed 1 voltage. The threshold voltages of the sense amplifiers should be adjusted to approximately the midpoint between the sum of the disturbed

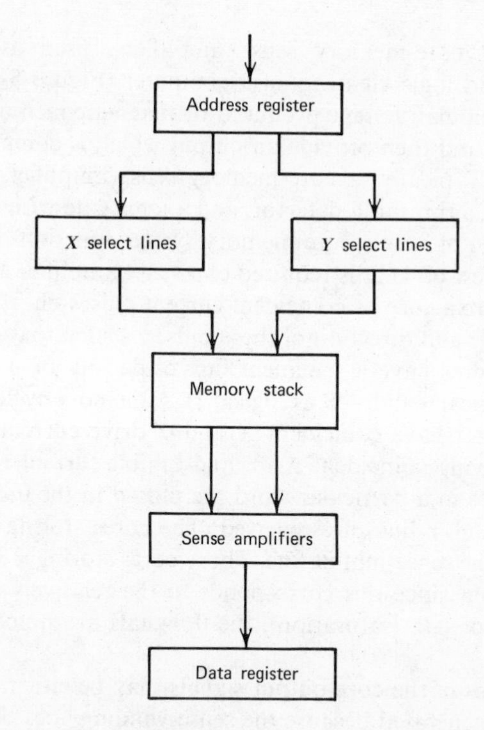

Figure 5-51. Typical digital computer-memory organization.

0 voltages and the minimum disturbed 1 voltage. This is illustrated in Figure 5-52. The region on either side of the threshold voltage in which the sense amplifier cannot detect the difference between a 0 and a 1 is called the uncertainty region. The uncertainty region of a sense amplifier is the sum of the variation of the input-threshold voltage and the differential-offset voltage. It must be minimized. Ideally, it would be 0, however, it should at least be smaller than the difference in voltage between the minimum 1 and the maximum 0.

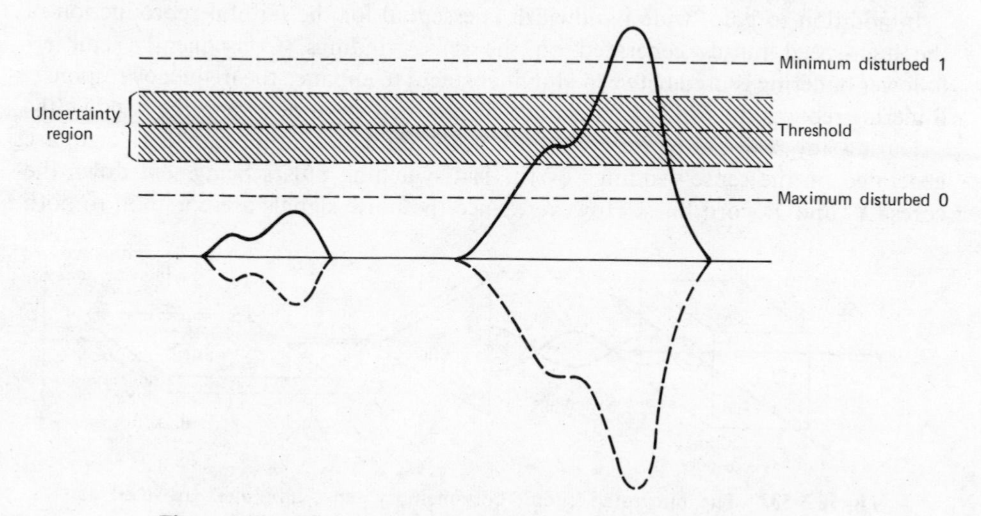

Figure 5-52. Typical core disturbed zero and disturbed one signals.

Integrated circuit core-memory sense amplifiers, used as interface circuitry between memory and logic elements of a computer (Figure 5-53), are designed to sense the low-level bipolar voltage produced by magnetic memories, determine their logic levels (0 or 1), and then provide an output which is compatible with the logic family being used. Typically, a core-memory sense amplifier consists of a linear amplifier, a rectifier, a threshold detector, and a logic gate with strobe (Figure 5-54).

A brief description of how a core-memory system functions should be helpful in getting an overall view of what is required of a sense amplifier and how it functions.

Data is written into a core by coincident current pulses on X, Y, and sense-inhibit lines. The magnitude and direction of these pulses is such that after the pulses have been applied, the cores have a remanent flux of density of $+B$ or $-B$ (typically, $+B$ is defined as logical 0 and $-B$ as logical 1). A memory cycle starts with all cores at 0. Those cores that have coincident X and Y drive currents are switched to 1, whereas those receiving coincident X, Y, and inhibit current remain at 0. To read the memory all cores in a particular word are pulsed in the 0 direction, that is, the currents in the X and Y lines are reversed. The cores storing a 1 switch to 0 and induce a signal on the sense-inhibit line. Those cores storing a 0 also induce a signal on the sense line, but since this corresponds to the relatively small change in flux density from $+B$ to $+B_s$ (saturation), the 0 signals are much smaller than the 1 signals.

The voltage pulses of the core-output signals may be either positive or negative. Bipolar signals are generated because the sense-winding goes through half the cores in one direction and through the remaining cores in the other direction.

Two differential amplifiers are used as the input stage of the sense amplifier to provide the gain necessary to make the low-magnitude (0 to 50 mV range) core-output signals compatible with the logic families being used (DTL or TTL). The differential-voltage gain of the amplifier is given by the equation: $A_{V\,\mathrm{diff}} = R_L/(r_e + R_e)$ where R_L is the load resistance, r_e is the emitter-diffusion resistance, and R_e is the emitter-degeneration resistance. Since the gain is a function of resistor ratios, rather than absolute values, fairly constant gain results. (A gain of 100 is sufficient for this type of sense amplifier.)

In addition to gain, wide bandwidth is essential for the faithful reproduction of the high-speed pulses generated on the sense-windings. Consequently, emitter-follower buffering is used between amplifier stages to enhance the frequency response. Buffering reduces the Miller capacitance effect which, in turn, widens the bandwidth.

During any read and write cycle both noise and core-output signals (1 or 0) are generated on the sense-windings by the fast-switching pulses being sent down the cores' X- and Y-word lines. However, since the noise signals are common to both

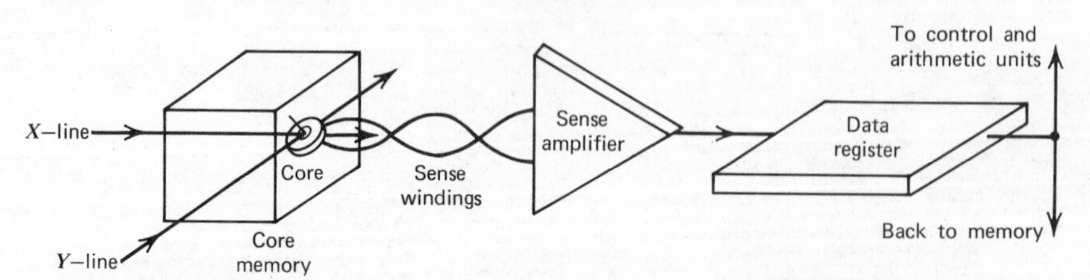

Figure 5-53. The integrated circuit core-memory sense amplifiers are used as interface circuitry between the core memory and the logic elements of a computer.

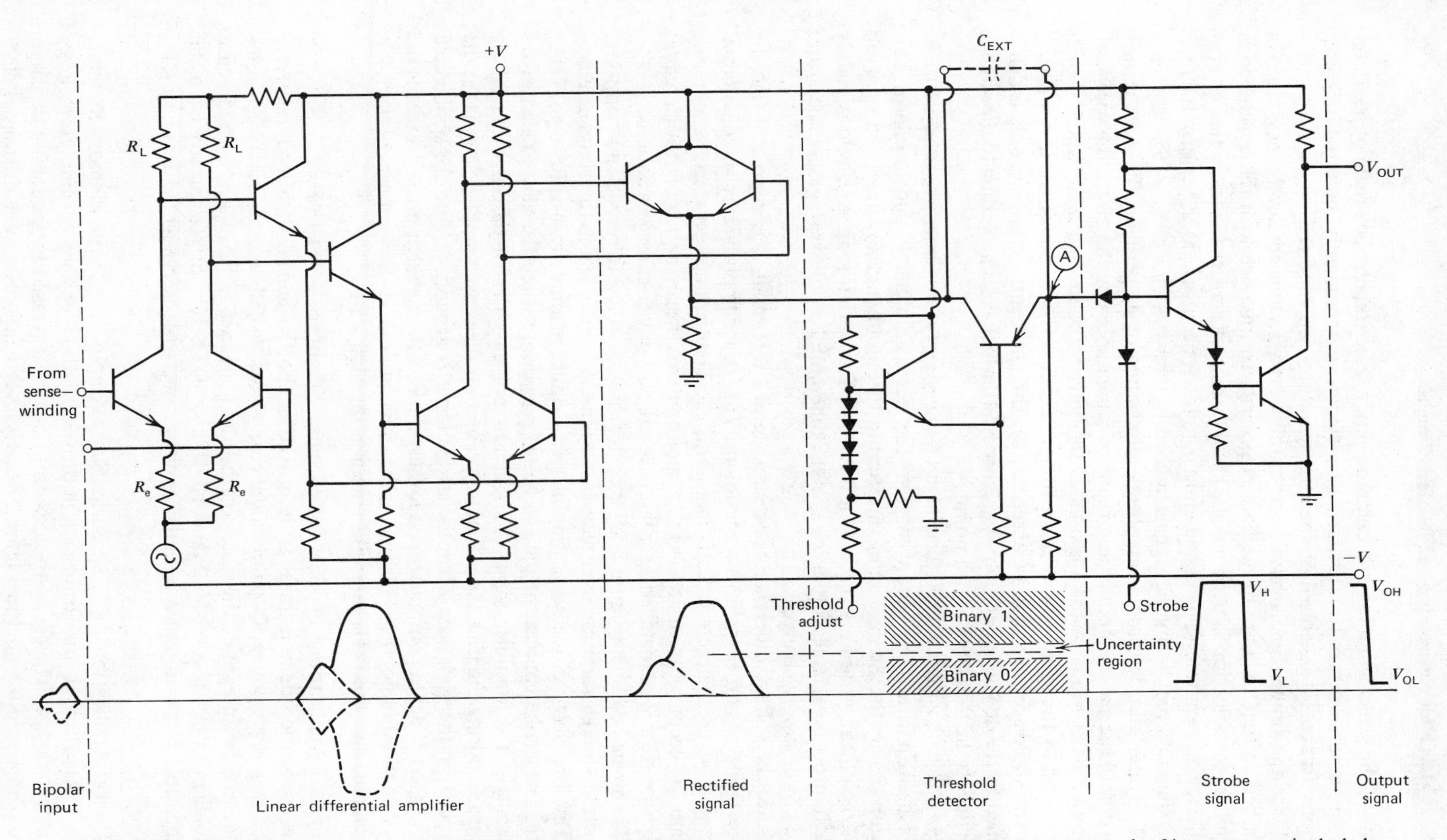

Figure 5-54. Simplified schematic of core-memory sense amplifier circuit. Sense of the typical signals present at each of its stages are included.

ends of the sense-windings, the common-mode rejection characteristic of the differential amplifier serves to attenuate these signals. Consequently, only the difference signals (0s and 1s) generated on the sense-windings are amplified.

Once the input signals have been amplified to a sufficient level, they must be rectified so that they are of the correct polarity to interface with the various logic families. Rectification is accomplished with a pair of emitter-followers tied at both collector and emitter. (The paired emitter-followers can be said to perform the logic "or" function for the analog signals applied to their bases.)

The threshold detector or amplitude discriminator is, perhaps, the most critical portion of the sense amplifier and requires capacitance coupling from the wire "or" output to the input of the logic gate. Its function is to determine which signals generated by the switching cores are 0s and which are 1s. Signals that do not exceed the threshold are a 0 and those exceeding the threshold are 1. The threshold is determined by the dc voltage at point A (Figure 5-54), which is set according to the voltage applied to the threshold adjust points.

As shown in Figure 5-54 there is an "uncertainty" region associated with the threshold in which the output state of the sense amplifier is undetermined. This region designates the maximum variation in threshold, and any amplified signal in this region can produce either a 1 or 0 output. The variation in threshold is mainly due to process variables, input offset of the differential-amplifier stages, temperature, and power-supply changes.

Since the voltage difference between a 1 and a 0 is small, it is important that the uncertainty region be held to a minimum. This is accomplished by the differential amplifier which characteristically maintains excellent balance between the amplifier inputs. By keeping the input offset to a minimum, the transistor pair of the differential amplifier stays well matched over temperature and power-supply variations.

The string of diodes (Figure 5-54) serve to offset or cancel the diode and transistor-junction voltage variation over temperature in the logic gate and threshold detector.

The final stage of the sense amplifier is a standard diode-transistor-logic (DTL) gate. The strobe enables the gate to be triggered only when the desired signals are present at the amplifier input. This action prevents the gate from being falsely triggered during the time there is noise present on the sense amplifier. Note that the output is in the high state when the strobe input is grounded. With the strobe held high (≥ 2.4 V dc) the output will go low (≤ 0.4 V) if the signal at point A exceeds the 1.4-V dc threshold of the DTL gate. If the signal at point A does not exceed the 1.4-V threshold, then the DTL gate will remain in the high state.

Typical Applications of 752X Series Sense Amplifiers—Small Memory Systems. This application demonstrates an improved method of sensing data from relatively small memory systems. Two individual core planes, usually consisting of 4096 cores each, can be integrated by each of the dual-channel SN7524 or SN7525 sense amplifiers as in Figure 5-55. Standard TTL or DTL ICs, driven directly from the compatible sense-amplifier outputs, may be selected to serve as the memory data register (MDR).

Typical Applications of 752X Series Sense Amplifiers—Large Memory Systems. This application demonstrates an improved method of sensing data from large memory systems. The signal-to-noise ratio can be increased by sectioning the large core planes as illustrated in Figure 5-56. Two segments, usually consisting of 4096

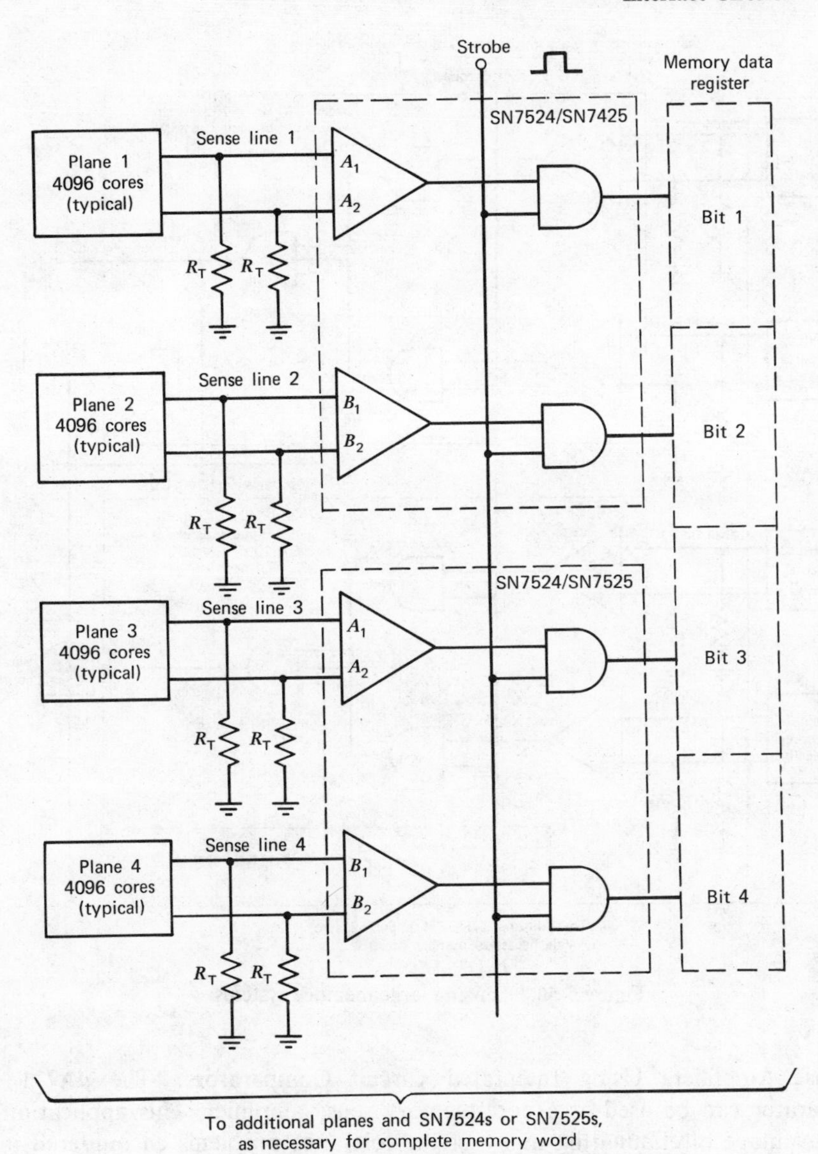

Figure 5-55. Sensing small memory systems.

cores each, can be interfaced by each of the dual-input channels of the SN7520/SN7521 or SN7522/SN7523 sense amplifiers. The cascaded output gates of the SN7520/SN7521 circuits may be connected to serve as the MDR. A number of SN7522/SN7523 sense amplifiers may be wire-AND connected to expand the input function of the MDR to interface all the segments of the plane. Complementary outputs, clear, and preset functions are provided for the MDR. Rules for combined fan-out and wire-AND capabilities must be observed.

Large Memory Systems with Selected Core Plane. Figure 5-57 depicts the application of the SN7520 and SN7522 sense amplifiers to large memory systems with selected core planes.

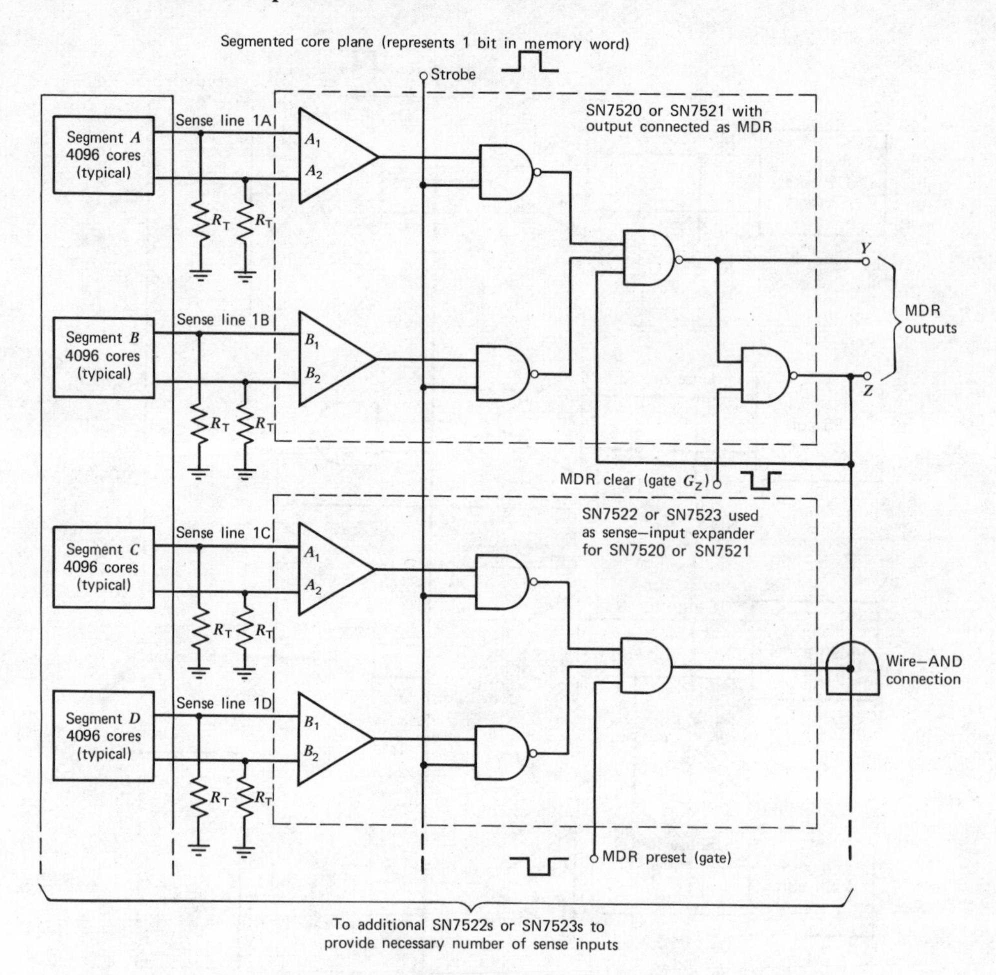

Figure 5-56. Sensing large memory systems.

Sense Amplifiers Using Integrated Circuit Comparators. The μA711 dual comparator can be used as a core-memory sense amplifier. This application has the advantage of eliminating many of the tolerance problems encountered in the design of a sense amplifier with the usual differential amplifier-threshold-detector combination. Using the dual comparator, the sense amplifier threshold is determined by external resistors and is practically independent of the characteristics of the IC. Excellent threshold stability over the full military temperature range is inherent in the design.

With this approach, a single IC can be used as a sense amplifier for practically all coincident current cores presently used. In addition, the circuit is ideally suited to positive 1, negative 0 memories such as bias and nondestructive read. Independent strobing of each comparator channel is provided, and pulse-stretching on the output is easily accomplished. Up to eight sense amplifiers can be OR'ed directly. In most applications, the μA711 can be used with only an external resistor network to determine threshold; for very demanding applications, the comparator can be combined with relatively inexpensive discrete parts to obtain the required performance.

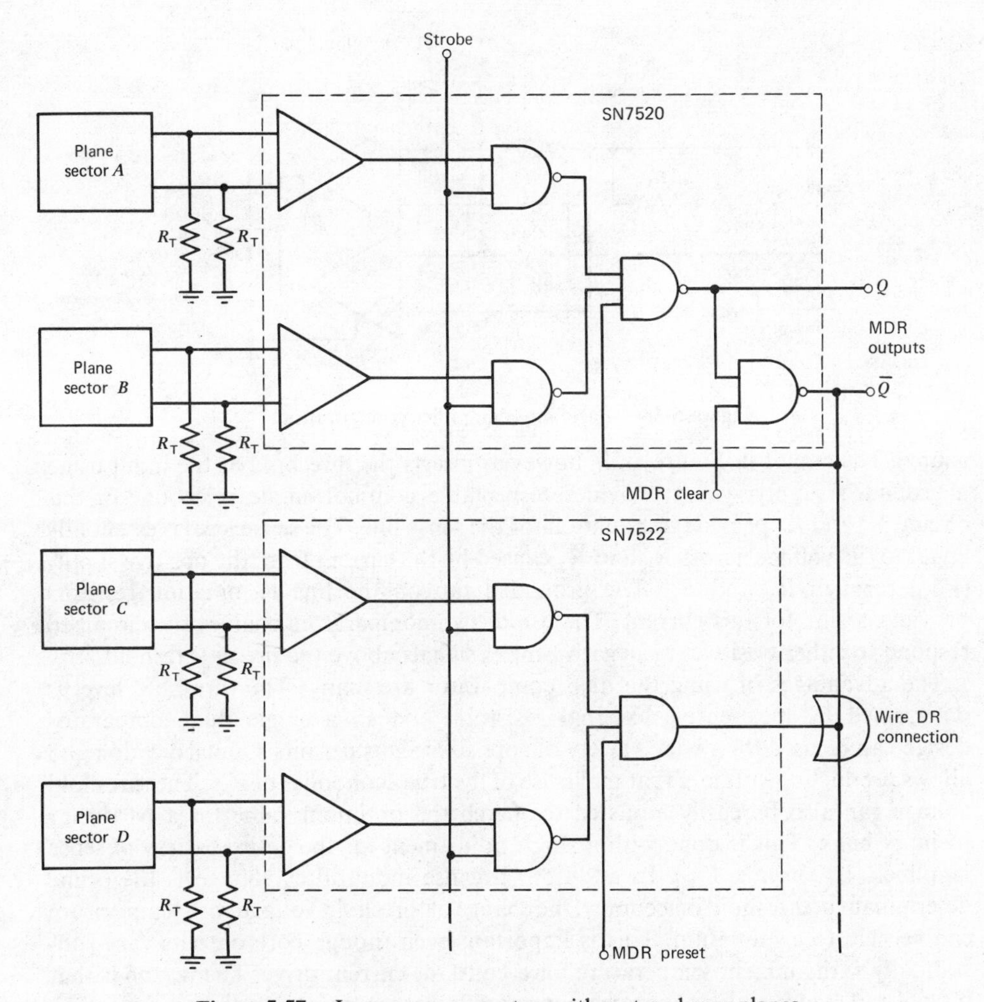

Figure 5-57. Large memory system with sectored core planes.

Conventional sense amplifiers are usually differential-input, differential-output amplifiers which amplify the output of the cores and eliminate the comparatively large common-mode signals present during read. A full-wave rectifier and sense-level threshold are inserted at the output of the amplifier to discriminate between the 0 and 1 outputs of the cores.

This approach has the disadvantage that the offset voltage, the differential gain, the output common-mode level, the rectifier offset, and the output threshold level must all be accurately controlled to get a precise input-referred threshold level. A much more satisfactory solution is to use a voltage-comparator and insert the threshold voltage at the input. In this case, the accuracy is only affected by the offset voltage of the comparator (which similarly affects the differential amplifier in the conventional approach). Variations in voltage gain and frequency response do not affect accuracy as long as these are high enough.

The problem in using a voltage-comparator as a sense amplifier has been that the threshold voltage is not easily inserted at the input when common-mode rejection is required. A straightforward approach to the problem requires a floating voltage

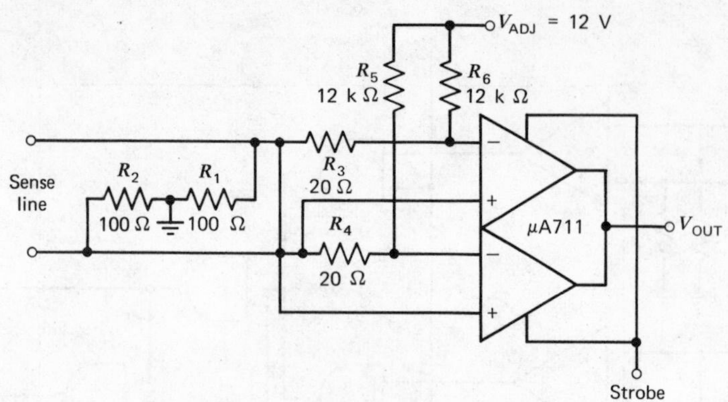

Figure 5-58. Basic sense amplifier configuration.

source. The circuit in Figure 5-58, however, inserts the threshold at the input using a grounded supply, yet it provides respectable common-mode rejection. In this circuit, R_1 and R_2 provide termination for the sense line. The sense level is essentially equal to the voltage across R_3 and R_4 caused by the current from the positive supply (V_{ADJ}) through R_5 and R_6. The grounded tap on the line termination resistors provides a sink for this current. The use of two comparators enables the circuit to respond to either positive- or negative-input signals above the preset threshold.

The advantages of using the dual comparator are many. The threshold level is determined by inexpensive external resistors, and so a single dual-comparator design can be used for a wide variety of core sizes. This permits standardization and allows production volumes that make use of the true economies of ICs. The threshold voltage can also be easily adjusted to match the optimum value for a particular memory bank. This is done with a single adjustment for the entire battery of sense amplifiers by varying V_{ADJ}. In addition, because input-offset voltage is the prime determinant of threshold inaccuracy, the change of threshold voltage with temperature can be held to a minimum. This is important even though core outputs vary substantially with ambient temperature for a constant-current drive. The reason is that if write and read functions are to be done at different temperatures, the core variation must be compensated with the current drive for the cores. If an attempt is made to compensate for core thermal variations with corresponding thermal changes in the sense amplifier threshold, it is not practical to write into the cores at one temperature and readout at another. The best situation is for the sense amplifier to maintain a constant threshold over the full temperature range.

With the circuit in Figure 5-58, the sense level will be affected by a common-mode signal, although only by approximately 8% for a 1-V signal with $V_{ADJ} = 12$ V. This is rarely a problem because in most memories the common-mode signal during read is less than 0.5 V. This sensitivity can be reduced by increasing R_5 and R_6 and using a higher voltage to set the threshold. In the unusual case where very large common-mode signals are present during read, the circuit shown in Figure 5-59 can be read. This circuit uses additional resistors to balance the input-bias network to the common-mode signal, and it provides at least an order of magnitude better common-mode rejection. This circuit is especially useful if the resistor network is purchased as a thin- (or thick-) film assembly, since the extra resistors add little complexity to the complete assembly and since the tolerance on the balancing resistors can be looser than the tolerance on the threshold-determining resistors.

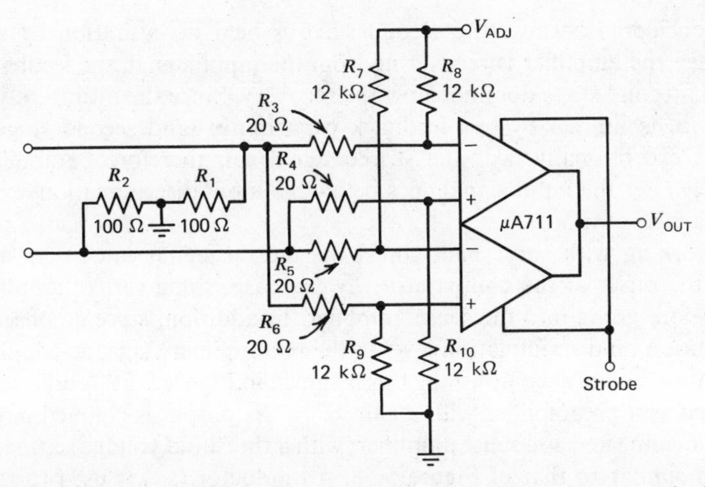

Figure 5-59. Sense amplifier with improved insensitivity to common-mode signals.

The high-frequency common-mode rejection can be severely degraded by unbalanced stray capacitances on the sense line and on the comparator input. For this reason, low-capacitance resistors should be used for R_5 and R_6; the physical layout should also be such as to minimize strays.

Because the comparator is directly coupled and employs low storage time devices, the recovery from either differential or common-mode overloads is less than 50 nsec—regardless of their amplitude. The direct coupling also removes the pattern sensitivity of threshold voltage seen with ac-coupled amplifiers.

The response of the sense amplifier to a typical input 1 is shown in Figure 5-60. With an input signal that is 7 mV above the discrimination level for 45 nsec, an output pulse that is above the logic threshold for 50 nsec is obtained.

In determining the recovery time of the sense amplifier to an overload or to a maximum 0 when time strobing is used, only the response time of the input stage

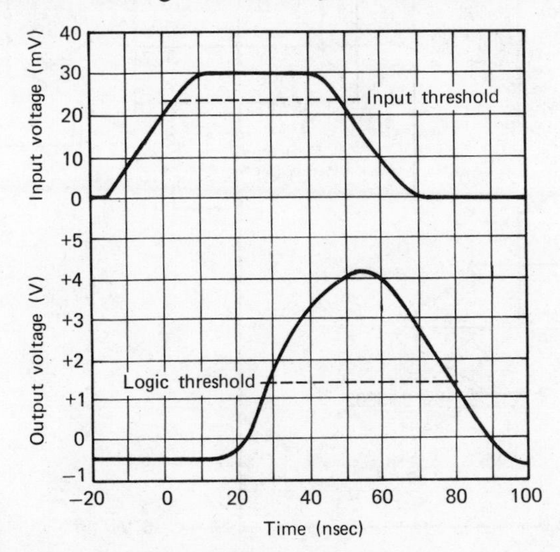

Figure 5-60. Response of the basic sense amplifier circuit, with a 23 mV threshold setting to a 30 mV input pulse.

need be considered because the second stage is held in saturation by the strobe. Hence, when the amplifier is recovering from the input signal, the feedback capacitance of the second stage does not slow the recovery. Since the overall response time of the device is limited by this feedback capacitance (and second stage storage), recovery is exceptionally fast; the strobe signal can therefore be applied almost immediately after the input signal goes below the input discrimination level without obtaining a false output.

When working with very small cores, the output signal can be so small as to approach the offset of the comparator. In this case, some sort of amplification is required before going into the sense amplifier. In addition, since dc offset will be a basic limitation on discrimination level in the preamplifier stage, ac-coupling is also required. A method of accomplishing this is shown in Figure 5-61. A differential-input stage is used as a preamplifier with a gain of 10. Its output is coupled into the dual comparator connected as a sense amplifier, with a threshold voltage setting of 50 mV, in a circuit similar to that of Figure 5-58. An inductor (L_1) is used to remove the offset of the input-stage transistors so that a matched pair does not have to be used. The differential preamplifier achieves temperature-compensated operation, within a few percent, over the entire -55 to $+125°C$ temperature range without the use of emitter-degeneration resistors. This is an advantage in that degeneration resistors deteriorate common-mode rejection and give nonoptimum high-frequency characteristics. The gain of the preamplifier, and therefore the equivalent input-referred threshold voltage, are most conveniently set by adjusting R_7.

The circuit as shown in Figure 5-61 has a sense threshold of 5 mV. Both the input

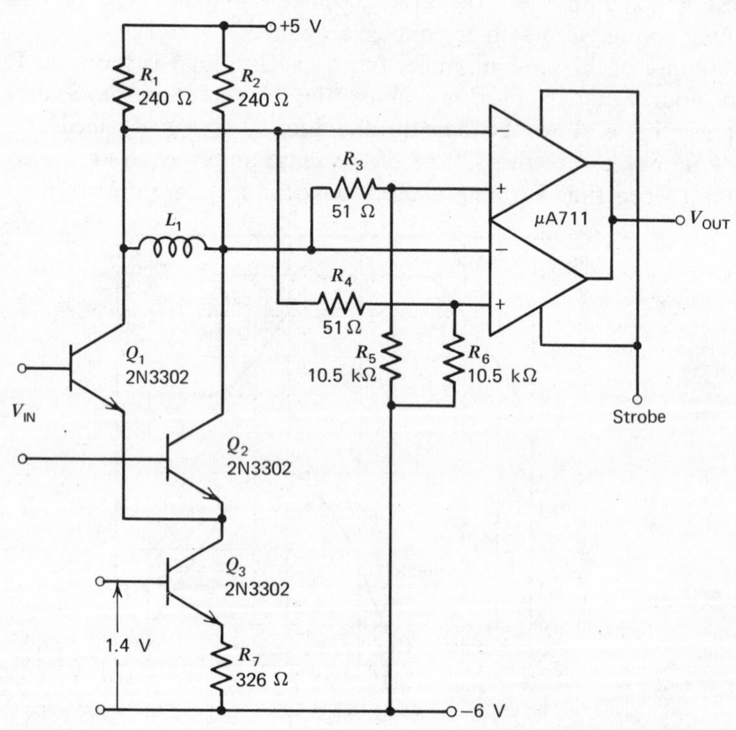

Figure 5-61. Sense amplifier circuit using temperature-compensated preamplifier for increased threshold sensitivity.

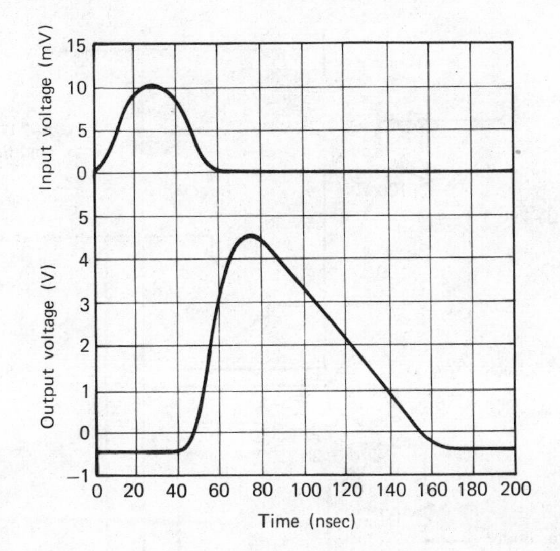

Figure 5-62. Response to 10-mV input pulse for 5-mV threshold.

stage and the dual-comparator circuit provide common-mode rejection. The response of the sense amplifier to a 10-mV input signal is shown in Figure 5-62; this can be improved by operating the preamplifier stage at a higher current than the nominal 1 mA in the indicated design.

A number of sense lines can be combined into a single sense amplifier using a circuit similar to that in Figure 5-61. Individual differential amplifiers and compensated current sources are used for each sense line. These differential amplifiers are combined into a single set of load resistors. If more than two (or possibly three) differential amplifiers are used, the inductor L_1 should be removed, and a pair of inductors should be connected across the load resistors R_1 and R_2 to prevent excessive dc voltage drop across them.

In the design of a memory, it is sometimes convenient to gate a number of sense lines into a single sense amplifier. A circuit which accomplishes this is shown in Figure 5-63. Differential preamplifiers are employed on each sense line, and are operated from a single temperature-compensated current source. The digital gate signal is applied to the center tap on the line termination resistors. The differential pair that has the highest input common-mode signal will operate, while the remainder will be cut off since their emitter-base junctions will be reverse-biased. The common collector outputs of the preamplifiers are fed to a μA711-connected as a sense amplifier as shown in Figure 5-61.

If the inductor in Figure 5-61 is used with the front end in Figure 5-63, the circuit should be gated on for a long time (as compared to the time constant of the inductor and the collector-load resistors) before trying to read. This is required to settle out the collector-current imbalance caused by offset between the differential transistors. The inductor, however, can be eliminated if matched input transistors (2N3729) are used, and if the offset of the input transistors is small by comparison to the desired threshold.

The logic drive circuitry should present a low impedance to the center tap on the line-termination resistors so that an excessive common-mode signal is not developed.

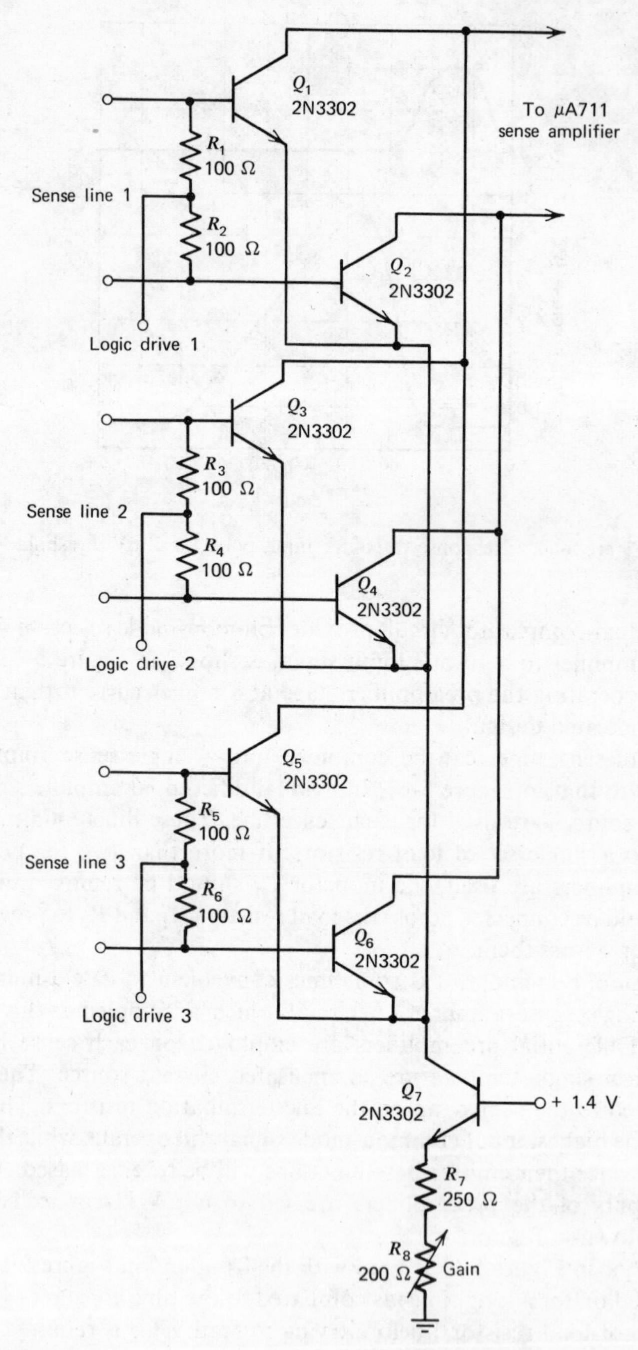

Figure 5-63. Three-channel gated preamplifier for sense amplifier.

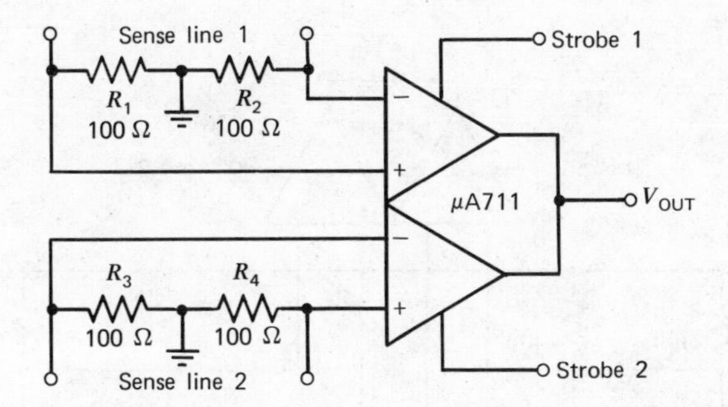

Figure 5-64. Dual sense amplifier for positive one, negative zero memory systems.

The gain of the preamplifiers and, therefore, the threshold setting are best adjusted by varying the emitter resistance in the current source as shown in Figure 5-63.

Certain types of memories (i.e., bias or nondestructive read) give an output of one polarity for a 1 and an output of the opposite polarity for a 0. In these applications, one-half of the μA711 can be used as a sense amplifier. No resistor network need be employed on the input to determine threshold since the comparator alone is designed to detect zero-crossing.

The complete μA711 can be used as a dual sense amplifier with common outputs, which can be separated (as far as the system logic is concerned) by using the independent strobe terminals. A circuit for this is shown in Figure 5-64.

Most often with positive 1, negative 0 memories, the output-pulse amplitude is comparable to the offset of the μA711. In this case, the circuit in Figure 5-64 cannot be used alone. A preamplifier with ac-coupling must be used to amplify the signal to a level that can be detected easily by the μA711. A circuit for doing this is shown in Figure 5-65. It is similar to the preamplifier in Figure 5-59 except that no attempt is made to control the gain since its actual magnitude is unimportant as long as it is high enough. As with the circuit in Figure 5-61, the inductors remove the offset of the preamplifier transistors so that matched pairs do not have to be used. Since the preamplifier shown has a gain of 10, the circuit will be able to discriminate a zero-crossing within 0.5 mV.

INTERFACE WITH MOS

Many high-performance MOS memories are built with input drive requirements and output levels not directly compatible with TTL. Interfacing between MOS and TTL can be accomplished with IC level shifters. The exact type of interface selected will depend on the system performance and input/output requirements.

MOS inputs are generally high-impedance and capacitive. This capacitive load will vary with the type and number of MOS circuits being driven. The voltage levels required range from 10 to 24 V to −20 V.

The SN75361 dual-channel and SN75365 quad-channel circuits are designed to drive MOS input circuits with positive level requirements. Inputs on the 75361 and 75365 are TTL-compatible, while the outputs have 100 mA continuous current capability to provide quick charge and discharge of capacitive loads.

MOS outputs are generally current-mode signals and represent a relatively small voltage or current swing between a logic 1 and a logic 0. The output sense amplifier or converter interface must be sensitive enough to detect and distinguish between

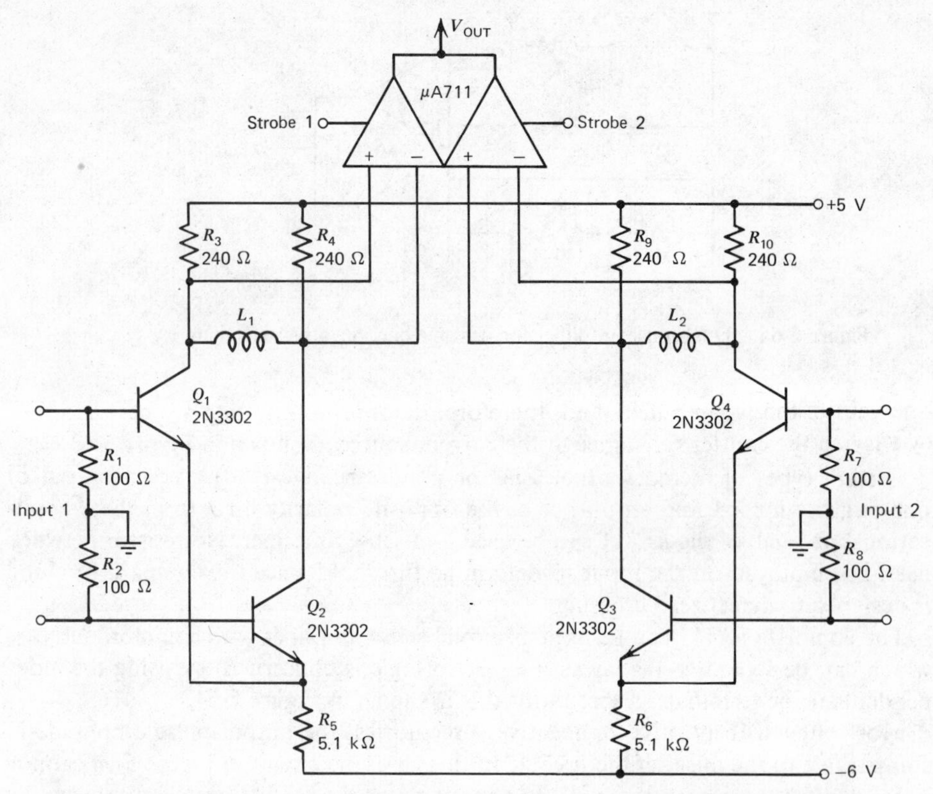

Figure 5-65. Dual sense amplifier for positive 1, negative 0 memory systems including preamplifiers.

these MOS logic levels. Most MOS memory chips have high-level outputs of 300 mV (logic 1) and low-level outputs of 5 mV (logic 0). To interface with this type of circuit, a sensitivity of ± 10 mV is desirable. For these applications, the SN75207/208 line receivers are recommended, since their sensitivity is $\leq \pm 10$ mV.

DISPLAY DRIVERS

Interface circuits play an important role in MOS calculator systems. For example, MOS-to-LED interfaces make quick recognition of calculator output data possible. The display driver exists solely because most digital readout devices are incompatible in one or more ways with the electronic circuits that control them. The incompatibility may be in voltage or current levels or in the respective number systems that logic and displays work with.

The definition of a display varies widely, even if you only consider those associated with electronics. A display can be anything from an indicator lamp to a CRT (cathode ray tube), and the nature and complexity of the associated drivers vary accordingly. In this section, display drivers and decoder/drivers for 7-segment readouts are discussed.

The ideal situation in any electronic system requiring a 7-segment display would be to have a display that is directly compatible with the controlling logic circuitry and available power-supply voltages so that no interface circuits are required.

Actually, progress is being made in that direction, and in the case of hand-held calculators, which typically use small LED (light-emitting diode), fluorescent or gas-discharge displays, the ideal situation is nearly at hand. The trend in LEDs is toward lower currents for a given light output so that direct segment driving from MOS is within the realm of possibility now. Digit driving, however, is still beyond the capabilities of MOS, thus discrete transistors or IC drivers must be used as interface elements.

However, most applications do require some type of interface circuitry between logic and display. The complexity depends on a number of factors such as the type of display selected, the format of data, the difference in voltage and/or current levels encountered and the number of digits to be driven.

The circuit of Figure 5-66 illustrates one of the simpler and more straightforward methods of a multiplexing system. Four 9312 8-input multiplexers select the BCD input data for the 9307 decoder. Address for the multiplexers and the scan decoder from the 9316 binary counter uses only the first three bits. Each 3-bit address selects the BCD input data to be displayed and the scan decoder energizes the correct display readout device. The counter progresses through all eight counts and repeats, activating each display unit for 1/8th of the scanning cycle. Higher operating voltage must be applied for the displays to have the same brightness as displays in a static mode (dc). The circuit as shown uses fluorescent diode displays. Other displays may be substituted with the appropriate drive circuitry.

LED Display Drivers. A couple of things are happening in 7-segment LED displays that affect display drivers. First, performance efficiency of the devices is steadily increasing. The newest 0.1-in. displays can be driven by 0.5 mA/segment average current in strobing applications, and manufacturers are talking about 0.1 to 0.2 mA/segment for small LEDs in the near future.

Second, most new 1/4-in. LED displays use only one diode/segment and rely on optocoupling to magnify and shape a dot of GaP or GaAsP material for uniform appearance over the whole segment. This means that GaAsP types have a 1.7-V drop instead of the 3.4-V drop of two diodes/segment displays, and GaP types have a 2.1-V drop instead of 4.2-V.

Display size is usually the key to selection and then you live with the problems dictated by size, such as average operating current, voltage drop, and electrical configuration; i.e., common cathode or common anode.

As shown in Figure 5-67*a* the common-cathode type has all cathodes tied to ground in a direct-drive application, or to a digit switch in a multiplexing application. Segment drive is through the individual anodes; therefore, a current source is required to drive the display.

Common-anode LEDs (Figure 5-67*b*) have all anodes tied to V_{CC} or a digit switch while segment drive is through the cathodes. In this case, current sinks are required. With either configuration it is important to remember that the LED is a diode and therefore requires current limiting in each segment to prevent uneven drive current and probable failure of the display.

LED technology demands that monolithic displays be common-cathode devices, thus most hand-held calculators and other systems using LEDs up to 3/16-in. high use the common-cathode configuration. At the 1/4-in. size, LEDs are available in either common-cathode or common-anode configuration since these sizes are of hybrid construction.

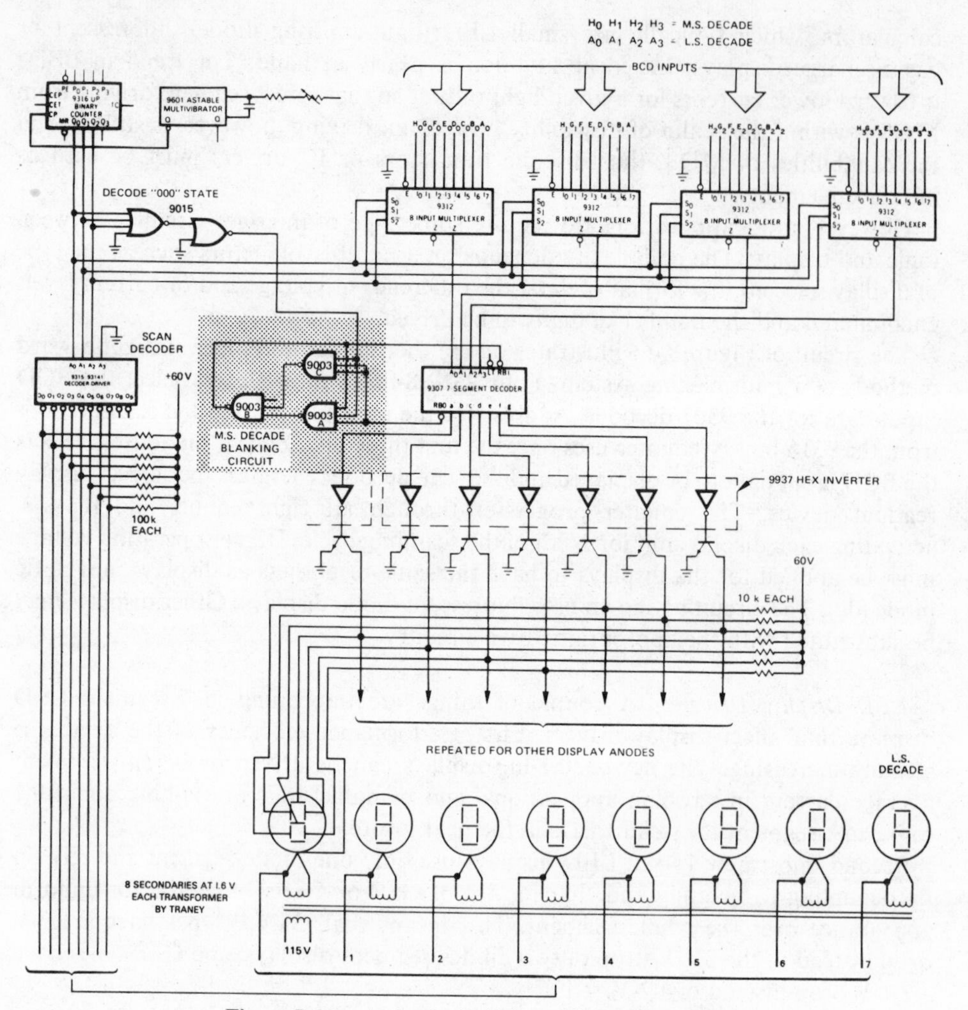

Figure 5-66. A simple, 8-digit multiplexed system.

A display driver for LEDs can be anything from a discrete transistor to an IC. In fact, the complete driver circuit may be a combination of both. The type of driver selected depends on the application and the current needs of the display.

Basically, there are three types of display interface circuits. The most simple is the IC driver, which is a circuit consisting of four or more individual drivers in a package that provides current gain for interfacing from circuits, like an MOS calculator chip, to the LED display. Then there is the decoder/driver which may or may not include a quad latch on the input for storage. This is usually a 16-*pin* DIP that accepts a BCD input and provides the 7-segment output.

The third type is the counter/decoder/driver. There are a number of bipolar and MOS counter/decoder/driver circuits on the market which may or may not require some type of interface between them and the display. Because the present topic is drivers, counter/decoder circuits will not be discussed.

A typical application for IC drivers is in interfacing between a calculator chip and an LED display. In this case, because the decoding is usually done on the chip,

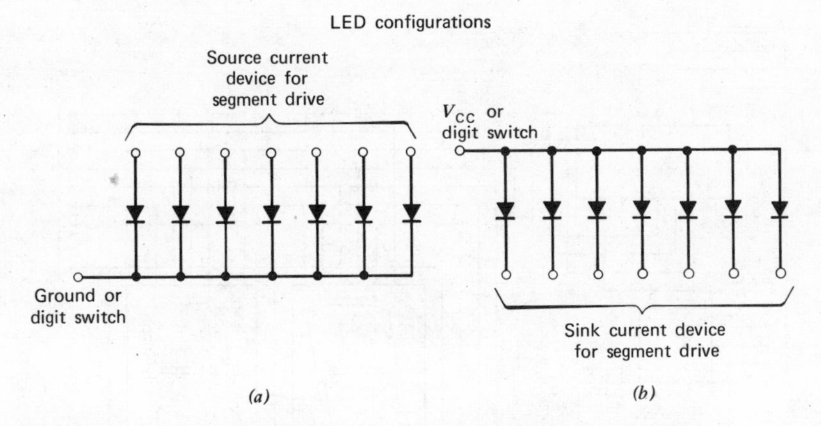

Figure 5-67. Monolithic LEDs are common-cathode devices requiring current sources for segment drive. Hybrid LEDs—1/4 in. and larger—can be either common-cathode (*a*) or common-anode (*b*) types.

the outputs are in 7-segment form. In addition, the digit and segment outputs are scanned or strobed, thus the function of the driver is to provide current gain between the chip and the display.

Since hand-held calculators usually use monolithic LEDs, they require anode segment drivers (current sources) and cathode digit drivers (current sinks). Drivers for calculators have typically been discrete components such as Darlington pairs or SCRs, but several manufacturers now have IC drivers designed specifically for this task.

Such is the case with the circuit shown in Figure 5-68. The multiplexing circuit shown uses two 9661 quad segment drivers. The four outputs of the first unit drive segments a, b, c, and d of each display. The four outputs of the second 9661 drive segments e, f, g, and the decimal point of each display.

In this circuit, the two 9661 segment drivers can be replaced with one 9660 7-segment decoder/driver when current required is less than 50 mA @ 3 V and BCD to 7-segment decoding is required.

Other segment/digit/lamp drivers can be used depending on display requirements. The SN75491 quad segment driver and SN75492 hex digit driver are used in a segment address and digit scan method of generating a multidigit LED display. This principle of time-sharing or multiplexing of the individual digits leads to significant cost reduction in the character generator logic and LED driver by eliminating redundant circuitry.

Basically these ICs are Darlingtons with the necessary biasing circuitry to eliminate the need for any other components except the segment current-limiting resistors. They are designed to operate from a 10 V maximum V_{SS}.

The peak currents required in the segment and digit drivers are inversely proportional to the number of displays being multiplexed. The SN75491 is capable of 50 mA peak current and the SN75492 will deliver 250 mA peak. This is adequate for most applications. Two SN75491s used with two SN75492s are all that are required to provide the drive for a large 12-digit display (Figure 5-69).

When designing with 1/4-in. and larger LEDs having a common-anode configuration, current sinks are required in the segment lines and current sources are

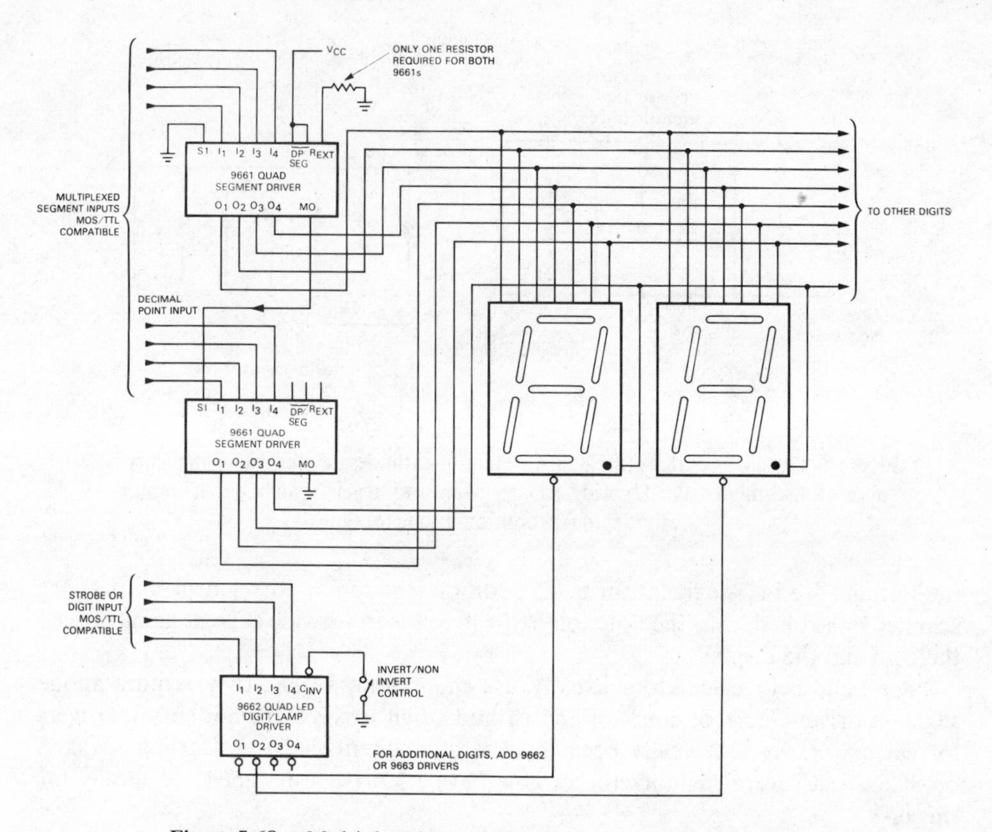

Figure 5-68. Multiplexed common cathode led calculator display.

required for digit select. Most digit drivers suitable for common-cathode operation can be used as segment drivers in common-anode configurations, and drivers like the 75491 can be used either as sources or sinks.

Integrated circuit current sources are limited to about a 75-mA output. Therefore, 1/4-in. and larger digit applications will probably require discrete components as digit drivers with common-anode LEDs.

When data is presented in a BCD format, then BCD-to-7-segment decoder/drivers come into the picture. There are a number of these devices available from most semiconductor manufacturers and from some display manufacturers in the 54/7400 and other TTL series.

Most that have been around for several years have open-collector outputs that require current-limiting resistors in series with each segment. Those that do have internal pull-up resistors have very limited source capability and are intended to drive interface circuits rather than drive an LED directly.

There have been some recent developments in decoder/drivers resulting in a number of changes over the previous TTL models. Many of the new devices have all or a combination of the following: constant-current outputs (fixed by the value of an internal or external resistor), ROM (read-only-memory) decoding, lower input current requirements, and quad latch or lamp test options (Figure 5-70).

Constant-current outputs eliminate the need for series resistors, but with present units, you have to live with the output current values offered. One way around that,

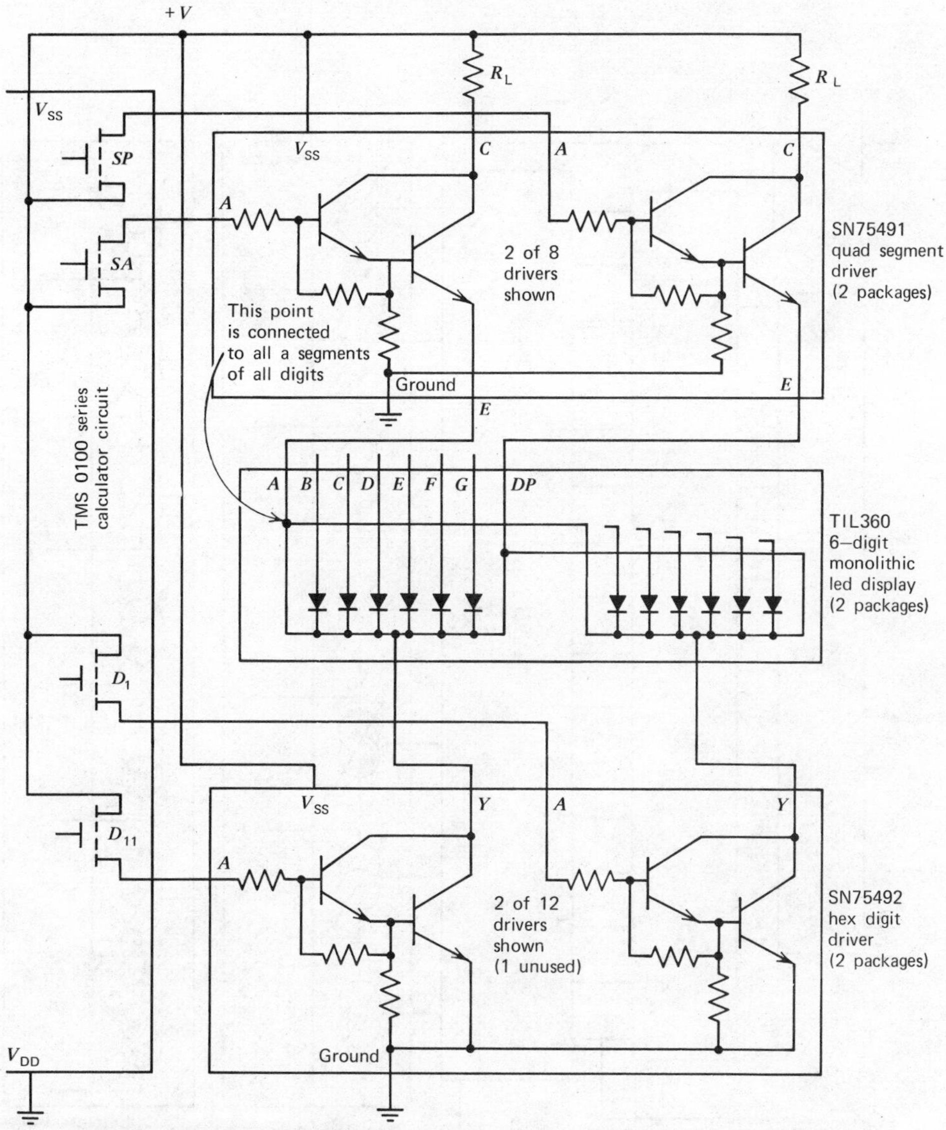

Figure 5-69. Time multiplexed to minimize display circuitry, up to 12 digits can be illuminated using two SN75491 quad segment drivers and two SN75492 hex digit drivers (Interface with MOS calculator.)

if you really must have a different value, is to use a driver with a higher output current than you want and strobe it to reduce the average output current. Another way is to request a different value if quantities justify a mask change.

Signetics has just introduced a family of 12 decoder/drivers that offer 5-, 20-, or 50-mA outputs in either source or sink devices with or without a quad latch.

These devices use ROM decoding which allows complete flexibility in decoding from four inputs to seven outputs, thus accommodating full hexidecimal and other decoding functions. However, a metallization mask change is required for other than BCD-to-7-segment decoding.

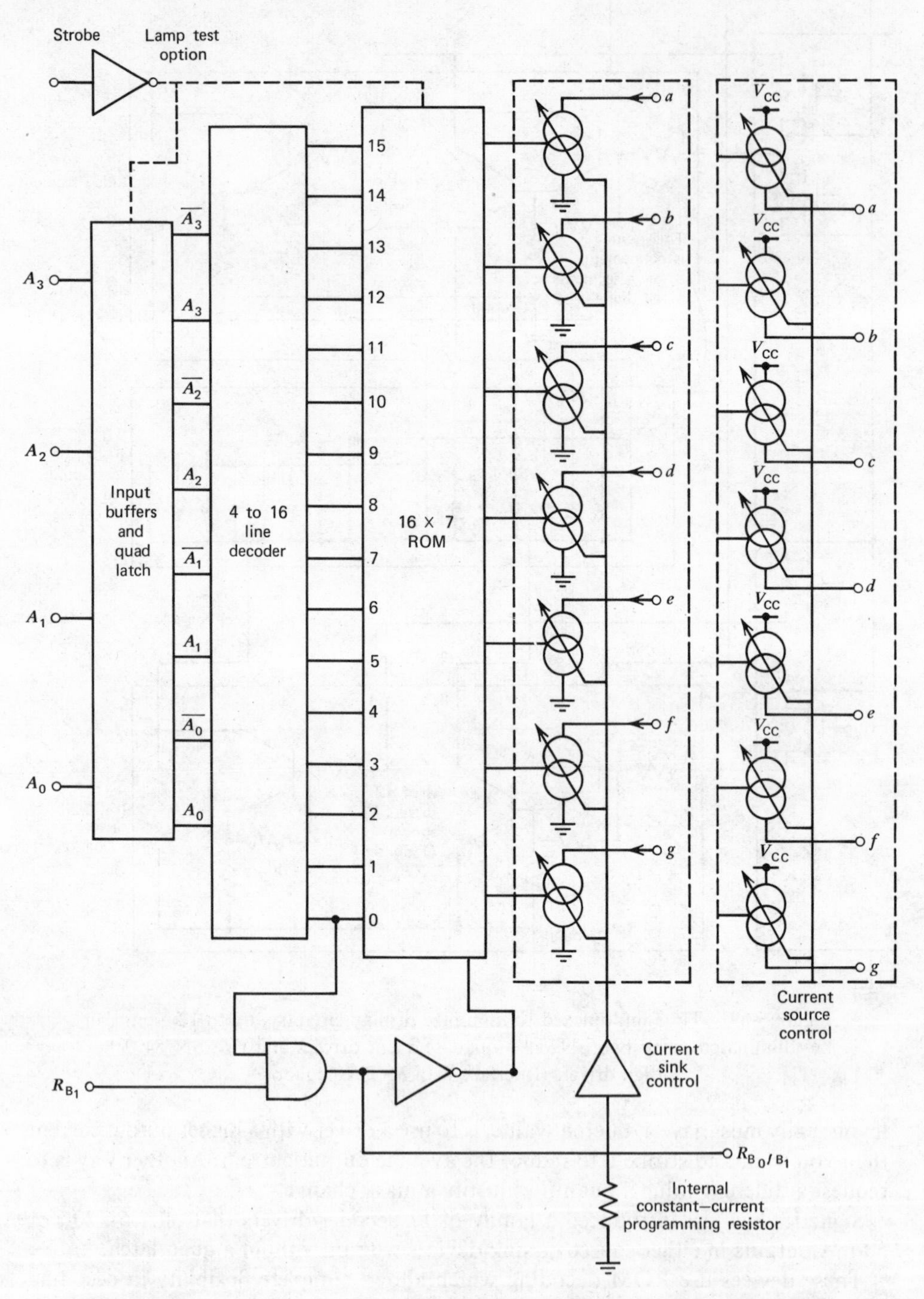

Figure 5-70. A new family of decoder/drivers from Signetics Corporation offers a quad latch or lamp test option, various level constant-current output, and ROM decoding.

Signetics' new family uses a nongold-doped low-power Schottky process to allow the use of the low-current *pnp* input structures shown in Figure 5-71. This reduces input current requirements to 125 μA, which is more than an order-of-magnitude reduction over older TTL types.

When a latch option is elected, lamp test is sacrificed on most decoder/drivers because of the need for a strobe input.

In addition to bipolar decoder/drivers for LEDs, CMOS versions are beginning to appear. Motorola offers a CMOS device (MC14511) with bipolar output drivers that sources up to 25 mA per segment for driving LEDs directly. It is a latch/decoder/driver that does not have constant-current outputs but does have provision for lamp test.

One thing to keep in mind when looking at current ratings of drivers is that you will require a higher rating for multiplexing applications than for direct drive. For example, a 20-mA output current is adequate for a typical 1/4-in. LED segment in direct drive, but it would not be adequate to drive 8-digit segments in a multiplexing situation. In that case each segment would carry an average current of 20 mA divided by 8 or 2.5 mA, which would not be adequate.

In a multiplexing mode, it is necessary for the driver to handle the peak current which is, of course, higher than the direct-drive current. Because of the LED's improved efficiency in pulsed mode and the eye's tendency to peak detect, it may not be necessary to have as high an average current in the multiplexed mode as you would in dc mode.

Gas-Discharge Display Interface Circuits. Interface circuits for mating gas-discharge displays to logic solve the problem of driving a high-voltage display from relatively low-level outputs. Typical anode supply voltages for these displays are 180 V, which is well beyond MOS capabilities and beyond conventional bipolar monolithic technology as well.

A number of companies have introduced drivers and decoder/drivers specifically designed to drive gas-discharge displays. Two basic monolithic processes are used to achieve the high breakdown voltages required.

Deep diffusion isolation, which is an extension of standard linear-processing

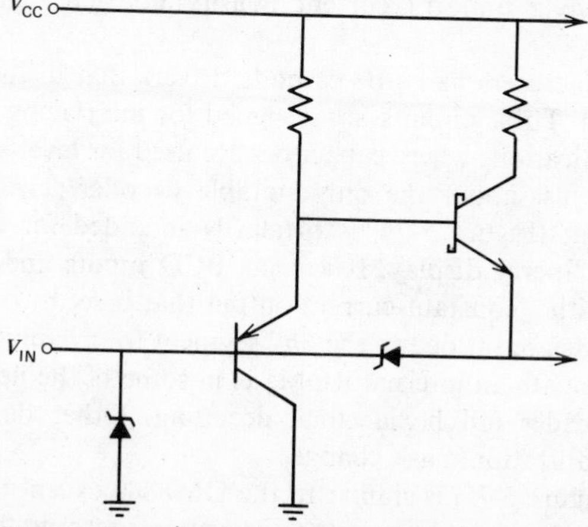

Figure 5-71. Low-current *pnp* input structures of Signetics Corporation's decoder/drivers is the result of a nongold-doped low-power Schottky process.

techniques, produces output breakdown voltages in the neighborhood of 150 V. Dielectric isolation is the other process, and it is capable of producing breakdowns above 200 V; however, not all devices made by that process have that high a spec.

It is not always necessary to be able to stand off the full anode supply voltage on the output of the driver. In direct-drive applications, all the switching is done in the cathode lines. With a 180-V anode supply and a 135-V drop across the display—which is typical in the newest gas-discharge models—the driver output only sees 45 V. Due to supply transients, the fact that the older display models have a 150-V drop, and the requirement for higher anode voltages when multiplexing, all the drivers designed for this type of display have significantly higher output breakdowns than 45 V.

In multiplexing applications, the combination of higher anode supply voltages and the necessity of switching in both the anode and cathode lines makes the job a bit more difficult.

There are various schemes for multiplexing so that it is not always necessary to use active devices for level-switching between the MOS output and anode or cathode drivers. Capacitors can also be used for level-switching; however, when transistors are used, they must be capable of handling the anode supply voltage without breakdown.

Besides the possibility of using discrete high-voltage transistors for level-shifting, there is an IC that contains eight dielectrically isolated transistors designed specifically for this job. Actually, this is one of a family of level-shifters with breakdown voltages ranging from 125 to 225 V, with 200 and 225-V units being used for this particular application.

In addition to level-shifters, there are a number of DIP IC anode and cathode drivers available for interfacing gas-discharge displays to MOS chips in multiplexing applications.

The DM8885 is a 16-*pin* DIP that has a constant-current output which is programmed between 0.2 and 1.5 mA by one external resistor. In order to drive all seven segments from one 16-*pin* package and still have pins for V_{CC}, ground and current programming, only six inputs are used. An internal decoder is used to provide the 7-segment outputs as shown in Figure 5-72. All output currents are ratioed to the "*b*" segment to give a uniform current density in each segment for optimum appearance.

Several IC manufacturers make IC cathode drivers that include a dc-restoration circuit (Figure 5-73). These circuits are designed for interfacing with MOS circuits in multiplexing applications where capacitors are used for level-shifting.

The DM8880 is just about the only suitable decoder/driver for gas-discharge displays. This circuit (Figure 5-74) is primarily intended for interfacing between TTL logic and the Sperry display. It accepts BCD inputs and drives the cathode segments directly with a constant-current output that is set by one external resistor. The segment currents are ratioed to the "*b*" segment to give uniform appearance.

Decoding is done with an internal ROM, as in some of the newest LED decoder/drivers, which provides full hexidecimal decoding. Other decoding formats are allowed with a metallization mask change.

The DM8884 (Figure 5-75) is similar to the DM8880 except that it has a decimal point and comma drive in addition to the 7-segment outputs. It does not have the ripple blanking input or the blanking input/ripple blanking output.

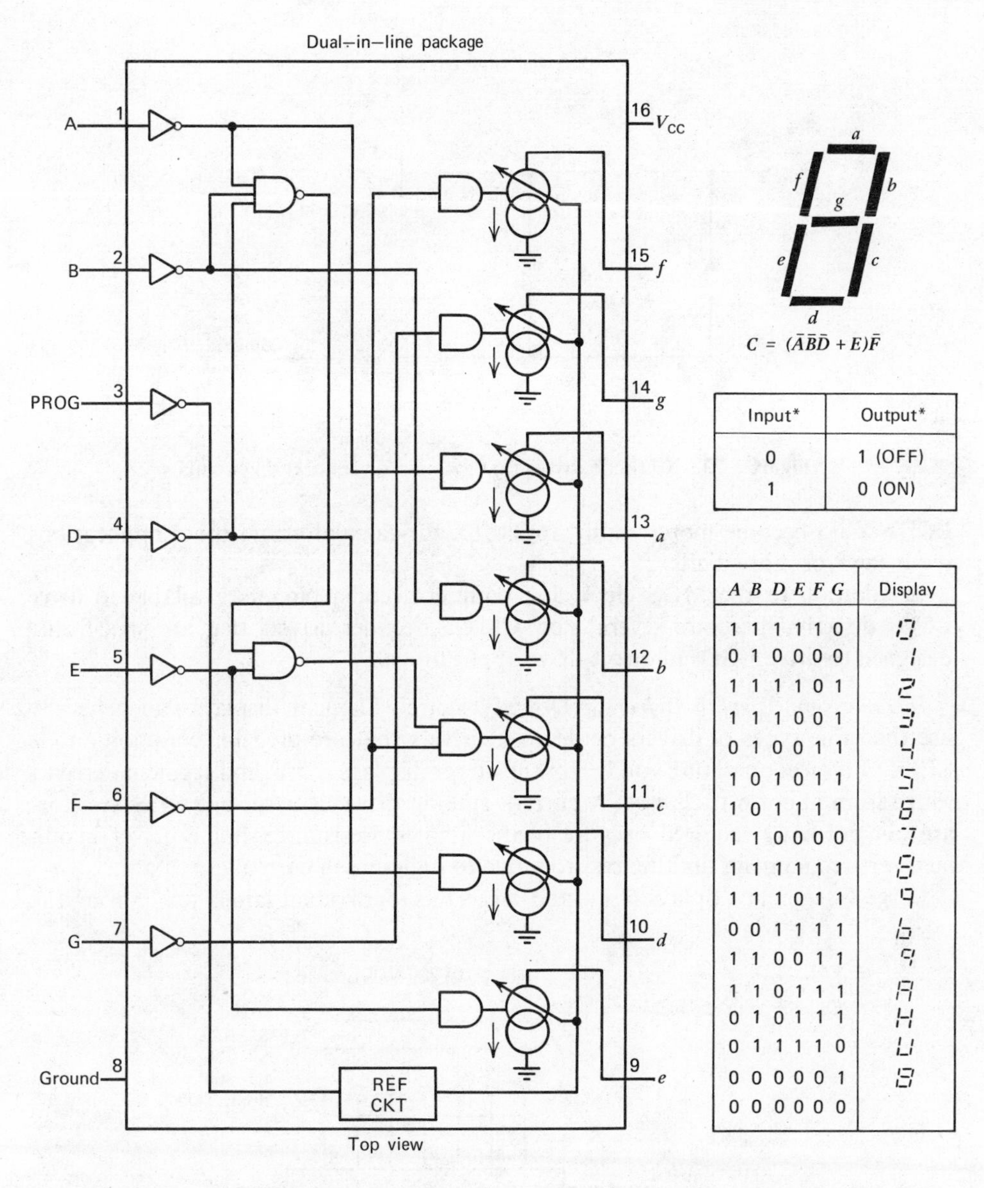

Input*	Output*
0	1 (OFF)
1	0 (ON)

A B D E F G	Display
1 1 1 1 1 0	0
0 1 0 0 0 0	1
1 1 1 1 0 1	2
1 1 1 0 0 1	3
0 1 0 0 1 1	4
1 0 1 0 1 1	5
1 0 1 1 1 1	6
1 1 0 0 0 0	7
1 1 1 1 1 1	8
1 1 1 0 1 1	9
0 0 1 1 1 1	b
1 1 0 0 1 1	[illegible]
1 1 0 1 1 1	A
0 1 0 1 1 1	H
0 1 1 1 1 0	U
0 0 0 0 0 1	[illegible]
0 0 0 0 0 0	

Figure 5-72. The DM8885 MOS-to-high-voltage cathode buffer. On-chip decoding allows 7-segment drive with six inputs [*note:* * positive logic].

Liquid-Crystal Drivers (*LCDs*). At the present time it is impractical to multiplex commercially available LCDs. This is primarily because of the inherent linearity of the device or lack of a sharp threshold to turn segments ON or OFF as in LEDs and gas-discharge displays. Optimization of turn-ON to turn-OFF time is also a problem for multiplexing.

These problems are in the process of being overcome, however. Several LCD manufacturers are multiplexing a large number of digits under laboratory conditions. When it becomes possible to produce production units with these capabilities, then

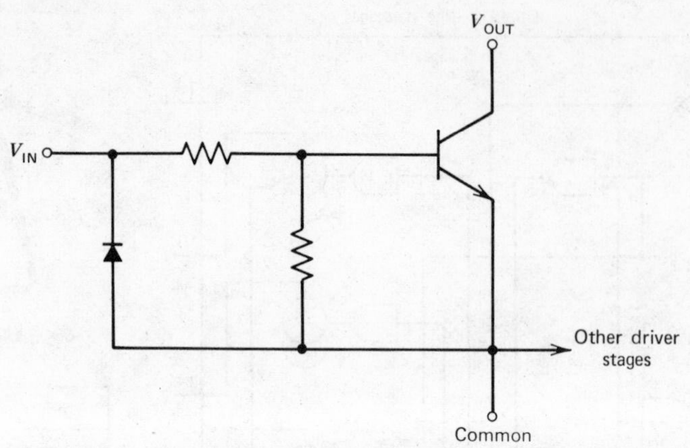

Figure 5-73. Cathode driver that includes dc restoration circuit.

LCDs could become more readily applicable for calculator and other applications requiring 8 or more digits.

In addition to the MOS clock and counter/decoder circuits available to drive LCDs directly, there are several new CMOS decoder/drivers that are specifically designed to drive LCDs in direct-drive applications.

Incandescent Display Drivers. Directly viewed filament displays can generally use the same types of drivers or decoder/drivers that are used for common-anode LEDs. A typical operating voltage for the newer displays is 5 V, and segment currents range from about 8 to 25 mA. A current-sinking driver is used, but series resistors are not normally required because of the filament resistance. Some driver manufacturers recommend limiting resistors due to high inrush currents.

Large 7-segment displays (1 in. and larger) use individual lamps for lighting the

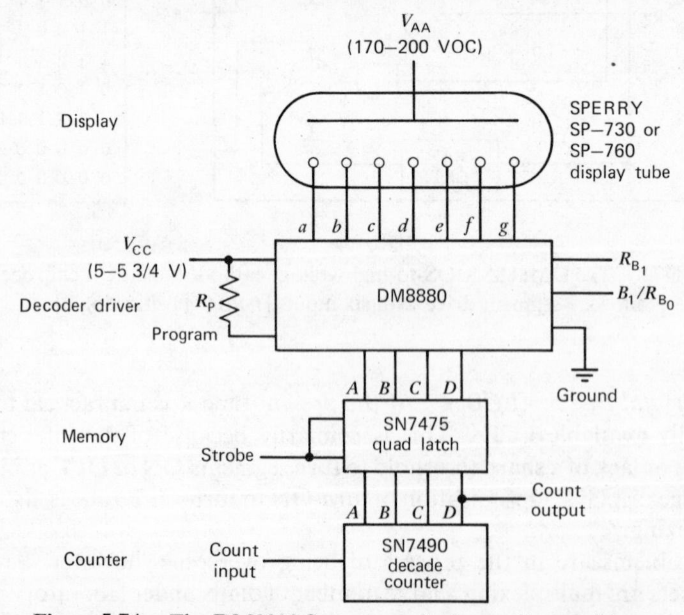

Figure 5-74. The DM8880 Sperry gas-discharge display driver.

Function	Decimal point	Comma	D	C	A	B	a	b	c	d	e	f	g
0	1	1	0	0	0	0	0	0	0	0	0	0	1
1	1	1	0	0	0	1	1	0	0	1	1	1	1
2	1	1	0	0	1	0	0	0	1	0	0	1	0
3	1	1	0	0	1	1	0	0	0	0	1	1	0
4	1	1	0	1	0	0	1	0	0	1	1	0	0
5	1	1	0	1	0	1	0	1	0	0	1	0	0
6	1	1	0	1	1	0	0	1	0	0	0	0	0
.7	1	1	0	1	1	1	0	0	0	1	1	1	1
8	1	1	1	0	0	0	0	0	0	0	0	0	0
9	1	1	1	0	0	1	0	0	0	0	1	0	0
10	1	1	1	0	1	0	1	1	0	0	0	1	1
11	1	1	1	0	1	1	1	1	0	0	0	1	0
12	1	1	1	1	0	0	0	0	1	1	1	0	0
13	1	1	1	1	0	1	0	1	1	0	0	0	0
14	1	1	1	1	1	0	1	1	1	1	1	1	0
15	1	1	1	1	1	1	1	1	1	1	1	1	1
*Decimal	0	1	X	X	X	X	X	X	X	X	X	X	X
point	0	0	X	X	X	X	X	X	X	X	X	X	X

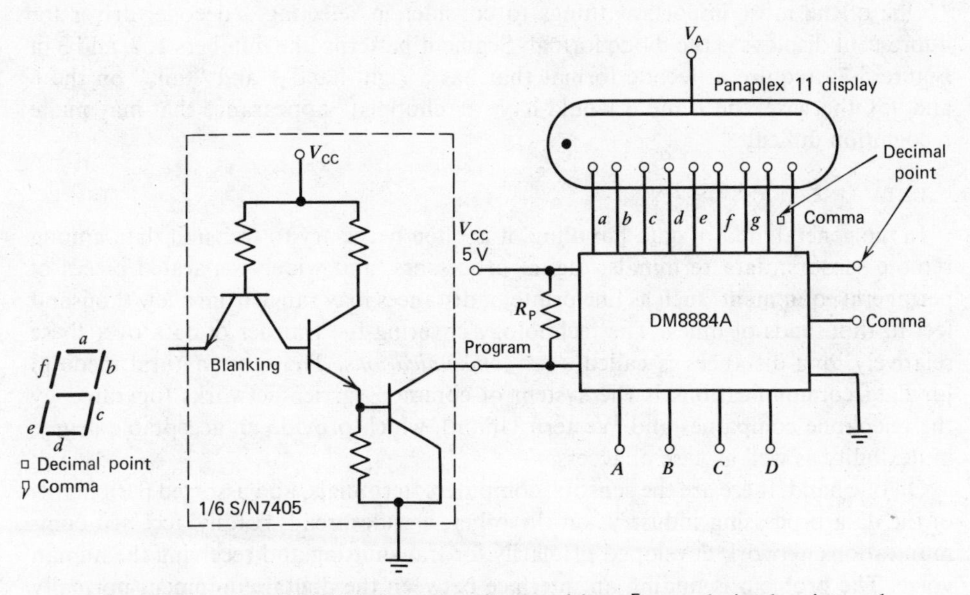

Figure 5-75. The DM8884 Panaplex display driver [*note:* decimal point and comma can be displayed with or without any numeral].

bar segments. Lamp voltages vary from about 3 to 28 V, and current requirements may exceed 100 mA in 4-in. displays.

When power requirements exceed the capabilities of monolithic devices, hybrid circuits are required.

Fluorescent Display Drivers. There does not seem to be a great deal of design activity with fluorescent displays in the U.S. and as a result, semiconductor manufacturers have not developed drivers specifically for this type of display.

With a few interface components, however, some of the LED or gas-discharge

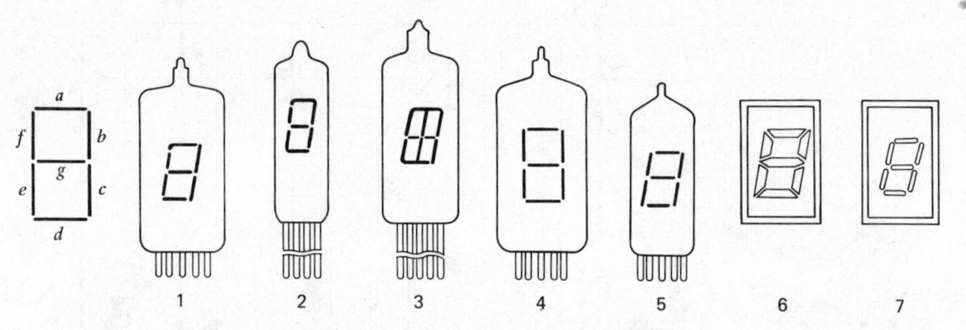

Figure 5-76. Decode format is an important consideration, particularly for fluorescent displays which may have segment patterns shown in 1, 2, and 3.

decoder/drivers can be used. Because of the voltage levels involved (as low as 10 V for dc operation of 0.3-in. high digits to over 50 V for multiplexing larger digits), the vacuum fluorescent displays can be directly driven from some MOS counter/decoder/drivers and calculator chips.

One of the most important things to consider in selecting a decoder/driver for fluorescent displays is the diode format. Segment patterns like numbers 1, 2, and 3 of Figure 5-76 require a decode format that has a right-hand 1 and "tails" on the 6 and 9. Otherwise, the 1 and 6 would have a "chopped" appearance that may make recognition difficult.

DATA LINE TRANSMISSION

In the general area of data handling, it is often necessary to transmit data among remote sensors, data terminals, digital processors, and widely separated pieces of peripheral equipment, such as line printers; distances may range from a few thousand feet to thousands of miles. The technology covering the transfer of data over these relatively long distances is called *data communications*. The most natural medium for data communications is the system of common-carrier networks (operated by the telephone companies and Western Union), which provide an acceptable degree of flexibility as well as ease of access.

On one hand, there are the sensors, computers, terminals, and assorted peripherals of the data-processing industry; on the other, a widespread, readily accessed communications network developed primarily for transmitting and receiving the human voice. The problem is finding an interface between the digital equipment normally used by the computer user, and the available analog communications networks. Such an interface is the *modem*.

The word modem is an acronym for *mo*dulator–*dem*odulator, whose primary function is to convert input digital data to a form compatible with basically analog transmission lines. *Modulation* is the D/A conversion process; *demodulation* is the reverse, whereby transmitted analog signals are reconverted to digital data compatible with the receiving data-handling equipment. Figure 5-77 summarizes the modem function.

In the process of carrying out this primary function, the modem must be compatible with the data communications equipment at its digital interface as well as with the telephone line at its analog interface. The former compatibility is governed by

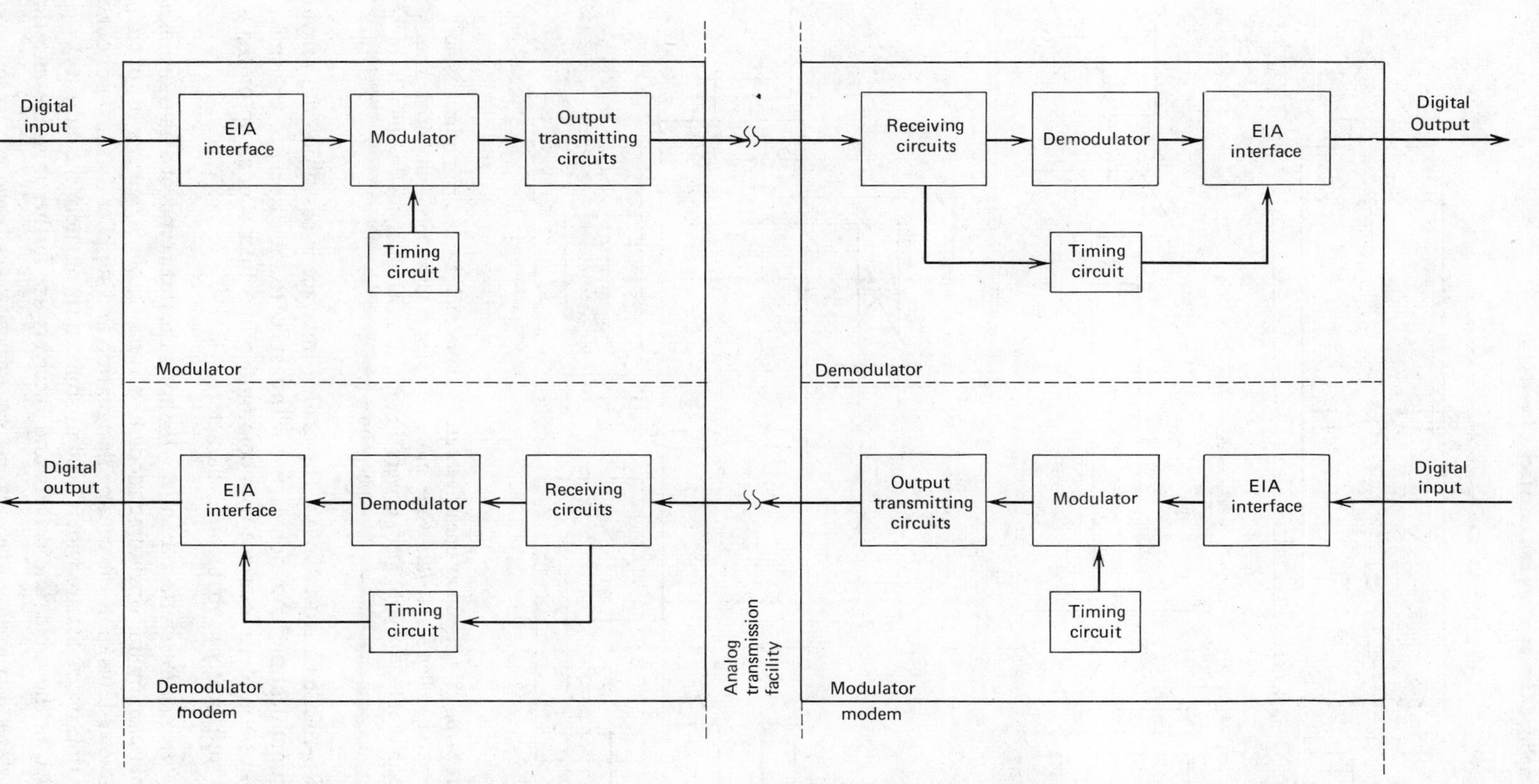

Figure 5-77. Simplified diagram showing modulation and demodulation functions. These devices interface digital data equipment with the analog transmission line.

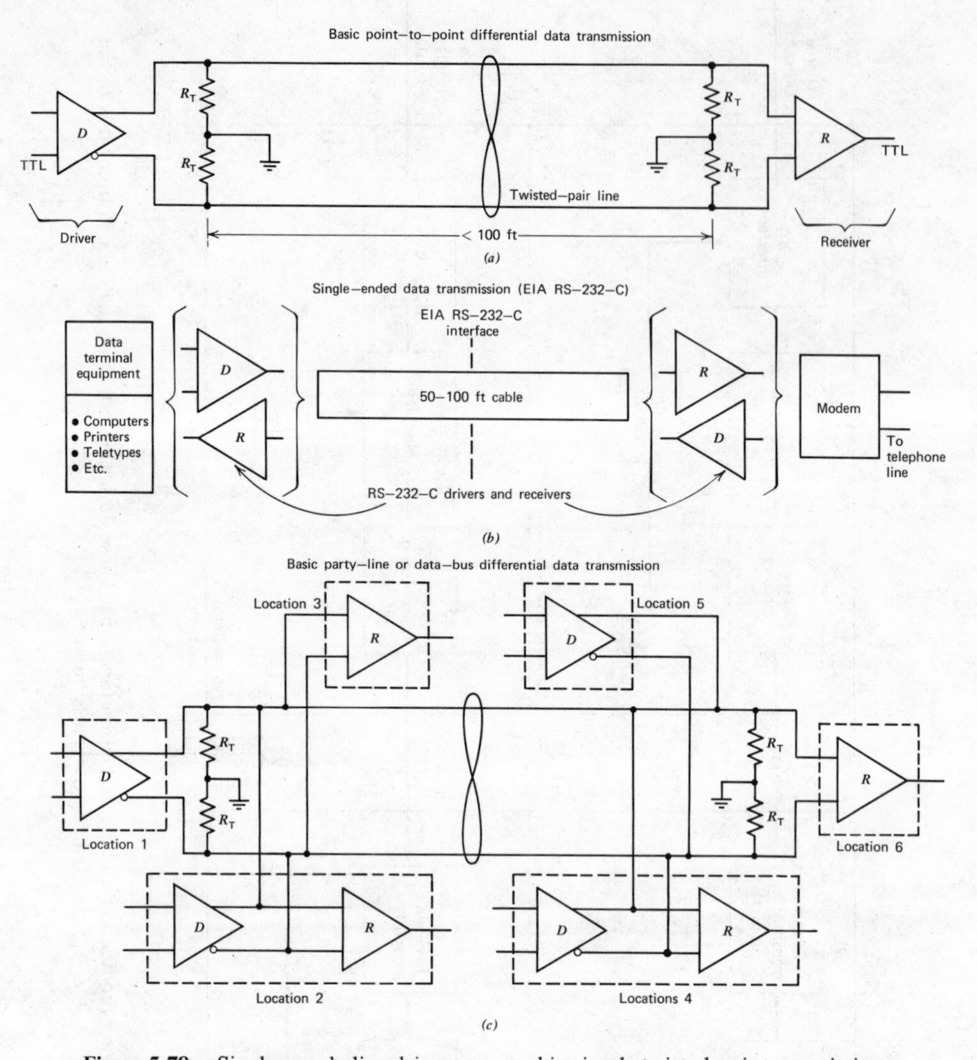

Figure 5-78. Single-supply line drivers are used in simple twisted-pair transmission systems of under 100 ft (*a*), while more complex multisupply units drive lines thousands of feet long. Modern drivers (*b*) with three-state outputs are suitable for party lines (*c*).

EIA Std RS-232Cc (August 1969).[2] The analog interface is based on data provided by the communications common carrier with respect to the associated transmission system. Thus a modern interface with communications lines is often optimized with respect to statistically probable line characteristics.

Line Drivers and Receivers. Figure 5-78 shows the three types of data transmission systems that employ line (including modem) drivers. Table 5-2 gives a list of circuits used in these systems, plus the catalog designation and supplier that first announced the device (linear circuits are widely second-sourced throughout the industry).

As shown in the table, line drivers and receivers are further divisible into three

[2] Interface between data terminal equipment and data communications equipment employing serial binary data interchange, Engineering Dept., Electronic Industries Association.

Table 5-2 Sample of line circuits available

Description	*Initial manufacturer's part number*	*Voltage supplies (V)*	*Comments*
Line drivers			
Dual differential	TI: SN55/75109	±5	Constant-current output (6 mA)
	TI: SN55/75110	±5	Constant-current output (12 mA)
	Fairchild: 9614	+5	Single supply
	National: DM7830	+5	Single supply
	Signetics: 8T13	+5	Party-line operation, about 75 mA source at 2.4 V
	Signetics: 8T23	+5	Short-circuit protection I_{OUT} = 59.3 mA at 3.11 V
Dual EIA	Signetics: 8T15	+5.5	Short-circuit protection, single supply
Dual EIA RS-232C	TI: SN75150	±12	±25-V output, short-circuit protection
Quad EIA RS-232C	Moto: MC1488	±15	Current-limited output
Quad bus	Signetics: 8T09	+5	Drives 25 loads on single bus, 40 mA sink per output
	Signetics: 8T26	+5	Tri-state outputs, 40 mA sink
Quad single/dual differential	National: DM7831	+5	V_{CC} clamp diode, tri-state outputs, 40 mA sink
	National: DM7832	+5	Same as 7831, but no clamp diodes
Line receivers			
Dual differential	TI: SN55/75107	±5	±25-mV input sensitivity, totem pole output
	TI: SN55/75108	±5	±25-mV input sensitivity, open collector output
	Fairchild: 9615	+5	Single supply
	National: DM7820	+5	Single supply
Dual single-ended	TI: SN75140	+5	Adjustable V_{REF} (1.5–3.5 V)
Dual EIA	National: 7822	+5	Tri-state outputs, independent strobe inputs
	Signetics. 8T16	+5	Single supply
Dual EIA RS-232C/Mil-Std 188	TI: SN75152	±12	Continuously adjustable hysteresis
Quad EIA RS-232C	TI: SN75154	+5 or +12	No external components required
	Moto: MC1489	+10	Built-in input hysteresis
Quad bus	Signetics: 8T26	+5	Tri-state outputs, 40 mA sink
Triple	Signetics: 8T24	+5	Fanout of 10 with TTL, high-speed, single supply

varieties: circuits operating from a single 5-V supply, circuits operating from multiple power supplies, and circuits that conform to the EIA RS-232 modem interface specification, including the largely development tri-state output circuits for party-line use.

The single-supply line driver and receiver circuits are used in medium performance twisted-pair transmission (Figure 5-78*a*). These are short data transmission lines for low data rates—say, up to 100 ft at less than 20 MHz.

In these systems, the transmitted data takes the form of a voltage level operating at fairly high currents—for most 5-V families, currents as high as 40 to 50 mA can be produced at standard TTL drive levels. The high power is needed to overcome line losses. If the line losses are low enough, the output capability can be relaxed—a few milliwatts will do in most cases. In addition, the receivers can have sensitivities in the 1-mV range.

The higher performing line drivers and receivers, requiring the multiple-supply circuits, are used on transmission lines up to several hundreds of feet at data rates often exceeding 20 MHz. However, the trade-off between line length and frequency must be made. If you want to go down low enough in data rate—say, into the kilohertz region—you can drive a transmission line a few thousand feet long; the telephone company does this every day.

The multiple-supply circuits are more complex, capable of outputs up to 100 mV, and with receivers capable of detecting signals as low as 1 mV. This performance is possible because the transmitted data takes the form of a constant current, generally kept at a low level of about 10 mA so that line losses can be minimized and line length maximized.

The line is terminated in its characteristic impedance, and because current and impedance both are low, its output-voltage levels are small, and it requires a more sensitive receiver. This means that the receivers need more than just the normal logic gate in the input. Often the same type of low-sensitivity front end that is built into many op amps is required here.

The third variety of line drivers and receivers—those meeting the EIA RS-232 specifications—is used as shown in Figure 5-78*b* to transmit data between terminal equipment such as teletypewriters and line printers. The RS-232 specification was developed when telephone users won the right in court to use their own peripheral driver and receiver equipment on a telephone company's lines.

All RS-232 circuits operate in conjunction with a modem—a modulator-demodulator unit that processes data for transmission over telephone lines. In effect, the modem set, usually the size of an office typewriter, works with data-processing equipment and converts bipolar (digital) signals to acoustic signals at the transmit end and then reconverts acoustic signals to bipolar at the receiving end, where another modem is located. Receiving and transmitting are normally accomplished over communication lines which require interface circuits that meet the EIA RS-232C specification.

The EIA specification defines the interface on the digital portion of the modem. Essentially this means that the RS-232 drivers must be capable of driving 50 to 100 ft of cable, while withstanding output short-circuit conditions of 25 V. An RS-232 receiver, in turn, must withstand +25-V signals excursions on the input.

To meet the +25-V input condition, the receivers have a cascaded input structure with an attenuator on the front end. To meet the short-circuit condition, the driver incorporates high voltage-breakdown transistors in a collapsing-circuit technique that shuts off the output current when a certain voltage level is exceeded.

Two of the newest line circuits to be developed are TI's dual-line driver, the SN75150, and its companion receiver, the SN75152. The driver has a dual-power supply, so that it can be operated either from a +12-V supply, normally used for telephone lines, or from a +5-V supply, normally used with TTL or DTL systems. In addition, the receiver has a 3- to 7-kΩ input resistance over the RS-232C voltage range, high enough for most peripheral applications, plus an active "pull-up" on the output for more symmetrical switching speeds.

Again, to meet the EIA RS-232 specification, outputs for the driver are current-limited for short-circuit protection to +25 V. Output voltage levels are nominally +6 V and are designed to drive capacitive loads greater than 2500 pF enabling the device to interface with all commercially available modems.

Because most modem sets use two receivers for each driver, the driver units are supplied as duals and the receivers as quads. This arrangement reduces the number of IC packages required in a modem design. Major applications for these types of drivers and receivers are computer time-sharing services, data-processing equipment, and other uses that call for relatively short single-line point-to-point data transmission and for level translators.

Being ideal for party line applications, line drivers and receivers with three-state outputs (discussed in detail in a later section) are growing in popularity. The basic party line or data bus differential transmission line, where six parties share a common transmission line, is shown in Figure 5-78*c*. By providing a high-impedance output state, in addition to the conventional low-impedance, high- and low-current states, equipment at any location can be switched into the high-impedance (OFF) state when not in operation.

National Semiconductor has pioneered in three-state output devices, for digital as well as linear applications. Quad drivers that are rapidly becoming the industry standard are National's DM7831/DM8831. Although intended primarily for computer logic manipulation, they can also drive transmission lines in a party line application. (A typical three-state party line configuration developed by the company for its 7831/8831 pair is illustrated in Figures 5-79*a* and *b*).

Computers and peripheral equipment are prime consumers of line drivers. This is because low-power logic circuits can transmit data only over very short distances—at most a few inches—and a system designer must go to independent line circuits if error-free data transmission over longer lines is required. Add to this the burden of noisy computer environments, and the problem of selecting suitable transmission line drivers and receivers becomes formidable. This is especially true when data is transmitted between computer consoles or between a computer and distant peripherals.

A typical computer interface system is diagramed in Figure 5-80. A balanced, twisted-pair transmission line is terminated in its characteristic impedance so that both lines are biased at a few hundred millivolts. When the driver converts input logic levels to voltage levels that control a current switch, the switch unbalances the voltage on the transmission lines, resulting in a difference voltage at the receiver input.

Usually the input of the receiver has a differential-input stage that exhibits high rejection of common-mode input signals. Intermediate stages convert the polarity of the input signal to the desired logic levels at the receiver output. Moreover, the

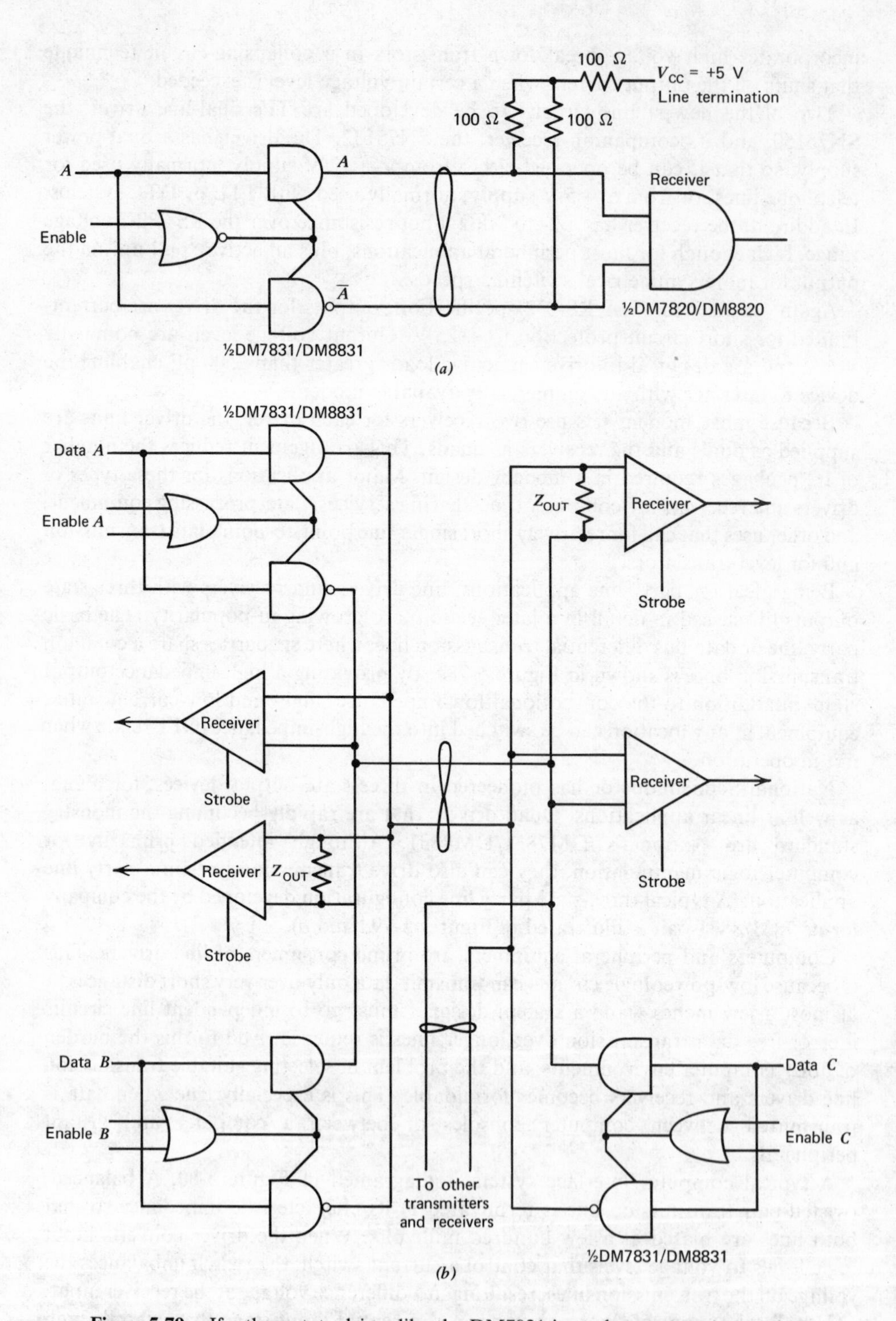

Figure 5-79. If a three-state driver like the DM7831 is used, many parties can be accommodated on a single transmission line. Line terminations shown in (*a*) have 100-Ω resistor networks. The 7831 also can be used for computer-logic manipulation.

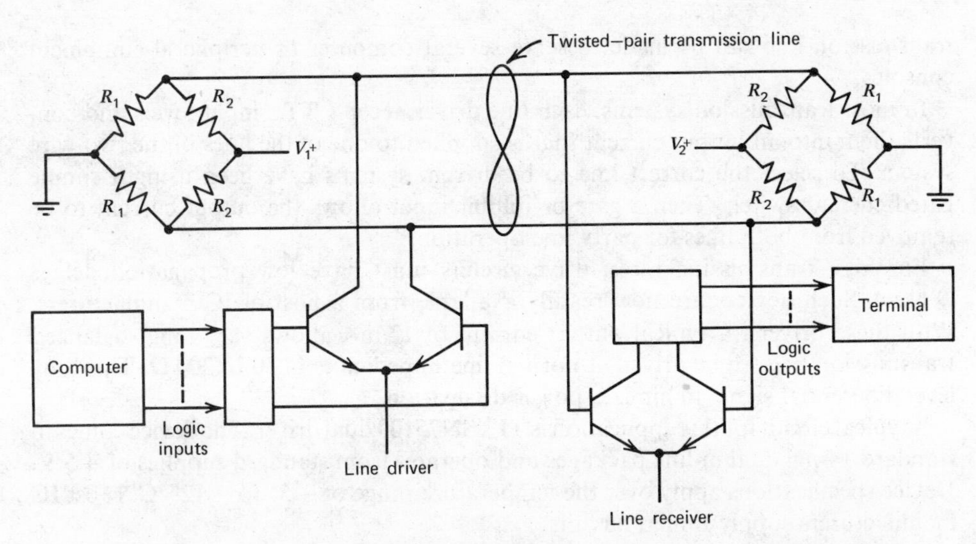

Figure 5-80. This computer interface system has a twisted-pair transmission line with both lines biased at a few hundred millivolts. The logic inputs control the driver which activates a switch to unbalance the two voltages and cause a voltage difference at the receiver.

driver output is capable of rejecting common-mode signals induced on the line, which also aids recovery of data.

An important feature of the system of Figure 5-80 is that provision can be made for removing the driver output current from both lines. Again, this is the party line application, since in this inhibit mode another driver may be used to transmit data over the line.

Adding strobes or gates to the receivers, moreover, will allow any driver to communicate with any or all enabled receivers, while other drivers are inhibited and other receivers are strobed OFF. This version of the system is shown in Figure 5-81. Line receivers and drivers may be connected anywhere along the line, so that a single

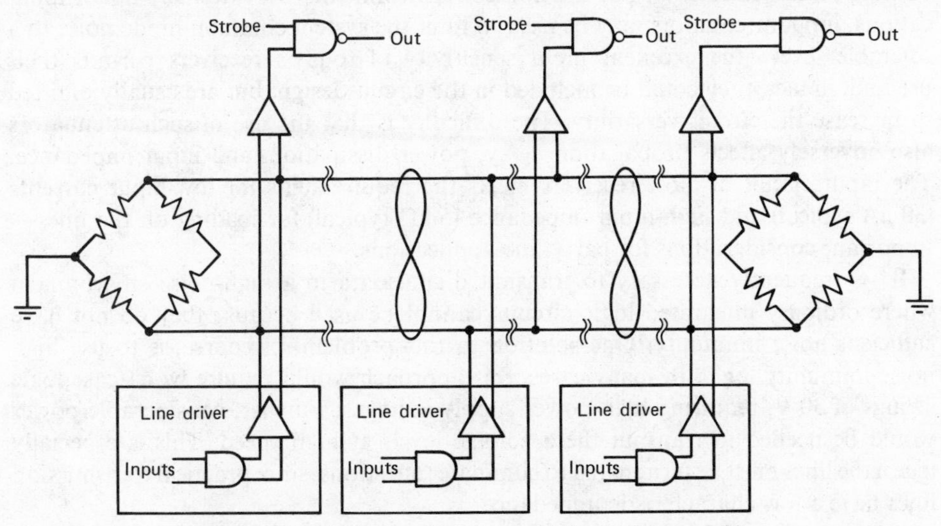

Figure 5-81. Addition of strobe capability to Figure 5-80.

transmission line can be made to serve several computer or peripheral-equipment consoles.

In most transmission systems, each line driver accepts TTL input levels and converts them into an output current that is supplied to one of the lines of the two-wire system. To select the correct line to be driven, systems have been using a simple gated current switch. Then a gate or inhibit input allows the output current to be removed from both lines for party line operation.

For high transmission rates, drive circuits must have low propagation delays (9 nsec). Such devices are now readily available from a host of IC manufacturers. With these drivers, a typical output current of 12 mA allows very long, balanced transmission lines to be driven at normal line impedances of 50 to 200 Ω. This low-level differential signal minimizes power dissipation.

A typical circuit for this application is TI's SN75109 dual driver. The device comes in standard 14-*pin* dual in-line packages and operates from standard supplies of +5 V. Device specifications apply over the temperature range of −55 to +125°C. Most IC manufacturers supply similar circuits.

Dual-line receivers are available to complement such drivers. A typical example is TI's SN75107, which is designed to detect differential-input signals as low as 20 mV in the presence of common-mode input noise in the range of ±3 V. The polarity of the low-level differential-input signal is then translated into high-level output logic levels compatible with TTL systems. Each receiver circuit also has TTL-compatible inputs used for strobing the receiver. One input allows independent strobing of the selected receiver, and the other, being common to both receivers in the package, allows simultaneous strobing of both receivers for increased logic versatility.

Most receivers have a nominal propagation delay of less than 20 nsec, making them ideal for use in high-speed systems and almost completely insensitive to overdrive voltages greater than 10 mV. Typically, circuits respond to input signals with repetition rates to 20 MHz.

The common-mode input-voltage range of most of these receivers is a high +3 V, making them useful in all but the noisiest environments. In extremely noisy applications, input attenuators may be used to limit the system common-mode noise to a tolerable level—the excellent input sensitivity of today's receivers permits their use and, in fact, they could be included in the circuit design, but are usually omitted to increase the circuit versatility. The difficulty is that the use of such attenuators also adversely affects propagation delay, power dissipation, and input impedance. The input circuit of most receivers meets the requirements for low-input currents (30 μA typical) and high-input impedance (5 kΩ typical) for loading on the lines—important considerations for party-line applications.

It[3] is frequently necessary to transmit digital data in a high-noise environment where ordinary integrated logic circuits cannot be used because they do not have sufficient noise immunity. One solution to this problem, of course, is to use high noise-immunity logic. In many cases, this approach would require worst case logic swings of 30 V, requiring high power-supply voltages. Further, considerable power would be needed to transmit these voltage levels at high speed. This is especially true if the lines must be terminated to eliminate reflections, since practical transmission lines have a low characteristic impedance.

[3] The material on the next few pages is excerpted from *IC's for Digital Data Transmission*, AN-22, National Semiconductor Corporation, Febuary 1969.

A much better solution is to convert the ground-referred digital data at the transmission end into a differential signal and transmit this down a balanced, twisted-pair line. At the receiving end, any induced noise, or voltage due to ground-loop currents, appears equally on both ends of the twisted-pair line. Hence, a receiver which responds only to the differential signal from the line will reject the undesired signals even with moderate voltage swings from the transmitter.

Figure 5-82 illustrates this situation more clearly. When ground is used as a signal return as in Figure 5-82*a*, the voltage seen at the receiving end will be the output voltage of the transmitter plus any noise voltage induced in the signal line. Hence, the noise immunity of the transmitter-receiver combination must be equal to the maximum expected noise from both sources.

The differential transmission scheme diagrammed in Figure 5-82*b* solves this problem. Any ground noise or voltage induced on the transmission lines will appear equally on both inputs of the receiver. The receiver responds only to the differential signal coming out of the twisted-pair line and delivers a single-ended output signal referred to the ground at the receiving end. Therefore, extremely high noise immunities are not needed; and the transmitter and receiver can be operated from the same supplies as standard integrated logic circuits.

The interconnection of the DM7830 line driver with the DM7820 line receiver is shown in Figure 5-83. With the exception of the transmission line, the design is rather straightforward. Connections on the input of the driver and the output or strobe of the receiver follow standard design rules for DTL or TTL integrated logic

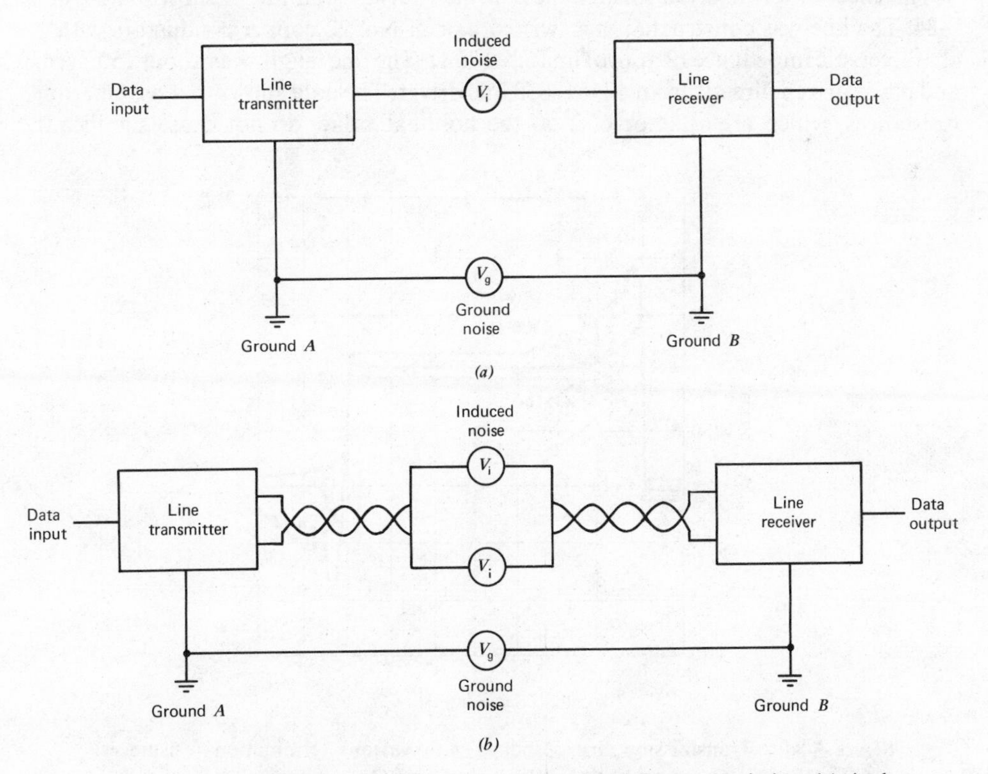

Figure 5-82. Comparing differential and single-ended data transmission: (*a*) single-end system; (*b*) differential system. (Courtesy National Semiconductor Corporation.)

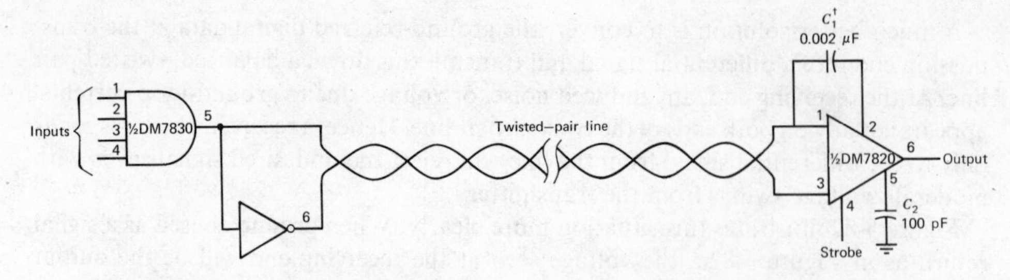

Figure 5-83. Interconnection of the line driver and line receiver (Courtesy National Semiconductor Corporation.)

circuits. The load presented by the driver inputs is equal to 3 standard digital loads, while the receiver can drive a worst-case fan-out of 2. The load presented by the receiver strobe is equal to one standard load.

The purpose of C_1 on the receiver is to provide dc isolation of the termination resistor for the transmission line. This capacitor can both increase the differential noise immunity, by reducing attenuation on the line, and reduce power dissipation in both the transmitter and receiver. In some applications, C_1 can be replaced with a short between *Pin*s 1 and 2, which connects the internal termination resistor of the DM7820 directly across the line; C_2 may be included, if necessary, to control the response time of the receiver, making it immune to noise spikes that may be coupled differentially into the transmission lines.

The effect of termination mismatches on the transmission line is shown in Figure 5-84. The line was constructed of a twisted pair of No. 22 copper conductors with a characteristic impedance of approximately 170 Ω. The line length was about 150 nsec. and it was driven directly from a DM7830 line driver. The data shows that termination resistances, which are a factor of 2 off the nominal value, do not cause significant

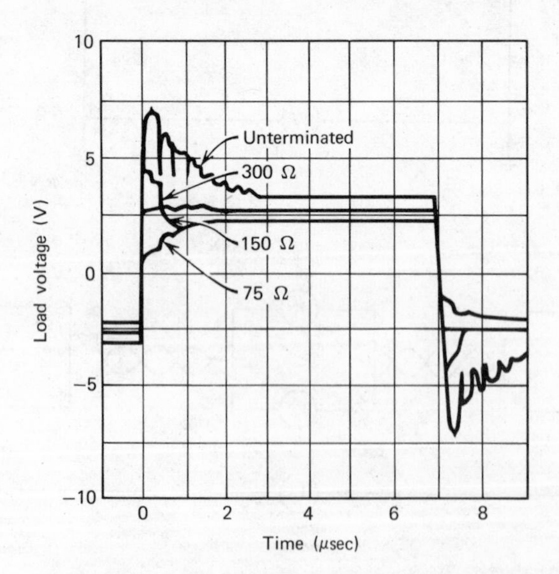

Figure 5-84. Transmission line response with various termination resistances. (Courtesy National Semiconductor Corporation.)

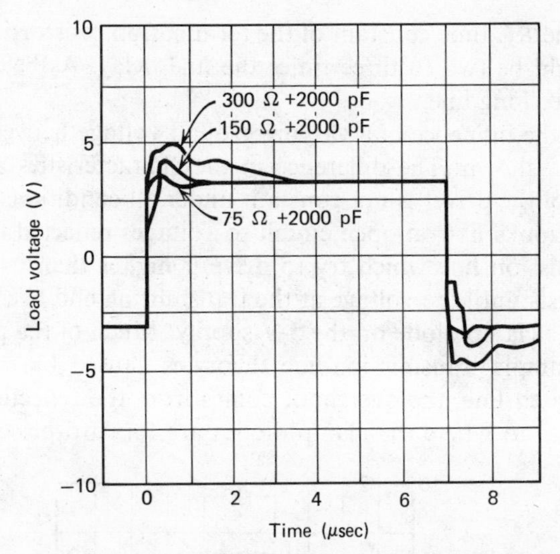

Figure 5-85. Line response for various termination resistances with a dc isolation capacitor. (Courtesy National Semiconductor Corporation.)

reflections on the line. The lower termination resistors do, however, increase the attenuation.

Figure 5-85 gives the line-transmission characteristics with various termination resistances when a dc isolation capacitor is used. The line is identical to that used in the previous example. It can be seen that the transient response is nearly the same as a dc-terminated line. The attenuation, on the other hand, is considerably lower, being the same as an unterminated line. An added advantage of using the isolation capacitor is that the dc signal current is blocked from the termination resistor which reduces the average power drain of the driver and the power dissipation in both the driver and receiver.

The effect of different values of dc isolation capacitors is illustrated in Figure 5-86.

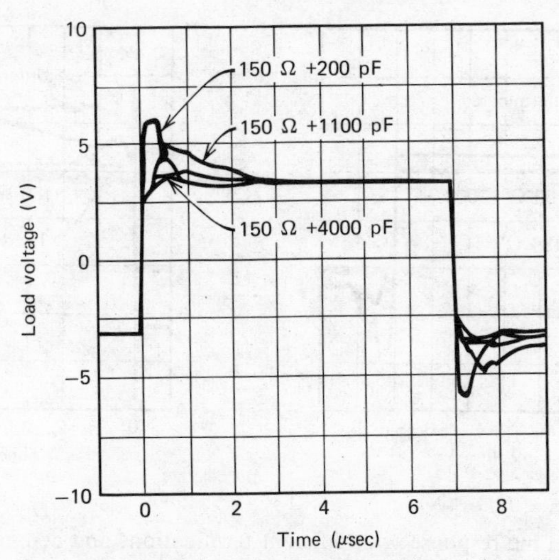

Figure 5-86. Response of terminated line with different dc isolation capacitors. (Courtesy National Semiconductor Corporation.)

This shows that the *RC* time constant of the termination resistor/isolation capacitor combination should be two to three times the line delay. As before, this data was taken for a 150 nsec long line.

In Figure 5-87, the influence of a varying ground voltage between the transmitter and the receiver is shown. The difference in the characteristics arises because the source resistance of the driver is not constant under all conditions. The high output of the transmitter looks like an open circuit to voltages reflected from the receiving end of the transmission line which try to drive it higher than its normal dc state. This condition exists until the voltage at the transmitting end becomes high enough to forward-bias the clamp diode on the 5-V supply. Much of the phenomena which does not follow simple transmission-line theory is caused by this. For example, with an unterminated line, the overshoot comes from the reflected signal charging the line capacitance to where the clamp diodes are forward-biased. The overshoot

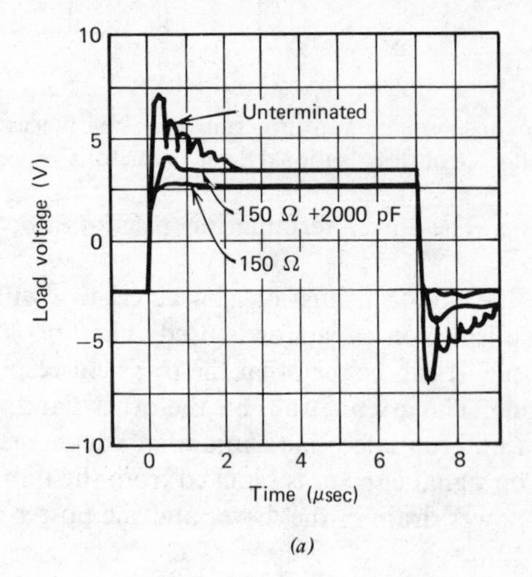

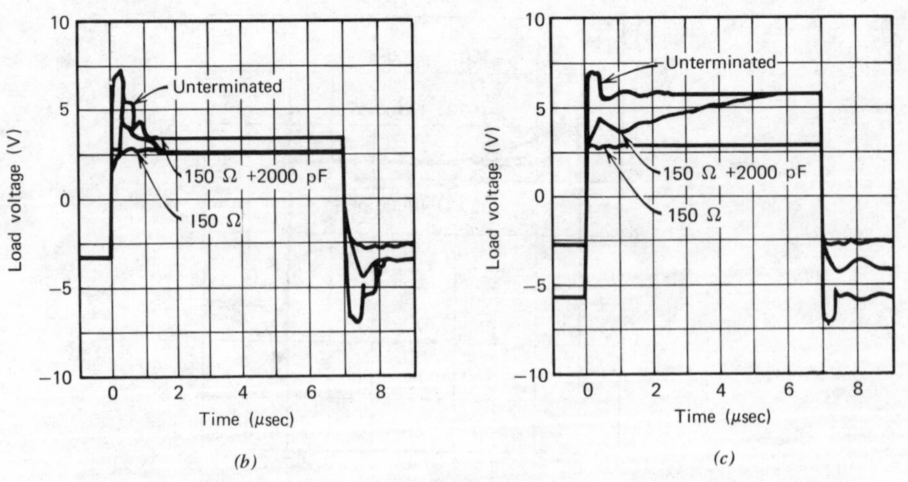

Figure 5-87. Line response with different terminations and common-mode input voltage: (*a*) $V_{CM} = 0$ V; (*b*) $V_{CM} = -15$ V; (*c*) $V_{CM} = 15$ V. (Courtesy National Semiconductor Corporation.)

then decays at a rate determined by the total line capacitance and the input resistance of the receiver.

When the ground on the receiver is 15-V more negative than the ground at the transmitting end, the decay with an unterminated line is faster, as shown in Figure 5-87*b*. This occurs because there is more current from the input resistor of the receiver to discharge the line capacitance. With a terminated line, however, the transmission characteristics are the same as for equal ground voltages because the terminating resistor keeps the line from getting charged.

Figure 5-87*c* gives the transmission characteristics when the receiver ground is 15 V more positive than the transmitter ground. When the line is not terminated, the differential voltage swing is increased because the high output of the driver will be pulled against the clamp diodes by the common-mode input current of the receiver. With a dc isolation capacitor, the differential swing will reach this same value with a time constant determined by the isolation capacitor and the input resistance of the receiver. With a dc-coupled termination, the characteristics are unchanged because the differential load current is large by comparison to the common-mode current so that the output transistors of the driver are always conducting.

The low output of the driver can also be pulled below ground to where the lower clamp diode conducts, giving effects which are similar to those described for the high output. However, a current of about 9 mA is required to do this, so it does not happen under normal operating conditions.

To summarize, the best termination is an *RC* combination with a time constant approximately equal to three times the transmission-line delay. Even though its value is not precisely determined, the internal termination resistor of the IC can be used because the line characteristics are not greatly affected by the termination resistor.

The only place that an *RC* termination can cause problems is when the data transmission rate approaches the line delay and the attenuation down the line (terminated) is greater than 3 dB. This would correspond to more than 1000 ft of twisted-pair cable with No. 22 copper conductors. Under these conditions, the noise margin can disappear with low duty-cycle signals. If this is the case, it is best to operate the twisted-pair line without a termination to minimize transmission losses. Reflections should not be a problem as they will be absorbed by the line losses.

The circuit shown in Figure 5-88 monitors the transmission line and will provide an inhibit control to the local driver when the line is busy. The local driver may be enabled when no data is being received from the line. If there is no data on the line, the retriggerable one-shot releases the line driver for operation. Before data is fed to the driver input, the local enable-control input must be taken high. The line sense amplifier is then inhibited to prevent interference with the local driver. The local driver may then transmit on the line.

Data transmission between racks of equipment in the same room may be accomplished relatively easy. Long-line transmission from a building or within a building of very large size may require signal-boosting. This can be accomplished simply by feeding through a receiver and driver to boost the signal. The problem becomes slightly more complex if bidirectional transmission is desired. The circuit of Figure 5-89 combines a SN7528 dual-channel sense amplifier, a SN7404 hex inverter, a SN75109 dual-channel driver, and a SN75107 dual-channel receiver to accomplish bidirectional boosting on a single transmission line. Adjustment of propagation

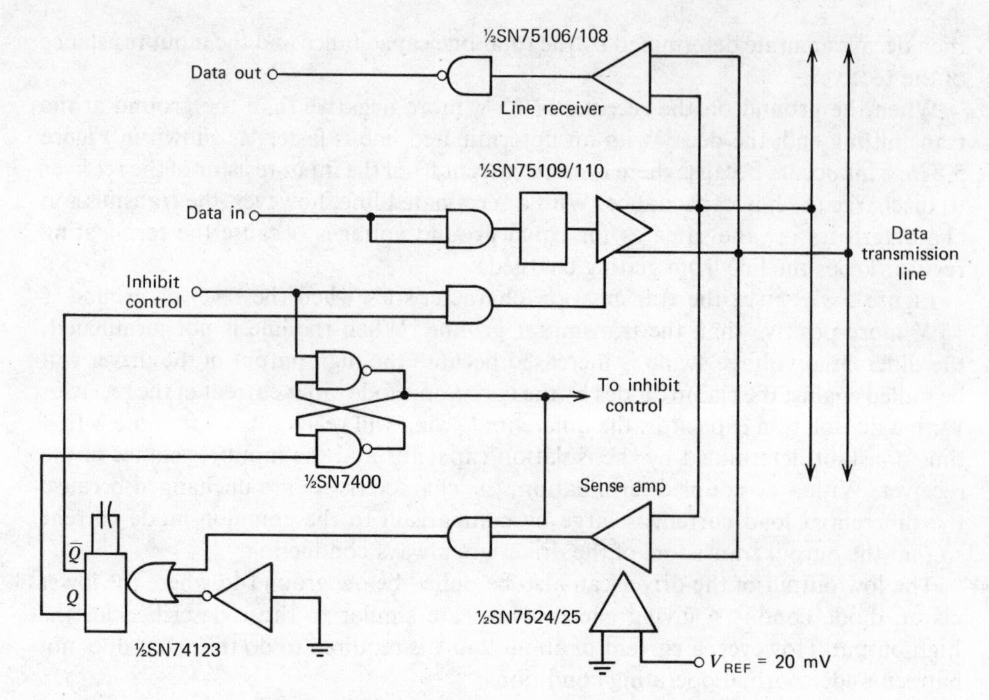

Figure 5-88. Data terminal with busy inhibit.

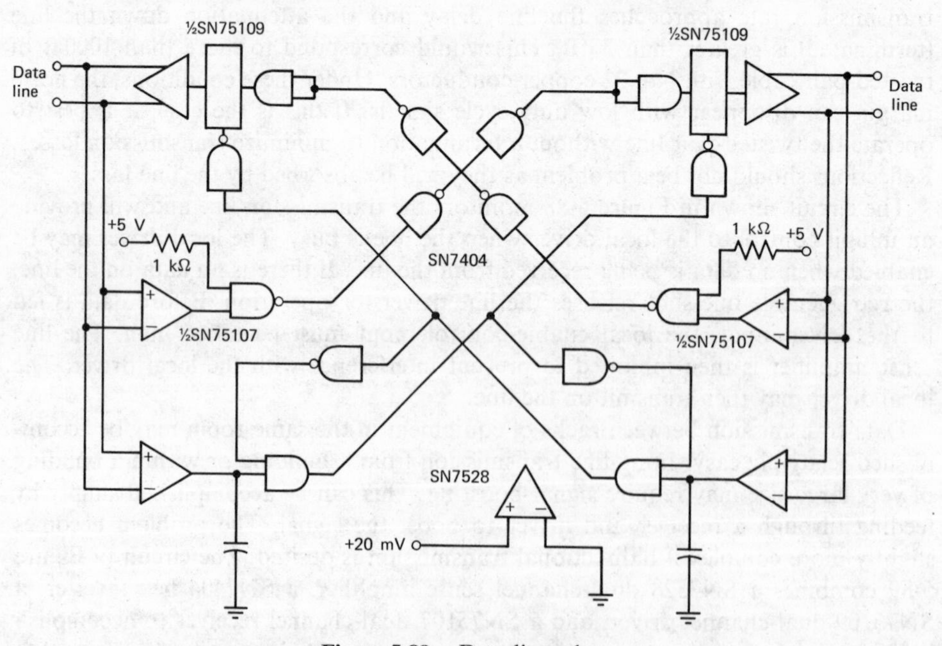

Figure 5-89. Data line relay.

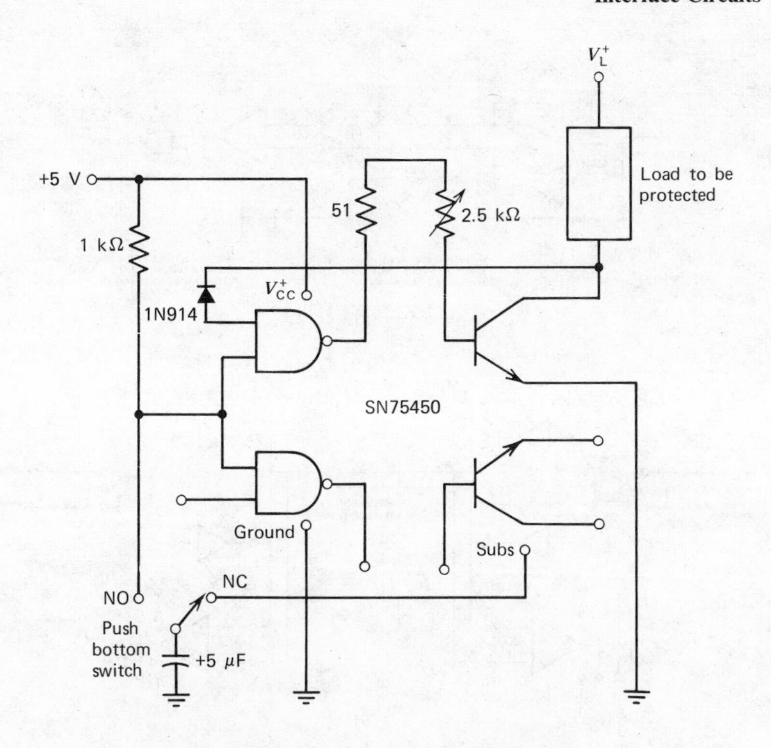

Figure 5-90. Electronic circuit breaker.

delays allows the receiving end driver to be held off while the other driver is allowed to transmit the data on.

A basic general-purpose driver has many applications in peripheral systems. The SN75450A operating as an instrumentation circuit breaker (Figure 5-90) is a good example of the versatility of such a device. The output-transistor saturation level locks the circuit on until the ON current increases to a point where the base drive is no longer adequate to maintain saturation.

As the output begins to pull out of saturation, its base drive is removed—completely opening the load path to ground. The limit at which the output-load current will result in shut-down is determined by the gate-to-base impedance control, which limits the base drive required to maintain saturation. This may be set for limit currents of from 50 to 250 mA. Figure 5-91 depicts several other line driver/receiver applications.

Tri-State Data-Switching Devices.[4] Data-routing is a natural tri-state-logic (TSL) function. The fact that the two-wide, bidirectional multiplexer of Figure 5-92 uses five fewer gates than a conventional design illustrates the efficiency of TSL in data-transfer applications. More complex switching chores are handled by the demultiplexer of Figure 5-93 and the multiplexer in Figure 5-94. The demultiplexer switches either of two inputs to any of its four outputs. Inputs may also be complemented while being switched. Examples of its bus-interchange applications are shown

[4] Excerpted from E. R. Hnatek, *A User's Handbook of Integrated Circuits*, Wiley, 1973.

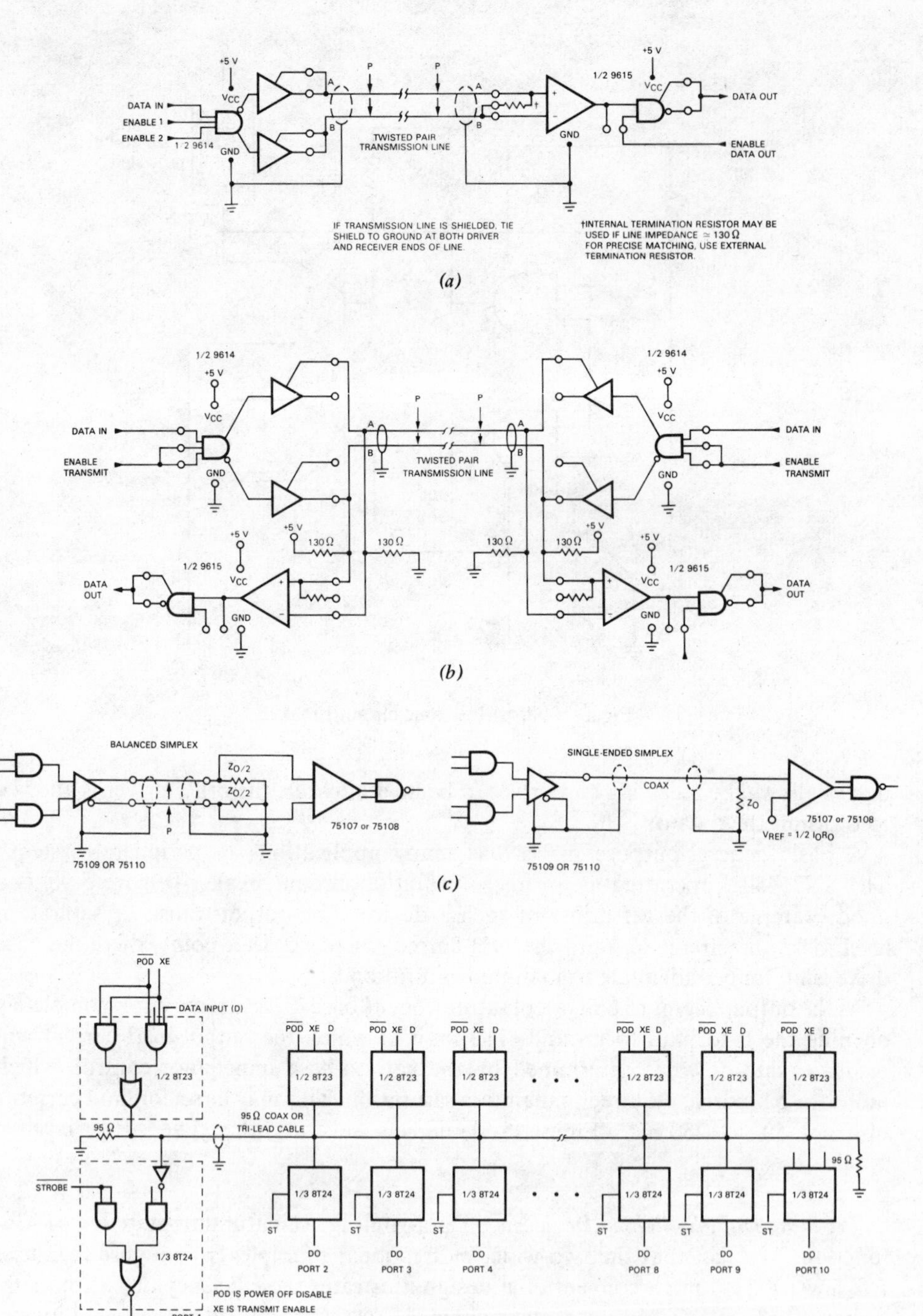

Figure 5-91. Transmission line interface element applications. (*a*) 9614/9615 simplex, balanced differential; (*b*) 9614/9615 half-duplex, differential; (*c*) simplex mode 75107 series; (*d*) IBM I/O interface application.

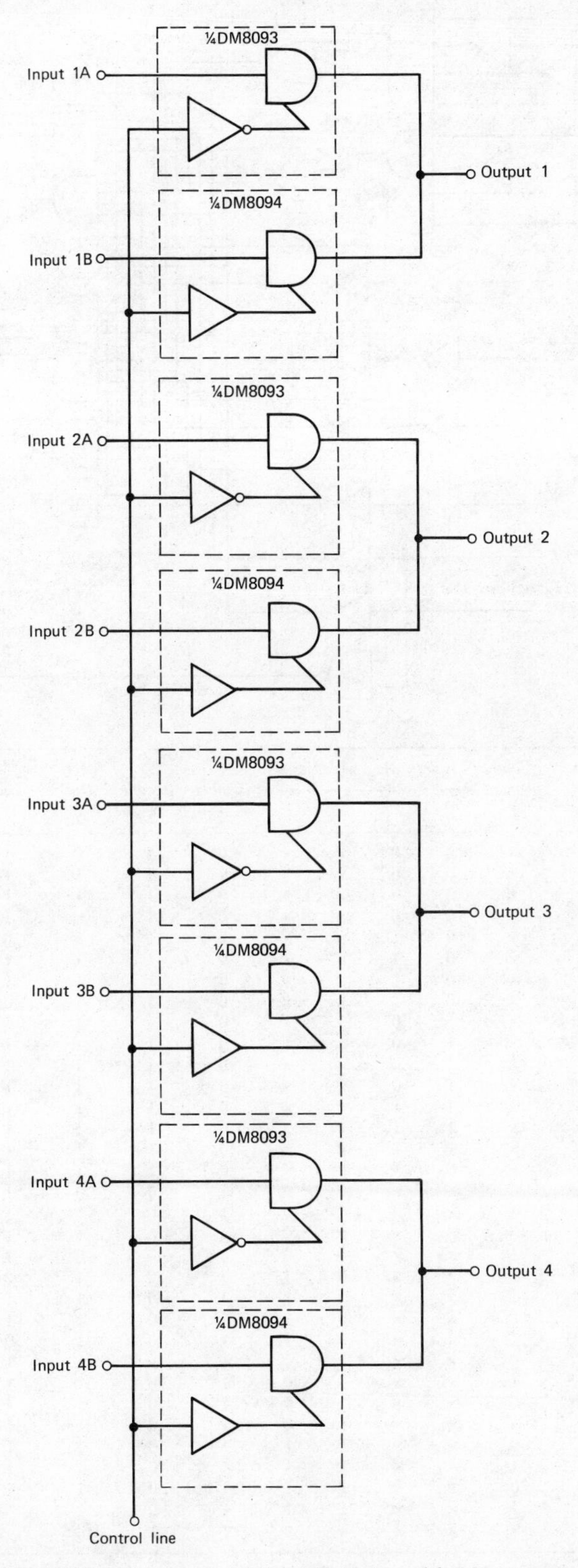

Figure 5-92. Two-wide 4-bit multiplexing with tri-state gates.

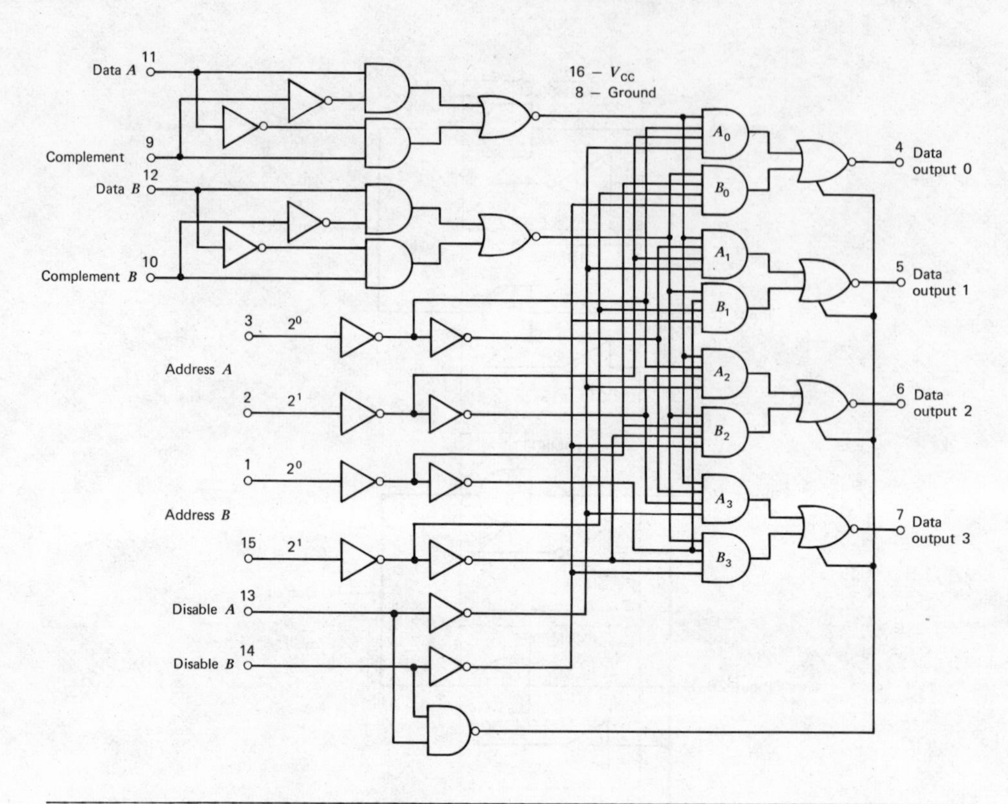

Data A	Complement A	Data B	Complement B	Address A 2^1	Address A 2^0	Address B 2^1	Address B 2^0	Disable A	Disable B	Out 0	Out 1	Out 2	Out 3
0	0	×	×	0	0	×	×	0	1	0	1	1	1
0	1	×	×	0	0	×	×	0	1	1	1	1	1
1	0	×	×	0	0	×	×	0	1	1	1	1	1
1	1	×	×	0	0	×	×	0	1	0	1	1	1
0	0	×	×	0	1	×	×	0	1	1	0	1	1
0	1	×	×	0	1	×	×	0	1	1	1	1	1
1	0	×	×	0	1	×	×	0	1	1	1	1	1
1	1	×	×	0	1	×	×	0	1	1	0	1	1
0	0	×	×	1	0	×	×	0	1	1	1	0	1
0	1	×	×	1	0	×	×	0	1	1	1	1	1
1	0	×	×	1	0	×	×	0	1	1	1	1	1
1	1	×	×	1	0	×	×	0	1	1	1	0	1
0	0	×	×	1	1	×	×	0	1	1	1	1	0
0	1	×	×	1	1	×	×	0	1	1	1	1	1
1	0	×	×	1	1	×	×	0	1	1	1	1	1
1	1	×	×	1	1	×	×	0	1	1	1	1	0
×	×	0	0	×	×	0	0	1	0	0	1	1	1
×	×	0	1	×	×	0	0	1	0	1	1	1	1
×	×	1	0	×	×	0	0	1	0	1	1	1	1
×	×	1	1	×	×	0	0	1	0	0	1	1	1
×	×	0	0	×	×	0	1	1	0	1	0	1	1
×	×	0	1	×	×	0	1	1	0	1	1	1	1
×	×	1	0	×	×	0	1	1	0	1	1	1	1
×	×	1	1	×	×	0	1	1	0	1	0	1	1
×	×	0	0	×	×	1	0	1	0	1	1	0	1
×	×	0	1	×	×	1	0	1	0	1	1	1	1
×	×	1	0	×	×	1	0	1	0	1	1	1	1
×	×	1	1	×	×	1	0	1	0	1	1	0	1
×	×	0	0	×	×	1	1	1	0	1	1	1	0
×	×	0	1	×	×	1	1	1	0	1	1	1	1
×	×	1	0	×	×	1	1	1	0	1	1	1	1
×	×	1	1	×	×	1	1	1	0	1	1	1	0
×	×	×	×	×	×	×	×	1	1	Hi-Z	Hi-Z	Hi-Z	Hi-Z

Figure 5-93. DM7230/DM8230 tri-state demultiplexer.

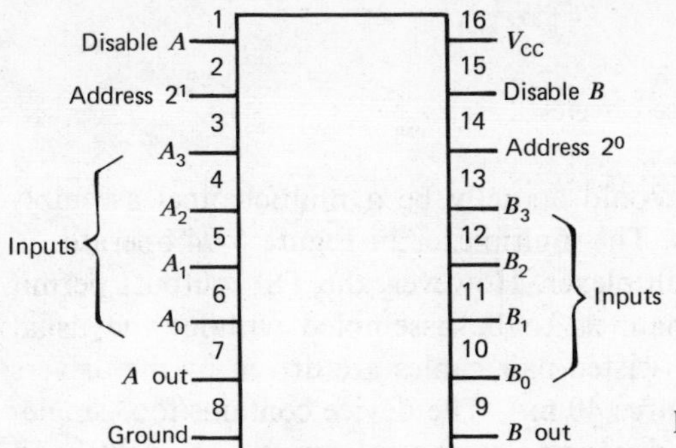

Figure 5-94. DM7214/DM8214 dual 4:1 or single 8:1 multiplexer.

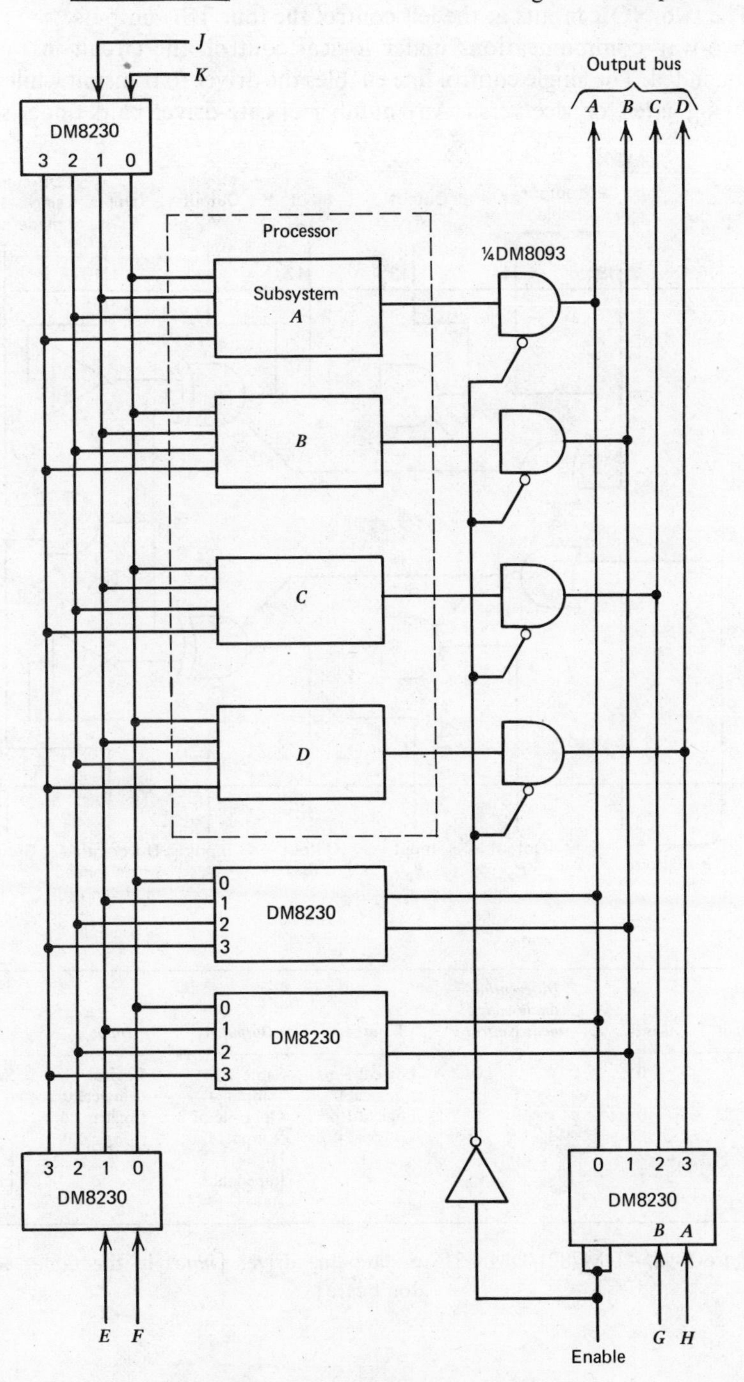

Figure 5-95. Bus-interchange examples.

in Figure 5-95. Each of the switches could actually be a multichannel assembly of many bus-connected demultiplexers. The multiplexer in Figure 5-94 operates in 8:1 or dual 4:1 modes like a TTL multiplexer. However, the TSL outputs permit multiplexers with as many as 512 channels to be assembled without the usual submultiplexers. Very long buses and twisted-pair cables are driven by the drivers in Figure 5-96. The outputs sink or source 40 mA. The device contains four single-ended drivers or two differential drivers, depending on control inputs to the right-hand NOR. The two NOR inputs at the left control the four TSL outputs.

For two-way communications under logical control, the circuit in Figure 5-97 is recommended. The single control line enables the driver to transmit while disabling the receiving gates, or vice versa. Any number of gate-driver pairs under single-line

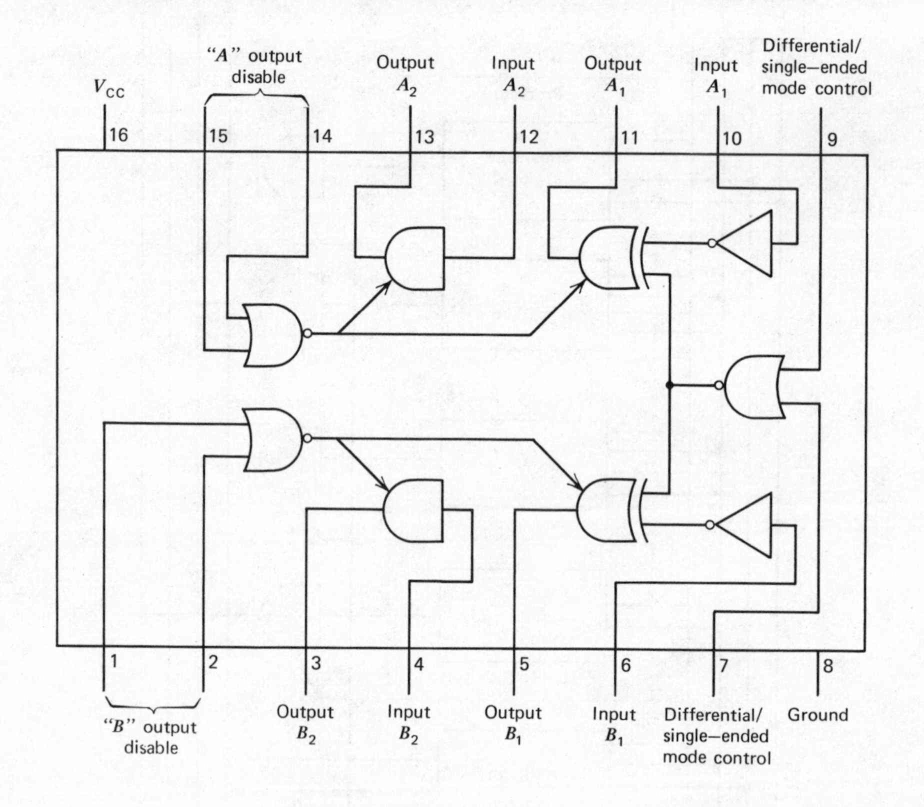

"*A*" *Output*	*Disable*	*Differential/ single-ended mode control*		*Input A_1*	*Output A_1*	*Input A_2*	*Output A_2*
0	0	0	0	Logical 1 or logical 0	Same as input A_1	Logical 1 or logical 0	Same as input A_2
0	0	×	1	Logical 1 or logical 0	Opposite of input A_1	Logical 1 or logical 0	Same as input A_2
		1	×				
1	×				High-impedance state		High-impedance state
×	1	×	×	×		×	

Figure 5-96. DM7821/DM8831 tri-state line driver [*note:* in the table, × = don't care].

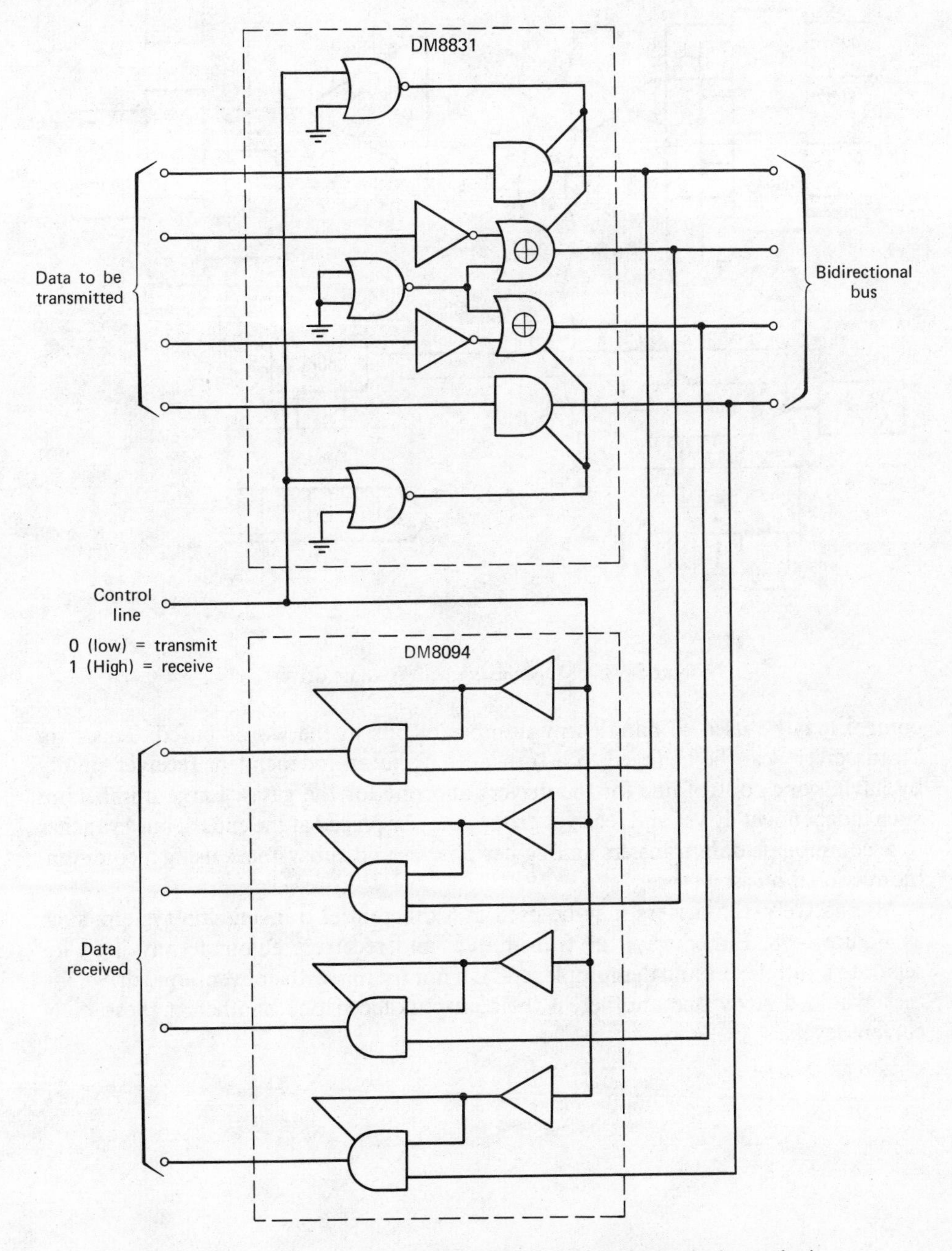

Figure 5-97. Bidirectional communication with bus line and tri-state logic.

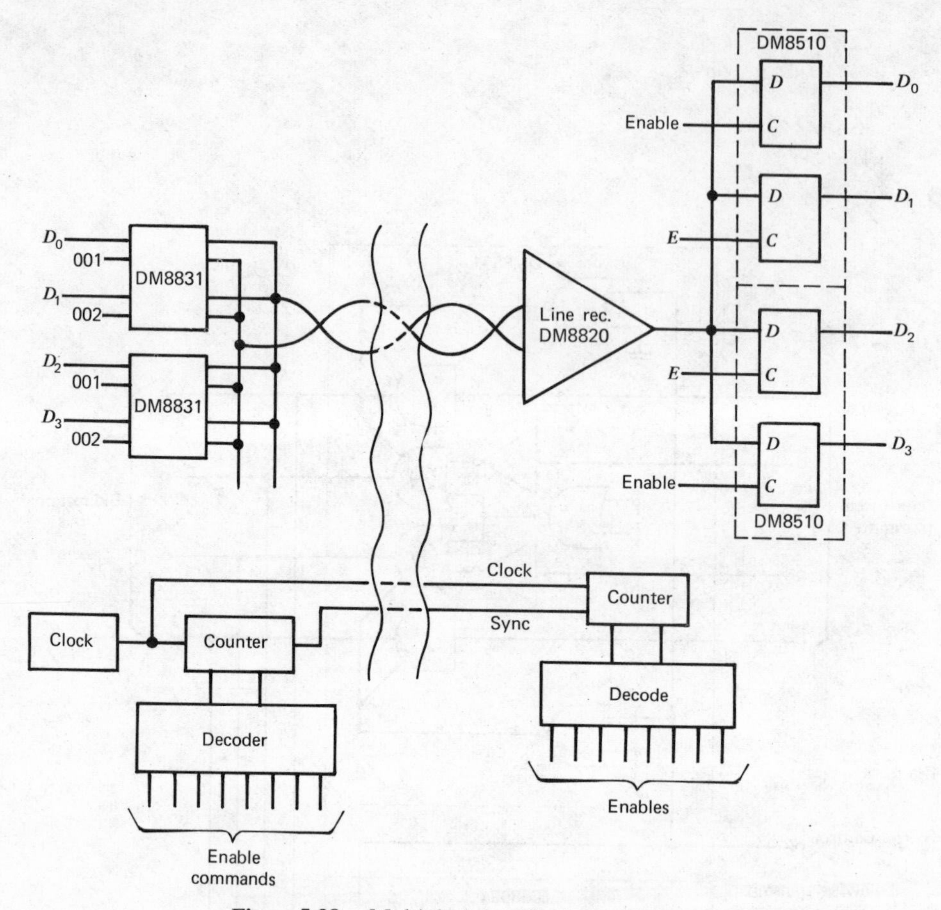

Figure 5-98. Multiplexing with tri-state drivers.

control may be used to handle any number of bits in the words bused. This data "transceiver" can be changed to a transmitter and an independent receiver simply by having one control line for the drivers and one for the gates. Large numbers of such independent driver and receiver groups may be placed at the ends of bus branches to accommodate data transfers among peripherals and subsystems, using a common, bidirectional bus structure.

Alternatively, the drivers may be used in a differential transmission system, such as Figure 5-98. Either way, the transmitters and receivers eliminate any need for separate multiplexers and demultiplexers. Do not try this with conventional drivers—they would destroy one another, if their outputs faced one another at these high-current levels.

Chapter 6

Monolithic Voltage-Regulators

INTRODUCTION

Next to the op amp, the voltage-regulator is the most common linear IC building block. The advent of the monolithic voltage-regulator has added a new dimension to regulation. The monolithic regulator offers the performance to replace discrete assemblies or portions of descrete assemblies in conventional central regulators. In addition, the small size, low cost, and ease of use uniquely fit the monolithic units for point-of-use regulation. With the many options available with IC regulators, it behooves an engineer to evaluate the possibility of employing such devices in his circuit.

The small size and low cost certainly make the monolithic regulator uniquely suited for local regulation. The extremely high-speed capability of these devices means that effective local regulation of communications equipment and computer boards is now practical. Hand-in-hand with these technical and price advantages is a significant reduction in the engineering effort necessary to design regulated power supplies.

Instead of a major engineering project, design of a power supply with ICs is literally a matter of minutes. The user need only determine the current and voltage requirements, in order to select the most economical unit that will meet his requirements. Once the right regulator and package have been chosen, the only remaining design effort involves calculating the appropriate values for the voltage-setting resistors (to provide the required output voltage) and for the current-limiting resistor (to protect against overloads). Conservatively, the first time a designer uses a monolithic regulator, this procedure may take as long as 10 min.

When higher current values are required, or if a switching regulator is desired, the design effort will be necessarily greater to select the external circuitry, but the time needed for the undertaking is still substantially less than the time required to start from scratch with a do-it-yourself regulator.

The primary function of any voltage-regulator is to hold the voltage in its output circuit at a predetermined value over the expected range of load currents. Working against the regulator are variations in load current, input voltage, and temperature.

The degree to which a regulator can maintain a constant voltage in the face of these variations is the basic figure of merit. Although there is a degree of interaction between these performance-degrading factors, especially between output current and temperature, it is most convenient to consider their effects separately.

Three families of integrated voltage-regulators have been developed: General-purpose positive and negative voltage-regulators (including a low-cost positive regulator for consumer products), and high-current, medium-current, and low-current series of fixed voltage three terminal regulators and dual tracking regulators.

The general-purpose types are designed to allow the user to set the output voltage and current levels by adding external components. They are complementary, so that precision positive/negative regulator subsystems can be assembled.

The high-, medium-, and low-current regulators supply a fixed output of 5, 6, 8, 12, 15, 18, and 24 V without external components, although external capacitors may be added to improve transient response and external resistors added to make the output adjustable or to make the device act as a current regulator.

DEFINITION OF TERMS

Load Regulation

Load regulation is defined as the percentage change in regulated output voltage for a change in load from minimum load current to the maximum load current specified. The general formula for load regulation is

$$\text{load regulation } (\%) = \frac{V_{\text{OUT(MIN)}} - V_{\text{OUT(MAX)}}}{V_{\text{OUT(MIN)}}} \times 100$$

where

$V_{\text{OUT(MIN)}}$ = output voltage with minimum rated load

$V_{\text{OUT(MAX)}}$ = output voltage with maximum rated load

A low value of load regulation is desirable.

Line Regulation

Line regulation, defined as the percentage change in regulated output voltage for a change in input voltage, is given by

$$\frac{\%}{V} = \left(\frac{\Delta V_{\text{OUT}}}{\Delta V_{\text{IN}} \times V_{\text{OUT}}}\right) \times 100$$

where

ΔV_{OUT} = change in output voltage V_{OUT} for a change in input voltage ΔV_{IN}

ΔV_{IN} = change in input voltage

V_{OUT} = nominal output voltage

A low value of line regulation is desirable.

Ripple Rejection

Ripple rejection is ac line regulation. Whereas line regulation is measured at dc, ripple rejection is defined as the line regulation for ac input signals at or above a given frequency with a specified value of bypass capacitor on the reference bypass terminal.

Standby (Quiescent) Current Drain

Standby current drain is the current that flows into the regulator and through to ground. It does not include any current drawn by the load or by external resistor networks. A low value of standby current drain is desirable.

Temperature Drift

Temperature drift is the percentage change in output voltage (or reference voltage) for a temperature variation from room temperature to either temperature extreme. Approximately 85 to 90% of this drift occurs in the regulator's internal reference circuitry. The remaining 10 to 15% is due to error amplifier or bias-current drift. Temperature drift is the major cause of output voltage change in an IC regulator. A low value of temperature drift is desirable.

Current-Limit Sense Voltage

The current-limit sense voltage is the voltage across the current-limit terminal required to cause the regulator to current-limit with a short-circuited output. This voltage is used to determine the value of the external current-limit resistor when external booster transistors are used.

Dropout Voltage

The dropout voltage is the input–output voltage differential at which the circuit ceases to regulate against further reductions in input voltage.

Feedback Sense Voltage

The feedback sense voltage is the voltage, referred to ground, on the feedback terminal of the regulator while it is operating in regulation.

Output–Input Voltage Differential

The voltage difference between the unregulated input voltage and the regulated output voltage for which the regulator will operate within specifications is called the output–input voltage differential.

BASIC REGULATOR TYPES

Voltage regulators are basically op amps combined on the monolithic chip with a voltage reference and a series-pass element.

Figure 6-1 depicts a typical circuit of a generalized series-dissipative regulator. An error amplifier converts the difference between the output sample and the reference voltage into an error signal that controls the voltage dropped by the series-pass transistor.

The general-purpose types require built-in frequency compensation to prevent oscillations. The compensation method must provide a high degree of rejection to

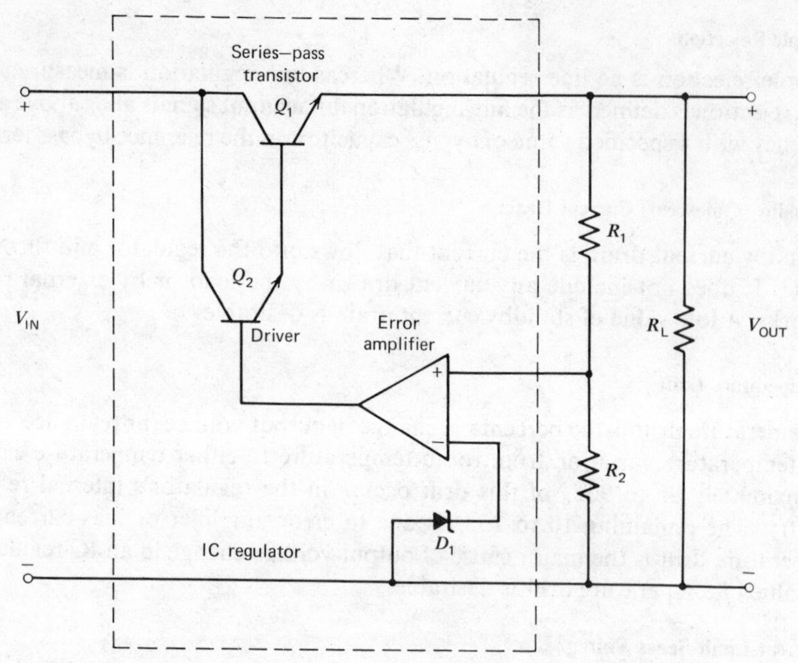

Figure 6-1. Series-dissipative regulator.

input-voltage transients, be stable to heavy reactive loads, and minimize the overshoot caused by large load and line transients.

In a series-dissipative regulator, the excess voltage is dissipated in the series transistor, which acts as a variable resistor. These losses result in a low circuit efficiency. The regulation of dc voltage is accomplished most efficiently by nondissipative or switching regulators. All high-efficiency regulators use a high-frequency switch (the higher the frequency, the smaller the components, but the more filtering required), a choke (energy-storage device), a filter capacitor, and a flyback diode to provide a path for the choke current during the period when the switch is OFF.

The nondissipative (switching) regulator has higher efficiency and lower weight than the dissipative regulator. It allows a designer to use smaller components and heat sinks to provide a specified output-power capability than he would need with a series regulator. The basic configuration of the switching regulator is presented in Figure 6-2.

Regulation occurs by controlling the duty cycle of Q_1. Transistor Q_1 serves as a switch and is either ON (saturated) or OFF, so that power dissipation is at a minimum. Freewheeling diode D_1 conducts during the time that Q_1 is cut OFF, thus maintaining current flow through inductor L_1. When Q_1 is ON, D_1 is reverse-biased and does not conduct. The load current I_L through L_1 increases according to the relationship

$$V_{IN} - V_{OUT} = \frac{(\Delta i_L)}{t_{ON}} \tag{6-1}$$

and it flows through the load and charges capacitor C_1. When V_{OUT} reaches V_{REF}, the voltage-comparator turns Q_1 OFF. The current through L_1 then decreases until D_1 is forward-biased. At this point the inductor current flows through D_1 and

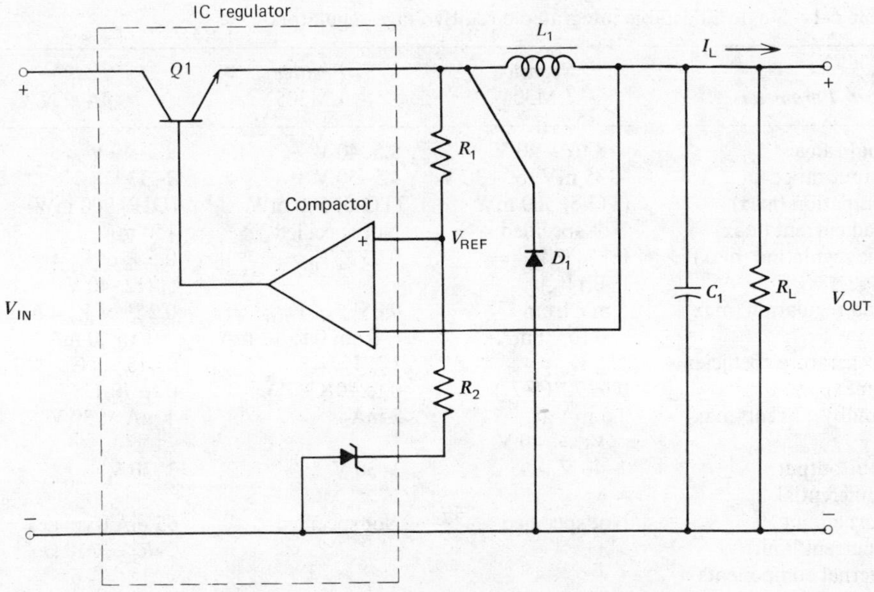

Figure 6-2. Basic switching regulator circuit.

decreases at a rate given by

$$V_{OUT} = L_1 \frac{(\Delta i_L)}{(t_{OFF})} \tag{6-2}$$

When the inductor current falls below the load current, the output capacitor begins to discharge and V_{OUT} decreases. When V_{OUT} decreases to slightly less than V_{REF}, the voltage-comparator turns Q_1 back ON and the cycle repeats itself. The output voltage is given by

$$V_{OUT} = \frac{V_{IN} \cdot t_{ON}}{t_{ON} + t_{OFF}} \tag{6-3}$$

The use of ICs in regulator circuits can enhance power-supply performance while minimizing cost and engineering time. Since only one IC is needed for a wide range of outputs, the part cost, board space, and purchasing problems are less when compared to discrete designs. Also engineering time is saved since typical and worst-case performance data, as well as application data, are available from the manufacturer before design is begun.

INTEGRATED CIRCUIT REGULAR FAMILIES[1]

The first entries into the IC regulator market were the single voltage adjustable types, such as the LM104, LM105, and μA723. Recently, three other classes have been developed: three-terminal fixed regulators. National's LM309, LM320, and LM340 lines, and Fairchild's μA7800, μA78M, and μA78L lines; adjustable dual-tracking regulators, Silicon General's SG3502 and Raytheon's RC4194; and fixed-dual regulators, Motorola's MC1468, Silicon General's SG3501 and Raytheon's RC4195.

[1] H. Gill, "IC Chips Put Voltage Regulation where It's Needed," *Products*, August 20, 1973.

Table 6-1 Single adjustable integrated circuit voltage-regulators

Parameter	*Negative* LM304	*Positive* LM305	*Positive* μA723C
Input range	−8 to −40 V	8.5–40 V	9.5–40 V
Output range	−35 mV to −30 V	4.5–30 V	2–37 V
Dissipation (max)	(TO-5) 500 mW	(TO-5) 500 mW	(DIP) 900 mW
Load current (max)	Not specified	Not specified	150 mA
Line regulation (max)	0.1% (ΔV_{IN} = 0.1V_{IN})	0.03%/V_{IN}	0.5% of V_{OUT} (12–40 V)
Load regulation (max)	5 mV from 0 to 20 mA	0.05% of V_{OUT} from 0 to 12 mA	0.2% of V_{OUT} from 1 to 50 mA
Temperature coefficient (max)	1% V_{OUT} 0 to 70°C (T_A)	1% V_{OUT} 0 to 70°C (T_A)	0.015%/°C 0 to 70°C
Standby current (max)	2.5 mA at V_{IN} = 40 V	2 mA	4 mA at 30 V
Input/output differential	2–40 V	3–30 V	3–30 V
Short circuit (current limit)	Not specified	Not specified	65 mA (typ) at R_{SC} = 10 Ω
External components	4	4	6

Which class of voltage-regulator you choose depends on your system requirements. The adjustable single regulator (Table 6-1) is for systems that use one nonstandard supply, such as 13.5 V, where voltage accuracies of better than ±2% are needed, or where a controlled short-circuit current is required.

The fixed output-voltage three-terminal regulators are good where standard output voltages can be used. These regulators have an output accuracy of ±4% and can deliver between 500 mA and 1.5 A depending on package style and the output-voltage requirement. The fixed regulators in this group need up to two external capacitors. And by adding external resistors, transistors, or op amps, they can control greater output voltages and currents than the basic regulator.

Adjustable dual-tracking regulators are designed for multiple-supply line systems using positive and negative voltages. Complementary voltages are controlled by building the regulator so the positive output tracks the negative output through a feedback loop connected between the negative output and the positive-error amplifier.

Using a dual-tracking regulator in place of two separate regulators reduces the external parts requirement significantly. Two μA723 regulators supplying a positive and negative voltage at 50 mA need nine resistors, two capacitors, one diode and one transistor. Load currents greater than 50 mA will increase the parts count still further. The same job can be done with one dual regulator, two resistors, and four capacitors. Dual-tracking regulators are available that provide 200 mA of load current per side.

Dual fixed ±15-V regulators are designed specifically for op amp circuits. Depending on the device type, these regulators require two to six external components and can deliver ±50 to ±100 mA of load current.

In this chapter we will be discussing the popular single voltage adjustable regulators and the three-terminal fixed regulators.

DISSIPATIVE VOLTAGE-REGULATORS

Positive Voltage-Regulators

In Figures 6-1 and 6-2 V_{OUT} is compared before integration with a reference voltage by an op amp driving the pass transistor; IC regulators include the amplifier and the pass transistor. Most regulators have an internal voltage reference, which may be supplemented in some applications by an external Zener diode.

There are two widely used positive voltage-regulators: The LM105 and the μA723. The basic connections using these regulators are shown in Figures 6-3, 6-4, and 6-5. The connection to give a stabilized output voltage greater than V_{REF} (those μA723 applications with outputs >7.5 V) is given in Figure 6-4, and that for output voltages less than V_{REF} (μA723 applications with output <6.5 V) is given in Figure 6-5. The equations governing the values of resistors R_1 and R_2 are given in the figure legends. In general, the ratio of R_1 and R_2 sets the division ratio of the output or reference voltage, Figures 6-3, 6-4, and 6-5, and the magnitude sets the impedance value of the tap. The inverting and noninverting input-bias currents are typically under 1 μA, and the input-offset current is typically under 200 μA.

Thus, an effective impedance of 2-kΩ at each input is quite satisfactory. For the μA723 in the configuration of Figure 6-5, a good rule of thumb is to set the current out of V_{REF} at 1 mA. This gives an effective impedance of less than 2 kΩ for all possible ratios (since $R_1 + R_2 \simeq 7.1$ kΩ), and does not significantly load V_{REF}. A current of at least 5 mA can be drawn from V_{REF}, but excessive current drain causes poor regulation at low V_{IN} and high dissipation at high V_{IN}, both undesirable. An upper limit of a 2-mA drain (in normal operation) is advisable.

The variation of V_{REF} from unit to unit is about $\pm 6\%$ for the μA723 at room temperature with an additional variation over the temperature range of about $\pm 1.5\%$. Although the full spread is not frequently seen, a worst-case design should certainly allow for such a spread. This variation will be directly reflected in the output voltage, and may necessitate the addition of a trim potentiometer as part of R_1 or R_2. In order to facilitate comprehension of the ease of designing with IC regulators, a practical example is presented.

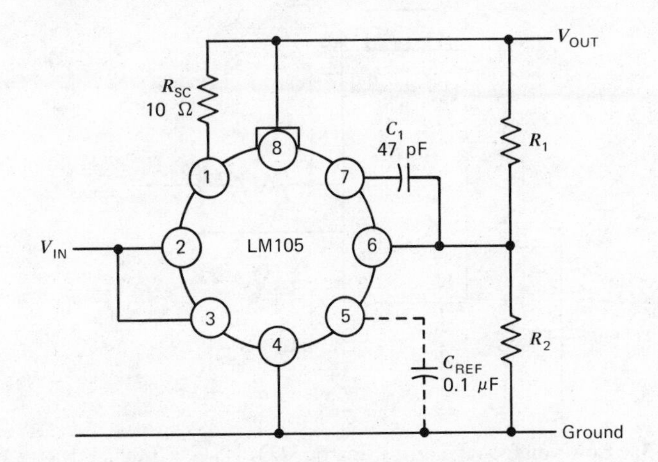

Figure 6-3. Basic LM105 regulator circuit.

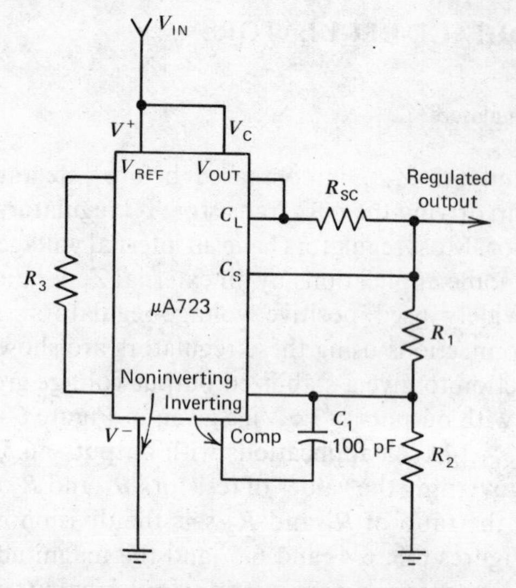

Figure 6-4. Basic μA723 regulator circuit [*note:* $V_{OUT} = V_{REF}[(R_1 + R_2)/R_2]$, $I_{SC} = V_{SENSE}\ R_{SC}$, $R_3 = R_1R_2/(R_1 + R_2)$ for minimum temperature drift: R_3 may be eliminated for minimum component count].

A linear regulator is desired which has the following specifications:

$$V_{IN} = 22 \text{ to } 30 \text{ V dc}$$
$$V_{OUT} = 12 \text{ V dc}$$
$$I_L = 10 \text{ mA}$$
$$\text{temperature range} = -55 \text{ to } +125°\text{C}$$
$$\text{current limit} = 25 \text{ mA}$$

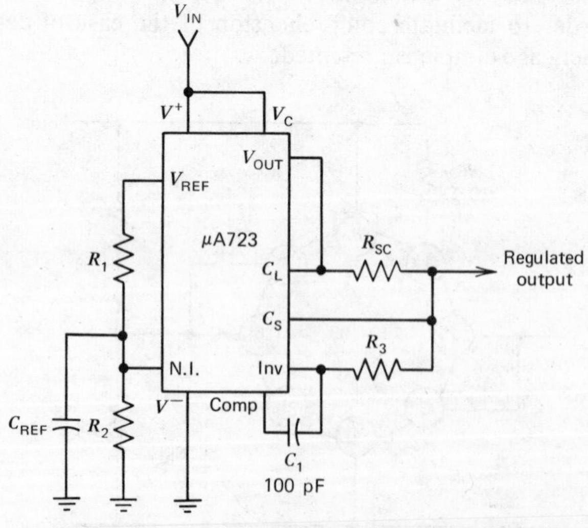

Figure 6-5. Low-voltage-regulator using the 723, $V_{OUT} = 2$ to 7 V [*note:* $V_{OUT} = V_{REF} \times R_2/(R_1 + R_2)$; $R_3 = R_1R_2/(R_2 + R_2)$ for minimum temperature drift].

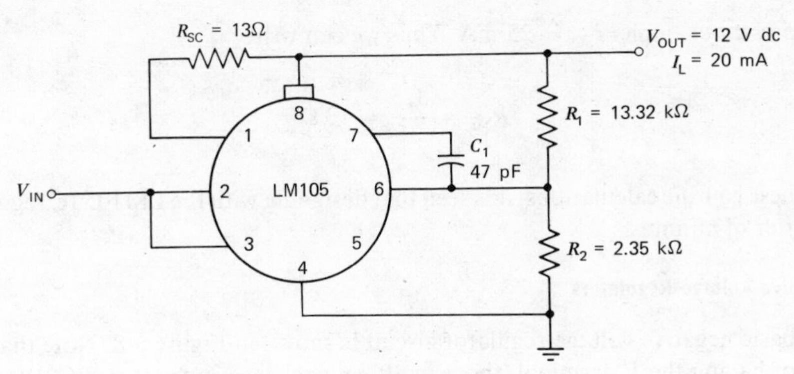

Figure 6-6. Basic LM105 regulator circuit.

These specifications dictate the choice of an LM105 type of IC regulator. The basic regulator circuit is diagrammed in Figure 6-6. In this circuit, the output voltage is set by R_1 and R_2. The resistor values are selected based on a feedback voltage of 1.8 V to *pin* 6 of the LM105. To keep thermal drift of the output voltage within specifications, the parallel combination of R_1 and R_2 should be about 2 kΩ. The values of R_1 and R_2 are easily found from Figure 6-7. For an output voltage of 12 V dc, we have

$$R_2 \simeq 2.35 \text{ k}\Omega$$

$$R_1 = 1.11 V_{OUT} = 1.11(12) = 13.32 \text{ k}\Omega$$

If we wish to use current-limiting in the circuit, resistor R_{SC} must be added. The value of R_{SC} is computed from the formula

$$R_{SC} = \frac{325}{I_{SC}}$$

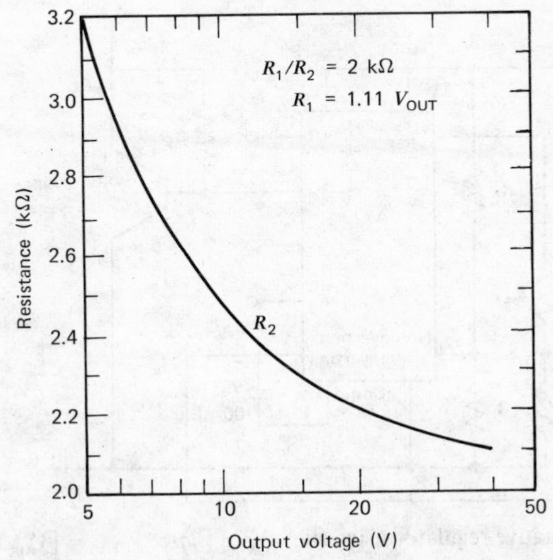

Figure 6-7. Optimum divider resistance values.

In the present example, $I_{SC} = 25$ mA. Thus we can write

$$R_{SC} = \frac{325}{25} = 13\ \Omega$$

From these routine calculations, it is seen that designing with the LM105 IC regulator is a matter of minutes.

Negative Voltage-Regulators

The basic negative voltage-regulator circuit is shown in Figure 6-8. Note that, for units not having the V_z terminal (those in 10-*pin* packages), an external 6.2-V Zener can be used. The inverting and noninverting input connections are reversed, and the pass transistor Q_1 acts as a level-shifted emitter-follower from V_{OUT} to drive the output. The regulator is driven from its own output, so the line regulation is excellent, the load regulation is controlled by the h_{FE} of Q_1 and the load regulation of the IC. Resistor R_5 must be of sufficient value to drive the maximum load current through Q_1 at the minimum input voltage, and large enough not to draw more than 10 mA through the internal Zener (the V_z terminal) at minimum load and maximum V_{IN}. This places a lower limit on the h_{FE} of Q_1, which for large currents may need to be a Darlington pair or equivalent.

Another limitation on this circuit is that the output voltage is the supply voltage for the regulator, thus the circuit as shown cannot be used for output voltages less than 8.5 or 9 V (depending on which part is used). The modification of this circuit as shown in Figure 6-9 overcomes this, limitation but assumes the presence of a positive supply. In general, this is satisfactory, since usually the need for a negative supply (i.e., with the incoming positive-supply terminal grounded) occurs only with symmetrical supplies.

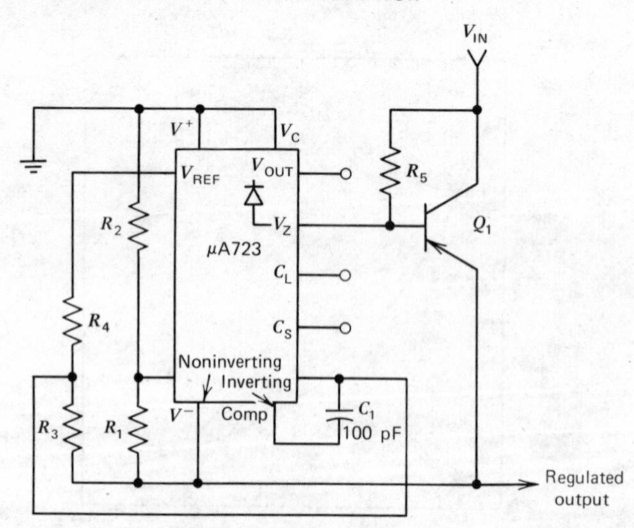

Figure 6-8. Negative regulator using the μA723 [*note:* $V_{OUT} = [V_{REF}R_3(R_1 + R_2)/R_1(R_3 + R_4)$; $R_3 \approx R_4$].

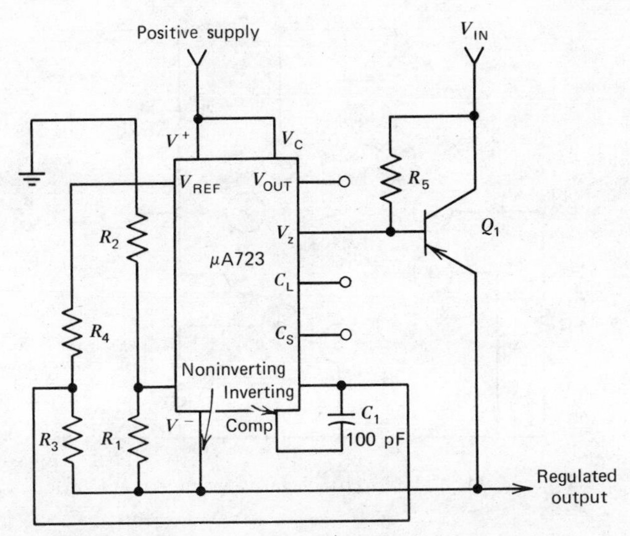

Figure 6-9. Modified negative regulator using the μA723.

The two resistors R_3 and R_4 in the μA723 circuits are needed to reduce the reference voltage; the V_{OUT} terminal will be about 5.5 V above the negative regulated line, and the inputs to the differential amplifier should not be more positive than V_{OUT}. Since V_{REF} is about 7.15 V, it must be divided down to below 5.5 V, or if a Darlington-type pass transistor is used, even less. Dividing by half gives a satisfactory level, with R_3 and R_4 being equal.

Since the amplifier is inverted, the built-in short-circuit protection components cannot be used. However, an external transistor connected as shown in Figure 6-10 will provide short-circuit protection in approximately the same manner. Note that the saturation voltage of Q_2 at the maximum current through R_5 must be less than the V_{BE} of Q_1. The other components are not affected. The diode and resistor shown in Figure 6-10 are frequently needed to avoid latchup and enable initial regulation to

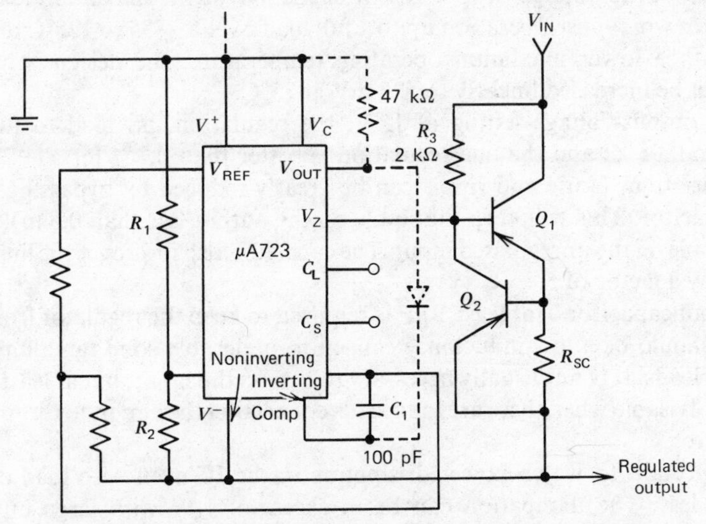

Figure 6-10. Negative voltage-regulator with short-circuit protection.

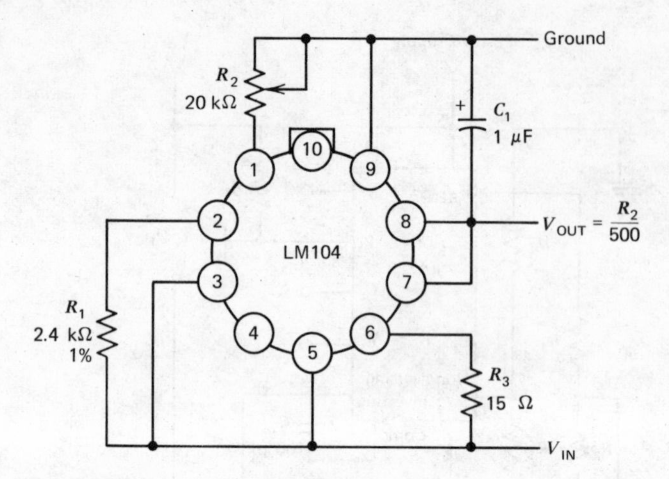

Figure 6-11. The LM104 negative voltage-regulator.

occur. The diode can be a very small signal device, since the forward current is only about 100 μA briefly when switch on occurs, and the reverse voltage is less than 2 V; the resistor should pass about 100 μA into V_{OUT} and the diode. The complex protection schemes developed in the overload protection section of this chapter can be applied to this configuration also.

Then there is a device such as the LM104 which was designed specifically as a negative voltage regulator. The basic circuit connection using the LM104 is shown in Figure 6-11.

This circuit can provide output voltages between 0 and -40 V at currents up to 25 mA. The output voltage is linearly dependent on the value of R_2, giving approximately 2 V for each 1 kΩ of resistance. The exact scale factor can be set up by trimming R_1. This should be done at the maximum output-voltage-setting in order to compensate for any mismatch in the internal divider resistors of the IC.

Short-circuit protection is provided by R_3. The value of this resistor should be chosen so that the voltage drop across it is 300 mV at the maximum load current. This ensures worst-case operation up to a full load over a -55 to 125°C temperature range. With a lower maximum operating temperature, the design value for this voltage can be increased linearly to 525 mV at 25°C.

For an output-voltage-setting of 15 V, the regulation, no load to full load, is better than 0.05%; and the line regulation is better than 0.2% for a $\pm 20\%$ input-voltage variation. Noise and ripple can be greatly reduced by bypassing R_2 with a 10 uF capacitor. This will keep the ripple on the output less than 0.5 mV for a 1-V, 120-Hz ripple on the unregulated input. The capacitor also improves the line-transient response by a factor of 5.

An output capacitor of at least 1 μF is required to keep the regulator from oscillating. This should be a low-inductance capacitor, preferably solid tantalum, installed with short leads. It is not usually necessary to bypass the input, but at least a 0.01-μF bypass is advisable when there are long leads connecting the circuit to the unregulated power source.

It is important to watch power dissipation in the IC even with load currents of 25 mA or less. The dissipation can be in excess of 1 W with large input-output voltage differentials, and this is above ratings for the device.

Overload Protection and Current-Limiting Circuits

The current-limiting characteristics of a regulator are important for two reasons: First, it is almost mandatory that a regulator be short-circuit-protected because the output is distributed to enough places that the probability of it becoming shorted is quite high; second, the sharpness of the limiting characteristics is not improved by the addition of external booster transistors. External transistors can increase the maximum output current, but they do not improve the load regulation at currents approaching the short-circuit current.

The simple short-circuit protection arrangement of Figure 6-4 gives the typical output characteristic shown in Figure 6-12. This figure shows the variation of the limit current with temperature, and also the sharp knee between the constant-voltage and constant-current regions. This method is adequate for most cases, and has the benefit of simplicity. The value of the required resistor is given by

$$\text{short-circuit current} = I_{SC} \simeq \frac{V_{SENSE}}{R_{SC}} \tag{6-4}$$

A typical plot of V_{SENSE} as a function of temperature is given in Figure 6-13 with some typical plots of I_{SC} values.

There are several variations of this basic circuit. Clearly, since the regulator is shut down when the voltage difference between C_L and C_S reaches V_{SENSE}, we can arrange a set of resistors to achieve this limit on some locus other than that given by Figures 6-4, 6-12, and 6-13. In particular, the arrangement of Figure 6-14 gives a limit relationship

$$I_{L_1}R_{SC_1} + I_{L_2}(R_{SC_1} + R_{SC_2}) < V_{SENSE} \tag{6-5}$$

This allows a higher limit on output 1 $[V_{SENSE}/R_{SC_1}$ if $I_{L_2} = 0]$ than on output 2 $[V_{SENSE}/(R_{SC_1} + R_{SC_2})$ if $I_{L_1} = 0]$, but protects both outputs to less than these limits if the other is carrying current. The disadvantage of this circuit is that only one output

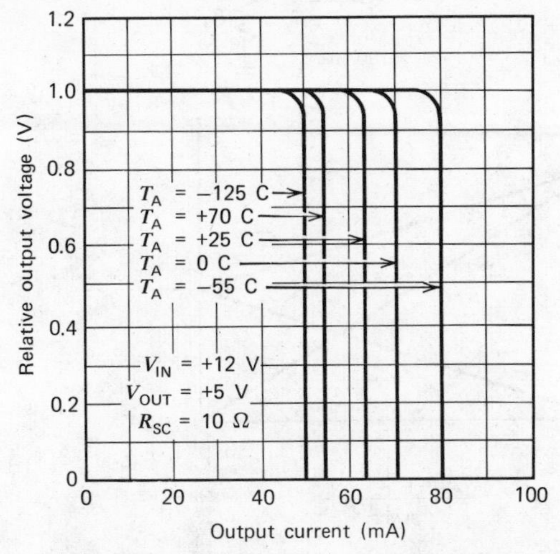

Figure 6-12. Current-limiting characteristics.

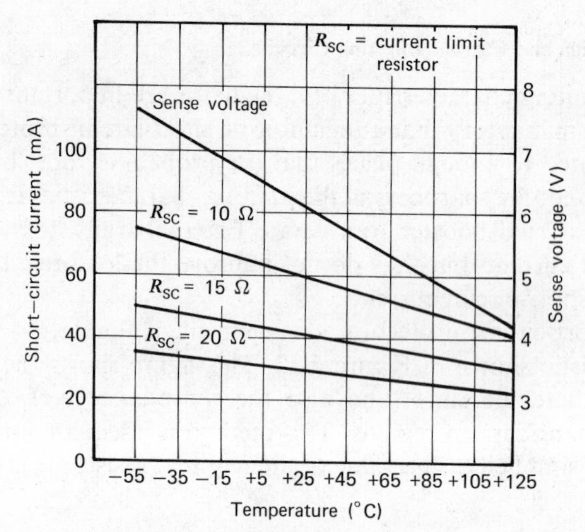

Figure 6-13. Sense voltage and short circuit current limit as a function of temperature.

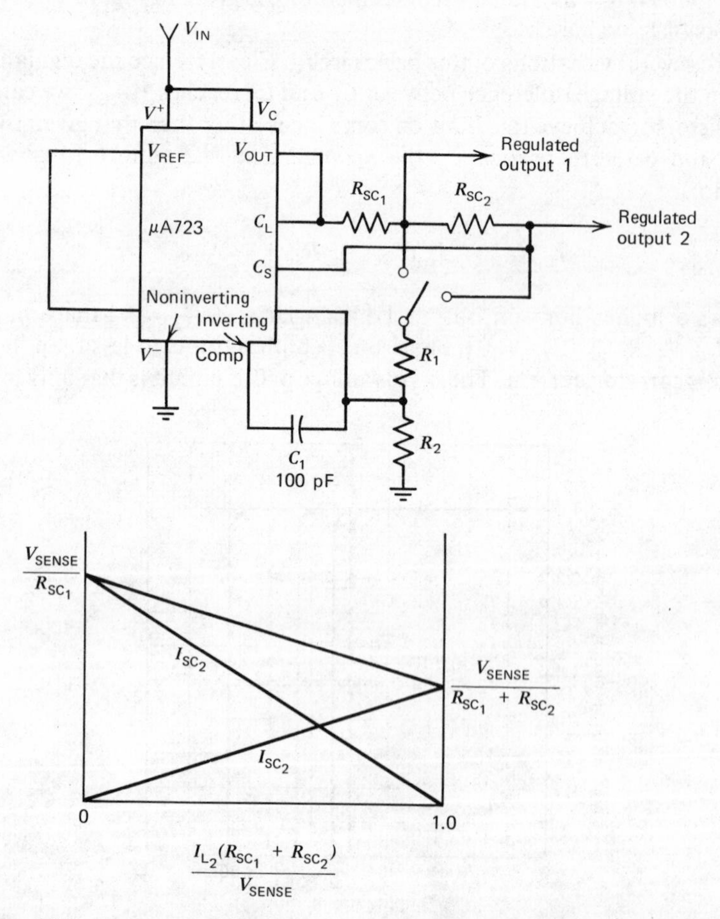

Figure 6-14. Current-limiting feature extended to provide two regulated outputs.

can be well regulated; the other output has an effective output impedance derived from R_{SC}. If output 1 is stabilized (by tying R_1 to it), then the output impedance of output 2 will be R_{SC_2} higher. If output 2 is stabilized, then output 1 has a low-output impedance, but a high mutual impedance from the load current I_{L_2}; R_{SC_2} is given by

$$R_{SC_2} \simeq \frac{\Delta V_{Q_1}}{\Delta I_{L_2}} \tag{6-6}$$

Another current-limiting arrangement is shown in Figure 6-15. Here, the current limits on each output are independent, but a similar output-impedance problem arises. Both of these applications would be better handled with two regulators.

In general, any linear combination of load currents and input or output voltages can be used to set the overload operating conditions. In extreme cases the use of an op amp summing amplifier to drive C_L or C_S could be considered. However, the simple limit scheme depicted in Figures 6-4 and 6-13 is adequate to cover most cases, and the resistor combination discussed here is generally only used in connection with a quite different problem, which is considered next.

The foremost limitation on the use of this simple protection arises from the power dissipation under short-circuit conditions. The power dissipation allowed depends on the package type. For simplicity, we will discuss the 10 lead TO-5 package values. At ambient temperatures below about 30°C, the power-dissipation limit is 800 mW, giving a junction-to-ambient temperature difference in the neighborhood of 120°C. Above this temperature, derating at 6.8 mW/°C, we find the following device power-dissipation capabilities: At an ambient of 50°C, 680 mW; at 75°C, 510 mW; at 100°C, 300 mW; and at 125°C, 170 mW.

Clearly, under short-circuit output conditions, the dissipation becomes the product of the full-input voltage and the short-circuit current. Consider a device with V_{IN} = 20 V, V_{OUT} = 15 V, and a current limit of 60 mA, operating at an ambient temperature

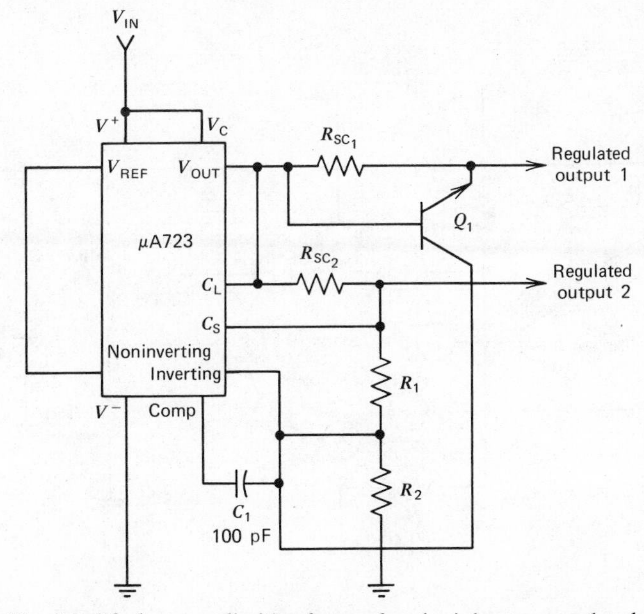

Figure 6-15. Modified current-limiting feature for obtaining two regulated outputs.

of T_A = 25°C. The standby current will be assumed to be 2 mA which gives a dissipation of 40 mW. The pass transistor dissipation just before overload is 20 V − 15 V or 5 V at 60 mA, or 300 mW, for a total of 340 mW, well within ratings.

But under short-circuit conditions, the initial-pass transistor dissipation is 20 V at 60 mA, or 1.2 W (a total of 1.24 W). This is significantly in excess of the allowed ratings, and will result in catastrophic failure of the IC regulator. The same problem arises in circuits using external-pass transistors, except that the dissipation is transferred to the pass transistor itself.

The solution to this problem is found in foldback current-limiting circuits. Foldback current-limiting, makes the output current under overload conditions decrease below the full-load current as the output voltage is pulled down. The short-circuit current can be made but a fraction of the full-load current.

FOLDBACK CURRENT-LIMITING IN POSITIVE REGULATORS

A high-current regulator using foldback-limiting is shown in Figure 6-16. A second booster transistor Q_1 has been added to provide a 2-A output current without causing excessive dissipation in the LM105. The resistor across its emitter-base junction bleeds off any collector-base leakage and establishes a minimum collector current for Q_2 to make the circuit easier to stabilize with light loads. The foldback characteristic is produced with R_4 and R_5. The voltage across R_4 bucks out the voltage dropped across the current-sense resistor R_3. Therefore, more voltage must be developed across R_3 before current-limiting is initiated. After the output voltage begins to fall, the bucking voltage is reduced, as it is proportional to the output voltage. With the output shorted, the current is reduced, as it is proportional to the output voltage.

With the output shorted, the current is reduced to a value determined by the current-limit resistor and the current-limit sense voltage of the LM105.

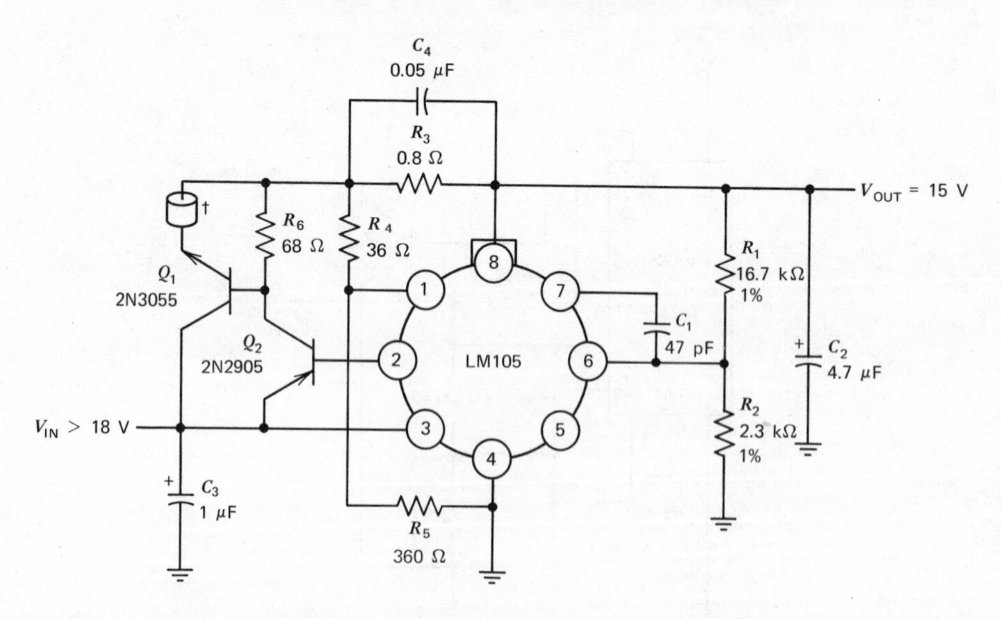

Figure 6-16. High-current regulator using foldback current-limiting.

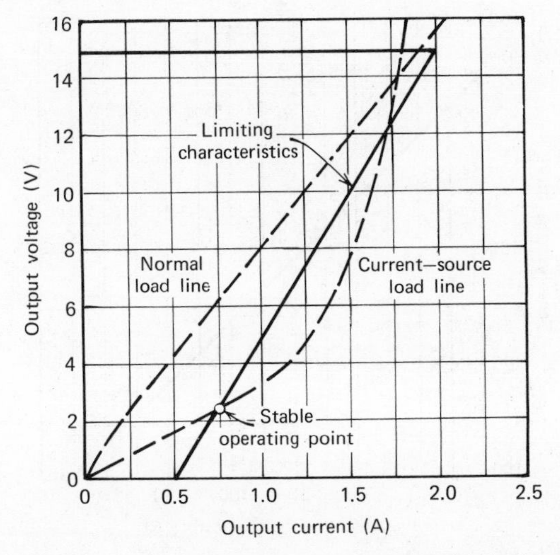

Figure 6-17. Limiting characteristics of regulator using foldback current-limiting.

Figure 6-17 illustrates the limiting characteristics. The circuit regulates for load currents up to 2-A. Heavier loads will cause the output voltage to drop, reducing the available current. With a short on the output, the current is only 0.5 A.

In design, the value of R_3 is determined from

$$R_3 = \frac{V_{\text{LIMIT}}}{I_{\text{SC}}} \tag{6-7}$$

where V_{LIMIT} is the current-limit sense voltage of the LM105, given in Figure 6-18, and I_{SC} is the design value of short-circuit current; R_5 is then obtained from

$$R_5 = \frac{V_{\text{OUT}} + V_{\text{SENSE}}}{I_{\text{BLEED}} + I_{\text{BIAS}}} \tag{6-8}$$

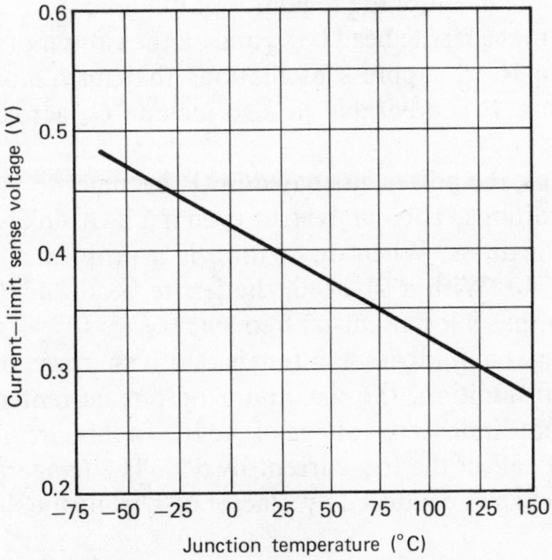

Figure 6-18. Current-limit sense voltage of the LM105 as a function of junction temperature.

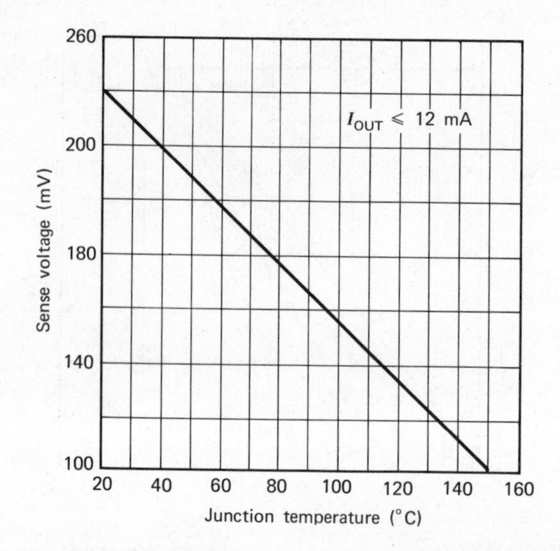

Figure 6-19. Maximum voltage drop across current-limit resistor at full load for worst-case load regulation of 0.1%.

where V_{OUT} is the regulated output voltage, V_{SENSE} is maximum voltage across the current-limit resistor for 0.1% regulation as indicated in Figure 6-19, I_{BLEED} is the preload current on the regulator output provided by R_5 and I_{BIAS} is the maximum current coming out of *Pin* 1 of the LM105 under full-load conditions. The current I_{BIAS} will be equal to 2 mA plus the worst-case base drive for the *pnp* booster transistor Q_2; I_{BLEED} should be made about ten times greater than I_{BIAS}. Finally, R_4 is given by

$$R_4 = \frac{I_{FL}R_3 - V_{SENSE}}{I_{BLEED}} \tag{6-9}$$

where I_{FL} is the output current of the regulator at full load.

It is recommended that a ferrite bead be strung on the emitter of the pass transistor, as shown in Figure 6-16 to suppress oscillations that may show up with certain physical configurations. It is advisable to also include C_4 across the current-limit resistor.

In some applications, the power dissipated in Q_2 becomes too great for a 2N2905 under worst-case conditions. This can be true even if a heat sink is used, as it should be in almost all applications. When dissipation is a problem, the 2N2905 can be replaced with a 2N3740. With a 2N3740, the ferrite bead and C_4 are not needed because this transistor has a lower cut-off frequency.

One of the advantages of foldback-limiting is that it sharpens the limiting characteristics of the IC. In addition, the maximum output current is less sensitive to variations in the current-limit sense voltage of the IC: In this circuit, a 20% change in sense voltage will only affect the trip current by 5%. The temperature sensitivity of the full-load current is likewise reduced by a factor of 4, while the short-circuit current is not.

Even though the voltage dropped across the sense resistor is larger with foldback-limiting, the minimum input-output voltage differential of the complete regulator is not increased above the 3 V specified for the LM105 as long as this drop is less than 2 V. This can be attributed to the low sense voltage of the IC by itself.

Figure 6-17 shows that foldback-limiting can only be used with certain kinds of loads. When the load looks predominately like a current source, the load line can intersect the foldback characteristic at a point where it will prevent the regulator from coming up to voltage, even without an overload. Fortunately, most solid-state circuitry presents a load line which does not intersect. However, the possibility cannot be ignored, and the regulator must be designed with some knowledge of the load.

With foldback-limiting, power dissipation in the pass transistor reaches a maximum at some point between full-load and short-circuited output. This is illustrated in Figure 6-20. However, if the maximum dissipation is calculated with the worst-case input voltage, as it should be, the power peak is not too high.

The basic μA723 foldback circuit is shown in Figure 6-21. Here, the combination of load current and output voltage, is such that

$$V_{SENSE} = (V_{OUT} + I_L \cdot R_{SC}) \frac{R_4}{R_3 + R_4} - V_{OUT} \tag{6-10}$$

Rearranging, we find

$$I_{KNEE} = \frac{V_{OUT} R_3}{R_{SC} R_4} + \frac{V_{SENSE}(R_3 + R_4)}{R_{SC} R_4} \tag{6-11}$$

In particular,

$$I_{SC} = \frac{V_{SENSE}(R_3 + R_4)}{R_{SC} R_4} \tag{6-12}$$

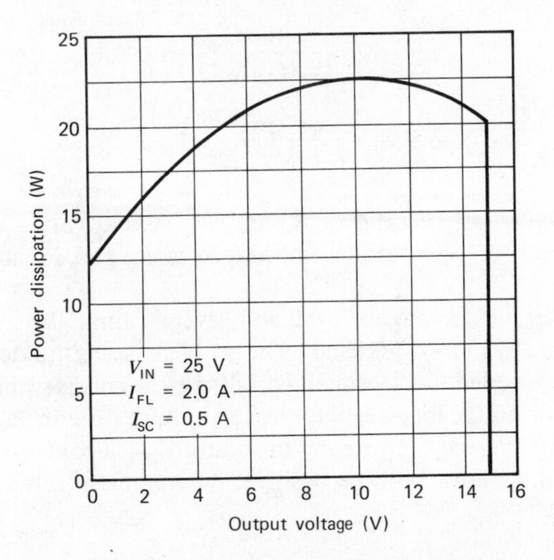

Figure 6-20. Power dissipation in series-pass transistors under overload conditions in regulator using foldback current-limiting.

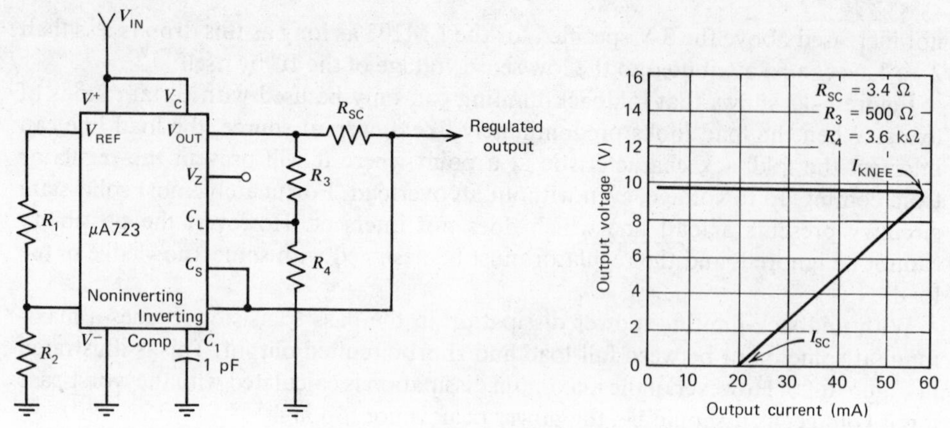

Figure 6-21. Foldback current-limiting regulator, $V_{OUT} = 2\text{–}7$ V [*note:*

$$I_{KNEE} = \frac{V_{OUT}R_3}{R_{SC}R_4} + \frac{V_{SENSE}(R_3 + R_4)}{R_{SC}R_4};$$

$$\frac{R_4}{R_3} = \frac{V_{OUT}I_{SC}}{V_{SENSE}(I_{KNEE} - I_{SC})} - 1;$$

$$V_{OUT} = V_{REF} \times \frac{R_1 + R_2}{R_2};$$

$$R_{SC} = \frac{V_{SENSE}}{I_{SC}}\left(1 + \frac{R_3}{R_4}\right);$$

$$I_{SC} = \frac{V_{SENSE}}{R_{SC}} \times \frac{R_3 + R_4}{R_4}\Bigg].$$

and for component values,

$$\frac{R_4}{R_3} = \frac{V_{OUT}I_{SC}}{V_{SENSE}(I_{KNEE} - I_{SC})} - 1 \tag{6-13}$$

$$R_{SC} = \frac{V_{SENSE}}{I_{SC}}\left(1 + \frac{R_3}{R_4}\right) \tag{6-14}$$

These equations give only a ratio of R_4 and R_3. Actual values may be chosen as follows. For the μA723, the parallel impedance of R_3 and R_4 should be less than 10 kΩ.

The foldback circuit for Figure 6-21 will severely limit the knee current if the difference between V_{IN} and V_{OUT} is small. The problem is accentuated by the relatively large value of R_{SC} needed (6-14), which will develop a considerable voltage drop at I_{KNEE}, and will worsen the load regulation. The voltage drop in R_{SC} is an addition to the minimum drop allowed between V_{IN} or V_C and V_{OUT}. From (6-13) we can also see that, since R_3 and R_4 must both be positive, the minimum ratio of short circuit to knee current is

$$\frac{I_{SC}}{I_{KNEE}} \geq \frac{V_{SENSE}}{V_{OUT} + V_{SENSE}} \tag{6-15}$$

For the voltage drop across R_{SC} to be no greater than V_{OUT} we need $R_4/R_3 > 1$, so (6-15) becomes

$$\frac{I_{SC}}{I_{KNEE}} > \frac{2V_{SENSE}}{V_{OUT} + V_{SENSE}} \tag{6-16}$$

For low V_{OUT} values, this can become a significant limitation. Thus, we can see that the limitations of the circuit of Figure 6-21 are that the highest I_{KNEE} value cannot always be obtained; the ratio of I_{SC}/I_{KNEE} is limited; and the voltage drop in R_{SC} can be excessive, adversely affecting the load regulation. A further disadvantage, not discussed before, is the more rapid change of shut-down currents with temperature. The benefits are easier assurance of start-up, and closer control of shut-down locus by resistor values against unit-to-unit variation.

Figure 6-22 shows a variation of Figure 6-21, with a nonlinear resistor for R_4 and using a Zener diode D_1. The Zener voltage V_{Z_1} should be less than the normal V_{OUT}. The overload characteristic is plotted for specific values, together with the same power-dissipation curve as before, for comparison. Once again, a good fit can be obtained. The currents can all be scaled by varying R_{SC}. The same criteria for selecting the values of R_3 and R_4 apply as they did for the circuit of Figure 6-21.

FOLDBACK CURRENT-LIMITING IN NEGATIVE REGULATORS

When the output of the LM104 is shorted, the dissipation in a series pass transistor can easily increase by a factor of 4. Therefore, foldback current-limiting, similar to that already described for positive regulators, is often desirable. A typical foldback current-limiting design for a negative regulator is shown in Figure 6-23.

Normally Q_3 is held in a nonconducting state by the voltage developed across R_4. However, when the voltage across the current-limit resistor R_7 increases to where it equals the voltage across R_4 (about 1 V), Q_3 turns ON and begins to rob base drive from the driver transistor Q_1. This causes an increase in the output current of the LM104, and it will go into current-limiting at a current determined by R_5. Since the base drive to Q_1 is clamped, the output voltage will drop with heavier loads. This reduces the voltage drop across R_4 and, therefore, the available output current. With the output completely shorted, the current will be about one-fifth the full-load current.

In design, R_7 is chosen so that the voltage drop across it will be 1 to 2 V under full-load conditions. The resistance of R_3 should be 1000 times the output voltage; R_4 is then determined from

$$R_4 \cong \frac{R_7 R_3 I_{FL}}{V_{OUT} + 0.5} \tag{6-17}$$

where I_{FL} is the load current at which limiting will occur. If it is desired to reduce the ratio of full load to short-circuit current, this can be done by connecting a resistance of 2 to 10 kΩ across the emitter-base of Q_3.

ADJUSTABLE CURRENT-LIMITING

In laboratory power supplies, it is often necessary to adjust the limiting current of a regulator. This, of course, can be done by using a variable resistance for the current-limit resistor. However, the current-limit resistor can easily have a value below that of

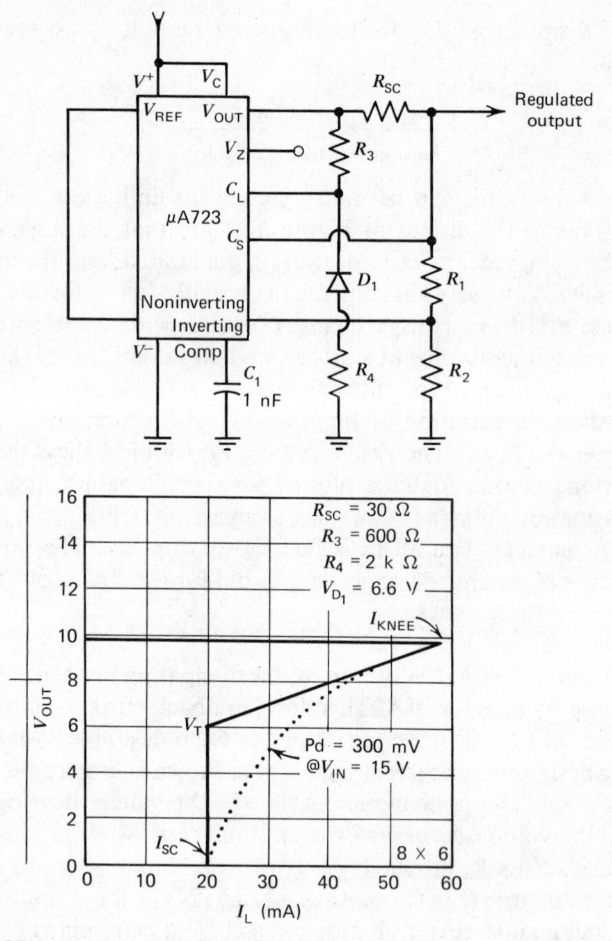

Figure 6-22. Another variation of Figure 6-21 using a nonlinear resistor and Zener diode [*note:*

$$R_{SC} = \frac{V_{SENSE}}{I_{SC}};$$

$$\frac{R_3}{R_4} = \frac{V_{SENSE}(I_{KNEE}/I_{SC} - 1)}{(V_{OUT} - V_T)};$$

$$I_{KNEE} = \left[\frac{R_3}{R_4}(V_{OUT} - V_T) + V_{SENSE}\right] \Big/ R_{SC};$$

$$V_{Z_1} = V_T + V_{SENSE}].$$

commercially available potentiometers. Discrete resistance values can be switched to vary the limiting current, but this does not provide continuously variable adjustment.

The circuit in Figure 6-24 solves this problem, giving a linear adjustment of limiting current over a five-to-one range. A silicon diode D_1 is included to reduce the current-limit sense voltage to approximately 50 mV. Approximately 1.3 mA from the reference supply is passed through a potentiometer R_4 to buck out the diode voltage. Therefore,

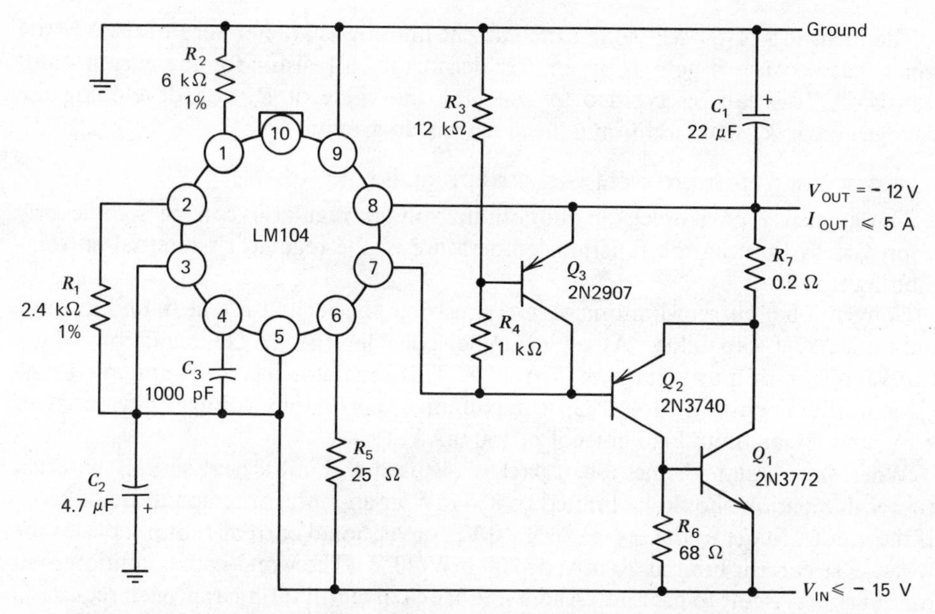

Figure 6-23. Foldback current-limiting using the LM104 negative regulator.

the effective current-limit sense voltage is nearly proportional to the resistance of R_4. The current through R_4 is fairly insensitive to changes in ambient temperature, and D_1 compensates for temperature variations in the current-limit sense voltage of the LM104. Therefore, the limiting current will not be greatly affected by temperature.

It is important that a potentiometer be used for R_4 and connected as shown. If a rheostat connection were used, it could open while it was being adjusted and momentarily increase the current-limit sense voltage to many times its normal value. This could destroy the series-pass transistors under short-circuit conditions.

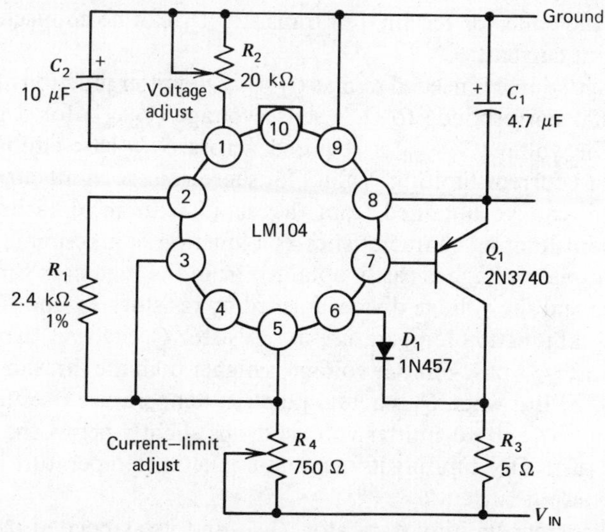

Figure 6-24. Adjustable current-limiting using the LM104 [*note:* $-40\text{ V} \leq V_{OUT} \leq 0$; $40\text{ mA} \leq I_{LIMIT} \leq 200\text{ mA}$].

The inclusion of R_4 will soften the current-limiting characteristics of the LM104 somewhat because it acts as an emitter-degeneration resistor for the current-limit transistor. This can be avoided by reducing the value of R_4 and developing the voltage across R_4 with additional bleed current to ground.

TEMPERATURE-LIMITING INCREASES OUTPUT CURRENT[2]

The efficiency of a precision monolithic voltage regulator can be significantly improved by limiting the junction temperature of the regulator's internal current-limiting transistor.

Conventional current-limiting severely restricts the regulator's peak and average output current capability. As an example, consider the μA723 regulator, which can supply an output voltage of 7 to 37 V. This regulator has a maximum storage (junction) temperature of 150°C, a maximum input/output voltage differential of 40 V, and a maximum load current of 150 mA.

When the regulator's metal-can package is used without a heat sink, its internal power dissipation should be limited to 800 mW at an ambient temperature of 25°C. If the input voltage to the regulator is 40 V, conventional current-limiting places the worst-case current limit at 20 mA, or 800 mW/40 V. (The worst-case condition is an output short circuit to ground.) And a foldback-current-limiting approach requires a limit knee setting of 24.2 mA, or 800 mW/(40 − 7) V.

Both of these approaches significantly limit the regulator's output current capability when the regulator must supply a load continuously at both intermediate- and high-output voltage levels, or when it must supply peak currents at any output voltage level. In contrast, temperature-limiting protects the regulator from burnout, while allowing it to provide the maximum possible output current (both continuous and pulsed), regardless of output voltage level, ambient temperature, and the amount of heat sinking.

In Figure 6-25, the regulator's own current-limiting transistor, Q_L, can be used to implement this temperature-limiting. The transistor's base-emitter junction, which has a temperature sensitivity of −1.8 mV/°C, can act as a temperature-sensor for the regulator. And the collector terminal of transistor Q_L can be connected to limit the regulator's output current.

A stable voltage source is needed to bias Q_L's base-emitter junction at the threshold voltage (V_{TH}) that corresponds to Q_L's sense voltage (V_{SENSE}) for a given junction temperature. The voltage V_{SENSE} is required across Q_L's base-emitter junction to implement output current-limiting. Values for sense voltage, limit current, and junction temperature can be obtained from the manufacturers' datasheet plot of the regulator's current-limiting characteristics as a function of junction temperature.

The threshold bias voltage is easily obtained from the regulator's internal voltage reference source and the voltage divider formed by resistors R_1 and R_2.

When the actual junction temperature of transistor Q_L is lower than the junction-temperature limit, Q_L's base-emitter voltage is higher than the threshold bias voltage, so that Q_L is OFF. But when Q_L's actual junction temperature rises to the junction-temperature limit, Q_L's base-emitter voltage drops slightly below the threshold bias voltage, and Q_L turns ON, limiting its maximum junction temperature by first limiting the regulator's output current.

The external current-limiting transistor, Q_{EXT} and its associated resistor, R_{SC}, are

[2] M. J. Shah, *Temperature Limiting Boosts Regulator Output Current*, *Electronics*, December 6, 1973.

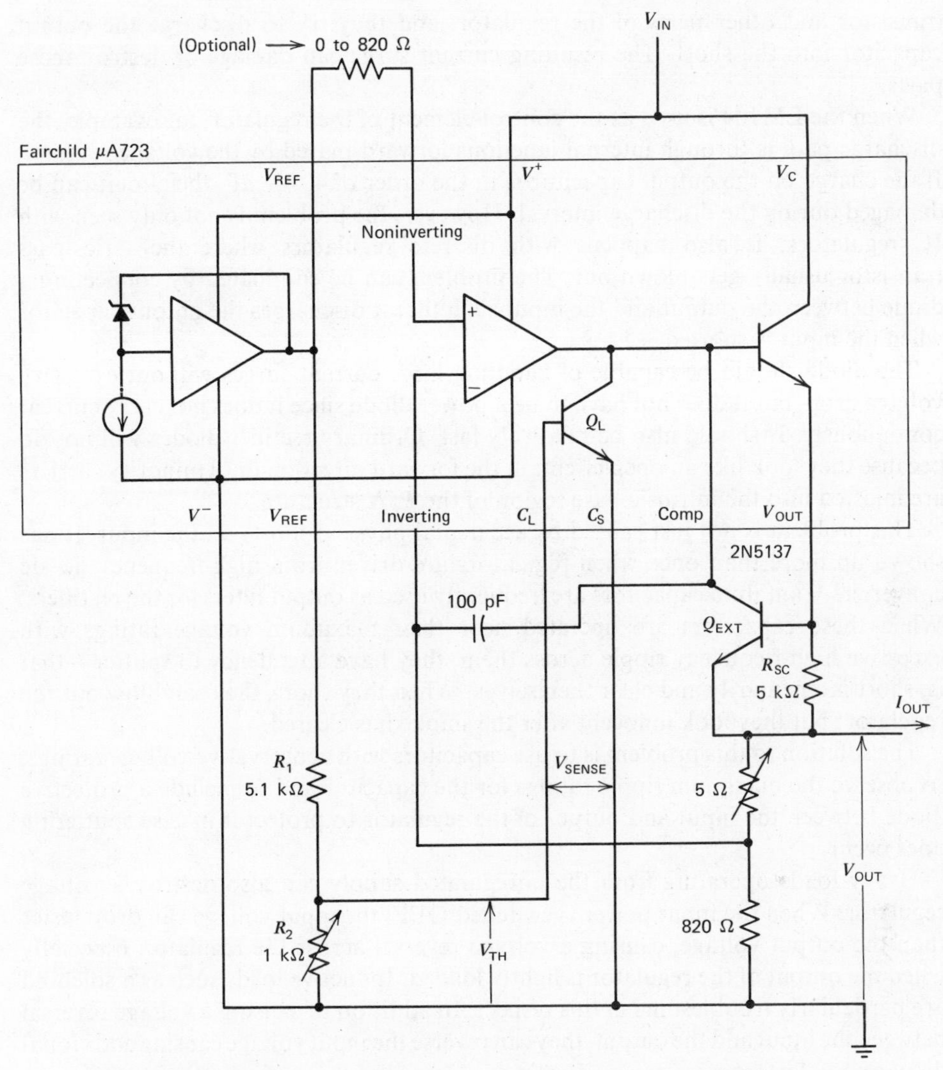

Figure 6-25. Using the μA723 voltage regulator internal current-limiting transistor to provide temperature-limiting. (Courtesy *Electronics*.)

needed to limit regulator output current below the 150-mA secondary breakdown limit of the regulator's internal output transistors. (The optional resistor can be included to minimize output voltage drift.) Figure 6-25 shows the regulator's output current capability over the full range of input/output differential voltage for three sense-voltage settings.

Using Protective Diodes

It is a little known fact that most voltage-regulators can be damaged by shorting out the unregulated input voltage while the circuit is operating—even though the output may have short-circuit protection. When the input voltage to the regulator falls instantaneously to zero, the output capacitor is still charged to the nominal output voltage. This applies voltage of the wrong polarity across the series-pass transistor and other parts of the regulator, and they try to discharge the output capacitor into the short. The resulting current surge can damage or destroy these parts.

When the LM104 is used as the control element of the regulator, for example, the discharge path is through internal junctions forward-biased by the voltage reversal. If the charge on the output capacitor is in the order of 40 V · μF, the circuit can be damaged during the discharge interval. However, the problem is not only seen with IC regulators. It also happens with discrete regulators where the series-pass transistor usually gets blown out. The problem can be eliminated by connecting a diode between the output and the input such that it discharges the output capacitor when the input is shorted.

The diode should be capable of handling large current surges without excessive voltage drop, but it does not have to be a power diode since it does not carry current continuously. It should also be relatively fast. Ordinary rectifier diodes will not do because they look like an open circuit in the forward direction until minority carriers are injected into the intrinsic base region of the *PIN* structure.

This problem is not just caused by accidental physical shorts on the input. It has shown up more than once when regulators are driven from high-frequency dc–dc converters. Tantalum capacitors are frequently used as output filters for the rectifiers. When these capacitors are operated near their maximum voltage ratings with excessive high-frequency ripple across them, they have a tendency to sputter—that is, short momentarily and clear themselves. When they short, they can blow out the regulator; but they look innocent after the smoke has cleared.

The solution to this problem is to use capacitors with conservative voltage ratings, to observe the maximum ripple ratings for the capacitor and to include a protective diode between the input and output of the regulator to protect it in case sputtering does occur.

Heavy loads operating from the unregulated supply can also destroy a voltage regulator. When the input power is switched OFF, the input voltage can drop faster than the output voltage, causing a voltage reversal across the regulator, especially when the output of the regulator is lightly loaded. Inductive loads such as a solenoid are particularly troublesome in this respect. In addition to causing a voltage reversal between the input and the output, they can reverse the input voltage causing additional damage.

In cases like this, it is advisable to use a multiple-pole switch or relay to disconnect the regulator from the unregulated supply separate from the other loads. If this cannot be done, it is necessary to put a diode across the input of the regulator to clamp any reverse voltages, in addition to the protective diode between the input and the output.

Yet another failure mode can occur if the regulated supply drives inductive loads. When power is shut off, the inductive current can reverse the output-voltage polarity, damaging the regulator and the output capacitor. This can be cured with a clamp diode on the output. Even without inductive loads, it is usually good practice to include this clamp diode to protect the regulator if its output is accidentally shorted to a negative supply.

A negative regulator with all these protective diodes is shown in Figure 6-26. Diode D_1 protects against output-voltage reversal, D_2 prevents a voltage reversal between the input and the output of the regulator, and D_3 prevents a reversal of the

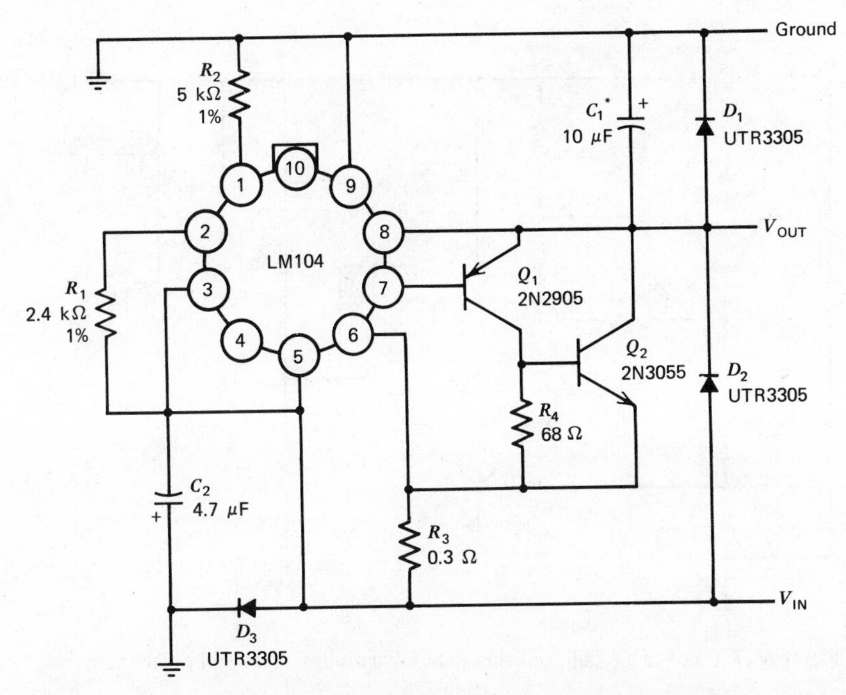

Figure 6-26. Negative regulator using protective diodes to prevent polarity-reversal damage [*note:* * solid tantalum].

input-voltage polarity. In many cases, D_3 is not needed if D_1 and D_2 are used, since these diodes will clamp the input voltage within two diode drops of ground. This is adequate if the input-voltage reversals are of short duration. The positive-regulator counterpart is shown in Figure 6-27.

Low-Power Regulator Circuits

If no external series-pass transistors are used, the regulators will provide output voltages in their specified ranges, at limited currents. The maximum current is about 20 to 100 mA for a positive regulator and about 25 mA for a negative regulator.

The circuits shown in Figures 6-3, 6-4, 6-5, 6-6, and 6-11 are typical low-power or bias-supply configurations. In the positive circuits R_{SC} is a current-sense resistor that makes the regulator exhibit a constant-current characteristic, C_1 is the compensation capacitor and C_{REF} can be added to bypass noise from the internal-voltage reference. The functions of these components will be further described in the next sections.

The output voltage of the negative regulator (LM104) in Figure 6-11 is linearly dependent on the value of R_2, giving approximately 2 V for each 1 kΩ of resistance. The exact scale factor can be set up by trimming R_1. This should be done at the maximum output voltage setting in order to compensate for any mismatch in the internal divider resistors of the IC.

External-Pass Transistor Circuits (High-Power Regulator)

The use of an external-pass transistor can readily extend the usable current range of an IC regulator to many amperes. The simplest circuit is that shown in Figure

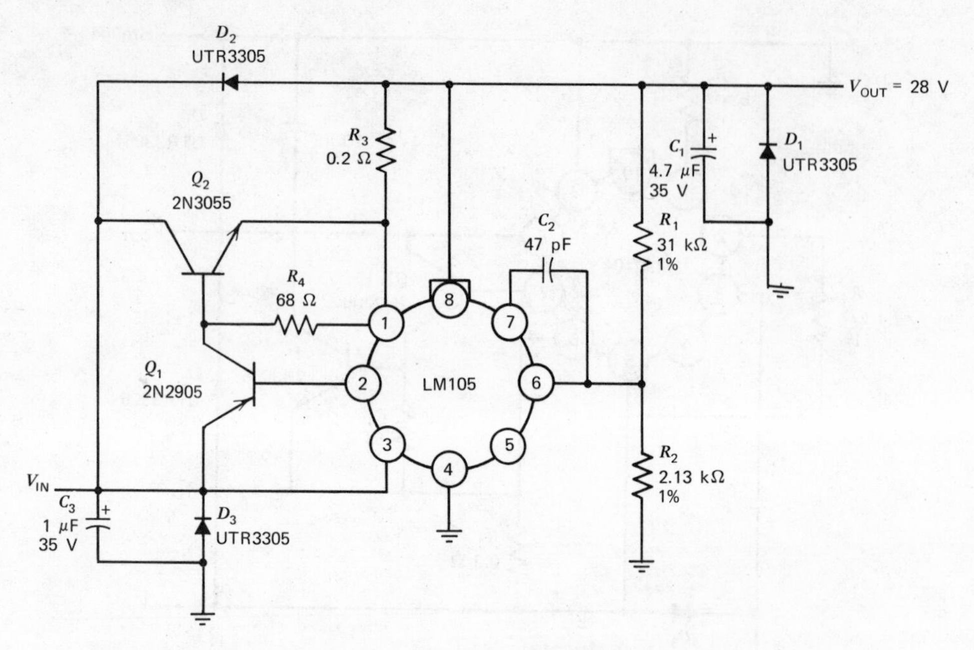

Figure 6-27. Positive regulator using protective diodes to prevent polarity reversal damage.

6-28*a*. Here, the internal Darlington-pass transistor configuration is extended to a triplet by the external *npn* transistor.

For further extension, this external device can be a Darlington, extending the string to a quad. This arrangement has the merit of economy, since *npn* transistors are generally cheaper than *pnp* transistors, but it has the disadvantage that the minimum differential voltage between input and output is increased by the V_{BE}s of the external transistors. The load regulation suffers somewhat as shown by the following equation:

$$V_{IN_2} \simeq \frac{I_L}{A_V}\left[R_{EXT} + \frac{(2+n)kT}{qI_L}\right] \tag{6-18}$$

where n is the number of external devices.

An alternative circuit avoiding these same disadvantages mentioned for the circuit of Figure 6-28*a* is shown in Figure 6-28*b*. This circuit uses a *pnp* transistor whose base drive is obtained from the V_C terminal. Resistor R_3 is used to ensure Q_1 turns OFF under no-load conditions, avoiding the excessive build-up of leakage current that can sometimes occur at high temperature. The effect of Q_1 is to multiply the h_{FE} of the output transistor in the IC, without increasing the V_{BE}; hence this circuit optimizes the load regulation also. Once again, this device can be replaced by a Darlington pair, but then the V_{BE} build-up begins to affect the differential-voltage limit again. The best arrangement is probably that shown in Figure 6-29. In this circuit the main power transistor Q_2 is an *npn* type, the V_{BE} build-up is minimized, and the load regulation is optimized. The short-circuit and overload protection techniques

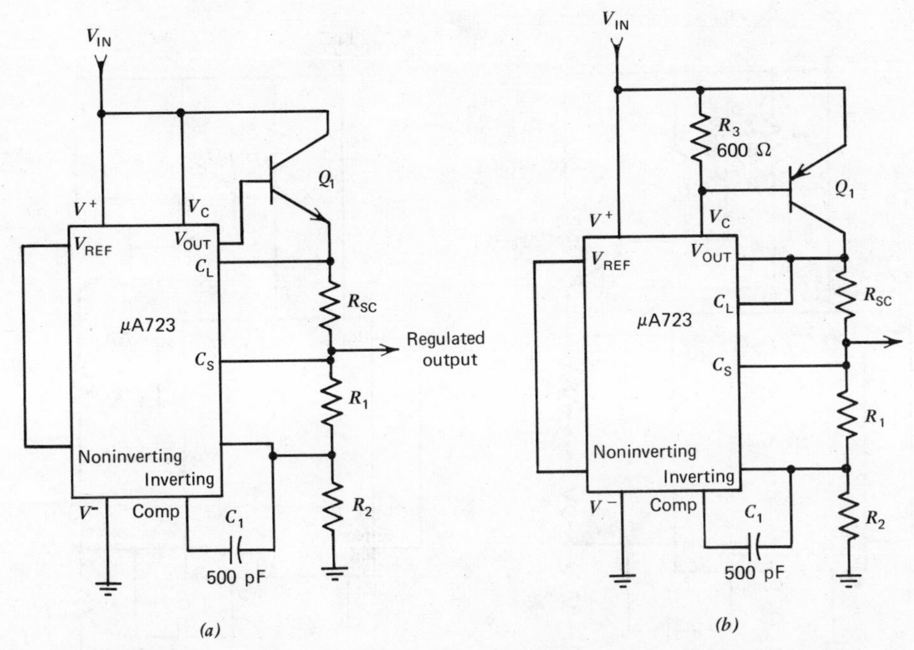

Figure 6-28. Variations of integrated circuit regulators using external-pass transistors to increase the load-current capability.

discussed previously can all be applied to these pass transistor circuits. The currents can be scaled by altering the value of R_{SC}.

This will protect the entire circuit against external overload, but not protect the regulator against pass transistor failure since the regulator IC may then be limited at currents of several amps, enough to destroy it. If this problem is serious, the circuits of Figures 6-28*b* and 6-29 may be protected additionally as shown in Figure 6-30. Here, R_{SC_1} performs the main short-circuit protection, but R_{SC_2} limits the IC output current to a safe value if the pass transistors fail; the normal voltage drop in R_{SC_2} should be small compared with the voltage drop in R_{SC_1}.

Figure 6-31 shows how an external-pass transistor is added to the LM105. The addition of an external *pnp* transistor does not increase the minimum input-output voltage differential. This would happen if an *npn* transistor were used in a compound emitter-follower connection with the *npn* output transistor of the IC. A single-diffused, wide-base transistor like the 2N3740 is recommended because it causes fewer oscillation problems than double-diffused, planar devices. In addition, it seems to be less prone to failure under overload conditions; and low-cost devices are available in power packages like the TO-66 or even TO-3.

When the maximum dissipation in the pass transistor is less than about 0.5 W, a 2N2905 may be used as a pass transistor. However, it is generally necessary to carefully observe thermal deratings and provide some sort of heat sink. In the circuit of Figure 6-31, the output voltage is determined by R_1 and R_2.

The resistor values are selected based on a feedback voltage of 1.8 V to *Pin* 6 of the LM105.

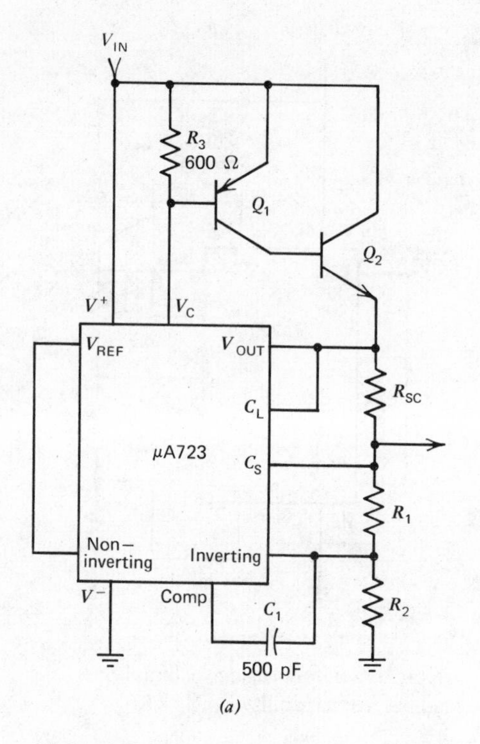

Figure 6-29. Integrated circuit regulator using external-pass transistors to increase the load-current capability.

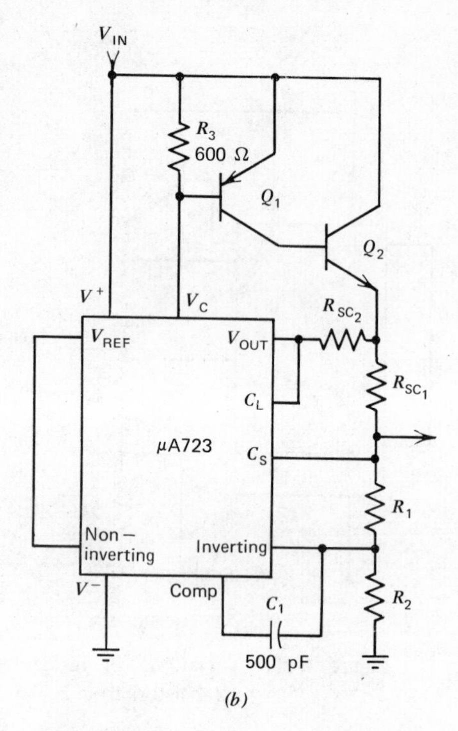

Figure 6-30. Another integrated circuit regulator using external-pass transistors to increase the load-current capability.

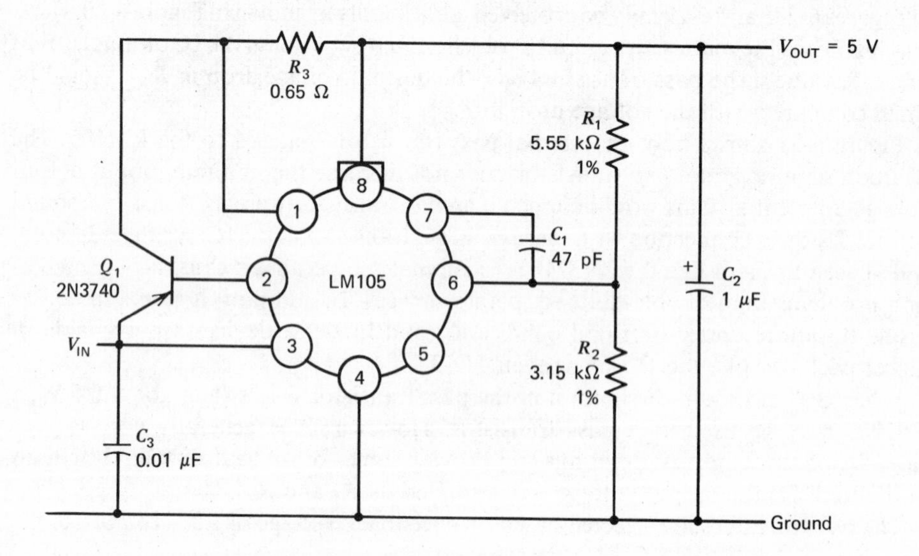

Figure 6-31. A 0.2-A regulator using the LM105 and an external-pass transistor.

The 1-μF output capacitor C_2 is required to suppress oscillations in the feedback loop involving the external booster transistor Q_1 and the output transistor of the LM105; C_1 compensates the internal regulator circuitry to make the stability independent for all loading conditions. Capacitor C_3 is not normally required if the lead length between the regulator and the output filter of the rectifier is short.

Current-limiting is provided by R_3. The current-limit resistor should be selected so that the maximum voltage drop across it, at full-load current, is equal to the voltage given in Figure 6-19 at the maximum junction temperature of the IC. This assures a no-load to full-load regulation better than 0.1% under worst-case conditions.

The short-circuit output current is also determined by R_3. Figure 6-18 shows the voltage drop across this resistor, when the output is shorted, as a function of junction temperature in the IC.

With the type of current-limiting used in Figure 6-31, the dissipation under short-circuit conditions can be more than three times the worst-case full-load dissipation.

Hence, the heat sink for the pass transistor must be designed to accommodate the increased dissipation if the regulator is to survive more than momentarily with a shorted output. It is encouraging to note, however, that the short-circuit current will decrease at higher ambient temperatures. This assists in protecting the pass transistor from excessive heating.

To regulate negative voltages, the circuit in Figure 6-32 is used. An LM104 contains the voltage reference and control circuitry while an external transistor is used to increase the power-handling capacity. A reference voltage is generated by driving a constant current, determined by R_1, through R_2. The voltage across this resistor is fed into an error amplifier. The error amplifier controls the output voltage at twice the voltage across R_2. The output voltage is resistor programmable with R_2 and adjustable down to zero.

Current limiting the LM104 is similar to the LM105. Voltage across R_3 turns ON an internal transistor that decreases drive to the output transistors. This current-limit

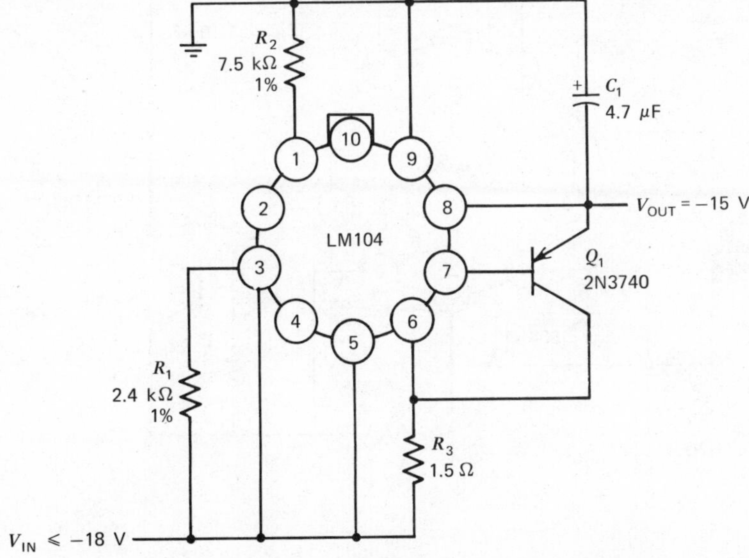

Figure 6-32. A 200-mA negative regulator using the LM104.

sense voltage is also temperature-dependent, decreasing from 0.65 V at 25°C to 0.45 V at 125°C.

Boosting the available output current from 200 mA is relatively simple. Figure 6-33 shows positive and negative 2-A regulators. An additional power transistor increases the current-handling capability of the regulator. Adding the boost transistors increases the output current without increasing the minimum input-output voltage differential.

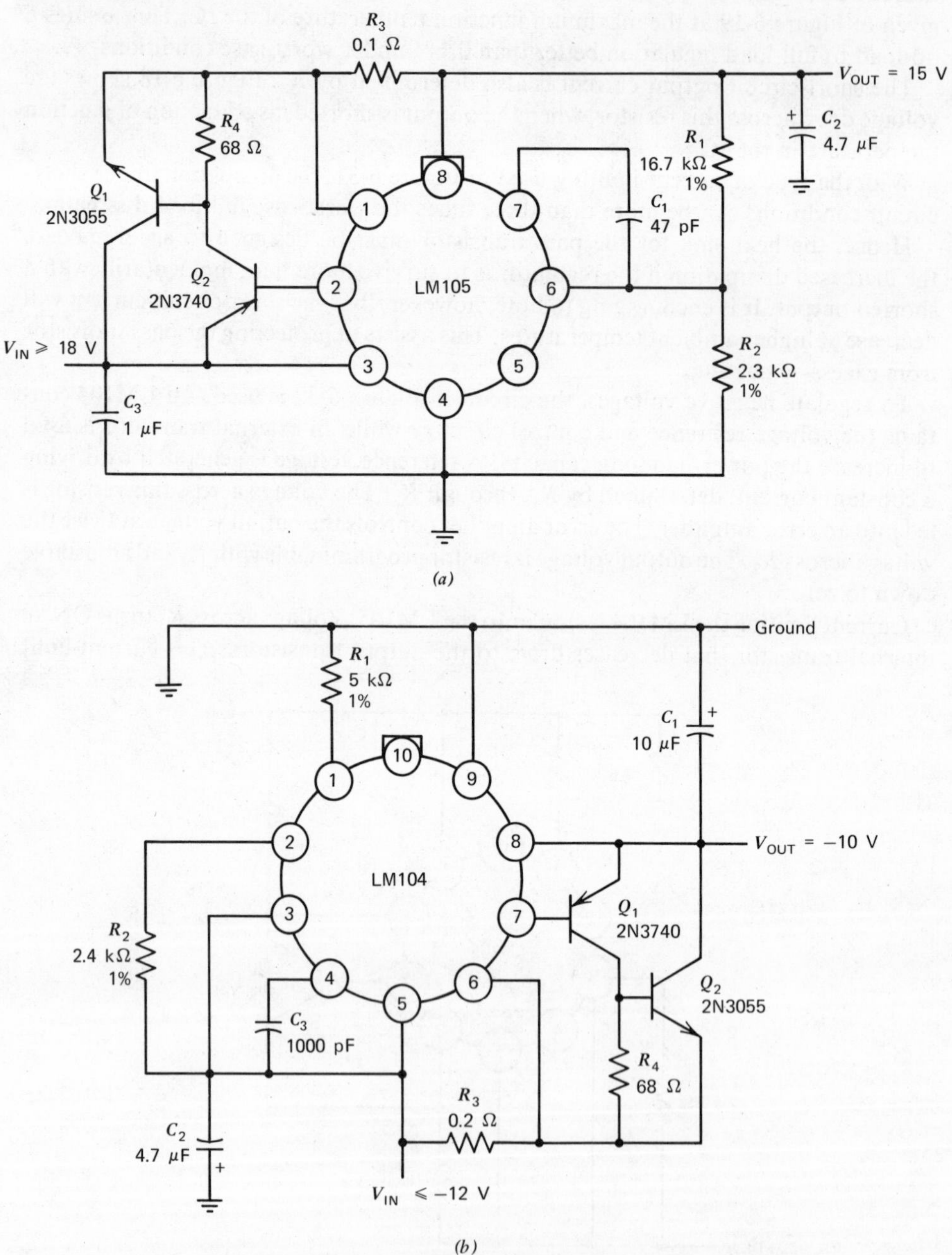

Figure 6-33. High-current (2-A) regulators using external-pass transistors.

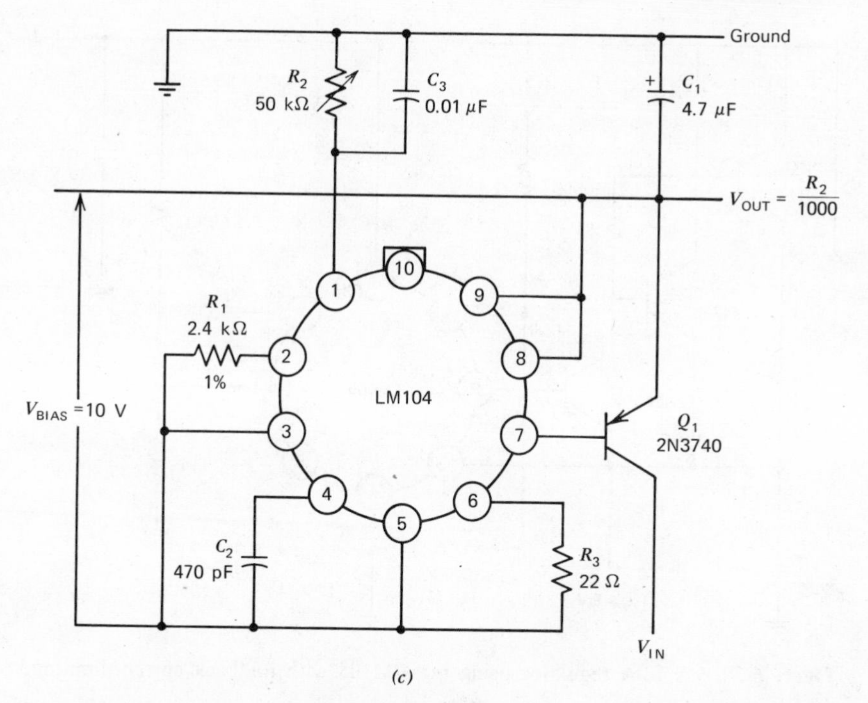

(c)

Figure 6-33. (*Continued.*)

The minimum differential will be 2 to 3 V, depending on the drive current required from the IC and operating temperature. Low input-output voltage differential allows more efficient regulation.

10-A POSITIVE REGULATOR

The output current of a regulator can be increased to any desired level by adding more booster transistors, increasing the effective current gain of the pass transistors. A circuit for a 10-A regulator is shown in Figure 6-34. A third *npn* transistor has been included to get higher current. A low-frequency device is used for Q_3 because it seems to better withstand abuse. However, high-frequency transistors must be used to drive it; Q_2 and Q_3 are both double-diffused transistors with good frequency response. This ensures that Q_3 will present the dominant lag in the feedback loop through the booster transistors, and back around the output transistor of the LM105. This is further ensured by the addition of C_3.

The circuit, as shown, has a full-load capability of 10 A. Foldback-limiting is used to give a short-circuit output current of 2.5 A. The addition of Q_3 increases the minimum input/output voltage differential, by 1 V, to 4 V.

External-Pass Transistors

Overstressing series-pass transistors has been the biggest cause of failures with IC regulators. This not only applies to the transistors within the IC, but also to the external booster transistors. Hence, in designing a regulator, it is of utmost importance to determine the worst-case power dissipation in all the driver and pass transistors.

Devices must then be selected which can handle the power. Further, adequate

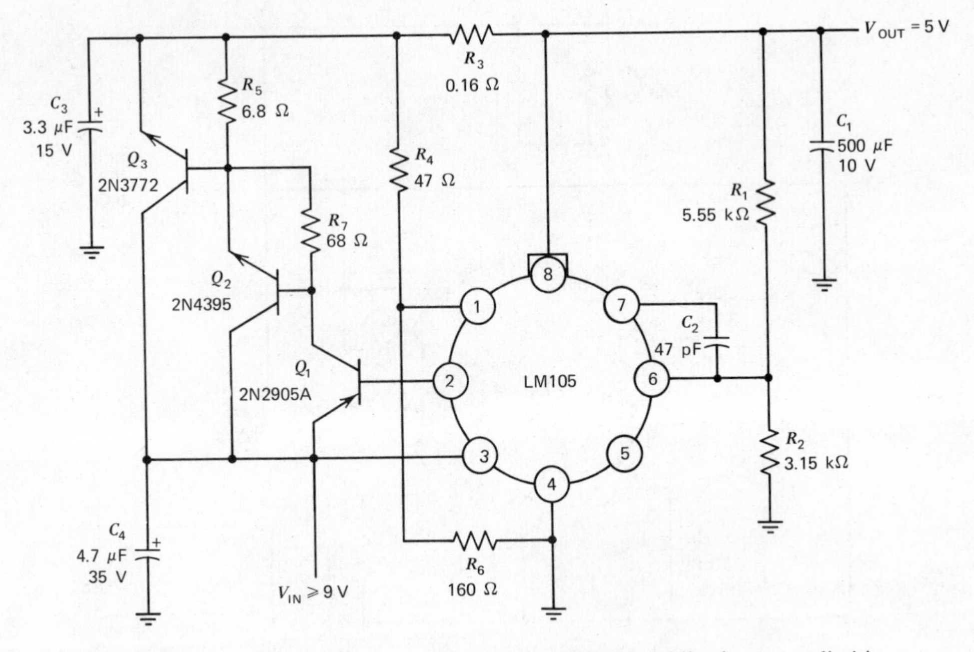

Figure 6-34. A 10-A regulator using the LM105 with foldback current-limiting.

heat sinks must be provided as even power transistors cannot dissipate much power by themselves.

Normally, the highest power dissipation occurs when the output of the regulator is shorted. If this condition requires heat sinks which are so large as to be impractical, foldback current-limiting can be used. With foldback-limiting, the power dissipated under short-circuit conditions can actually be made less than the dissipation at full load.

Figure 6-35 makes the point about dissipation limitations more strongly. It gives

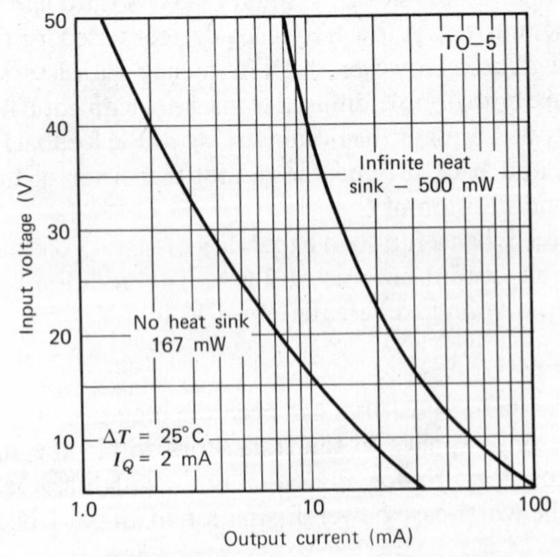

Figure 6-35. Dissipation-limited short-circuit output current for an IC regulator in a TO-5 package.

the maximum short-circuit output current for a positive IC regulator in a TO-5 package, assuming a 25°C temperature rise between the chip and ambient and a quiescent current of 2 mA. Dual-in-line or flat packages give results which are, at best, slightly better, but are usually worse. If the short-circuit current is not of prime concern, Figure 6-35 can also be used to give the maximum output current as a function of the input-output voltage differential. However, the increased dissipation due to the quiescent current flowing at the maximum input voltage must be taken into account. In addition, the input-output differential must be measured with the maximum expected input voltages.

As another example, the worst-case input voltage for Figure 6-32 can be 25 V. With a shorted output at 125°C, the current through the pass transistor will be 300 mA; and the dissipation in it will be 7.5 W. This clearly establishes the need for an efficient heat sink.

For lower power operation, a 2N2905 with a clip on heat sink can be used for the external-pass transistor. However, when the worst-case dissipation is above 0.5 W, it is advisable to employ a power device such as the 2N3740 with a good heat sink.

The current-limit resistor in Figure 6-32 is chosen so that the voltage drop across it is 300 mV, with maximum load current, for operation to 125°C. With lower maximum ambients this voltage drop could be increased by 2.2 mV/°C. If possible, a fast-acting fuse rated about 25% higher than the maximum load current should be included in series with the unregulated input.

When a booster transistor is used, the minimum input-output voltage differential of the regulator will be increased by the emitter-base voltage of the added transistor. This establishes the minimum differential at 2 to 3 V, depending on the base drive required by the external transistor.

Adding a driver transistor increases current gain, of course, and thus increases the output current capability. It is advisable to use the 2N3740 for output currents above 1 A, because it has a good heat sink. However, its frequency response is poorer than the 2N2905s, so a 1000-pF compensation capacitor should also be added. The need for a good heat sink on the other transistor is obvious.

Experience shows that a single-diffused transistor such as a 2N3055 (or a 2N3772 for higher currents) is preferred over a double-diffused, high-frequency transistor for the series-pass element. The slower, single-diffused devices are less prone to secondary breakdown and oscillations in linear regulator applications. The resistor across the emitter-base junction fixes the minimum collector current of the driver to minimize oscillation problems with light loads. It is still possible to experience oscillations with certain physical layouts, but these can almost always be eliminated by stringing a ferrite bead, such as a Ferrox-cube K5-001-00/3B, on the emitter lead of Q_2.

The low-resistance values required for the current-limit resistor R_3 are sometimes not readily available. A suitable resistor can be made using a piece of resistance wire or even a short length of kovar lead wire from a standard TO-5 transistor.

The current-limit sense voltage can be reduced to about 400-mV by inserting a germanium diode (or a diode-connected germanium transistor) in series with the current-limit terminal of the regulator. This diode will also compensate the sense voltage and make the short-circuit current essentially independent of temperature.

With high-current regulators it is especially important to use a low-inductance capacitor on the output. The lead length on this capacitor must also be made short.

Otherwise, the capacitor leads can resonate with smaller bypass capacitors (like 0.1-μF ceramic) which may be connected to the output. These resonances can lead to oscillations. With short leads on the output capacitor, the Q of the tuned circuit can be made low enough so that it cannot cause trouble.

Series-Dissipative Integrated Circuit Regulator Applications

CONSTANT-CURRENT SOURCE

The circuit in Figure 6-36 utilizes the LM105 as a 1-A constant-current source. Here the LM105 regulates the emitter current of a Darlington-connected transistor, and the output current is taken from the collectors. The use of a Darlington connection for Q_1 and Q_2 improves the accuracy of the circuit by minimizing the base-current error between the emitter and collector current.

The output of the LM105, which drives the control transistors, must be protected against short circuits with R_6 to limit the current when Q_2 saturates. The minimum load current for the IC is provided by R_7; D_1 is included to absorb the kickback of inductive loads when power is shut OFF. The output current of the circuit is adjusted with R_2. When used as a current source, the LM105 is capable of driving servo control motors.

TEMPERATURE CONTROLLER

A circuit for an oven-temperature controller using the LM105 is given in Figure 6-37. Temperature changes in the oven are sensed by a thermistor. This signal is fed to the LM105, which controls power to the heater by switching the series-pass transistor Q_2 ON and OFF. Since the pass transistor will be nearly saturated in the

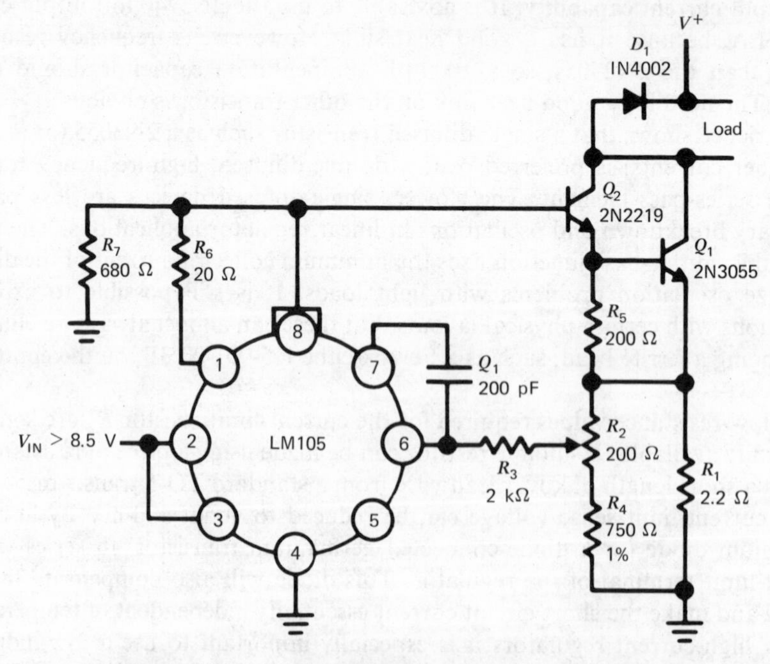

Figure 6-36. A 1-A current source using the LM105.

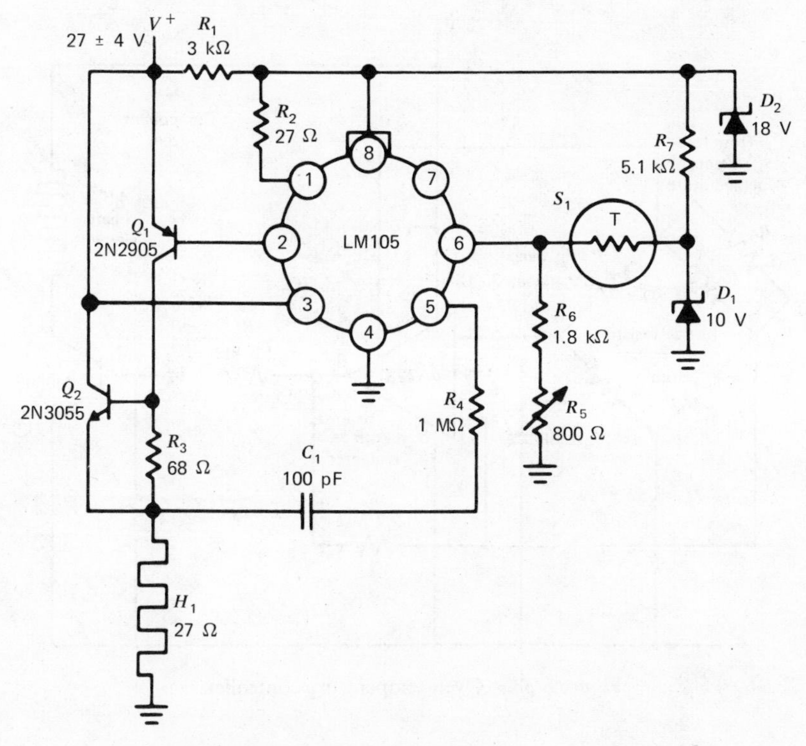

Figure 6-37. Switching-temperature controller.

ON condition, its power dissipation is minimized. In operation, if the oven temperature should try to increase, the thermistor resistance will drop, increasing the voltage on the feedback terminal of the regulator. This action shuts OFF power to the heater. The opposite would be true if the temperature dropped.

Variable duty-cycle switching action is obtained by applying positive feedback around the regulator from the output to the reference-bypass terminal (which is also the noninverting input to the error amplifier) through C_1 and R_4. When the circuit switches ON or OFF, it will remain in that state for a time determined by this RC time constant.

Additional details of the circuit are that base drive to Q_1 is limited, to a value determined by R_2, by the internal current-limiting circuitry of the LM105. Moreover, D_2 provides a roughly regulated supply for D_1 in addition to fixing the output level of the LM105 at a level which properly biases the internal transistors. The reference diode for the thermistor sensor D_1 need not be a temperature-compensated device as long as it is put in the oven with the thermistor. Finally, the temperature is adjusted with R_5.

Figure 6-38 is another circuit for an oven-temperature controller using the μA723 voltage regulator. This circuit will control the temperature of an oven within ± 1°C over an ambient swing of -50°C to the temperature of the oven. In addition, if R_2 is selected to provide adequate drive to Q_1, the input voltage can vary from 10 to 37 V. Because transistor Q_1 is saturated when ON, no heat sink is required. Resistor R_1 sets the trip-point hysteresis at 2°C with the circuit values indicated. If the heater

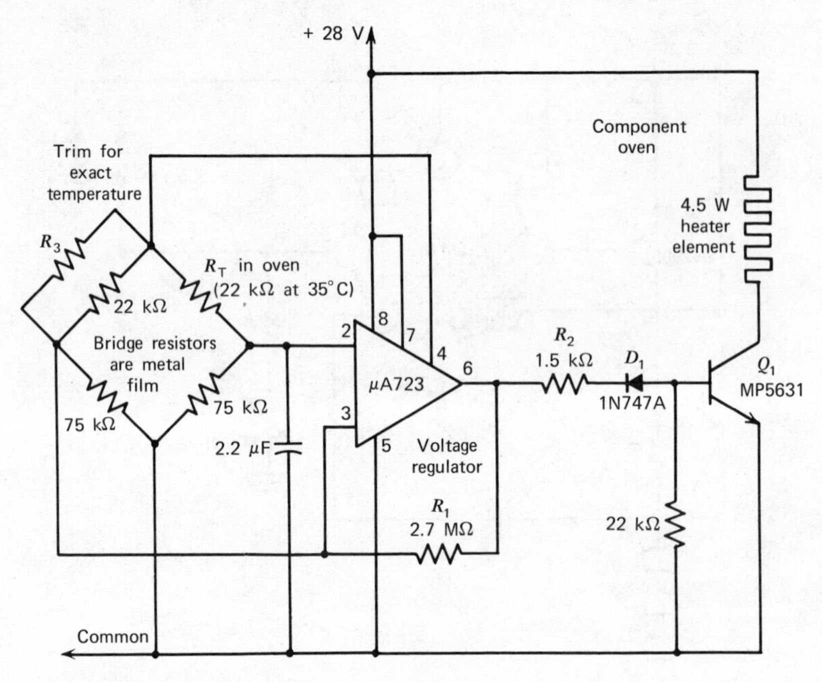

Figure 6-38. Oven-temperature controller.

current is less than 150 mA, it can be supplied directly from the μA723 by connecting the element between *pin* 6 and the common line.

POWER AMPLIFIER

In Figure 6-39 the LM105 is used as a high-gain amplifier and connected to a quasicomplementary power-output stage. Feedback around the entire circuit stabilizes the gain and reduces distortion. In addition, the regulation characteristics of the LM105 are used to stabilize the quiescent output voltage and to minimize ripple feedthrough from the power supply.

The LM105 drives the output transistors Q_5 and Q_6 for positive-going output signals whereas Q_1, operating as a current source from the 1.8 V on the reference terminal of the LM105, supplies base drive to Q_3 and Q_4 for negative-going signals. Transistor Q_2 eliminates the dead zone of the class-B output stage, and it is bypassed by C_5 to present a lower driving impedance to Q_3 at high frequencies. The voltage drop across Q_2 will be a multiple of its emitter-base voltage determined by R_9 and R_{10}. These resistors can therefore be selected to give the desired quiescent current in Q_4 and Q_6. It is important that Q_2 be mounted on the heat sink with the output and driver transistors to prevent thermal runaway.

Output current-limiting is obtained with D_2 and D_3. Diode D_2 clamps the base drive of Q_3 when the voltage drop across R_6 exceeds one diode drop, and D_3 clamps the base of Q_5 when the voltage across R_1 becomes greater than two diode drops. When D_3 becomes forward-biased, R_{11} is needed to limit the output current of the LM105. The power-supply ripple is peak-detected by D_1 and C_1 to obtain increased positive-output swing by operating the LM105 at a higher voltage than Q_5 and Q_6 during the troughs of the ripple. This also reduces the ripple seen by the LM105.

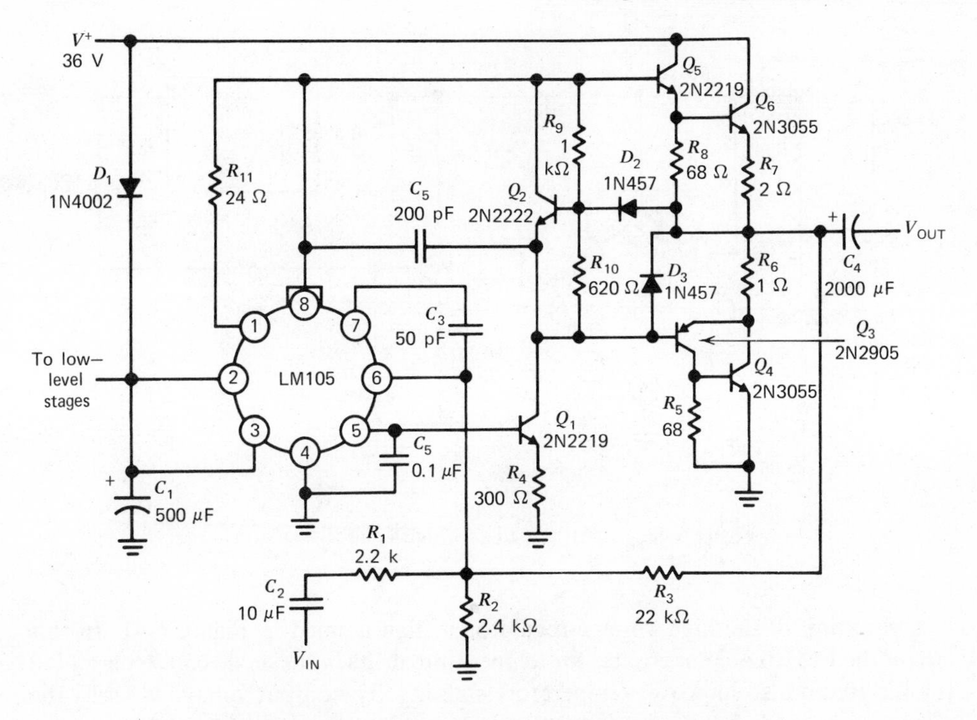

Figure 6-39. Power amplifier with current-limiting.

Capacitor C_5 bypasses any Zener noise on the reference terminal of the LM105 that would otherwise be seen on the output.

The quiescent output voltage is set with R_2 and R_3 in the same way as with a voltage regulator. The ac voltage gain is determined by the ratio of R_1 and R_3, since the circuit is connected as a summing amplifier.

HIGH VOLTAGE-REGULATOR

Integrated circuit regulators were designed primarily for applications with output voltages below 80 V. However, they can be used as high voltage-regulators under certain circumstances. An example of this, a circuit regulating the output of a 2-kV supply, appears in Figure 6-40. The LM105 senses the output of the high-voltage supply through a resistive divider and varies the input to a dc/dc converter, which generates the high voltage. Hence, the circuit regulates without having any high voltages impressed across it.

Under ordinary circumstances, the feedback terminal of the LM105 wants to operate from a 2-kΩ divider impedance. Satisfying this condition on a 2-kV regulator would require that about 2 W be dissipated in the divider. This, however, is reduced to 40 mW by the addition of Q_1, which acts as a buffer for a high-impedance divider, operating the LM105 from the proper source resistance. The other half of the transistor Q_2, compensates for the temperature drift in the emitter-base voltage of Q_1, so that it is not multiplied by the divider ratio. The circuit does have an uncompensated drift of 2 mV/°C; but since this is added directly to the output, and not multiplied by the divider ratio, it will be insignificant with a 2-kV regulator.

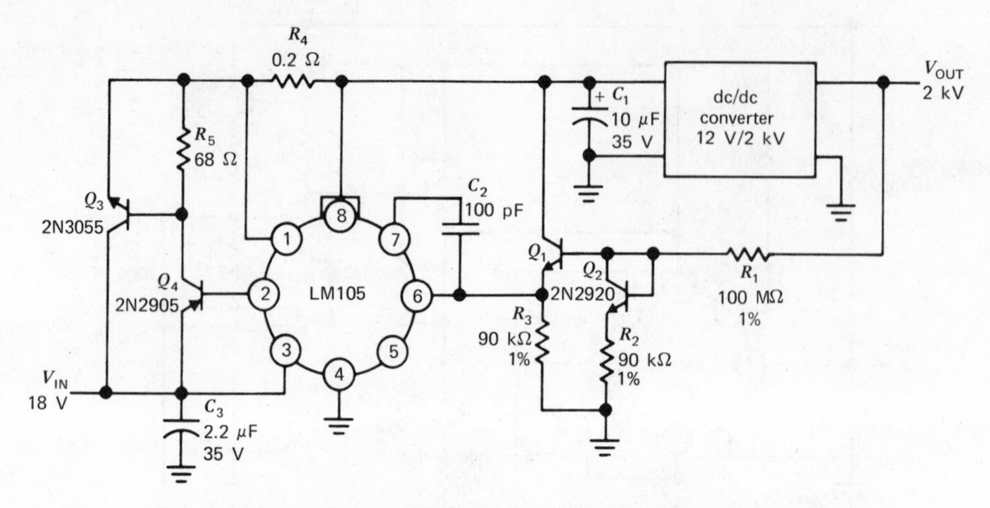

Figure 6-40. LM105 used as a high voltage-regulator.

A variation of the high voltage-regulator is diagrammed in Figure 6-41. In this circuit the LM101A compensates for temperature drifts in the dc/dc converter and in the LM105; it also improves temperature stability. The output voltage is set by the values of R_1 and R_2. If an adjustable output voltage is desired, a potentiometer could be inserted between R_1 and R_2. If a 10-kΩ potentiometer were inserted, the output could be adjusted from about 400 to about 1500 V. For an output voltage of -500 V, the line and load regulation are better than 0.1% for line voltage from 10 to 24 V and load currents to 10 mA. Temperature stability is 0.2%/°C over the temperature range -55 to $+125$°C.

LIGHT-INTENSITY REGULATOR

Figure 6-42 gives the circuit for a light-intensity regulator using the LM105. A phototransistor senses the light level and drives the feedback terminal of the LM105 to control current flow into an incandescent bulb; R_1 serves to limit the in-rush current to the bulb when the circuit is first turned ON.

The current gain of the phototransistor Q_2 is fixed at 10 (to make it less temperature-sensitive) by R_3 and the temperature-compensating diode D_1. A photodiode, such as the 1N2175, could be substituted for the phototransistor if it had sufficient light-sensitivity, and R_3 and D_1 could be eliminated. The input voltage does not have to be regulated because the sensitivity of a phototransistor or photodiode is not greatly affected by the voltage drop across it. A photoconductor can also be used in place of the phototransistor, except that input voltage would have to be regulated.

TRACKING REGULATORS

In systems that have more than one regulated-output voltage, it is desirable to adjust all supplies with a single potentiometer. In Figure 6-43 a single potentiometer controls both outputs of the 5- and 15-V regulators. To ensure that both regulators operate with the same reference voltage, their internal reference voltages are tied together (*pin* 5). Resistors R_2 and R_2' are fixed at 2 kΩ. Resistor R_1 is calculated for an output voltage using 1.6 V as a reference voltage. The following expression is used

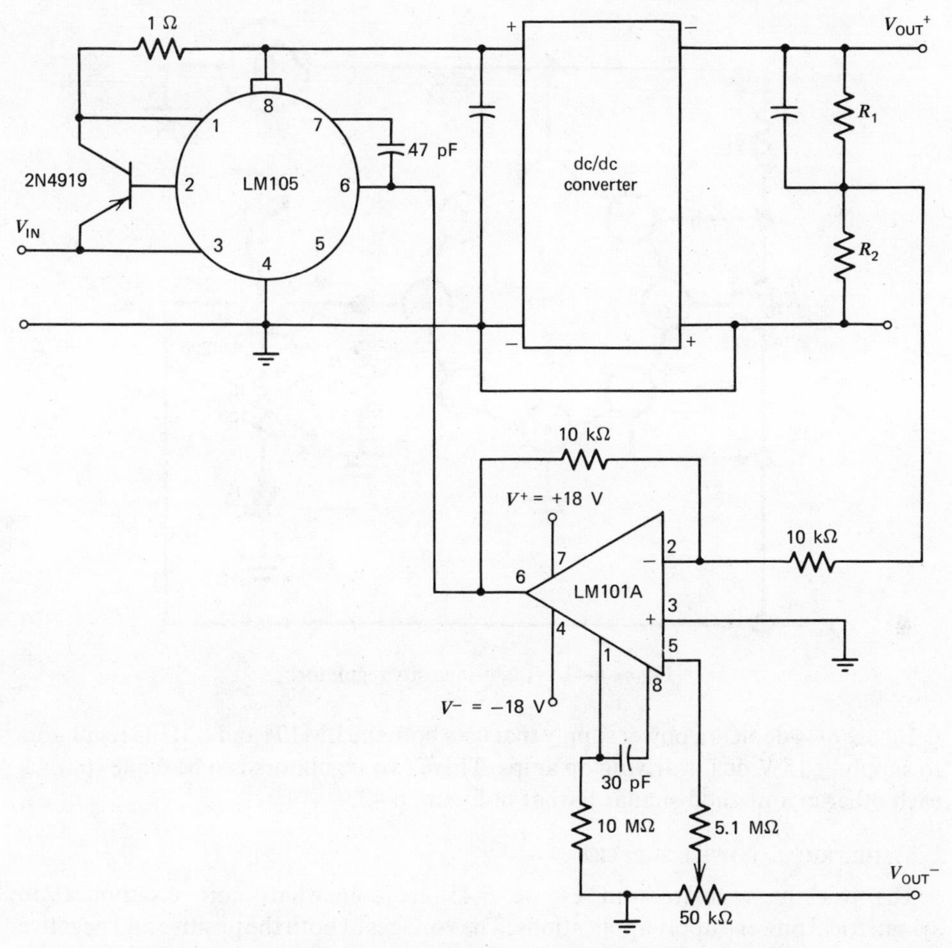

Figure 6-41. High voltage-regulator using the LM101A and LM105.

to determine R_1:

$$R_1 = \frac{(V_{OUT} - 1.6 \text{ V})2000 \ \Omega}{1.6 \text{ V}} \tag{6-19}$$

Potentiometer R_5 adjusts both regulators within 2% of the desired output for reference variations from 1.6 to 2 V. If the reference is 2 V, R_5 is 324 Ω and the output voltages are 5.1 and 14.9 V. If the reference is approximately 1.8 V, both outputs are within 1% nominal.

Potentiometer R_5, connected between the lower section of the output-voltage divider network and ground, compensates for any variations from the nominal 1.8-V reference of the regulators. Note that the wiper of R_5 is connected to one side of the potentiometer. If a rheostat connection were used, the arm might open-circuit during adjustment, causing large transients on the output.

If 1% resistors are used, there is an additional worst-case error of 2% of each regulator. Resistor errors are inherent in any type of tracking regulator system, even if the adjustment is proven to be exact theoretically.

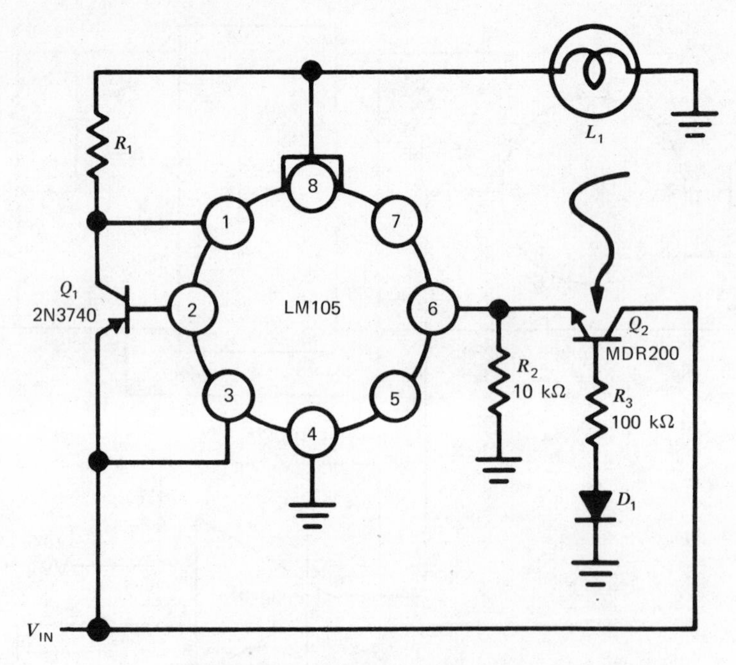

Figure 6-42. Light-intensity regulator.

Figure 6-44 depicts a power supply that uses both the LM104 and LM105 regulators to supply ±15 V dc for driving op amps. These two regulators can be made to track each other in a method similar to that of Figure 6-43.

SYMMETRICAL POWER SUPPLIES

The tracking regulators in Figure 6-45 are somewhat more economical in symmetrical power-supply applications. The voltages at both the positive and negative outputs can be set within ±1.5% by a single adjustment.

The positive voltage is regulated by an LM105, while an LM104 regulates the negative supply. The unusual feature is that the two regulators are interconnected by

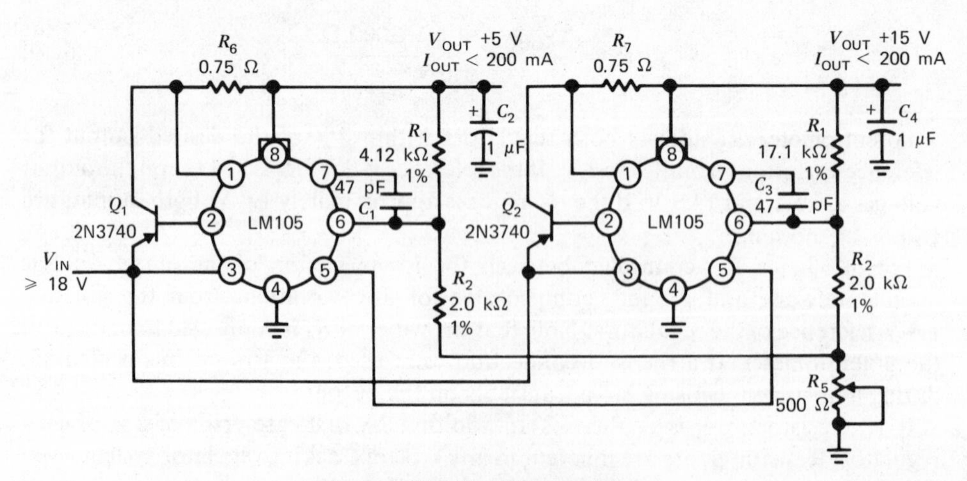

Figure 6-43. Tracking regulator.

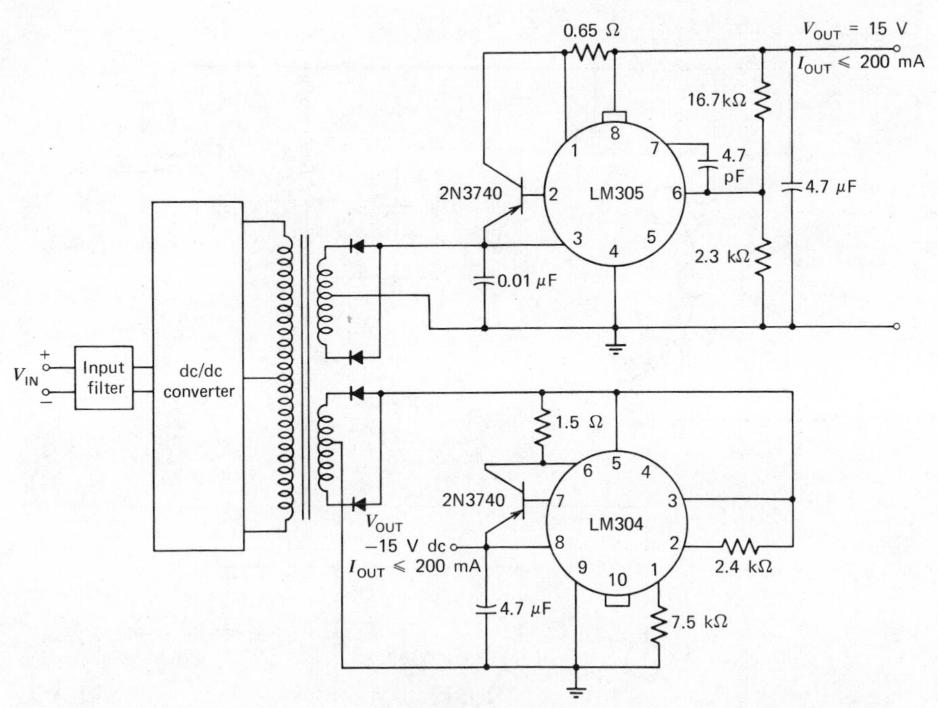

Figure 6-44. LM104 and LM105 voltage-regulators used on a ±15-V operational amplifier power supply.

R_3. This not only eliminates one precision resistor, but the reference current of the LM104 stabilizes the LM105 so that a ±10% variation in its reference voltage is only seen as a ±3% change in output voltage. This means that in many cases the output voltage of both regulators can be set up with sufficient accuracy by trimming a single resistor R_1. The line regulation and temperature drift of the circuit is determined primarily by the LM104, so both output voltages will tend to track.

Output ripple can be reduced by about a factor of 5 to less than 2 mV/V by bypassing *Pin* 1 of the LM104 to ground with a 10-μF capacitor. A center-tapped transformer with a bridge rectifier can be used for the unregulated power source.

Since positive voltage-regulators can be used as negative voltage-regulators, both halves of the power supply can be built with the same type of regulator if desired. Figures 6-46 and 6-47 show two such designs. In the first, one LM105 is made to operate as a negative regulator by grounding the normal output terminal (*pin* 8) and connecting the ground terminal (*pin* 4) to the output. Hence, it regulates the voltage between the output and ground terminals. A *pnp* booster Q_2 is connected in the normal manner to drive a *npn* series-pass transistor Q_3. The additional components (Q_4, R_7, R_8, R_9, and R_{10}) provide current-limiting. In Figure 6-47, the output *pin* is grounded and a transformer's split secondaries create a floating voltage source for the negative regulator.

Many negative regulator applications occur in symmetrical supplies, for example, ±15 V for most IC op amps. Figure 6-48 shows a simple method to ensure the negative supply follows the positive supply, even in shutdown. When using the μA723 IC regulator, the circuit uses the positive-supply output as a reference for the negative

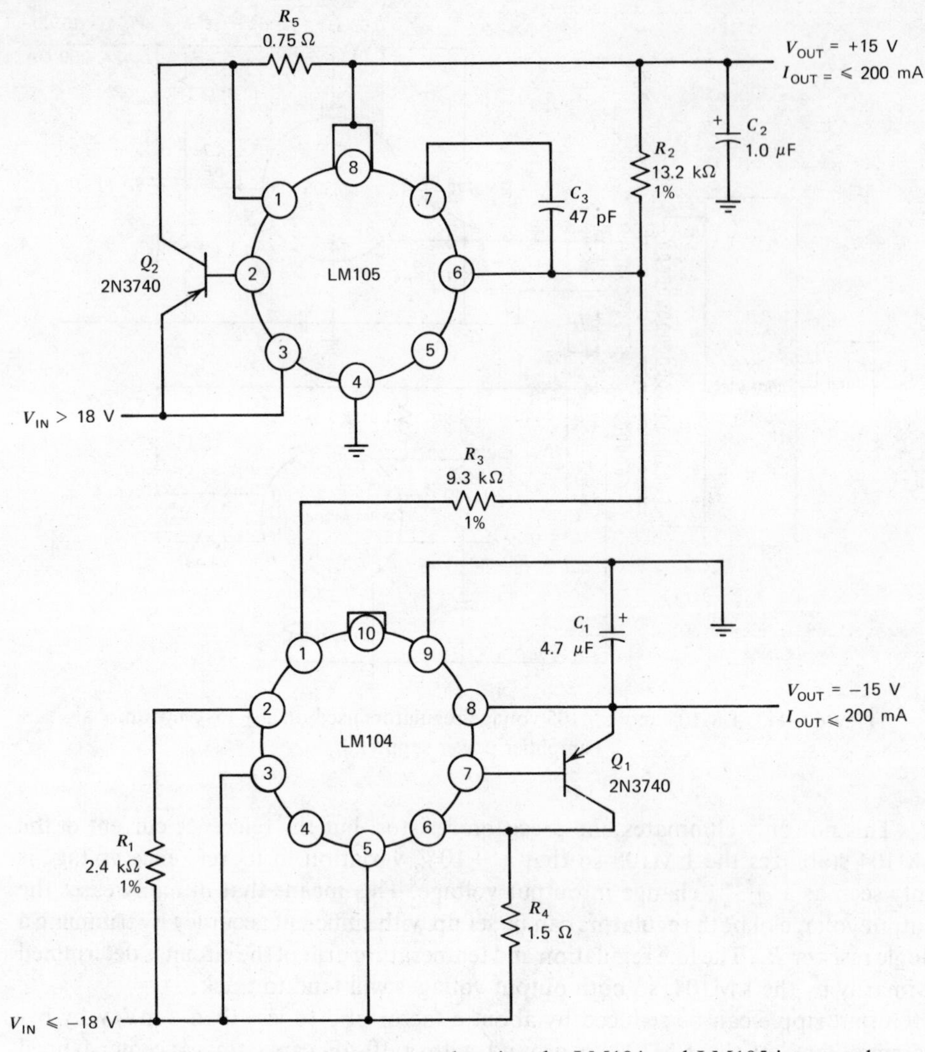

Figure 6-45. Symmetrical power supply using the LM104 and LM105 integrated circuit regulators.

regulator. The negative regulator can be provided with its own protection, as shown. However, the negative regulator would shutdown without affecting the positive supply.

HIGH-EFFICIENCY REGULATOR

Another high-power regulator is shown in Figure 6-49. This circuit is useful when low-output voltages are required. Here, the series-pass transistor Q_2 and the regulator are operated from separate supplies. The series-pass transistor is run off of a low-voltage main supply which minimizes the input-output differential for increased efficiency. The regulator, on the other hand, operates from a low-power bias supply with an output greater than 8.5 V.

With this circuit, care must be taken that Q_2 never saturates. Otherwise, Q_1

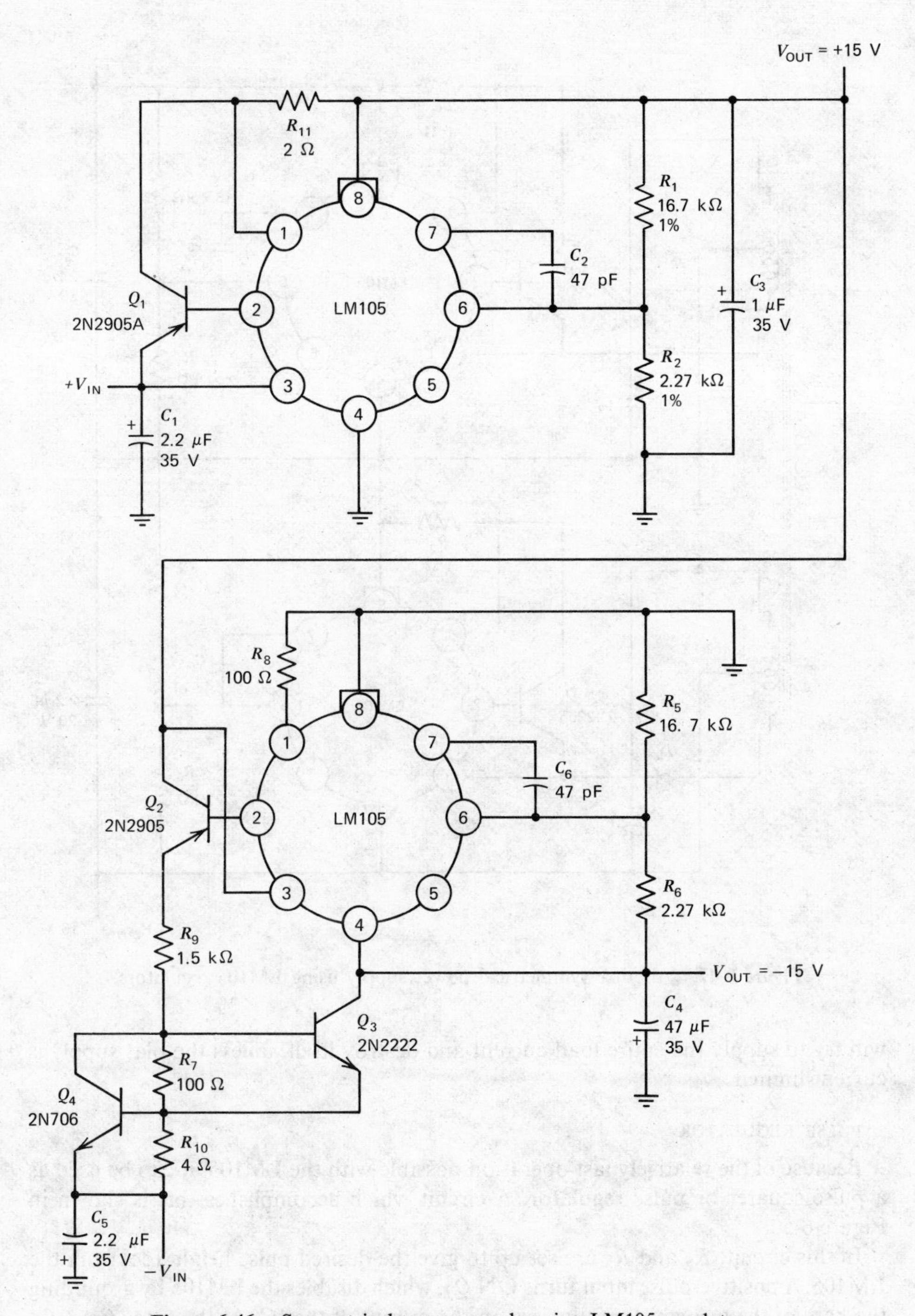

Figure 6-46. Symmetrical power supply using LM105 regulators.

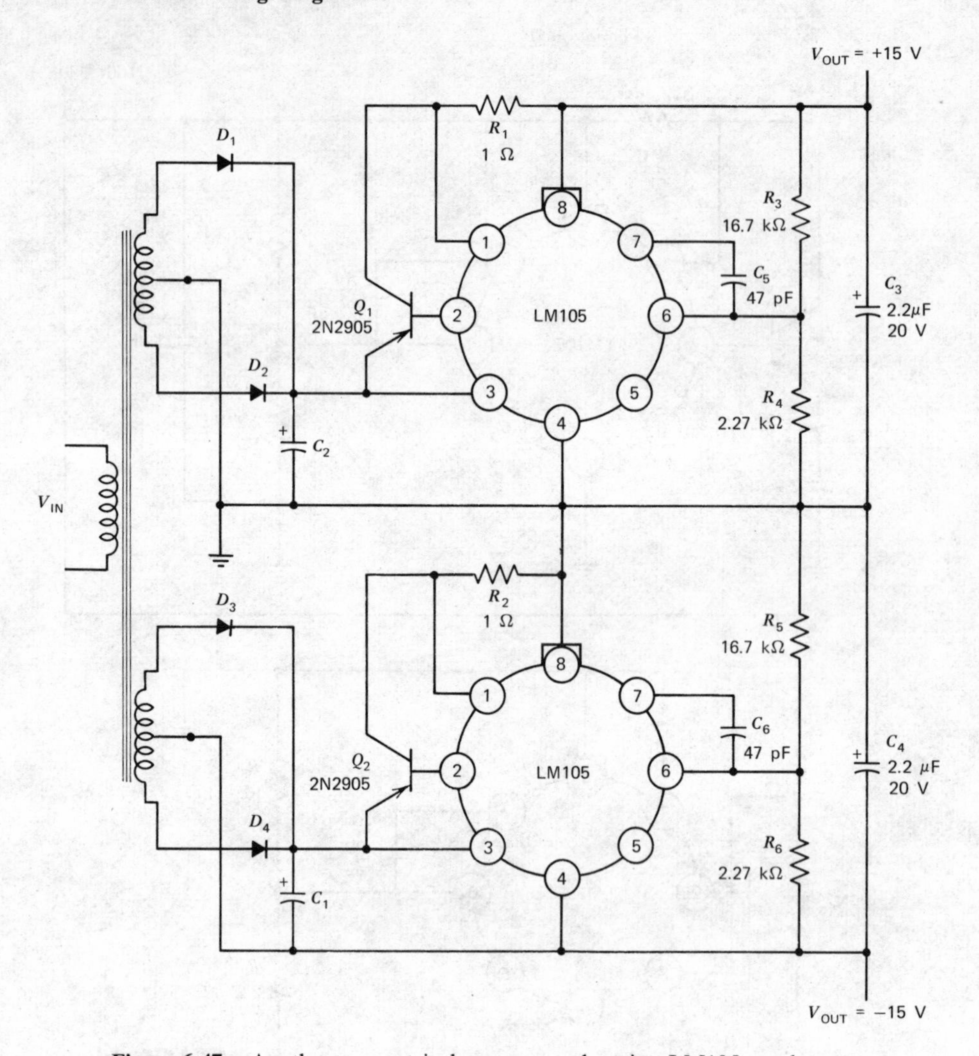

Figure 6-47. Another symmetrical power supply using LM105 regulators.

will try to supply the entire load current and destroy itself, unless the bias supply is current-limited.

PULSE REGULATOR

Because of the relatively fast operation possible with the LM105, it can be used as a pulse squarer or pulse regulator. A circuit which accomplishes this is shown in Figure 6-50.

In this circuit, R_2 and R_3 are set up to give the desired pulse height (dc) from the LM105. A positive-pulse input turns ON Q_1, which disables the LM105 by grounding the base of the *npn* emitter-followers on the output of the IC. At the same time, Q_2 grounds the regulator output, providing current-sinking capability.

If additional output-current drive is needed, an *npn* buffer should be used on the LM105 in place of a *pnp* because of the difficulties encountered in stabilizing the *pnp* circuit without capacitance on the output.

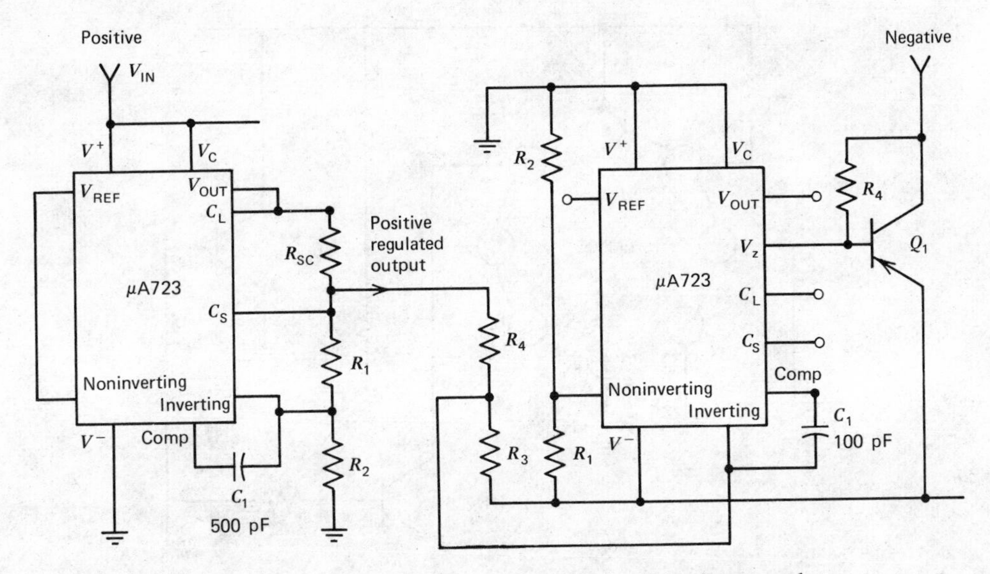

Figure 6-48. Symmetrical power supply using the 723 voltage-regulator.

This method of pulsing the circuit ON and OFF, that is, pulling *Pin* 7 down within one diode drop of ground, can be used as an electrical shut-off for voltage or current regulators.

FLOATING REGULATORS

The voltage limitations of IC regulators can be overcome by means of floating regulators. The limits on V_{IN} can be overcome by using a preregulator to feed the $V+$ and V_C lines, but the V_{OUT} limitation requires greater sophistication. The circuits of

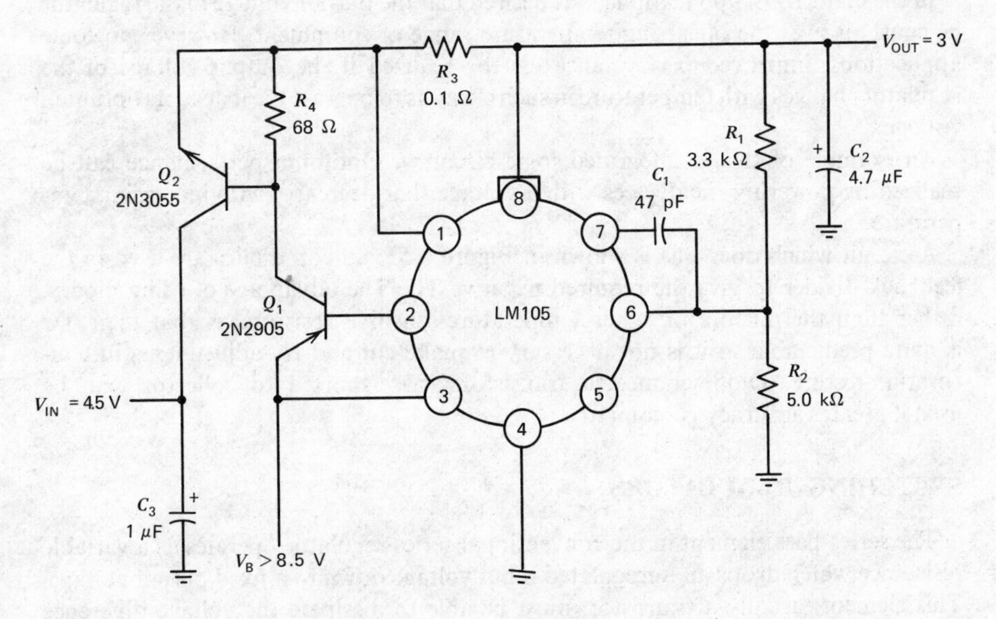

Figure 6-49. High-efficiency regulator.

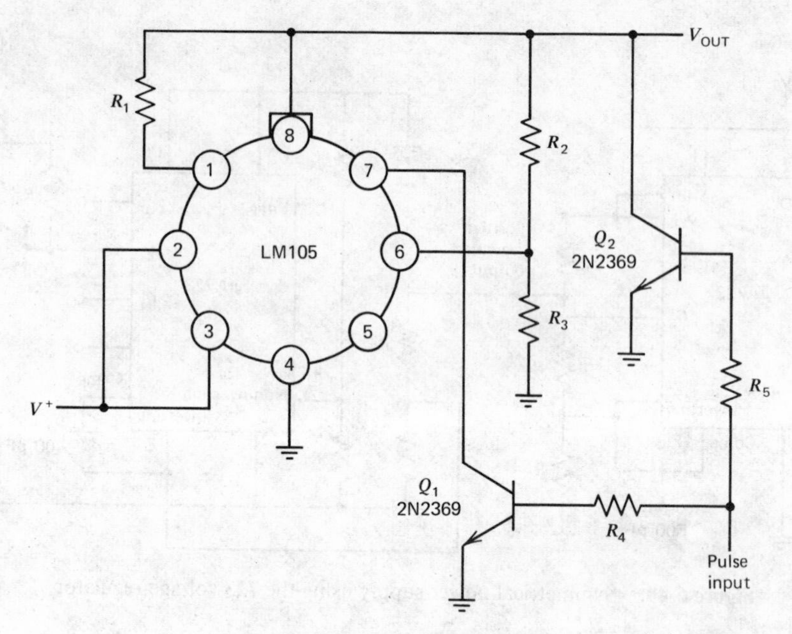

Figure 6-50. Pulse regulator.

Figures 6-51 and 6-52 show two techniques that can be used to give output voltages well outside the range of the IC device. In both circuits, R_5 and D_1 provide a low-voltage supply to the regulator itself. The circuit of Figure 6-51 depicts a positive-floating regulator and Figure 6-52 shows a negative-floating regulator.

TEMPERATURE-COMPENSATING REGULATORS

In the majority of applications, it is desired that the output voltage of the regulator be constant over the operating temperature range of equipment. However, in some applications, improved performance can be realized if the output voltage of the regulator changes with temperature in such a way as to operate the load at its optimum voltage.

An example of this is integrated logic circuitry. Optimum performance can be realized by powering the devices with a voltage that decreases with increasing temperature.

A circuit which does this is shown in Figure 6-53. Silicon diodes are used in the feedback divider to give the required negative TC. The advantage of using diodes, rather than thermistors or other temperature-sensitive resistors, is that their TC is quite predictable so it is not necessary to make cut-and-try adjustments in temperature-testing. Diode-connected transistors (base-shorted to collector) can be used if greater accuracy is required.

SWITCHING REGULATORS

The series-pass element in the conventional series regulator operates as a variable resistance which drops an unregulated input voltage down to a fixed output voltage. This element, usually a transistor, must be able to dissipate the voltage difference between the input and output at the load current. The power generated can become

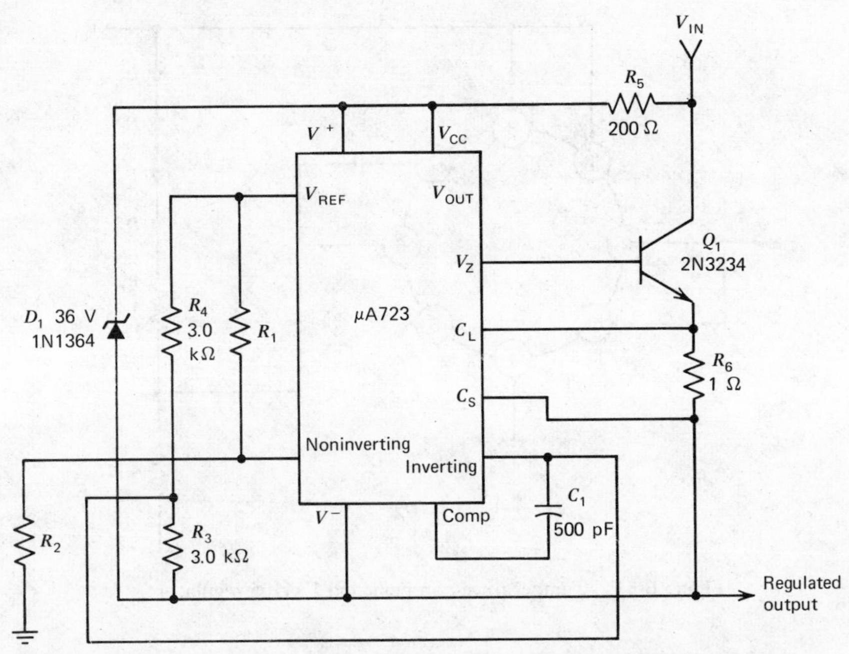

Figure 6-51. Positive-floating regulator [*note:* typical performance—regulated output voltage +50 V, line regulation (V_{IN} = 20 V) 15 mV, load regulation (I_L = 50 mA) 20 mV].

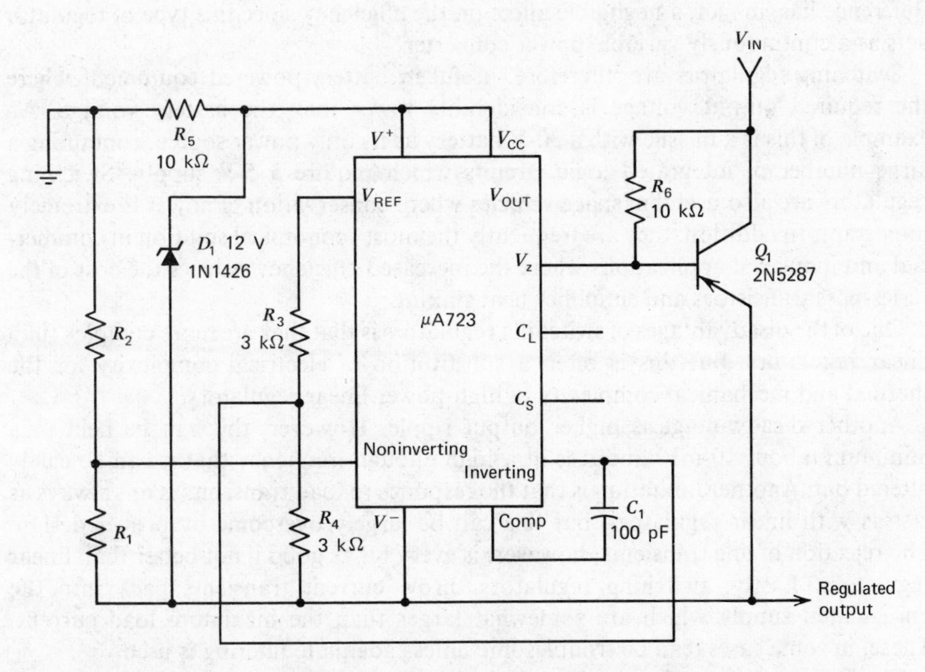

Figure 6-52. Negative-floating regulator [*note:* typical performance—regulated output voltage −100 V, line regulation (V_{IN} = 20 V) 30 mV, load regulation (I_L = 100 mA) 20 mV].

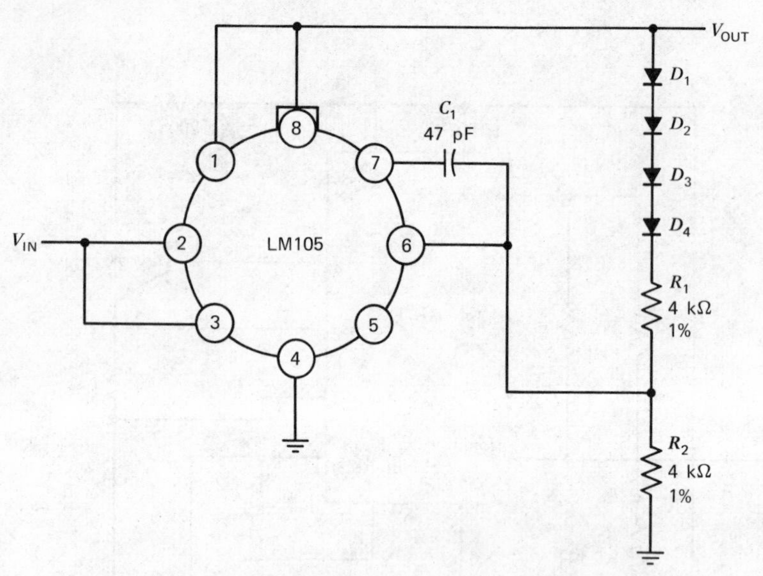

Figure 6-53. Temperature-compensated LM105 regulator.

excessive, particularly when the input voltage is not well regulated and the difference between the input and output voltages is large.

Switching regulators, on the other hand, are capable of high-efficiency operation even with large differences between the input and output voltages. The voltage difference has, in fact, a negligible affect on the efficiency since this type of regulator acts as a continuously variable power converter.

Switching regulators are, therefore, useful in battery-powered equipment where the required output voltage is considerably lower than the battery voltage. An example of this is a missile with a 30-V battery as its only power source, containing a large number of integrated logic circuits which require a 5-V supply. Switching regulators are also useful in space vehicles where conservation of power is extremely important. In addition, they are frequently the most economical solution in commercial and industrial applications where the increased efficiency reduces the cost of the series-pass transistors and simplifies heat sinking.

One of the disadvantages of switching regulators is that they are more complex than linear regulators, but this is often a substitution of electrical complexity for the thermal and mechanical complexity of high-power linear regulators.

Another disadvantage is higher output ripple. However, this can be held to a minimum (about 10-mV) and it is at a high enough frequency that it can be easily filtered out. Another limitation is that the response to load transients is not always as fast as with linear regulators, but this can be largely overcome by proper design. The rejection of line transients, however, is every bit as good if not better than linear regulators. Lastly, switching regulators throw current transients back into the unregulated supply which are somewhat larger than the maximum load current. These, in some cases, can be troublesome unless adequate filtering is used.

The basic principle of operation of a self-oscillating regulator is shown in Figure 6-54 where the switch S_1 is a transistor switched or driven from a threshold detector with hysteresis. When the switch is closed, the voltage across L_1 causes the current to

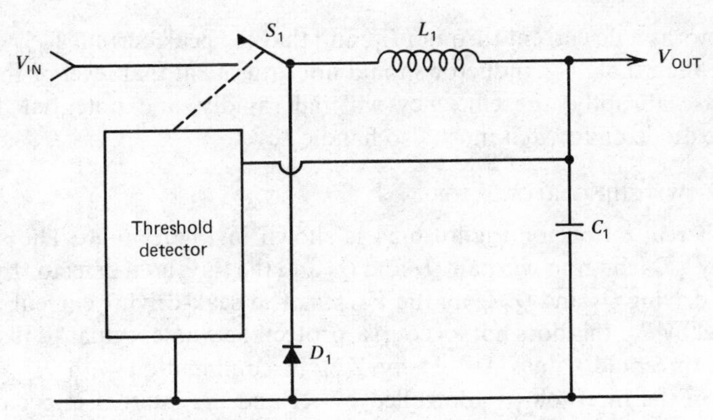

Figure 6-54. Basic switching regulator block diagram.

increase almost linearly. This current, less the load current, charges up C_1 until the upper threshold of the threshold detector V_{UT} is reached. At this point, S_1 opens, and the voltage across D_1 reverses so that it conducts, in the forward direction, the current through L_1 decreases, eventually discharging C_1 to the lower threshold V_{LT}, at which point S_1 closes again. The basic waveforms involved are sketched in Figure 6-55 for two different duty cycles. The detailed theory and design of switching regulators is complex, but the basic operation assumes a frequency significantly above the resonant frequency of L and C. The harmonics of the input square wave are attenuated, and the phase-shift of the lowest frequency component is close to 180°. This minimizes the ripple to the difference in the threshold levels ($V_{UT} - V_{LT}$).

The frequency of operation is a function of load level, input voltage, and threshold levels. It is possible to build a synchronous regulator, which switches in sychronism with an external drive signal, but the EMI (electromagnetic interference) problem generally requires that switching regulators be well shielded. The choice of operating frequency is in general a compromise between the inductor size becoming uneconomical at low frequencies and switching losses in D_1 and S_1 becoming excessive at high frequencies. The optimum switching frequency for switching regulators has been determined to be between 20 and 100 kHz. At lower frequencies, the core becomes unnecessarily large; and at higher frequencies, switching losses in S_1 (a transistor) and D_1 become excessive. It is important, in this respect, to minimize switching losses by assuring that both S_1 and D_1 are fast-switching devices. Note that the load

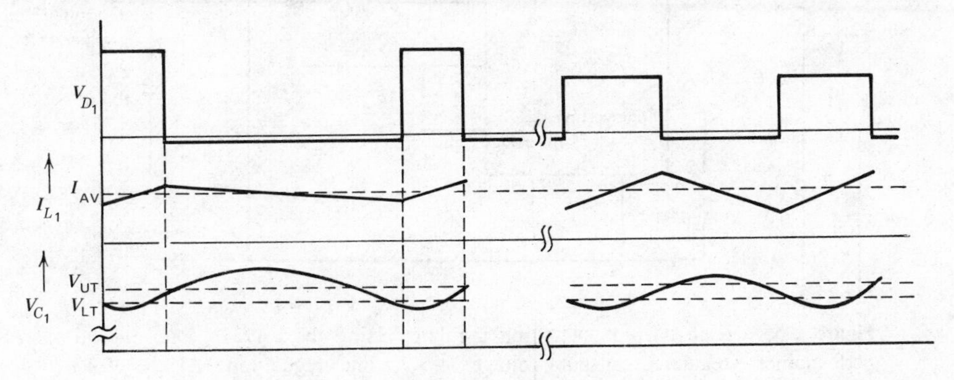

Figure 6-55. Basic switching regulator waveforms.

current flows as a dc current through L_1, and that the peak current I_{MAX} can be quite a bit more than that. The inductor should not saturate at that level or the frequency will increase abruptly, the efficiency will fall rapidly, and potential disaster will ensue. The diode and switch must also handle I_{MAX}.

POSITIVE-SWITCHING REGULATOR

A circuit realization for Figure 6-54 is shown in Figure 6-56. The switch S_1 is replaced by a Darlington *pnp* pair Q_1 and Q_2 and the threshold detector by the μA723 regulator, driving Q_1 and Q_2 from the V_{CC} *pin*. The peak-driving current through the μA723 is set by R_6; this does not, of course, protect the whole regulator, just the μA723 itself. The threshold values are set by V_{REF} in combination with R_1 and R_2; the hysteresis in the threshold is controlled by R_4 and the internal impedance of V_{REF} (about 2 kΩ), and derived from the voltage across D_1.

Figure 6-57 depicts the basic switching regulator circuit using the LM105 regulator. In this circuit, R_3 determines the base drive for the switch transistor Q_1, providing enough drive to saturate it with maximum load current. Capacitor C_2 minimizes output ripple; C_3 removes fast-rise time transients, and R_4 produces positive feedback. Losses are kept low by Q_1 and D_1, which are both fast-switching devices.

The following equations are used to calculate the values of R_4, L_1, and C_1, given the switching frequency and the output ripple. These equations provide a starting point for designing a switching regulator to fit a given application.

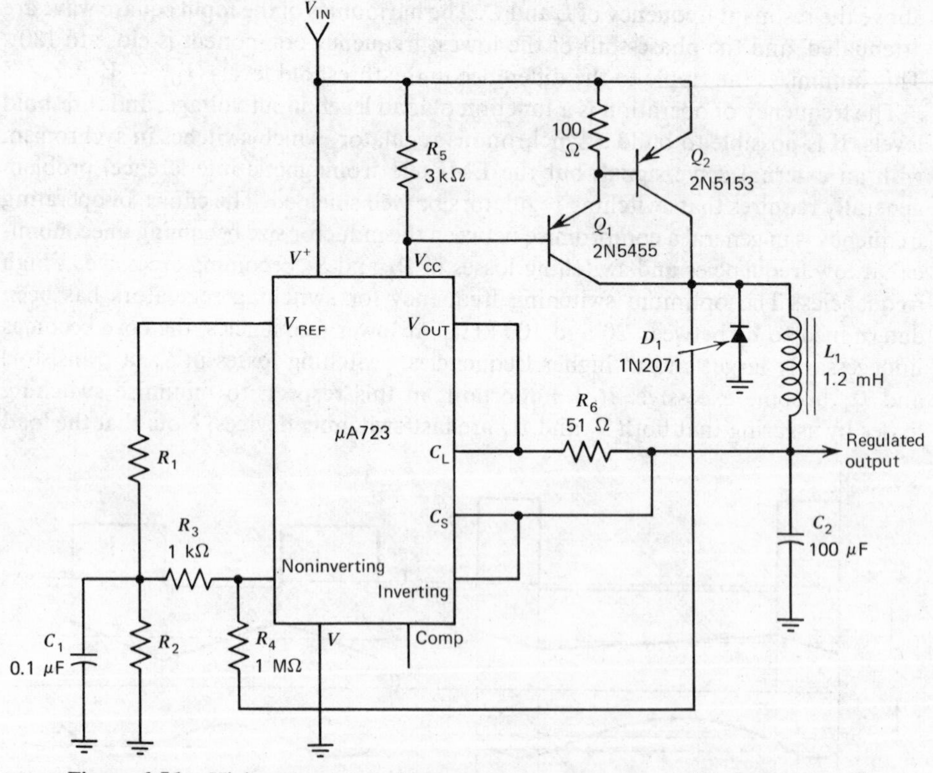

Figure 6-56. High current switching regulator using the μA723 [*note:* typical performance—regulated output voltage +5 V, line regulation ($\Delta V_{IN} = 30$ V) 10 mV, load regulation ($\Delta I_L = 2$ A) 80 mV].

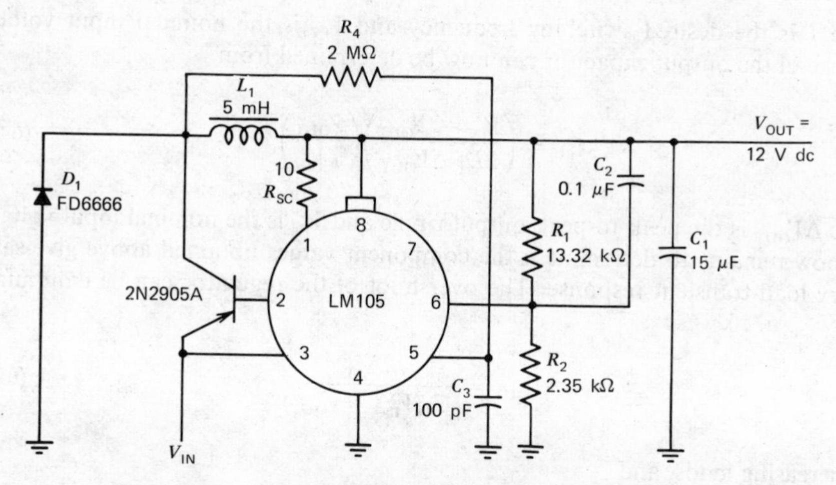

Figure 6-57. A 500-mA switching regulator using the LM105.

The output ripple of the regulator at the switching frequency is mainly determined by R_4. It should be evident from the description of circuit operation that the peak-to-peak output ripple will be nearly equal to the peak-to-peak voltage fed back to *pin* 5 of the LM105. Since the resistance looking into *pin* 5 is approximately 1000 Ω, this voltage will be

$$\Delta V_{REF} \simeq \frac{1000 V_{IN}}{R_4} \tag{6-20}$$

In practice, the ripple will be somewhat larger than this. When the switch transistor shuts OFF, the current in the inductor will be greater than the load current; thus the output voltage will continue to rise above the value required to shut OFF the regulator.

The values of the components of Figures 6-56 and 6-57 are determined from the following equations and discussions: An important consideration in choosing the value of the inductor is that it be large enough so that the current through it does not change drastically during the switching cycle. If it does, the switch transistor and catch diode must be able to handle peak currents which are significantly larger than the load current. The change in inductor current can be written as

$$\Delta I_L \simeq \frac{V_{OUT} t_{OFF}}{L} \tag{6-21}$$

In order for the peak current to be about 1.2 times the maximum load current, it is necessary that

$$L_1 = \frac{2.5 V_{OUT} t_{OFF}}{I_{OUT(MAX)}} \tag{6-22}$$

A value for t_{OFF} can be estimated from

$$t_{OFF} = \frac{1}{f}\left(1 - \frac{V_{OUT}}{V_{IN}}\right) \tag{6-23}$$

where f is the desired switching frequency and V_{IN} is the nominal input voltage. The size of the output capacitor can now be determined from

$$C_1 = \left(\frac{V_{IN} - V_{OUT}}{2L_1\, \Delta V_{OUT}}\right)\left(\frac{V_{OUT}}{f\, V_{IN}}\right)^2 \tag{6-24}$$

where ΔV_{OUT} is the peak-to-peak output ripple and V_{IN} is the nominal input voltage.

It now remains to determine if the component values obtained above give satisfactory load-transient response. The overshoot of the regulator can be determined from

$$\Delta V_{OUT} = \frac{L_1(\Delta I_L)^2}{C_1(V_{IN} - V_{OUT})} \tag{6-25}$$

for increasing loads, and

$$\Delta V_{OUT} = \frac{L_1(\Delta I_L)^2}{C_1 V_{OUT}} \tag{6-26}$$

for decreasing loads, where ΔI_L is the load-current transient. The recovery time is

$$t_r = \frac{2L_1\, \Delta I_L}{V_{IN} - V_{OUT}} \tag{6-27}$$

$$t_r = \frac{2L_1\, \Delta I_L}{V_{OUT}} \tag{6-28}$$

for increasing and decreasing loads, respectively.

In order to improve the load transient response, it is necessary to allow larger peak to average current ratios in the switch transistor and catch diode. Reducing the value of inductance given by (6-22) by a factor of 2 will reduce the overshoot by four times and halve the response time. This, of course, assumes that the output capacitance is doubled to maintain a constant switching frequency.

The above equations outline a design procedure for determining the value for R_4, L_1, and C_1, given the switching frequency and the output ripple. These equations are not exact, but they do provide a starting point for designing a regulator to fit a given application.

As an example, this design method will be applied to a LM105 regulator which must deliver 15 V at a maximum current of 300 mA from a 28-V supply. To start, a 40-kHz switching frequency will be selected along with an output ripple of 14 mV, peak-to-peak.

Resistor R_4 is calculated to be 2 MΩ. In determining L_1, t_{OFF} is found to be 11.6 μsec from (6-23). Inserting this into (6-22) gives a value of 1.45 μH for L_1. The value of C_1 obtained from (6-24) is then 57.5 μF.

In the actual circuit of Figure 6-58 a standard value of 47 μF is used for C_1; and L_1 is adjusted to 1.7 μH. The switching frequency obtained experimentally on this circuit is 60 kHz and the peak-to-peak output ripple is 20 mV. The fairly large disagreement between the calculated and experimental values is not alarming since

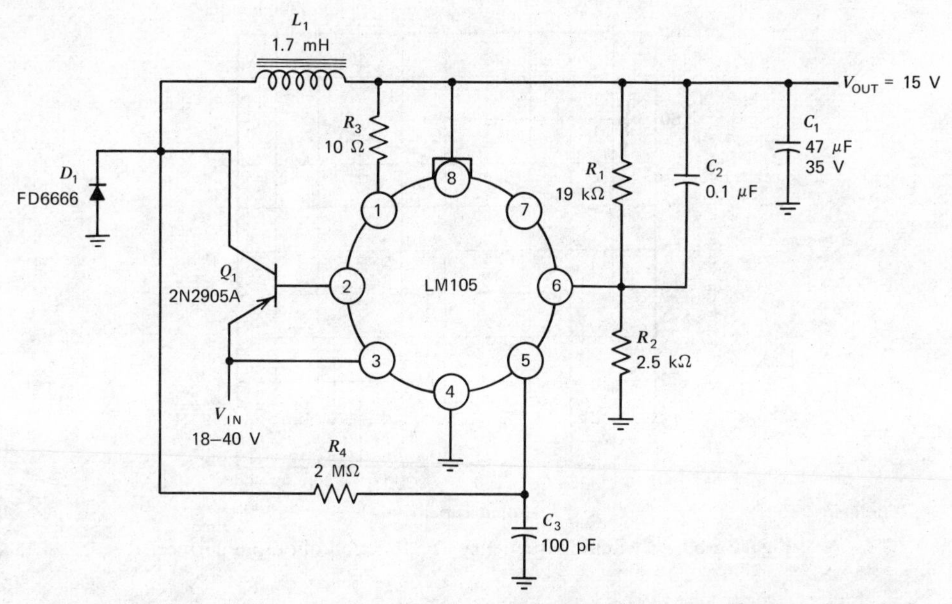

Figure 6-58. Switching regulator using the LM105 [*note:* * biasing diagram is top view; 125 turns #22 on Arnold Engineering A262123-2 molybdenum permalloy core].

many simplifying assumptions were made in the derivation of the equations. They do, however, provide a convenient method of handling a large number of mutually dependent variables to arrive at a working circuit. More exact expressions would involve a design procedure which is too cumbersome to be of practical value.

The variation of switching frequency with input voltage and load current is shown in Figures 6-59 and 6-60. The sharp rise in frequency at low output currents happens because the output transistor of the LM105 begins to supply an appreciable portion of the load current directly.

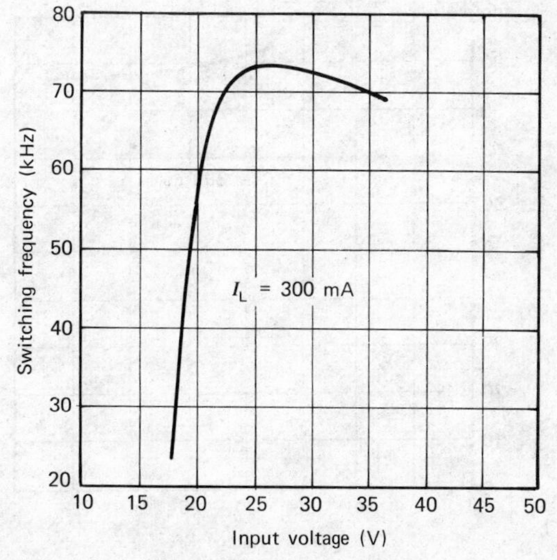

Figure 6-59. Switching frequency as a function of input voltage.

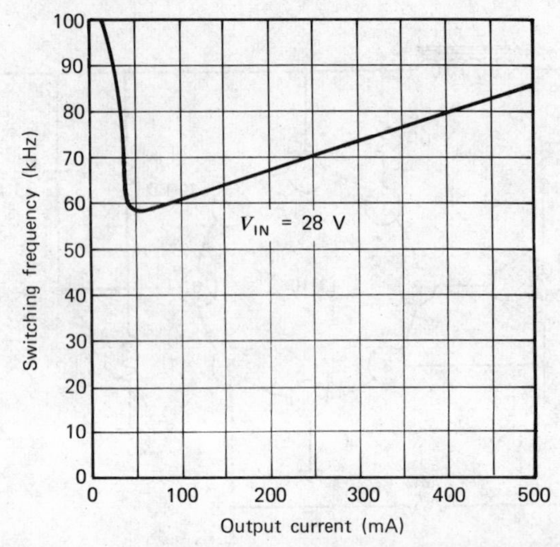

Figure 6-60. Switching frequency as a function of output current.

The efficiency of the regulator over a wide range of input voltages and output currents is given in Figures 6-61 and 6-62. In applications where the frequency of operation is an important parameter (e.g., to avoid EMI problems or for synchronizing of several regulators), the regulator may be locked to an external signal by means of the circuit shown in Figure 6-63. The incoming signal is converted to a triangular wave by R_3 and C_3 and added to the reference voltage. If the drive signal is already a triangular wave, C_3 can be omitted and R_3 increased to about 100 kΩ; but for square-wave drive of 5-V amplitude, 25 kΩ is correct.

The amplitude of the triangular wave at the reference *pin* must exceed the amplitude of the ripple on V_{OUT}; 50 mV is suitable for normal use. The switching control waveforms are sketched in Figure 6-64. The duty cycle of the switching pulse is controlled by the values of V_{OUT} and V_{REF}, thus ensuring the required regulation.

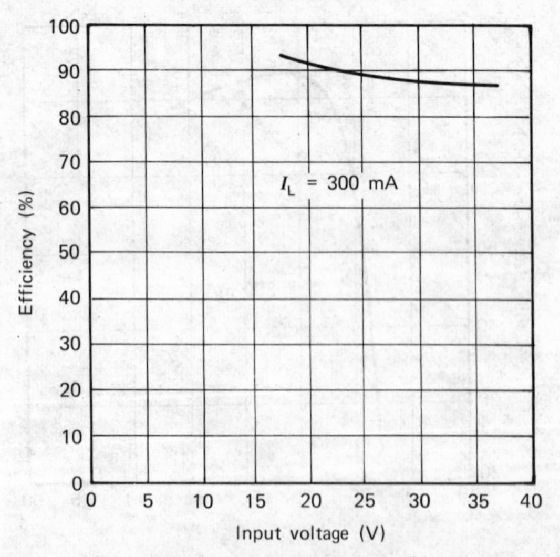

Figure 6-61. Efficiency as a function of input voltage.

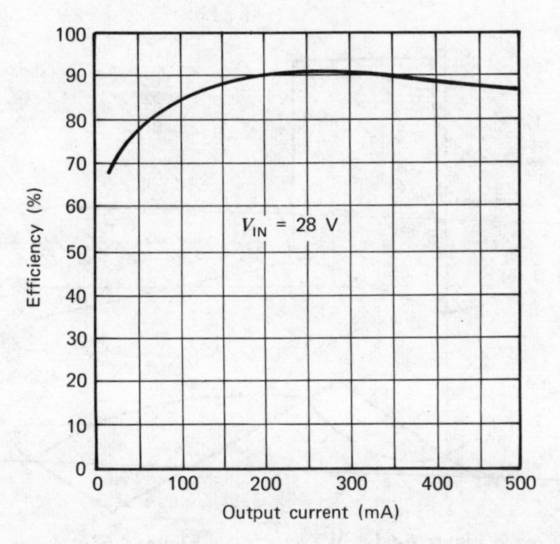

Figure 6-62. Efficiency as a function of output current.

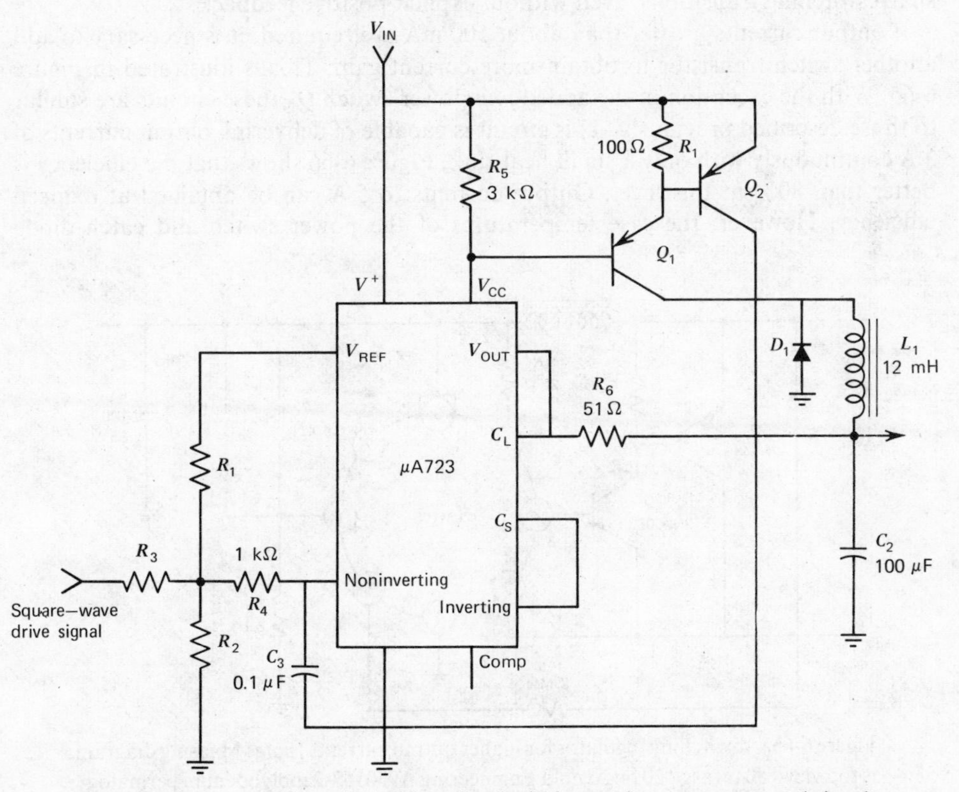

Figure 6-63. Connection of μA723 used to lock the regulator to an external signal.

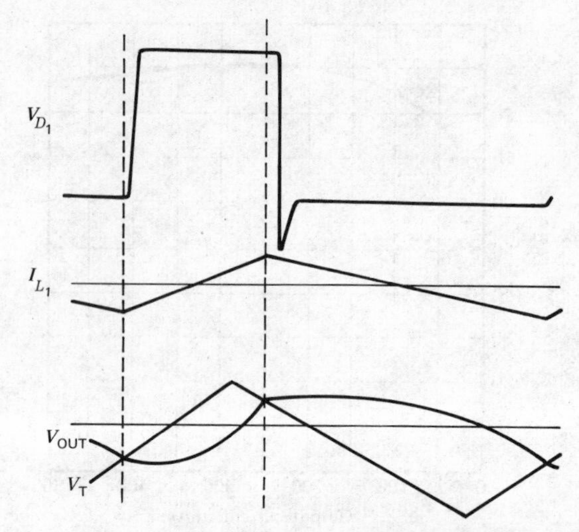

Figure 6-64. Waveforms of Figure 4-63.

Note that neither switching point is controlled in phase relative to the controlling waveform. Thus, spikes from other controlled regulators could cause false switching, if inadequately shielded from each other. Clearly, the line regulation is closely related to the amplitude of the triangular wave. The gain of the amplifier is adequate to give sharp switching transitions, even without explicit positive feedback.

If output currents greater than about 500 mA are required, it is necessary to add another switch transistor to obtain more current gain. This is illustrated in Figure 6-65. With the exception of the added *pnp* power switch Q_2 these circuits are similar to those described previously. This circuit is capable of delivering output currents of 3 A continuously with only a small heat sink. Figure 6-66 shows that the efficiency is better than 80% at this level. Output currents to 5 A can be obtained at reduced efficiency. However, the case temperatures of the power switch and catch diode

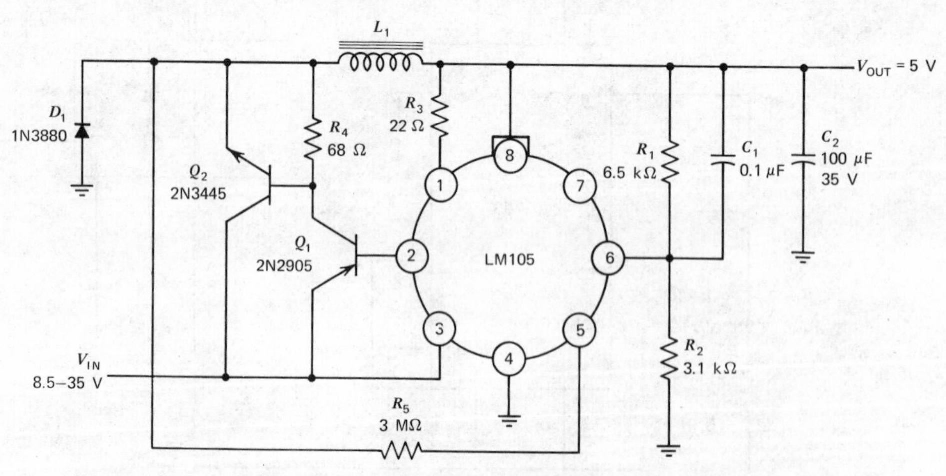

Figure 6-65. Switching regulator for higher output currents [*note:* * biasing diagram is top view; 50 turns #20 on Arnold Engineering A930157-2 molybdenum permalogy core].

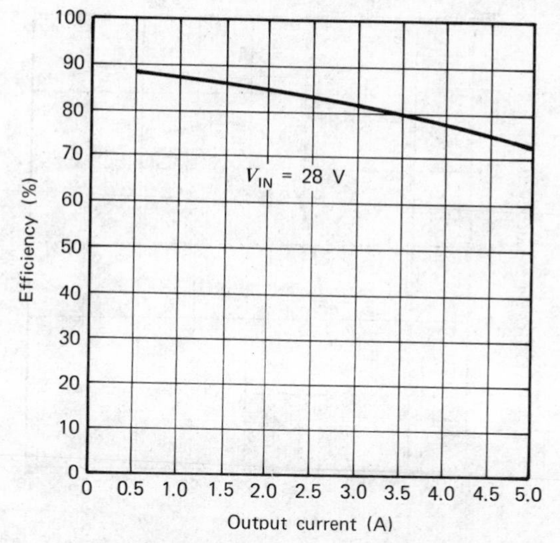

Figure 6-66. Switching regulator efficiency as a function of output current.

approach 100°C under this condition, so continuous operation is not recommended unless more heat sinking is provided.

Figure 6-67 shows that the efficiency is not significantly affected by input voltage. In Figure 6-68 it can be seen that the switching frequency is fairly constant over a wide range of input voltages. Figure 6-69 shows that the switching frequency increases with increasing load current. The higher dc current through the inductor reduces the incremental inductance causing the frequency to go up. Figure 6-70, illustrates the line regulation of the device. This can be improved by putting a small capacitor (0.01 μF) in series with the positive-feedback resistor R_3 to isolate the reference terminal from the dc input-voltage changes.

At low-output currents the inductor current can drop to zero at some time after the switch transistor turns OFF. When this happens, ringing occurs on the switching waveform. This is perfectly normal and causes no ill effects.

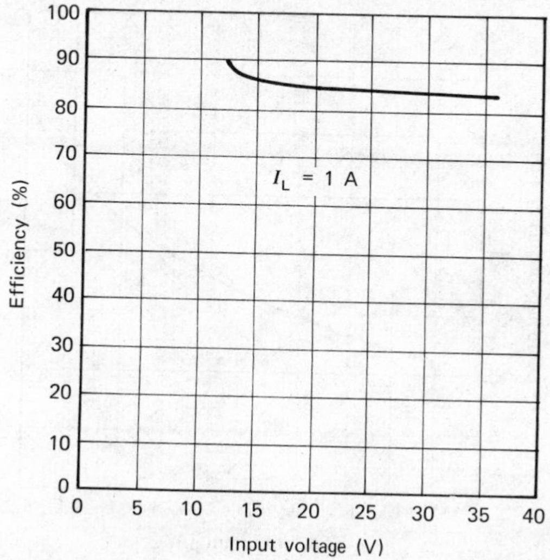

Figure 6-67. Switching regulator efficiency as a function of input voltage.

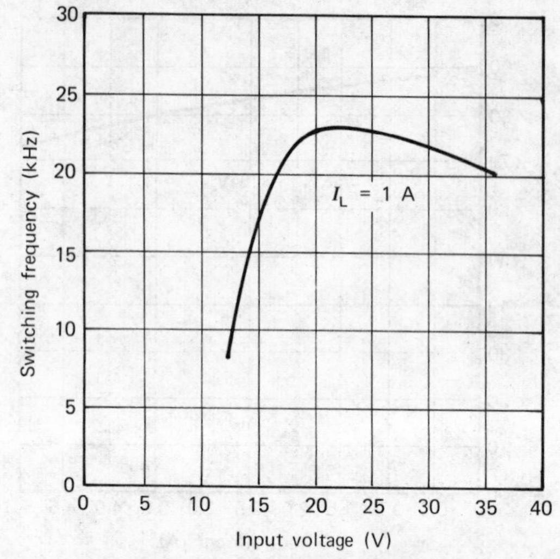

Figure 6-68. Variation of switching frequency with input voltage.

The use of solid tantalum capacitors for C_1 and C_3 is recommended when the regulator is expected to perform over the full military temperature range. The reason for using 35-V capacitors on the output, even though the output voltage is only 10-V, is that the 40-mV peak-to-peak ripple on the output would, for example, exceed the ratings of a 100-μF, 15-V capacitor.

Aluminum electrolytic capacitors have been used successfully over a limited temperature range. And there is basically no reason why wet foil or wet slug tantalums could not be used as long as their equivalent series resistance is low enough so that they behave like capacitors with the high-frequency switched-current waveform. It is also important that manufacturer's data be consulted to ensure that they can withstand the high-frequency ripple.

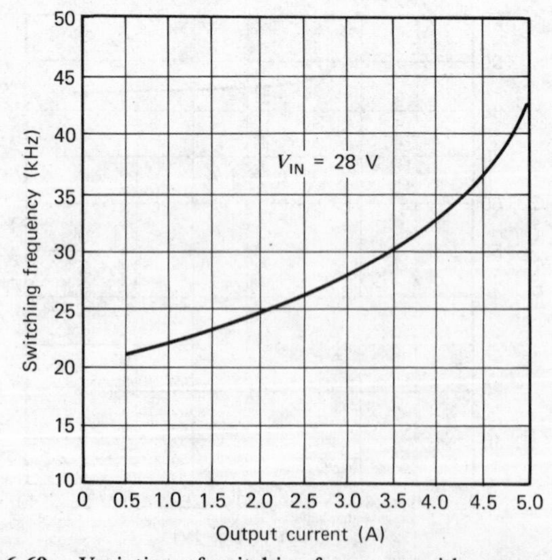

Figure 6-69. Variation of switching frequency with output current.

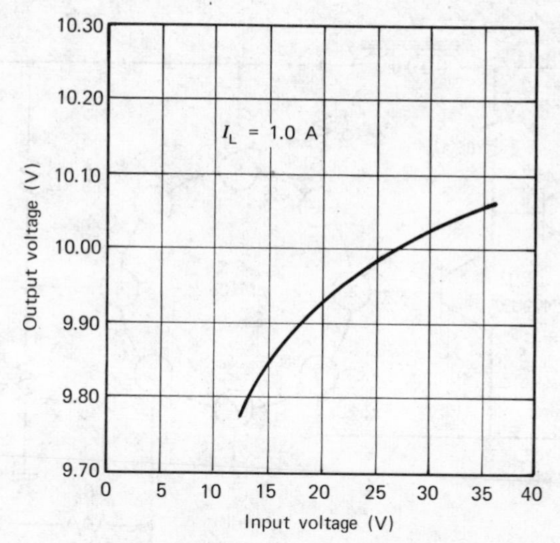

Figure 6-70. Line regulation.

As was mentioned with the low-current regulators, it is necessary to use fast-switching diodes and transistors in these circuits. Ordinary silicon rectifiers or low-frequency power transistors will operate at drastically reduced efficiencies and will quickly overheat in these circuits.

DRIVEN SWITCHING REGULATOR

When a number of switching regulators are used together in a system it is sometimes desirable to synchronize their operation to more uniformly distribute the switched-current waveforms on the input line. Synchronous operation is also wanted when a switching regulator is operated in conjunction with a power converter.

A circuit for synchronizing the switching regulator with a square-wave drive signal is shown in Figure 6-71. In this circuit, positive feedback is not used. Instead, the square-wave drive signal is integrated; and the resulting triangular wave (about 40 mV peak-to-peak) is applied to the reference bypass terminal of the LM105. This triangular wave will cause the regulator to switch since its gain is so high that the waveform overdrives it. The duty cycle of the switched waveform is controlled by the voltage on the feedback terminal *Pin* 6.

If this voltage goes up, the duty cycle will decrease since it is picking off a smaller portion of the triangular wave on *Pin* 5. By the same token, the duty cycle will decrease if the voltage on *Pin* 6 drops.

This action produces the desired regulation: If the output voltage starts to go up, it will raise the voltage on *Pin* 6 such that smaller portion of the triangular wave is picked off. This reduces the duty cycle, counteracting the output voltage increase.

In order for this circuit to work properly, the ripple voltage on *Pin* 6 should be less than a quarter of the peak-to-peak amplitude of the triangular wave. If this condition is not satisfied, the regulator will try to oscillate at its own frequency. Further, since the resistance looking into *Pin* 5 is about 1 kΩ, the integrating capacitor, C_3 should have a capacitive reactance of less than 100 Ω at the drive frequency. The value of R_3 is determined so that the amplitude of the triangular wave on *Pin* 5 is about 40 mV.

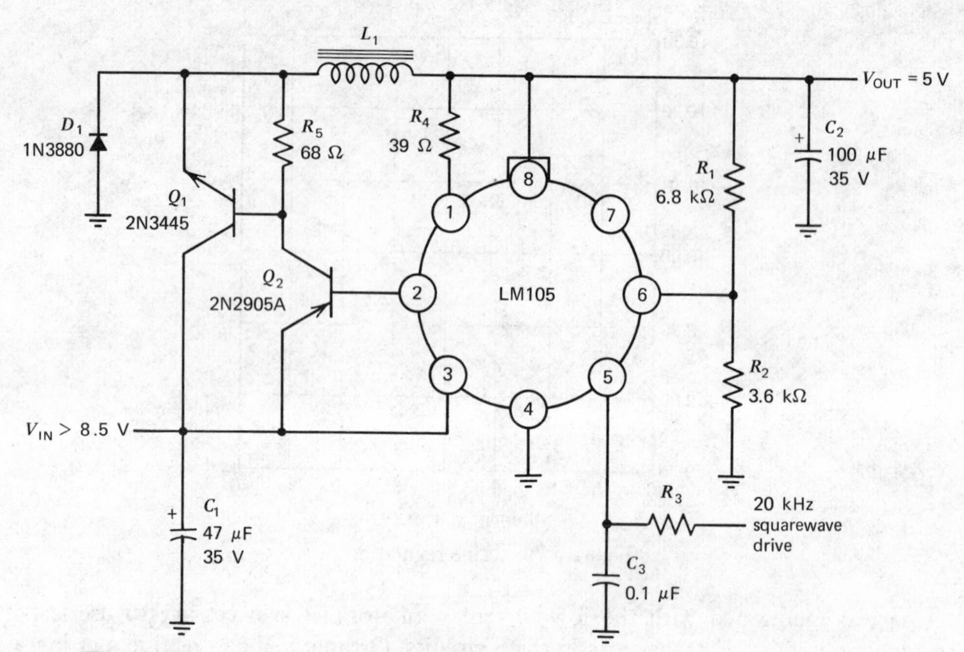

Figure 6-71. Driven switching regulator [*note:* * biasing diagram is top view; † 100 turns #22 on Arnold Engineering A930157-2 molybdenum permalloy core].

Driven regulators also have other advantages. For one, it is possible to design the *LC* filter independent of switching frequency considerations. Hence, lower output ripple and better transient response can be realized. A second advantage is the frequency stability. In a self-oscillating regulator, the switching frequency is controlled by a relatively large number of factors. As a result, it is not well determined when normal tolerances are taken into account. With low- and medium-power regulators, this is not usually a problem since the efficiency does not vary greatly with frequency. However, high-power regulators tend to be more frequency-sensitive and it is desirable to operate them at constant frequency.

CURRENT-LIMITING

In the circuits described previously, the regulator is not protected from overloads or a short-circuited output. Providing short-circuit protection is no simple problem, since it is necessary to keep the regulator switching when the output is shorted. Otherwise, the dissipation will become excessive even though the current is limited.

A circuit that does this is shown in Figure 6-72. The peak current through the switch transistor is sensed by R_8. When the voltage drop across this resistor becomes large enough to turn ON Q_3, the output voltage begins to fall since current is being supplied to the feedback terminal of the regulator from the collector of Q_3 so less has to be supplied from the output through R_1. Furthermore, the circuit will continue to oscillate, even with a shorted output, because of positive feedback through R_6 and the relatively long discharge time constant of C_2.

It is necessary to put a resistor R_7 in series with the base of Q_3 to ensure that excessive current will not be driven into the base. In addition, a capacitor C_4 must be added across the input of Q_3 so that it does not turn ON prematurely on the large

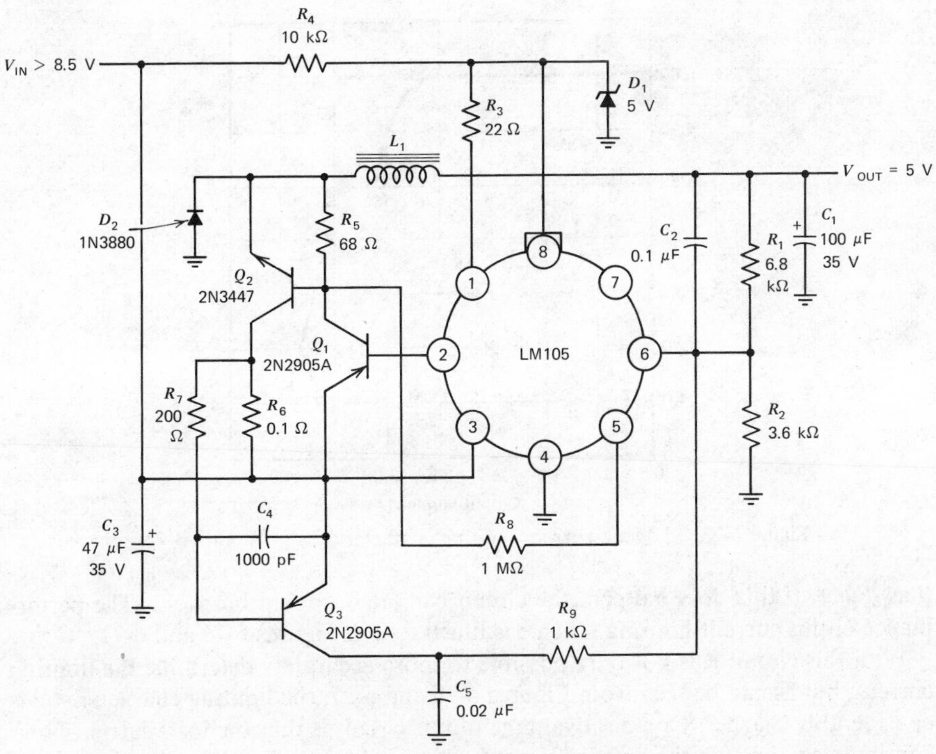

Figure 6-72. LM105 switching regulator with current-limiting [*note:* * biasing diagram is top view; 70 turns #20 on Arnold Engineering A930157-2 molybdenum permalloy core].

current spike (about twice the load current) through the switch transistor caused by pulling the stored charge out of the catch diode. A Zener diode bias supply must also be used on the output of the LM105 since the current-limiting will not work if the voltage on this point drops below about 1 V. The current-limiting characteristics of this circuit are shown in Figure 6-73. Figure 6-74 shows how the average input current actually drops off as the circuit goes into current-limiting.

This current-limiting scheme protects the switching transistors from overload or short-circuited output. However, the dropout current and short-circuit current are not well controlled, so it is difficult to prove that the circuit will sustain a continuous short circuit under worst-case conditions. This is particularly true with high-current regulators where the required amount of overdesign can become quite expensive.

Figure 6-75 shows a circuit which is more easily designed for continuous short-circuit protection under worst-case conditions. In this circuit, the current-sensing resistor is located in series with the inductor. Therefore, the peak-limiting current can be more precisely determined since the current spike generated by pulling the stored charge out of the catch diode does not flow through the sense resistor.

Operation of this circuit is essentially the same as the previous one in that an *npn* transistor Q_4 senses the overcurrent condition and turns ON Q_3 which supplies the current-limit signal to the feedback terminal. The Zener diode D_3 is required on the feedback terminal to guarantee that this terminal cannot go more than 0.5 V higher

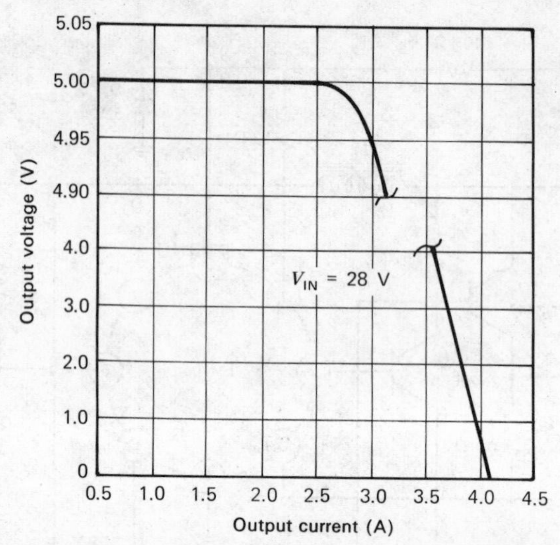

Figure 6-73. LM105 current-limiting characteristics for Figure 6-72.

than *Pin* 1. If this does happen, the circuit can latch up and burn out. The performance of this current-limiting scheme is illustrated in Figures 6-76 and 6-77.

With this circuit it is not only possible to more accurately determine the limiting current, but as can be seen from Figures 6-76 and 6-77, the limiting characteristic is considerably sharper. One disadvantage of this circuit is that the load current flows continuously through the current-sense resistor, reducing efficiency. As an example, with a 5-V regulated output the efficiency will be reduced by 10% at full load.

Figure 6-78 depicts a switching regulator using the μA723 that is, immune to overloads and short circuits. This circuit is designed for a nominal load current of 0.5 A, but it can deliver from 150 mA to 1 A maximum—at which point it current-limits. Input voltages can range from 9.5 to 40 V.

The output current is limited by Q_2, which senses the voltage across R_7. The

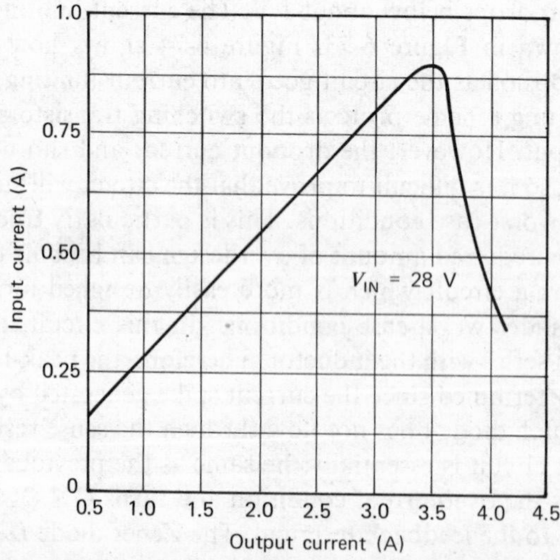

Figure 6-74. Input current drop as regulator goes into limiting for Figure 6-72.

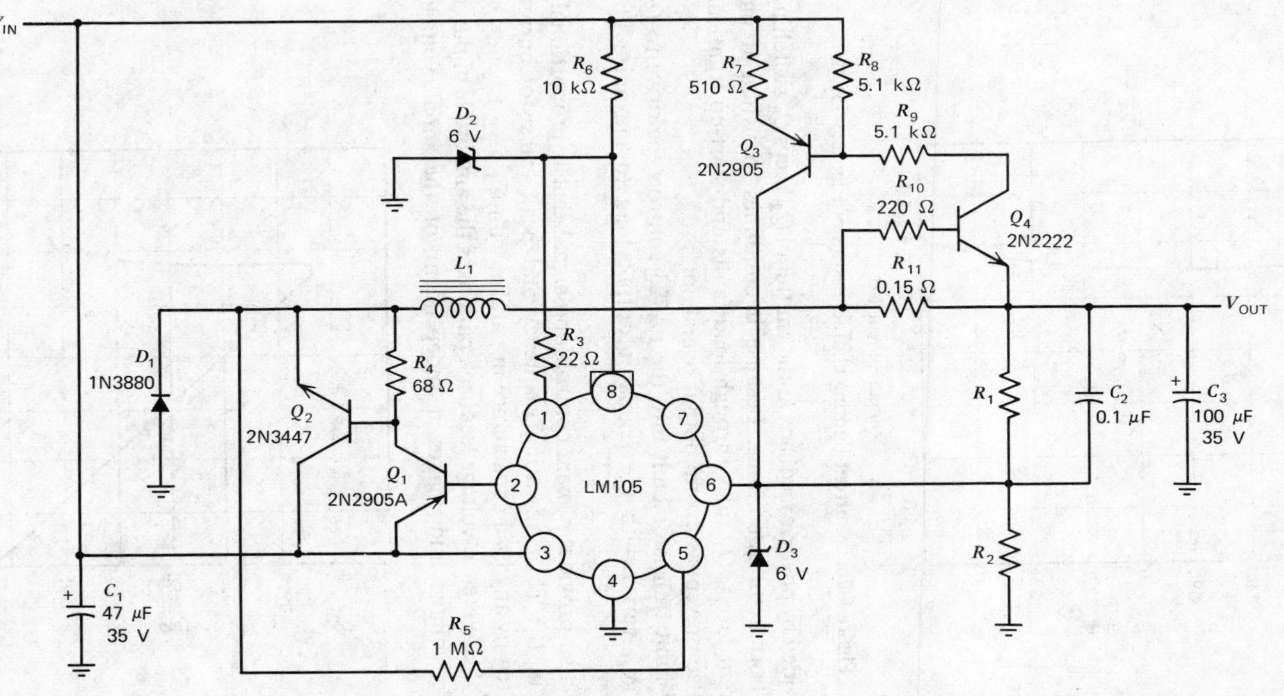

Figure 6-75. Switching regulator with continuous short-circuit protection [*note:* * biasing diagram is top view; 70 turns #20 on Arnold Engineering A930157-2 molybdenum permalloy core].

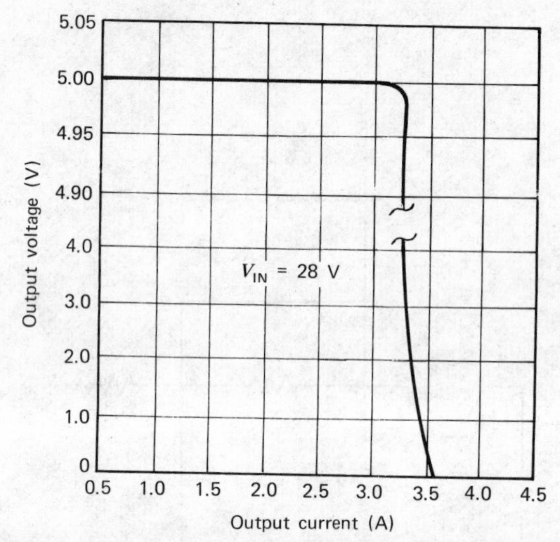

Figure 6-76. Current-limiting characteristics of Figure 6-75.

hysteresis provided by positive feedback at the emitter of Q_2 keeps Q_1 switching under overload or shortcircuit conditions. Because it continues to switch at moderate rates and current levels, Q_1 rides through load faults and start-up transients. The current limit also prevents start-up output overshoot.

The current-limit transistor built into the μA723 controls the drive to Q_1. This drive, set by R_7, varies with the regulator output load to maximize operating efficiency.

Resistor R_7 also provides a main control-loop signal that is proportional to the current ramp in L_1. Thus operating frequency and peak transistor currents are almost unaffected by any load capacitance in parallel with C_3.

Diode D_2 in the reference-voltage feedback path fixes the amplitude of the control-loop hysteresis voltage and makes it independent of the input voltage. The

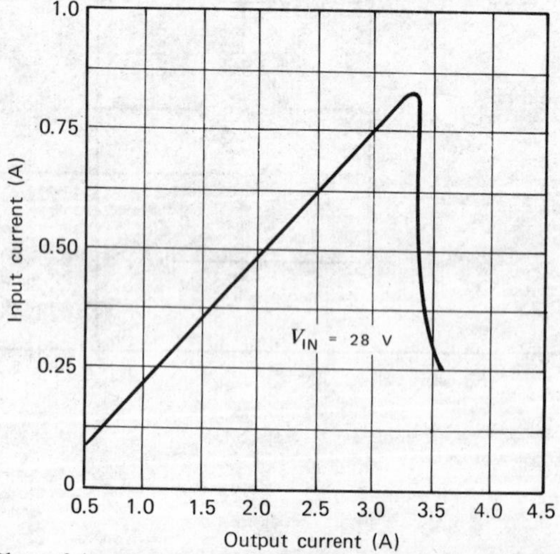

Figure 6-77. Plot of input current as regulator goes into current-limiting for Figure 6-75.

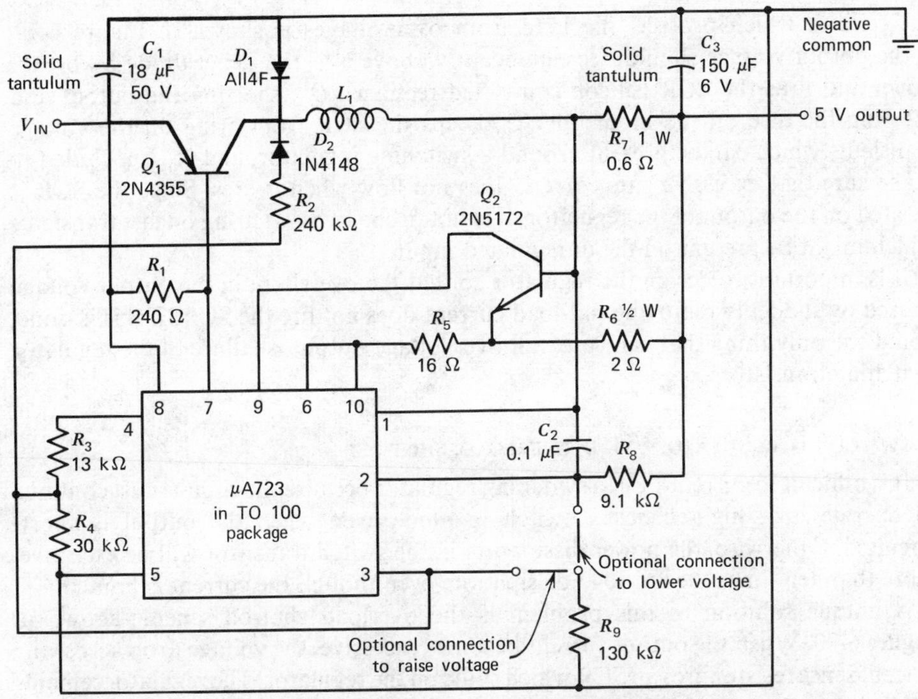

Figure 6-78. Switched regulator with current-limiting can deliver half an amp using a low-cost plastic-packaged output transistor. Output voltage can be adjusted by hand-selecting resistor R_9 during final testing [*note:* * keep heavy lines close together to minimize magnetic interference; ** L_1 is 170 turns of #26 wire on mag-inc. 55204-A2 moly-permalloy core].

combination of this fixed hysteresis with the inductor current feedback through C_2 stabilizes the operating frequency. As a result, the frequency changes by less than 2 to 1, while the input voltage varies by more than 4 to 1.

At any given input voltage, the operating frequency is an inverse function of the inductance of L_1. The inductor may be modified to vary the operating frequency if sufficient current-carrying capacity is provided. The unit described here operates between 4 and 8 kHz. The inductor begins to saturate above 500-mA output current; therefore the frequency will increase for heavy loading. Saturation is gradual, so the current-limit circuit retains control. Winding the inductor with fewer turns of larger gauge wire will result in better high-current performance, but at the expense of higher operating frequency and reduced efficiency at lower currents.

The resistor values shown provide a nominal output of 5 V. Resistor tolerances and individual differences in μA723 ICs can cause an unacceptable regulated voltage. The optional resistor R_9 can be connected to raise or lower the output voltage, as required.

OVERVOLTAGE PROTECTION

One objection brought up against switching regulators is that they can fail with the output voltage going up to the unregulated input voltage which is frequently several times the regulated output voltage. This can destroy the equipment that the regulator is supplying.

A circuit which protects the load from overvoltages is shown in Figure 6-79. If the output voltage should rise significantly above 6 V, the Zener diode D_2 breaks down and fires the SCR (silicon-controlled rectifier), Q_3, shorting the output and blowing the fuse on the input line; C_3 keeps the SCR from firing on the voltage transients which can be present around a switching regulator, and R_7 is included to make sure that excessive gate current does not flow when it fires. Since the SCR is located on the output of the regulator, it is not prone to dV/dt firing on fast transients which might be present on the unregulated input.

It is important to design the regulator so that the overshoot in the output voltage caused by suddenly removing full-load current does not fire the SCR. If this is done, about the only thing that can cause an overvoltage output is failure of the regulator switching transistors.

SWITCHING REGULATOR WITH OVERLOAD SHUT-OFF

It is difficult to current-limit a switching regulator because the circuit must continue to operate in a high-efficiency switching mode even when the output is short-circuited. Otherwise, the power dissipation in the switch transistor will be excessive, more than ten times the full-load dissipation, even though the current is limited.

A unique solution to this problem is the overload shut-off scheme shown in Figure 6-80. When the output current becomes excessive, the voltage drop across the current sense resistor fires an SCR which shuts off the regulator. The regulator remains OFF, dissipating practically no power, until it is reset by removing the input voltage.

In the actual circuit, complementary transistors Q_1 and Q_2 replace the SCR since it is difficult to find devices with a low enough holding current (about 50 μA). When the voltage drop across R_4 rises to about 0.7 V, Q_2 turns ON, removing the base drive to the output transistors on the LM105 through *Pin* 7. Then Q_1 latches Q_2, holding the regulator OFF until the input voltage is removed. It will then start when power is applied if the overload has been removed.

CONSTANT-CURRENT SOURCE USING A SWITCHING REGULATOR

Figure 6-81 depicts a switching regulator circuit using the μA723 IC regulator that provides a 1 A constant-current output and has a peak-to-peak ripple of 28 mA. The μA723 functions as a reference source and a comparator. Resistors R_1 and R_2 reduce the internal reference from 7.15 to 3 V and drive the IC's noninverting input, while resistors R_3 and R_4 drive the IC's inverting input. When the IC's comparator terminals are balanced, approximately 1 V appears across shunt resistor R_5.

A hysteresis voltage of approximately 28 mV is applied to the IC's noninverting input via R_6. This sets the circuit's minimum output ripple at 28 mA *p-p*. However, this ripple current will be greater if the storage time of output transistor Q_1 is significant.

Transistor Q_1 is a current booster, while inductor L_1 and capacitor C_1 filter the switched waveform when the circuit's feedback loop requires a current increase; the output stage of the IC regulator conducts and a pulse current of approximately 12 mA flows to V_C. The size of the current pulse is determined by R_7. This current pulse drives Q_1. Diode D_1 biases the output stage of the IC regulator, while D_2 operates as a free-wheeling diode. The circuit's maximum frequency is a function of the load, and is typically 20 kHz.

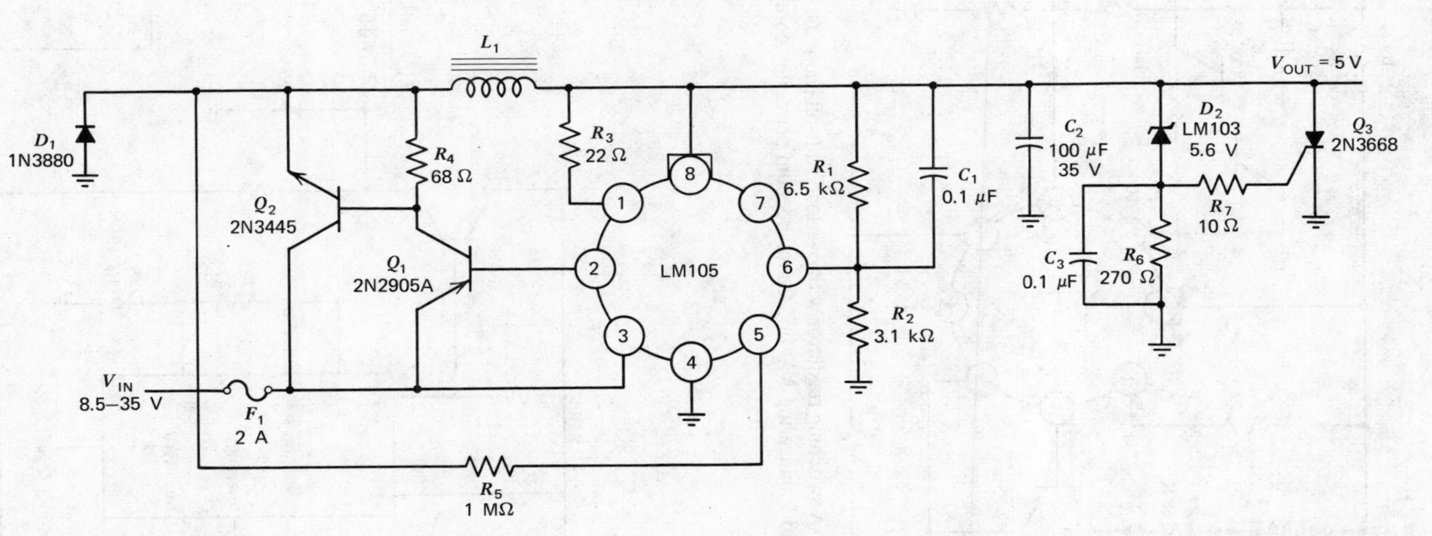

Figure 6-79. Switching regulator with crowbar over voltage protection [*note:* 60 turns #20 on Arnold Engineering A930157-2 molybdenum permalloy core].

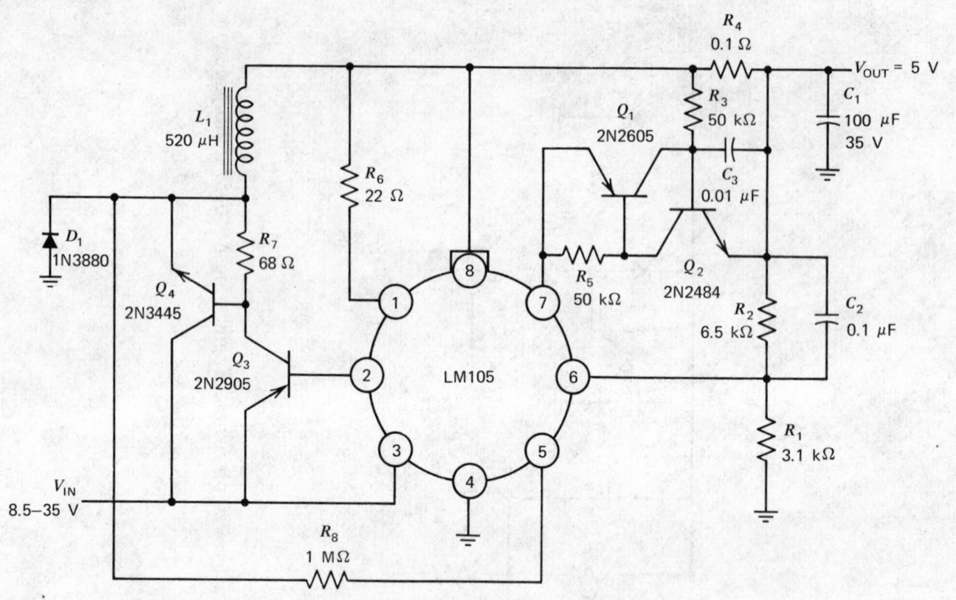

Figure 6-80. A 3-A switching regulator with overload shut-off [*note:* 60 turns #20 on Arnold Engineering A930157-2 molybdenum permalloy core].

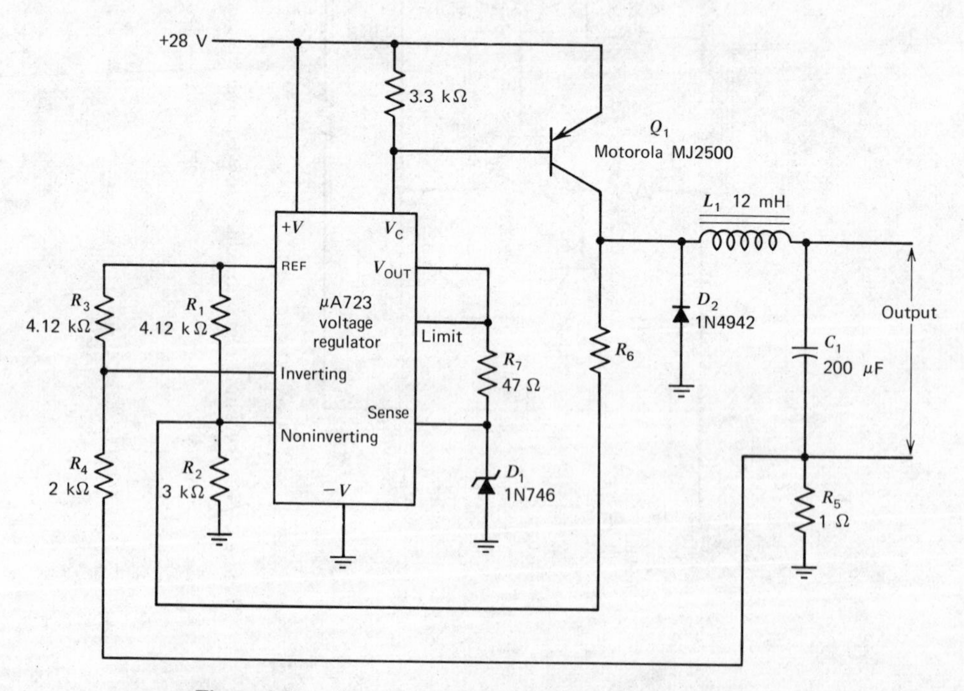

Figure 6-81. Constant-current source switching regulator.

NEGATIVE-SWITCHING REGULATORS

A negative-switching regulator using the μA723 IC may be constructed as shown in Figure 6-82. The same basic restrictions apply here as in the simple negative regulator of Figure 6-8 (the output voltage must be at least 8.5 V unless a positive supply is available, as for Figure 6-9). The basic operation is the same as before.

Negative switching regulators built with the LM104 show efficiencies approaching 90% even when the regulated output voltage is only a fraction of the input voltage. With proper design, transient response and ripple can also be made quite acceptable.

The LM104 self-oscillating switching regulator in Figure 6-83 operates in much the same way as a linear regulator. The reference current is set up at 1 mA with R_1, and R_2 determines the output voltage in the normal fashion. The circuit is made to oscillate by applying positive feedback through R_5 to the noninverting input on the error amplifier of the LM104. When the output voltage is low, the internal-pass transistor of the IC turns ON and drives Q_1 into saturation. The current feedback through R_5 then increases the magnitude of the reference voltage developed across R_2; Q_1 will remain ON until the output voltage comes up to twice this reference voltage. At this point, the error amplifier goes into linear operation, and the positive feedback makes the circuit switch OFF. When this happens, the reference voltage is lowered by feedback through R_5, and the circuit will stay OFF until the output voltage drops to where the error amplifier again goes into linear operation. Hence, the circuit regulates with the output voltage oscillating about the nominal value with a peak-to-peak ripple of around 40 mV.

The power conversion from the input voltage to a lower output voltage is obtained by the action of the switch transistor Q_1, the catch diode D_1, and the LC filter. The

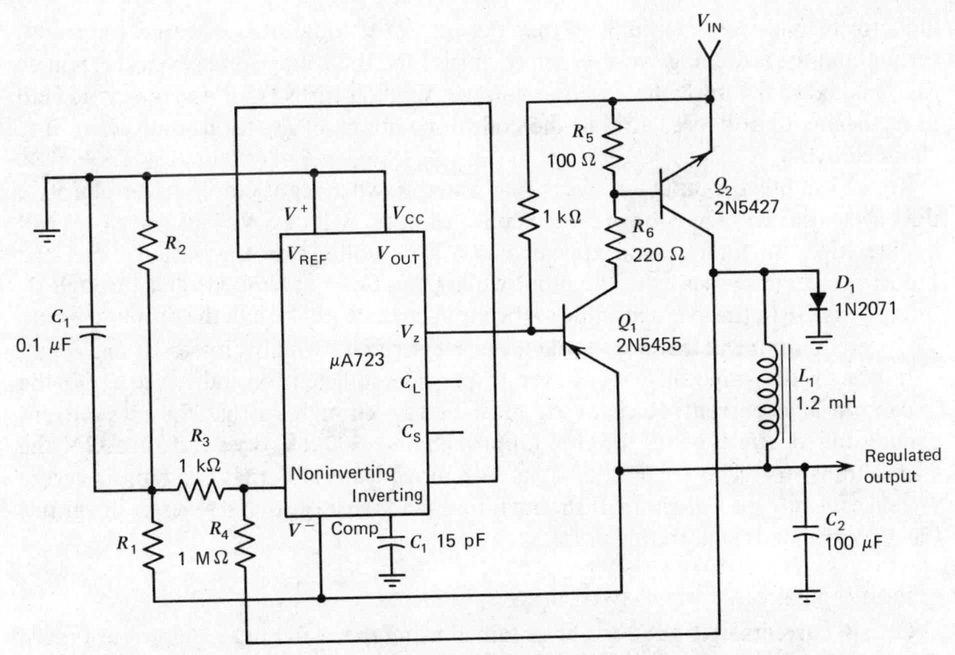

Figure 6-82. Negative-switching regulator using the μA723 [*note:* typical performance—regulated output voltage −15 V, line regulation (ΔV_{IN} = 20 V) 8 mV, load regulation (ΔI_L = 2 A) 6 mV].

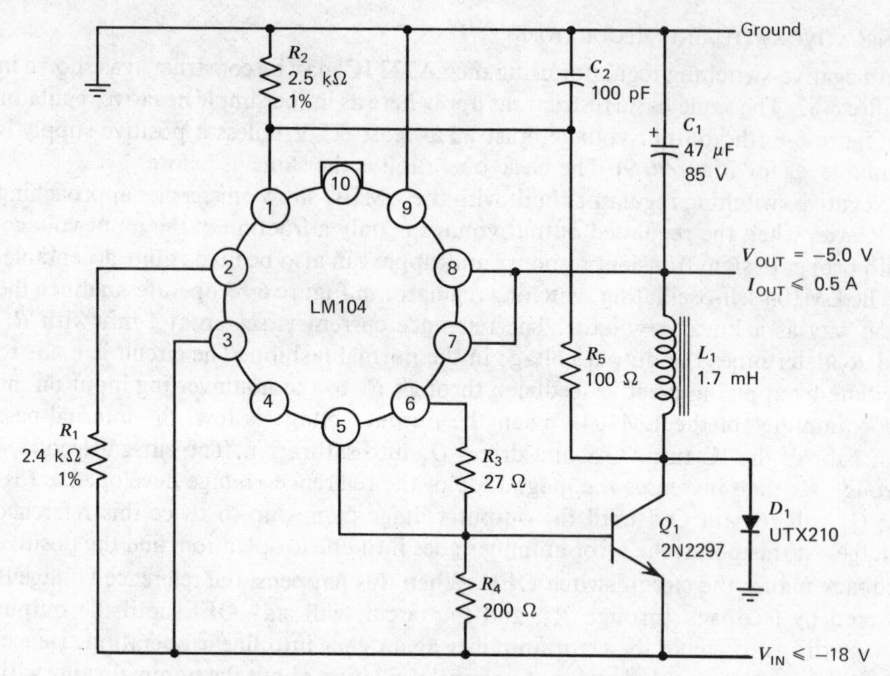

Figure 6-83. Negative-switching regulator using the LM104 [*note:* 125 turns #22 on Arnold Engineering A262123-2 molybdenum permalloy core].

inductor is made large enough so that the current through it is essentially constant throughout the switching cycle. When Q_1 turns ON, the voltage on its collector will be nearly equal to the unregulated input voltage. When it turns OFF, the magnetic field in L_1 begins to collapse, driving the collector voltage of Q_1 to ground where it is clamped by D_1.

If, for example, the input voltage is 10 V and the switch transistor is driven at a 50% duty cycle, the average voltage on the collector of Q_1 will be 5 V. This waveform will be filtered by L_1 and C_1 and appear as a 5 V dc voltage on the output. Since the inductor current comes from the input while Q_1 is ON but from ground through D_1 while Q_1 is OFF, the average value of the input current will be half the output current. The power output will therefore equal the input power if switching losses are neglected.

In design, the value of R_3 is chosen to provide sufficient base drive to Q_1 at the maximum load current. Resistor R_4 must be low enough so that the bias current coming out of *Pin* 5 of the LM104 (approximately 300 μA) does not turn ON the switch transistor. The purpose of C_2 is to remove transients that can appear across R_2 and cause erratic switching. It should not be made so large that it severely integrates the waveform fed back to this point.

HIGH-CURRENT NEGATIVE-SWITCHING REGULATORS

Output currents up to 3 A can be obtained using the switching regulator in Figure 6-84. The circuit is identical to the one described previously (Figure 6-83), except that Q_2 has been added to increase the output-current capability by about an order of magnitude. It should be noted that the reference-supply terminal is returned to the

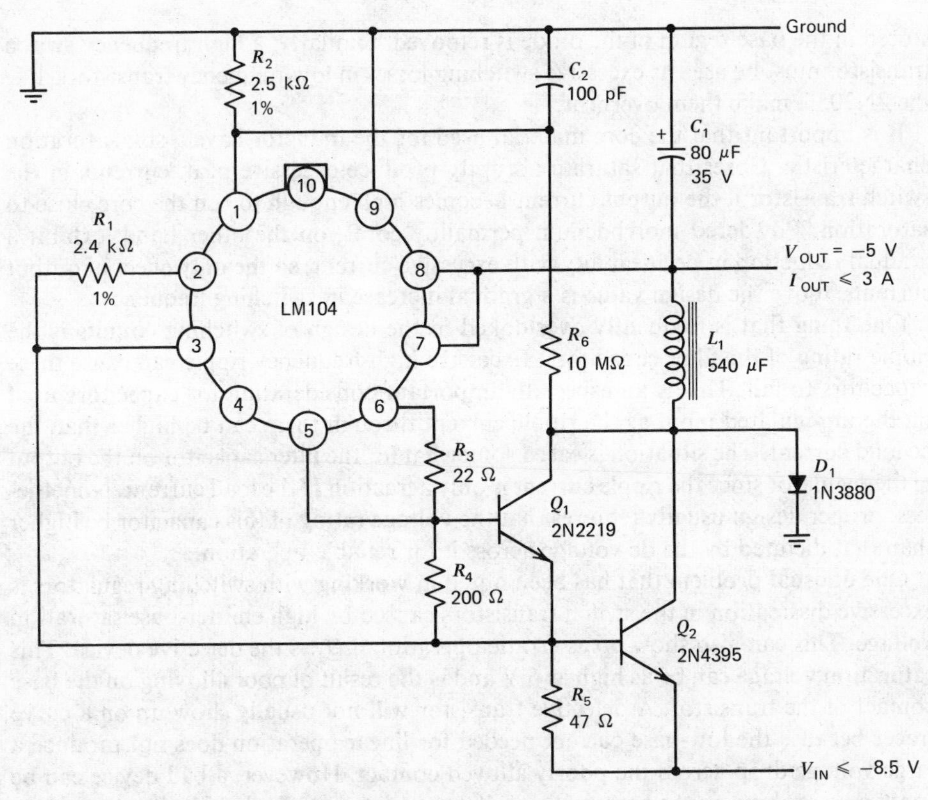

Figure 6-84. High-current switching regulator using the LM104 [*note:* 60 turns #20 on Arnold Engineering A930157-2 molybdenum permalloy core].

base of Q_2, rather than the unregulated input. This is done because the LM104 will not function properly if *Pin* 5 gets more than 2 V more positive than *Pin* 3. The reference current, as well as the bias currents for *Pins* 3 and 5, is supplied from the unregulated input through R_5, so its resistance must be low enough so that Q_2 is not turned ON with about 2 mA flowing through it.

The line regulation of this circuit is worsened somewhat by the unregulated input voltage being fed back into the reference for the regulator through R_6. This effect can be eliminated by connecting a 0.01-μF capacitor in series with R_6 to remove the dc component of the feedback.

There are a number of precautions that should be observed with all switching regulators, although they are more inclined to cause problems in high-current applications: For one, fast-switching diodes and transistors must be used. If D_1 is an ordinary junction rectifier, voltages in the order of 10-V can be developed across it in the forward direction when the switch transistor turns OFF. This happens because low-frequency rectifiers are usually manufactured with a *PIN* structure which presents a high forward impedance until enough minority carriers are injected into the diode-base region to increase its conductance.

This not only causes excessive dissipation in the diode, but the diode also presents a short circuit to the switch transistor when it first turns ON, until all the charge stored in the base region of the diode is removed. Similarly, a high-frequency switch transistor must be used as excessive switching losses in low-frequency transistors, like the 2N3055, make them overheat.

It is important that the core material used for the inductor have a soft saturation characteristic. Cores that saturate abruptly produce excessive peak currents in the switch transistor if the output current becomes high enough to run the core close to saturation. Powdered molybdenum permalloy cores, on the other hand, exhibit a gradual reduction in permeability with excessive current, so the only effect of output currents above the design value is a gradual increase in switching frequency.

One thing that is frequently overlooked in the design of switching circuits is the ripple rating of the filter capacitors. Excessive high-frequency ripple can cause these capacitors to fail. This is an especially important consideration for capacitors used on the unregulated input as the ripple current through them can be higher than the dc load current. The situation is eased somewhat for the filter capacitor on the output of the regulator since the ripple current is only a fraction of the load current. Nonetheless, proper design usually requires that the voltage rating of this capacitor be higher than that dictated by the dc voltage across it for reliable operation.

One unusual problem that has been noted in working with switching regulators is excessive dissipation in the switch transistors caused by high emitter-base saturation voltage. This can also show up as erratic operation if Q_1 is the defective device. This saturation voltage can be as high as 5 V and is the result of poor alloying on the base contact of the transistor. A defective transistor will not usually show up on a curve tracer because the low-base current needed for linear operation does not produce a large voltage drop across the poorly alloyed contact. However, a bad device can be spotted by probing on the bases of the switch transistors while the circuit is operating.

It is necessary that the catch diode D_1 and any bypass capacitance on the unregulated input be returned to ground separately from the other parts of the circuit. These components carry large current transients and can develop appreciable voltage transients across even a short length of wire. If C_1, C_2, or R_2 have any common ground impedance with the catch diode or the input-bypass capacitor, the transients can appear directly on the output.

NEGATIVE-SWITCHING REGULATOR WITH CURRENT-LIMITING

The LM104 switching regulator circuits described previously are not protected from overloads or a short-circuited output. The current-limiting of the LM104 is used to limit the base drive of the switch transistor, but this does not effectively protect the switch transistor from excessive current. A circuit which provides current-limiting and protects the regulator from short circuits is shown in Figure 6-85.

The current through the switch transistor produces a voltage drop across R_9. When this voltage becomes large enough to turn ON Q_3, current-limiting is initiated. This occurs because Q_3 takes over as the control transistor and regulates the voltage on *pin* 8 of the LM104. This point is the feedback terminal of the error amplifier and is separated from the actual output of the regulator by not shorting the regulated output and booster output terminals of the IC. Hence, with excessive output current, the circuit still operates as a switching regulator with Q_3 regulating the voltage fed back to the error amplifier as the output voltage falls off.

A resistor R_7 is included so that excessive base current will not be driven into the base of Q_3. Capacitor C_4 ensures that Q_3 does not turn ON from the current spikes

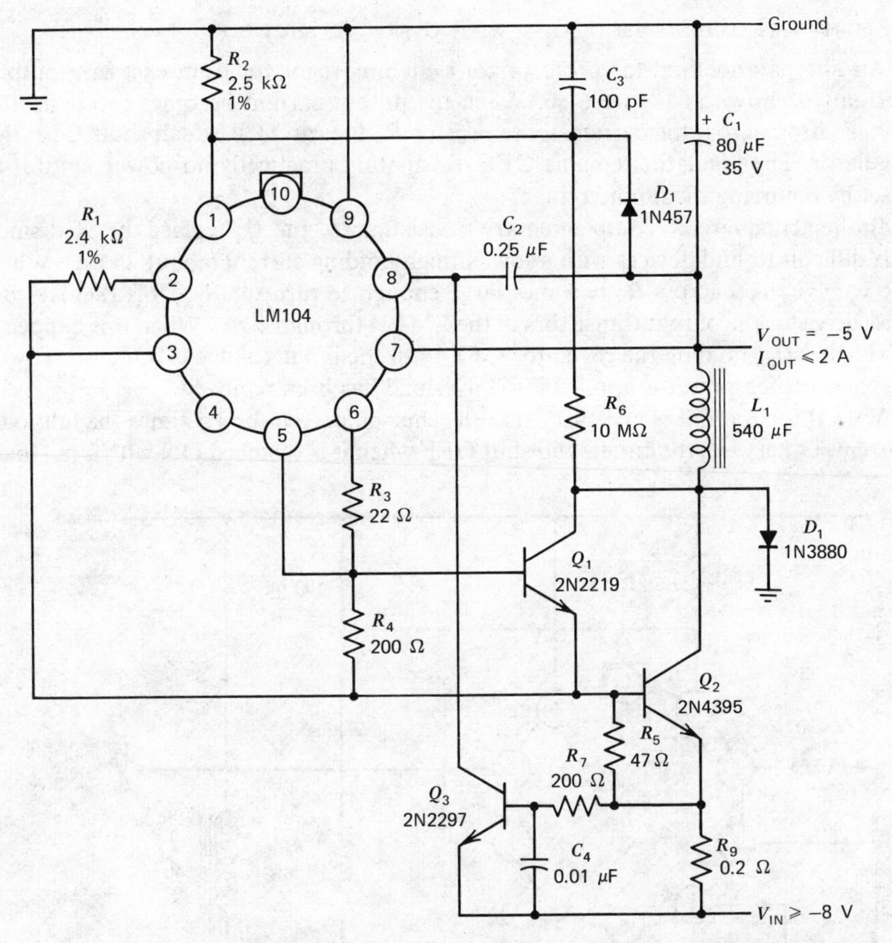

Figure 6-85. LM104 negative-switching regulator with current-limiting [*note:* 60 turns #20 on Arnold Engineering A930157-2 molybdenum permalloy core.

through the switch transistor caused by pulling the stored charge out of the catch diode (these are about twice the load current.) This capacitor also operates in conjunction with C_2 to produce sufficient phase delay in the feedback loop so that the circuit will oscillate in current-limiting. However, C_4 should not be made so large that it appreciably integrates the rectangular waveform of the current through the switch transistor.

As the output voltage falls below half the design value, D_1 pulls down the reference voltage across R_2. This permits the current-limiting circuitry to keep operating when the unregulated input voltage drops below the design value of output voltage, with a short on the output of the regulator.

A transistor with good high-current capability was chosen for Q_3 so that it does not suffer from secondary breakdown effects from the large peak currents (about 200 mA) through it. With a shorted output, these peak currents occur with the full-input voltage across Q_3. The average dissipation in Q_3 is, however, low.

NEGATIVE-SWITCHING REGULATOR WITH OVERLOAD SHUT-OFF

An alternate method for protecting a switching regulator from excessive output currents is shown in Figure 6-86. When the output current becomes too high, the voltage drop across the current-sense resistor R_8 fires an SCR which shuts OFF the regulator. The regulator remains OFF, dissipating practically no power, until it is reset by removing the input voltage.

In the actual circuit, complementary transistors Q_3 and Q_4 replace the SCR since it is difficult to find devices with a low enough holding current (about 25 μA). When the voltage drop across R_8 becomes large enough to turn ON Q_4, this removes the base drive for the output transistors of the LM104 through *Pin* 4. When this happens, Q_3 latches Q_4, holding the regulator OFF until the input voltage is removed. It will then start when power is applied if the overload has been removed.

With this circuit, it is necessary that the shut-off current be 1.5 times the full-load current. Otherwise, the circuit will shut OFF when it is switched ON with a full load

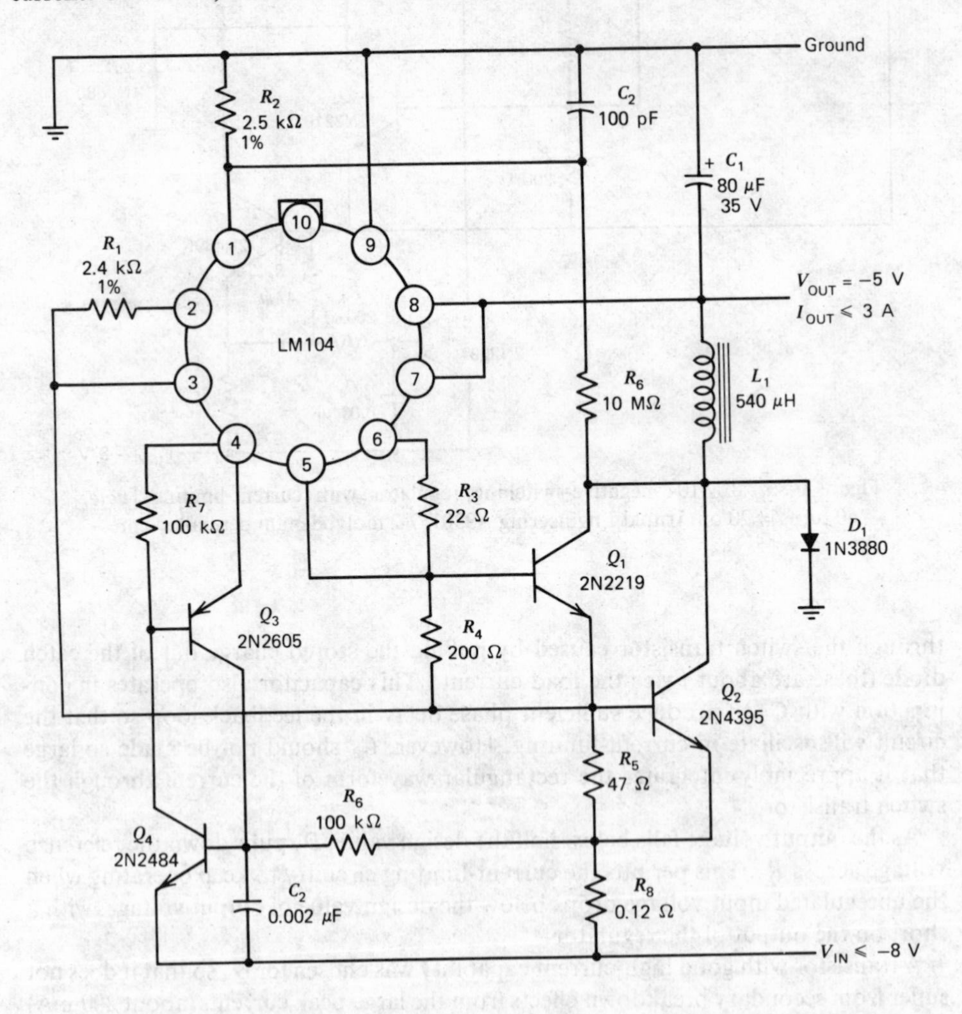

Figure 6-86. LM104 negative-switching regulator with overload shut-off [*note:* 60 turns #20 on Arnold Engineering A930157-2 molybdenum permalloy core.

because of the excess current required to charge the output capacitor. The shut-off current can be made closer to the full-load current by connecting a 10-μF capacitor across R_2 which will limit the charging current for C_1 by slowing the rise time of the output voltage when the circuit is turned ON.

However, this capacitor will also bypass the positive feedback from R_6 which makes the regulator oscillate. Therefore, it is necessary to put a 270-Ω resistor in the ground end of the added capacitor and provide feedback to this resistor from the collector of Q_1 through a 1-MΩ resistor.

High-Efficiency Switching Regulator

A switching regulator normally experiences losses in its switching element control network. A solution to this problem is a circuit design in which substantially all of the current applied to the control circuit is fed to the load through an inductive network (Figure 6-87).

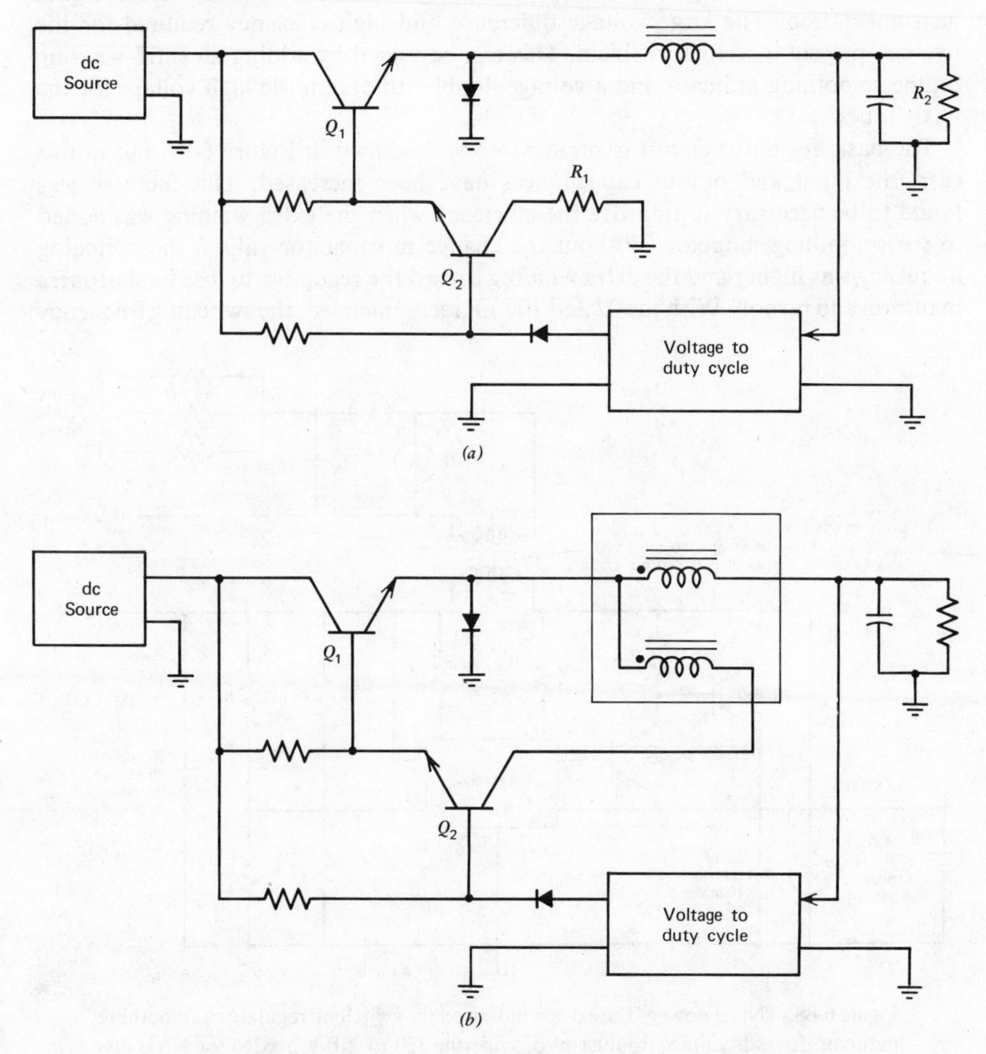

Figure 6-87. Block diagrams of high-efficency switching regulators.

In Figure 6-87*a*, the driver transistor Q_2 derives its collector current I_c through R_1, which causes a considerable power loss I_c^2R.

In Figure 6-87*b* collector current I_c is obtained from secondary winding, or tap, on inductance L_1 which is otherwise unchanged. In this way current I_c is obtained from a low-impedance circuit with resultant low I_c^2R losses. In addition I_c contributes to the load current flowing through R_2 because of the transformer action of the modified inductance L as the load current increases. Thus the base driving current Q_1 increases as the load current is increased, and this causes Q_1 saturation voltage to remain low.

This approach raises the efficiency of a regulator from 83.8% to 94.5%. With virtually no increase in complexity or cost, appreciable reduction in size is achieved by elimination of the heat-sink requirement.

Figure 6-88 depicts a practical example of this principle in which a switching regulator using an IC regulator is used for driving both ICs and Nixie tubes in digital instrumentation. The large voltage difference and high efficiency required for the two can present a serious problem. This can be solved by adding an extra winding on the smoothing inductor and a voltage doubler to obtain the high voltage for the Nixie tubes.

The basic regulator circuit is similar to one described in Figure 6-87, but in this case, the input and output capacitances have been increased. This increase was found to be necessary to improve the efficiency when the extra winding was added to the smoothing inductor. Without the change in capacitor values, the switching frequency was higher and the extra winding caused the regulator to lose its sharpness in turnon and turnoff. With the 22 and 100 μF recommended, the switching frequency

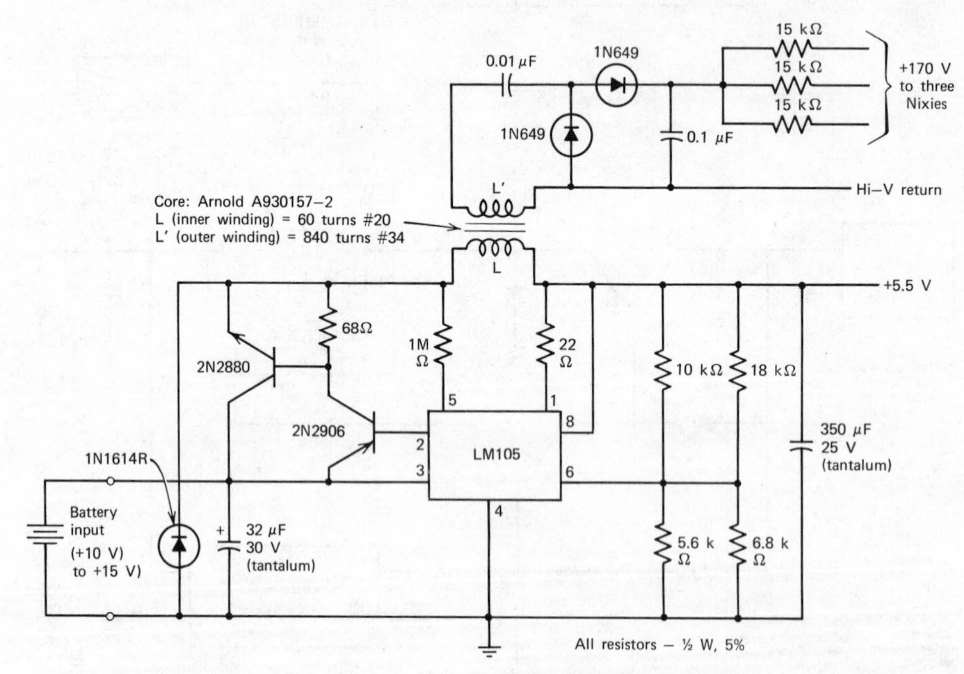

Figure 6-88. Nixie power. The extra winding on the switching regulator's smoothing inductor drives a voltage doubler to provide the 150 to 210 V needed for Nixie display tubes.

would be between 30 and 75 kHz. With the values shown—32 and 350 μF—the frequency drops to about 9.8 kHz, and efficiency is in the 78% to 85% range, depending on the low-voltage load.

The circuit produces two separate output voltages for an input voltage of from 10 to 15 V. The TTL logic circuits are supplied 5.5 V, regulated to within $\pm 1\%$, for load currents from 0.1 to 2 A. Maximum noise voltage is 0.5 V peak-to-peak. Driving the three Nixie tubes takes from 150 to 210 V at about 6 mA.

LOW RADIO-FREQUENCY NOISE-SWITCHING REGULATOR

Figure 6-89 is the complete schematic diagram of a low radio-frequency (r-f) noise-switching regulator. This circuit is more complex than the previous switching regulator circuits because of the stringent r-f requirements. Also, because of the added components and r-f requirements, the performance has degraded somewhat from that of a conventional switching regulator. Capacitors C_4 and C_5 were added to the circuit of Figure 6-57 to reduce high-frequency switching transients by slowing down the switching speeds of the output transistors. They also protect the semiconductor devices from transients. Capacitor C_4 protects the emitter-base junction of Q_2, whereas C_5 protects both the IC and the collector-base junction of Q_1 by controlling the rise times of the switching voltages.

In addition, capacitors C_6 and C_7 suppress transients on the unregulated input, C_8 minimizes the input impedance seen by the regulator, and C_3 improves filtering of the switching noise in the regulated output. The other capacitors have the same function as before, except that C_1 is much larger. This circuit oscillates in the range of 20 to 40 kHz. Resistors R_1 and R_2 still determine the output-voltage level, but R_6 was added to ensure sufficient positive feedback. The feedback is now mainly through R_6.

Inductor L_1 has the same function as before, but the lower V_{OUT} and the added inductive resistance provided by R_4 allow L_1 in this case to be only 550 μH; L_2 and L_3 are high-frequency input and output filters. The diodes on the collector of Q_2 are used to prevent reverse-input voltage polarity and heavy negative-going input transients from damaging the regulator.

Although I_{MAX} is only 500 mA, the higher dissipation of this slower switching circuit and the need to operate the regulator at relatively high temperatures required the use of two external transistors. Pass transistor Q_2 is an *npn* type for the same reason that Q_1 is a *pnp*—so that cascade connections on the IC booster output could be used. Finally, R_5 was added to terminate Q_2 and ensure low leakage in the OFF state.

DEVELOPMENT OF A 5-V, 1-A MONOLITHIC REGULATOR

The trend in IC voltage regulators is for more foolproof, high- and medium-current circuits with a minimum of external components, whether they be active or passive. One of the drawbacks of earlier regulators was the considerable number of external parts they required, which led the user to think twice about changing from discretes to ICs. Additionally, because of the high-current requirements of digital systems, single-point regulation creates many problems. Heavy power buses must be used to distribute the regulated voltage. With low voltages and currents of many amperes, voltage drops in connectors and conductors can cause an appreciable change in the load voltage. This situation is further complicated because TTL draws transient

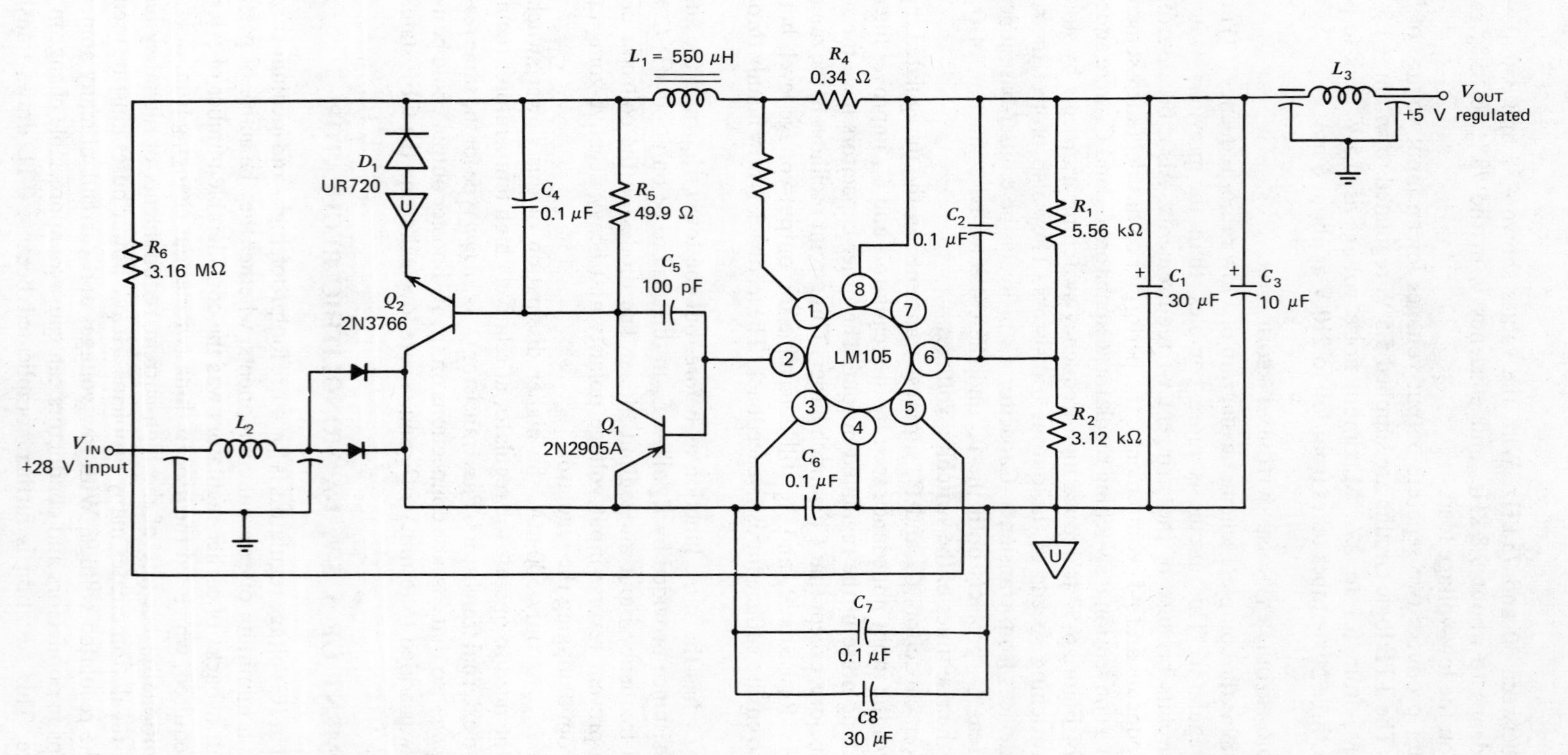

Figure 6-89. Low noise-switching regulator with 5-V output regulated to 1% by LM105 integrated circuit regulator.

currents many times the steady-state current when it switches. These current transients can cause false operation unless large bypass capacitors are used.

These problems have led to the development of the LM109 on-card monolithic regulator. It is quite simple to use in that it requires no external components. The IC has three active leads—input, output, and ground—and can be supplied in standard transistor power packages. Output currents in excess of 1 A can be obtained. Furthermore, no adjustments are required to set up the output voltage, and with the overload protection provided, it is virtually impossible to destroy the regulator. The simplicity of the regulator, coupled with low-cost fabrication and improved reliability of monolithic circuits, now makes the regulator quite attractive.

The design of the LM109 regulator is based on several break throughs (in the absence of which progress previously had been retarded) in the development of high-current IC regulators: On-chip thermal shutdown in conjunction with current-limiting, on-chip series-pass transistor, and a stable internal-voltage reference. Because of their uniqueness, each item is discussed.

Thermal shutdown limits the maximum junction temperature and protects the regulator regardless of input voltage, type of overload, or degree of heat sinking. With an external-pass transistor, there is no convenient way to sense junction temperature; thus it is much more difficult to provide thermal-limiting. Thermal protection is, in itself, a very good reason for putting the pass transistor on the chip. When a regulator is protected by current-limiting alone, it is necessary to limit the output current to a value substantially lower than is dictated by dissipation under normal operating conditions, in order to prevent excessive heating when a fault occurs.

Thermal-limiting provides virtually absolute protection for any overload condition. Hence the maximum output current under normal operating conditions can be increased. This tends to make up for the limitation that an IC has a lower maximum junction temperature than a discrete transistor.

When a regulator works with a relatively low voltage across the IC, the internal circuitry can be operated at comparatively high currents without causing excessive dissipation. Both the low voltage and the larger internal currents permit higher junction temperatures. This can also reduce the heat sinking required—especially for commercial temperature-range parts. Figure 6-90 is the schematic diagram of the LM109 1.5-A, 5-V monolithic voltage-regulator, which utilizes thermal shutdown and current-limiting techniques.

Thermal protection of the LM109 limits the maximum chip temperature to 175°C, thereby protecting the regulator from overheating, regardless of input voltage, type of overload, or degree of heat sinking. Having the temperature sensor on the chip, close to the series-pass transistor, enables the regulator to shut down in a matter of milliseconds, should an overload occur. Once the overload is removed, and the chip cools to 165°C, the regulator safely turns back ON, resuming normal operation.

The output current of the regulator is limited when the voltage across R_{14} becomes large enough to turn ON Q_{14}. This ensures that the output current cannot get high enough to cause the pass transistor to go into secondary breakdown or to damage the aluminum conductors on the chip.

Furthermore, when the voltage across the pass transistor exceeds 7 V, the current through R_{15} and D_3 reduces the limiting current, again to minimize the chance of secondary breakdown. The performance of this protective circuitry is illustrated in Figure 6-91.

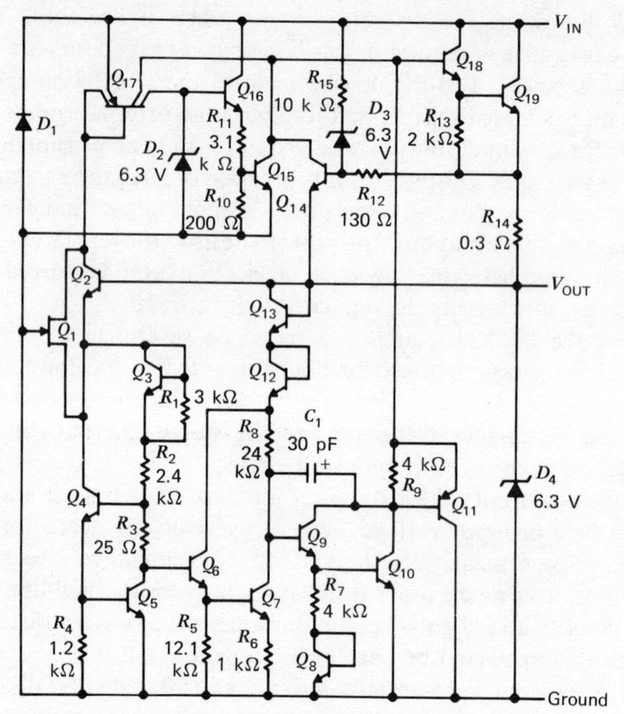

Figure 6-90. Detailed schematic of the LM109 regulator.

Even though the current is limited, excessive dissipation can cause the chip to overheat. In fact, the dominant failure mechanism of solid-state regulators is excessive heating of the semiconductors, particularly the pass transistor. Thermal protection attacks the problem directly by putting a temperature regulator on the IC chip. Normally, this regulator is biased below its activation threshold and does not affect circuit operation. If for any reason, however, the chip approaches its maximum

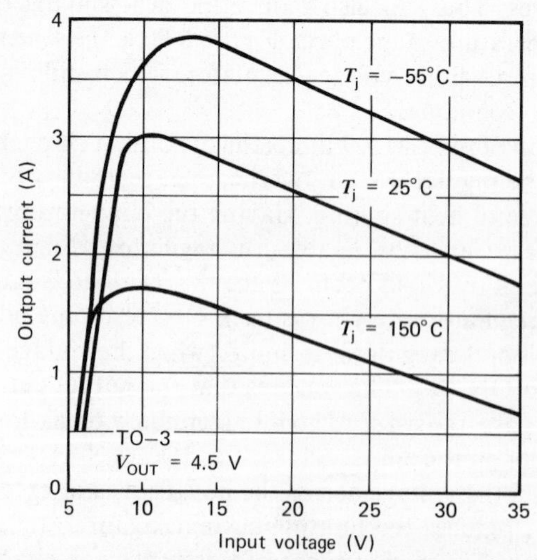

Figure 6-91. LM109 current-limiting characteristics.

operating temperature, the temperature regulator turns ON and reduces internal dissipation to prevent any further increase in chip temperature.

The thermal-protection circuitry develops its reference voltage with a conventional Zener diode D_2. Transistor Q_{16} is a buffer that feeds a voltage-divider, delivering about 300-mV to the base of Q_{15} at 175°C. The emitter-base voltage Q_{15} is the actual temperature sensor because, with a constant voltage applied across the junction, the collector current rises rapidly with increasing temperature. Although some form of thermal protection can be incorporated in a discrete regulator, ICs have a distinct advantage: The temperature-sensing device detects increases in junction temperature within milliseconds.

Schemes that sense case or heat-sink temperature take several seconds, or longer. With the longer response times, the pass transistor usually blows out before thermal-limiting has come into effect.

Another protective feature of the regulator is the crowbar clamp on the output. If the output voltage tries to rise, D_4 will break down and limit the voltage to a safe value. If the rise is caused by failure of the pass transistor such that the current is not limited, the aluminum conductors on the chip will fuse, disconnecting the load. Although this destroys the regulator, it does protect the load from damage. The regulator is also designed so that it is not damaged if the unregulated input is shorted to ground when there is a large capacitor on the output. Furthermore, if the input voltage tries to reverse, D_1 will clamp this for currents up to 1 A.

Figure 6-92 is a photomicrograph of the regulator chip. It can be seen that the pass transistors, which must handle more than 1 A, occupy most of the chip area. The output transistor is actually broken into segments. Uniform current distribution is ensured by also breaking the current-limit resistor into segments and using them to equalize the current and thus the chip temperature.

With many regulators that have a large capacitor on the output, an accidental

Figure 6-92. Photomicrograph of the LM109 regulator shows that high-current pass transistor takes more area than control circuitry.

short on the input of the regulator could destroy the series-pass transistor by discharging the output capacitor through it in the reverse direction. With the LM109 operating in the fixed 5-V output configuration, this type of failure is impossible. Also, the regulator is protected from input voltage reversals, provided the current is limited to some value less than 1 A. Larger currents will result in the melting of the aluminum interconnections on the chip itself.

The internal-voltage reference for this regulator is probably the most significant departure from standard design techniques. Temperature-compensated Zener diodes are normally used for the reference. However, these have breakdown voltages between 7 and 9 V, which puts a lower limit on the input voltage to the regulator. For low-voltage operation, a different kind of reference is needed.

The reference in the LM109 does not use a Zener diode. Instead, it is developed from the highly predictable emitter-base voltage of the transistors. In its simplest form, the reference developed is equal to the energy-band-gap voltage of the semiconductor material. For silicon, this is 1.205 V; thus, the reference need not impose minimum input-voltage limitations on the regulator. An added advantage of this reference is that the output voltage is well determined in a production environment, eliminating the need for individual adjustment of the regulators.

A simplified version of this reference is shown in Figure 6-93. In this circuit, Q_1 is operated at a relatively high-current density. The current density of Q_2 is about ten times lower, and the emitter-base voltage differential (ΔV_{BE}) between the two devices appears across R_3. If the transistors have high-current gains, the voltage across R_2 will also be proportional to ΔV_{BE}. Here Q_3 is a gain stage that will regulate the output at a voltage equal to its emitter-base voltage plus the drop across R_2. The emitter-base voltage of Q_3 has a negative TC, whereas the ΔV_{BE} component across R_2 has a positive TC. It will be shown that the output voltage is temperature-compensated when the sum of the two voltages is equal to the energy-band-gap voltage.

Conditions for temperature compensation can be derived starting with the equation

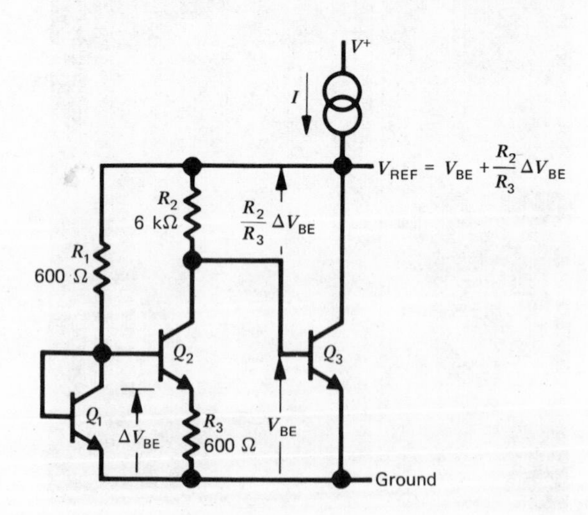

Figure 6-93. The LM109 low-voltage reference in one of its simplest forms.

for the emitter-base voltage of a transistor, which is

$$V_{BE} = V_{g0}\left(1 - \frac{T}{T_0}\right) + V_{BE0}\left(\frac{T}{T_0}\right) + \frac{nkT}{q}\log_e \frac{T_0}{T} + \frac{kT}{q}\log_e \frac{I_C}{I_{C0}} \tag{6-29}$$

where V_{g0} is the extrapolated energy-band-gap voltage for the semiconductor material at absolute zero, q is the charge of an electron, n is a constant depending on how the transistor is made (about 1.5 for double-diffused *npn* transistors), k is Boltzmann's constant, T is absolute temperature, I_C is collector current, and V_{BE0} is the emitter-base voltage at T_0 and I_{C0}.

The emitter-base voltage differential between two transistors operated at different current densities is given by

$$\Delta V_{BE} = \frac{kT}{q}\log_e \frac{J_1}{J_2} \tag{6-30}$$

where J is current density. Referring to (6-30), we see that the last two terms are quite small and are made even smaller by making I_C vary as absolute temperature. At any rate, they can be ignored for now because they are the same order as errors caused by nontheoretical behavior of the transistors that must be determined empirically.

If the reference is composed of V_{BE} plus a voltage proportional to V_{BE}, the output voltage is obtained by adding (6-29) in its simplified form to (6-30):

$$V_{REF} = V_{g0}\left(1 - \frac{T}{T_0}\right) + V_{BE0}\left(\frac{T}{T_0}\right) + \frac{kT}{q}\log_e \frac{J_1}{J_2} \tag{6-31}$$

Differentiating with respect to temperature yields

$$\frac{\partial T_{REF}}{\partial T} = -\frac{V_{g0}}{T_0} + \frac{V_{BE0}}{T_0} + \frac{kT}{q}\log_e \frac{J_1}{J_2} \tag{6-32}$$

For zero temperature drift, this quantity should equal zero, giving

$$V_{g0} = V_{BE0} + \frac{kT_0}{q}\log_e \frac{J_1}{J_2} \tag{6-33}$$

The first term on the right is the initial emitter-base voltage; the second is the component proportional to emitter-base voltage differential. Hence, if the sum of the two is equal to the energy-band-gap voltage of the semiconductor, the reference will be temperature compensated.

In Figure 6-90 the ΔV_{BE} component of the output voltage is developed across R_8 by the collector current of Q_7. The emitter-base voltage differential is produced by operating Q_4 and Q_5 at high densities while operating Q_6 and Q_7 at much lower current levels. The extra transistors improve tolerances by making the emitter-base voltage differential larger. In addition, R_3 serves to compensate the transconductance of Q_5, so that the ΔV_{BE} component is not affected by changes in the regulator output voltage or the absolute value of components.

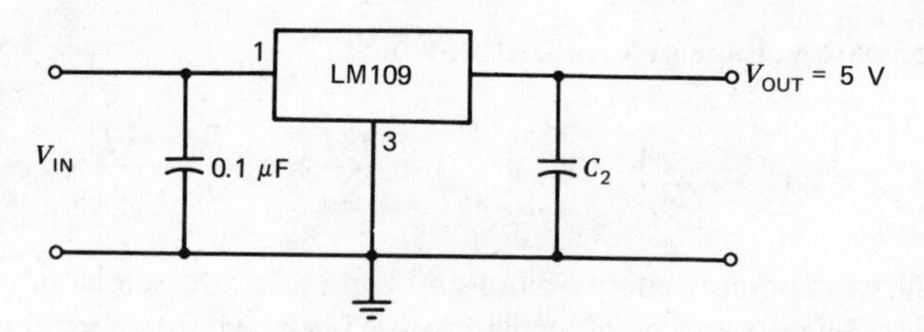

Figure 6-94. Fixed 5-V regulator—0.1 pF required if regulator is located an appreciable distance from the power-supply filter; although C_2 is not required, it gives improved transient response if this is desired.

Thus the LM109 performs a complete regulation function on a single silicon chip, requiring no external components. It was basically designed to provide a +5-V supply for TTL and DTL circuits. However, the design breakthroughs achieved in the LM109 have led to the widespread development of many high-current IC regulators by all the major IC manufacturers. (National's LM320 and LM340 series, Fairchild's μA7800 series, μA78M series, and μA78L series, and so on).

Figures 6-94 and 6-95 depict the ease of use of the LM109 and the minimal number of external components required to operate the device. The LM109 is very simple to use in that it requires no external components and it has three active leads (input, output, and ground); output currents in excess of 1 A can be obtained, no adjustments are required to set up the output voltage, and the device has internal overload and thermal shutdown protection—making it virtually blowoutproof. The internal-frequency compensation of the regulator permits it to operate with or without a bypass capacitor on the output. However, an output capacitor improves the transient response and reduces the high-frequency output impedance.

The LM109 can be made to deliver a voltage higher than the specified output by adding two resistors to the output (Figure 6-96*a*). The concept is simple but problems can arise since the quiescent current (I_Q) for the LM109 (or μA7800 Series) is not constant for either line, load, or temperature variations. The output voltage developed is

$$V_{OUT} = 5\text{ V} + R_2\left(I_Q + \frac{5\text{ V}}{R_1}\right) \qquad (6\text{-}34)$$

The LM109 data sheet indicates that I_Q decreases by 750 μA from $T_J = 25°C$ to $T_J = 150°C$, and it will also increase by 200 μA due to a 20% increase in line voltage. Although the ambient temperature remains constant, T_J can easily jump 100°C just by varying the load current. The output voltage is also dependent on the absolute

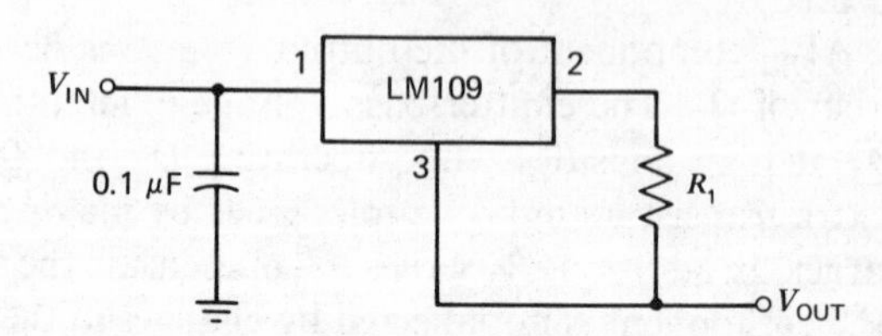

Figure 6-95. LM109 used as a current regulator, where R_1 determines the output current.

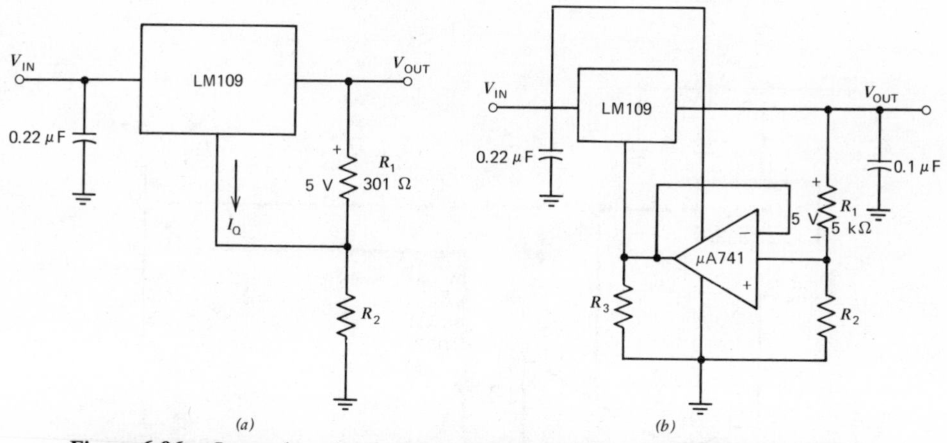

Figure 6-96. Increasing the output-voltage capability of an integrated circuit regulator may (*a*) involve a simple resistor network, or (*b*) employ a 741 to buffer the effects of quiescent current (I_Q) variations [*note:* (*a*) and (*b*) $V_{OUT} = 5\text{ V} + R_2(I_Q + 5\text{ V}/R_1)$; (*b*) $R_3 = (V_{OUT} - 5\text{ V})/I_{Q(typ)}$].

value of I_Q. Therefore, variations in I_Q from unit to unit will require different values for R_2, depending on the output-voltage accuracy requirement (not a good situation for mass production).

This technique for increasing the regulator's output voltage capability is not practical when accuracies of better than 3% are needed. A much improved scheme (Figure 6-96*b*) uses an additional resistor (R_3) and a μA741 op amp connected as a voltage-follower. The μA741 buffers the output resistor divider network from variations in I_Q. The equation for V_{OUT} can still be used for determining the output voltage. A minimum output of 7 V is required of this circuit since the μA741 will not operate in the linear region when its input voltage comes within 2 V of the negative supply (ground), R_3 is used to reduce the μA741 output drive requirements to less than ±2 mA and is given by $R_3 = V_{OUT} - 5\text{ V}/I_Q$ (typical). Adding the μA741 buffer allows the circuit's load and line regulation specs to equal those of an unmodified LM109 or μA7800.

PRECISION REGULATORS

Precision regulators can be built using an IC op amp as the control amplifier and a discrete Zener as a reference, where the performance is determined by the reference. Figure 6-97 presents the circuit schematic of a simple regulator. It is capable of providing better than 0.01% regulation for worst-case changes of line, load, and temperature. Typically, the line rejection is 120 dB to 1 kHz; and the load regulation is better than 10 μV for a 1-A change. Temperature is the worst source of error; however, it is possible to achieve less than a 0.01% change in the output voltage over a −55 to +125°C range.

The operation of the regulator is straightforward. An internal-voltage reference is provided by a high-stability Zener diode. The LM108A op amp compares a fraction of the output voltage with a reference. The output of the amplifier controls the ground terminal of an LM109 regulator through source follower Q_1. Frequency compensation for the regulator is provided by both the R_1–C_2 combination and output capacitor C_3.

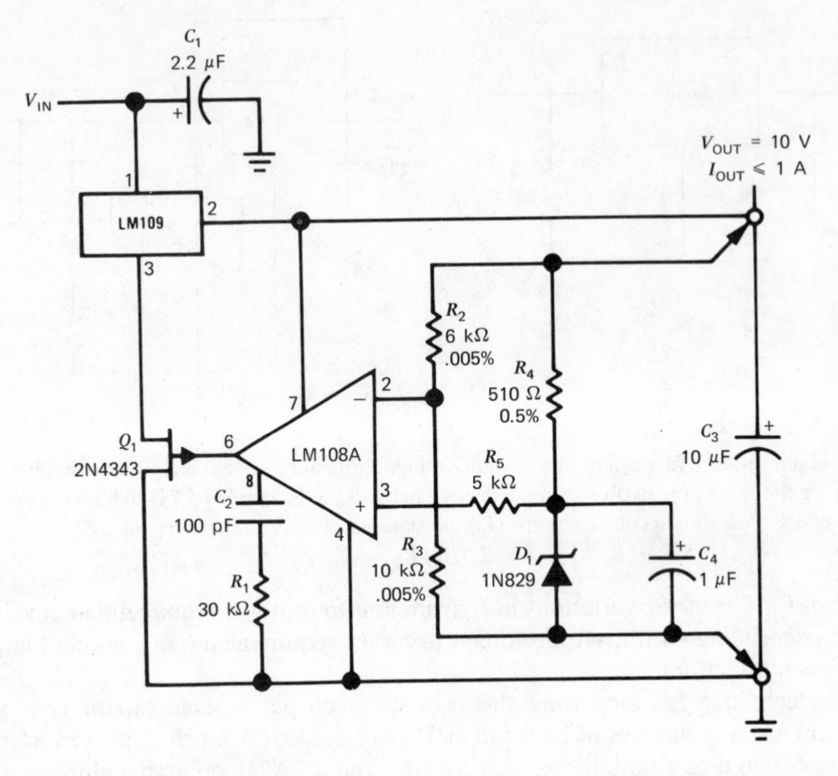

Figure 6-97. High-stability positive regulator.

The use of an LM109 instead of discrete power transistors has several advantages. First, the LM109 contains all the biasing and current-limit circuitry needed to supply a 1-A load. This simplifies the regulator. Second, and probably most important, the LM109 has thermal-overload protection, making the regulator virtually burnout-proof.

Although the regulator is relatively simple, some precautions must be taken to eliminate possible problems. A solid tantalum output capacitor must be used. Unlike electrolytics, solid tantalum capacitors have low-internal impedance at high frequencies. Low impedance is needed both for frequency compensation and to eliminate possible minor loop oscillations.

Some unusual problems are encountered in the construction of a high-stability regulator. Component choice is most important, since resistors, amplifiers, and Zeners can contribute to temperature drift. Also, good circuit layout is needed to eliminate the effect of lead drops, pick-up, and thermal gradients.

The resistors must be low TC wirewound or precision metal film. Ordinary 1% carbon film, tin oxide, or metal-film units are not suitable because they may drift as much as 0.5% over temperature. The resistor accuracy need not be 0.005% as in Figure 6-97, however, they should track better than 1 ppm/°C. Additionally, wire-wound resistors usually have lower thermoelectric effects than film types. The resistor driving the Zener is not quite as critical, but it should change less than 0.2% over temperature.

The 1N829 diode is representative of the better Zeners available. However, it still has a TC of 0.005 %/°C or a maximum drift of 0.05 % over a −55 to 125°C temperature range. The drift of the Zener is usually linear with temperature and may be varied by changing the operating current from its nominal value of 7.5 mA. The TC changes by about 50 μV/°C for a 15 % change in operating current. Therefore, the temperature drift of the regulator may be minimized by adjusting the Zener current.

Good construction techniques are important. It is necessary to use remote sensing at the load, as shown in the schematic. Even an inch of wire will degrade the load regulation. The voltage-setting resistors, the Zener, and the amplifier should also be shielded. Board leakages or stray capacitance can easily introduce 100 μV of ripple or dc error into the regulator. Generally, short wire length and single-point are helpful in obtaining proper operation.

AC/DC CONVERTER

The ac/dc converter of Figure 6-98 is completely foolproof. Both the LM309 and LM320 voltage-regulators have internal current-limiting to limit peak-output currents to a safe value. Also, thermal shutdown is provided to keep each IC safe from overheating. Each regulator is thus blowoutproof.

The μA741 op amp is also a foolproof device. It has internal overload protection on the inputs and output; it does not latch up when the common-mode range is exceeded; and it is also free from oscillation.

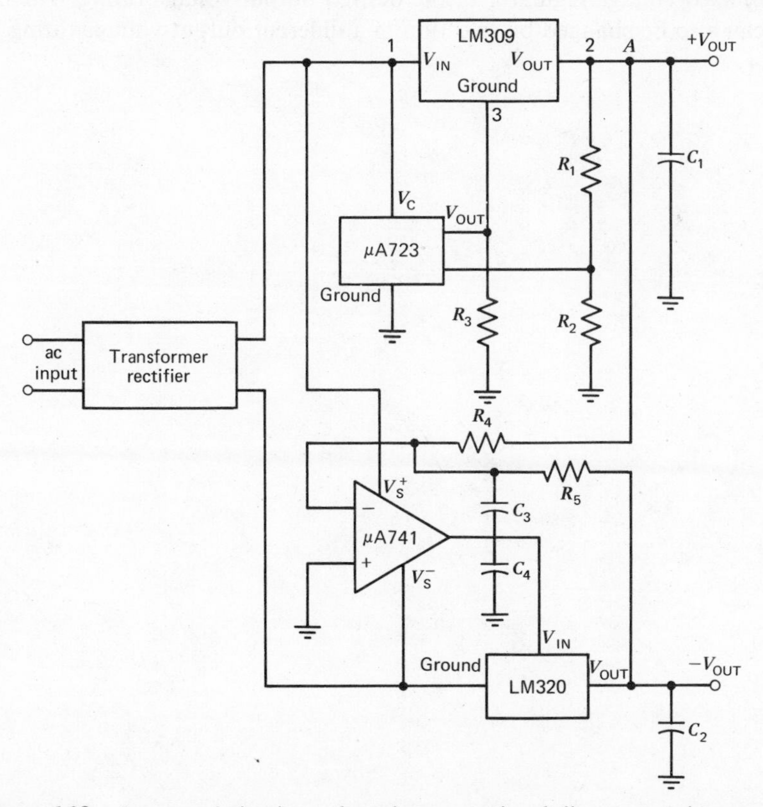

Figure 6-98. Integrated circuits replace the conventional discrete transistors and diodes in this ac/dc converter.

The circuit operates in the classical manner: an ac voltage is applied to a transformer and rectified to obtain raw, but usable, positive and negative dc voltages. These voltages are fed, respectively, to the LM309 and LM320 series regulators to obtain the desired output voltages. In this circuit, both the LM309 and LM320 are used as positive- and negative-regulator pass transistors due to their high-current capabilities and foolproof protection features. The μA723 regulator is used only as a precision voltage reference which regulates the ground terminal of the LM309.

Resistors R_1 and R_2 are used to set up the positive-output voltage and as a supply to the μA723 reference. The μA741 is used to invert the positive voltage to a negative voltage for the LM320. The connection point A on the positive output is fed back through R_4 to the inverting input of the μA741 to enable the negative output to track the positive output. Capacitors C_1 and C_2 are used to smooth the output voltage. With the following component values the circuit provides a +15-V, 1-A output and a −15-V, 1-A output with a total circuit efficiency of 80%:

$R_1 = 5.6\ \text{k}\Omega \qquad R_2 = 5.6\ \text{k}\Omega \qquad R_3 = 300\ \Omega \qquad R_4 = 20\ \text{k}\Omega \qquad R_5 = 20\ \text{k}\Omega$

$C_1 = C_2 = 47\ \mu\text{F} \qquad C_3 = 0.05\ \mu\text{F} \qquad C_4 = 0.03\ \mu\text{F}$

The positive output can be adjusted to any value (within the LM309's operating specifications) by means of R_1 and R_2 or by replacing the LM309 with an LM340 (μA7800) fixed-voltage regulator of the desired output-voltage rating. The negative output can also be changed by selection of a different output-voltage rating LM320 regulator.

Chapter 7

The Integrated Circuit Timer

INTRODUCTION

The introduction of the first IC timer by Signetics in 1972 added a new dimension to the field of linear ICs. Since then these versatile timing circuits have replaced both thermal relays and electro mechanical devices in a variety of timing functions. The 555 monolithic timing circuit is a highly stable controller capable of producing accurate time delays or oscillation. Additional terminals are provided for triggering or resetting if desired. In the time-delay mode of operation, the time is precisely controlled by one external resistor and capacitor. For stable operation as an oscillator, the free-running frequency and the duty cycle are both accurately controlled with two external resistors and one capacitor. The circuit may be triggered and reset on falling waveforms, and the output structure can source or sink up to 200 mA or drive TTL circuits.

A functional diagram of the timer is shown in Figure 7-1. Basically it consists of two comparators controlling the state of a flip-flop which drives a discharge transistor and an output stage. Comparator II sets the initial state of the flip-flop. It is controlled by an external-threshold signal (referenced to one-third V_{CC}), that is applied to *pin* 2. The flip-flop in turn, controls the state of the output. A negative-going pulse (less than one-third V_{CC}) applied to the trigger input, sets comparator II and the flip-flop.

The various *pin* functions are described below:

PIN 1 Ground—usually connected to ground. The voltage should be the most negative of any voltage appearing at the other *pins*.

PIN 2 Trigger—level-sensitive point to 1/3 V_{CC}. When the voltage at this *pin* is brought below 1/3 V_{CC} the flip-flop is set causing *pin* 3 to produce a high state. Allowable applied voltage is between V_{CC} (*pin* 8) and ground (*pin* 1).

PIN 3 Output—level here is normally low and goes high during the timing interval. Since the output stage is active in both directions, it can source or sink up to 200 mA.

PIN 4 Reset—when voltage at this *pin* is less than 0.4 V, the timing cycle is interrupted, returning the timer to its nontriggered state. This is an overriding function so that the

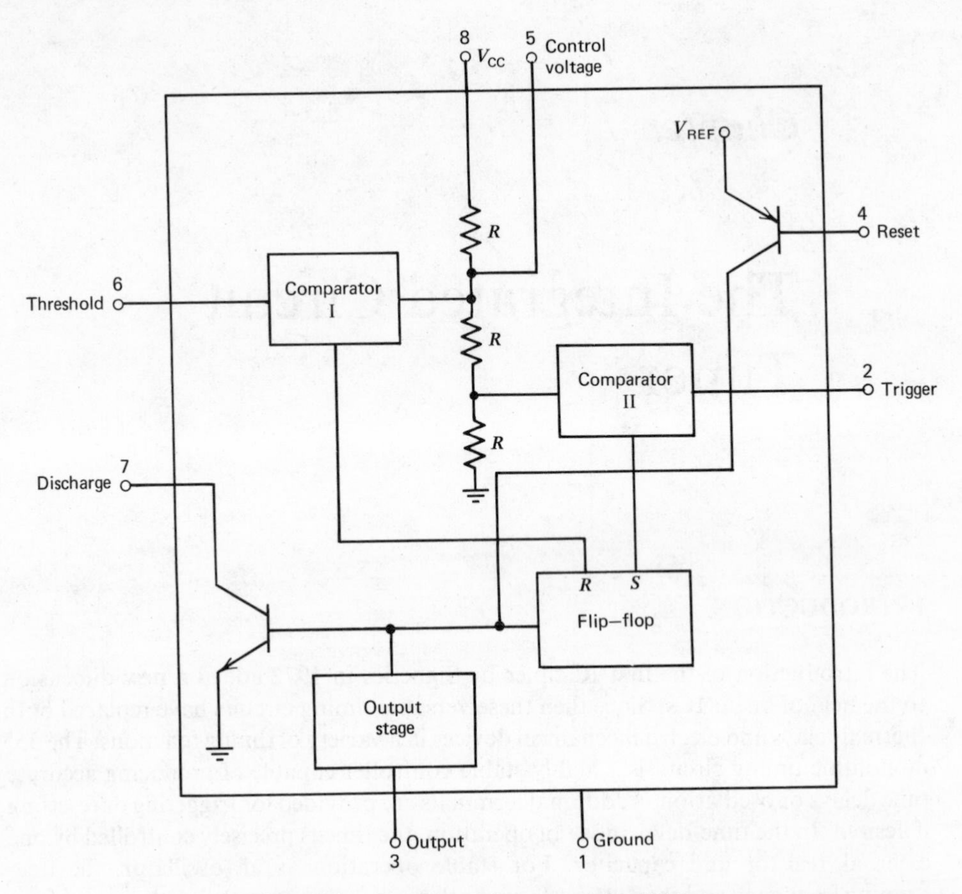

Figure 7-1. Functional diagram of the 555 timer.

timer can not be triggered unless reset is released (*pin* 4 > 1.0 V). Other useful functions include oscillation gating in the astable mode and state assurance for power-supply start-up. Allowable applied voltage is between ground (*pin* 1) and V_{CC} (*pin* 8). When not used, connect to V_{CC}.

PIN 5 Control voltage—internally derived 2/3 V_{CC} point. A resistor-to-ground or an external voltage may be connected to *pin* 5 to change the comparator reference points. When not used for this purpose, a capacitor-to-ground greater than or equal to 0.01 μF is recommended for all applications. The purpose is to filter power-supply noise spikes from causing inconsistant timing. Allowable applied voltage is between ground (*pin* 1) and V_{CC} (*pin* 8).

PIN 6 Threshold—level-sensitive point to 2/3 V_{CC}. When the voltage at this *pin* is brought greater than 1/3 V_{CC} the flip-flop is reset causing *pin* 3 to produce a low state. Allowable applied voltage is between ground (*pin* 1) and V_{CC} (*pin* 8).

PIN 7 Discharge—collector of a transistor switch to ground (*pin* 1). It is normally used to discharge the timing capacitor. Since the collector current is limited, it can accommodate very large capacitors (in excess of 1,000 μF) without damage.

PIN 8 V_{CC} power-supply voltage connected here can range from 4.5 to 16 V with respect to ground (*pin* 1). Timing is relatively independent of this voltage. The timing error due to power-supply variation is typically less than 0.05%/V.

APPLICATION INFORMATION

Monostable Operation

In this mode of operation, the timer functions as a one-shot. Referring to Figure 7-2 the external capacitor is initially held discharged by the discharge transistor inside the timer.

Upon application of a negative trigger pulse to *pin* 2 the flip-flop is set which releases the short circuit across the external capacitor and drives the output high. The voltage across the capacitor, now, increases exponentially with the time constant $t = R_A C$. When the voltage across the capacitor equals 2/3 V_{CC}, the comparator resets the flip-flop which in turn discharges the capacitor rapidly and drives the output to its low state. Figure 7-3 shows the actual waveforms generated in this mode of operation.

The circuit triggers on a negative-going input signal when the level reaches 1/3 V_{CC}. Once triggered, the circuit will remain in this state until the set time is elapsed, even if it is triggered again during this interval. The time that the output is in the high state is given by $t = 1.1 R_A C$ and can easily be determined by Figure 7-4. Notice that since the charge rate, and the threshold level of the comparator are both directly proportional to supply voltage, the timing interval is independent of supply. If the reset terminal (*pin* 4) and trigger terminal (*pin* 2) are connected together and held at ground, the timing cycle will commence on the positive edge.

Practical Considerations

Timing Resistor. The minimum value for R_A is established from the fact that the initial charging current i must not be so large as to prevent the discharge transistor from doing its job (see *pin* 7 description above). Initial charging i therefore, should be limited to 5 mA. Normally R_A would be chosen to be as large as possible to allow for a small value of C. The maximum value of R_A is determined by the threshold current required at *pin* 6. Initial minimum charging current is 1 μA. For any power-supply voltage, then, the timing can be varied by 5000 to 1 by changing only R_A!

Timing Capacitor. The minimum practical value for C is 100 pF. The criteria for this value is to keep C much larger than the parasitic nonlinear capacitance at *pins*

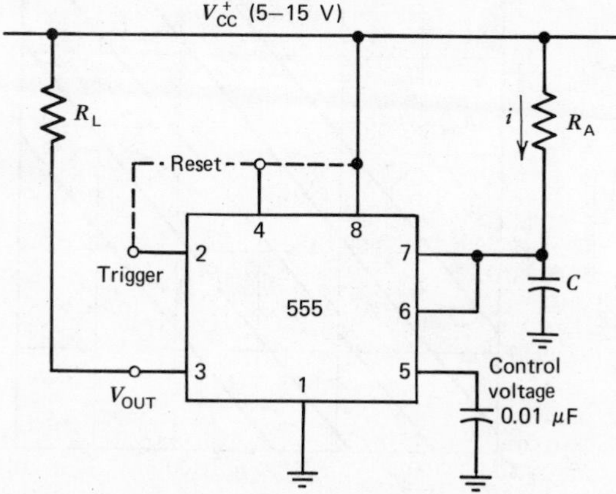

Figure 7-2. Monostable operation [*note:* maximum initial i = 5 mA, minimum initial i = 1 μA].

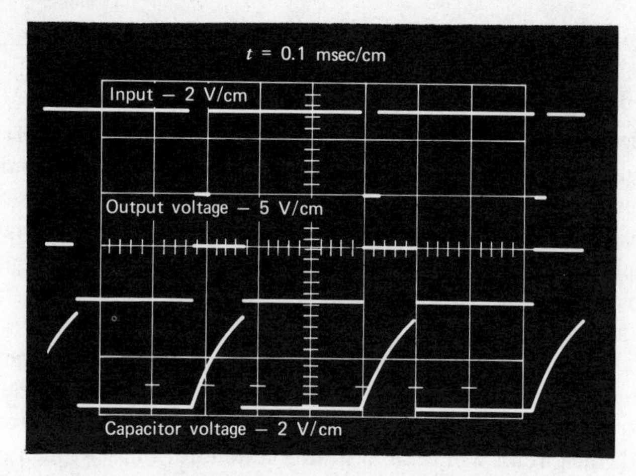

Figure 7-3. Output waveforms for monostable operation [*note:* $R_A = 9.1\ k\Omega$, $C = 0.01\ \mu F$; $R_L = 1\ k\Omega$].

6 and 7. The maximum practical value for C is usually determined by capacitor leakage currents. When very long times are required (approaching 1 hr) R_A is chosen to produce the minimum initial charging current of 1 μA. The capacitor-leakage current must then be less than 0.1 μA. The best choice for large capacitors with low leakage is the tantalum sintered anode type such as Sprague 109D. Another helpful fact is the decrease of leakage current with operating voltage. When the operating voltage is 1/2 of the rated working voltage, the leakage current is typically 1/5 of its rating.

Output-Load Considerations. The 555 timer provides an extremely useful time-delay output function capable of sinking or sourcing up to 200 mA. In addition to the large sink/source capability, typical rise and fall times are 100 nsec. To accomplish this type of driving capability, an output structure which differs significantly from standard logic totem poles must be utilized. Referring to the output fall time in Figure 7-5, note the possible problem of driving TTL gates.

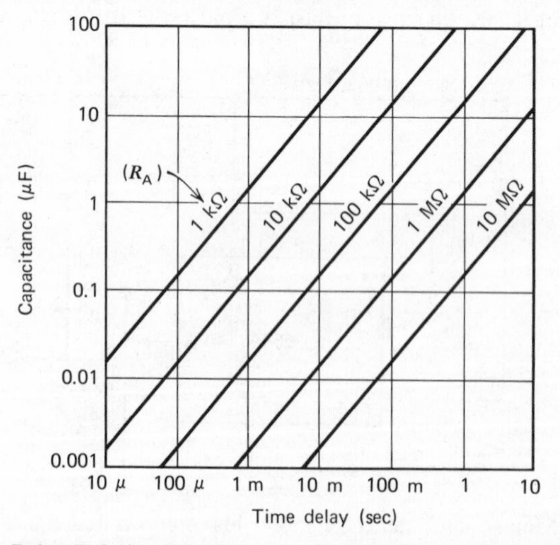

Figure 7-4. Relationships between timing elements and time delay.

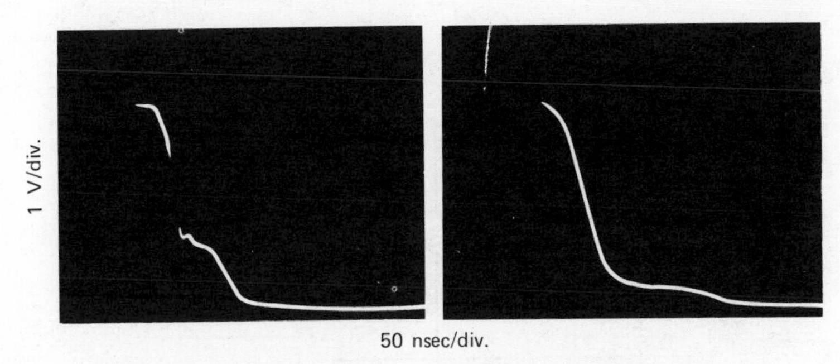

Figure 7-5. Nonlinearity in output-fall time.

The discontinuity occurs 2 V_{BE} above ground which is right at TTL threshold. Thus, a double output transition or unnecessary delay in transition may occur, depending upon the level of the glitch and the threshold of the particular gate. However, if the 555 is required to drive logic, a 100-pF capacitor to ground from *pin* 3 will move the glitch to about 0.5 V, which is well out of the threshold area of the gate. This eliminates any problems which might have occurred at the logic output.

A negative voltage at *pin* 3 can cause a latch-up. This condition normally occurs when driving relays to ground as shown in Figure 7-6. During the timing cycle the relay is energized. At the end of the timing cycle when the output goes low, the back emf due to relay-coil inductance can become very large. Diode D_1 should be included in all cases where an inductive load is involved. Even with D_1, however, voltage V_1 can become negative by one diode drop which could latch the 555. The solution is to add diode D_2 that prevents *pin* 3 of the 555 from becoming negative. Diode D_2 is not necessary when driving relays to V_{CC}.

Trigger Level. Triggering of the timer occurs on the negative-going edge of the trigger pulse. The threshold which must be attained is below 1/3 V_{CC}. There is, however, one significant restriction. The trigger level must not remain below this threshold longer than the timing cycle (1.1 $R_A C$). This condition can cause erratic time intervals. In applications where this condition can exist, the input-differentiation circuit of Figure 7-7 should be used. In differentiating the trigger input it is possible to produce a positive spike which exceeds the power supply in amplitude. Should this happen, the 555 will trigger again, causing an output for both leading and

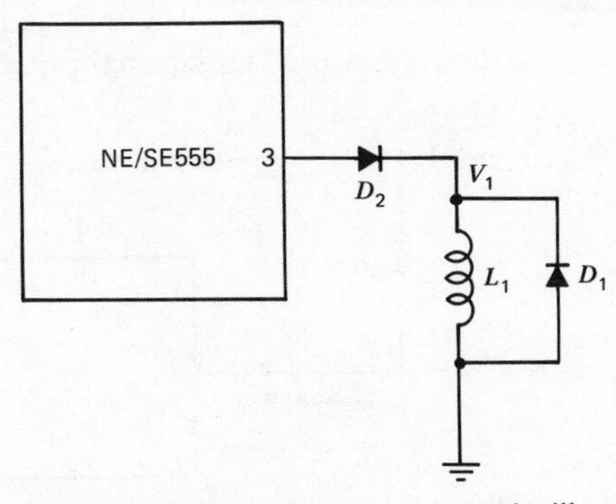

Figure 7-6. A diode in series with relay to ground will prevent latch-up.

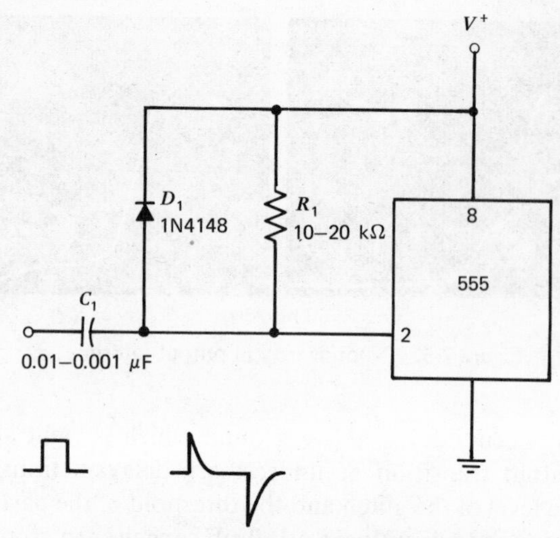

Figure 7-7. Trigger input differentiator.

trailing edges of the trigger waveform. If this occurs, the solution is to clamp the spike to a level less than V_{CC} with diode D_1 as shown.

A small bias current is required at the trigger terminal (*pin* 2). Normally this current is supplied by the trigger source. In applications where the trigger has essentially an infinite source impedance, a 10-MΩ resistor from *pin* 2 to V_{CC} will supply the necessary bias current (Figure 7-8).

Control Voltage. Although the control-voltage level can vary between ground (*pin* 1) and V_{CC} (*pin* 8), operation is not guaranteed. Modulation level extremes at *pin* 5 to ensure operation must be held to within 2 V_{BE} of ground to 1 V_{BE} lower than V_{CC}. This voltage level may be changed by connecting a voltage source at *pin* 5 or using a resistor to ground.

Comparator-reference voltages are changed proportional to the applied signal at the control-voltage terminal. The threshold will be the applied voltage and the trigger level will be 1/2 of the applied voltage. This point is useful then for voltage control of the timing interval or oscillator frequency when used in astable operation.

Astable Operation

If the timer is connected as shown in Figure 7-9 (*pins* 2 and 6 connected), it will trigger itself and free-run as a multivibrator. The external capacitor charges through

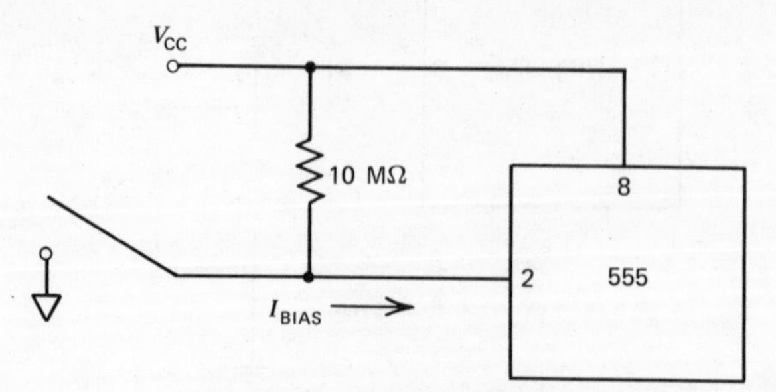

Figure 7-8. Trigger bias current.

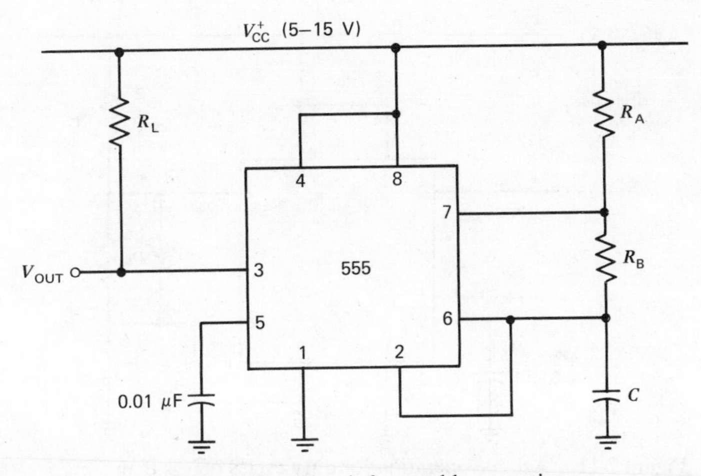

Figure 7-9. Circuit for astable operation.

R_A and R_B and discharges through R_B only. Thus the duty cycle may be precisely set by the ratio of these two resistors. In this mode of operation, the capacitor charges and discharges between 1/3 V_{CC} and 2/3 V_{CC}. As in the triggered mode, the charge and discharge times, and therefore the frequency are independent of the supply voltage.

Figure 7-10 shows actual waveforms generated in this mode of operation. The charge time (output high) is given by

$$t_1 = 0.693(R_A + R_B)C \tag{7-1}$$

and the discharge time (output low) by

$$t_2 = 0.693(R_B)C \tag{7-2}$$

Thus the total period is given by

$$T = t_1 + t_2 = 0.693(R_A + 2R_B)C \tag{7-3}$$

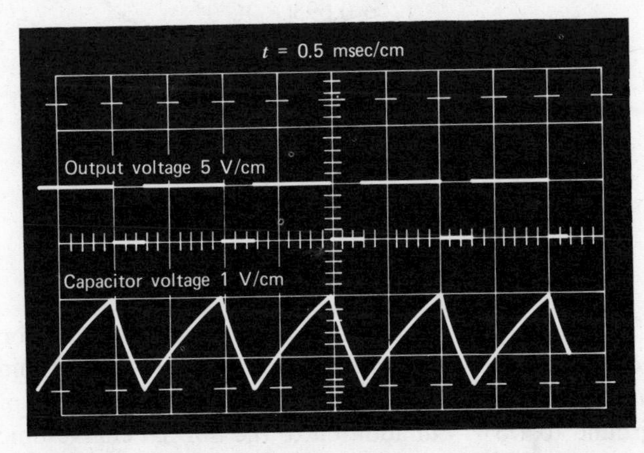

Figure 7-10. Output waveforms for astable operation [*note:* $R_A = 4\ k\Omega$, $R_B = 3\ k\Omega$, $R_L = 1\ k\Omega$].

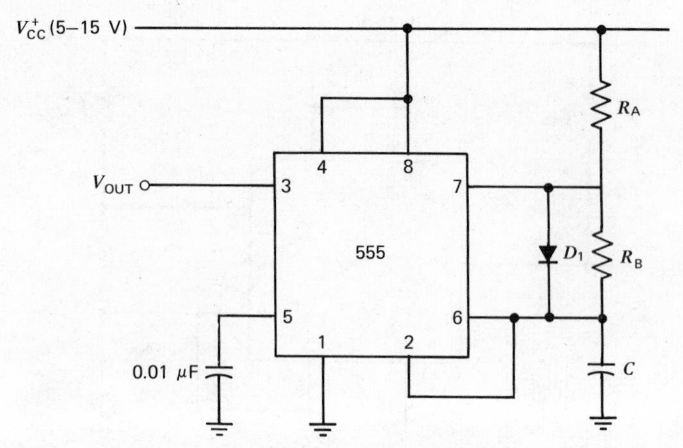

Figure 7-11. Astable operation for duty cycles of less than 50%.

The frequency of oscillation is then

$$f = \frac{1}{T} = \frac{1.44}{(R_A + 2R_B)C} \tag{7-4}$$

The duty cycle is given by $D = (R_A + R_B)/(R_A + 2R_B)$. With the addition of a diode in parallel with R_B as shown in Figure 7-11, duty cycles of less than 50% can be achieved.

Capacitor-charging current is now through R_A and D_1 with a discharge path through R_B. In this configuration the output high and low times are a function of independent resistors. Output high time is given by

$$t_1 \approx 0.693(R_A)C \tag{7-5}$$

Output low time is

$$t_2 \approx 0.693(R_B)C \tag{7-6}$$

The frequency and duty cycles are given by

$$f \approx \frac{1.44}{(R_A + R_B)C} \tag{7-7}$$

$$D \approx \frac{R_A}{R_A + R_B} \tag{7-8}$$

Reset control (*pin* 4) may be grounded to stop oscillation thus forming a gated oscillator. When the standard configuration (Figure 7-9) is used, a timing discrepancy exists for the first cycle of oscillation. This occurs because the capacitor voltage must charge from ground to 1/3 V_{CC} in addition to the normal charge time (between 1/3 and 2/3 V_{CC}). This effect can be minimized by using the circuit in Figure 7-11).

Very small duty cycles will cause a total time discrepancy of less than 40% of that duty cycle when using the 555 as a gated oscillator. For example, if $R_A = 1\ k\Omega$ and $R_B = 1\ M\Omega$ the total first cycle error is less than 0.04%. When diode D_1 is not used, R_B should be chosen greater than 2 kΩ. Restrictions on other component values are the same as discussed under monostable operation above.

Power-Up Time Delay

For applications where a time delay is needed with power turn-ON, the circuit of Figure 7-12 can be used. By tying the trigger to *pin* 6, the timing capacitance holds the trigger low while power is coming up. Triggering is then guaranteed and timing occurs in the normal manner; V_{CC} must be removed from this circuit before retriggering can occur.

Missing-Pulse Detector (Retriggerable Monostable Operation)

Retriggerable operation can be achieved by using an external transistor to discharge the timing capacitor before its voltage crosses the 2/3 V_{CC} threshold. One application of retriggerable operation is a missing-pulse detector as shown in Figure 7-13. The timing cycle is continuously reset by the input-pulse train. A decrease in frequency or a missing pulse allows the completion of the timing cycle which causes a change in the output level. For this application the time delay should be set to be slightly longer than the normal time between pulses. Figure 7-14 shows the actual waveforms seen in this mode of operation.

Frequency Division

The basic circuit for monostable operation (Figure 7-2) can be used to perform frequency division. If the input frequency is known, the timer can easily be used as a frequency divider by adjusting the length of the timing cycle. Figure 7-15 shows the waveforms of the timer when used as a divide by three circuit. This application makes use of the fact that this circuit cannot be retriggered during the timing cycle.

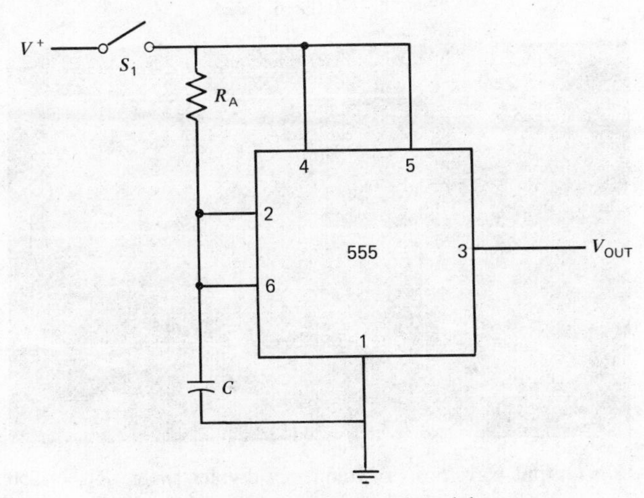

Figure 7-12. Power-up time delay.

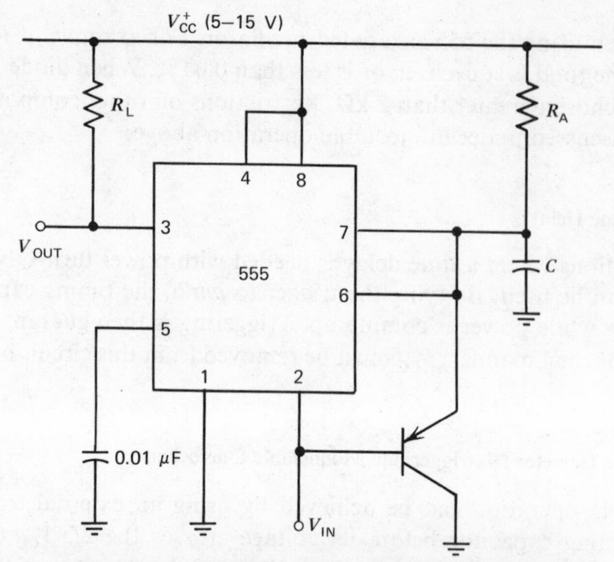

Figure 7-13. Retriggerable monostable operation.

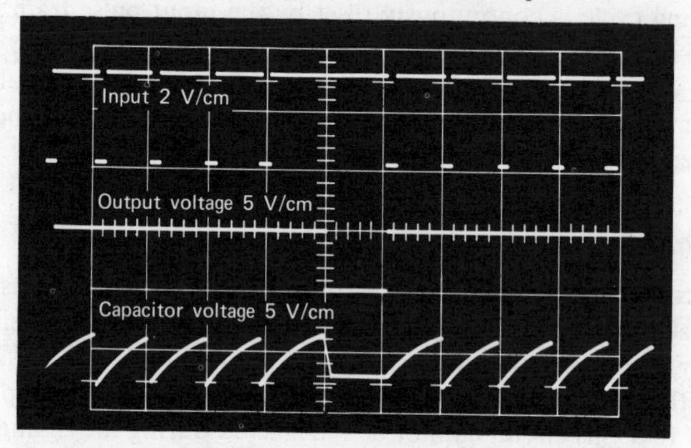

Figure 7-14. Output waveforms for missing-pulse detector [*note:* $R_A = 1\ k\Omega$, $C = 0.09\ \mu F$].

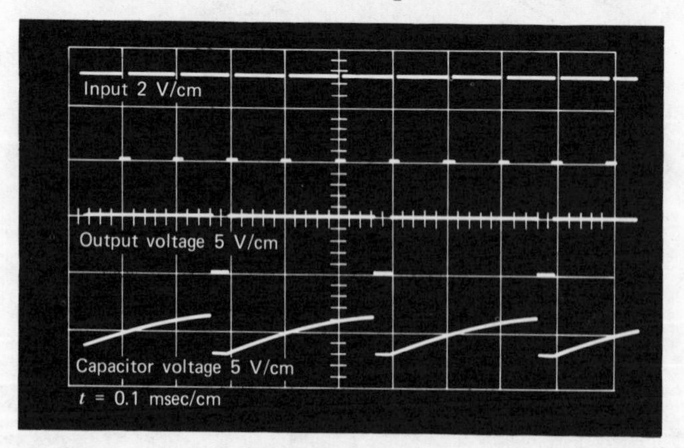

Figure 7-15. Output waveforms of frequency divider [*note:* $R_A = 1250\ \Omega$, $C = 0.02\ \mu F$, $R_L = 1\ k\Omega$].

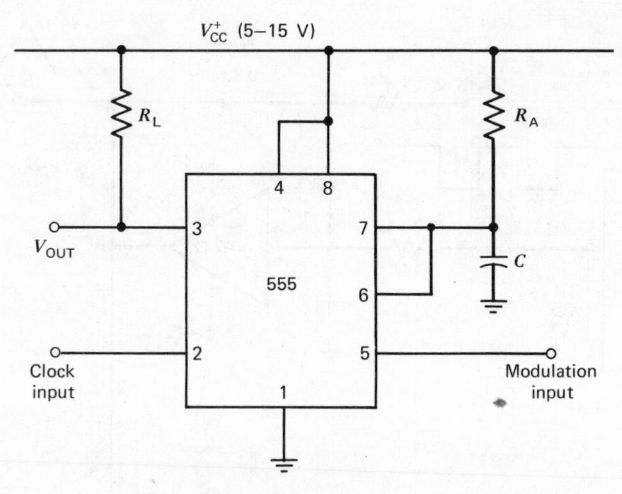

Figure 7-16. Pulse-modulation circuit.

Pulse-Width Modulation

In this application, the timer is connected in the monostable mode as shown in Figure 7-16. The circuit is triggered with a continuous pulse train and the threshold voltage is modulated by the signal applied to the control-voltage terminal (*pin* 5). This has the effect of modulating the pulse width as the control voltage varies. Figure 7-17 shows the actual waveforms generated with this circuit.

One should keep in mind, however, that the pulse width (using the circuit of Figure 7-16) is not a linear function of the modulating voltage applied at *pin* 5. The timing-capacitor voltage is given by

$$V_{CAP} = V_{CC}\left(1 - V - \frac{t}{RC}\right) \qquad (7\text{-}9)$$

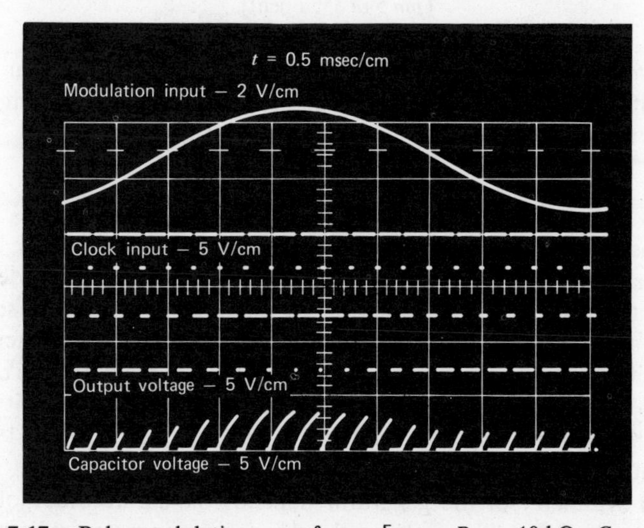

Figure 7-17. Pulse-modulation waveforms [*note:* R_A = 10 kΩ, C = 0.02 μF].

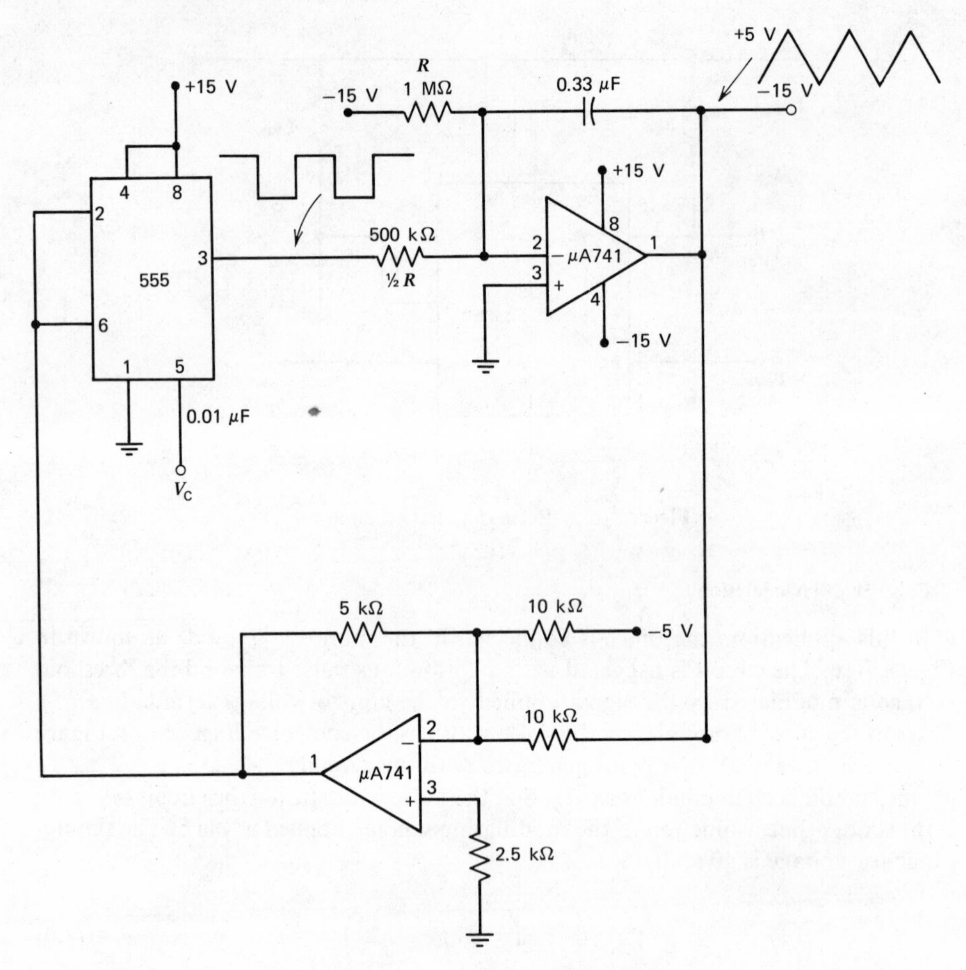

Figure 7-18. VCO using 555 timer as a schmitt trigger [*note:* $f = 0.75/RC$ (*pin* 5 of 555 open)].

This voltage would have to be a linear function of time to result in a linear relationship of control voltage versus pulse width. A method of achieving a triangular (linear) voltage at the timing capacitor is shown in Figure 7-18.

Square-Wave Generator

An inexpensive square-wave generator can be put together quickly by using the 555-type timer. With only one external resistor, this circuit can be made to generate square waves (Figure 7-19). The generator of Figure 7-19*a* produces square waves because the output voltage is essentially 180° out of phase with capacitor voltage, making capacitor-voltage change in a direction that forces the output to change state. The circuit of Figure 7-19*b* works identically and may be used where it is necessary to eliminate the slight ringing that occurs with the circuit of Figure 7-19*a* just prior to positive-output transitions.

The output symmetry of the generator depends on the accuracy of the timer's internal resistor string which produces the device's comparator-reference voltages.

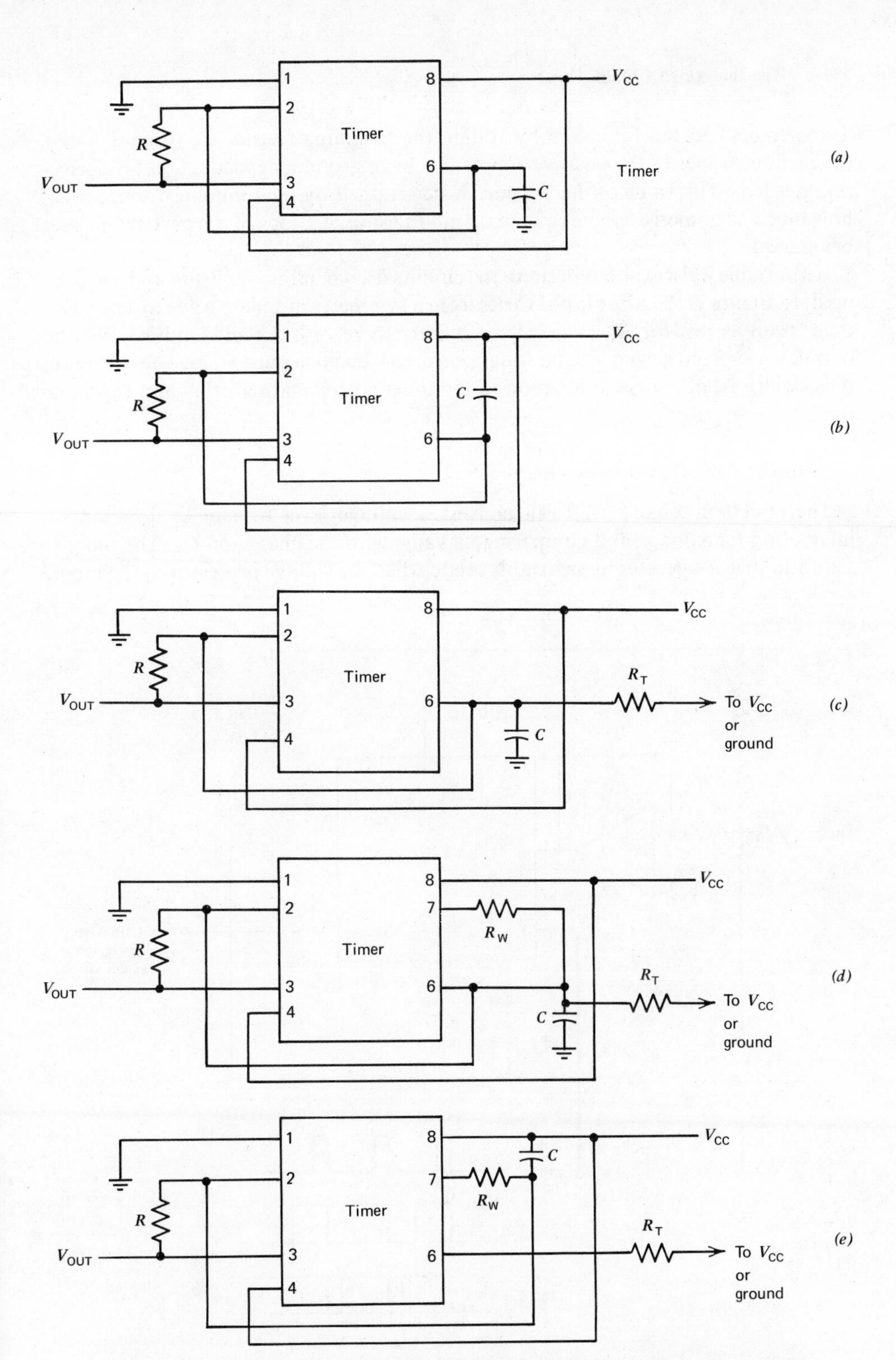

Figure 7-19. Square-wave generator. Timer integrated circuit (*a*) generates symmetrical square-wave output. Slight output ringing can be eliminated with circuit (*b*). Output symmetry can be adjusted with external trimmer (*c*), or output pulse width made variable, as in circuits (*d*) and (*e*).

These errors can be eliminated by adding the trimming resistor R_T (Figure 7-19*c*). (The trimmer goes to the positive-supply line or to ground, depending on the correction needed.) The value of the trimmer is determined by the timing resistance used, how much asymmetry can be tolerated, and the specifications of the particular timer being used.

If a variable pulse width is desired, the circuits drawn in Figures 7-19*d* and *e* may be used. In Figure 7-19*d*, the output varies from a symmetrical square wave to a negative pulse train as resistor R_W ranges from infinity to zero. In circuit Figure 7-19*e*, the output varies from a symmetrical square wave to a positive pulse train. The minimum pulse width is, of course, a function of the timer's propagation delay and capacitor size.

Voltage-to-Pulse-Duration Converter[1]

The circuit in Figure 7-20 can convert a voltage level to a pulse duration by integrating the voltage and comparing its value with the charge on C_1. The timer is set up so that it operates in an astable mode when no voltage is present at the input.

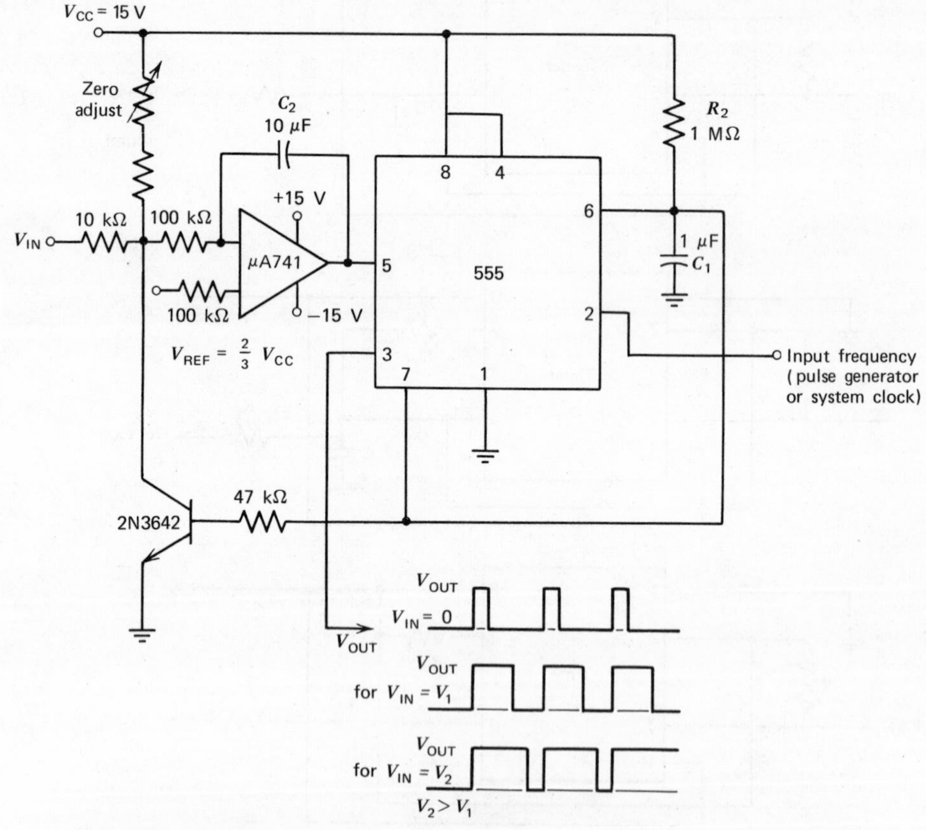

Figure 7-20. Voltage levels can be converted to pulse durations by combining an operational amplifier and a timer integrated circuit [*note:* V_{IN} is limited to ± 2 diode drops within ground or below V_{CC}].

[1] E. R. Hnatek and J. Wyland, "Unconventional Uses for IC Timers," *Electronic Design* **12,** June 7, 1973.

When the input voltage increases, the width of the output pulses increases, but the frequency of the pulse train remains the same (Figure 7-20). Basically this circuit is a dual-slope integrator and has an accuracy of better than 1%. Possible uses occur in data-acquisition and telemetry systems.

Expanded Duty Cycle

The duty-cycle range of the 555-type timer IC can be extended by providing independent charging and discharging paths for the timing capacitor. With the circuit of Figure 7-21*a* suggested by the manufacturer for an astable rectangular-pulse generator, the duty cycle can be adjusted from about 0.01 to almost 50%. The charging path for the capacitor is the series network of resistors R_A and R_B to supply V_{CC}; the discharge path is through resistor R_B to ground.

The addition of two diodes, as done in Figure 7-21*b*, makes the capacitor's charging and discharging paths independent of each other so that the timer's duty cycle can be extended to more than 99%. The charging path is now through resistor R_A and diode D_1 to the supply, while the discharging path is through R_B and D_2 to ground.

For the component values shown, the timer's duty cycle can be adjusted from less than 10% to greater than 90% with only a 1% variation in the output period, which, in this case, is 1 msec. The output pulse frequency is 1 kHz.

Voltage-to-Frequency Converter

A simple and accurate voltage-to-frequency converter can be built with a 555 timer, a LM301-A op amp and a few additional parts. Applications for a voltage-to-frequency circuit are numerous—in A/D conversion, analog-data transmission and voltage-controlled ramp generators, to list a few.

The op amp, connected as an integrator, generates a positive-going ramp from a negative-input voltage (Figure 7-22*a*). When the ramp reaches two-thirds of V_{CC}, the timer triggers, bringing the output voltage at *pins* 3 and 7 close to zero. Transistor Q_1 then turns ON, rapidly discharging C_1. Since the discharge time is constant, the linearity of the converter is limited at high frequencies and is controlled by the R_5C_3 time constant. Resistor R_5 and capacitor C_3 slow the retrace slope at the input (*pin* 2) of the 555. When the voltage at this input reaches one-third of V_{CC}, the timer resets, the outputs at *pins* 3 and 7 return to high levels, Q_1 is turned OFF and the next cycle begins.

The retrace time for the voltage-to-frequency converter is about 1 μsec. This results in less than a 0.2% nonlinearity deviation from the best straight line over the frequency range of 0 to 10 kHz. At frequencies for which the retrace time can be neglected, the conversion constant is given by

$$f = 3\frac{V_{IN}}{2(V_{CC}R_1C_1)} \tag{7-10}$$

In the circuit shown, $f = (10^3)(V_{IN})$.

If the V_{CC} supply is not sufficiently regulated, the timer's reference voltage at *pin* 5 can be stabilized if a Zener diode is connected to ground and a resistor to V_{CC} and if the resistor is adjusted for the nominal Zener current. The voltage at *pin* 5 then stays at the Zener voltage. Capacitor C_1 should be a polycarbonate or polystyrene type

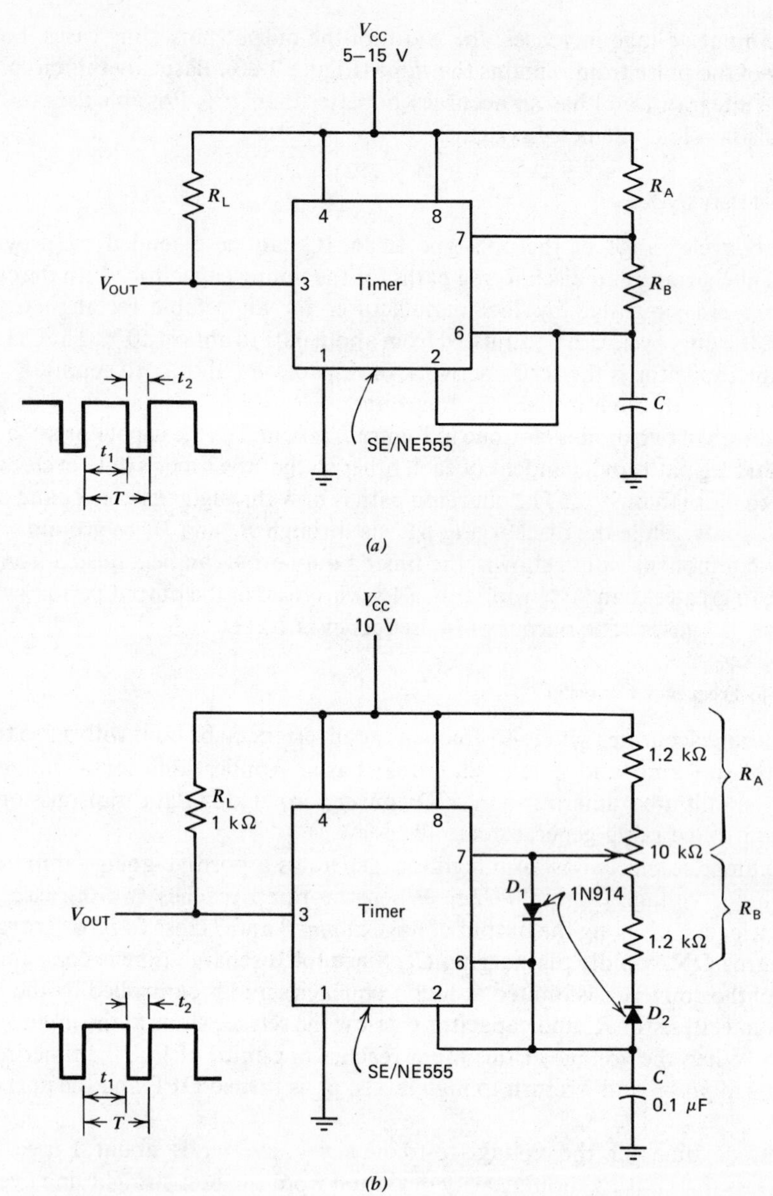

Figure 7-21. Increased duty cycle. Usual configuration for astable multivibrator (*a*) limits output duty cycle to about 50%. Adding two diodes, as shown in (*b*), separates the capacitor's charging and discharging paths, allowing duty cycles of greater than 99% to be achieved [*note:* (*a*) and (*b*), $t_1 = 0.685R_AC$, $t_2 = 0.685R_BC$, $T = 0.685(R_A + R_B)C$].

for best thermal stability. The zero adjust is set by R_3 so that the output frequency with a shorted input is about 0.1 Hz. Worst-case shift of this adjustment with a 20°C temperature change is 1.2°C, corresponding to a 1.2-Hz frequency shift.

The output at *pin* 7 of the 555 is directly compatible with DTL and TTL logic circuits. If it is necessary to design for a positive input, the modified version of the circuit shown in Figure 7-22*b* can be used. The only drawback here is that the circuit

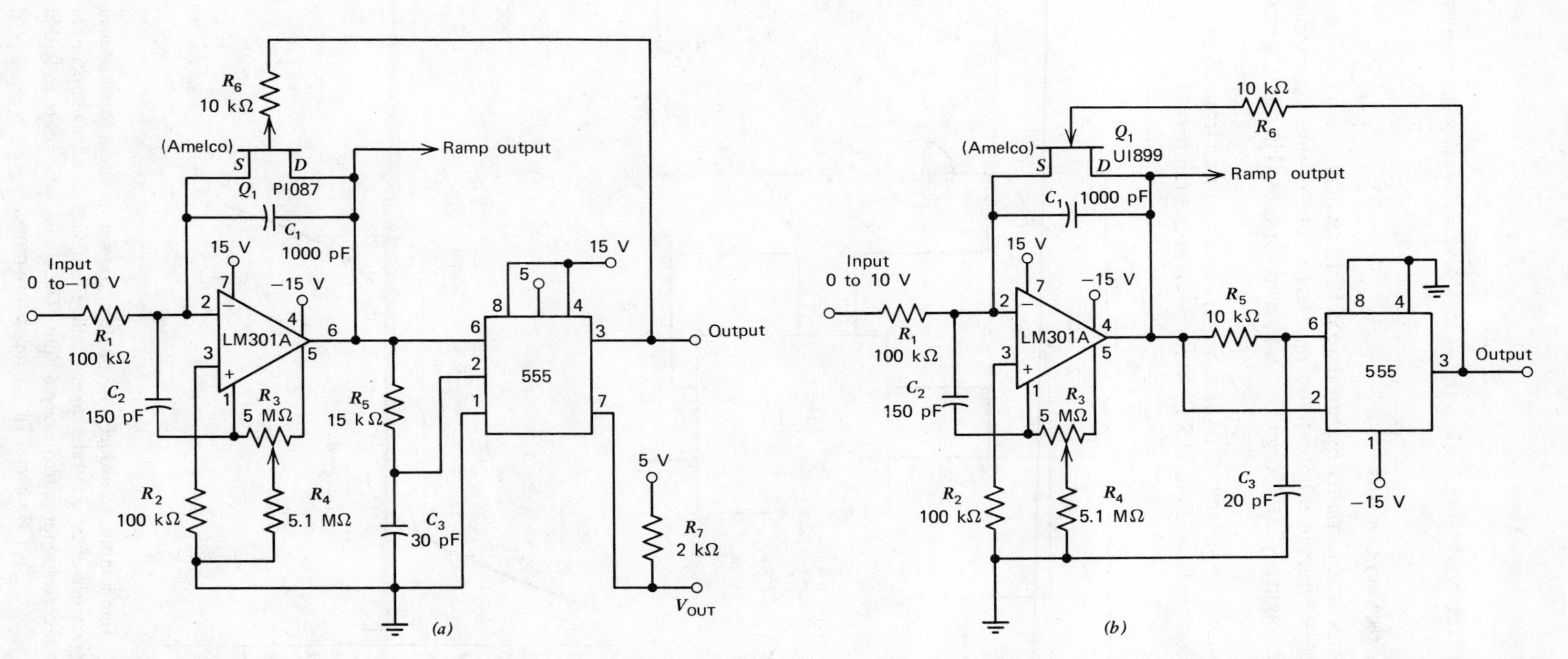

Figure 7-22. Linear voltage-to-frequency converter (*a*) achieves good linearity over 0 to −10 V. Its mirror image (*b*) provides the same linearity over the 0 to +10 V range but is not DTL/TTL compatible.

has no direct logic compatibility, unless the −15-V level is used as the common for the logic supply.

Temperature-to-Frequency Converter

When wired as an astable multivibrator, the 555 IC timer may be used to generate a square-wave output voltage whose frequency has a one-to-one correspondence with temperature (Figure 7-23). A negative TC thermistor is used in the IC's charging network.

The circuit's output frequency varies in a nearly linear manner from 38 to 114 Hz as temperature changes from 37 to 115°F. At no point in this temperature range does

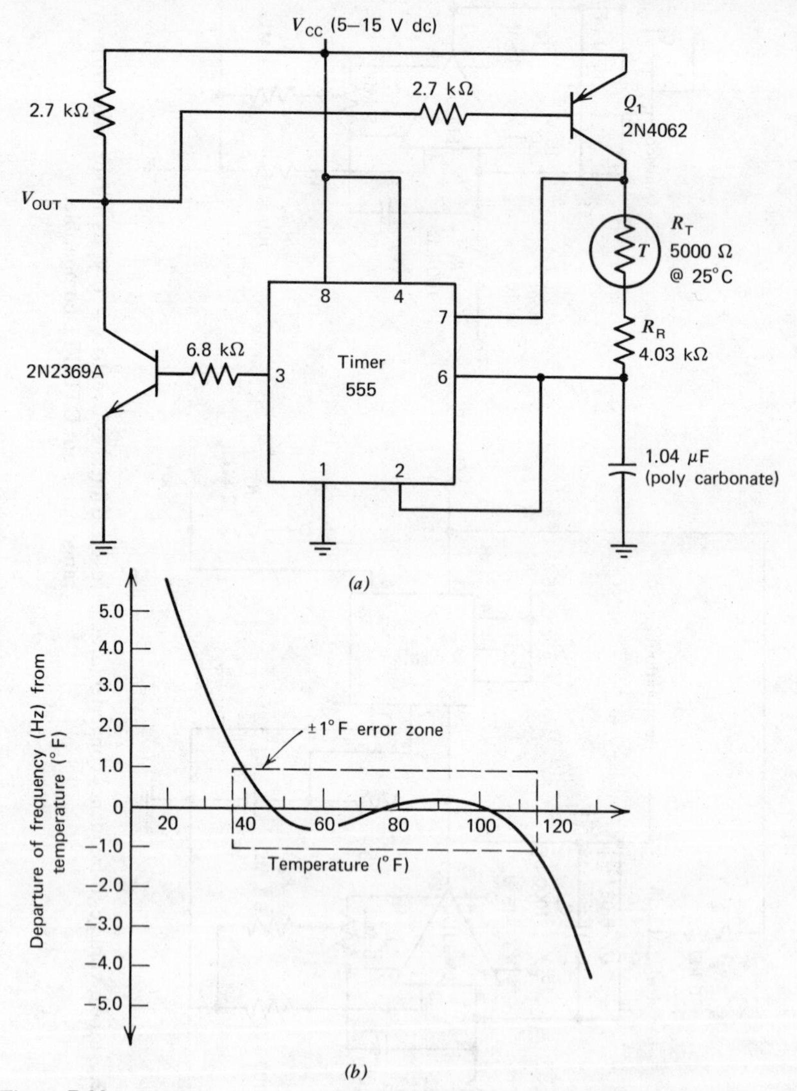

Figure 7-23. Temperature transducer. A couple of transistors and a thermistor in the charging network of the 555-type timer enable this device to sense temperature and produce a corresponding frequency output. The circuit is accurate to within ±1 Hz over a 78°F temperature range.

the frequency count differ by more than ±1 Hz from the corresponding temperature. Due to the small parts count, low cost, and low-power requirements (9.3 mA at 10 V dc), this temperature-to-frequency converter makes an inexpensive temperature transducer that can be used for telemetry applications.

The conventional astable configuration for the 555-type timer employs two fixed resistors. In place of one of these, the converter circuit uses a thermistor/resistor series combination. The other fixed resistor is replaced by transistor Q_1, which is turned ON during the charging interval and OFF during the discharging interval. The transistor's near-zero on-resistance and very large off-resistance result in equal charge and discharge intervals that depend on only R_T and R_B. Operating frequency can then be given by

$$f = \frac{1}{2(R_T + R_R)C \ln(2)} \quad \text{or} \quad f = \frac{k}{R_T + R_R} \tag{7-11}$$

Frequency variation with temperature, therefore, is similar to the voltage variation of a thermistor/resistor divider network. (This type of divider is often used in a bridge arrangement to produce a linearized voltage output with temperature.) The divider's output voltage can be expressed as

$$V_{OUT} = \frac{R_B}{R_R + R_T} V_S \tag{7-12}$$

Since the demoninators of this equation and the frequency equation are the same, the frequency/temperature relationship of the converter circuit will have the same shape and degree of linearity as that of the voltage output of a conventional thermistor/resistor divider.

When a thermistor having an R_0 value of 5000 Ω at 25°C and a resistance ratio of 906:1 over the temperature range of 0 to 50°C is used, the converter circuit produces a linearity error of less than ±1 over a 78°F range. Figure 7-23 contains a plot of this temperature/frequency performance.

It is purely coincidental that the frequency count of the circuit is the same as the useful fahrenheit temperature range (37 to 115°F) for which the circuit is nominally designed. In general, the frequency will be linear with respect to temperature in any interval of interest, but the frequency count will probably be different from the absolute value of the temperature being sensed.

To minimize circuit error, it may be necessary to use temperature-stable polycarbonate capacitors. For this circuit, off-the-shelf capacitors having nominal ±5% tolerances were employed, with the final capacitance being a number of parallel capacitors hand-selected to give the correct frequency count at a given temperature. The IC timer itself contributes negligible error to the frequency output over temperature. Without adequate power-supply bypassing, the circuit is somewhat sensitive to supply-voltage variations.

Servo System Controller[2]

In the motor controller shown (Figure 7-24), the transmitter consists of a timer connected as a variable duty-cycle oscillator. Diode D_1 and a potentiometer R_1

[2] *Ibid.*

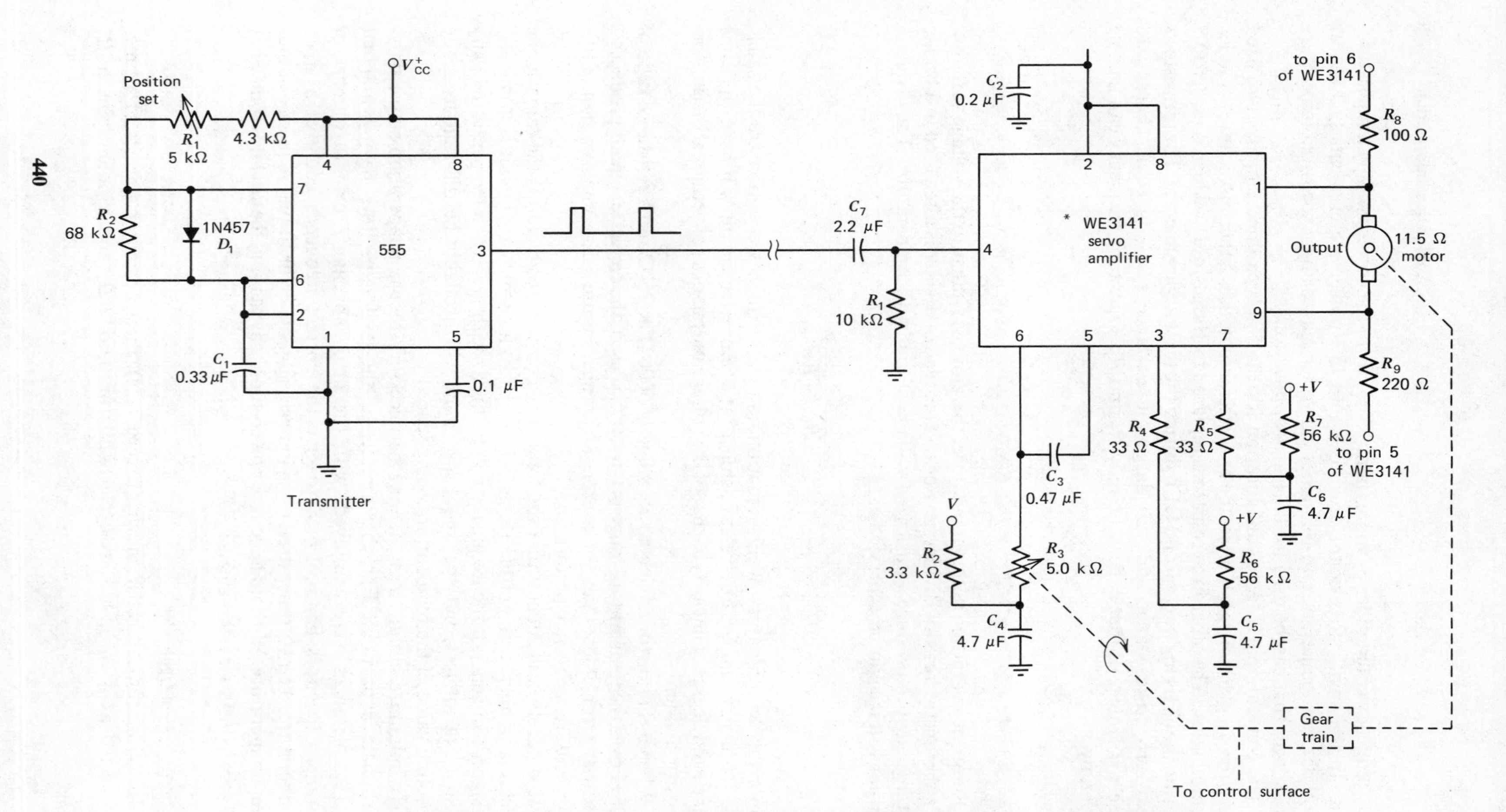

Figure 7-24. To remotely control a servo motor, the 555 needs only six extra components.

provide the charge current to C_1, which sets the duration of the positive portion of the output cycle. Potentiometer R_1 can vary the time duration from 1 to 2 msec. During the negative portion of the cycle, D_1 is back-biased and a discharge time of 16 msec is set by resistor R_2.

Servo drive is generated by the servo amplifier WE3141. This IC receives the pulse-width modulation from the transmitter and compares the 1 to 2 msec pulse width with the duration of an internally generated pulse. If the pulse widths are not of equal duration, the difference is stretched and applied to the output stage. Depending upon whether the input pulse is longer or shorter than the generated pulse, the motor will be driven either clockwise or counterclockwise to adjust the internal pulsewidth to match that of the transmitter.

Resistors R_4 and R_5 set the null point of the amplifier. The 33-Ω values allow a null period of about 4 to 5 μsec. This hysteresis is necessary to prevent the system from hunting. This type of circuit is useful in a wide range of remote-control systems.

Touch Switch[3]

A touch switch for security or convenience purposes can be constructed from the 555 monolithic timer and just a few additional components. Some of the applications of the touch switch include: Switchless keyboards, thief annunciators, activators for the physically handicapped, bounceless electronic switches (with no moving parts), and novelty controls.

The trigger input on the device (Figure 7-25) is the key feature in touch-switch applications. Requiring only 500 nA to fire at $(1/3)V_{CC}$ supply voltage (referenced to circuit ground), the device is easily triggered by the voltage differential found between a floating (nongrounded) human body and the circuit itself. This is 20 V or more, depending on static build-up. The touch plate can be any conducting material with virtually no size limitations.

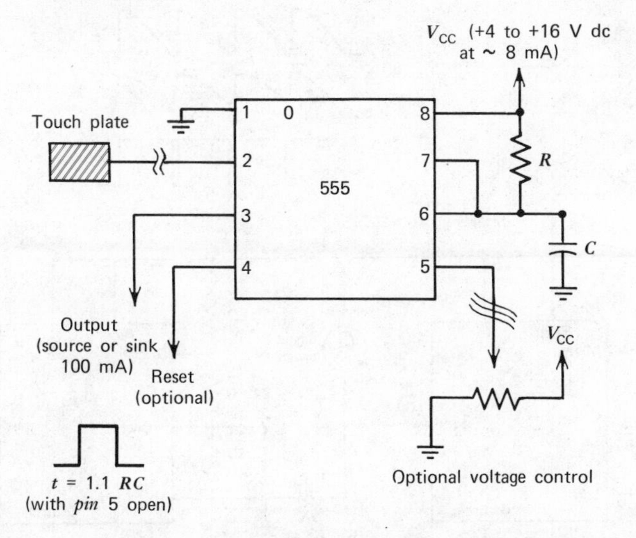

Figure 7-25. A touch plate at the trigger input converts the 555 timer into a versatile touch switch.

[3] J. C. Heater, "Monolithic Timer Makes Convenient Touch Switch," *EDN*, Dec. 1, 1972.

Once triggered, the device cannot be retriggered and it will time out. However, if the duration of human contact exceeds the *RC* time constant of the timer, random spikes occur in the output after the time out. This can be avoided by making fairly large time constants, so that the device will not time out before contact is removed.

The duration of the output pulse is controlled by both the *RC* time constant and the control voltage input (*pin* 5). By varying the voltage at *pin* 5, the timing can be changed by about one decade. If the entire *RC* network is omitted and *pin* 7 is connected directly to *pin* 6, the circuit will latch ON when "touched."

Tone–Burst Generator

With very few external components, two IC timers can be made to function as a tone-burst generator that is useful for radio and telephone applications. In the circuit shown in Figure 7-26, one timer controls the tone burst, and the other generates its frequency.

TIMER_1 is set up as an astable oscillator. But its threshold inputs are kept high by the additional *RC* network (R_2 and C_2) for longer than it takes the timer's discharge circuit to completely discharge the main *RC* network (R_1 and C_1). This assures that the output period of TIMER_1 remains almost constant, no matter if the burst is the first one or the last one.

The period that TIMER_1's output remains high can be approximated by the standard equation for delay-mode operation:

$$T_{\text{ON}} = 1.1R_1(C_1 + C_2) \tag{7-13}$$

The burst output time (when the output is low) can be adjusted to the desired value by the R_2C_2 network. This period is approximated by the equation for astable-mode operation:

$$T_{\text{OFF}} = 0.693R_2C_2 \tag{7-14}$$

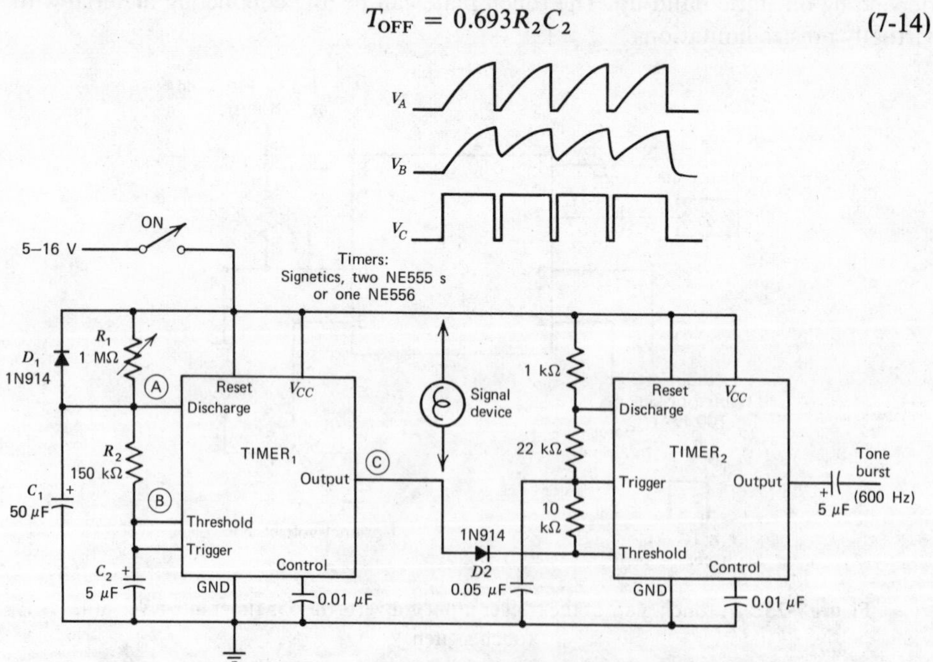

Figure 7-26. Tone–burst generator using IC timer.

When the added time period (burst length) approaches or exceeds the main time period, the two timing networks interact.

For this circuit, the output of $TIMER_1$ remains high for 1 minute and goes low for a half-second. The best way to activate the circuit is to switch the V_{CC} supply lead for the entire circuit. Diode D_1 assures that capacitor C_1 will be discharged after any partial periods.

The control timer ($TIMER_1$) can provide the output for a lamp, bell, buzzer, or other signaling device. (This timer's output must be used to sink the signaling device, which must also be wired to the supply line.) $TIMER_2$ operates as the tone oscillator, determining the frequency of the tone burst. The manner in which $TIMER_2$ is keyed eliminates the need for an intermediate device to invert the output of $TIMER_1$ to operate the reset lead of $TIMER_2$.

This simple tone-burst generator can be used as an audible timing reminder for long-distance telephone calls or for radio repeaters that have 3-minute shutdown timers. The same arrangement can be used to generate sampling pulses for a sample-and-hold circuit or for a serial-to-parallel data converter for Ascii-character detectors.

Solid-State Thermostat

Although it is really intended for timing applications, the 555- IC timer makes an economical and versatile solid-state thermostat when used with a negative TC thermistor. The timer's internal resistive divider establishes reference voltage at $(1/3)V_{CC}$ and $(2/3)V_{CC}$ for each of the timer's comparators. When an external voltage applied to the threshold input (*pin* 6) exceeds $(2/3)V_{CC}$, an output is generated by the threshold comparator that toggles the flip-flop. This turns ON the discharge transistor and results in a low-output signal from the timer's driver-amplifier output stage.

In most applications, as in this one, the turn-on of the timer's discharge transistor lowers the voltage at the threshold input to less than $(2/3)V_{CC}$. If the trigger input then drops below $(1/3)V_{CC}$, the trigger comparator generates a pulse that retoggles the flip-flop, drives the discharge transistor OFF, and causes the output stage to return to its high output level.

This circuit action lends itself nicely to temperature-control applications, particularly those normally reserved for thermostats that must maintain an environment within a bounded temperature range. A voltage that is directly proportional to temperature will rise (along with temperature) until threshold voltage $(2/3)V_{CC}$ is reached. The timer's output stage will then change state, so that a refrigeration unit can be turned ON or an oven can be turned OFF. Temperature will then drop until $(1/3)V_{CC}$ exists at the trigger input, causing the output stage to return to its first state—with the refrigerator OFF or the oven ON.

For the thermostat in Figure 7-27, thermistor/resistor divider networks produce the voltage that is directly proportional to temperature. When temperature is rising (high-output state, discharge transistor OFF), the threshold-input voltage is determined by the division between the combination of $(R_T + R_1)$ and R_2, and increases as the value of R_T decreases.

When R_T is equal to the thermistor resistance at the hot setpoint temperature, R_{TH}, the divider relationship needed to establish $(2/3)V_{CC}$ at the threshold input is

$$\frac{R_{TH} + R_1}{R_{TH} + R_1 + R_2} = \frac{1}{2} \qquad (7\text{-}15)$$

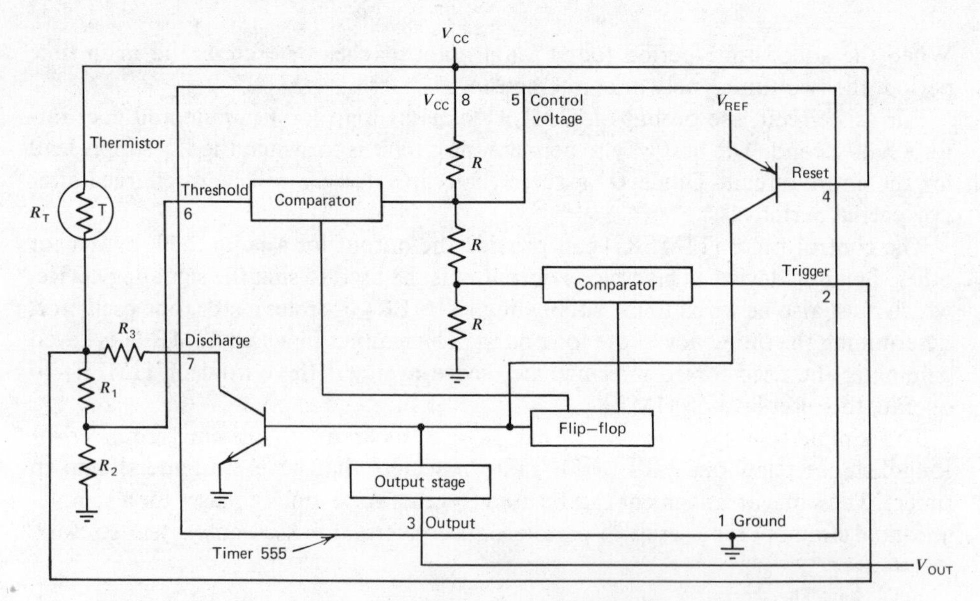

Figure 7-27. Temperature controller. Voltage divider formed by the thermistor and fixed resistors converts IC timer to solid-state thermostat. Upper and lower temperature limits are set by the switching voltages of the threshold comparator and the trigger comparator, respectively.

After an input to the threshold comparator reaches this level, the discharge transistor is switched ON, effectively placing R_3 in parallel with $(R_1 + R_2)$.

As the temperature drops, R_T increases in value, and the division is between R_T and $[R_3 \| (R_1 + R_2)]$. When R_T is equal to the resistance at the cold setpoint temperature R_{TC}, the divider must produce $(1/3)V_{CC}$ at the trigger input. The divider relationship becomes

$$\frac{R_3 \| (R_1 + R_2)}{R_{TC} + [R_3 \| (R_1 + R_2)]} = \frac{1}{2} \tag{7-16}$$

Therefore, the impedance level of the thermistor-resistor dividers is effectively changed in different ways, depending on whether the thermostat is in the rising temperature portion of its operating cycle or the cooling portion. This is necessary since a thermistor's resistance varies quasi-exponentially with temperature and may exhibit a two- or threefold change over a narrow temperature range. That is, the thermistor's cold setpoint resistance R_{TC} may be several times larger than its hot setpoint resistance R_{TH}.

If a standard thermistor is used and its resistance as a function of temperature is known, a straightforward design approach applies. When R_{TC} exceeds R_{TH} by a factor of 2 or more, let $R_2 = R_{TC}$ and let $K = R_{TC}/R_{TH}$ where K is a constant. For proper divider ratios

$$R_1 = \left(\frac{K}{2 - 1}\right) R_{TH} \tag{7-17}$$

$$R_2 = K R_{TH} \tag{7-18}$$

$$R_3 = \frac{3K^2 - 1}{4K - 2} R_{TH} \tag{7-19}$$

However, if the setpoint resistance ratio, R_{TC}/R_{TH} is less than 2, then let $R_1 = 0$ and $R_2 = 2R_{TH}$, so that

$$R_3 = \frac{2R_{TH}R_{TC}}{2R_{TH} - R_{TC}} \tag{7-20}$$

(For this analysis it is assumed that the timer's trigger and threshold inputs do not load the dividers.)

Thermistor power dissipation must be kept as low as possible to maintain the accuracy of the thermostat's setpoints. By operating the timer from the lowest possible supply voltage—5 V—thermistor self-heating can be minimized. But at high temperature setpoints, where thermistor resistance may be quite low (only a few hundred ohms), this approach may not be practical. On the other hand, at very cold temperatures, the thermistor–resistor divider impedance levels must be evaluated in terms of the timer's threshold and trigger input impedance levels.

To prevent noise signals from causing premature state changes, the timer's trigger and threshold inputs should be bypassed with capacitors. This is particularly important when divider impedance levels are high, the environment is noisy, or long leads are used to connect the thermistor to the circuit.

DC/DC Converter

Figure 7-28 depicts a +5 to ±15 V dc/dc converter using the 555 timer and two NE550 voltage-regulators that can be used to power op amps from either TTL logic supplies or 6-V batteries.

The principle of operation is generally applicable to most dc/dc converters. In this circuit the oscillator frequency was chosen to be approximately 17 kHz for best operation of transformer T_1. The frequency is given by

$$f = \frac{1.44}{(R_1 + 2R_2)C_2} \tag{7-21}$$

Regulation is accomplished by using two NE550 voltage-regulators. The output voltage of the positive regulator is given by

$$V_{OUT}^{+} = 1.63\,\frac{R_5 + R_6}{R_6} \tag{7-22}$$

and the output voltage for the negative regulator by

$$V_{OUT}^{-} = 1.63\,\frac{R_7 + R_8}{R_8} \tag{7-23}$$

where R_4 and R_{11} are current-sense resistors. They limit the output current to an acceptable value and thus provide short-circuit protection.

With this circuit, good ground connections must be provided to keep the rms noise below 500 μV. The experimental converter shown was constructed on a double-sided printed circuit board with one side for circuit traces and the other side a ground plane.

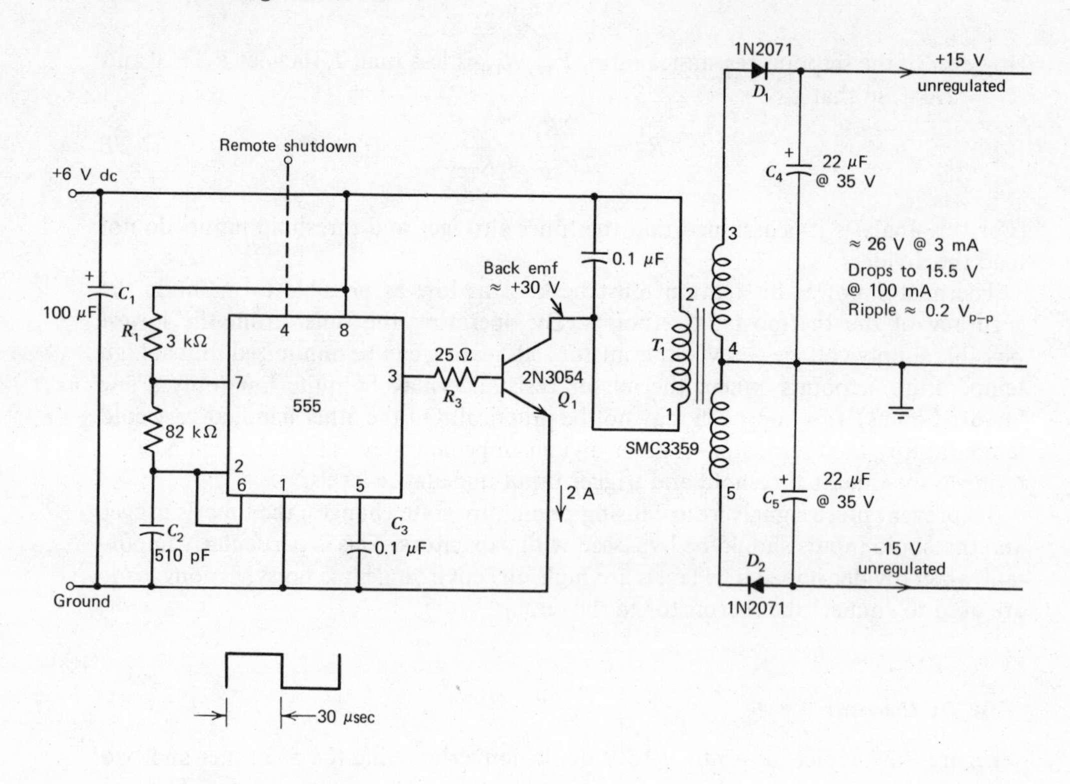

Figure 7-28. A +5 to ±15 V dc converter. Power efficiency at full load

Transistor Q_1 should be selected with appropriate breakdown voltages. Diodes D_1 and D_2 have peak inverse-voltage ratings of greater than 130 V.

The circuit of Figure 7-27 as constructed provides an output current of 100 mA; line and load regulation of 0.1%; power efficiency at full load is better than 75%; recovery from sudden application of load is <10 μsec and the transformer back emf at Q_1 is 30 V.

Transformerless DC/DC Converter[4]

A compact transformerless dc/dc converter, using the 555 IC timer, derives a negative-supply voltage from a positive one. The technique allows dual-supply op amps to operate from a single-supply line and still deliver bipolar outputs.

A square wave is generated by the 555 timer (Figure 7-29) plus four external passive components. Five additional passive components are required to derive the negative supply from the clock.

The circuit has component values chosen to give a 2-kHz pulse-repetition frequency, with the coupling and filter capacitors selected to minimize ripple under heavy loads. Since the timer is insensitive to variations in supply voltage and has good output drive

[4] B. Pearl, "Positive Voltage Changed into Negative, and No Transformer Is Required," *Electronic Design* **11**, May 24, 1973.

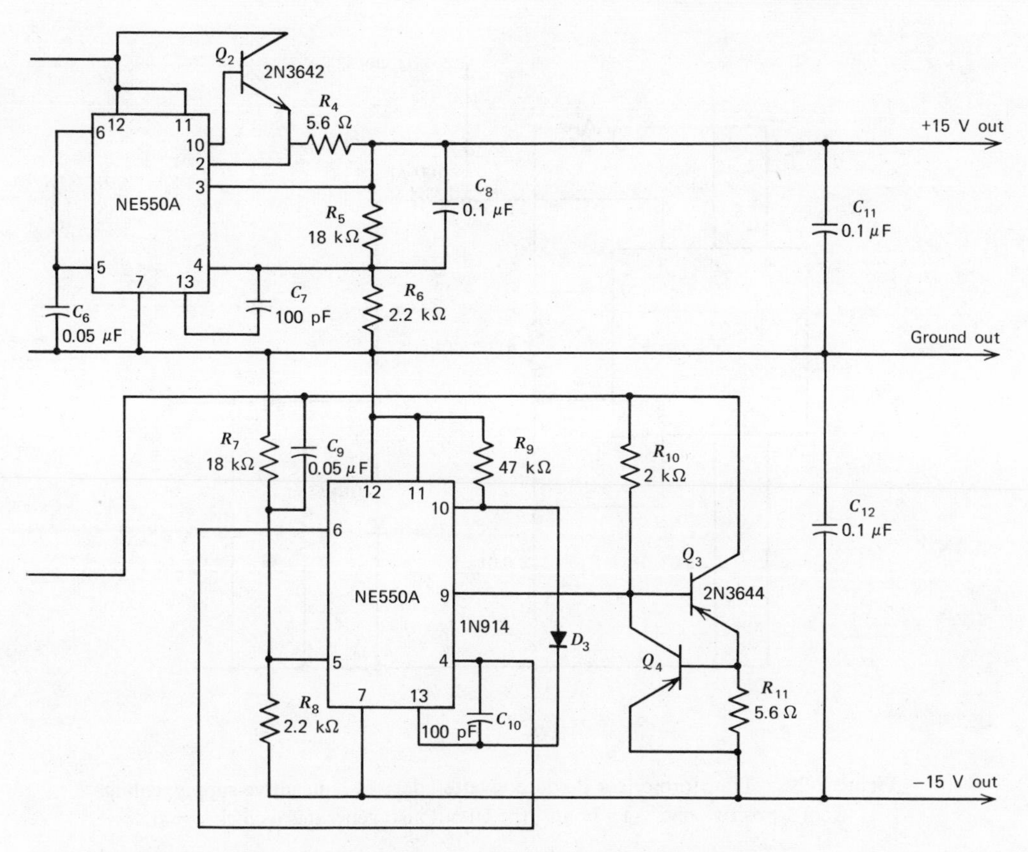

is >75%. Approximately 700 mW quiescent power drain from 5-V supply.

capability, it makes an excellent system clock. For a specific application, the capacitor values used depend on required clock frequency load and ripple rejection.

With a 500-Ω load, typically equivalent to ten μA741s, the negative-output voltage tracks the positive supply, but its absolute value is always about 3 V lower (Figure 7-30). Output regulation for a constant +10-V input is approximately 10% for a change from no load to a load of 500 Ω. Usable outputs are available with load impedances as low as 50 Ω which would represent about 70 μA741-type op amps.

Automobile Applications

AUTOMOBILE TACHOMETER[5]

Pulses generated by the opening and closing of the distributer points are fed into the input of the tachometer circuit (Figure 7-31) and are shaped and clamped by R_1 and D_1. They are then passed on the trigger terminal of the timer by C_1. Triggering of *pin* 2 causes the output of the timer to go high for a period determined by the expression

$$T = 1.1(R_4C_2) \tag{7-24}$$

[5] E. R. Hnatek and J. Wyland, "Inconventional Uses for IC Timers," *Electronic Design* **12**, June 7, 1973.

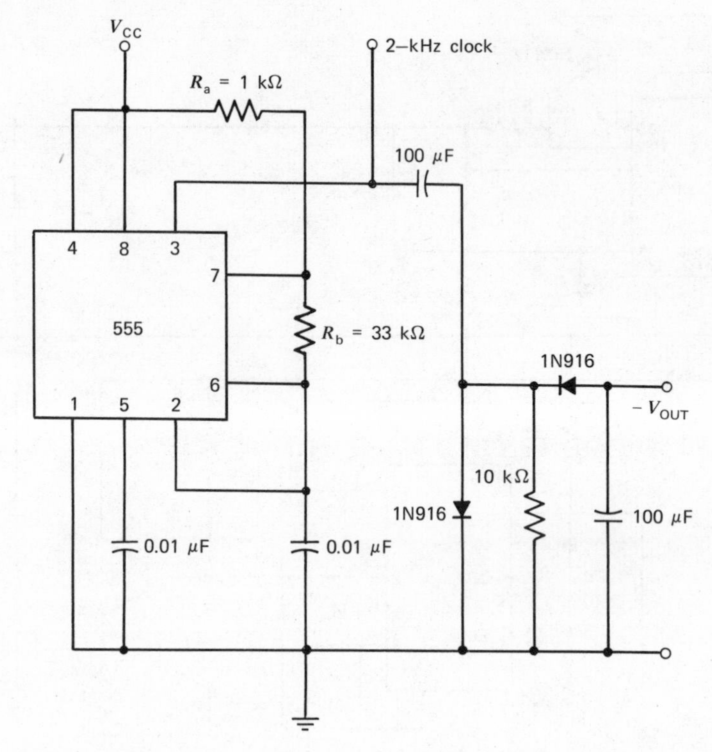

Figure 7-29. Transformerless dc-dc converter derives a negative-supply voltage from a positive one. As a bonus, the circuit also generates a clock signal.

During this time diode D_2 is back-biased and resistors R_5 and R_6 provide a calibrated current to the meter. After the time duration elapses, *pin* 3 goes low, shunting all current around the meter. The ratio of the time for which current flows through the meter to the time for which it is shunted to ground provides an accurate motor reading of engine reductions per minute. For a V-8 engine, the frequency of pulses at the ignition points is four times the engine revolutions per minute, since the points close eight times per revolution of the camshaft and the engine runs at twice the speed of the distributer shaft. A constant current must be applied to the meter during the one-shot period. This is supplied by the vehicle's electrical system via R_7, C_3, and a 9-V Zener D_3.

SPEED-WARNING DEVICE[6]

To obtain speed information, an electronic pick-up is mounted on the brake backing plate. A magnet, small enough to avoid upsetting the wheel balance, is attached to the brake drum. The magnet induces trigger pulses in the pick-up coil. Each time the wheel rotates, a pulse triggers the first of two timers connected as missing-pulse detectors. If the pulses occur at a low enough frequency (Figure 7-32*a*), the output from the first timer, IC_1, will alternate between its high and low states. The second timer IC_2 is then driven from the output of the first (Figure 7-32*b*). If pulses continue to appear at the input from IC_1, the second detector's output remains high. When the speed setting is exceeded, pulses occur too rapidly for the *RC* time

[6] *Ibid.*

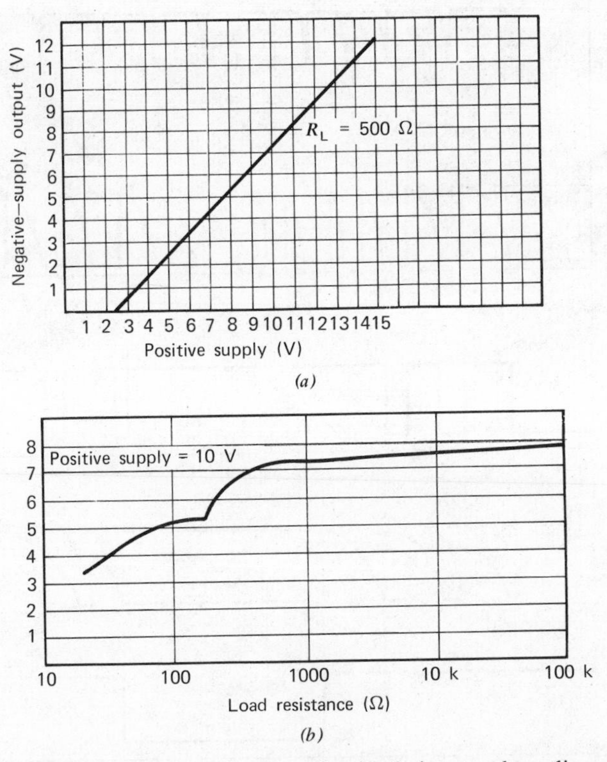

Figure 7-30. Negative output-voltage tracks the dc input voltage linearly (*a*) but its magnitude is about 3 V lower. Application of a 500-Ω load (*b*) causes a 10% change from the no-load value.

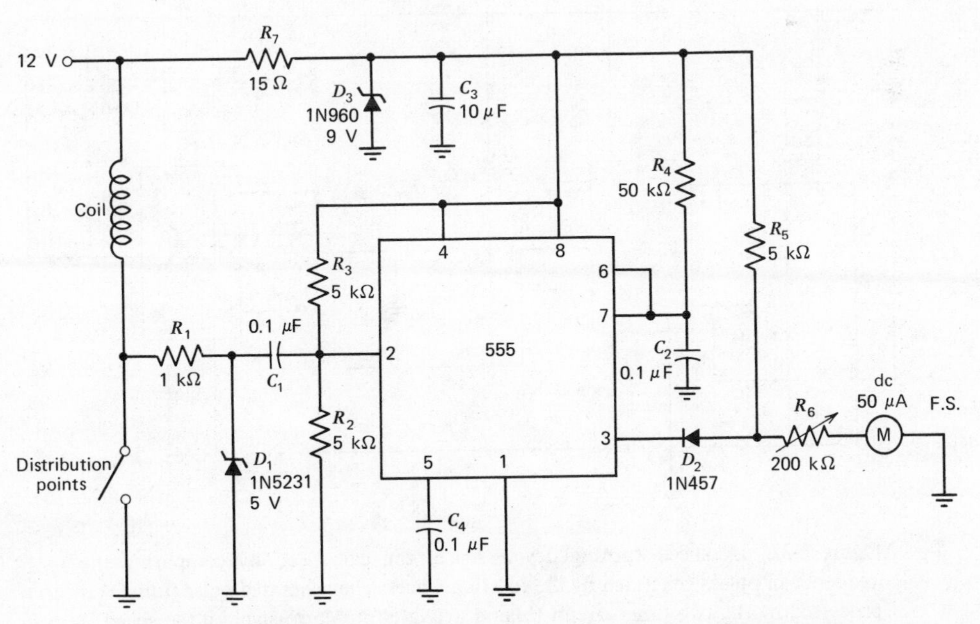

Figure 7-31. A car tachometer can be replaced by this lower power circuit.

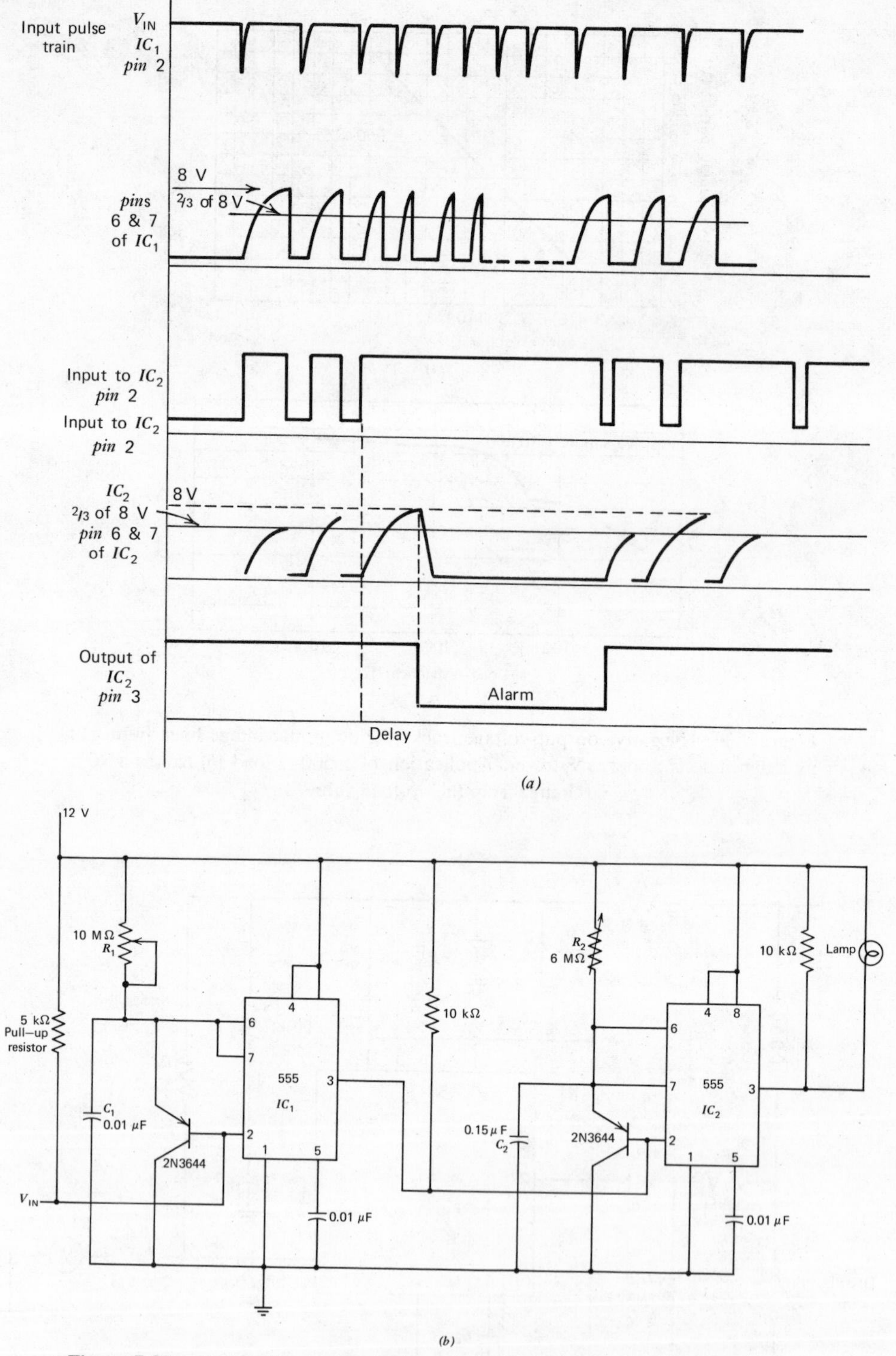

Figure 7-32. A speed-warning device senses the car speed by comparing the frequency of pulses generated by the rotating wheel. The generated pulse train (*a*) is processed by the two timer circuit (*b*) and activates an alarm signal if the speed is too high.

constant of IC_1 to react, and its output remains in the high state, after a delay. This in turn causes the output of IC_2 to go to its low state and turns ON a warning indicator. The speed, and hence the necessary time constants for the timers, can be calculated from tire size. Given a standard tire size of 14 in., the approximate outer diameter of the tire will be 25 in. Therefore each revolution of the tire corresponds to πD, or 78.5 in. Since 1 mile per hour (mph) is equal to 1.467 ft/sec, each revolution can be equated to miles per hour. The time constant of IC_1 is then set to the time necessary for the rotation of the tire to equal a speed of 5 mph, or multiples thereof. Resistor R_1 can either be a potentiometer that has been calibrated for several speeds or several predetermined fixed resistors and a rotary switch. A 556 dual timer can be used in place of two single 555 timers.

AUTOMOBILE BURGLAR ALARM

An inexpensive car burglar alarm system can be built with only two low-cost 555-IC timers or one 556 dual timer. The timers are connected as shown in Figure 7-33.

Timer A serves two purposes: It provides a time delay (roughly 1.1 $R_A C_A$) for alarming the system and allowing the driver to exit the car, and it also permits the driver to enter the car and disarm the alarm. This time delay eliminates the need for an inconvenient and vulnerable arming switch on the outside of the car. The on/off switch for the alarm can be hidden somewhere under the car's dashboard.

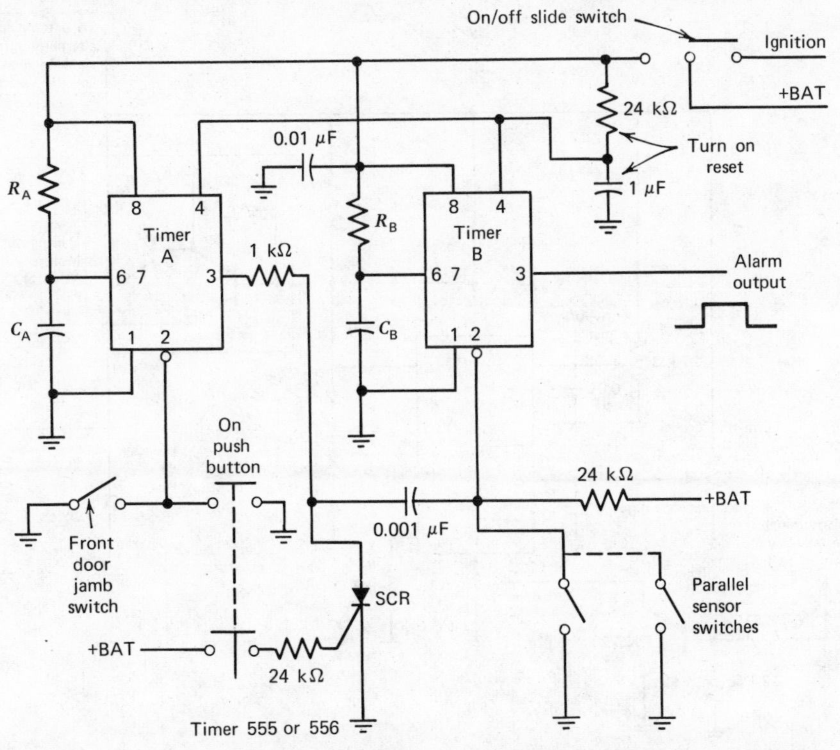

Figure 7-33. Auto burglar alarm. Timer A produces a safeguard delay, allowing driver to disarm alarm and eliminating vulnerable outside control switch. The SCR prevents timer A from triggering timer B, unless timer B is triggered by strategically located sensor switches.

When the alarm goes OFF, timer B is triggered ON by the falling edge of the output from timer A. After the initial turn-ON, however, the SCR prevents timer B from triggering until one of the grounding-type sensor switches fires this timer.

AUTOMOBILE VOLTAGE-REGULATOR[7]

A 555 IC timer, in combination with a power Darlington transistor pair, can provide low-cost automotive voltage-regulation. Such a regulator can even make it easier to start a car in cold weather.

As shown in Figure 7-34, the circuit requires very few parts. The value of resistor R_1 is chosen to prevent the timer's quiescent current, when the timer is OFF (output, *pin* 3, low), from turning ON the Darlington pair.

If battery voltage becomes too low, the timer turns ON, driving its output high and drawing a current of about 60 mA through resistor R_2. This causes a sufficient biasing voltage to be developed across resistor R_1, and the Darlington turns ON, supplying the energizing current to the field coil of the car's alternator. Diode D_1 suppresses the reverse voltage of the field coil when the Darlington pair is turned OFF.

The regulator's low-voltage turn-ON point is fixed by setting the voltage at the timer's trigger input (*pin* 2) to approximately half the reference voltage existing at its control-voltage input (*pin* 5). The high-voltage turn-OFF point is set by making the voltage at the timer's threshold input (*pin* 6) equal to the reference voltage at *pin* 5.

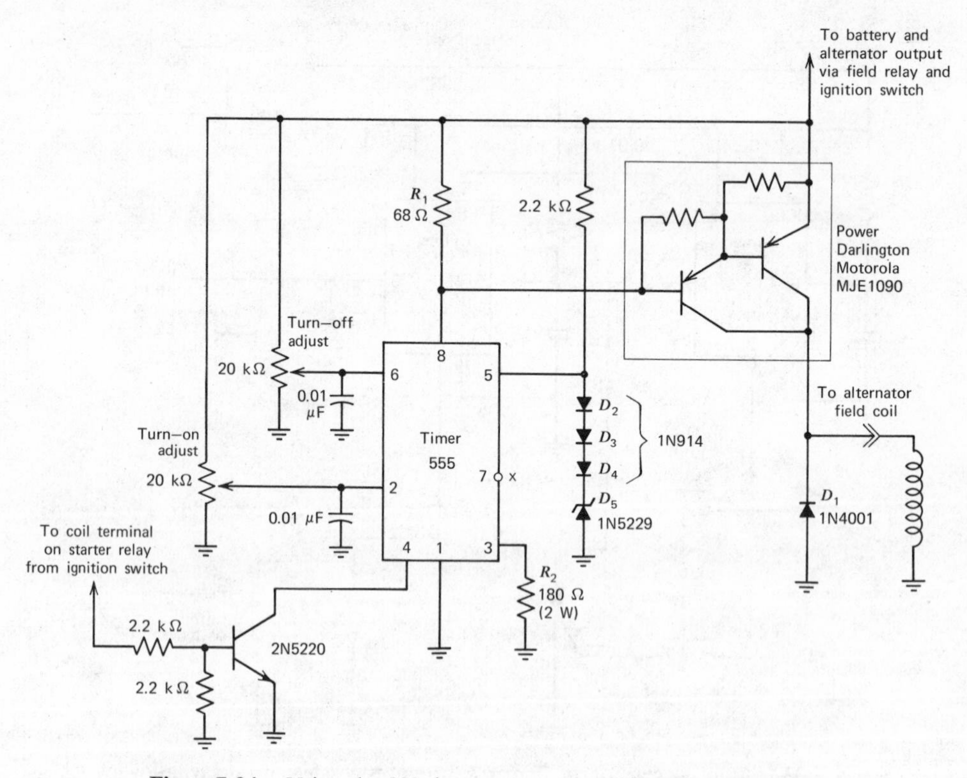

Figure 7-34. Using the 555 timer as an automobile voltage regulator.

[7] T. J. Fusar, "IC Timer Makes Economical Automobile Voltage Regulator," *Electronics*, Feb. 21, 1974.

At 77°F, the turn-ON voltage is typically 14.4 V, and the turn-OFF voltage is typically 14.9 V. These voltage levels, of course, should be set to match the charging requirement of a given car's specific battery-alternator combination.

The value of the reference voltage is established by the diode string, D_2 through D_5; here, it is approximately 5.9 V. The output voltage has a negative temperature coefficient of -11 mVolts/°F.

A transistor and a couple of resistors can be added to the circuit for better cold-weather starting. During starting, the transistor holds the timer in its OFF state, lightening the load on the car's cranking motor. (And to prevent radio interference, a 10-μF capacitor can be connected from the Darlington emitter to ground.)

BATTERY CHARGER

The 555 IC timer can conveniently function as the heart of an automatic battery charger as shown in Figure 7-35. The circuit is intended to maintain a full charge on a standby battery supply for an instrument that is always connected to the ac power line, whether or not in use. This charger uses the timer's two on-chip comparators, its set-reset flip-flop, and its high-current driver amplifier.

The Zener diode, D_1, provides a reference voltage for both comparators through the timer's internal resistive divider network. The output of the timer (*pin* 3) switches between 0 and 10 V.

The circuit is calibrated by substituting a variable dc-power supply for the nickel-cadmium batteries. The OFF adjustment potentiometer is then set for the desired

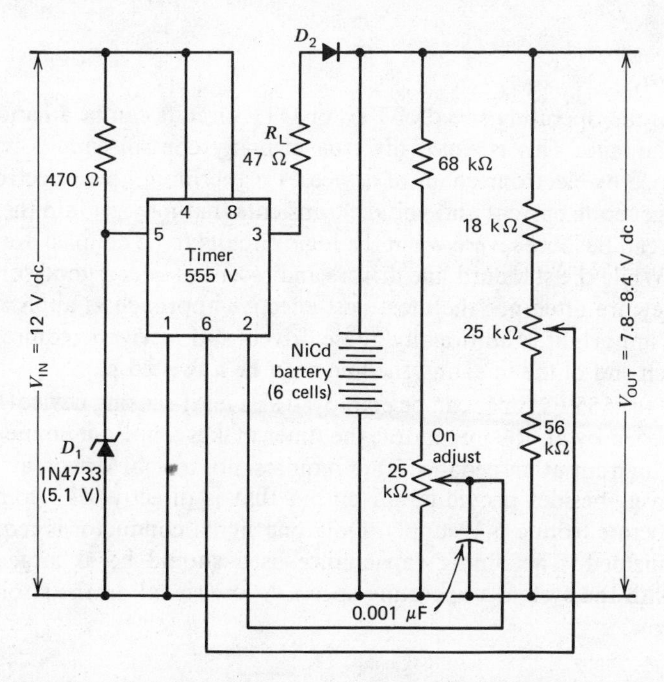

Figure 7-35. Battery charger. Integrated timer functions as the command post for automatic battery-charging circuit. The Zener diode sets the reference voltages for the timer's on-chip comparators. Desired turn-on and turn-off battery voltages are determined by the potentiometers.

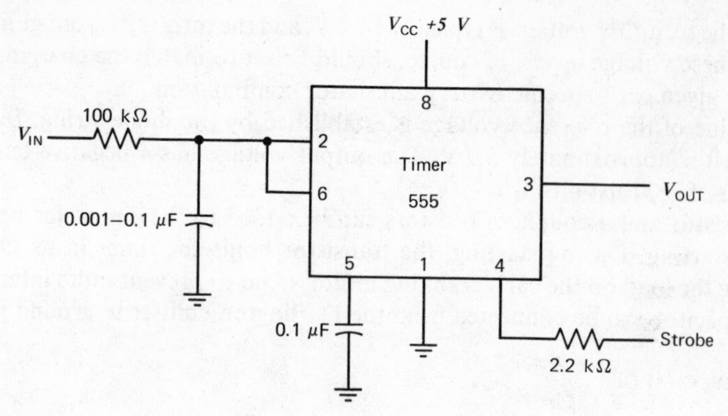

Figure 7-36. Interface circuit. Timer makes excellent line receiver for control application involving relatively slow electromechanical devices. It can work without special drivers over single unshielding lines.

battery cut-off voltage, which is typically 1.4 V per cell: and the ON adjustment potentiometer is set for the desired turn-on voltage, around 1.3 V per cell.

Resistor R_L limits the circuit's operating current to less than 200 mA under all conditions. Diode D_2 prevents the battery from discharging through the timer when the timer is in its OFF state. The capacitor stops oscillation during the circuit's OFF transition. The feedback divider can be decoupled for better load transient immunity, if desired.

Line Receiver

Sometimes the operating speed of TTL or DTL circuits can be a handicap, rather than an advantage. This is especially true in many control circuits where system speed is limited by electromechanical devices. Furthermore, these electromechanical devices can generate current and voltage transients that may get into the logic paths.

Problems can become severe when the logic circuits to be coupled are not close to each other. While the standard line drivers and receivers offer a good solution to this problem, they are often not the most cost-effective approach in applications where speed is not important. Additionally, these drivers and receivers require an interface device at each end of the line; and the line must be a twisted pair.

However, the 555 IC timer can be employed as a level-sensing device (Figure 7-36). When preceeded by an *RC* integrator, the timer makes a noise-immune line receiver that has a high-input impedance and requires no special driver at the sending end. Moreover, besides providing an output that is directly TTL-compatible, the timer can operate from a 5-V supply. Only one signal conductor is required, and it can be unshielded. The timing capacitance used should be as large as possible, consistent with the system's operating speed. A low signal on the strobe line holds the output low.

Chapter 8

Audio Circuits

In this chapter, we will discuss three categories of audio circuits: Audio amplifiers and preamplifiers, FM detectors/limiters, and phase-locked loops. All of these types of ICs are achieving increased popularity due to their versatility, availability, and low cost.

AUDIO AMPLIFIERS

Introduction

The specifications for an audio amplifier state that a required amount of power must be delivered to a specified load (loudspeaker) from a specified signal source (normally the output of a detector) as shown in the block diagram of Figure 8-1. Thus the problem of audio-amplifier design consists essentially of the design of the series of amplifier stages necessary to match the signal to the load. Each of these stages may be thought of as being an individual power amplifier working at a level intermediate to the stages preceding and following it. If the power requirement is 1 or 2 W, then all that is required is the use of one of the readily available IC audio amplifiers. If the power requirement is greater than 1 to 2 W, an IC preamplifier could be used, followed by a discrete output stage, perhaps class B.

The circuit shown in Figure 8-2 is an example of a monolithic IC preamp with a discrete output stage transformer-coupled to a speaker load. Balanced biasing is used, with Q_1 and Q_2 operating at $I_C = 2.2$ mA. The major portion of the gain is obtained from Q_2. The 2-kΩ resistors help to reduce signal flow through Q_1. Transistor Q_3 is a buffer that provides high-input impedance and low-output impedance to emitter-follow Q_4 and the discrete output transistor Q_5. Transformer T_1 provides an impedance match between Q_5 and the speaker. For compensation, negative feedback can be externally connected, as shown in the figure.

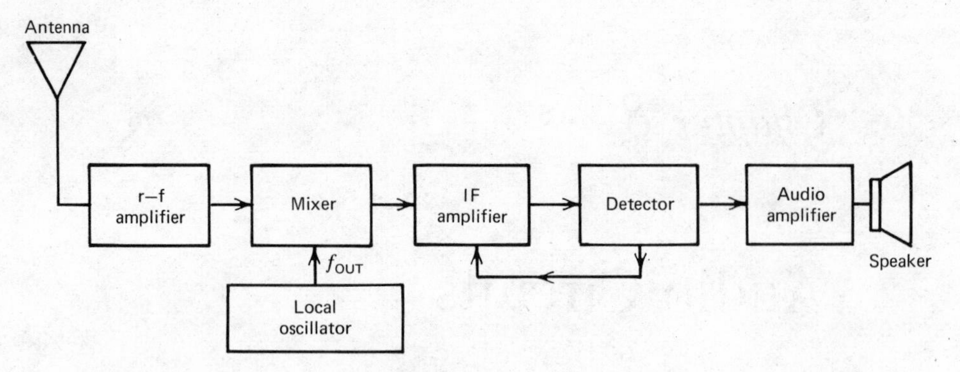

Figure 8-1. Block diagram of superhetrodyne receiver for audio communications.

Audio-Amplifier Applications

PHONOGRAPH AMPLIFIER[1]

The audio-amplifier application presented here use the versatile LM380 as the IC audio amplifier circuit for illustration purposes. The principles, however, apply to most IC audio amplifiers.

One of the most basic circuits using an IC audio amplifier is as a phonograph

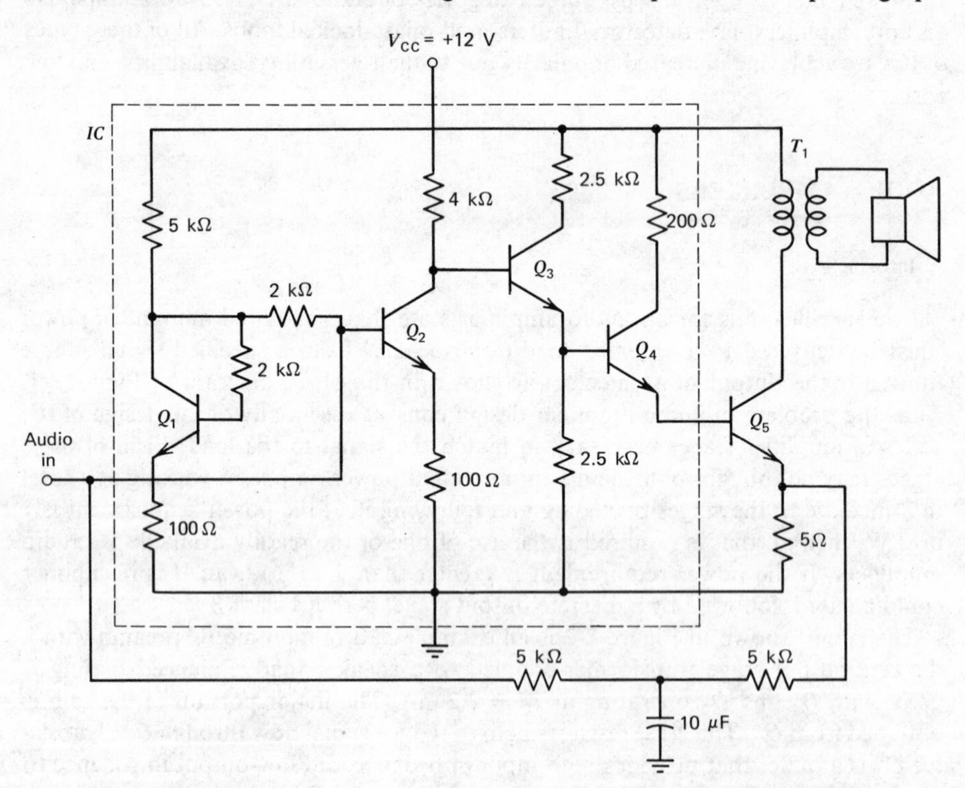

Figure 8-2. Integrated circuit audio amplifier with discrete output transistor Q_5.

[1] Portions of this section were taken from *LM380 Power Audio Amplifier AN69*, National Semiconductor Corporation, Dec. 1972.

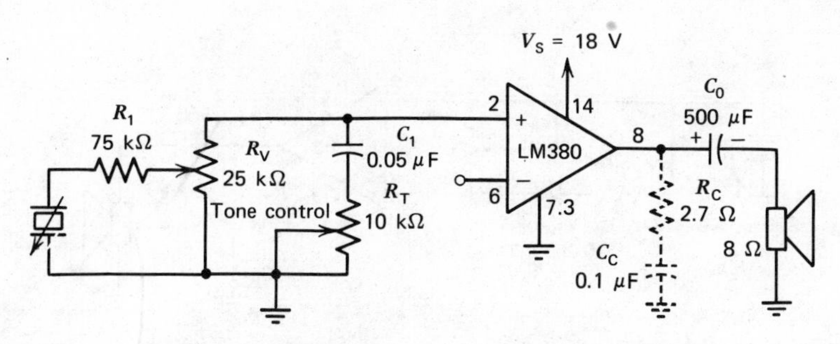

Figure 8-3. Phonograph amplifier. (Courtesy National Semiconductor Corporation.)

amplifier. Figure 8-3 shows the LM380 with a voltage-divider volume control and a high-frequency roll-off tone control. When maximum input impedance is required or the signal attenuation of the voltage divider volume control is undesirable, a "common-mode" volume control may be used as seen in Figure 8-4. With this volume control, the source-loading impedance is only the input impedance of the amplifier when in the full-volume position. This reduces to one-half the amplifier-input impedance at the zero volume position. Equation (8-1) describes the output voltage as a function of the potentiometer setting:

$$V_{OUT} = 50V_{IN}\left(1 - \frac{150 \times 10^3}{K_1R_V + 150 \times 10^3}\right) \quad 0 \le K_1 \le 1 \quad (8\text{-}1)$$

This common-mode volume control can be combined with a common-mode tone control as seen in Figure 8-5.

This circuit has a distinct advantage when transducers of high-source impedance are used, in that, the full-input impedance of the amplifier is realized. It also has an advantage with transducers of low-source impedance since the signal attenuation of the input voltage divider is eliminated. The transfer function of the circuit of Figure 8-5 is given by

$$\frac{V_{OUT}}{V_{IN}} = 50\left(1 - \frac{150 \times 10^3}{150 \times 10^3 + \dfrac{K_1R_TK_2R_V + (K_2R_V/j2\pi fC_1)}{K_1R_T + K_2R_V + (1/j2\pi fC_1)}}\right) \quad \begin{matrix} 0 \le K_1 \le 1 \\ 0 \le K_2 \le 1 \end{matrix} \quad (8\text{-}2)$$

Figure 8-6 shows the response of the circuit of Figure 8-5.

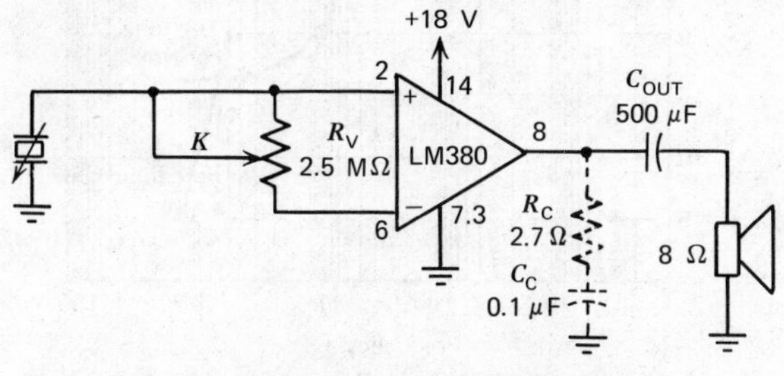

Figure 8-4. Common-mode volume control. (Courtesy National Semiconductor Corporation.)

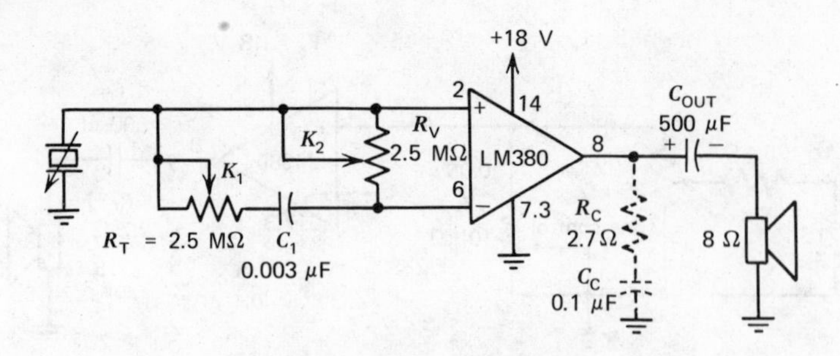

Figure 8-5. Common-mode volume and tone control. (Courtesy National Semiconductor Corporation.)

Most phonograph applications require frequency response-shaping to provide the RIAA (Recording Industries Association of America) equalization characteristic. When recording, the low frequencies are attenuated to prevent large undulations from destroying the record groove walls. (Bass tones have higher energy content than high-frequency tones.) Conversely, the high frequencies are emphasized to achieve greater signal-to-noise ratio. Therefore, when played back, the phono amplifier should have the inverse-frequency response as shown in Figure 8-7. This response is achieved with the circuit of Figure 8-8. The midband gain, between frequencies f_2 and f_3 (Figure 8-7) is established by the ratio of R_1 to the input resistance of the amplifier (150 kΩ).

$$\text{midband gain} = \frac{R_1 + 150\ \text{k}\Omega}{150\ \text{k}\Omega} \tag{8-3}$$

Capacitor C_1 sets the corner frequency f_2 where $R_1 = X_{C_1}$.

$$C_1 = \frac{1}{2\pi f_2 R_1} \tag{8-4}$$

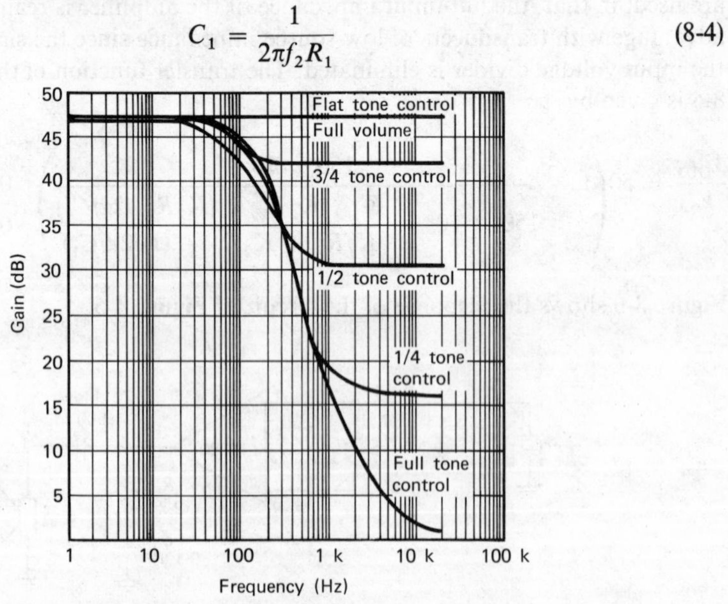

Figure 8-6. Tone-control response. (Courtesy National Semiconductor Corporation.)

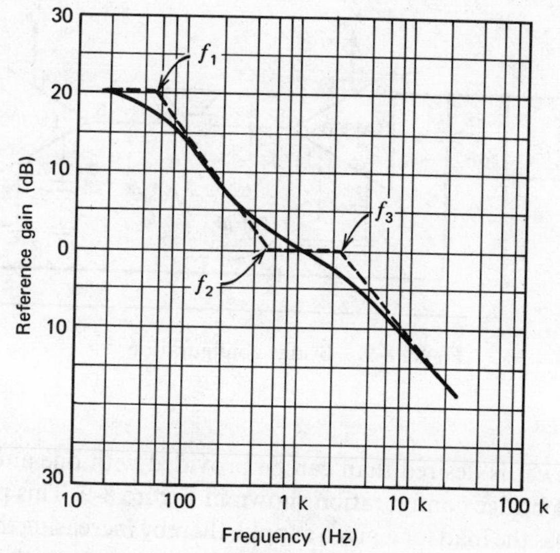

Figure 8-7. RIAA playback equalization. (Courtesy National Semiconductor Corporation.)

Capacitor C_2 establishes the corner frequency f_3 where X_{C_2} equals the impedance of the inverting input. This is normally 150 kΩ. However, in the circuit of Figure 8-8 negative feedback reduces the impedance at the inverting input as

$$Z = \frac{Z_0}{1 + A_0\beta} \tag{8-5}$$

where Z_0 = impedance at node 6 without external feedback (150 kΩ), A_0 = gain without external feedback (50), β = feedback transfer function $\beta = (A_0 - A)/A_0A$, and A = closed-loop gain with external feedback. Therefore,

$$C_2 = \frac{1}{2\pi f_3(Z_0/1 + A_0\beta)} = \frac{1}{2\pi f_3(150 \times 10^3/1 + 50\beta)} \tag{8-6}$$

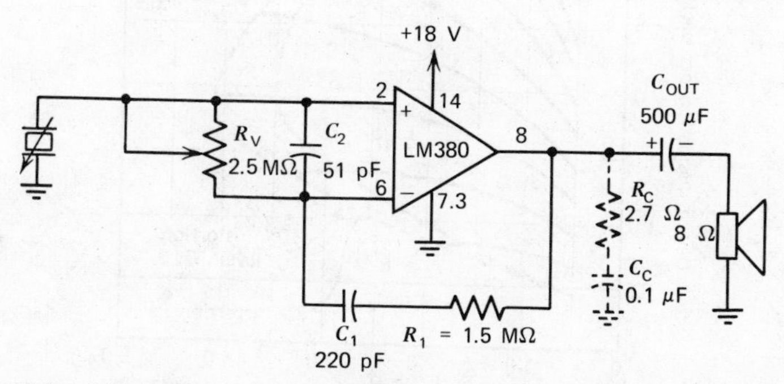

Figure 8-8. RIAA phonograph amplifier. (Courtesy National Semiconductor Corporation.)

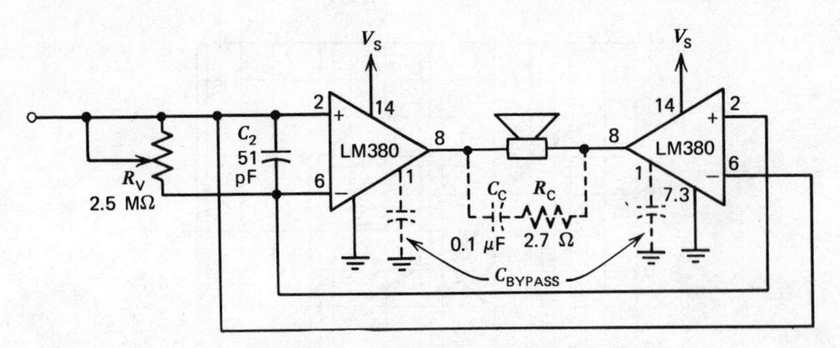

Figure 8-9. Bridge configuration.

BRIDGE AMPLIFIER

Where more power is desired than can be provided with one amplifier, two amps may be used in the bridge configuration shown in Figure 8-9. This provides twice the voltage swing across the load for a given supply, thereby increasing the power capacity by a factor of 4 over the single amplifier. However, in most cases the package dissipation will be the first parameter limiting power delivered to the load. When this is the case, the power capability of the bridge will be only twice that of the single amplifier.

Figures 8-10 and 8-11 show output versus device package dissipation for both 8- and 16-Ω loads in the bridge configuration. The 3 and 10% harmonic distortion contours double back due to the thermal-limiting of the LM380. Different amounts of heat-sinking will change the point at which the distortion contours bend.

The quiescent output voltage of the LM380 is specified at 9 ± 1 V with an 18-V supply. Therefore, under the worst-case condition, it is possible to have 2 V dc across the load.

With an 8-Ω speaker, this is 0.25 A which may be excessive. Three alternatives are available: (1) care can be taken to match the quiescent voltages, (2) a nonpolar capacitor may be placed in series with the load, (3) the offset-balance control of

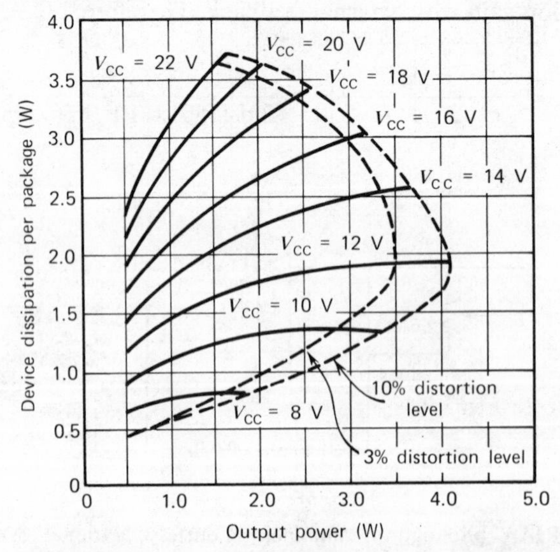

Figure 8-10. An 8-Ω load in the bridge configuration.

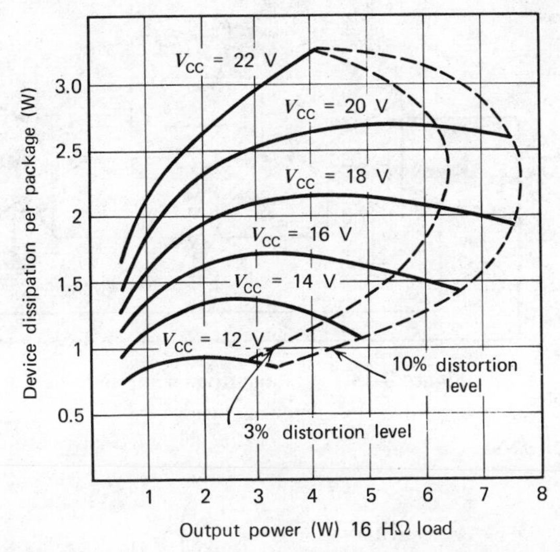

Figure 8-11. A 16-Ω load in the bridge configuration.

Figure 8-12 may be used. The circuits of Figures 8-9 and 8-12 employ the common-mode volume control as shown before. However, any of the various input connection schemes discussed previously may be used. Figure 8-13 shows the bridge configuration with the voltage divider input. As will be discussed later in the "Biasing" section, the undriven input may be ac- or dc-grounded. If V_s is an appreciable distance from the power-supply (>3 in.) filter capacitor, it should be decoupled with a 1-μF tantalum capacitor.

HIGH-INPUT IMPEDANCE CIRCUIT

The junction FET isolation circuit shown in Figure 8-14 raises the input impedance to 22 MΩ for low-frequency input signals. The gate to drain capacitance (2-pF maximum for the KE4221 shown) of the FET limits the input impedance as frequency increases.

At 20 kHz the reactance of this capacitor is approximately $-j4$ MΩ giving a net input-impedance magnitude of 3.9 MΩ. The values chosen for R_1, R_2, and C_1 provide an overall circuit gain of at least 45 for the complete range of parameters specified for the KE4221. When using another FET device, the relevant design

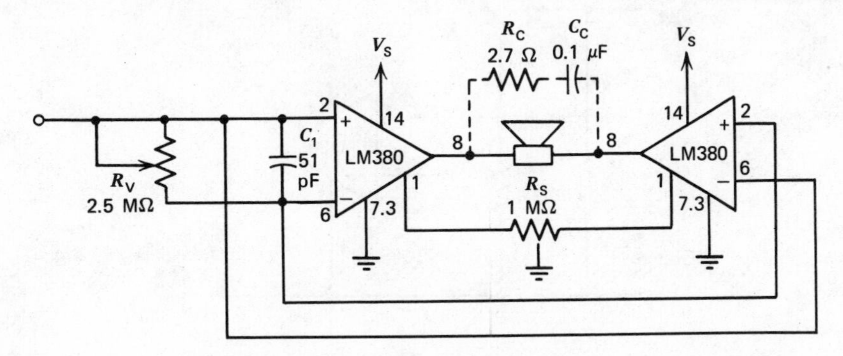

Figure 8-12. Quiescent balance control.

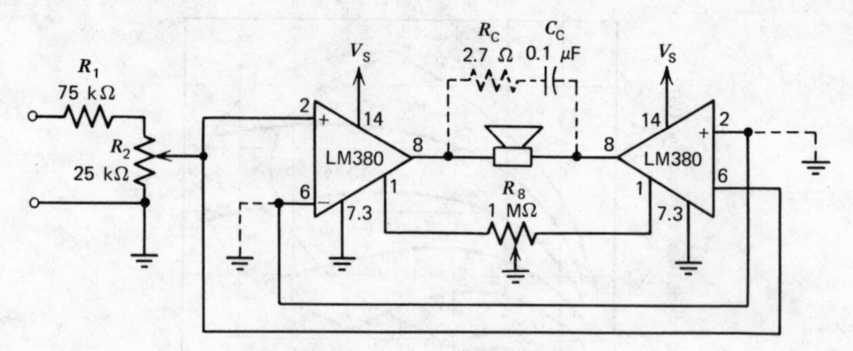

Figure 8-13. Voltage-divider input.

equations are as follows:

$$A_V = \left(\frac{R_1}{R_1 + 1/g_m}\right) \tag{8-7}$$

$$g_m = g_{mo}\left(1 - \frac{V_{GS}}{V_p}\right) \tag{8-8}$$

$$V_{GS} = I_{DS}R_1 \tag{8-9}$$

$$I_{DS} = I_{DSS}\left(1 - \frac{V_{GS}}{V_p}\right)^2 \tag{8-10}$$

The maximum value of R_2 is determined by the product of the gate reverse leakage I_{GSS} and R_2. This voltage should be 10 to 100 times smaller than V_p. The output impedance of the FET source-follower is

$$R_0 = \frac{1}{g_m} \tag{8-11}$$

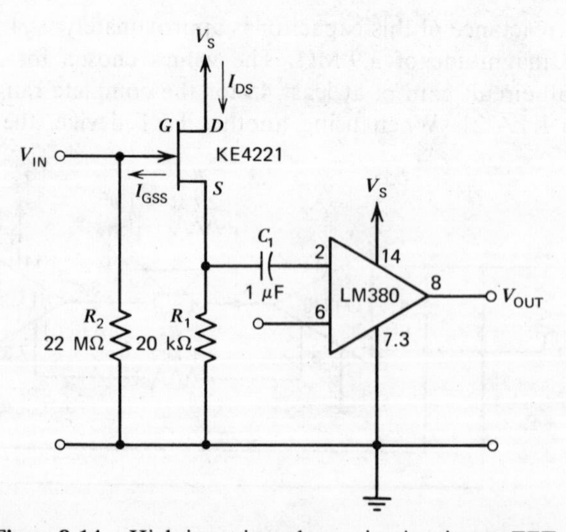

Figure 8-14. High-input impedance circuit using an FET.

so that the determining resistance for the interstage RC time constant is the input resistance of the LM380.

BOOSTED GAIN USING POSITIVE FEEDBACK

For applications requiring gains higher than the internally set gain of 50, it is possible to apply positive feedback around the LM380 for closed-loop gains of up to 300. Figure 8-15 shows a practical example of an LM380 in a gain of 200 circuit. The equation describing the closed-loop gain is

$$A_{V_{CL}} = \frac{-A_{V(w)}}{1 - [A_{V(w)}/(1 + (R_1/R_2))]} \tag{8-12}$$

where $A_{V(w)}$ is complex at high frequencies but is nominally the 40 to 60 specified on the data sheet for the pass band of the amplifier. If $1 + R_1/R_2$ approaches the value of $A_{V(w)}$, the denominator of (8-12) approaches zero, the closed-loop gain increases toward infinity, and the circuit oscillates. This is the reason for limiting the closed-loop gain values to 300 or less. Figure 8-16 shows the loaded and unloaded bode plot for the circuit shown in Figure 8-15.

The 24-pF capacitor C_2 shown in Figure 8-15 was added to give an overdamped square-wave response under full-load conditions. It causes a high-frequency roll-off of

$$f_2 = \frac{1}{2\pi R_2 C_2} \tag{8-13}$$

The circuit of Figure 8-17 will have a very long (1000 sec) turn on time if R_L is not present, but only a 0.01 sec turn on time with an 8-Ω load.

OSCILLATION

Normal power-supply-decoupling precautions should be taken when using the LM380. If V_S is more than 2 or 3 in. from the power-supply filter capacitor, it should be decoupled with a 0.1-μF disk ceramic capacitor at the V_S terminal of the IC.

The R_C and C_C dotted line components of Figure 8-17 and in the previous figures will suppress a 5 to 10 mHz small-amplitude oscillation which can occur during the

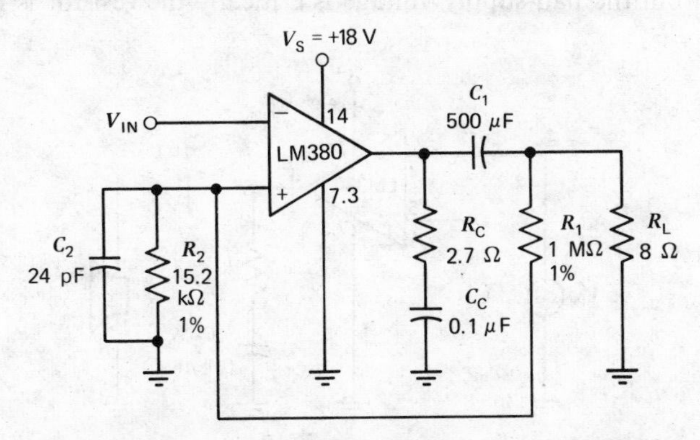

Figure 8-15. Boosted gain of 200 using positive feedback.

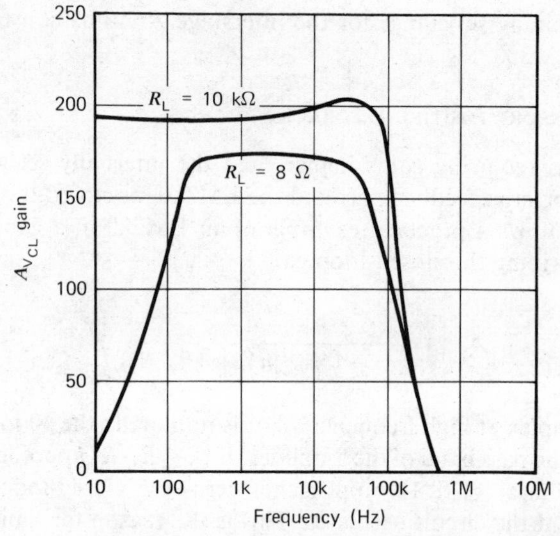

Figure 8-16. Boosted-gain bode plot.

negative swing into a load that draws high current. The oscillation is too high of a frequency to pass through a speaker, but it should be guarded against when operating in an r-f-sensitive environment.

BIASING

The design of the LM380 enables input transducers which are referenced to ground to be direct-coupled to either the inverting or noninverting inputs of the amplifier. The unused input may be either: (1) left-floating; (2) returned to ground through a resistor or capacitor; or (3) shorted to ground. In most applications where the noninverting input is used, the inverting input is left-floating. When the inverting input is used and the noninverting input is left-floating, the amplifier may be found to be sensitive to board layout since stray coupling to the floating input is positive feedback. This can be avoided by employing one of the following: (1) ac grounding the unused input with a small capacitor—this is preferred when using high-source impedance transducers; (2) returning the unused input to ground through a resistor—this is preferred when using moderate to low dc source-impedance transducers and when output offset from the half-supply voltage is critical—the resistor is made equal to

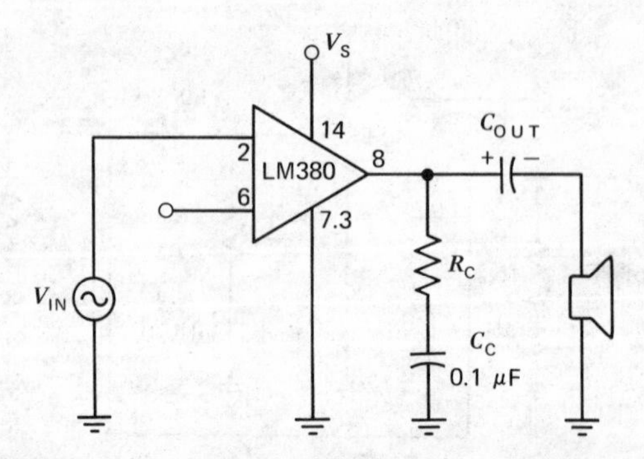

Figure 8-17. Minimum component configuration with oscillation suppression.

the resistance of the input transducer, thus maintaining balance in the input-differential amplifier and minimizing output offset; (3) shorting the unused input to ground—this is used with low dc source-impedance transducers or when output-offset voltage is noncritical.

STEREO TAPE-CARTRIDGE PLAYBACK SYSTEM[2]

The PA237 is a 2-W IC audio amplifier with the basic connection configuration shown in Figure 8-18. This circuit (in dotted lines) uses a quasicomplementary push-pull output-circuit connection comprising Q_4, Q_5, Q_6, Q_7, and Q_8. Three transistors are required for the composite *pnp* transistor since the h_{FE} of Q_6 (lateral *pnp*) is about 1. Transistors Q_6 and Q_7 are equivalent to a discrete *pnp* driver. This arrangement of push-pull composite transistors provides high-current amplification so that Q_2 can operate at low current ($\simeq 0.5$ mA). This enables R_3 to be relatively large (18 kΩ) and to enhance the voltage amplification. Transistors Q_1 and Q_2 form a differential amplifier where Q_1 operates as an emitter-follower and drives Q_2 as a quasicommon-base stage. Transistor Q_2 does operate common-base when R_6 is 0 (i.e., feedback removed from the base of Q_2). This circuit then has no phase inversion of the audio

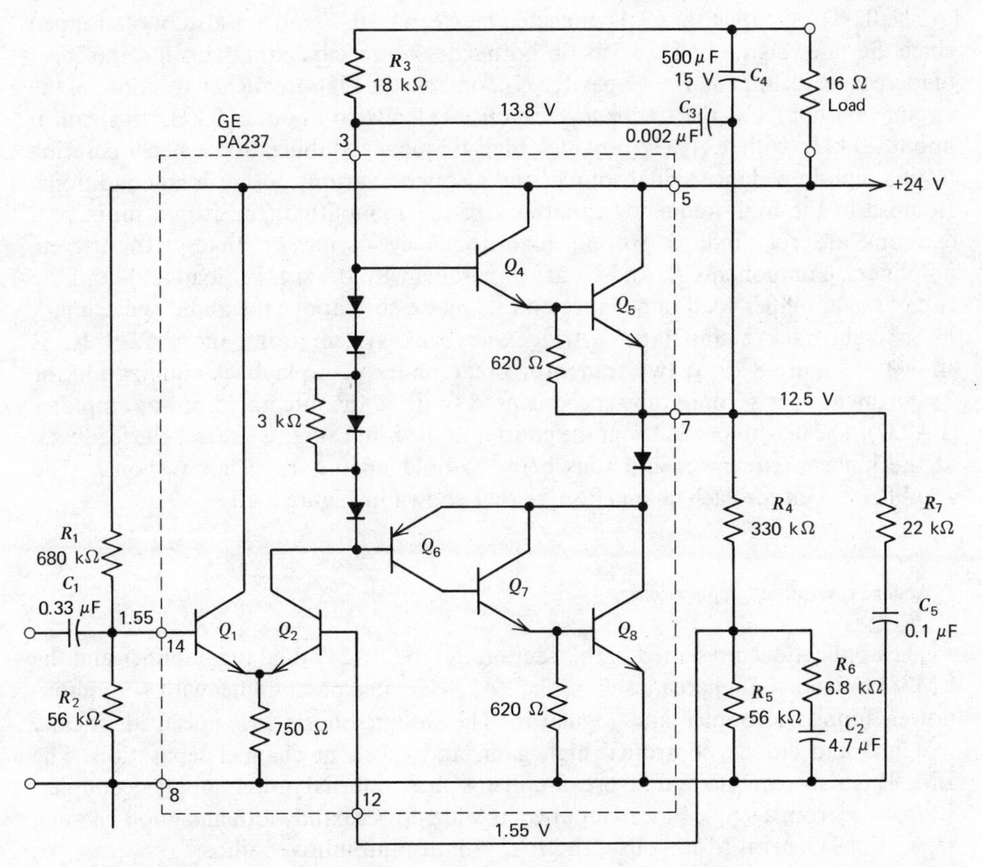

Figure 8-18. Basic connection diagram for the PA237. (Courtesy John Wiley & Sons.)

[2] D. V. Jones and R. F. Shea, *Transistor Audio Amplifiers*, Wiley, 1968.

signal from input to output except within the composite *pnp* transistor. The feedback signal applied to the base of Q_2 is in phase, as required with this connection of the differential amplifier.

The differential amplifier as shown in Figure 8-18, with voltage-source base bias, provides good bias stability against variations in h_{FE} for Q_1 and Q_2, and also against temperature variations. It is important that the dc bias voltage at lead 7 be stable if maximum power output at low distortion is to be maintained. The differential amplifier with dc feedback to the base of Q_2 provides this stability.

The diode string between the bases of Q_4 and Q_6 provides the necessary dc bias voltage for the quasicomplementary push-pull output to minimize distortion at low-signal levels. The value of R_3 can be reduced to maintain 0.5 mA through the series diodes at lower supply voltage. The power-output diode in series with Q_5 and Q_8 assists in the bias stabilization of this direct-coupled circuit, and it also provides local ac feedback in the effective emitter of the composite *pnp* transistor.

The output load is returned to the power supply, which is an ac ground, so that the load-coupling capacitor can also serve to bootstrap resistor R_3. This positive feedback increases the open-loop (i.e., $R_6 = 0$) voltage amplification. The negative feedback with $R_6 = 6.8\ \text{k}\Omega$ is 24 dB.

The 0.002-μF capacitor (C_3) connected between leads 7 and 3 is also bootstrapped since the same signal phase exists on both sides of the capacitor. This has the effect of increasing its impedance. Capacitor C_3 controls the high-frequency response of the circuit. Without C_3, the frequency response extends to about 100 kHz (instead of about 20 kHz with C_3); C_3 provides high-frequency stabilization when operating from a high-impedance signal source, and also with various output-load conditions. Because of the high-frequency capability of the monolithic transistors, more precautions are required to prevent high-frequency oscillation than with discrete amplifiers. Components R_7 and C_5 are in parallel with the speaker load and limit the rise of the amplifier-load impedance and its phase-shift above the audio spectrum.

An eight-track stereo tape-cartridge playback system using the PA237 IC is shown in Figure 8-19. A two-transistor preamplifier with playback equilization for $3\frac{3}{4}$ ips (inches per second) tape speed is used to drive the integrated power amplifier (PA237). The 0.5-μF capacitor at the emitter of the input stage decreases the feedback at the higher frequencies and thus helps to hold up the frequency response. The amplifier circuit for each channel is like that shown in Figure 8-18.

Audio Preamplifier Applications[3]

The applications presented in this section use the PA239 dual preamplifier and the LM381 low-noise dual preamplifier. The PA239 is a dual preamplifier with a common-power supply decoupler and regulator. The design boosts low noise, identically matched and closely controlled high gain, and excellent channel separation. The LM381 is also a low-noise dual preamplifier with an internal power-supply decoupler. It operates from a single power supply, has wide power bandwidth and good channel separation. General Motors uses the PA239 in its automotive radios.

[3] Portions of this section were taken from *Low Noise Dual Preamplifiers AN64*, National Semiconductor Corporation, May 1972.

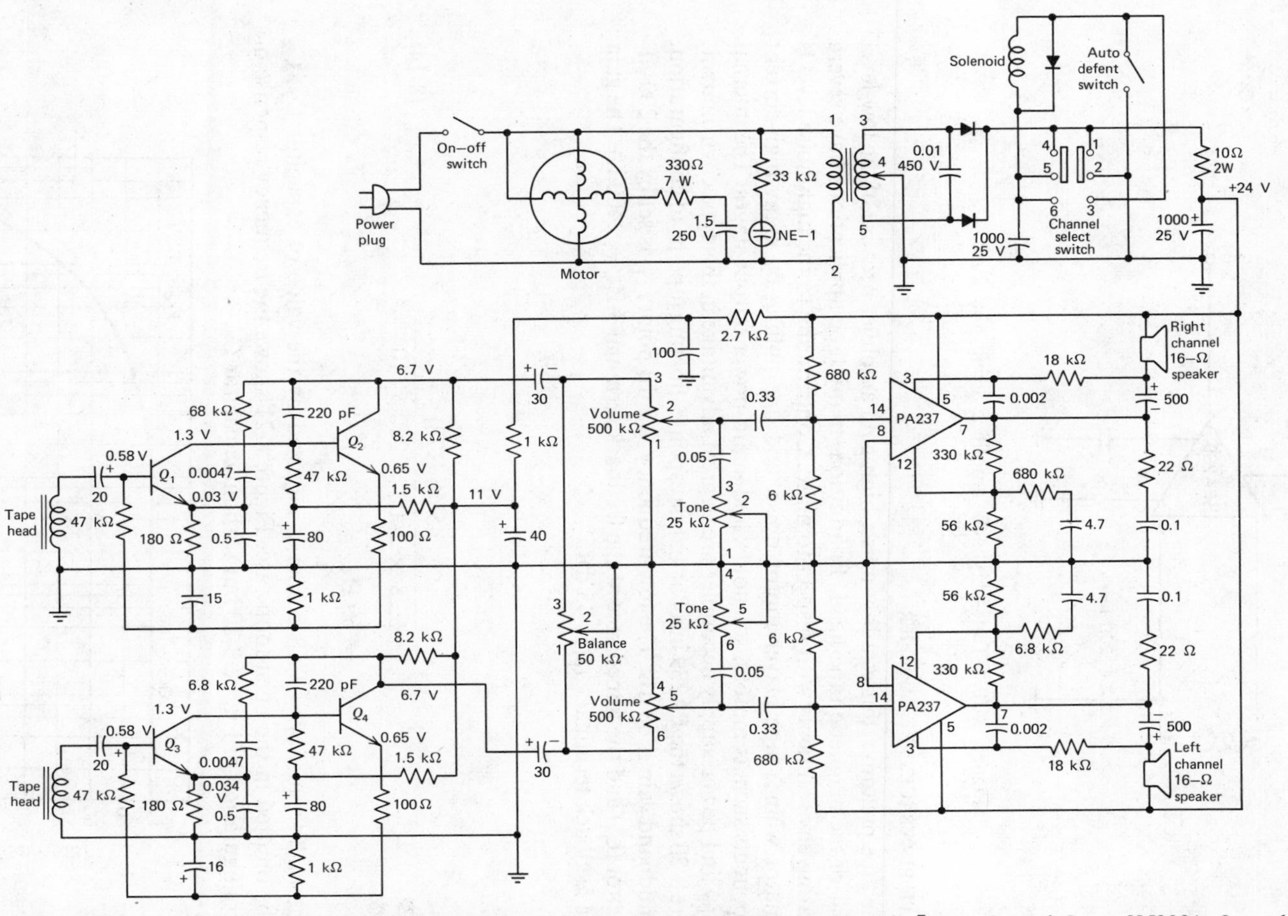

Figure 8-19. An 8-track, 4-W stereo cartridge-tape system with integrated power supply [*note:* Q_1 and Q_3 = 2N3901, Q_2 and Q_4 = 2N2925; all capacitors in microfarads unless otherwise noted]. (Courtesy John Wiley & Sons.)

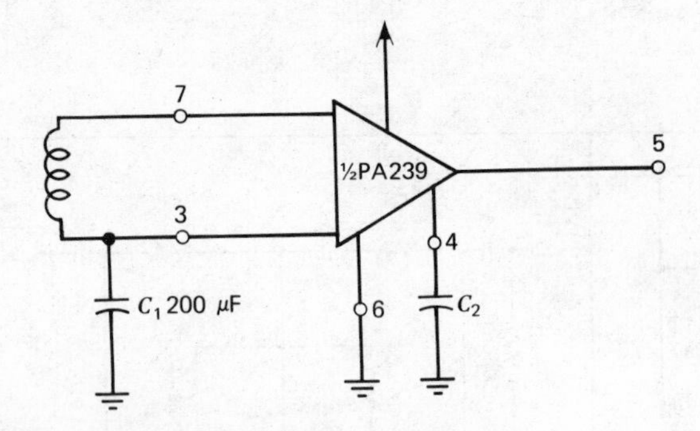

Figure 8-20. Tape-playback amplifier using the PA239.

TAPE-PLAYBACK PREAMPLIFIER

The tape preamp requires a low-noise, high-gain amplifier because of the low-level output of the magnetic tape head. In addition, many tape applications have severe operating environments with respect to noise. Contributing to the high noise level is alternator whine, tape player motor noise, and line pulling by many high-current components such as motors, solenoids, and audio-power output stages. The internal biasing and power-supply decoupler are of great advantage in this type environment. Figure 8-20 shows the PA239 in a flat-band response, floating tape-head configuration. The midband gain is 68 dB. The low-frequency −3 dB point (f_1) is set by the 200-μF capacitor (C_1) and the impedance at the feedback terminal 3(Z_3 = 2400 Ω). The gain at the feedback terminal (A_F) is 45 dB.

$$f_1 = \frac{A_F}{2\pi Z_3 C_1}$$

$$= \frac{45 \text{ db } (178)}{6.28 \times 2400 \times 200 \times 10^{-6}} \tag{8-14}$$

$$= 60 \text{ Hz}$$

The high-frequency −3 dB corner is adjusted by the value of capacitor C_2; C_2 can be omitted in the open-loop case. Figure 8-21 shows the frequency response of the configuration of Figure 8-20 but without capacitor C_2.

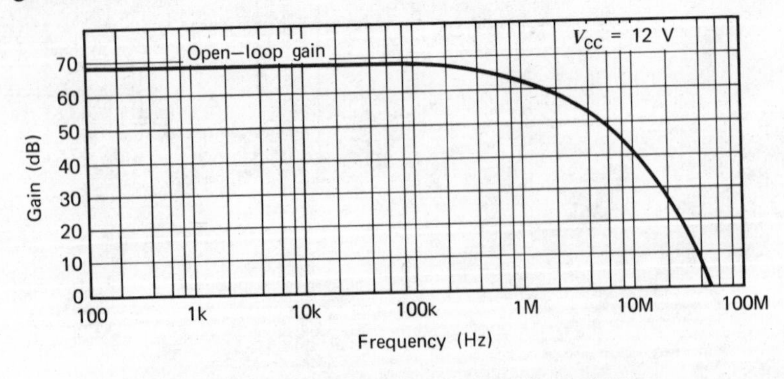

Figure 8-21. Frequency response of Figure 8-20.

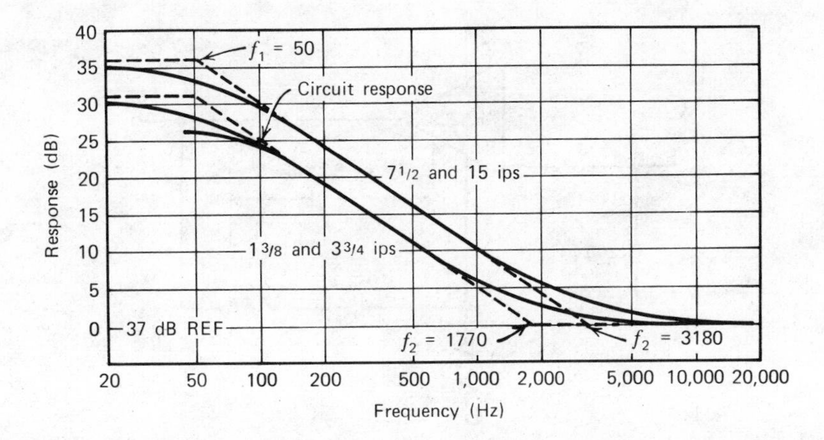

Figure 8-22. NAB equalization characteristic.

For music applications, compensation is required to provide the NAB (National Association of Broadcasting) standard tape playback equalization. Figure 8-22 shows the NAB equalization characteristic.

The NAB response is achieved for a system with a tape-head sensitivity of 800 μV at 1 kHz in the circuit configuration of Figure 8-23. This signal strength is the NAB standard recording level. The preamp gain is shaped so that all standard-level frequencies are reproduced at the output of the preamplifier at a level of 100 mV_{rms}. (Reference gain is 42 dB at 1 kHz.) The high-frequency gain of the NAB standard equalization is 37 dB. Therefore, stabilization of the amplifier must be accomplished. In compliance with (8-14) C_2 is chosen as 800 pF. This rolls off the frequency response above 10 kHz and provides unity gain at $\approx$70 kHz:

$$f_{3\,\text{dB}} = \frac{1}{2\pi C_2 R}$$

$$= \frac{1}{6.28 \times 800 \times 10^{-12} \times 2 \times 10^4}$$

$$= 10 \text{ kHz} \tag{8-15}$$

$$f_{\text{OUT}} = \frac{10^{A/20}}{2\pi C_2 R}$$

$$= \frac{10^{37/20}}{6.28 \times 800 \times 10^{-12} \times 2 \times 10^4} = \frac{70}{100 \times 10^{-6}} = 70 \text{ kHz}$$

where $f_{3\,\text{dB}}$ is the 3-dB corner frequency and f_{OUT} is the output frequency.

For other sensitivity levels and response characteristics, one may adjust the feedback and roll-off components. Figure 8-24 and (8-14) through (8-18) show the general case. The amplifier gain is controlled by the ratio

$$A_{\text{F}} = \frac{Z_{\text{F}}}{R_1} \qquad R_1 \ll Z_{\text{F}} \tag{8-16}$$

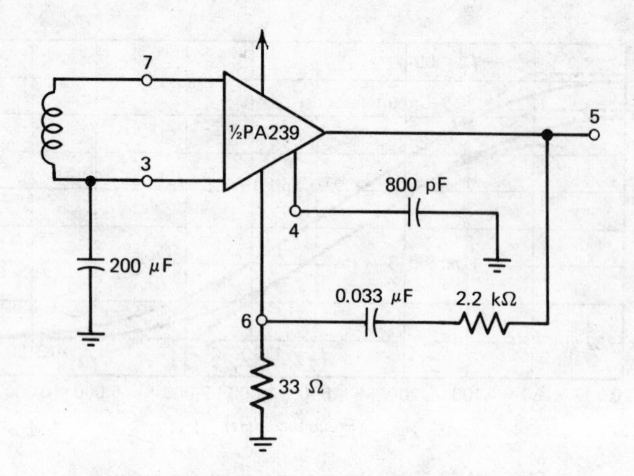

Figure 8-23. NAB system.

The high-frequency gain (0-dB reference, Figure 8-22) is controlled by the real part of the feedback impedance:

$$A_{\mathrm{F(0\ db\ ref)}} = \frac{R_F}{R_1} \tag{8-17}$$

The base boost (f_2 point, Figure 8-22) is determined by R_F and C_F,

$$f_2 = \frac{1}{2\pi R_F C_F} \tag{8-18}$$

As shown before in (8-14), the low-frequency roll-off is controlled by capacitor C_1:

$$f_1 = \frac{A_F}{2\pi Z_3 C_1}$$

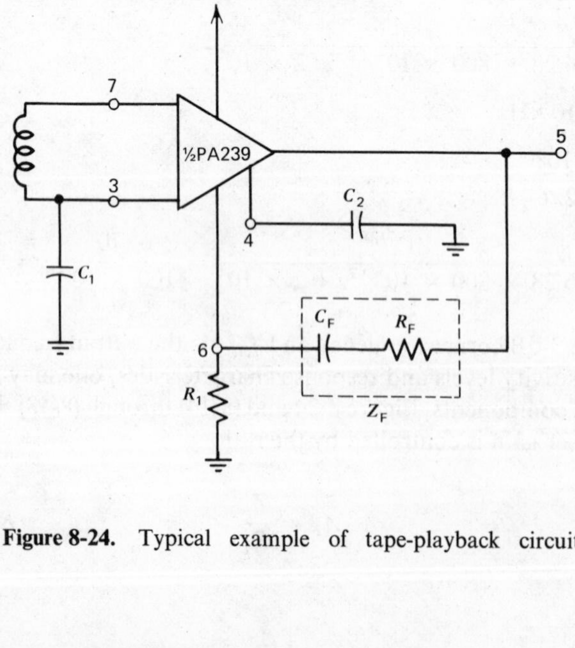

Figure 8-24. Typical example of tape-playback circuit.

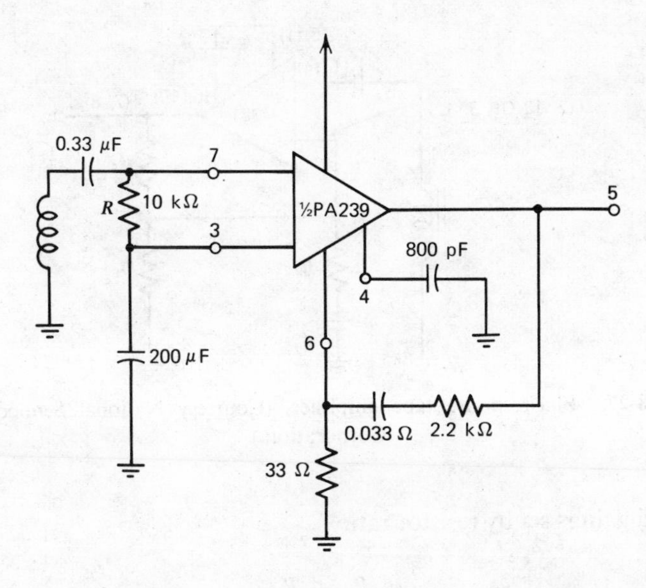

Figure 8-25. PA239 grounded head configuration.

The previous circuits have shown the PA239 with a floating head, that is, the dc bias current (3-μA maximum) passes through the head. The circuit of Figure 8-25 shows the PA239 in a grounded head configuration. Resistor *R* is limited to 10-kΩ maximum to provide adequate dc feedback. For greater than 10-kΩ input impedance, the circuit of Figure 8-26 is recommended.

Figure 8-27 shows the LM381 in a flat response tape-playback configuration.

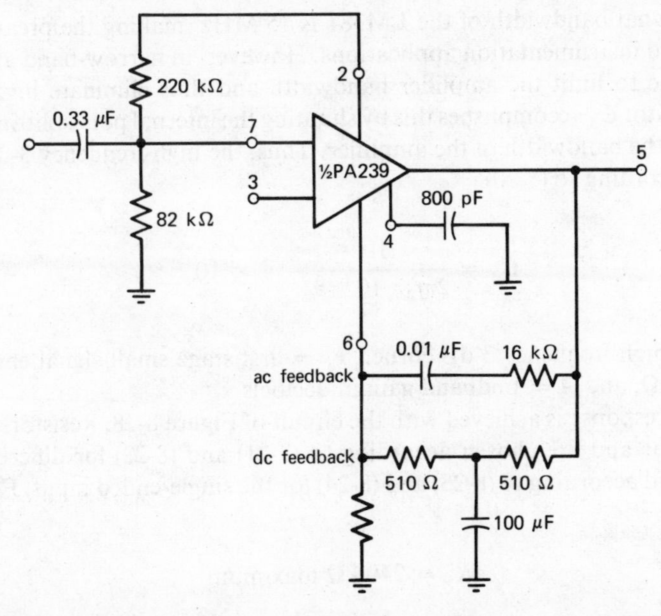

Figure 8-26. PA239 grounded head configuration for greater than 10-kΩ input impedance.

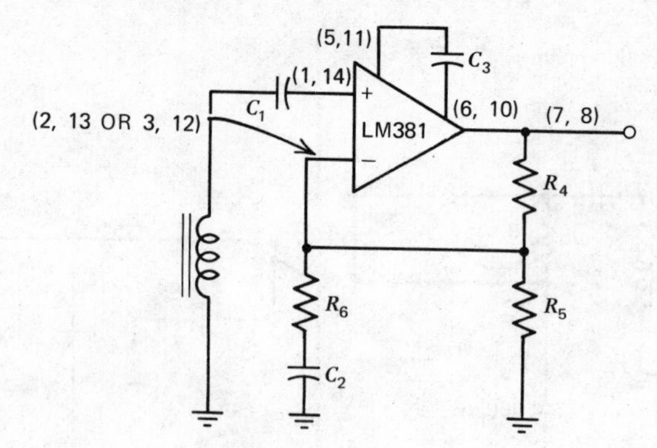

Figure 8-27. Flat-response tape amplifier. (Courtesy National Semiconductor Corporation.)

The midband gain is set by resistor ratio

$$\frac{R_4 + R_6}{R_6}$$

Capacitor C_2 sets the low-frequency 3-dB corner where $X_{C_2} = R_6$:

$$C_2 = \frac{1}{2\pi f_{OUT} R_6} \tag{8-19}$$

The small-signal bandwidth of the LM381 is 15 MHz, making the preamp suitable for wide-band instrumentation applications. However, in narrow-band applications, it is desirable to limit the amplifier bandwidth and thus eliminate high-frequency noise. Capacitor C_3 accomplishes this by shunting the internal pole-splitting capacitor C_1, limiting the bandwidth of the amplifier. Thus, the high-frequency 3-dB corner is set by C_3 according to (8-20).

$$C_3 = \frac{1}{2\pi f_3 r_e 10^{A/20}} - 4 \times 10^{-12} \tag{8-20}$$

where f_3 = high-frequency 3-dB corner, r_e = first stage small-signal emitter resistance $\approx 2.6\ \text{k}\Omega$, and A = midband gain in decibels.

The NAB response is achieved with the circuit of Figure 8-28. Resistors R_4 and R_5 set the dc bias and are chosen according to (8-21) and (8-22) for differential-input operation, and according to (8-23) and (8-24) for the single-ended input. For differential input,

$$R_5 = 240\ \text{k}\Omega \text{ maximum} \tag{8-21}$$

$$R_4 = \left(\frac{V_{CC}}{2.4} - 1\right) R_5 \tag{8-22}$$

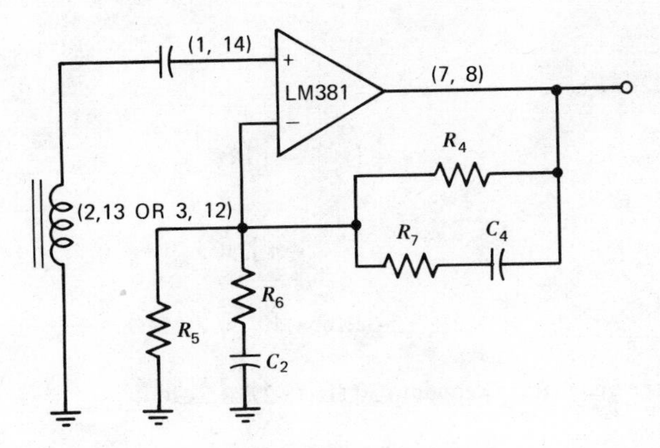

Figure 8-28. NAB tape preamplifier using the LM381. (Courtesy National Semiconductor Corporation.)

For single-ended input,

$$R_5 = 1200\ \Omega \text{ maximum} \tag{8-23}$$

$$R_4 = \left(\frac{V_{CC}}{1.2} - 1\right) R_5 \tag{8-24}$$

The reference gain of the preamp, above corner frequency f_2 (Figure 8-22) is set by the ratio

$$\text{reference gain}_{0\text{ dB}} = \frac{R_7 + R_6}{R_6} \tag{8-25}$$

The corner frequency f_2 (Figure 8-22) is determined where $X_{C_4} = R_7$ and is given by

$$f_2 = \frac{1}{2\pi C_4 R_7} \tag{8-26}$$

Corner frequency f_1 is determined where $X_{C_4} = R_4$:

$$f_1 = \frac{1}{2\pi C_4 R_4} \tag{8-27}$$

The low-frequency 3-dB roll-off point f_{OUT} is set where $X_{C_2} = R_6$

$$f_{OUT} = \frac{1}{2\pi C_2 R_6} \tag{8-28}$$

Example. Design a NAB-equalized preamp for a tape player requiring 0.5 V rms output from a head sensitivity of 800 μV at 1 kHz, $3\frac{3}{4}$ ips. The power-supply voltage is 24 V and the differential-input configuration is used.

1. Let $R_5 = 240\ \text{k}\Omega$.
2. From (8-22) we have

$$R_4 = \left(\frac{V_{CC}}{2.4} - 1\right)R_5$$

$$= \left(\frac{24}{2.4} - 1\right)2.4 \times 10^5$$

$$= 2.16 \times 10^5 \approx 2.2\ \text{M}\Omega$$

3. For a corner frequency, f_1 equal to 50 Hz, (8-27) is used:

$$C_4 = \frac{1}{2\pi f_1 R_4} = \frac{1}{6.28 \times 50 \times 2.2 \times 10^6}$$

$$= 1.44 \times 10^{-9}$$

$$\approx 1500\ \text{pF}$$

4. From Figure 8-22 the corner frequency $f_2 = 1770$ Hz at $3\frac{3}{4}$ ips; resistor R_7 is found from (8-26):

$$C_4 = \frac{1}{2\pi f_2 R_7}$$

$$R_7 = \frac{1}{6.28 \times 1770 \times 1.5 \times 10^{-9}} = 6 \times 10^4$$

$$\approx 62\ \text{k}\Omega$$

5. The required voltage gain at 1 kHz is

$$A_V = \frac{0.5\ V_{\text{rms}}}{800\ \mu V_{\text{rms}}} = 6.25 \times 10^2\ \text{V/V} = 56\ \text{dB}$$

6. From Figure 8-22 we see the reference frequency gain above f_2 is 5 dB down from the 1-kHz value or 51 dB (355 V/V); and from (8-25),

$$\text{reference gain}_{0\ \text{dB}} = \frac{R_6 + R_7}{R_6} = 355$$

$$R_6 = \frac{R_7}{355 - 1} = \frac{62\ \text{k}\Omega}{354} = 175$$

$$\approx 180\ \Omega$$

7. For low-frequency corner $f_{\text{OUT}} = 40$ Hz, then (8-28) becomes

$$C_2 = \frac{1}{2\pi f_0 R_6} = \frac{1}{6.28 \times 40 \times 180} = 2.21 \times 10^{-5}$$

$$\approx 20\ \mu\text{F}$$

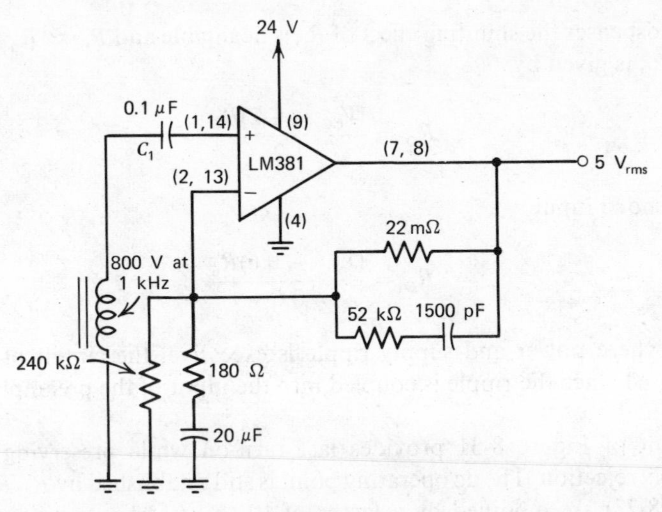

Figure 8-29. Typical tape-playback amplifier. (Courtesy National Semiconductor Corporation.)

This circuit is shown in Figure 8-29 and requires approximately 5 sec to turn on for the gain and supply voltage chosen in the example. Turn-on time can closely be approximated by

$$t_{ON} \approx -R_4 C_2 \ln\left(1 - \frac{2.4}{V_{CC}}\right) \tag{8-29}$$

As seen by (8-29), increasing the supply voltage decreases turn-on time. Decreasing the amplifier gain also decreases turn-on time by reducing the $R_4 C_2$ product.

Where the turn-on time of the circuit of Figure 8-28 is too long, the time may be shortened by using the circuit of Figure 8-30. The addition of resistor R_D forms a voltage divider with R_6'. This divider is chosen so that zero dc voltage appears across C_2. The parallel resistance of R_6' and R_D is made equal to the value of R_6 found by

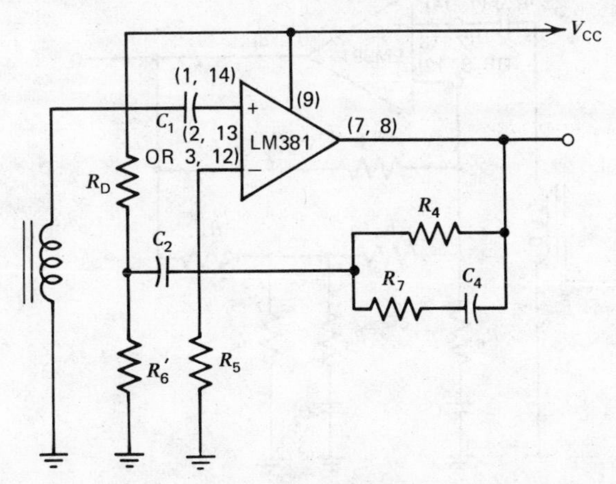

Figure 8-30. Fast turn-on NAB tape preamplifier. (Courtesy National Semiconductor Corporation.)

(8-25). In most cases the shunting effect of R_D is negligible and $R_6' \approx R_6$. For differential input, R_D is given by

$$R_D = \frac{(V_{CC} - 1.2)R_6'}{1.2} \tag{8-30}$$

For single-ended input,

$$R_D = \frac{(V_{CC} - 0.6)R_6'}{0.6} \tag{8-31}$$

In cases where power and supply ripple is excessive, the circuit of Figure 8-30 cannot be used since the ripple is coupled into the input of the preamplifier through the divider.

The circuit of Figure 8-31 provides fast turn-on while preserving the 120-dB power-supply rejection. The dc operating point is still established by R_4/R_5. However, (8-21) and (8-23) are modified by a factor of 10 to preserve dc bias stability. For differential input, (8-21) is modified as

$$R_5 = \frac{2V_{BE}}{100I_{Q_2}} = \frac{1.2}{50 \times 10^{-8}}$$
$$= 24\ \text{k}\Omega \text{ maximum}$$

For single-ended input, (8-23) becomes

$$R_5 = \frac{V_{BE}}{50I_{FB}} = \frac{0.6}{50 \times 10^{-4}}$$
$$= 120\ \Omega \text{ maximum}$$

Equations (8-25), (8-26), and (8-28) describe the high-frequency gain and corner frequencies f_2 and f_{OUT} as before. Frequency f_1 now occurs where X_{C_+} equals the

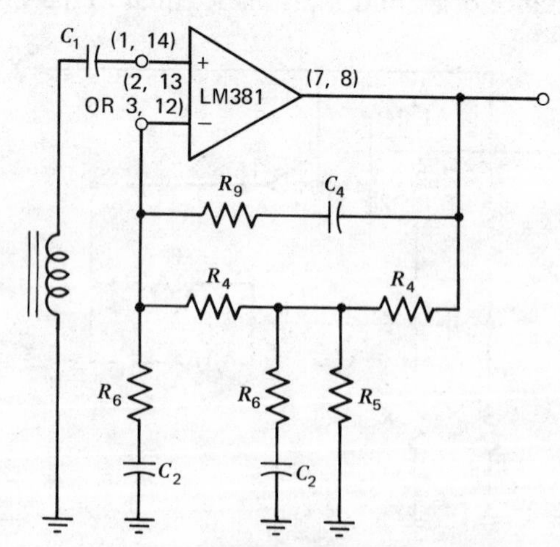

Figure 8-31. Two-pole fast turn-on NAB tape preamplifier. (Courtesy National Semiconductor Corporation.)

composite impedance of the R_4, R_6, C_2 network as given by (8-32),

$$C_4 = \frac{1}{2\pi f_1 R_6([(R_4 + R_6)/R_6])^2 - 1)} \tag{8-32}$$

The turn-on time becomes

$$t_{ON} \approx -2\sqrt{R_4 C_2} \ln\left(1 - \frac{2.4}{V_{CC}}\right) \tag{8-33}$$

TAPE-RECORD PREAMPLIFIER

For tape-recording, the frequency response must be the mirror image of the NAB playback equalization so that the composite record and playback response is flat. Figure 8-32 shows the record characteristic superimposed on the NAB playback response.

Figure 8-33, curve A, shows the response characteristics of a typical laminated core, quarter-track head. Curve B shows the amplifier response required to make the composite of the two provide the proper NAB recording characteristic. This response is obtained by the circuit of Figure 8-34. The gain is established by (8-34) and (8-35):

$$A = \frac{R_2 + Z_F}{Z_F} \tag{8-34}$$

$$\begin{aligned} Z_F &= \frac{R_1(R_4 + 1/2\pi f C_4)}{R_1 + (R_4 + 1/2\pi f C_4)} \\ &\approx R_1 \parallel R_4 \text{ high frequencies} \\ &\approx R_1 \text{ low frequencies} \end{aligned} \tag{8-35}$$

Capacitor C_4 establishes the corner frequency f_2:

$$C_4 = \frac{1}{2\pi f_2(2.42R_1)} \tag{8-36}$$

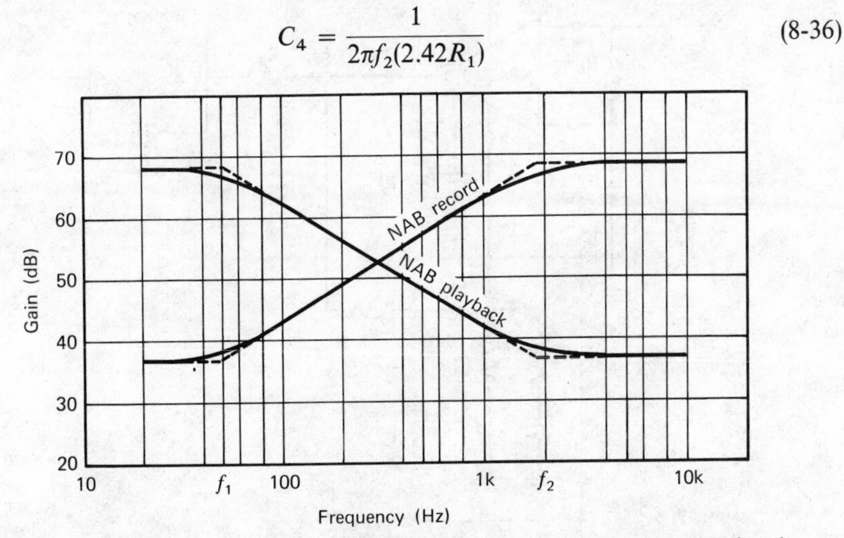

Figure 8-32. NAB record and playback characteristics. (Courtesy National Semiconductor Corporation.)

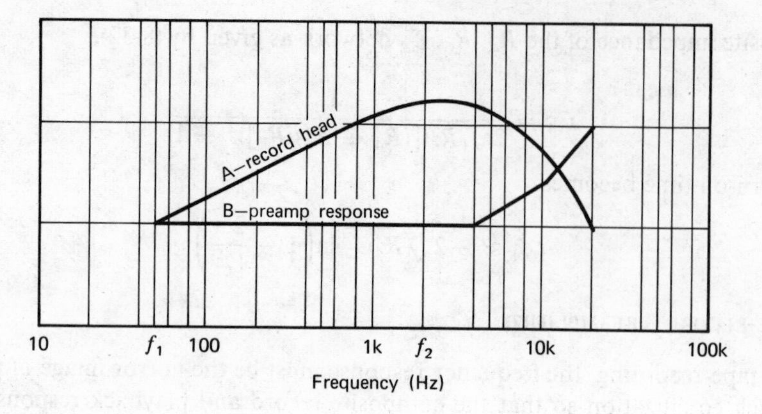

Figure 8-33. Record head and preamplifier response.

Capacitor C_1 is chosen to give the desired low-frequency response f_1:

$$C_1 = \frac{A_F}{2\pi f_1 Z_3} \tag{8-37}$$

where A_F = gain to feedback terminal 3 (45-dB), f_1 = low-frequency cut-off, and Z_3 = impedance at terminal 3 (2400 Ω).

Example. A recording head requiring 30 μA is used with a microphone of 2-mV output. The 30-μA current source is simulated by a 1-V_{rms} output driving through 33-kΩ resistance:

$$R = \frac{1\ V_{rms}}{30 \times 10^{-6}} = 33\ \text{k}\Omega$$

Therefore, the gain required is 1 V/2 mV = 500.

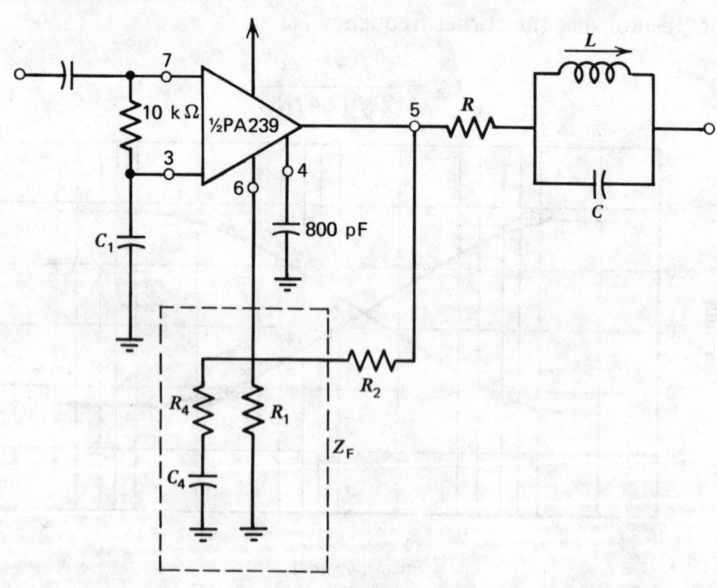

Figure 8-34. Circuit for obtaining NAB response.

From (8-34) and (8-35), we see the low-frequency gain (below f_2, Figure 8-33) is

$$A = \frac{R_2 + R_1}{R_1} = 500 = 54 \text{ dB}$$

Let $R_1 = 100$, so that $R_2 = 499R_1 \approx 51 \text{ k}\Omega$. From Figure 8-33 we see that the tape-head response falls off 24 dB from 4 to 20 kHz. The open-loop gain of the PA239 is 68 dB and the gain below f_2 is set to 54 dB. Therefore only 14 dB of treble boost can be obtained from the amplifier. This will result in flat response out to $\approx$14 kHz, beyond which the response will roll-off. Therefore, there is no need to set the high-frequency gain, and resistor R_4 may be eliminated. From (8-36),

$$C_4 = \frac{1}{2\pi(4 \times 10^3)2.42(100)}$$

$$= 0.16\ \mu\text{F}$$

From (8-37),

$$C_1 = \frac{A_F}{2\pi f_1 Z_3}$$

$$= \frac{45 \text{ dB } (178)}{6.28 \times 50 \times 2400} = 200\ \mu\text{F}$$

The resulting circuit is shown in Figure 8-35, where L and C provide a bias trap to present a high impedance to the bias frequency, thus preventing intermodulation distortion.

The response to provide the NAB recording characteristic using the LM381 preamplifier is shown in Figure 8-36. Resistors R_4 and R_5 set the dc bias as before

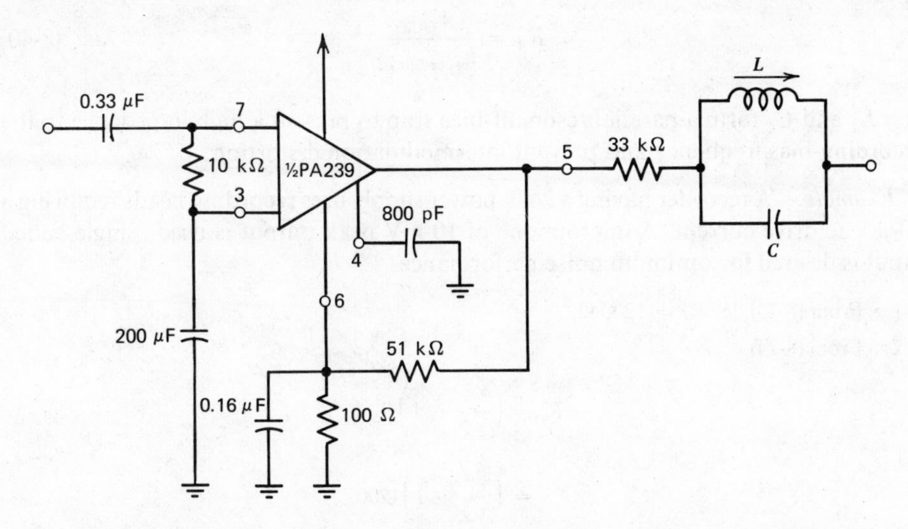

Figure 8-35. Typical example of tape-record preamplifier circuit.

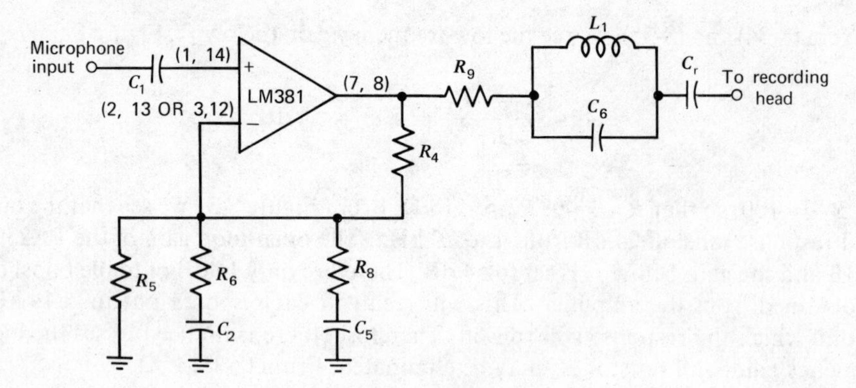

Figure 8-36. Tape-recording preamplifier. (Courtesy National Semiconductor Corporation.)

using (8-21) and (8-22) for the differential input and (8-23) and (8-24) for the single-ended input. Resistor R_6 and capacitor C_2 set the midband gain as before (8-19). Capacitor C_5 sets the high-frequency 3-dB point f_3 (Figure 8-23) as

$$f_3 = \frac{1}{2\pi C_5 R_6} \tag{8-38}$$

The preamp gain increases at 6 dB/octave above f_3 until $R_8 = X_{C_5}$.

$$R_8 = \frac{1}{2\pi f_4 C_5} \tag{8-39}$$

where f_4 = desired high-frequency cut-off. Resistor R_9 is chosen to provide the proper recording head current,

$$R_9 = \frac{V_{\text{OUT}}}{i_{\text{record head}}} \tag{8-40}$$

and L_1 and C_6 form a parallel resonant-bias trap to present a high impedance to the recording-bias frequency and prevent intermodulation distortion.

Example. A recorder having a 24-V power supply uses recording heads requiring a 30-μA ac drive current. A microphone of 10-mV peak output is used. Single-ended input is desired for optimum noise performance.

1. From (8-23), let $R_5 = 1200\ \Omega$.
2. From (8-24),

$$R_4 = \left(\frac{V_{CC}}{1.2} - 1\right) R_5$$

$$= \left(\frac{24}{1.2} - 1\right) 1200$$

$$= 2.28 \times 10^4 \approx 22\ \text{k}\Omega$$

3. The maximum output of the LM381 is $(V_{CC} - 2\text{ V})_{p-p}$. For a 24-V power supply, the maximum output is 22 V_{p-p} or 7.8 V_{rms}. Therefore, an output swing of 6 V_{rms} is reasonable. From (8-40),

$$R_9 = \frac{V_{OUT}}{i_{OUT\ \text{record head}}}$$

$$= \frac{6\text{ V}}{30\ \mu\text{A}} = 200\text{ k}\Omega$$

4. Let the high-frequency cut-off $f_4 = 16$ kHz (Figure 8-33). The recording-head frequency response begins falling off at approximately 4 kHz. Therefore, the preamp gain must increase at this frequency to obtain the proper composite characteristic. The slope is 6 dB/octave for the two octaves between f_3 (4 kHz) and the cut-off frequency f_4 (16 kHz). Therefore, the midband gain lies 12 dB below the peak gain. We are allowing 6 V_{rms} output-voltage swing. Therefore,

$$\text{peak gain} = \frac{6\text{ V}}{10\text{ mV}} = 600 \text{ or } 55.6\text{ dB}$$

$$\text{midband gain} = 43.6\text{ dB or } 150$$

5. Since the midband gain $= (R_4 + R_6)/R_6 = 150$, then

$$R_6 = \frac{R_4}{149} = \frac{22 \times 10^3}{149} = 147.7$$

$$R_6 \approx 150\ \Omega$$

6. From (8-19),

$$C_2 = \frac{1}{2\pi f_{OUT} R_6}$$

$$= \frac{1}{6.28 \times 50 \times 150}$$

$$= 2.12 \times 10^{-5}$$

$$\approx 20\ \mu\text{F}$$

7. From (8-38),

$$C_5 = \frac{1}{2\pi f_3 R_6}$$

$$= \frac{1}{6.28 \times 4 \times 10^3 \times 150}$$

$$= 266 \times 10^{-7}$$

$$\approx 0.27\ \mu\text{F}$$

8. From (8-39),

$$R_8 = \frac{1}{2\pi f_4 C_5}$$

$$= \frac{1}{6.28 \times 16 \times 10^3 \times 2.7 \times 10^{-7}}$$

$$= 36.8$$

$$\approx 33\ \Omega$$

The realization of this example is shown in Figure 8-37.

PHONOGRAPH PREAMPLIFIER

Crystal and ceramic phonograph cartridges provide output signals of 100 mV to 1 V and therefore do not require preamplification. Magnetic cartridges, however, provide much lower outputs and require the use of a preamplifier such as the PA239. Table 8-1 shows the rated outputs of several magnetic cartridges.

Output voltage is specified for a given modulation velocity. Since the magnetic pick-up is a velocity device, output is proportional to velocity. For example, a cartridge producing 5 mV at 5 cm/sec will produce 1 mV at 1 cm/sec and the cartridge could be equivalently specified as having a sensitivity of 1 mV/cm/sec.

In order to transform cartridge sensitivity into useful preamp design information, we need to know typical and maximum modulation velocity limits found on high-quality stereo records. The RIAA recording characteristic (which has to do with magnetic phono playback equalization) is used almost universally today; it establishes a limit on maximum recording velocity of 25 cm/sec in the range from 800 to 2500 Hz. Typical, good-quality records are normally recorded at a velocity of 3 to 5 cm/sec. It follows then that the worst-case overload condition is at 25 cm/sec and 800 Hz.

Using some typical design parameters, we can design a cartridge preamp circuit as follows:

1. Assume cartridge sensitivity of 0.5 mV/cm/sec.
2. Maximum output from this cartridge at 25 cm/sec is 12.5 mV_{rms}.

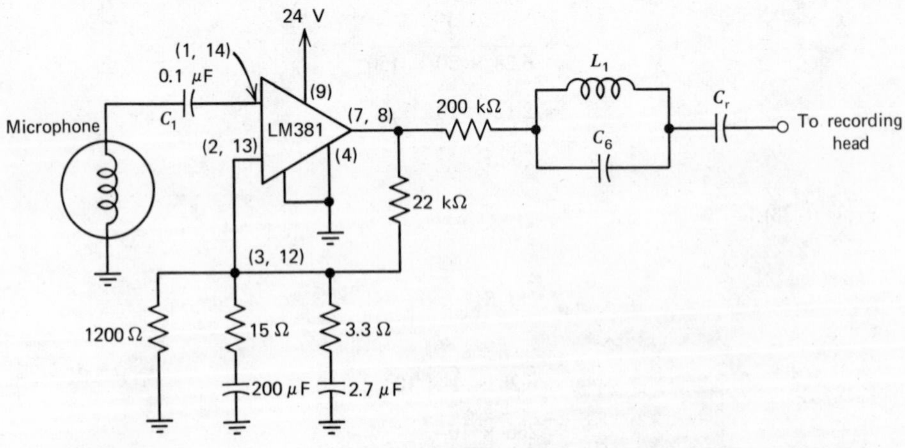

Figure 8-37. Typical tape-recording amplifier. (Courtesy National Semiconductor Corporation.)

Table 8-1

Manufacturer	*Model*	*Output at 5 cm/sec (mV)*
Empire Scientific	999	5
	888	8
Shure	V-15	3.5
	M91	5
	M31	10.5
Pickering	V-15 AM3	6
	V-15 AT3	5

3. Maximum "undistorted" output swing for the PA239 is approximately 1.25 V_{rms} with 20 dB of feedback gain reduction at 800 cycles. This determines the gain at 800 Hz to be as follows:

$$\text{maximum preamp gain} = \frac{\text{maximum preamp swing capability}}{\text{maximum head output}} = \frac{1.25\ V_{rms}}{12.5\ \text{mV}} = 40\ \text{dB}$$

A circuit diagram and the resulting preamp response curve for the above application is shown in Figures 8-38 and 8-39. Figure 8-40 depicts the schematic diagram of a phono preamp using the LM380, and Figure 8-41 shows the response curve for this circuit. Referring to Figure 8-40, resistors R_4 and R_5 set the dc bias (8-21 and 8-22; or 8-23 and 8-24). The 0-dB reference gain is set by the ratio

$$\text{reference gain}_{0\ \text{dB}} = \frac{R_{10} + R_6}{R_6} \tag{8-41}$$

The corner frequency f_1 (Figure 8-41) is established where $X_{C_7} = R_4$ or

$$C_7 = \frac{1}{2\pi f_1 R_4} \tag{8-42}$$

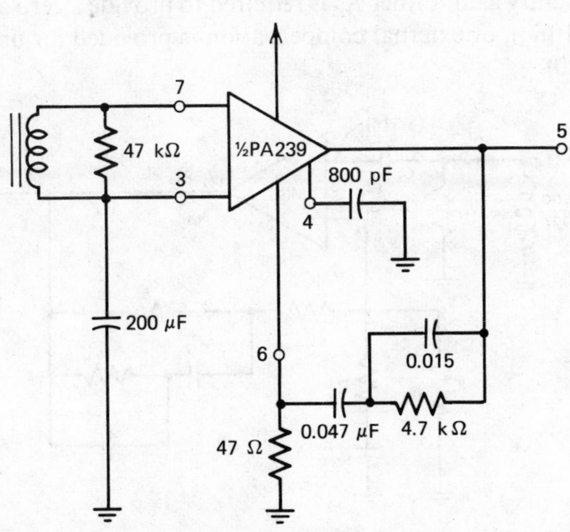

Figure 8-38. Typical phonograph preamplifier circuit schematic.

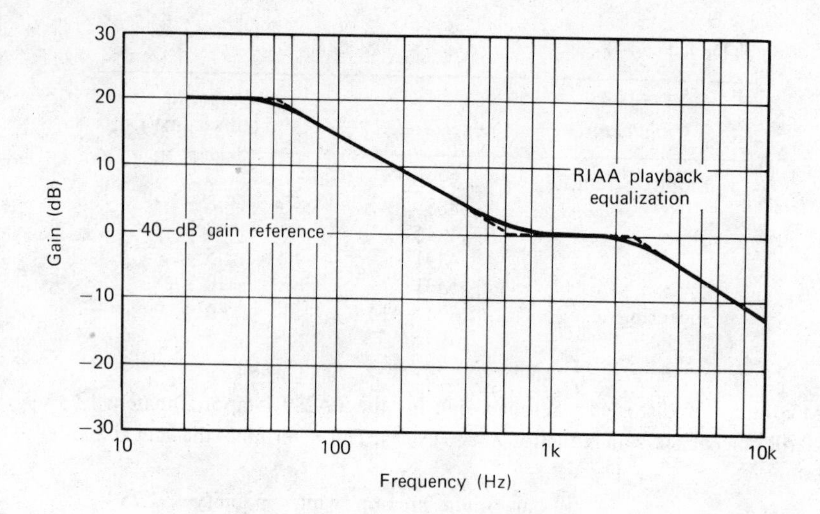

Figure 8-39. Response for circuit of Figure 8-38.

Likewise, frequency f_2 occurs where $X_{C_7} = R_{10}$ or

$$C_7 = \frac{1}{2\pi f_2 R_{10}} \tag{8-43}$$

The third corner frequency f_3 is determined where $X_{C_8} = R_{10}$:

$$C_8 = \frac{1}{2\pi f_3 R_{10}} \tag{8-44}$$

Resistor R_Z is used to insert a zero in the feedback loop since the LM381 is not compensated for unity gain. Either R_Z is required to provide a zero at or above a gain of 20 dB ($R_Z = 10R_6$), or external compensation is provided for unity gain stability according to (8-20).

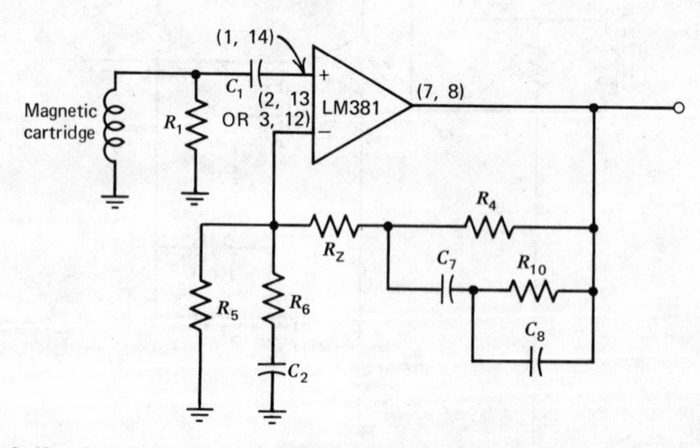

Figure 8-40. RIAA phonograph preamplifier. (Courtesy National Semiconductor Corporation.)

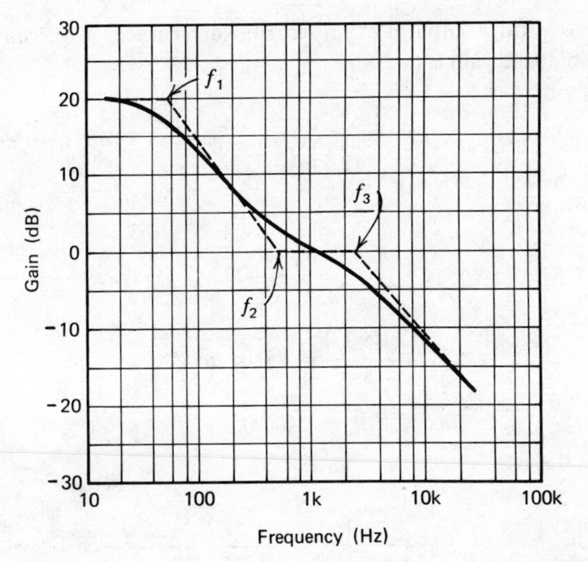

Figure 8-41. RIAA playback equalization. (Courtesy National Semiconductor Corporation.)

Example. Design a phono preamp operating from a 30-V supply, with a cartridge of 0.5 mV/cm/sec sensitivity, to drive a power amplifier of 5-V_{rms} input-overload limit.

1. From (8-21), let $R_5 = 100\ \text{k}\Omega$.
2. From (8-22)

$$R_4 = \left(\frac{V_{CC}}{2.4} - 1\right)R_5 = \left(\frac{30}{24} - 1\right)10^5$$

$$= 11.5 \times 10^5 \approx 1.2\ \text{M}\Omega$$

3. From (8-42),

$$C_7 = \frac{1}{2\pi f_1 R_4}$$

$$= \frac{1}{6.28 \times 50 \times 1.2 \times 10^6}$$

$$= 2.65 \times 10^9$$

$$\approx 0.003\ \mu\text{F}$$

4. From (8-43),

$$C_7 = \frac{1}{2\pi f_2 R_{10}}$$

$$R_{10} = \frac{1}{6.28 \times 500 \times 3 \times 10^{-9}} = 1.03 \times 10^5$$

$$\approx 100\ \text{k}\Omega$$

5. The maximum cartridge output at 25 cm/sec is (0.5 mV/cm/sec) × (25 cm/sec) = 12.5 mV; the required midband gain is therefore

$$\frac{5\ \text{V}_{\text{rms}}}{12.5\ \text{mV}_{\text{rms}}} = 400$$

6. Equation (8-41) gives

$$\text{reference gain}_{0\ \text{dB}} = \frac{R_{10} + R_6}{R_6} = 400$$

$$R_6 = \frac{100\ \text{k}\Omega}{399} = 251 \approx 240\ \Omega$$

$$R_Z = 10R_6 = 2400\ \Omega$$

7. Equation (8-19) gives

$$C_2 = \frac{1}{2\pi f_{\text{OUT}} R_6} = \frac{1}{6.28 \times 40 \times 240} = 1.7 \times 10^{-5}$$

$$\approx 20\ \mu\text{F}$$

8. From (8-44), we have

$$C_8 = \frac{1}{2\pi f_3 R_{10}} = \frac{1}{6.28 \times 2200 \times 6.8 \times 10^4}$$

$$= 7.23 \times 10^{-10}$$

$$\approx 0.001\ \mu\text{F}$$

The completed design is shown in Figure 8-42 where a 47-kΩ input resistor has been included to provide the RIAA standard cartridge load.

Figure 8-43 shows one channel of a practical preamp for a stereo phonograph. The preamp is complete with RIAA equalization, bass and treble tone control, balance control, and volume control.

AUDIO MIXER

In many audio applications, it is desirable to provide a mixer to combine or select several inputs. Such applications include public address systems where more than one microphone is used: Tape recorders, high-fidelity phonographs, guitar amplifiers, etc.

Figure 8-44 shows the LM381 in a mixer configuration. Inputs at A, B, C, and N can be selected and combined (summed) with potentiometers R_A, R_B, R_C, R_N. Resistors R_4 and R_5 establish the dc quiescent point in accordance with (8-21) and (8-22). (Only the differential-input configuration is used in the mixer application since the high-source inpedance of the input potentiometers would negate any advantage of the single-ended input.) Input-bias current is supplied through resistor R_F. Therefore, an upper limit of R_F should be established to avoid output offset-voltage problems. A safe upper limit is to let

$$R_F = R_4 \text{ maximum} \tag{8-45}$$

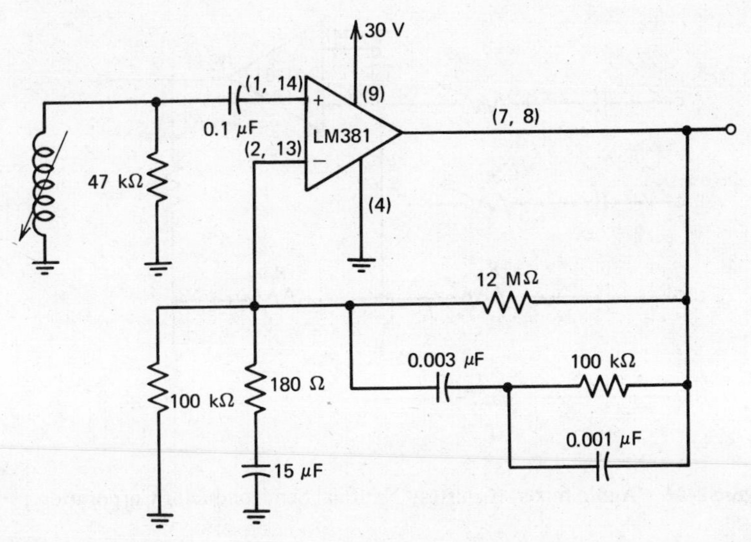

Figure 8-42. Typical magnetic phonograph preamplifier. (Courtesy National Semiconductor Corporation.)

The voltage gain of the mixer is

$$|A_{V_{\mathrm{A,B,C}}}| = \frac{R_4R_\mathrm{F} + R_4R_5 + R_5R_\mathrm{F}}{R_5(R_{\mathrm{A,B,C}} + R_{\mathrm{S_{A,B,C}}})} \tag{8-46}$$

Where the values of R_F and the source impedance R_S are such that the gain of the circuit of Figure 8-44 is inadequate, the configuration of Figure 8-45 may be used. The voltage gain of the mixer is now

$$|A_V| = \frac{R_\mathrm{F}}{R_{\mathrm{A,B,C}} + R_{\mathrm{S_{A,B,C}}}} \tag{8-47}$$

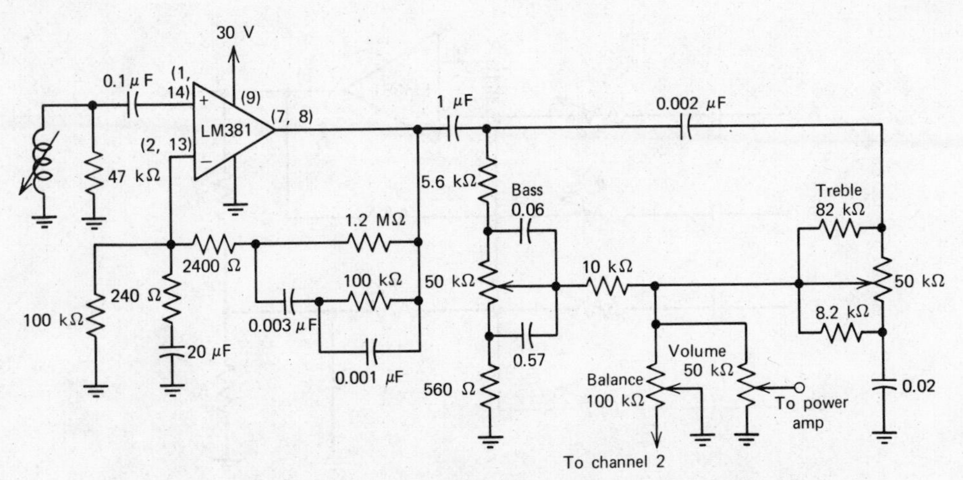

Figure 8-43. Single channel of complete phonograph preamplifier. (Courtesy National Semiconductor Corporation.)

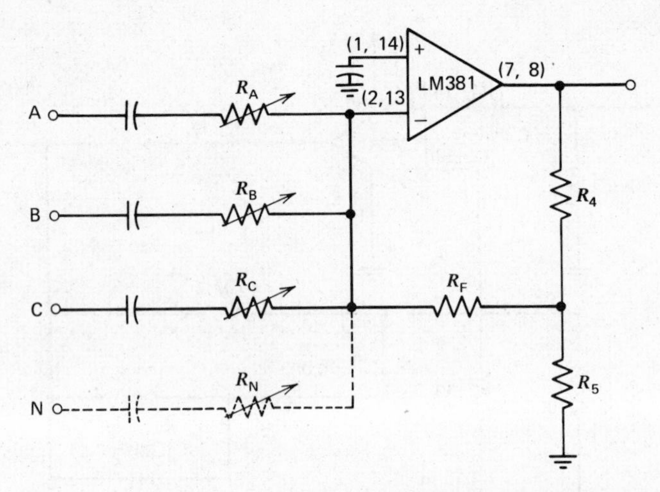

Figure 8-44. Audio mixer. (Courtesy National Semiconductor Corporation.)

Since Resistor R_F is no longer required to supply the input-bias current, it does not have the upper limit as in the previous circuit. Therefore, the open-loop gain of the LM381 can be realized. Capacitor C_1 shunts the ac feedback of the $R_4 - R_5$ network and is found by

$$C_1 = \frac{10^{A/20}}{2\pi f_{OUT} R_V}$$

$$A_0 = \text{amplifier open-loop gain in dB}$$

$$f_{OUT} = \text{low-frequency 3-dB corner}$$

Example. Design a microphone mixer for use with 600-Ω dynamic microphones with an output level of 10 mV. The mixer should operate from a 24-V supply and deliver a 5-V output. A dynamic range of 80 dB is desired.

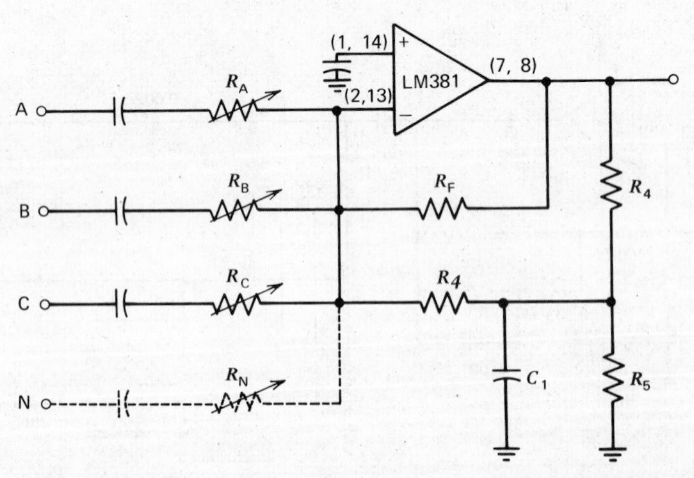

Figure 8-45. High-gain version of Figure 8-44. (Courtesy National Semiconductor Corporation.)

1. From (8-2), $R_5 = 24\ \text{k}\Omega$.
2. Equation (8-22) gives

$$R_4 = \left(\frac{V_{CC}}{2.4} - 1\right)R_5$$

$$= \left(\frac{24}{2.4} - 1\right)24 \times 10^3$$

$$= 2.16 \times 10^5 \approx 220\ \text{k}\Omega$$

3. For 5-V output,

$$\text{gain} = \frac{5\ \text{V}}{10\ \text{mV}} = 500$$

4. For a 80-dB dynamic range

$$\text{attenuation} = \frac{500}{80\ \text{dB}} = 5 \times 10^2$$

5. From (8-46),

$$|A_V| = \frac{R_4R_F + R_4R_5 + R_5R_F}{R_5(R_{A,B,C} + R_S)}$$

$$R_F = \frac{|A_V|\ R_5(R_{A,B,C} + R_S) - R_4R_5}{R_4 + R_5}$$

At maximum volume, $R_{A,B,C} = 0$ and gain $= 500$, therefore

$$R_F = \frac{500 \times 2.4 \times 10^4(0 + 600) - (2.2 \times 10^5)(2.4 \times 10^4)}{2.2 \times 10^5 + 2.4 \times 10^4}$$

$$= 7.87\ \text{k}\Omega \approx 8.2\ \text{k}\Omega$$

At maximum attenuation,

$$R_{A,B,C} = \frac{R_4R_F + R_4R_5 + R_5R_F - |A_V|\ R_5R_S}{A_VR_5}$$

$$= 5.99 \times 10^6 \approx 5\ \text{M}\Omega$$

The resulting circuit is shown in Figure 8-46.

ULTRASONIC TRANSFORMERLESS REMOTE CONTROL RECEIVER

In cascade, the two PA239 amplifier sections will provide up to 136 dB typical voltage gain in the 40-kHz range with excellent stability and noise performance. Perhaps most unique about the PA239 as an ultrasonic receiver, is the fact that the high selectivity required in such a receiver can be achieved by simple *LC* tuning at the high-impedance 20-kΩ interstage node. Other ICs without high-impedance levels must resort to more costly impedance transformers. The external component count to perform this type receiver function with the PA239 is, in general, much less than it is with other ICs recommended for this service. A circuit showing the rudiments of an ultrasonic receiver is shown in Figure 8-47.

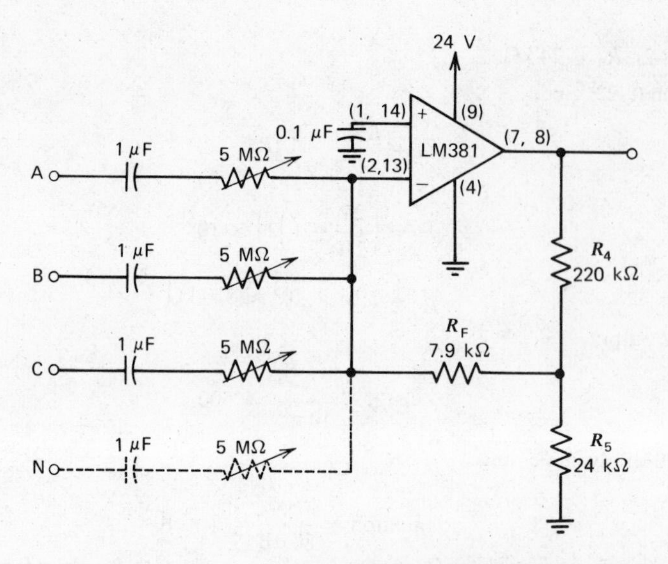

Figure 8-46. Example of audio mixer circuit. (Courtesy National Semiconductor Corporation.)

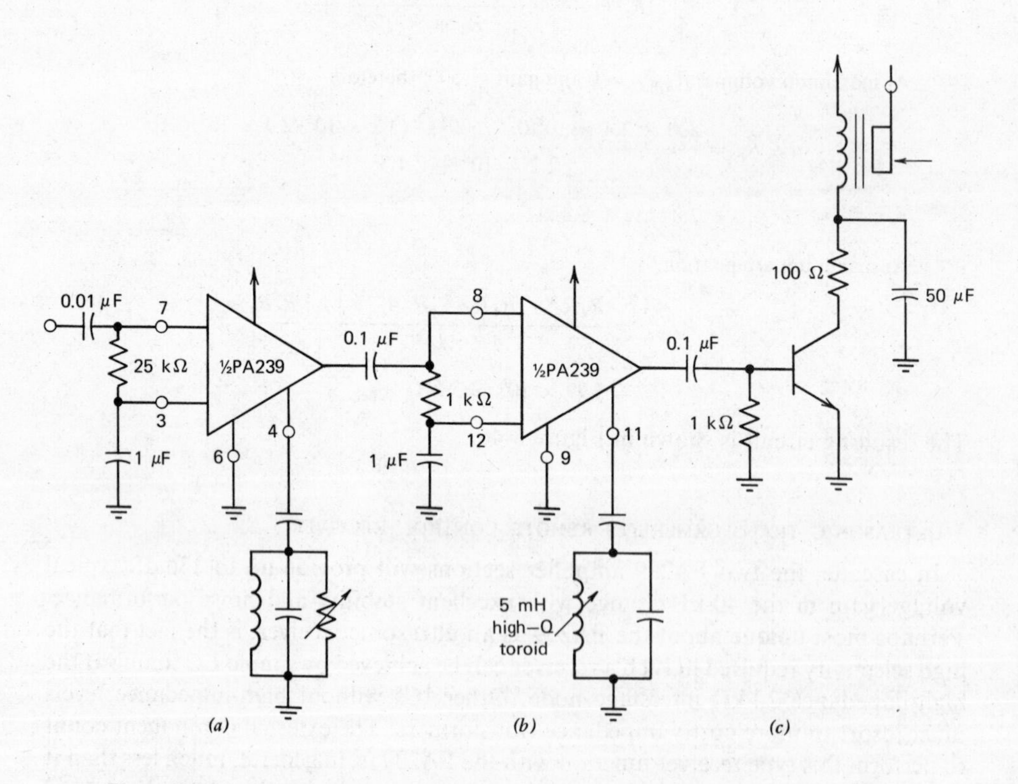

Figure 8-47. Circuit schematic diagram of ultrasonic receiver: (*a*) low selectivity fixed-tuned first stage; (*b*) high selectivity second stage; (*c*) control relay driver.

FM DETECTOR-LIMITER

Introduction

The purpose of a detector stage (shown in Figure 8-1) in a communications receiver is to separate the modulation from the IF (intermediate frequency), discard the IF, and pass the information on to stages that will provide further gain, such as audio amplifiers for radio reception.

The circuits discussed in this section utilize the ULN2111A FM detector-limiter. The main feature of its design is a quadrature detection scheme in which a simple single adjustment of the phase-shift network is all that is required to get satisfactory performance. This circuit is utilized in General Motors' automotive radios.

Applications

APPLICATION IN FM CIRCUITRY

The basic design philosophy behind the ULN2111A is that the function of balance should be placed in the active portion of the system rather than in the passive portion. The major difference proposed by this philosophy to preceding design approaches is that balance need not be adjustable and, consequently, restrictions associated in the design and adjustment of balanced coil detectors (such as the ratio detector) is greatly relieved.

In any system using a balanced discriminator, the major figures of merit include: Audio recovery, AMR (AM rejection, extraneous noise that FM must pass), dynamic range, and THD (third harmonic distortion). Of these characteristics, only audio recovery must be adjusted to the circuit externally.

Audio recovery is effected primarily by the resonant simple *LC* network and the injection value at the detector input. To obtain a proper value for audio recovery, specific characteristics of the S curve must be selected. Primarily, these characteristics resolve to a choice of slope of the S curve giving a specified value for $\Delta V/\Delta F$ which is the basic expression of audio recovery. Two values must be chosen:

1. Peak-to-peak separation desired
2. Peak-to-peak voltage required at the nodes of the S curve

The peak-to-peak separation is usually defined by the service for which the system is intended. For FM, this value is 550 kHz. Once the choice is made, and the center frequency, as well as the peak-to-peak separation is known, an approximation of the circuit Q can be made, by using

$$Q = \frac{F}{F_0} \qquad Q = \frac{F_0}{\Delta F} \tag{8-48}$$

Although the desired circuit Q is fairly low (indicating a high-L network), it is desirable to use a high-C network. The input to the detector will introduce some variable capacitance. This can be minimized through the use of at least 100 pF as the C part of the resonant circuit. The inductor chosen along with this capacitance value will yield a network with a Q somewhat higher than desired. This should be reduced through use of a parallel resistor across the network. Figure 8-48 shows complete

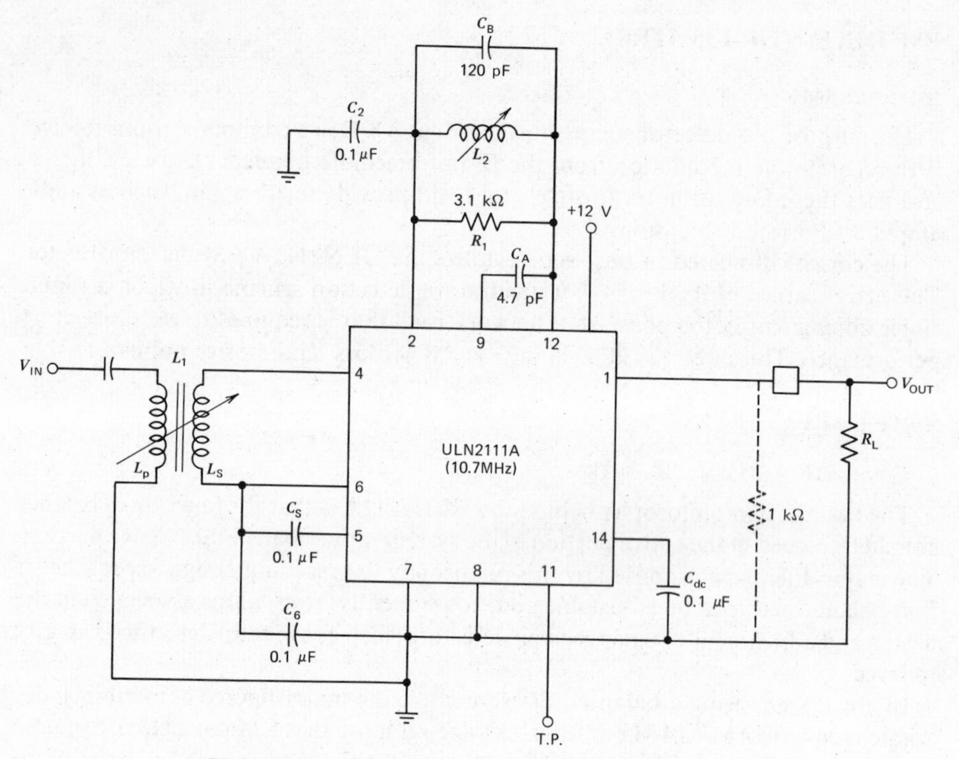

Figure 8-48. Quadrature network for FM receivers.

quadrature networks for FM circuitry. Optimum component values are given in the figure.

The final component required is a small decoupling capacitor placed between the network and the low amplifier output. To ensure linear-detector operation, the reactance of this capacitor should be substantially large as compared to the impedance of the tuned circuit at resonance.

The other factor governing the audio output is the injection value at the input to the detector. The optimum value is 60 mV rms at the resonant frequency of the network.

To complete the circuit, it is necessary to select a de-emphasis capacitor and several bypass capacitors. The de-emphasis capacitor should have a time constant of 75 μsec. Satisfactory results will be obtained with a value in the range of 0.01 to 0.05 μF. The bypasses are not critical and any value from 0.05 to 0.01 μF will do. As with any bypass capacitor, the objective is to keep the reactance as low as possible. A noninductive, ceramic capacitor is best in this application. The amplifier and detector inputs are designed to operate at dc bias levels provided by a part of the diode-divider chain in the device. It is intended that a coil, or part of a transformer, be used as a low dc resistance path between the bias source and the respective input connection, thus providing the bias level required (Figure 8-48). For best operation, the dc resistance value should be as low as possible. Values of 100 to 300 Ω are quite satisfactory.

The amplifier will tolerate a considerable dynamic range at the input. The minimum level of input is approximately 1.0 mV for good AMR and effective clipping. The maximum input swing is limited by the collector-base diode of the input transistor.

In the circuit of Figure 8-48, the maximum peak swing should be limited to about 1.5 V. The dynamic tolerable range will then be in excess of 60 dB, which is more than adequate for satisfactory performance.

The detector-output-loading is not critical. The source resistance is quite low (on the order of 200 Ω). To prevent clipping at the output when the a-c load impedance is less than 2000 Ω, a 1000-Ω resistor should be added between the output and the ground.

Tuning the device in a system is quite simple. Two alternatives are available: (1) Tune the receiver to a strong station and then tune the quadrature coil to maximum audio output, or (2) apply an FM generator to the amplifier input through a small decoupling capacitor, and tune as above.

The circuit layout should be carried out with reasonable care inasmuch as the device has an inherent high gain and high band-pass characteristics. It is important to separate the input and output components to prevent undesired coupling which may cause oscillation.

The ULN2111A is a simple circuit to use. Balance and symmetry are a function of the monolithic circuit, thus relieving the system designer of many restrictions normally found in other methods of detection.

TWO-BLOCK FM IF STRIP

An IF strip for use in FM receiver applications can be designed by using a pair of ULN2111A ICs. The availability of inexpensive ceramic filters has generated considerable interest in fixed tuned nonadjustable IF strips. The advantages of such IF strips are simplicity, smaller space requirements, and lower production costs.

Figure 8-49 shows a diagram for a two-block IF stage. The second block is hooked up as a standard quadrature detector. De-emphasis of 75 μsec is supplied by a 0.001-μF capacitor at *pin* 14. Filtering at this point will be required even for stereo applications to prevent the strip from oscillating. For stereo, however, this capacitor may be reduced to 100 pF which is sufficiently small to maintain satisfactory band-pass.

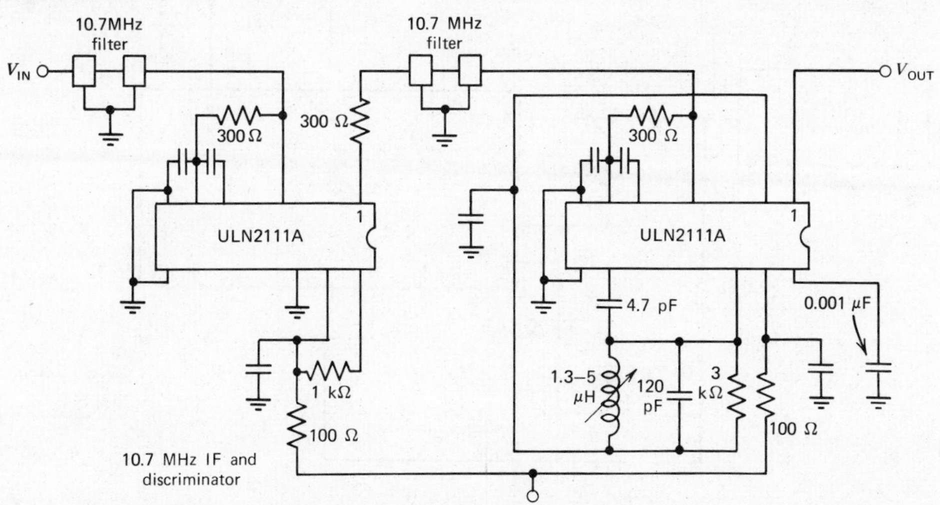

Figure 8-49. Diagram of two-block IF strip [*note:* all capacitors = 0.01 μF ceramic filters–Murata 10.7 SF].

No provision has been made in the strip for AGC. Due to the limiting action of the amplifier and the wide dynamic input range of permissible signal range, no AGC is required for FM receiver service.

SUPPRESSED CARRIER MODULATION

Suppressed carrier modulation can be provided using the ULN2111A. The test circuit for this application is shown in Figure 8-50. With a carrier frequency of 50 kHz, the modulator frequency is 3 kHz. At this low frequency, the feedback capacitor is changed to 6.7 μF as shown in Figure 8-50. This value is necessary so that a 55-dB gain is realized in the amplifier section at low frequencies.

At higher frequencies, the carrier frequency is 5 MHz and the modulating frequency is 5 kHz. At these frequencies, the value of C_f can be 0.05 μF.

In the circuit diagram in Figure 8-50, the modulating signal input is at *pin* 12. A phase-shift network is not a requirement. Instead, a 50-Ω resistor is placed between *pins* 2 and 12 to ensure proper bias to the input of the gate detector. By adjusting the dc bias of the gate where the modulation is applied, full-carrier suppression is achieved. The 0.1-μF capacitor, placed between *pin* 2 and ground, decouples the limiter supply. Since the input to the limiter is floating, consequently requiring a dc reference, a 50-Ω resistor is used between *pins* 4 and 6. This resistor can be any value from 50 to 200 Ω. This dc reference is provided by the diode supply line which must be decoupled by an external capacitor from *pin* 6 to ground. This avoids regeneration between bases through insufficient base-source impedance.

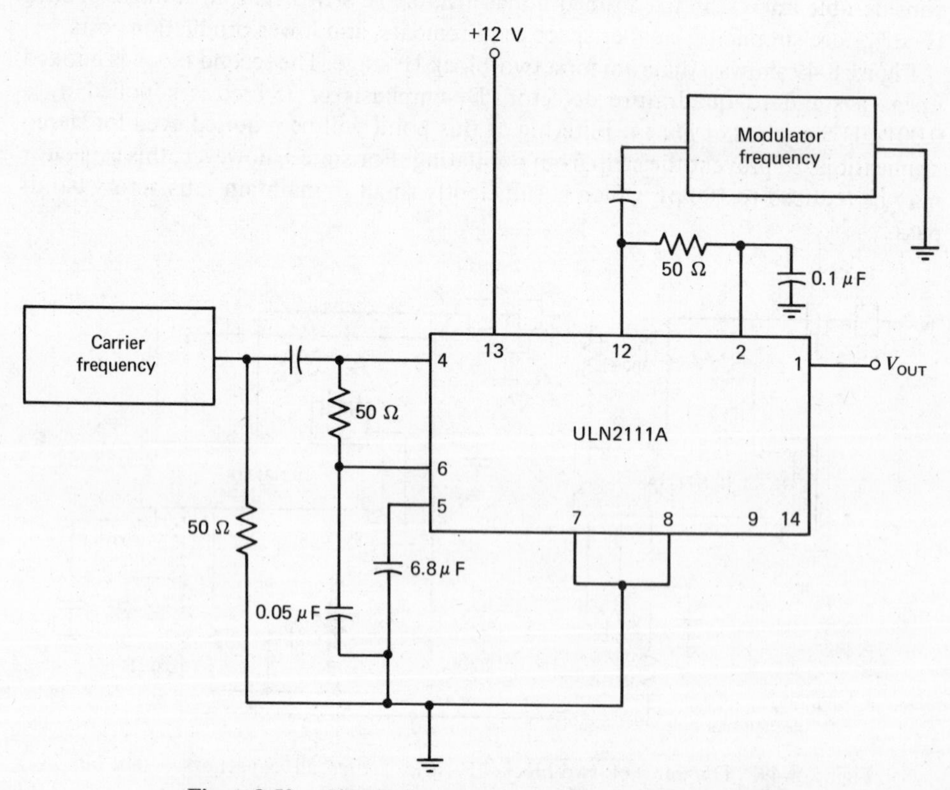

Figure 8-50. Circuit diagram for suppressed carrier modulator.

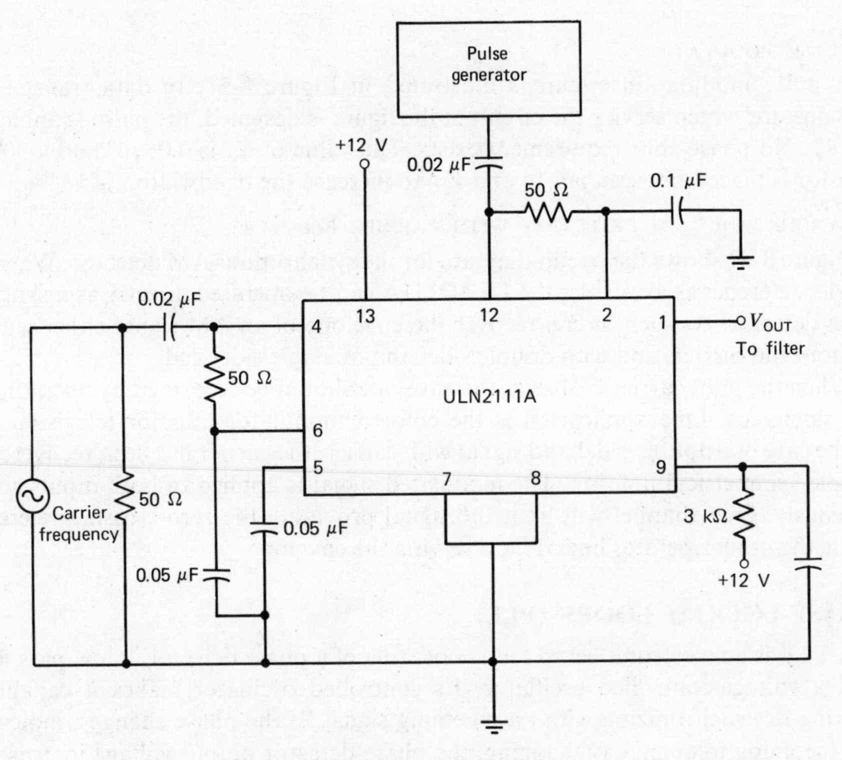

Figure 8-51. Pulse modulation circuit.

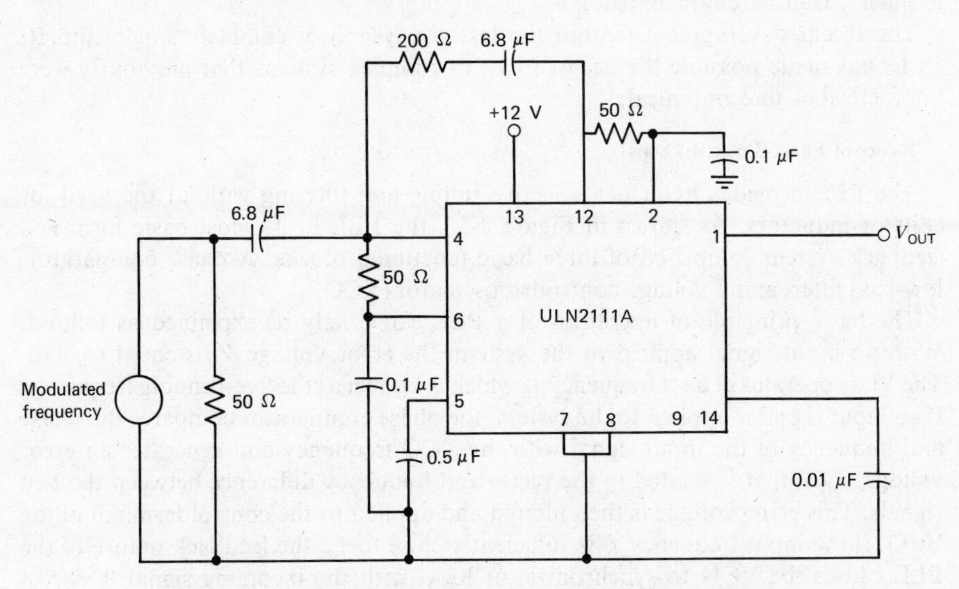

Figure 8-52. Synchronous AM detection with carrier frequency.

PULSE MODULATION

A pulse modulation circuit is illustrated in Figure 8-51. In data transmission systems, for which service the circuit in the figure is designed, the pulse is applied to *pin* 12. No phase-shift requirement exists. The value of C_f is 0.05 μF and a 3000-Ω resistor is placed between *pin* 14 and V_{CC} to increase the bandwidth.

SYNCHRONOUS AM DETECTION WITH CARRIER RECOVERY

Figure 8-52 shows the circuit diagram for the synchronous AM detector. When the carrier reference is available, the ULN2111A can be operated directly as a synchronous detector. As such, it can recover the envelope of an AM signal either with or without the carrier, and with double sideband or single sideband.

When the pilot carrier is absent, the envelope should be recovered by operating on the sidebands. One application is the color demodulator in color television sets. In the case of a double-sideband signal with carrier, the carrier has been recovered by simple symmetrical limiting. The modulated signal is applied to both inputs simultaneously. One channel will limit the signal providing the zero-crossing reference, while the other operates linearly, preserving the envelope.

PHASE-LOCKED LOOPS (PLL)

A PLL is an electronic servo loop consisting of a phase detector, a low-pass filter, and a voltage-controlled oscillator. Its controlled oscillator makes it capable of locking or synchronizing with an incoming signal. If the phase changes, indicating the incoming frequency is changing, the phase-detector output voltage increases or decreases just enough to keep the oscillator frequency the same as the incoming frequency, preserving the locked condition. Thus, the average voltage applied to the controlled oscillator is a function of the frequency of the incoming signal. In fact, the low-pass filter voltage is the demodulated output when the incoming signal is frequency-modulated (provided the controlled oscillator has a linear voltage-to-frequency transfer characteristic).

The recent development (within the last two years) of complete single-chip IC PLLs has made possible the use of PLLs in complex systems that previously were impractical or uneconomical.

Basics of Phase-Locked Loops

The PLL provides frequency selective tuning and filtering without the need for coils or inductors. As shown in Figure 8-53, the PLL in its most basic form is a feedback system comprised of three basic functional blocks: A phase comparator, low-pass filter, and a voltage-controlled oscillator (VCO).

The basic principle of operation of a PLL can briefly be explained as follows: With no input signal applied to the system, the error voltage V_d is equal to zero. The VCO operates at a set frequency f_0 which is known as the free-running frequency. If an input signal is applied to the system, the phase comparator compares the phase and frequency of the input signal with the VCO frequency and generates an error voltage, $V_e(t)$, that is related to the phase and frequency difference between the two signals. This error voltage is then filtered and applied to the control terminal of the VCO. If the input frequency f_s is sufficiently close to f_0, the feedback nature of the PLL causes the VCO to synchronize, or lock, with the incoming signal. Once in lock, the VCO frequency is identical to the input signal, except for a finite phase difference.

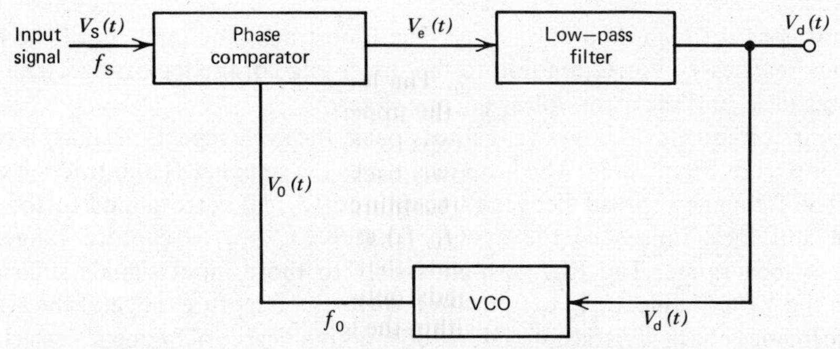

Figure 8-53. The basic phase-locked loop consists of three functional blocks.

Two key parameters of a PLL system are its lock and capture ranges. They can be defined as follows:

Lock Range. Range of frequencies in the vicinity of f_0, over which the PLL can maintain lock with an input signal. It is also known as the tracking or holding range. Lock range increases as the overall gain of the PLL is increased.

Capture Range. Band of frequencies in the vicinity of f_0 where the PLL can establish or acquire lock with an input signal. It is also known as the acquisition range. It is always smaller than the lock range, and is related to the low-pass filter bandwidth. It decreases as the filter bandwidth is reduced.

The lock and capture ranges of a PLL can be illustrated with reference to Figure 8-54, which shows the typical frequency-to-voltage characteristics of a PLL. In the figure, the input is assumed to be swept slowly over a broad frequency range. The vertical scale corresponds to the loop-error voltage.

In the upper part of Figure 8-54, the loop frequency is being gradually increased. The loop does not respond to the signal until it reaches a frequency f_1, corresponding to the lower edge of the capture range. Then, the loop suddenly locks on the input, causing a negative jump of the loop-error voltage. Next, V_d varies with frequency with a slope equal to the reciprocal of the VCO voltage-to-frequency conversion

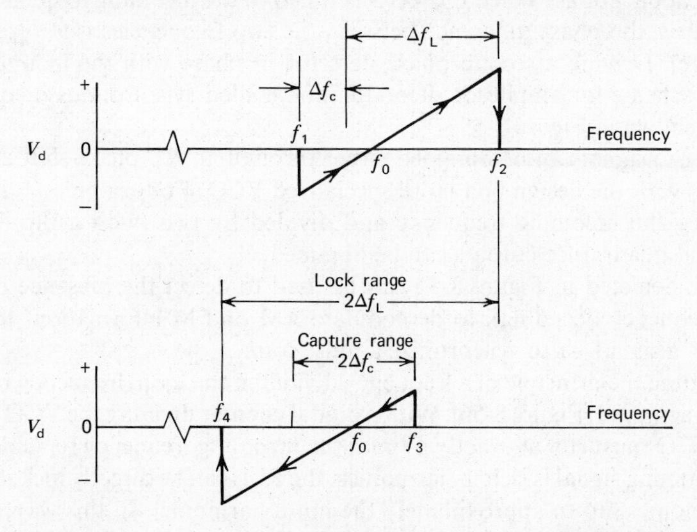

Figure 8-54. Typical phase-locked loop frequency-to-voltage transfer characteristic.

gain, and goes through zero as $f_s = f_0$. The loop tracks the input until the input frequency reaches f_2, corresponding to the upper edge of the lock range. The PLL then loses lock, and the error voltage drops to zero.

If the input frequency is now swept slowly back, the cycle repeats itself as shown in the lower part of Figure 8-54. The loop recaptures the signal at f_3 and traces it down to f_4. The frequency spread between (f_1, f_3) and (f_2, f_4) corresponds to the total capture and lock ranges of the system; that is, $f_3 - f_1$ = capture range and $f_2 - f_4$ = lock range. The PLL responds only to those input signals sufficiently close to the VCO frequency f_0 to fall within the lock or capture range of the system. Its performance characteristics, therefore, offer a high degree of frequency selectivity, with the selectivity characteristics centered about f_0.

If an incoming frequency is far removed from that of the VCO, so that their difference exceeds the pass band of the low-pass filter, it will simply be ignored by the PLL. Thus, the PLL is a frequency-selective circuit.

Three features of the PLL are particularly important for ICs:

1. The center frequency of the resulting filter is represented by the free-running frequency of the VCO. Since stable oscillators can be designed in integrated form requiring only one resistor and one capacitor to determine frequency, a stable and accurate filter is possible without thin-film elements.
2. The selectivity of the PLL is determined by the low-pass filter and the loop gain. With a single *RC* low-pass filter, the selectivity of a six to twelve-pole *LC* filter can be achieved (the shape of the low-pass filter curve is translated into the frequency range of the VCO). This minimizes the number of external (or internal) capacitors. Also, center-frequency and band-pass characteristics can be adjusted separately.
3. The linearity of the phase-locked loop as an FM discriminator depends only on the voltage-frequency conversion characteristics of the VCO. Such oscillators can be built with a linearity of better than 0.1% with the help of integrated current sources.

The basic PLL is useful only for FM. Adding a second phase detector greatly extends its range cf applications. When the VCO is tuned to the incoming frequency and the loop is locked, the phase difference between the two frequencies is 90° (resulting in 0 dc voltage). Driving a second phase detector in phase with the incoming signal provides a scheme for amplitude demodulation (called synchronous or quadrature detection), shown in Figure 8-55.

The 0° VCO signal can be obtained either through an *RC* phase-shift network or (preferably) with the design of a more specialized VCO. For example, if the VCO is run at twice the incoming frequency and divided by two (with a flip-flop), both in-phase and quadrature outputs can be obtained.

The PLL depicted in Figure 8-55 can be used to detect the presence of a signal (tone or frequency decoding), to demodulate AM or FM information, to track or reconstitute a signal, or to synchronize a pulse train.

By extending the principle of a frequency divider in the loop, frequency multiplication can be achieved (Figure 8-56). With a digital counter dividing the VCO frequency by n, the VCO must run at exactly n times the incoming frequency to achieve lock.

If the incoming signal is rich in harmonics, the PLL can be directly locked onto one of the harmonics (up to approximately the ninth harmonic). In this way, frequency multiplication is achieved without the use of a divider.

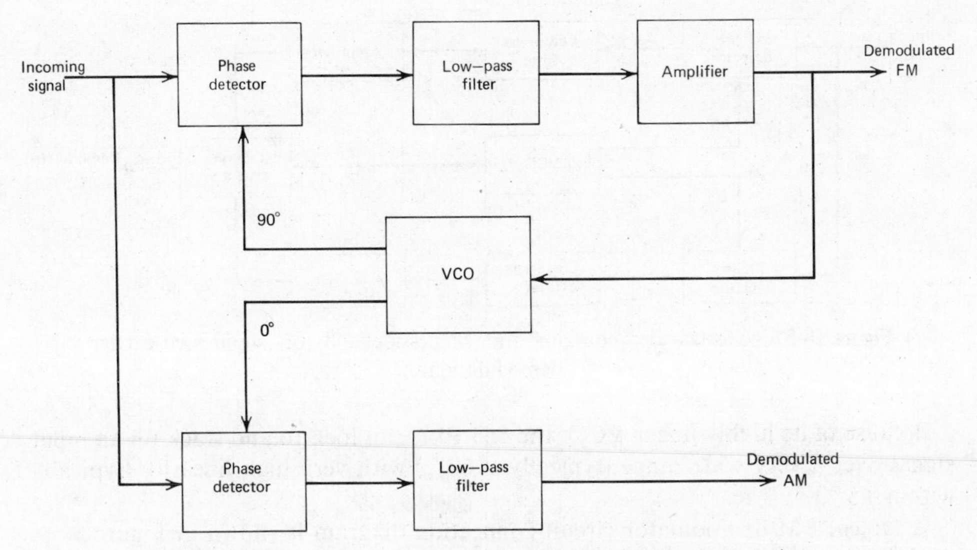

Figure 8-55. Phase-locked loop extended to amplitude detection.

In all of these applications, the PLL can operate only if a carrier is present. A configuration called the Costas demodulator allows the selective reception and demodulation of single-sideband, suppressed carrier, or vestigial sideband signals (Figure 8-57). In this method, the two detected signals (90° apart in phase) are mixed in a third phase detector, and their difference is fed to the control terminal of the VCO. The PLL is free-running until the modulation appears. The time required to achieve lock can, however, be made short enough so that little or no information is lost.

The NE565 PLL is a general-purpose circuit designed for highly linear FM demodulation. During lock, the average dc level of the phase-comparator output signal is directly proportional to the frequency of the input signal. As the input frequency shifts, it is this output signal which causes the VCO to shift its frequency to match that of the input. Consequently, the linearity of the phase-comparator output with frequency is determined by the voltage-to-frequency transfer function of the VCO.

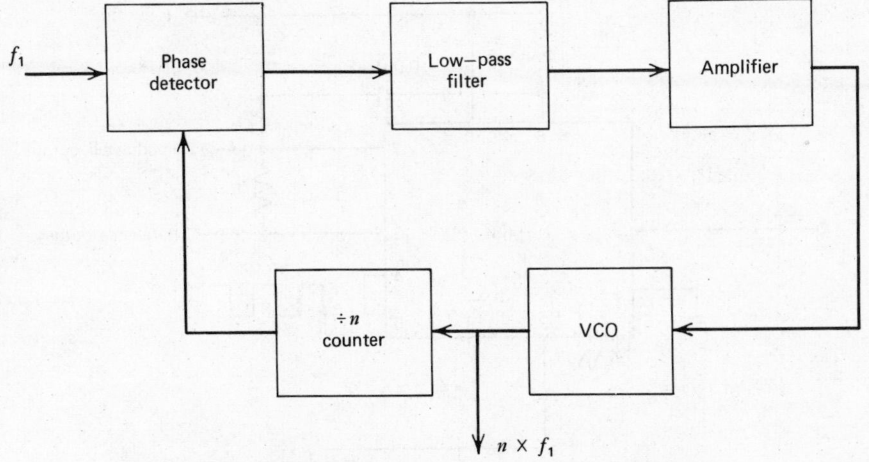

Figure 8-56. The phase-locked loop as a frequency multiplier.

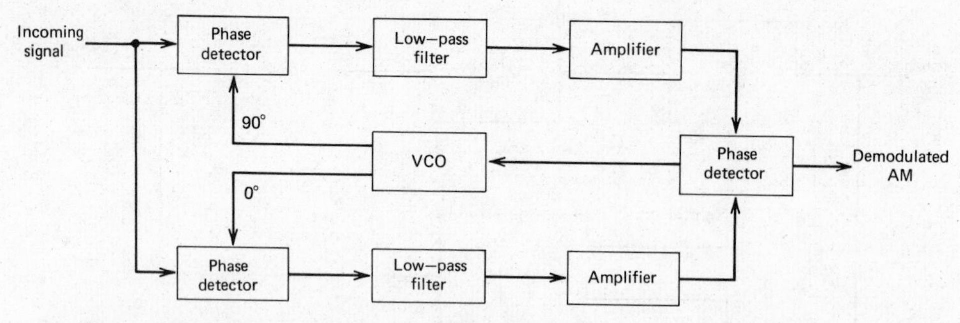

Figure 8-57. Costas demodulator for single-sideband or suppressed-carrier demodulation.

Because of its highly linear VCO, the 565 PLL can lock to and track on an input signal over a very wide range (typically +60%) with very high linearity (typically within 0.5%).

A typical FM demodulator circuit-connection diagram is shown in Figure 8-58. The VCO free-running frequency is given by

$$f_0 = \frac{1}{4R_1C_1} \tag{8-49}$$

and should be adjusted to be at the center of the input signal frequency range; C_1 can be any value, but R_1 should be within the range of 2000 to 20,000 Ω with an optimum value on the order of 4000 Ω. The source can be direct-coupled if the dc resistances seen from *pins* 2 and 3 are equal and there is no dc voltage difference applied between the *pins*. A short between *pins* 4 and 5 connects the VCO to the phase comparator; *pin* 6 provides a dc reference voltage that is close to the dc potential of the demodulated output (*pin* 7). Thus, if a resistance (R_2 in Figure 8-58) is connected between *pins* 6 and 7, the gain of the output stage can be reduced with little change in the dc voltage level at the output. This allows the lock range to be decreased without affecting the free-running frequency. In this manner, the lock range can be decreased from ±60% of f_0 to approximately ±20% of f_0 at ±6 V.

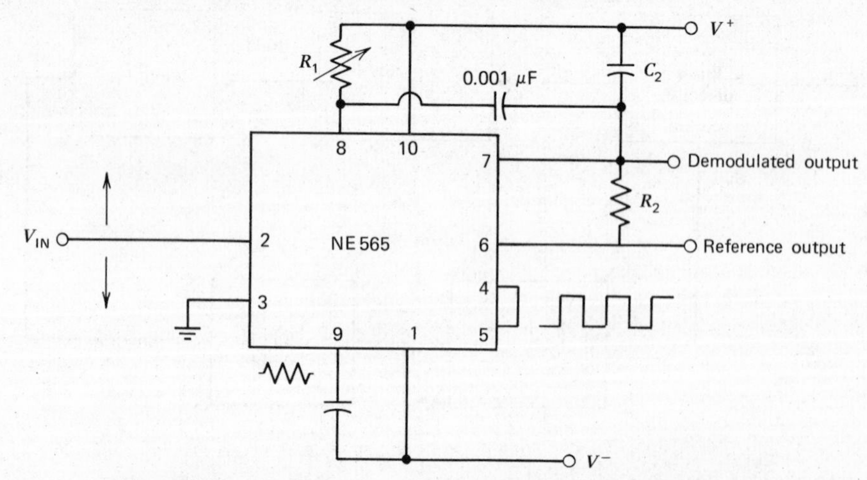

Figure 8-58. Typical connection diagram for FM demodulation.

A small capacitor, typically 0.001 μF, should be connected between *pins* 7 and 8 to eliminate possible oscillation in the voltage-controlled current source. The input-signal level can be between 1 mV and several volts peak-to-peak. However, if the loop is to track wide frequency deviations, it is recommended that the input level be greater than 100 mV_{p-p}.

The dc level at the inputs (*pins* 2 and 3) should be equal and must be in the lower half of the total supply-voltage range to allow proper operation of the phase comparator. If, for example, supply voltages of ± 60 V are used with a ground reference in the center, the dc voltage at the inputs should be approximately between ground and -4 V. For cases in which only one supply voltage is used, the inputs can be referenced to an appropriate voltage obtained from a voltage divider connected between supply and ground.

The control voltage for the VCO (demodulated at *pin* 7) is referenced to the positive supply as is the voltage-controlled current source in the VCO. When the VCO is operating at its free-running frequency, this voltage has typically a value of 0.125 V_{CC} below positive supply where V_{CC} is the total supply voltage applied across the circuit. Again, for example, if supply voltages of ± 6 V are used, the output voltage at *pin* 7 would typically be 1.5 V below positive supply (or 4.5 V above ground).

The PLL is directly analogous to conventional feedback or servo systems and is characterized by the same equations. The open-loop transfer function for any PLL is

$$H(s) = \frac{K_0 K_d A F(s)}{s} \tag{8-50}$$

where K_0 is the VCO conversion gain in radians per second per volt, K_d is the phase-detector gain factor in volts per radian, A is the gain of the amplifier, and $F(s)$ is the transfer function of the filter. When the loop is closed, the transfer function becomes

$$H(s) = \frac{K_0 K_d A F(s)}{s + K_0 K_d A F(s)} \tag{8-51}$$

If the filter is omitted entirely, $F(s) = 1$, the loop transfer function becomes

$$H(s) = \frac{K_0 K_d A}{s + K_0 K_d A} \tag{8-52}$$

This is the characteristic equation for a first-order loop. A loop of this type has a pole, initially at the origin, and a zero at ∞. This results in the root locus shown in Figure 8-59 as the loop gain is varied. A root locus of this type predicts a flat response with a 6-dB/octave roll-off which is typical of a first-order loop. The output, however, contains both sum and difference frequency components and must be filtered outside the loop to obtain the desired signal. An important characteristic of the first-order loop, however, is that the capture and lock ranges are equal.

A second-order loop is obtained by the use of a single-pole filter with a transfer function of the form

$$F(s) = \frac{1}{s\tau + 1} \tag{8-53}$$

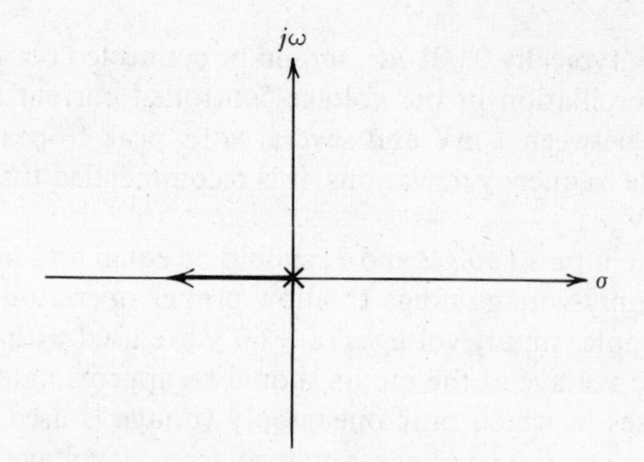

Figure 8-59. First-order loop.

This results in a closed-loop transfer function of

$$H(s) = \frac{K_0 K_d A/\tau}{s^2 + (s/\tau) + K_0 K_d A/\tau} \tag{8-54}$$

This equation can be written as

$$H(s) = \frac{K_0 K_d A/\tau}{s^2 + 2\zeta\omega_n s + \omega_n^2}$$

where

$$\zeta = \frac{1}{2}\left[\frac{1}{K_0 K_d A \tau}\right]^{1/2} \tag{8-55}$$

and

$$\omega_n = \left[\frac{K_0 K_d A}{\tau}\right]^{1/2} \tag{8-56}$$

in which ω_n is the "natural" frequency of the loop and ζ is the damping factor.

The root locus for a second-order loop is shown in Figure 8-60. The loop has poles at $s = 0$, and $s = -1/\tau$ initially, and two zeros at $s = \infty$. As the loop gain is increased,

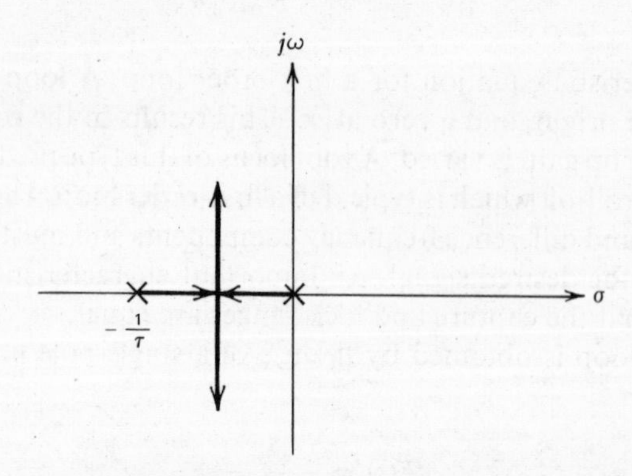

Figure 8-60. Second-order loop.

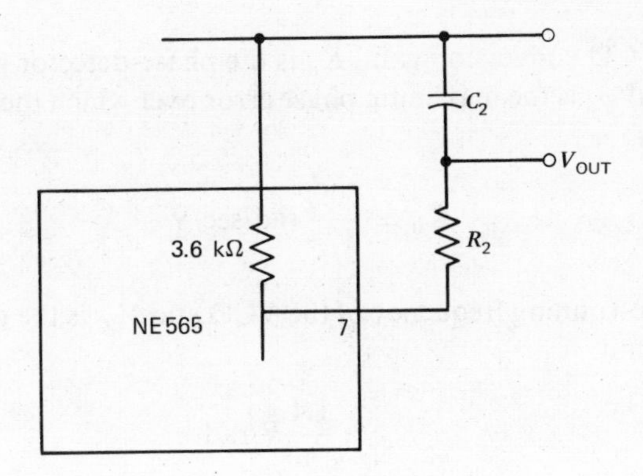

Figure 8-61. Lead-lag filter connection.

the poles move together and break into the complex plane. If the loop gain and the filter-time constant are large, as is usually the case, the poles will be highly complex, and the loop will be badly underdamped with a poor transient response. This also results in a capture range which is much smaller than the lock range.

A scheme to improve the response in this case is shown in Figure 8-61. Insertion of a resistor R_2 in a series with the filter capacitor C_2 forms a lag-lead filter which basically allows the loop to operate in a mode similar to that of a first-order loop.

The lead-lag filter increases damping, yet still achieves filtering since the output is taken across the capacitor C_2. This type of filter has the form

$$F(s) = \frac{s\tau_2 + 1}{s(\tau_1 + \tau_2) + 1} \tag{8-57}$$

which introduces a zero at $s = -1/\tau_2$ as well as a pole initially at $s = -1/(\tau_1 + \tau_2)$. The root locus for a loop using this type of filter is shown in Figure 8-62. Here it is seen that for large values of loop gain, the poles are no longer complex, but lie on the real axis to give a flat overdamped response.

The lock range, over which the loop will track the input signal, can be calculated from the expression

$$\omega_L = \pm K_0 K_d A \phi_d \tag{8-58}$$

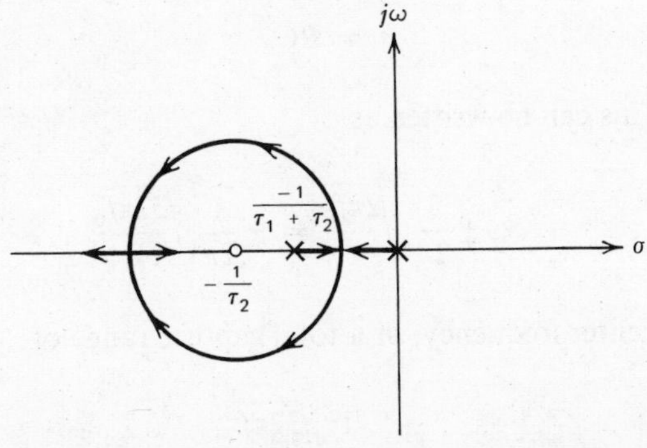

Figure 8-62. Second-order loop with lead-lag filter.

where K_0 is the VCO conversion gain, K_d is the phase-detector gain factor, A is the amplifier gain, and ϕ_d is the maximum phase error over which the loop can remain in lock. For the 565,

$$K_0 \simeq \frac{50 f_0}{V_{CC}} \text{ rad/sec/V} \tag{8-59}$$

where f_0 is the free-running frequency of the VCO and V_{CC} is the total supply voltage applied to the circuit.

$$K_d \simeq \frac{1.4}{\pi} \text{ V/rad} \tag{8-60}$$

$$A \simeq 1.4 \tag{8-61}$$

$$\phi_d \simeq \pm \frac{\pi}{2} \text{ rad} \tag{8-62}$$

The lock range for the 565 then becomes

$$f_L = \frac{\omega_L}{2\pi} \simeq \pm \frac{8 f_0}{V_{CC}} \text{ Hz} \tag{8-63}$$

to each side of the center frequency, or a total range of

$$f_L \simeq \frac{16 f_0}{V_{CC}} \text{ Hz} \tag{8-64}$$

The capture range, over which the loop can acquire lock with the input signal is given approximately by

$$\omega_C \simeq \pm \sqrt{\frac{\omega_L}{\tau}} \tag{8-65}$$

where ω_L is the lock range,

$$\omega_L = 2\pi f_L \tag{8-66}$$

and τ is the time constant of the loop filter,

$$\tau = RC_2 \tag{8-67}$$

with $R \simeq 3.6\ \text{k}\Omega$. This can be written as

$$f_c \simeq \pm \frac{1}{2\pi} \sqrt{\frac{2\pi f_L}{\tau}} = \pm \frac{1}{2\pi} \sqrt{\frac{32\pi f_0}{\tau V_{CC}}} \tag{8-68}$$

to each side of the center frequency, or a total capture range of

$$f_c \simeq \frac{1}{\pi} \sqrt{\frac{32\pi f_0}{\tau V_{CC}}} \tag{8-69}$$

This approximation works well for narrow capture ranges ($f_c = (1/3)f_L$) but becomes too large as the limiting case is approached ($f_c = f_L$).

The effects of the time constant τ on the capture range f_c can be seen from the typical lock-and-capture range characteristic shown in Figure 8-63. Here the capture range decreases as the value of the filter capacitor C_2 (and hence τ) is increased while the lock range remains constant.

For applications where both a narrow lock range and a large output-voltage swing are required, it is necessary to inject a constant current into *pin* 8 and increase the value of R_1. One scheme for this is shown in Figure 8-64. The basis for this scheme is the fact that the output voltage controls only the current through R_1 while the current through Q_1 remains constant. Thus, if most of the charging current is due to Q_1, then the total current can be varied only a small amount due to the small change in current through R_1. Consequently, the VCO can track the input signal over a small frequency range, yet the output voltage of the loop (control voltage of the VCO) will swing its maximum value.

Diode D_1 is a Zener diode used to allow a large voltage drop across R_A than would otherwise be available; D_4 is a diode which should be matched to the emitter-base junction of Q_1 for temperature stability. In addition, D_1 and D_2 should have the same breakdown voltages, and D_3 and D_4 should be similar so that the voltage seen across R_B and R_C is the same as that seen across *pins* 10 and 1 of the PLL. This causes the frequency of the loop to be insensitive to power-supply variations. The center frequency can be found by

$$f_O \simeq \frac{2R_B}{(R_B + R_C)R_A C_1} + \frac{1}{4R_1 C_1} \text{ Hz} \tag{8-70}$$

and the lock range is given by

$$\begin{aligned} f_L &\simeq \frac{1.4V_D f_O}{(V_1 - V_2 - V_Z - V_D)[R_B R_1/(R_B + R_C)R_A + 1/8]} \\ &\simeq \frac{11.2V_D(R_B + R_C)R_A f_O}{(V_1 - V_2 - V_Z - V_D)[8R_B R_1 + R_A(R_B + R_C)]} \text{ Hz} \end{aligned} \tag{8-71}$$

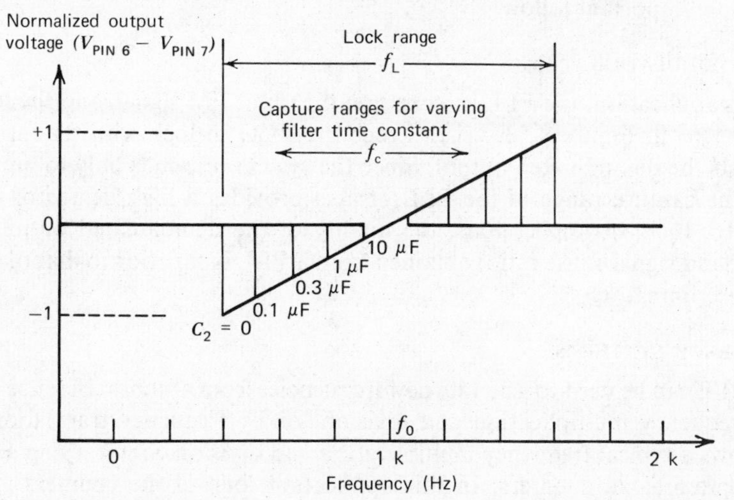

Figure 8-63. Typical lock-and-capture range characteristics.

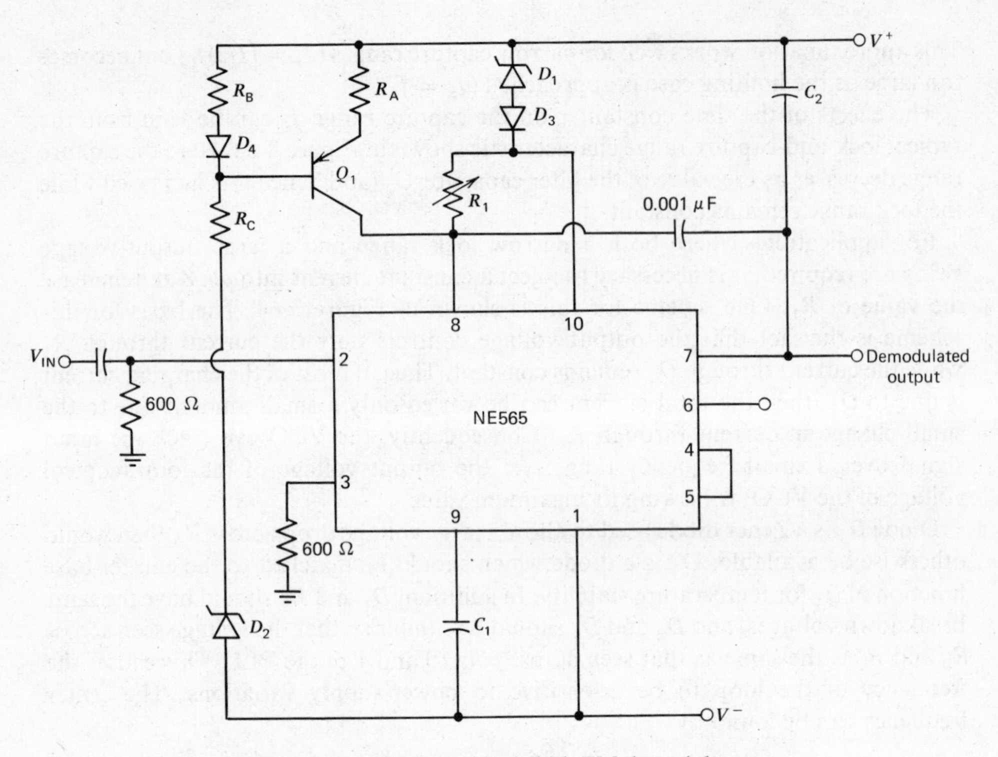

Figure 8-64. Narrow bandwidth FM demodulator.

where V_D = forward-biased diode voltage $\simeq$ 0.7 V, V_Z = Zener diode breakdown voltage, V_1 = positive-supply voltage, V_2 = negative-supply voltage, and f_0 = free-running VCO center frequency.

Applications for Phase-Locked Loops

As a versatile building block, the PLL covers a wide range of applications. Some of the more important follow.

FM DEMODULATION

In this application, the PLL is locked on the input FM signal, and the loop-error voltage, $V_d(t)$ in Figure 8-53, which keeps the VCO in lock with the input signal, represents the demodulated output. Since the system responds only to input signals within the capture range of the PLL, it also provides a high degree of frequency selectivity. In most applications, the quality of the demodulated output (i.e., its linearity and signal/noise ratio) obtained from a PLL is superior to that of a conventional discriminator.

FREQUENCY SYNTHESIS

The PLL can be used to generate new frequencies from a stable reference source by either frequency multiplication and division, or by frequency translation. Figure 8-65 shows a typical frequency multiplication and division circuit, using a PLL and two programmable counters. In this application, one of the counters is inserted between the VCO and phase comparator and effectively divides the VCO frequency

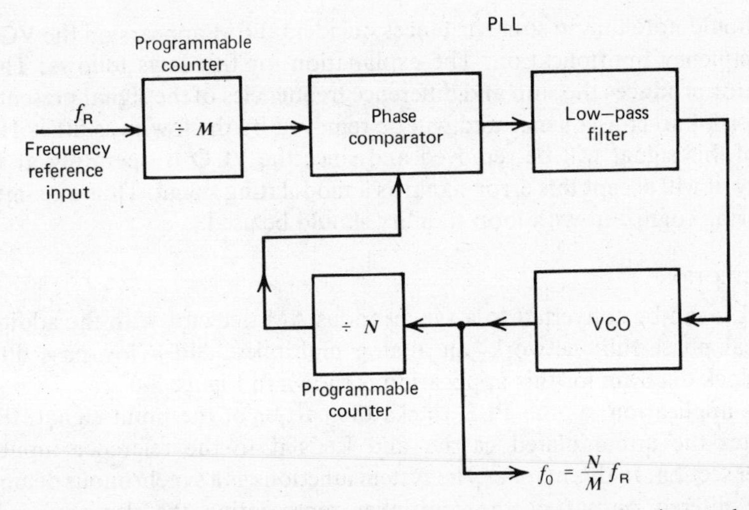

Figure 8-65. A frequency multiplier-divider can be constructed using a phase-locked loop.

by the counter's modulus N. When the system is in lock, the VCO output is related to the reference frequency f_R by the counter moduli M and N as

$$f_O = \frac{N}{M} f_R \tag{8-72}$$

By adding a multiplier and an additional low-pass filter to a PLL (Figure 8-66), one can form a frequency translation loop. In this application, the VCO output is shifted from the reference frequency f_R by an amount equal to the offset frequency, f_1, that is, $f_O = (f_R + f_1)$.

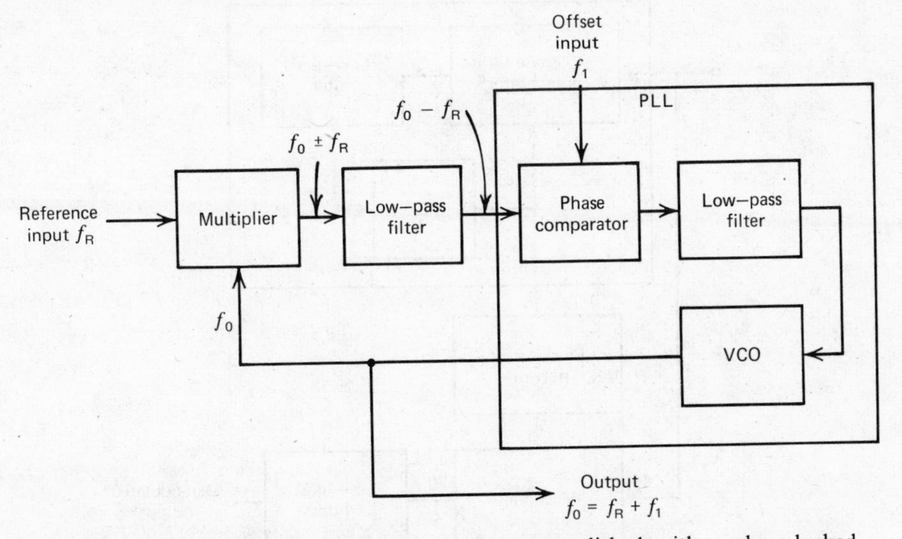

Figure 8-66. Frequency translation can be accomplished with a phase-locked loop by adding a multiplier and an additional low-pass filter to the basic phase-locked loop.

One should note that in some instances, incidental FM appears on the VCO when doing frequency multiplication. The explanation for this is as follows: The phase comparator produces the sum and difference frequencies of the signal presented to it. The sum is of no consequence and will be removed by the low-pass filter. However, not all of this signal will be removed and since the VCO is operating at a higher frequency, it will accept this error signal as a modulating signal. Thus maximum low-pass filtering cognizant with loop stability should be used.

AM DETECTION

The PLL can be converted to a synchronous AM detector with the addition of a noncritical phase-shift network, an analog multiplier, and a low-pass filter. The system block diagram for this application is shown in Figure 8-67.

In this application, as the PLL tracks the carrier of the input signal, the VCO regenerates the unmodulated carrier and feeds it to the reference input of the multiplier section. In this manner, the system functions as a synchronous demodulator with the filtered output of the multiplier representing the demodulated audio information.

TONE DETECTION

In this application, the PLL is again connected as shown in Figure 8-67. When a signal tone is present at the input within a frequency band corresponding to the capture range of the PLL, the output dc voltage is shifted from its tone-absent level. This shift is easily converted to a logic signal by adding a threshold detector with logic-compatible output levels.

STEREO DECODING

In commercial FM broadcasting, suppressed carrier AM modulation is used to superimpose the stereo information on the FM signal. To demodulate the complex

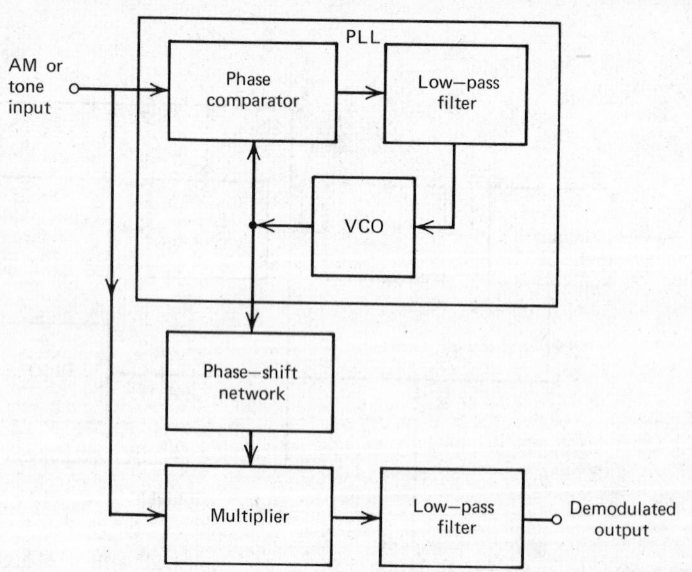

Figure 8-67. AM and tone detection are possible by adding three functional blocks to the basic phase-locked loop.

stereo signal, a low-level pilot tone is transmitted at 19 kHz (1/2 of actual carrier frequency). The PLL can be used to lock onto this pilot tone, and regenerate a coherent 38-kHz carrier which is then used to demodulate the complete stereo signal. A number of highly specialized monolithic circuits have been developed for this application. Typical examples of monolithic stereo decoder circuits using the PLL principle are the MC1310 and CA3090E, manufactured by Motorola and RCA, respectively.

Stereo decoders normally incorporate *LC* filters, both to provide 19-kHz selectivity and as part of a doubling circuit to generate the required 38-kHz subcarrier. One of the problems with these decoders has been system alignment, since it is necessary to adjust the coils very accurately to obtain maximum stereo separation. Even after obtaining the correct alignment, the problem of aging of the coils and temperature and humidity variations still exists.

A solution to the problem is the PLL stereo decoder shown in Figure 8-68. The MC1310 requires only one adjustment, which is controlled by a potentiometer. The correct procedure for alignment is to open the input *pin* 2 and monitor *pin* 10 with a frequency counter. The potentiometer is adjusted until a 19.00-kHz frequency reading is obtained.

For those having limited test equipment, a very simple procedure may be followed which will result in separation of within a few decibels of optimum. This method consists of simply listening to a stereo broadcast and adjusting the potentiometer until the stereo pilot lamp turns on. Then the center of the lock-in range is found by

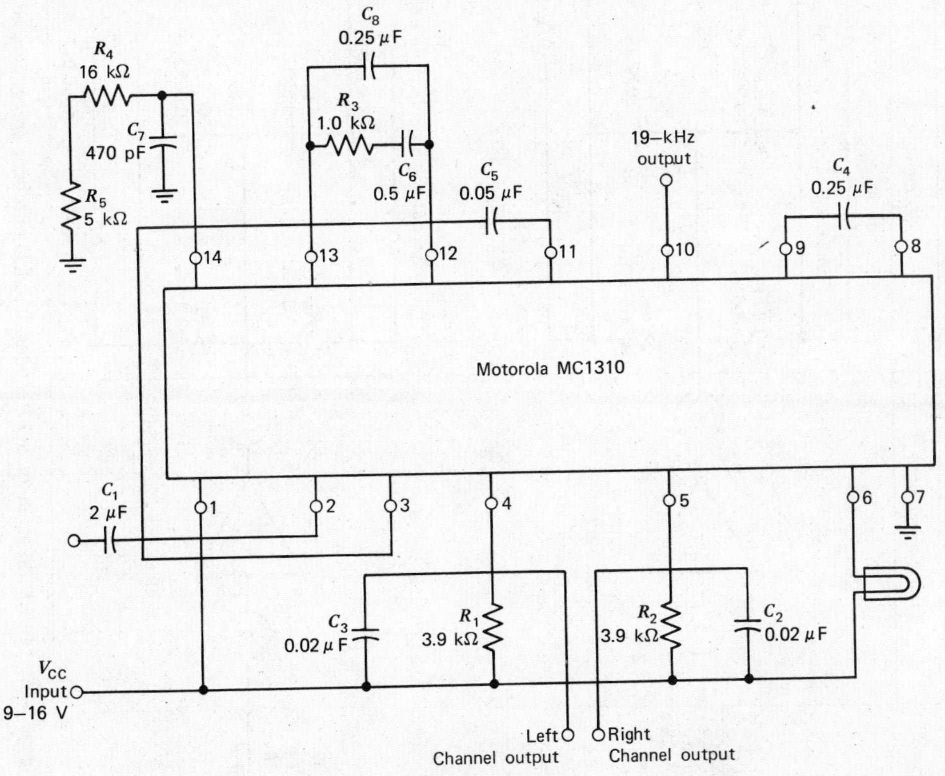

Figure 8-68. All adjustments for the stereo decoder are made with R_5.

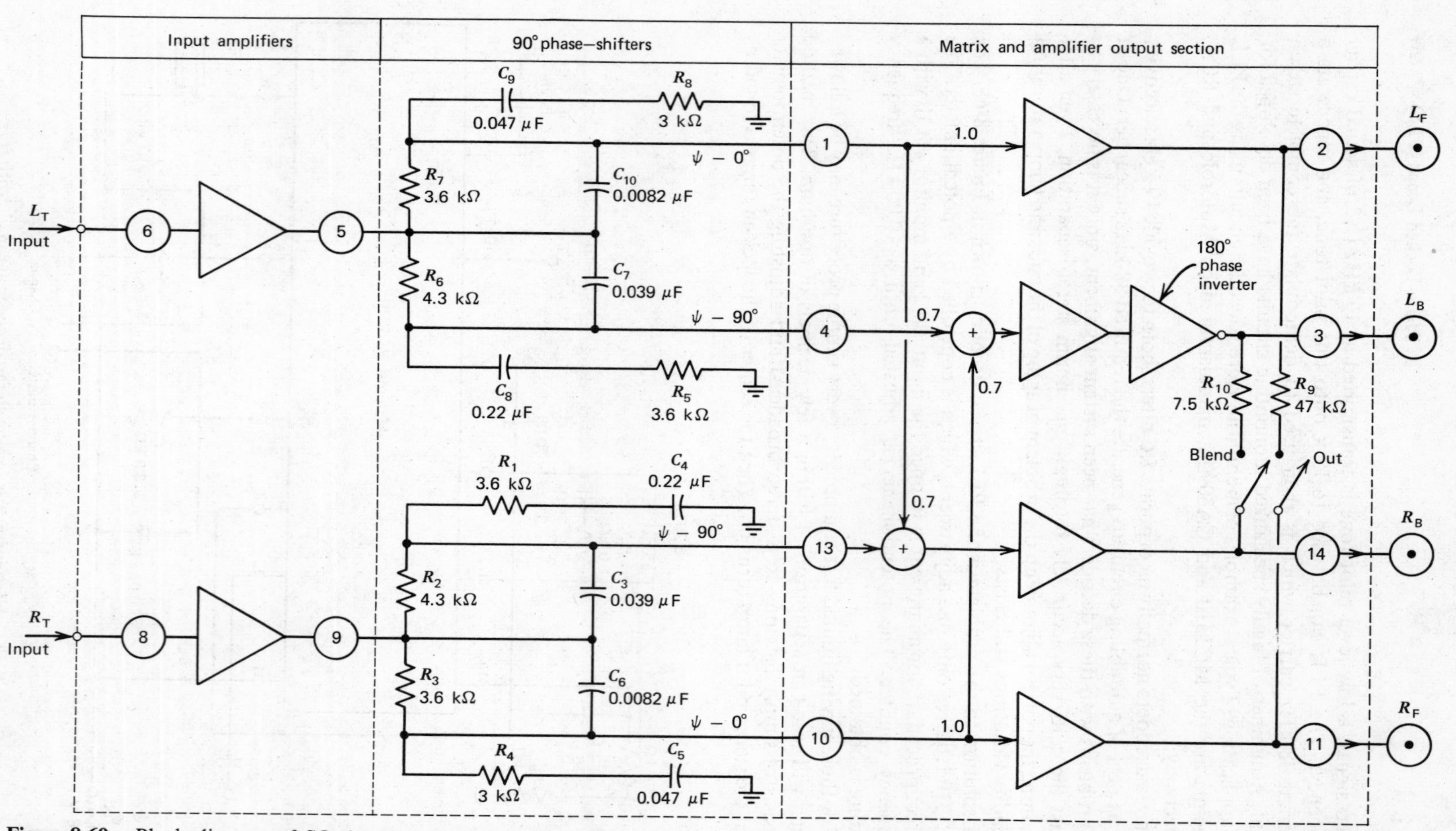

Figure 8-69. Block diagram of SQ decoder. Circled numbers are integrated circuit connections [*note:* circled numbers are integrated circuit connections]. (Courtesy *Popular Electronics*.)

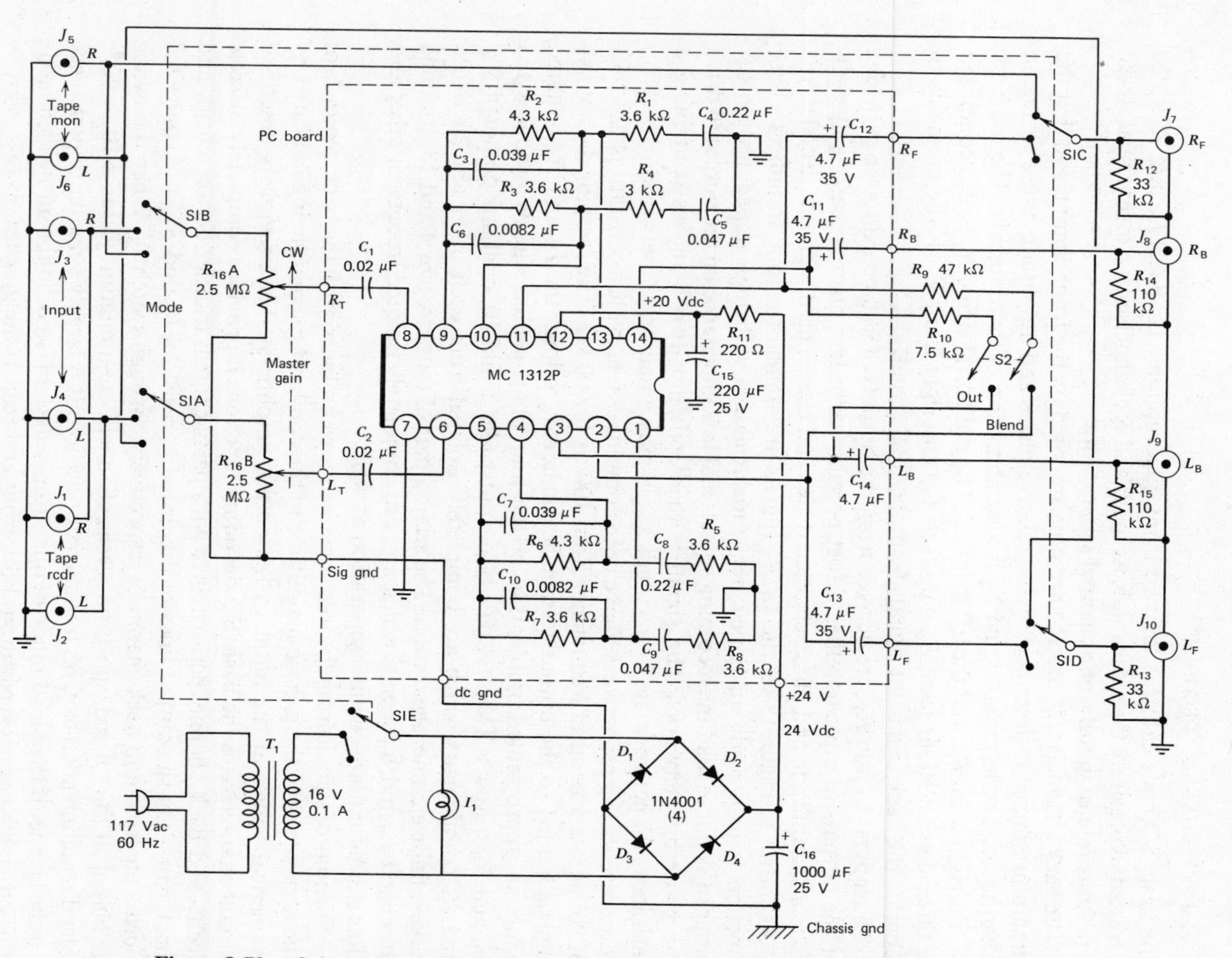

Figure 8-70. Schematic diagram of complete SQ decoder. (Courtesy *Popular Electronics*.)

rotating the potentiometer back and forth until the center of the lamp ON range is found. This method produces separation of 40 dB (typical) with total harmonic distortion typically 0.3%.

SQ FOUR-CHANNEL DECODER[4]

The basic SQ code calls for the two front quadraphonic channels L_F and R_F to be recorded in precisely the same mode as the *L* and *R* channels of a conventional stereo disk, thus retaining full front-channel separation.

The back channels, L_B and R_B, are superimposed on the front channels so that the quadrature image is described by the relative phase and amplitude of the signals in the grooves. The left total signal L_T is a combination of three signals: The left back L_B, the right back R_B, and the left front L_F. The right total signal R_T is also composed of three signals: Right back R_B, left back L_B, and right front R_F. It then remains to separate the back and front signals from the total signals.

The inputs, L_T and R_T, are derived from the outputs of a standard stereo cartridge or the outputs of a preamplifier. Most popular cartridges possess reasonably well-matched amplitude and phase characteristics and can be used with this decoder.

As shown in Figure 8-69, the L_T/R_T signals are applied to input amplifiers whose function is to provide an impedance transformation with high impedance on the input so as not to load the incoming signals and low impedance on the output to drive the phaseshift networks. The signal is then split into two branches containing a reference psi-network ($\psi - 0°$) and a psi-plus quadrature network ($\psi - 90°$), respectively. The psi networks have been computed to provide a constant phase-shift of 90° across a band of frequencies from 100 to 10,000 Hz. The merging L_T and R_T signals are fed to the output terminals, unaltered, to form the L_F and R_F outputs, while an appropriate combination of the four phase-shifted signals produces the L_B and R_B outputs. The first pair of outputs, *pins* 1 and 10, contains dominant L_F and R_F components which are completely isolated from each other and therefore have infinite channel separation. The second pair of outputs, *pins* 4 and 13, contains dominant L_B and R_B signals which are also completely isolated from each other and thus exhibit infinite channel separation as well.

Portions of signals from the front channels are combined with the back channels of the opposite side (the precise magnitude of the combining signals being indicated by numerical values at the input to the output amplifiers). The previously described circuit characterizes the basic SQ decoding function; it provides completely discrete front- and back-channel performance with partial signal transfer between front and back pairs. On the output lines of the decoder, resistors R_9 and R_{10} are connected across the front and back channels, respectively, through switch S_2. When the switch is placed in the "blend" position, cross-channel contamination yields a 10 to 40% blend recommended by CBS for matrix operation. It is largely a matter of preference to the listener if the blend mode of operation is desired, since in the unblended mode the left/right channel separation, both front and rear. is the greatest.

The realization of Figure 8-69 is shown in Figure 8-70. The entire decoder has been reduced to an IC that operates in conjunction with external phase-shift networks.

[4] M. Esformes, "SQ Four-channel Decoder," *Popular Electronics*, July 1973.

Index

Absolute value generator, 131, 132
AC coupled amplifier, 98, 100
AC coupled comparator, 269, 270, 271
AC/DC converter, 123-130, 132, 419-420
Active component simulation, 25-27
Active filters, 39, 44, 45, 61, 62, 65, 87, 95, 131, 132, 157, 160, 167-202, 235-237, 496-500, 507, 508
Active pull-up resistors, 300, 313
A/D converter, 85, 100, 101, 103, 147, 207, 238, 246, 265-270, 435
Adjustable current limiting, 351-354
Adjustable voltage reference, 135-136
AGC (Automatic Gain Control), 494
All-pass filter, 167, 237
AM detection, 508
AMR (AM Rejection), 491, 492
Analog commutator buffers, 102-103
Analog memory, 65
Analog switch, 102
Antilog generator, 213-215, 218
Astable multivibrator, 426-429, 434-435, 438-439, 442-443
Audio amplifier design guidelines, 463-465
Audio amplifier frequency response, 468
Audio amplifiers, 87, 455-490
Audio amplifier stabilization, 469
Audio circuits, 435-512
Audio mixer, 486-490
Audio recovery, 491
Automobile applications, 447-454

Balanced discriminator, 491
Bandpass filter, 167, 170-185, 189, 190, 191, 194, 201, 236-237
Band reject filter, 167, 237
Base boost, 470
Base-emitter voltage differential, 70, 134, 256, 354, 414-416
Base resistor, 7, 10, 11, 118, 201
Battery charger, 453-454
Bias circuits, 13-17, 31
Bias current compensation, 66-70, 110, 204, 205, 250, 257
Bilateral current sources, 104, 105
Binary-weighted resistor network, 269, 270
Biquadratic filter, 194-202
Bridge amplifier, 460-461
Bridge-T filter, 177, 189
Burglar alarm, 451-452
Buried layer, 10
Bus signal line, 280, 281
Butterworth filter, 169
Bypass circuits, 64, 74, 78, 79, 82, 83, 95, 97, 101, 116, 189, 220, 247, 248, 342, 361, 365, 366, 369, 389, 390, 403, 404, 407, 409, 411, 415, 445

Capacitance multiplication, 26, 27, 90-91
Capture range, 497, 498, 504, 505, 506
Characteristic impedance, 280, 312, 313, 316
Chopper stabilized operational amplifiers, 44-49, 57, 58, 102, 116, 117, 154, 155, 207, 208
Clipping amplifier, 125, 126, 160
CMOS circuits, 303, 306
CMRR (Common Mode Rejection Ratio), 37, 49, 51, 72, 73, 109, 110, 111, 112, 113, 114, 116
Common Mode Rejection, 29, 37, 41, 80, 110, 112, 123, 230, 253, 260, 286, 289, 290, 291, 293, 313, 315
Common mode slew rate, 251
Common mode voltage range, 51, 52, 56, 60, 61, 108, 109, 114, 136, 250, 255, 262, 290, 293, 313, 316, 419
Comparator DC accuracy, 239, 240, 241
Comparator differential bias current, 250, 251
Comparator differential input voltage, 241
Comparator error, 243, 245, 246, 250, 251, 253, 255, 256, 257
Comparator input bias current, 239, 240, 241, 242, 247, 250, 251, 253, 256, 257, 270, 271
Comparator input offset current, 238, 240, 241, 242, 247, 250, 251, 278
Comparator input offset voltage, 238, 241, 243, 250, 251, 290
Comparator input resistance, 240, 241, 250, 263, 278
Comparator input voltage range, 239, 247
Comparator logic compatibility, 245
Comparator output leakage current, 239
Comparator protection circuitry, 248
Comparator response time, 239, 242-245, 291, 292, 293
Comparators, 10, 62-63, 156, 159, 160, 162, 163, 193, 226, 227, 228, 229, 238-279, 288-296
Comparator saturation voltage, 239
Comparator strobe circuitry, 243, 244, 247, 265, 286, 292, 315, 316, 454
Comparator strobe on current, 239
Comparator strobe release time, 243, 244, 245, 454

Comparator strobe voltage, 244
Comparator supply current, 239
Comparator transfer functions, 239, 240, 247, 249, 252, 253, 258
Comparator useage guidelines, 247, 248, 270
Comparator voltage gain, 239
Complementary output stage, 2-17
Computers, 279, 280, 282, 283, 308, 313, 316
Constant current outputs, 297-305
Constant current source voltage regulator, 366, 398, 400
Core circuits, 281-295
Core memory, 282
Core memory stacks, 281-282
Cost, 49, 50, 95, 109, 112, 230, 265, 331, 335, 454
Costas demodulator, 494
Coupling/decoupling capacitors, 20
Crosstalk, 280, 281
Crystal oscillator, 263, 264
Cube generator, 213, 214
Current amplifier, 30, 31
Current limiting, 224, 225, 271, 339, 343-354, 358, 359, 361, 363, 364, 368-369, 376, 392-397, 404-407, 411, 414, 415, 418, 419, 445
Current monitor, 121, 122
Current reference, 14
Current regulators, 107
Current source loads, 17-19, 22, 23, 233
Current sources, 104-108
Current-to-voltage converter, 61-62, 120

D/A converter, 85, 102, 147, 154, 155, 269
Darlington pair, 39, 41-44, 67
Data communications, 308-330
Data delay line, 321, 322
Data transmission, 308-330, 496
DC/DC converter, 369, 370, 371, 373, 445-448
DC restoration circuit, 304, 306
Decoder/driver, 299-308
Deemphasis capacitor, 492, 493
Deep isolation diffusion, 303-304
Dielectric isolation, 6, 8, 10, 304
Difference amplifier, 57
Differential amplifier, 17, 87, 92, 109-119, 260, 284, 285, 286, 466
Differential input audio amplifier, 472, 473, 476
Differential input instrumentation amplifier, 115-117
Differential input stage, 292, 313
Differential transmission line, 317
Differentiator, 57-59, 193, 204, 205
Diffusion, 1, 3, 10, 11, 40, 303, 304
Diffusion furnace, 1, 2
Display drivers, 296-308
Dissipative voltage regulators, 223-225, 331-337, 409-420, 445-447, 452, 453
Distortion, 249, 460, 461
Double ended limit detector, 264, 265, 266
Driven switching regulator, 391-392
DTL (Diode-Transistor-Logic), 101, 238, 250, 259, 273, 274, 276, 278, 286, 313, 317, 436, 437
Dual comparators, 243, 244, 245, 247, 255, 257, 261, 262, 264, 265-267, 288-296
Dual operational amplifiers, 50, 446
Dual timers, 451
Dual tracking regulators, 230-233, 335, 336
Dual voltage regulators, 230-233, 335, 336, 370-374

ECL (Emitter-Coupled-Logic), 260-261
Efficiency, 225, 227, 228, 334, 374, 376, 377, 380, 386, 387, 389, 401, 408, 409, 446
Electronic circuit breaker, 323
Elliptic filters, 190, 191, 192
EMI (Electromagnetic Interference), 227, 381, 386
Emitter resistors, 11, 201
Epitaxial process, 4, 6
Epitaxy, 1, 4, 6
Error amplifier, 224, 361, 404
Error output voltage, 85, 114, 115, 126, 141, 150
Error sources, 53, 54, 55, 56, 57, 62, 64-66, 72, 85, 91, 105, 106, 109, 110, 113, 126, 141, 149, 150, 151, 158, 206, 207, 209, 212, 215, 217, 218, 219, 220, 222, 223, 231, 271, 417
External pass transistor circuits, 357-366, 369-370, 374, 375, 377, 382, 386-387, 388, 391-394, 395, 398, 399, 400, 401, 402-407

Feedforward compensation, 74, 82-86, 88, 95, 126, 128, 159, 207, 212, 220
Ferrite core memory, 282-286
FET (Field-Effect-Transistor), 7, 8, 9, 11, 13, 14, 39-44, 49, 77, 103, 106, 130, 131, 148, 151, 152, 153, 154, 156, 162, 177, 189, 190, 193, 194, 216, 219
Fixed output voltage regulator, 332, 335, 336, 409-420
Floating regulator, 377-379
Fluorescent display drivers, 297, 307, 308
FM circuitry, 491, 492-496, 498, 499, 506
FM demodulator, 496, 497, 500, 501, 505, 506
FM detector/limiter, 491-496
FM if strip, 493-494
Foldback current limiting, 225, 346-352, 363, 364
Four channel stereo decoder, 511-512
Free running multivibrator, 156, 228, 271-273
Frequency compensation, 51, 54, 57, 61, 63, 73-77, 82-89, 106, 126, 132, 138, 141, 156, 212, 214, 273, 333
Frequency divider, 429-430
Frequency doubler, 271, 273
Frequency response, 10, 31-34, 36, 37, 38, 51, 59, 61, 62, 76, 83, 85, 95, 99, 102, 204, 205, 458, 466, 471, 472, 473, 474, 477, 478, 479, 481, 484, 501, 506
Frequency roll-off, 51, 52, 75, 76, 77, 86, 87, 204, 469, 470, 473, 479, 501-506
Frequency synthesis, 506-508
Full wave rectifier, 128, 129, 131, 132

Gain magnitude response, 178, 179, 182, 183, 184, 187, 200
Gas discharge display driver, 297, 303-305, 306, 307, 308
General purpose voltage regulators, 332, 335, 336, 337-409
Grounding and lead inductance, 79-82, 109, 110, 220

Halfwave rectifier, 128, 129, 284, 285
Harmonic distortion, 460
High efficiency regulator, 374, 376, 377, 407-409
High gain amplifiers, 57, 58
High impedance amplifier, 39, 54-55, 98, 99, 101
High impedance audio amplifier, 461-463
High impedance comparator, 275, 278
High pass filter, 167, 172, 196, 199, 200, 201, 235
High stability voltage reference, 136-138
High voltage regulator, 233-234, 368-369
High voltage technology, 303-304
Hystersis, 229, 249, 253, 254, 255, 278, 279, 380, 396, 397, 441

IC processing steps, 3, 4-6
Incandescent display drivers, 306-307
Inductance simulation, 26, 27
Input circuits, 38-44
Integrators, 39, 59-61, 65, 67, 83, 85, 149, 163, 164, 166, 193, 204-209, 271, 435, 437
Instrumentation amplifiers, 109-119, 132-135
Interface circuits, 238-330, 454
Internally compensated operational amplifiers, 273
Internal voltage reference, 414-416
Inverting amplifier, 32, 53-54, 56, 71, 72, 89, 115, 128, 129, 132, 149
I/O bus, 279
Ion implantation, 11

Junction capacitor, 7, 12, 202
Junction isolation, 4-8

Ladder network, 100, 154, 265, 266, 268
Lamp driver, 63, 165, 264, 265
Lateral *pnp* transistor, 7, 9, 10, 24, 25, 82, 86, 95, 465
LED (Light Emitting Diode) driver, 143, 296-303
Level detector, 245, 246, 251-256
Level shifting, 19-22, 295, 304
Level shifting amplifier, 103-104, 108
Light intensity regulator, 370, 372
Linear components, 7, 8, 9
Linear phase filter, 172
Line driver, 280, 281, 310-330, 454
Line printer, 312
Line receiver, 238, 260, 261, 280, 281, 310-330, 454
Line regulation, 224, 232, 334, 340, 342, 373, 388, 389, 391, 403, 446
Liquid crystal display driver, 305-306
Load current, 223, 233, 234, 336, 354, 355, 366, 402
Load regulation, 223, 233, 234, 342, 350, 351, 358, 361, 370, 381, 382, 385, 387, 417, 446, 447
Lock range, 497, 498, 500, 503, 504, 505
Logarithmic amplifier, 65
Logarithmic converter, 203, 204, 209-220
Logarithmic multiplier/divider, 134
Long time comparator, 271, 273
Low pass filter, 132, 157, 167, 168, 169, 190, 196, 200, 201, 235, 496-500, 507, 508
Low power regulator circuit, 357
Low RF switching regulator, 409-410
Low temperature coefficient current sources, 107-108

Magnetic data register, 281, 286
Magnetic transducer, 250-251
Masks, 1, 3, 9, 301
Metallization mask, 3, 301
Miller capacitance, 284
Miller integrator, 271
Missing pulse detector, 429-430
Modem, 308, 309, 312
Monostable (one-shot) multivibrator, 156, 157, 245, 423, 424, 429-432
MOS calculator, 296, 297, 299, 301, 306, 308
MOS interface circuits, 295-296, 293, 303, 304, 308
MOS memories, 261, 262, 263, 295, 296
Multiple collector transistor, 10, 18, 19
Multiple feedback filter, 177, 178, 190
Multipler/divider, 203, 215-217, 218-219, 220
Multistage limiter, 249
Multivibrator/oscillators, 145, 156-162, 245, 262-264, 423, 424, 429-432

NAB (National Association of Broadcasters), equalization, 469, 470, 472, 473, 475, 476, 477, 478, 479, 480
Nanoammeter, 141-143
Negative regulators, 335, 336, 340-342, 351, 353, 354, 356-357, 361, 362, 363, 373, 374, 375, 376, 377, 379, 401-407
Noninverting amplifier, 32, 54-55, 66, 67, 68, 72, 73, 123, 148
Noninverting logarithmic generator, 212-213
Nonlinear amplifier, 203-223
Notch filter, 186, 187, 188, 189
NPN transistor, 7, 8, 9

Offset voltage balancing, 71-73
Offset voltage compensation, 70-71
Offset voltage drift, 65, 66, 70
One-shot multivibrator, 156, 157, 245, 423, 424, 429-432
Operational amplifier, 8, 13, 15, 28-237, 271, 273-279, 331, 337, 417-419, 420, 432, 434, 435, 437, 447
Operational amplifier frequency response, 10, 31-34, 36, 37, 38, 51, 59, 61, 62, 76, 83, 85, 95, 99, 102, 204, 205, 458, 466, 471, 472, 473, 474, 477, 478, 479, 481, 484, 501, 506
Operational amplifier input bias current, 15, 16, 31, 34, 35, 41, 49, 53, 56, 79, 95, 121, 204, 206, 207, 213, 216
Operational amplifier input impedance, 29, 30, 31, 35, 36, 37, 49, 54, 55, 56, 57, 78, 89, 105, 109, 110, 115, 116, 123, 145
Operational amplifier input offset current, 15, 34, 35, 40, 49, 79, 206, 216
Operational amplifier input offset voltage, 14, 35, 39, 41, 44, 49, 53, 70, 71-73, 89, 216
Operational amplifier output impedance, 29,

30, 35, 104, 145, 177, 185, 232
Operational amplifier protection circuitry, 25, 50, 79, 91-95, 96, 97, 126, 130, 134, 156, 216
Operational amplifier specifications, 34-38, 49, 50
Operational amplifier stability, 77, 78, 86, 94, 109, 138
Optical coupling, 260
Oscillators, 65, 87, 145, 156-162, 164, 245, 262-264, 423, 424, 429-432
Output current swing, 63, 64, 97, 104, 126, 138, 175, 176
Output stages, 22-25
Overload protection, 224, 225, 339, 343-354, 358, 359, 361, 363, 364, 368, 369, 376, 392-397, 398, 400, 404-407, 411, 414, 415, 418, 419, 445
Overvoltage protection, 397, 398
Oxide capacity, 7, 12, 202

Panaplex display driver, 304, 307
Party line applications, 310, 312, 313, 315, 316
Peak detector, 256-257
Peak-to-peak separation, 491
Phase detector, 496-498
Phase lock loop (PLL), 8, 496-512
Phase lock loop control voltage, 501
Phase lock loop error voltage, 496, 497
Phase lock loop frequency, 497-506
Phase lock loop linearity, 498, 500
Phase lock loop selectivity, 498, 506, 509
Phase lock loop synchronization (lock), 490, 499
Phase lock loop tracking (holding) range, 497
Phase lock loop transfer function, 501-506
Phase lock loop useage hints, 501, 503, 505
Phase response, 178, 179, 182, 183, 184, 187, 200
Phase shift, 54, 83, 84, 86, 191, 193, 204, 220, 225, 494
Phonograph amplifier, 456-460, 482-486, 487
Photocell amplifier, 120-131
Photocell, 218, 219
Photodiode/phototransistor, 370, 372
Photodiode sensor, 119-120, 256, 257
Photoresist, 1, 4
Piezoelectric accelerometer, 119
Pinch resistor, 7, 11, 201
Playback equalization, 458, 459, 469, 477, 482, 484, 485, 486
Positive voltage regulator, 335-340, 346-351, 358-379, 382-400
Power amplifier, 368-369
Power bandwidth, 74, 86, 87
Power booster, 108-109, 138, 140
Power dissipation, 342, 345, 346, 348, 349, 354, 359, 361, 364, 367, 369, 378, 398, 403, 404, 405, 409, 414, 415
Power supply bypassing, 247, 248
Power supply rejection, 79, 247
Power supply useage considerations, 78-82, 83, 97
Precision AC/DC converters, 123-130, 132, 419-420
Precision clamp, 97, 125, 126, 157, 158
Precision current source, 105-106
Precision rectifier, 126-130
Precision voltage regulator, 417-419
Programmable operational amplifier, 141-142
Pull-up resistor, 250, 280
Pulse height discriminator, 238
Pulse regulator, 376, 377, 378
Pulse shape restorer, 246
Pulse Width Modulator (PWM), 431, 432, 495-496
Punch-through transistors, 11, 39-44

Quadrature network, 492, 498
Quiescent current, 23

Radio Frequency (RF), 409-410, 464
Read-Only-Memory (ROM), 300, 301, 302, 304
Rectifiers, 126-130, 131, 132, 284, 285
Regulators, 8, 66, 223-234, 271, 273, 331-420, 445-447, 452, 453
Relaxation oscillator, 145
Resistance multiplication, 89-90
RIAA equalization, 458, 482, 484, 485, 486
Ripple voltage, 232
RMS detector, 131-134
Roll-off curve, 51, 52, 75, 76, 77, 86, 87, 204, 469, 470, 473, 479, 501-506
Root extractor, 217-218

Sample-and-hold accuracy, 150-152
Sample-and-hold circuit, 39, 45, 65, 67, 95, 143, 147-155, 204, 443
Sample-and-hold stability, 152-154
Schmitt trigger, 238, 249, 255
Schottky barrier diode, 10, 11, 128, 262
Schottky process, 8, 303
Self-tuned filter, 191-194
Sense amplifier, 246, 261, 262, 263, 282, 283, 284, 286-296, 321, 322
Separation, 491, 509
Series dissipative voltage regulators, 223-234, 333-334, 337, 378, 409-420, 445-447, 452, 453
Series pass element, 223, 225, 337, 348, 349, 353, 355, 356, 363-366, 367, 374, 378, 380, 413, 414, 420
Servo preamplifier, 222-223
Servo system, 501-503
Servo system controller, 439-441
Settling time, 52, 155
Seven segment display drivers, 296-303
Short circuit protection, 225, 234, 341, 342, 343-354, 355, 358, 359, 365, 392-397, 445
Silicon, 1-12, 39
Sine wave oscillators, 87, 157-162, 164
Single eneded input audio amplifier, 472, 473, 476, 480
Single ended transmission line, 317, 329-330
Single supply voltage reference, 139-141
Slew rate, 15, 36, 37, 38, 39, 41, 51, 52, 74, 82, 84, 96, 97, 128, 175, 176, 177, 207, 279
Solid state thermostat, 443-445
Speed warning device, 448, 450, 451
Square wave generator, 432-434
State variable filter, 171, 194-202

Stero decoding, 508-512
Stereo tape cartridge playback system, 465-467, 468-477
Summing amplifier, 39, 56, 66, 68, 85, 132, 203
Superbeta transistors, 11, 39-44
Supergain transistors, 11, 39-44
Suppressed carrier modulation, 494-495
Switching circuit drivers, 100-112
Switching regulators, 225-229, 271, 273, 274, 331, 334, 378-409
Switching regulator design hints, 226-227
Switching regulator frequency, 225, 228, 273, 334, 335, 381, 382, 383, 384, 385, 386, 389, 390, 392, 397, 398, 408, 409, 445
Symmetrical power supplies, 372-374, 375, 376, 377
Synchronized regulator, 386, 391
Synchronous AM detection, 495-496, 498

Tachometer, 145-147, 447, 448, 449
Tape record preamplifier, 477-482
Teletypewriter, 317
Temperature compensated sense amplifier, 292
Temperature compensating regulators, 378, 380
Temperature controllers, 366-368
Temperature-to-frequency converter, 438-439
Temperature limiting, 354-355, 411-414, 416, 418, 419
Termination mismatches, 318, 320
THD (Third Harmonic Distortion), 491
Thermal shutdown, 411-414, 416, 418, 419
Thermal stability, 436
Thermometer, 143-144
Thermostat, 443-445
Thick film resistors, 201, 290
Thin film capacitors, 204
Thin film resistors, 201, 290
Threshold detector, 163, 164, 166, 238, 246, 284, 285, 286, 380, 382, 508
Threshold voltage, 289, 290, 292, 295
Timer, 65, 143, 163, 421-454
Timer application hints, 423-426, 435, 436, 445
Timer control voltage, 426
Timer errors, 434, 439
Timer output load considerations, 424-425
Timer timing elements, 423-424
Timer trigger level, 423, 425, 426, 429, 441-442
Timer waveforms, 424, 425, 427, 430, 431, 442, 450
Tone burst generator, 442-443
Tone control response, 458, 486
Tone detector, 508
Touch switch, 441-442
Tracking regulator, 230-233, 335, 336, 370-374
Transducers, 119, 120, 250, 251, 438-440, 443-445
Transistor compensation, 67-69
Transistion time, 36, 37, 96
Transmission line, 229, 280, 308-338
Transmission line noise, 316, 317, 320
Transmission line response, 318-321
Triangular wave generator, 163-166, 168
Tri-State Logic (TSL), 310, 311, 313, 314, 323-330
TTL (Transistor-Transistor-Logic), 101, 102, 147, 238, 250, 255, 259, 260, 273, 274, 276, 278, 281, 286, 295, 300, 303, 312, 313, 316, 317, 409, 416, 421, 424, 425, 437, 445, 454
Tunable filter, 186, 187, 188, 189
Twin T filter, 177, 185, 186, 187
Twisted pair transmission line 312, 313, 315, 317, 318, 321
Two-pole compensation, 74, 87-88

Ultrasonic remote control receiver, 489-490
Unity gain buffer, 39, 49, 55, 56, 66, 67, 72, 73, 79, 95-109, 115, 116, 125, 137, 145, 149, 150, 152, 153, 158, 185, 186, 189, 217, 218, 268

VCO (Voltage Controlled Oscillator), 168, 432, 496-512
Vertical *pnp* transistor, 7, 8, 9, 10, 23, 25
Voltage comparators, 10, 62-63, 156, 159, 160, 162, 163, 193, 226, 227, 228-229, 238-279, 288-296
Voltage controlled amplifier, 130-131
Voltage doubler, 128
Voltage follower, 39, 49, 55, 56, 66, 67, 72, 73, 95-109, 115, 116, 125, 137, 145, 149-150, 152, 153, 158, 159, 185, 186, 189, 209, 213, 229, 230, 266-270
Voltage-to-frequency converter, 435-438
Voltage gain, 14, 17, 28, 29, 30, 32-34, 79, 207
Voltage-to-pulse duration converter, 434-435
Voltage references, 13, 14, 15, 17, 135-141, 333, 334, 335, 337, 341, 370, 371
Voltage regulator, 8, 66, 223-234, 271, 273, 274, 331-420, 445-447, 452, 453
Voltage regulator bypass capacitor, 342, 361, 365, 366, 369, 389, 390, 403, 404, 407, 409, 411, 415, 416
Voltage regulator current limiting, 224, 225, 271, 339, 343-354, 358, 359, 361, 363, 364, 368-369, 376, 392-397, 404-407, 411, 414, 415, 418, 419, 445
Voltage regulator current limit sense voltage, 339, 347, 348, 353, 362, 365
Voltage regulator design hints, 226, 227, 342, 355-357, 361, 365, 366, 369, 376, 381, 382-391, 398-400, 403, 404, 409, 418, 419
Voltage regulator dropout voltage, 333
Voltage regulator efficiency, 225, 227, 228, 334, 374, 376, 377, 380, 386, 387, 389, 401, 408, 409, 446
Voltage regulator error signal, 231
Voltage regulator feedback sense voltage, 333, 339, 343
Voltage regulator filters, 225, 227, 382-391, 409, 410
Voltage regulator input-output voltage differential, 333, 348, 362, 363
Voltage regulator line regulation, 233, 332, 340, 342, 373, 388, 389, 391, 403, 446
Voltage regulator load current, 336, 354, 355, 366, 402

Voltage regulator load regulation, 233, 332, 340, 342, 350, 351, 358, 361, 370, 417, 446, 447
Voltage regulator output ripple, 381, 382, 383
Voltage regulator overload shutoff, 225, 227, 228, 334, 374, 376, 377, 380, 386, 387, 389, 398, 400, 406, 407, 408, 409, 446
Voltage regulator overvoltage protection, 397-398
Voltage regulator power dissipation, 342, 345, 346, 348, 349, 354, 359, 361, 364, 367, 369, 378, 398, 403, 404, 414, 415
Voltage regulator protective diodes, 355-357
Voltage regulator ripple rejection, 333, 342, 368, 373
Voltage regulator short circuit protection, 341, 342, 343-354, 355, 358, 359, 365, 392-397, 445
Voltage regulator specifications, 332-333
Voltage regulator standby current drain, 333, 337, 365, 368
Voltage regulator temperature drift, 333, 337, 348, 367, 370, 371, 373, 417, 418, 419
Voltage regulator temperature limiting, 354-355, 411-414, 416, 418, 419
Voltage regulator transients, 271, 334, 365, 378, 380, 384, 392, 401, 409, 411
Voltage regulator voltage reference, 333, 334, 335, 337, 341, 370, 371
Voltage splitter, 229-230
Volume control, 457, 460, 461, 486

Wafers, 1, 2, 3
Weinbridge oscillator, 159, 161-162, 164
Window comparator, 257-259

Zener reference, 13, 14, 107, 135-141
Zero crossing detector, 238, 249-251, 270, 274